D1674632

Springer Series in Computational Mathematics

14

Springer
Berlin
Heidelberg
New York
Barcelona
Budapest
Hong Kong
London
Milan
Paris
Santa Clara
Singapore
Tokyo

E. Hairer   G. Wanner

Solving Ordinary Differential Equations II

Stiff and Differential-Algebraic Problems

Second Revised Edition

With 137 Figures

Ernst Hairer
Gerhard Wanner
Université de Genève
Section de Mathématiques, C.P. 240
2–4 rue du Lièvre
CH-1211 Genève 24
Switzerland

Cataloging-in-Publication Data applied for

Die Deutsche Bibliothek - CIP-Einheitsaufnahme

Solving ordinary differential equations / E. Hairer ; G. Wanner. - Berlin ; Heidelberg ; New York ; Barcelona ; Budapest ; Hong Kong ; London ; Milan ; Paris ; Santa Clara ; Singapore ; Tokyo : Springer.
Bd. 1 verf. von E. Hairer, S. P. Norsett und G. Wanner
NE: Hairer, Ernst; Norsett, Syvert P.; Wanner, Gerhard
2. Stiff and differential algebraic problems. - 2., rev. ed. - 1996
(Springer series in computational mathematics ; 14)
ISBN 3-540-60452-9
NE: GT

Mathematics Subject Classification (1991): 65Lxx, 34A50

ISBN 3-540-60452-9 Springer-Verlag Berlin Heidelberg New York

ISBN 3-540-53775-9 1.Auflage Springer-Verlag Berlin Heidelberg New York

Printed in Germany

Typesetting: Camera-ready copy produced from the authors
SPIN 10472348 41/3143-5 4 3 2 1 0 – Printed on acid-free paper

To Evi and Myriam

From the Preface to the First Edition

"Whatever regrets may be, we have done our best."
(Sir Ernest Shackleton, turning back on 9 January 1909 at 88° 23' South.)

Brahms struggled for 20 years to write his first symphony. Compared to this, the 10 years we have been working on these two volumes may even appear short.

This second volume treats stiff differential equations and differential algebraic equations. It contains three chapters: Chapter IV on one-step (Runge-Kutta) methods for stiff problems, Chapter V on multistep methods for stiff problems, and Chapter VI on singular perturbation and differential-algebraic equations.

Each chapter is divided into sections. Usually the first sections of a chapter are of an introductory nature, explain numerical phenomena and exhibit numerical results. Investigations of a more theoretical nature are presented in the later sections of each chapter.

As in Volume I, the formulas, theorems, tables and figures are numbered consecutively in each section and indicate, in addition, the section number. In cross references to other chapters the (latin) chapter number is put first. References to the bibliography are again by "author" plus "year" in parentheses. The bibliography again contains only those papers which are discussed in the text and is in no way meant to be complete.

It is a pleasure to thank J. Butcher, G. Dahlquist, and S.P. Nørsett (coauthor of Volume I) for their interest in the subject and for the numerous discussions we had with them which greatly inspired our work. Special thanks go to the participants of our seminar in Geneva, in particular Ch. Lubich, A. Ostermann and M. Roche, where all the subjects of this book have been presented and discussed over the years. Much help in preparing the manuscript was given by J. Steinig, Ch. Lubich and A. Ostermann who read and re-read the whole text and made innumerable corrections and suggestions for improvement. We express our sincere gratitude to them. Many people have seen particular sections and made invaluable suggestions and remarks: M. Crouzeix, P. Deuflhard, K. Gustafsson, G. Hall, W. Hundsdorfer, L. Jay, R. Jeltsch, J.P. Kauthen, H. Kraaijevanger, R. März, and O. Nevanlinna. ... Several pictures were produced by our children Klaudia Wanner and Martin Hairer, the one by drawing the other by hacking.

The marvellous, perfect and never failing TEX program of D. Knuth allowed us to deliver a camera-ready manuscript to Springer Verlag, so that the book could be produced rapidly and at a reasonable price. We acknowledge with pleasure the numerous remarks of the planning and production group of Springer Verlag concerning fonts, style and other questions of elegance.

March, 1991 The Authors

Preface to the Second Edition

The preparation of the second edition allowed us to improve the first edition by rewriting many sections and by eliminating errors and misprints which have been discovered. In particular we have included new material on

- methods with extended stability (Chebyshev methods) (Sect. IV.2);
- improved computer codes and new numerical tests for one- and multistep methods (Sects. IV.10 and V.5);
- new results on properties of error growth functions (Sects. IV.11 and IV.12);
- quasilinear differential equations with state-dependent mass matrix (Sect. VI.6).

We have completely reorganized the chapter on differential-algebraic equations by including three new sections on

- index reduction methods (Sect. VII.2);
- half-explicit methods for index-2 systems (Sect. VII.6);
- symplectic methods for constrained Hamiltonian systems and backward error analysis on manifolds (Sect. VII.8).

Our sincere thanks go to many persons who have helped us with our work:

- all readers who kindly drew our attention to several errors and misprints in the first edition, in particular C. Bendtsen, R. Chan, P. Chartier, T. Eirola, L. Jay, P. Kaps, J.-P. Kauthen, P. Leone, S. Maset, B. Owren, and L.F. Shampine;
- those who read preliminary versions of the new parts of this edition for their invaluable suggestions: M. Arnold, J. Cash, D.J. Higham, P. Kunkel, Chr. Lubich, A. Medovikov, A. Murua, A. Ostermann, and J. Verwer.
- the staff of the Geneva computing center and of the mathematics library for their constant help;
- the planning and production group of Springer-Verlag for numerous suggestions on presentation and style.

All figures have been recomputed and printed, together with the text, in Postscript. All computations and text processings were done on the SUN workstations of the Mathematics Department of the University of Geneva.

April 1996 The Authors

Contents

Chapter IV. Stiff Problems – One-Step Methods

Chapter V. Multistep Methods for Stiff Problems

Chapter VII. Differential-Algebraic Equations of Higher Index

Chapter IV. Stiff Problems – One-Step Methods

This chapter introduces stiff (styv (Swedish first!), steif (German), stìf (Islandic), stijf (Dutch), raide (French), rígido (Spanish), rígido (Portuguese), stiff (Italian), kankea (Finnish), δύσκαμπτο (Greek), merev (Hungarian), rigid (Rumanian), tog (Slovenian), čvrst (Serbo-Croatian), tuhỳ (Czecho-Slovakian), sztywny (Polish), jäik (Estonian), stiegrs (Latvian), standus (Lithuanian), stign (Breton), zurrun (Basque), sert (Turkish), жесткий (Russian), твърд (Bulgarian), קשיח (Hebrew), سَاق (Arabic), بنگک (Urdu), سخت (Persian), कठिण (Sanscrit), कड़ा (Hindi), 刚性 (Chinese), 硬い (Japanese), cứng (Vietnamese), ngumu (Swaheli) ...) differential equations. While the intuitive meaning of stiff is clear to all specialists, much controversy is going on about it's correct mathematical definition (see e.g. p.360-363 of Aiken (1985)). The most pragmatical opinion is also historically the first one (Curtiss & Hirschfelder 1952): *stiff equations are equations where certain implicit methods, in particular BDF, perform better, usually tremendously better, than explicit ones.* The eigenvalues of the Jacobian $\partial f/\partial y$ play certainly a role in this decision, but quantities such as the dimension of the system, the smoothness of the solution or the integration interval are also important (Sections IV.1 and IV.2).

Stiff equations need new concepts of stability (A-stability, Sect. IV.3) and lead to mathematical theories on order restrictions (order stars, Sect. IV.4). Stiff equations require implicit methods; we therefore focus in Sections IV.5 and IV.6 on implicit Runge-Kutta methods, in IV.7 on (semi-implicit) Rosenbrock methods and in IV.9 on semi-implicit extrapolation methods. The actual efficient implementation of implicit Runge-Kutta methods poses a number of problems which are discussed in Sect. IV.8. Section IV.10 then reports on some numerical experience for all these methods.

With Sections IV.11, IV.12 and IV.13 we begin with the discussion of contractivity (B-stability) for linear and nonlinear differential equations. The chapter ends with questions of existence and numerical stability of the implicit Runge-Kutta solutions (Sect. IV.14) and a convergence theory which is independent of the stiffness (B-convergence, Sect. IV.15).

IV.1 Examples of Stiff Equations

... Around 1960, things became completely different and everyone became aware that the world was full of stiff problems.
(G. Dahlquist in Aiken 1985)

Stiff equations are problems for which explicit methods don't work. Curtiss & Hirschfelder (1952) explain stiffness on one-dimensional examples such as

$$y' = -50(y - \cos x). \tag{1.1}$$

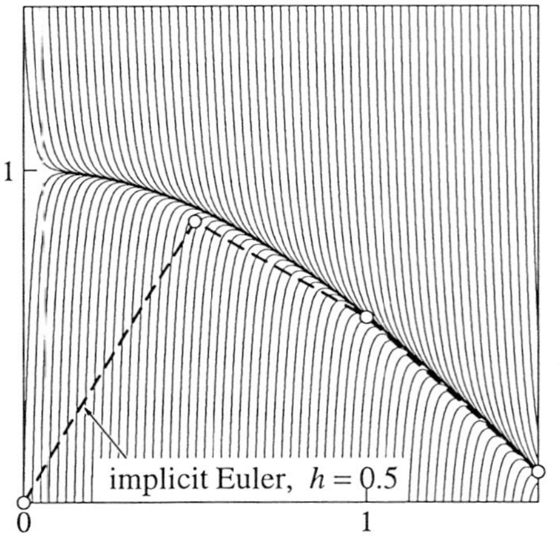

Fig. 1.1. Solution curves of (1.1) with implicit Euler solution

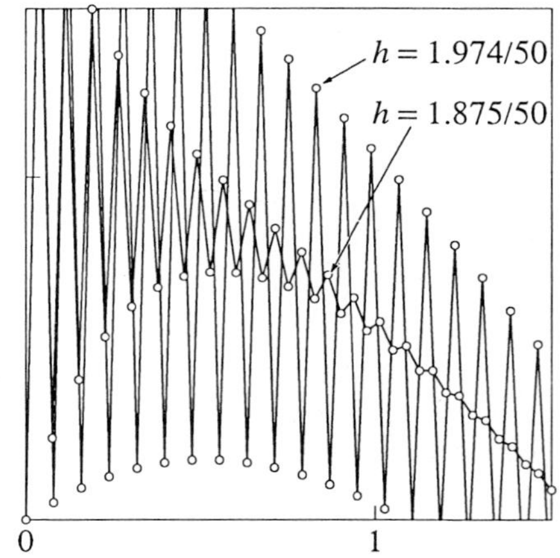

Fig. 1.2. Explicit Euler for $y(0) = 0$, $h = 1.974/50$ and $1.875/50$

Solution curves of Equation (1.1) are shown in Fig. 1.1. There is apparently a smooth solution in the vicinity of $y \approx \cos x$ and all other solutions reach this one after a rapid "transient phase". Such transients are typical of stiff equations, but are neither sufficient nor necessary. For example, the solution with initial value $y(0) = 1$ (more precisely $2500/2501$) has *no* transient. Fig. 1.2 shows Euler polygons for the initial value $y(0) = 0$ and step sizes $h = 1.974/50$ (38 steps) and $h = 1.875/50$ (40 steps). We observe that whenever the step size is a little too large (larger than 2/50), the numerical solution goes too far beyond the equilibrium and violent oscillations occur.

Looking for better methods for differential equations such as (1.1), Curtiss and Hirschfelder discovered the BDF method (see Sect. III.1): the approximation

$y \approx \cos x$ (i.e., $f(x,y)=0$) is only a crude approximation to the smooth solution, since the derivative of $\cos x$ is not zero. It is much better, for a given solution value y_n, to search for a point y_{n+1} where the slope of the vector field is directed towards y_n, hence

$$\frac{y_{n+1}-y_n}{h}=f(x_{n+1},y_{n+1}). \tag{1.2}$$

This is the implicit Euler method. The dotted line in Fig. 1.1 consists of three implicit Euler steps and demonstrates impressively the good stability property of this method. Equation (1.1) is thus apparently "stiff" in the sense of Curtiss and Hirschfelder.

Extending the above idea "by taking higher order polynomials to fit y at a large number of points" then leads to the BDF methods.

Chemical Reaction Systems

> When the equations represent the behaviour of a system containing a number of fast and slow reactions, a forward integration of these equations becomes difficult. (H.H. Robertson 1966)

The following example of Robertson's (1966) has become very popular in numerical studies (Willoughby 1974):

$$\begin{array}{cccl} A & \xrightarrow{0.04} & B & \text{(slow)} \\ B+B & \xrightarrow{3\cdot 10^7} & C+B & \text{(very fast)} \\ B+C & \xrightarrow{10^4} & A+C & \text{(fast)} \end{array} \tag{1.3}$$

which leads to the equations

$$\begin{array}{llll} \text{A:} & y_1' = -0.04y_1 + 10^4 y_2 y_3 & & y_1(0)=1 \\ \text{B:} & y_2' = 0.04y_1 - 10^4 y_2 y_3 & -3\cdot 10^7 y_2^2 & y_2(0)=0 \\ \text{C:} & y_3' = & 3\cdot 10^7 y_2^2 & y_3(0)=0. \end{array} \tag{1.4}$$

After a bad experience with explicit Euler just before, let's try a higher order method and a more elaborate code for this example: DOPRI5 (cf. Volume 1). The numerical solutions obtained for y_2 with $Rtol=10^{-2}$ (209 steps) as well as with $Rtol=10^{-3}$ (205 steps) and $Atol=10^{-6}\cdot Rtol$ are displayed in Fig. 1.3. Fig. 1.4 presents the step sizes used by the code and also the local error estimates. There, all rejected steps are crossed out.

We observe that the solution y_2 rapidly reaches a quasi-stationary position in the vicinity of $y_2'=0$, which in the beginning ($y_1=1$, $y_3=0$) is at $0.04\approx 3\cdot 10^7 y_2^2$, hence $y_2\approx 3.65\cdot 10^{-5}$, and then very slowly goes back to zero again. The numerical method, however, integrates this smooth solution by thousands of apparently unnecessary steps. Moreover, the chosen step sizes are more or less independent of the chosen tolerance. Hence, they seem to be governed by stability

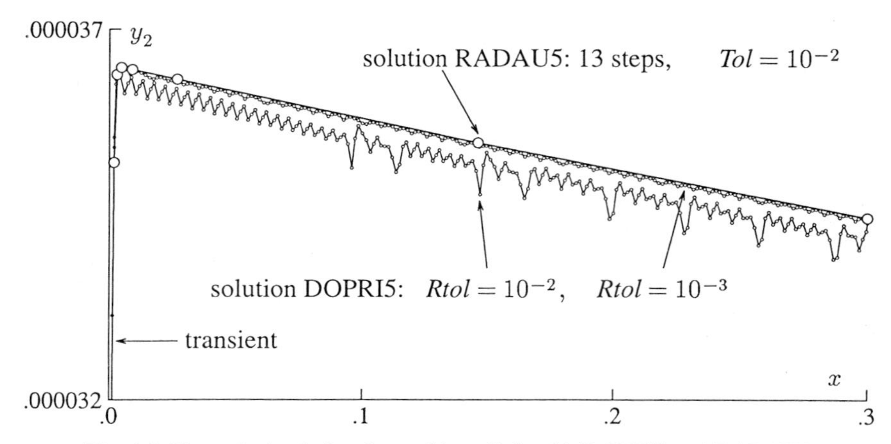

Fig. 1.3. Numerical solution for problem (1.4) with DOPRI5 and RADAU5

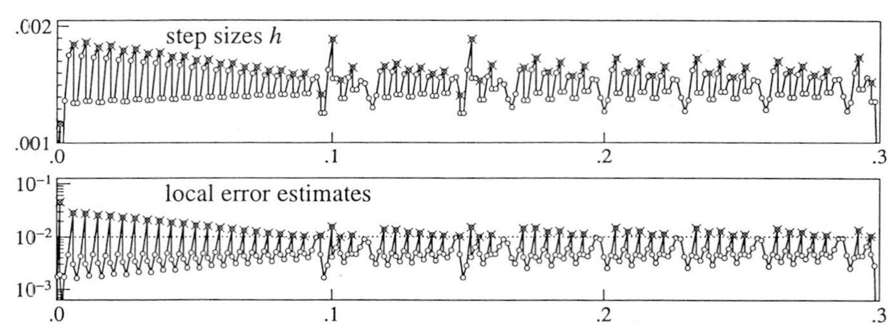

Fig. 1.4. Step sizes and local error estimates of DOPRI5, $Tol = 10^{-2}$

rather than by precision requirements. It can also be seen that an implicit Runge-Kutta code (such as RADAU5 described in Sections IV.5 and IV.8) integrates this equation without any problem.

Electrical Circuits

> This behavior is known, at least in part, to any experienced worker in the field. (G. Hall 1985)

One of the simplest nonlinear equations describing a circuit is van der Pol's equation (see Sect. I.16)

$$\begin{aligned} y_1' &= y_2 & y_1(0) &= 2 \\ y_2' &= \mu(1-y_1^2)y_2 - y_1 & y_2(0) &= 0. \end{aligned} \tag{1.5}$$

We have seen in Chapter II that this equation is easily integrated for moderate values of μ. But we now choose $\mu = 500$ and suspect that the problem might

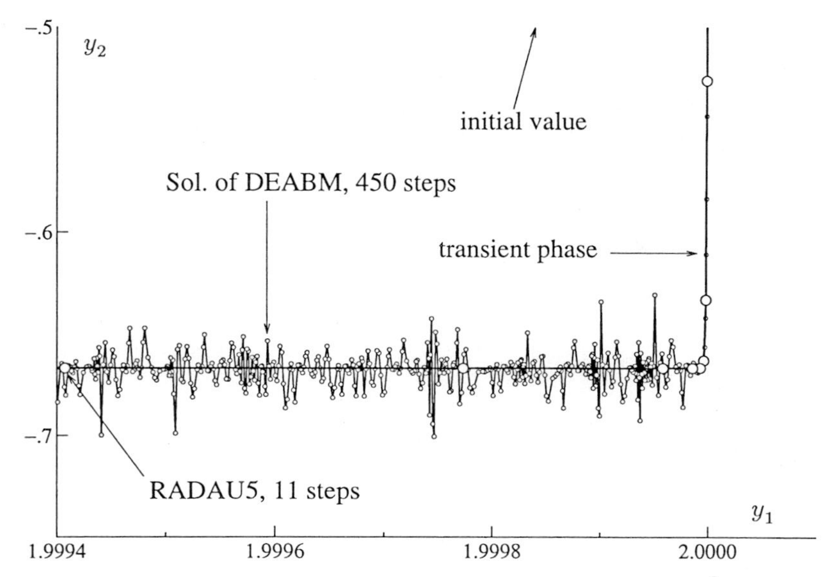

Fig. 1.5. Numerical solution for DEABM at equation (1.5'), $Rtol = 10^{-2}$, $Atol = 10^{-7}$

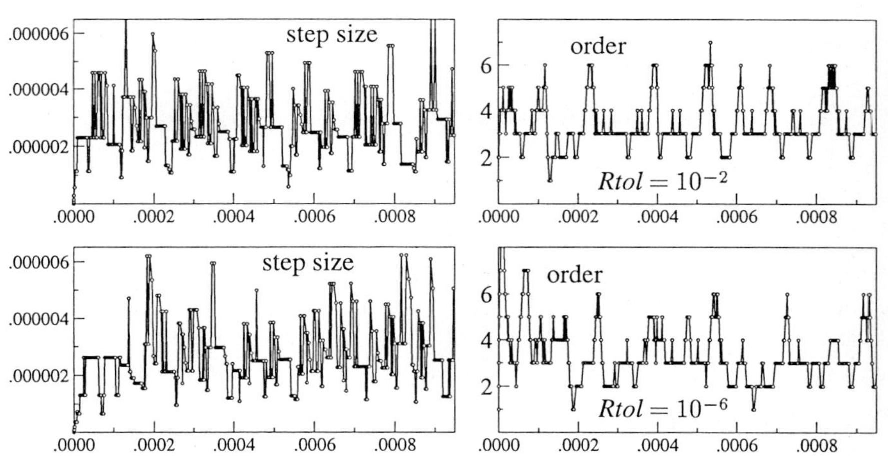

Fig. 1.6. Step sizes and orders for DEABM, $Rtol = 10^{-2}$, 10^{-6}, $Atol = 5 \cdot 10^{-8}$

become difficult. It turns out that the period of the solution increases with μ. We therefore rescale the solutions and introduce $t = x/\mu$, $z_1(t) = y_1(x)$, $z_2(t) = \mu y_2(x)$. In the resulting equation the factor μ^2 multiplies the entire second line of f. Substituting again y for z, x for t and $\mu^2 = 1/\varepsilon$ we obtain

$$\begin{aligned} y_1' &= y_2 \\ y_2' &= \mu^2\big((1-y_1^2)y_2 - y_1\big) \end{aligned} \qquad \text{or} \qquad \begin{aligned} y_1' &= y_2 \\ \varepsilon y_2' &= (1-y_1^2)y_2 - y_1 . \end{aligned} \tag{1.5'}$$

The steady-state approximation (see Vol. I, Formula (I.16.5)) then becomes independent of μ.

Why not try a multistep code this time? For example the predictor-corrector Adams code DEABM of Shampine & Watts. Figures 1.5 and 1.6 show the numerical solution, the step sizes and the orders for the first 450 steps. Eventually the code stops with the message *Idid* $= -4$ ("the problem appears to be stiff"). The implicit Runge-Kutta code RADAU5 integrates over the same interval in 11 steps.

Diffusion

Stalling numerical processes must be wrong.
(A "golden rule" of Achi Brandt)

Another source of stiffness is the translation of diffusion terms by divided differences (method of lines, see Sect. I.1) into a large system of ODE's. We choose the Brusselator (see (16.12) of Sect. I.16) in one spatial variable x

$$\begin{aligned}\frac{\partial u}{\partial t} &= A + u^2 v - (B+1)u + \alpha\frac{\partial^2 u}{\partial x^2}\\ \frac{\partial v}{\partial t} &= Bu - u^2 v + \alpha\frac{\partial^2 v}{\partial x^2}\end{aligned} \tag{1.6}$$

with $0 \le x \le 1$, $A = 1$, $B = 3$, $\alpha = 1/50$ and boundary conditions

$$\begin{aligned}u(0,t) &= u(1,t) = 1, & v(0,t) &= v(1,t) = 3,\\ u(x,0) &= 1 + \sin(2\pi x), & v(x,0) &= 3.\end{aligned}$$

We replace the second spatial derivatives by finite differences on a grid of N points $x_i = i/(N+1)$ $(1 \le i \le N)$, $\Delta x = 1/(N+1)$ and obtain from (1.6)

$$\begin{aligned}u_i' &= 1 + u_i^2 v_i - 4u_i + \frac{\alpha}{(\Delta x)^2}(u_{i-1} - 2u_i + u_{i+1}),\\ v_i' &= 3u_i - u_i^2 v_i + \frac{\alpha}{(\Delta x)^2}(v_{i-1} - 2v_i + v_{i+1}),\\ u_0(t) &= u_{N+1}(t) = 1, \quad v_0(t) = v_{N+1}(t) = 3,\\ u_i(0) &= 1 + \sin(2\pi x_i), \quad v_i(0) = 3, \quad i = 1,\ldots,N.\end{aligned} \tag{1.6'}$$

Table 1.1. Results for (1.6') with ODEX for $0 \le t \le 10$

N	*Tol*	accepted steps	rejected steps	function calls
10	10^{-4}	21	3	365
20	10^{-4}	81	25	1138
30	10^{-4}	167	45	2459
40	10^{-4}	275	62	4316
40	10^{-2}	266	59	3810

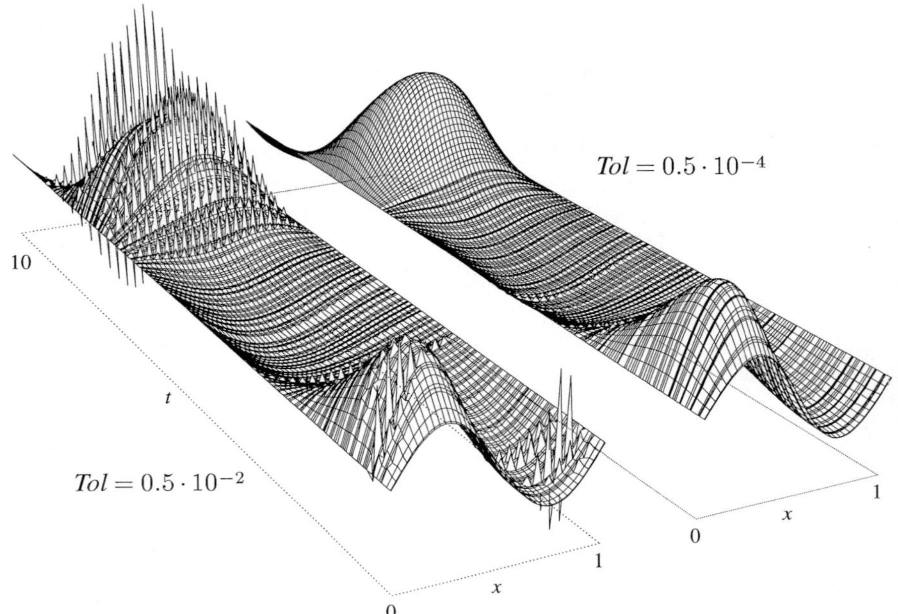

Fig. 1.7. Solution $u(x,t)$ of (1.6') with $N = 40$ using ODEX

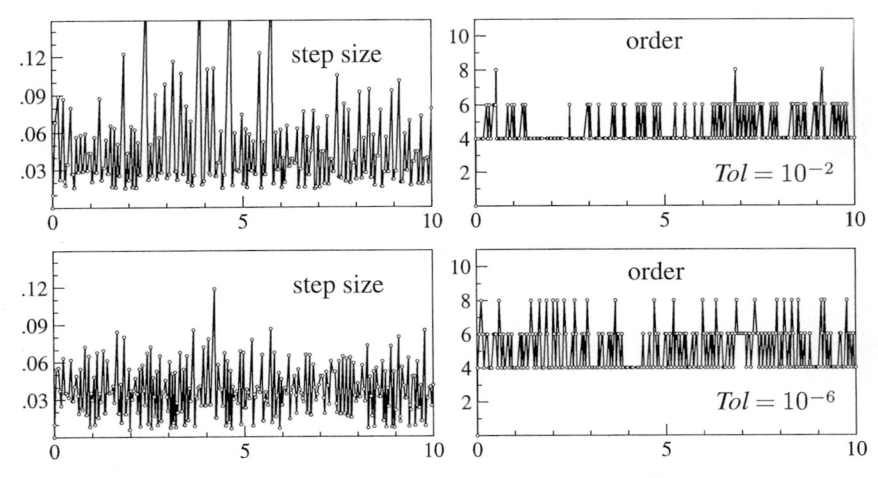

Fig. 1.8. Step size and order of ODEX at (1.6') with $N = 40$

This time we try the extrapolation code ODEX (see Volume I) and integrate over $0 \le t \le 10$ with $Atol = Rtol = Tol$. The number of necessary steps increases curiously with N, as is shown in Table 1.1. Again, for N large, the computing time is nearly independent of the desired tolerance, the computed solutions, however, differ considerably (see Fig. 1.7). Even the smooth 10^{-4}-solution shows curious stripes which are evidently unconnected with the behaviour of the solution. Fig. 1.8 shows the extremely ragged step size and order changes which take place in this example.

We again have all the characteristics of a "stiff" problem, and the use of an implicit method promises better results. However, when applying such a method, one must carefully take advantage of the banded or sparse structure of the Jacobian matrix. Otherwise the numerical work involved in the linear algebra would increase with N^3, precisely as the work for the explicit method (N^2 for the number of steps and N for the work per step).

A "Stiff" Beam

> Although it is common to talk about "stiff differential equations," an equation per se is not stiff, a particular initial value problem for that equation may be stiff, in some regions, but the sizes of these regions depend on the initial values and the error tolerance.
> (C.W. Gear 1982)

Let us conclude our series of examples by a problem from mechanics: the motion of an elastic beam. We suppose the beam inextensible of length 1 and thin. So we neglect shearing forces and rotatory inertia. We further want to allow it arbitrarily large movements. Thus, the most natural coordinate system to use is the angle θ as a function of arc length s and time t. We further suppose the beam clamped at $s=0$ and a force $\vec{F}=(F_x,F_y)$ acting at the free end $s=1$. The beam is then described by the equations

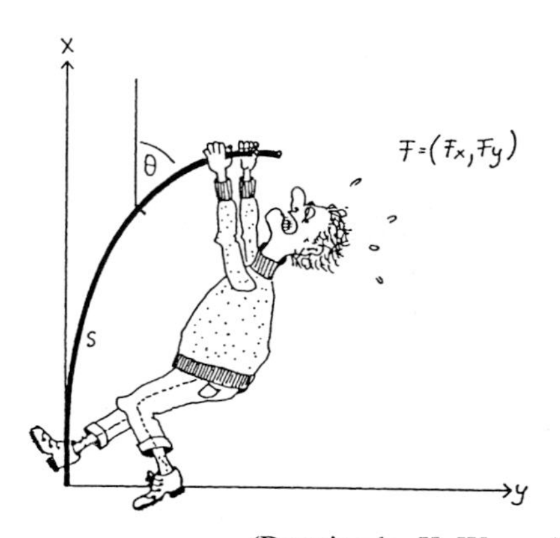

(Drawing by K. Wanner)

$$x(s,t)=\int_0^s \cos\theta(\sigma,t)d\sigma, \qquad y(s,t)=\int_0^s \sin\theta(\sigma,t)d\sigma. \tag{1.7}$$

In order to obtain the equations of motion for this problem, we apply Lagrange theory (Lagrange 1788). This requires that we form $L=T-U$ where T is the kinetic and U the potential energy. For the first of these we have simply

$$T=\frac{1}{2}\int_0^1 \Big((\dot{x}(s,t))^2+(\dot{y}(s,t))^2\Big)ds. \tag{1.8}$$

The potential energy is made up of energy from bending (depending on the curvature) and from exterior forces as follows:

$$U = \frac{1}{2}\int_0^1 (\theta'(s,t))^2\,ds - F_x(t)x(1,t) - F_y(t)y(1,t). \tag{1.9}$$

Here dots and primes denote derivatives with respect to t and s respectively. The equations of motion are now obtained by a "trivial" calculation (we are grateful to our colleague J. Descloux for having shown us how this must be done!) using the Hamilton principle which leads to (see Exercise 2)

$$\begin{aligned}\int_0^1 & G(s,\sigma)\cos\big(\theta(s,t)-\theta(\sigma,t)\big)\ddot{\theta}(\sigma,t)d\sigma \\ &= \theta''(s,t) + \cos\theta(s,t)F_y(t) - \sin\theta(s,t)F_x(t) \\ &\quad - \int_0^1 G(s,\sigma)\sin\big(\theta(s,t)-\theta(\sigma,t)\big)\big(\dot{\theta}(\sigma,t)\big)^2 d\sigma, \qquad 0\le s\le 1\end{aligned} \tag{1.10}$$

$$\theta(0,t)=0, \qquad \theta'(1,t)=0 \tag{1.11}$$

where

$$G(s,\sigma) = 1-\max(s,\sigma) \tag{1.12}$$

is Green's function for the problem $-w''(s)=g(s)$, $w'(0)=w(1)=0$. If we discretize the integrals with the help of the midpoint rule

$$\int_0^1 f(\theta(\sigma,t))d\sigma = \frac{1}{n}\sum_{k=1}^n f(\theta_k), \qquad \theta_k = \theta\Big(\big(k-\frac{1}{2}\big)\frac{1}{n},t\Big), \qquad k=1,\ldots,n \tag{1.13}$$

Equations (1.10) become

$$\begin{aligned}\sum_{k=1}^n a_{lk}\ddot{\theta}_k &= n^4\Big(\theta_{l-1}-2\theta_l+\theta_{l+1}\Big) + n^2\Big(\cos\theta_l\,F_y - \sin\theta_l\,F_x\Big) \\ &\quad - \sum_{k=1}^n g_{lk}\sin(\theta_l-\theta_k)\dot{\theta}_k^2, \qquad l=1,\ldots,n\end{aligned} \tag{1.10'}$$

$$\theta_0 = -\theta_1, \qquad \theta_{n+1}=\theta_n \tag{1.11'}$$

where

$$a_{lk} = g_{lk}\cos(\theta_l-\theta_k), \qquad g_{lk} = n+\frac{1}{2}-\max(l,k). \tag{1.14}$$

> Integration without preparation is frustration.
> (Reverend Leon Sullivan)

Numerical integration of (1.10') seems quite tedious, since the acceleration $\ddot{\theta}$ is only given implicitly. The computation of $\ddot{\theta}_k$ requires the solution of a linear system $A\ddot{\theta} = v$. Due to the special structure of A, this can be done efficiently, since with $B = (b_{lk})$, $b_{lk} = g_{lk} \sin(\theta_l - \theta_k)$, we have

$$A + iB = \operatorname{diag}\Big(e^{i\theta_1}, \ldots, e^{i\theta_n}\Big) G \operatorname{diag}\Big(e^{-i\theta_1}, \ldots, e^{-i\theta_n}\Big). \tag{1.15}$$

The matrix $G = (g_{lk})$ has the beautiful inverse

$$G^{-1} = \begin{pmatrix} 1 & -1 & & & \\ -1 & 2 & -1 & & \\ & -1 & \ddots & \ddots & \\ & & \ddots & 2 & -1 \\ & & & -1 & 3 \end{pmatrix}, \tag{1.16}$$

a positive definite tridiagonal matrix (a natural coincidence: G^{-1} represents the second order difference operator, and G comes from the Green function for a second order integration problem). Now

$$(A + iB)^{-1} = C + iD = \operatorname{diag}\Big(e^{i\theta_1}, \ldots, e^{i\theta_n}\Big) G^{-1} \operatorname{diag}\Big(e^{-i\theta_1}, \ldots, e^{-i\theta_n}\Big)$$

and

$$AC - BD = I, \qquad AD + BC = 0 \tag{1.17}$$

lead to $A^{-1} = C + DC^{-1}D$. We can also simplify the term $-\sum g_{lk} \sin(\theta_l - \theta_k)\dot{\theta}_k^2$, which in vector notation is $-B\dot{\theta}^2$, with the formula $A^{-1}B = -DC^{-1}$ (from (1.17)). The accelerations $\ddot{\theta}_k$ are now obtained from (1.10') as follows.

a) Let $v_l = n^4(\theta_{l-1} - 2\theta_l + \theta_{l+1}) + n^2(\cos\theta_l F_y - \sin\theta_l F_x)$,

b) Compute $w = Dv + \dot{\theta}^2$ (D is bidiagonal);

c) Solve the tridiagonal system $Cu = w$,

d) Compute $\ddot{\theta} = Cv + Du$.

Thus the evaluation of (1.10') reduces to $\mathcal{O}(n)$ operations (instead of $\mathcal{O}(n^3)$). We choose the initial conditions

$$\theta(s,0) = 0, \qquad \dot{\theta}(s,0) = 0 \tag{1.18}$$

and apply the exterior forces

$$F_x = -\varphi(t), \quad F_y = \varphi(t), \qquad \varphi(t) = \begin{cases} 1.5 \cdot \sin^2 t & 0 \le t \le \pi \\ 0 & \pi \le t\,. \end{cases} \tag{1.19}$$

The resulting system of ODE's is then integrated for $0 \le t \le 5$ by the code DOP853 of Volume I, although strictly speaking, the code is of too high an order for such a problem. The results are summarized in Table 1.2.

We observe the same phenomenon as before, the number of necessary steps increases like $\mathcal{O}(n^2)$ (the numerical work like $\mathcal{O}(n^3)$), and is more or less independent of the chosen tolerance. The numerical solution for $n = 40$ is displayed in Fig. 1.9. Only each 20th of the nearly 9000 steps is drawn (otherwise the picture would just be completely black). The computed solution looks perfectly smooth and there is no apparent reason for the need of *so* many steps. In fact due to lack

Table 1.2. Results for the beam (1.10') with DOP853

n	*Tol*	accepted steps	rejected steps	function calls
5	10^{-7}	142	35	2091
10	10^{-7}	383	26	4884
20	10^{-7}	1397	273	19769
40	10^{-7}	6913	1347	97775
20	10^{-3}	1486	450	22784
20	10^{-5}	1967	266	26532
20	10^{-7}	1397	273	19769

of stability, the numerical method produces small vibrations which are invisible for $Tol = 10^{-7}$, and which force the integrator to such small step sizes. If we relax the high precision requirement, these oscillations become visible (Fig. 1.10).

High Oscillations

Let us now choose slightly perturbed initial values in the beam equation (1.10'). Instead of (1.18) we put

$$\theta_1 = \ldots = \theta_{n-1} = 0,\ \theta_n = 0.4, \quad \dot\theta_1 = \ldots = \dot\theta_n = 0. \tag{1.18'}$$

This time, the *correct* solution for $n = 10$ of (1.10') computed with $Tol = 10^{-6}$ and more than 2000 steps is displayed in Fig. 1.11.

The solution is highly oscillatory, no damping wipes out the fast vibrations since the system is conservative. Hence also an implicit method, if required to follow all these oscillations, would need the same number of steps and there would of course be no advantage in using it. So we see that the decision whether a problem should be regarded as stiff or nonstiff ("... that is the question"), may also depend on the chosen initial conditions. On the other hand, we shall see in Sect. IV.2 that whenever these high oscillations are not desired, implicit methods are a marvellous instrument for wiping them out.

Exercises

1. (Curtiss & Hirschfelder 1952). "It is interesting to notice that this method of integration (the implicit Euler) may be used in either direction". Integrate equation (1.1) *backward* with step size -0.5 and initial value $y(1.5) = 0$ in three steps. Observe that the numerical solution remains stable and follows the smooth solution.

2. Derive the equations of motion (1.10) for the elastic beam from (1.8) and (1.9).

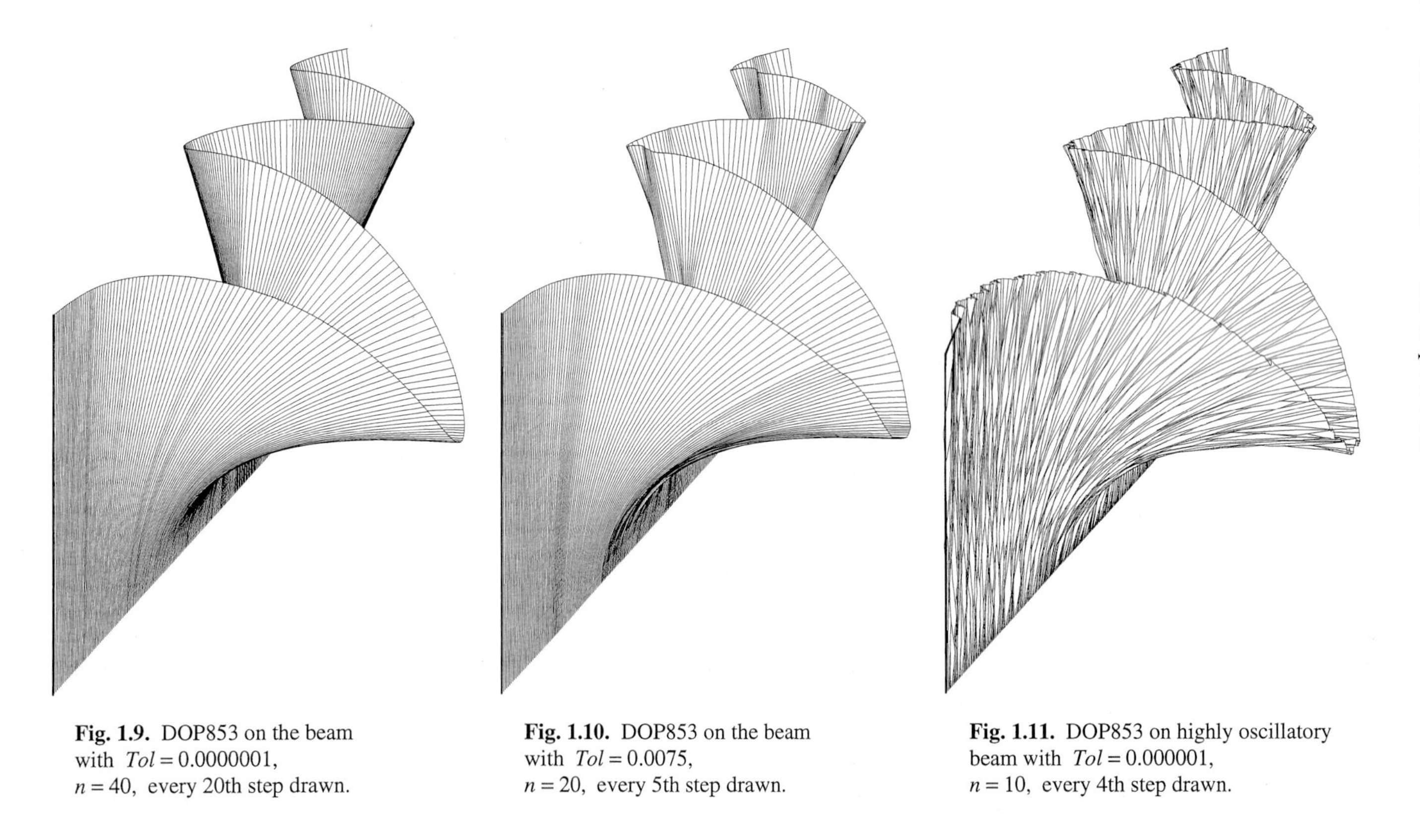

Fig. 1.9. DOP853 on the beam with $Tol = 0.0000001$, $n = 40$, every 20th step drawn.

Fig. 1.10. DOP853 on the beam with $Tol = 0.0075$, $n = 20$, every 5th step drawn.

Fig. 1.11. DOP853 on highly oscillatory beam with $Tol = 0.000001$, $n = 10$, every 4th step drawn.

Hint. If you want to avoid differentiation in function spaces, then discretize the beam as, say,

$$x_j = \Delta s \sum_{k=1}^{j} \cos\theta_k, \quad y_j = \Delta s \sum_{k=1}^{j} \sin\theta_k, \quad j = 1,\ldots,n, \quad \Delta s = \frac{1}{n} \tag{1.20}$$

$$T = \frac{\Delta s}{2} \sum_{j=1}^{n} \Big(\dot x_j^2 + \dot y_j^2\Big) = \frac{\Delta s}{2} \sum_{j=1}^{n} \dot z_j \dot{\bar z}_j, \qquad z_j = \Delta s \sum_{k=1}^{j} e^{i\theta_k}$$

$$U = \frac{\Delta s}{2} \sum_{j=1}^{n} \Big(\frac{\theta_j - \theta_{j-1}}{\Delta s}\Big)^2 - F_x \Delta s \sum_{k=1}^{n} \cos\theta_k - F_y \Delta s \sum_{k=1}^{n} \sin\theta_k,$$

form the Lagrange function $L = T - U$ and apply n-dimensional Lagrange theory (Lagrange (1788), Vol. II, Sect. VII and VIII, a very clear derivation can be found in Sommerfeld (1942), Vol. I, §36)

$$\frac{d}{dt}\Big(\frac{\partial L}{\partial \dot\theta_k}\Big) - \frac{\partial L}{\partial \theta_k} = 0$$

or

$$\sum_{l=1}^{n} L_{\dot\theta_k \dot\theta_l} \ddot\theta_l = L_{\theta_k} - L_{\dot\theta_k t} - \sum_{l=1}^{n} L_{\dot\theta_k \theta_l} \dot\theta_l. \tag{1.21}$$

3. Apply an explicit code to the Oregonator (Chapter I, Equation (16.15))

$$\begin{aligned} y_1' &= 77.27\Big(y_2 + y_1(1 - 8.375\times 10^{-6} y_1 - y_2)\Big) \\ y_2' &= \frac{1}{77.27}(y_3 - (1+y_1)y_2) \\ y_3' &= 0.161(y_1 - y_3) \end{aligned} \tag{1.22}$$

and study its performance.

4. a) Compute the equations of motion of the *hanging rope* (Fig. 1.12) of length 1 by using the results of Exercise 2. The potential energy has to be replaced by

$$U = -\int_0^1 x(s,t)ds.$$

Result.

$$\begin{aligned} &\int_0^1 G(s,\sigma)\cos\big(\theta(s,t) - \theta(\sigma,t)\big)\ddot\theta(\sigma,t)d\sigma \\ &= -\int_0^1 G(s,\sigma)\sin\big(\theta(s,t) - \theta(\sigma,t)\big)\,(\dot\theta(\sigma,t))^2 d\sigma - (1-s)\sin\theta(s,t) \end{aligned} \tag{1.23}$$

for $0 \le s \le 1$, or, when discretized

$$\sum_{k=1}^{n} a_{lk}\ddot{\theta}_k = -\sum_{k=1}^{n} b_{lk}\dot{\theta}_k^2 - n\Big(n+\frac{1}{2}-l\Big)\sin\theta_l. \tag{1.23'}$$

b) Do numerical computations with DOPRI5 or DOP853. Choose as initial position a hanging rope in equilibrium which is then released at one end.

Hint. The hanging rope in equilibrium satisfies, in the usual coordinates,

$$\int_{x_0}^{x_1} y\sqrt{1+(y')^2}\,dx = \min \qquad \text{with} \qquad \int_{x_0}^{x_1} \sqrt{1+(y')^2}\,dx = 1,$$

which, using a Lagrange multiplier, becomes

$$\int_{x_0}^{x_1} (y-\lambda)\sqrt{1+(y')^2}\,dx = \text{ stat }.$$

Applying (2.6) of Sect. I.2 yields $y-\lambda = K\sqrt{1+(y')^2}$ with solution

$$y = \lambda + K\cosh\Big(\frac{x+\alpha}{K}\Big).$$

Suitable choices of the parameters and change of coordinates ($K=1/2$, $\lambda = -K\cosh(\alpha/K)$, $x\to y$, $y\to -x$) then lead to

$$\theta(s,0) = \pi/2 - \arctan(\sinh(2\alpha)-2s). \tag{1.24}$$

Result. DOP853 has computed the solution for $0\le t\le 5$, $n=60$ and $Tol = 10^{-5}$, $\alpha = 0.6$, in 203 steps (Fig. 1.12). The number of steps increases here like $\mathcal{O}(n)$, so the rope is — evidently — less stiff than the beam.

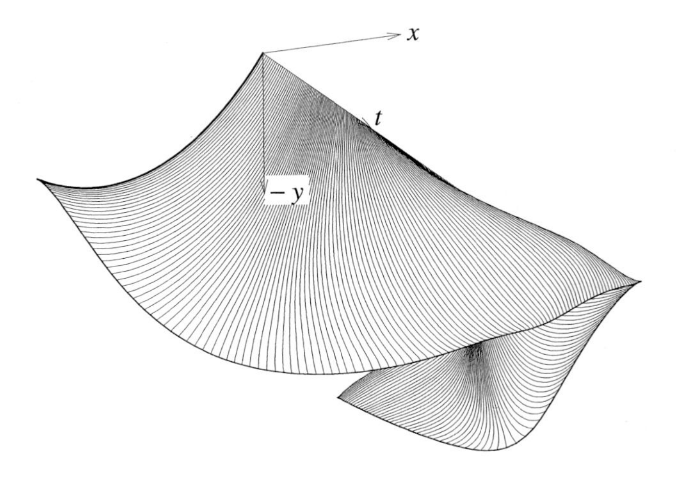

Fig. 1.12. Movement of hanging rope, every step drawn

IV.2 Stability Analysis for Explicit RK Methods

> ... werden wir bei dem Anfangswertproblem hyperbolischer Gleichungen erkennen, dass die Konvergenz allgemein nur dann vorhanden ist, wenn die Verhältnisse der Gittermaschen in verschiedenen Richtungen gewissen Ungleichungen genügen.
>
> (Courant, Friedrichs & Lewy 1928)

The first analysis of instability phenomena and step size restrictions for hyperbolic equations was made in the famous paper of Courant, Friedrichs & Lewy (1928). Later, many authors undertook a stability analysis, very often independently, in order to explain the phenomena encountered in the foregoing section. An early and beautiful paper on this subject is Guillou & Lago (1961).

Stability Analysis for Euler's Method

Let $\varphi(x)$ be a smooth solution of $y' = f(x,y)$. We linearize f in its neighbourhood as follows

$$y'(x) = f\big(x,\varphi(x)\big) + \frac{\partial f}{\partial y}\big(x,\varphi(x)\big)\big(y(x)-\varphi(x)\big) + \ldots \tag{2.1}$$

and introduce $y(x)-\varphi(x) = \overline{y}(x)$ to obtain

$$\overline{y}'(x) = \frac{\partial f}{\partial y}\big(x,\varphi(x)\big)\cdot\overline{y}(x) + \ldots = J(x)\overline{y}(x) + \ldots\,. \tag{2.2}$$

As a first approximation we consider the Jacobian $J(x)$ as constant and neglect the error terms. Omitting the bars we arrive at

$$y' = Jy. \tag{2.2'}$$

If we now apply, say, Euler's method to (2.2'), we obtain

$$y_{m+1} = R(hJ)y_m \tag{2.3}$$

with

$$R(z) = 1 + z. \tag{2.4}$$

The behaviour of (2.3) is studied by transforming J to Jordan canonical form (see Sect. I.12). We suppose that J is diagonalizable with eigenvectors $v_1,\ldots,v_n$ and write y_0 in this basis as

$$y_0 = \sum_{i=1}^{n}\alpha_i v_i\,. \tag{2.5}$$

Inserting this into (2.3) we obtain

$$y_m = \sum_{i=1}^{n} \bigl(R(h\lambda_i)\bigr)^m \alpha_i \cdot v_i \,, \tag{2.6}$$

where the λ_i are the corresponding eigenvalues (see also Exercises 1 and 2). Clearly y_m remains bounded for $m \to \infty$, if for all eigenvalues the complex number $z = h\lambda_i$ lies in the set

$$S = \Bigl\{ z \in \mathbb{C};\, |R(z)| \le 1 \Bigr\} = \Bigl\{ z \in \mathbb{C};\, |z - (-1)| \le 1 \Bigr\}$$

which is the circle of radius 1 and centre -1. This leads to the explanation of the results encountered in Example (1.1). There we have $\lambda = -50$, and $h\lambda \in S$ means that $0 \le h \le 2/50$, in perfect accordance with the numerical observations.

Explicit Runge-Kutta Methods

An explicit Runge-Kutta method (Sect. II.2, Formula (2.3)) applied to (2.2') gives

$$\begin{aligned} g_i &= y_m + hJ \sum_{j=1}^{i-1} a_{ij} g_j \\ y_{m+1} &= y_m + hJ \sum_{j=1}^{s} b_j g_j \,. \end{aligned} \tag{2.7}$$

Inserting g_j repeatedly from the first line, this becomes

$$y_{m+1} = R(hJ) y_m$$

where

$$R(z) = 1 + z \sum_j b_j + z^2 \sum_{j,k} b_j a_{jk} + z^3 \sum_{j,k,l} b_j a_{jk} a_{kl} + \ldots \tag{2.8}$$

is a polynomial of degree $\le s$.

Definition 2.1. The function $R(z)$ is called the *stability function* of the method. It can be interpreted as the numerical solution after one step for

$$y' = \lambda y \,, \qquad y_0 = 1 \,, \qquad z = h\lambda \,, \tag{2.9}$$

the famous *Dahlquist test equation.* The set

$$S = \Bigl\{ z \in \mathbb{C} \,;\; |R(z)| \le 1 \Bigr\} \tag{2.10}$$

is called the *stability domain* of the method.

Theorem 2.2. *If the Runge-Kutta method is of order p, then*

$$R(z) = 1 + z + \frac{z^2}{2!} + \ldots + \frac{z^p}{p!} + \mathcal{O}(z^{p+1}) \,.$$

Proof. The exact solution of (2.9) is e^z and therefore the numerical solution $y_1 = R(z)$ must satisfy

$$e^z - R(z) = \mathcal{O}(h^{p+1}) = \mathcal{O}(z^{p+1}) . \tag{2.11}$$

Another argument is that the expressions in (2.8) appear in the order conditions for the "tall" trees $\tau, t_{21}, t_{32}, t_{44}, t_{59}, \ldots$ (see Table 2.2 of Sect. II.2, p. 148). They are therefore equal to $1/q!$ for $q \le p$. □

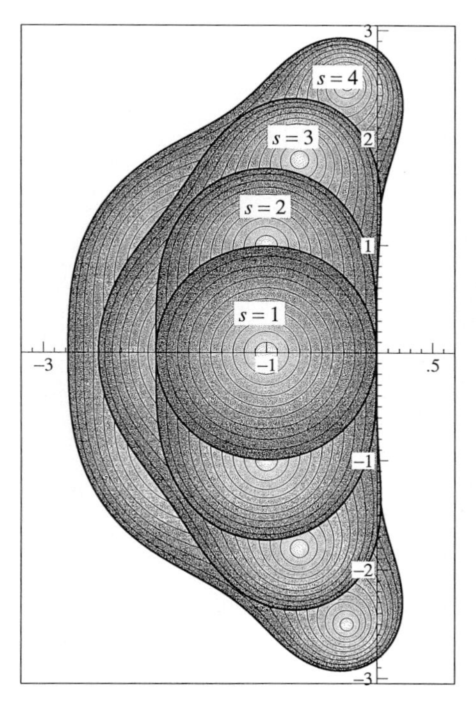

Fig. 2.1. Stability domains for explicit Runge-Kutta methods of order $p = s$

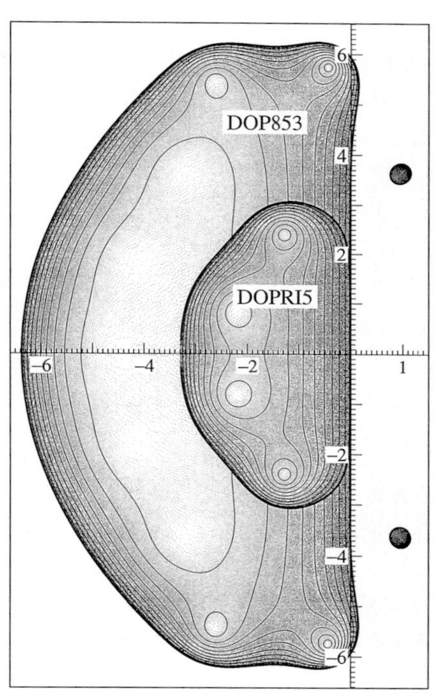

Fig. 2.2. Stability domains for DOPRI methods

As a consequence, all explicit Runge-Kutta methods with $p = s$ possess the stability function

$$R(z) = 1 + z + \frac{z^2}{2!} + \ldots + \frac{z^s}{s!}. \tag{2.12}$$

The corresponding stability domains are represented in Fig. 2.1.

The method of Dormand & Prince DOPRI5 (Sect. II.5, Table 5.2) is of order 5 with $s = 6$ (the 7th stage is for error estimation only). Here $R(z)$ is obtained by direct computation. The result is

$$R(z) = 1 + z + \frac{z^2}{2} + \frac{z^3}{6} + \frac{z^4}{24} + \frac{z^5}{120} + \frac{z^6}{600}. \tag{2.13}$$

For DOP853 (Sect. II.5, Fig. 5.3), $R(z)$ becomes

$$R(z) = \sum_{j=0}^{8} \frac{z^j}{j!} + 2.6916922001691 \cdot 10^{-6} z^9 + 2.3413451082098 \cdot 10^{-7} z^{10} + 1.4947364854592 \cdot 10^{-8} z^{11} + 3.6133245781282 \cdot 10^{-10} z^{12}. \tag{2.14}$$

The stability domains for these two methods are given in Fig. 2.2.

Extrapolation Methods

The GBS-algorithm (see Sect. II.9, Formulas (9.10), (9.13)) applied to $y' = \lambda y$, $y(0) = 1$ leads with $z = H\lambda$ to

$$\begin{aligned} y_0 &= 1, \qquad y_1 = 1 + \frac{z}{n_j} \\ y_{i+1} &= y_{i-1} + 2\,\frac{z}{n_j}\,y_i \qquad\qquad i = 1, 2, \ldots, n_j \\ T_{j1} &= \frac{1}{4}(y_{n_j-1} + 2y_{n_j} + y_{n_j+1}) \\ T_{j,k+1} &= T_{j,k} + \frac{T_{j,k} - T_{j-1,k}}{(n_j/n_{j-k})^2 - 1}\,. \end{aligned} \tag{2.15}$$

The stability domains for the diagonal terms T_{22}, T_{33}, T_{44}, and T_{55} for the harmonic sequence

$$\{n_j\} = \{2, 4, 6, 8, 10, \ldots\}$$

(the one which is used in ODEX) are displayed in Fig. 2.3. We have also added those for the methods *without* the smoothing step (II.9.13c), which shows some difference for negative real eigenvalues.

Analysis of the Examples of IV.1

The Jacobian for the Robertson reaction (1.3) is given by

$$\begin{pmatrix} -0.04 & 10^4 y_3 & 10^4 y_2 \\ 0.04 & -10^4 y_3 - 6 \cdot 10^7 y_2 & -10^4 y_2 \\ 0 & 6 \cdot 10^7 y_2 & 0 \end{pmatrix}$$

which in the neighbourhood of the equilibrium $y_1 = 1$, $y_2 = 0.0000365$, $y_3 = 0$ is

$$\begin{pmatrix} -0.04 & 0 & 0.365 \\ 0.04 & -2190 & -0.365 \\ 0 & 2190 & 0 \end{pmatrix}$$

with eigenvalues

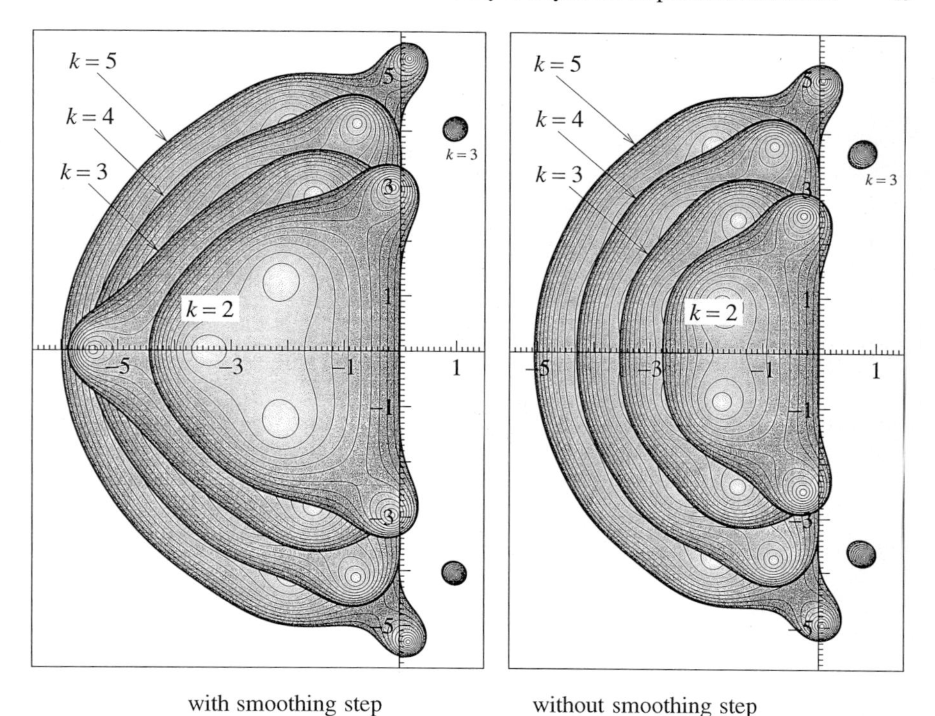

Fig. 2.3. Stability domains for GBS extrapolation methods

$$\lambda_1 = 0, \quad \lambda_2 = -0.405, \quad \lambda_3 = -2189.6.$$

The third one produces stiffness. For stability we need (see the stability domain of DOPRI5 in Fig. 2.2) $-2190h \geq -3.3$, hence $h \leq 0.0015$. This again confirms the numerical observations.

The Jacobian of example (1.6') (Brusselator reaction with diffusion) is a large $2N \times 2N$ matrix. It is composed of reaction terms and diffusion terms:

$$J = \begin{pmatrix} \operatorname{diag}(2u_i v_i - 4) & \operatorname{diag}(u_i^2) \\ \operatorname{diag}(3 - 2u_i v_i) & \operatorname{diag}(-u_i^2) \end{pmatrix} + \frac{\alpha}{(\Delta x)^2} \begin{pmatrix} K & 0 \\ 0 & K \end{pmatrix} \tag{2.16}$$

where

$$K = \begin{pmatrix} -2 & 1 & & & \\ 1 & -2 & 1 & & \\ & 1 & \ddots & \ddots & \\ & & \ddots & -2 & 1 \\ & & & 1 & -2 \end{pmatrix}. \tag{2.17}$$

The eigenvalues of K are known (see Sect. I.6, Formula (6.7b)), namely

$$\mu_k = -4\Big(\sin \frac{\pi k}{2N+2}\Big)^2, \tag{2.18}$$

and therefore the double eigenvalues of the right hand matrix in (2.16) are

$$-\frac{4\alpha}{(\Delta x)^2}\Big(\sin\frac{\pi k}{2N+2}\Big)^2 = -4\alpha(N+1)^2\Big(\sin\frac{\pi k}{2N+2}\Big)^2, \tag{2.19}$$

and are located between $-4\alpha(N+1)^2$ and 0. Since this matrix is symmetric, its eigenvalues are well conditioned and the first matrix on the right side of (2.16) with much smaller coefficients can be regarded as a small perturbation. Therefore the eigenvalues of J in (2.16) will remain close to those of the unperturbed matrix and lie in a stripe neighbouring the interval $[-4\alpha(N+1)^2, 0]$. Numerical computations for $N=40$ show for example that the largest negative eigenvalue of J varies between -133.3 and -134.9, while the unperturbed value is $-4\cdot 41^2\cdot\sin^2(40\pi/82)/50 = -134.28$. Since most stability domains for ODEX end close to -5.5 on the real axis (Fig. 2.3), this leads for $N=40$ to $h\le 0.04$ and the number of steps must be ≥ 250 (compare with Table 1.1).

In order to explain the behaviour of the beam equation, we linearize it in the neighbourhood of the solution $\theta_k = \dot\theta_k = 0$, $F_x = F_y = 0$. There (1.10') becomes

$$G\ddot\theta = n^4\begin{pmatrix} -3 & 1 & & & \\ 1 & -2 & 1 & & \\ & 1 & \ddots & \ddots & \\ & & \ddots & -2 & 1 \\ & & & 1 & -1 \end{pmatrix}\theta, \tag{2.20}$$

since for $\theta = 0$ we have $A = G$ and $B = 0$. We now insert G^{-1} from (1.16) and observe that the matrices involved are, with the exception of two elements, equal to $\pm K$ of (2.17). We therefore approximate (2.20) by

$$\ddot\theta = -n^4K^2\theta. \tag{2.21}$$

This second order equation was integrated in IV.1 as a first order system

$$\begin{pmatrix}\theta\\ \dot\theta\end{pmatrix}^{\cdot} = \begin{pmatrix}0 & I\\ -n^4K^2 & 0\end{pmatrix}\begin{pmatrix}\theta\\ \dot\theta\end{pmatrix} = E\begin{pmatrix}\theta\\ \dot\theta\end{pmatrix}. \tag{2.22}$$

By solving

$$\begin{pmatrix}0 & I\\ -n^4K^2 & 0\end{pmatrix}\begin{pmatrix}y\\ z\end{pmatrix} = \lambda\begin{pmatrix}y\\ z\end{pmatrix}, \tag{2.23}$$

we find that λ is an eigenvalue of E iff λ^2 is an eigenvalue of $-n^4K^2$. Thus Formula (2.18) shows that the eigenvalues of E are situated on the imaginary axis between $-4n^2i$ and $+4n^2i$. We see from Fig. 2.2 that the stability domain of DOP853 covers the imaginary axis between approximately $-6i$ and $+6i$. Hence for stability we need $h\le 1.5/n^2$ and the number of steps for the interval $0\le t\le 5$ must be larger than $\approx 10n^2/3$. This, again, was observed in the numerical calculations (Table 1.2).

Automatic Stiffness Detection

Neither is perfect, but even an imperfect test can be quite useful, as we can show from experience ... (L.F. Shampine 1977)

Explicit codes applied to stiff problems are apparently not very efficient and the remaining part of the book will be devoted to the construction of more stable algorithms. In order to avoid that an explicit code waste too much effort when encountering stiffness (and to enable a switch to a more suitable method), it is important that the code be equipped with a cheap means of detecting stiffness. The analysis of the preceding subsection demonstrates that, whenever a nonstiff code encounters stiffness, the product of the step size with the dominant eigenvalue of the Jacobian lies near the border of the stability domain. We shall show two manners of exploiting this observation to detect stiffness.

Firstly, we adapt the ideas of Shampine & Hiebert (1977) to the Dormand & Prince method of order 5(4), given in Table II.5.2. The method possesses an error estimator $err_1 = y_1 - \widehat{y}_1$ which, in the nonstiff situation, is $\mathcal{O}(h^5)$. However in the stiff case, when the method is working near the border of the stability domain S, the distance $d_1 = y_1 - y(x_0 + h)$ to the smooth solution is approximately $d_1 \approx R(hJ)d_0$, where J denotes the Jacobian of the system, $R(z)$ is the stability function of the method, and $d_0 = y_0 - y(x_0)$. Here we have neglected the local error for an initial value on the smooth solution $y(x)$. A similar formula, with R replaced by $\widehat{R}$, holds for the embedded method. The error estimator satisfies $err_1 \approx E(hJ)d_0$ with $E(z) = R(z) - \widehat{R}(z)$. The idea is now to search for a second error estimator $\widetilde{err}_1$ (with $\widetilde{err}_1 \approx \widetilde{E}(hJ)d_0$) such that

i) $|\widetilde{E}(z)| \le \theta |E(z)|$ on $\partial S \cap \mathbb{C}^-$ with a small $\theta < 1$;

ii) $\widetilde{err}_1 = \mathcal{O}(h^2)$ for $h \to 0$.

Condition (i) implies that $\|\widetilde{err}_1\| < \|err_1\|$ when $h\lambda$ is near ∂S (the problem is possibly stiff), and condition (ii) will lead to $\|\widetilde{err}_1\| \gg \|err_1\|$ for step sizes which are determined by accuracy requirements (when the problem is not stiff). If $\|\widetilde{err}_1\| < \|err_1\|$ occurs several times in succession (say 15 times) then a stiff code might be more efficient.

For the construction of $\widetilde{err}_1$ we put $\widetilde{err}_1 = h(d_1 k_1 + d_2 k_2 + \ldots + d_s k_s)$, where the $k_i = f(x_0 + c_i h, g_i)$ are the available function values of the method. The coefficients d_i are determined in such a way that

$$\sum_{i=1}^{s} d_i = 0, \qquad \sum_{i=1}^{s} d_i c_i = 0.02 \tag{2.24}$$

(so that (ii) holds) and that θ in (i) is minimized. A computer search gave values which have been rounded to

$$\begin{aligned} d_1 &= -0.08536, \quad d_2 = 0.088, \quad d_3 = -0.0096, \\ d_4 &= 0.0052, \quad d_5 = 0.00576, \quad d_6 = -0.004. \end{aligned} \tag{2.25}$$

The factor 0.02 in (2.24) has been chosen such that θ in (i) is close to 0.3 on large parts of the border of S, but $|\widetilde{E}(z)/E(z)|$ soon becomes larger than 1 if z approaches the origin.

In Fig. 2.4 we present the contour lines $|\widetilde{E}(z)/E(z)| = Const$ ($Const = 4, 2, 1, 0.5, 0.3, 0.2, 0.14, 0.1$) together with the stability domain of the method. A numerical experiment is illustrated in Fig. 2.5. We applied the code DOPRI5 (see the Appendix to Volume I) to the van der Pol equation (1.5') with $\varepsilon = 0.003$. The upper picture shows the first component of the solution, the second picture displays the quotient $\|\widetilde{err}_1\|/\|err_1\|$ for the three tolerances $Tol = 10^{-3}, 10^{-5}, 10^{-7}$. The last picture is a plot of $h|\lambda|/3.3$ where h is the current step size and λ the dominant eigenvalue of the Jacobian and 3.3 is the approximate distance of ∂S to the origin.

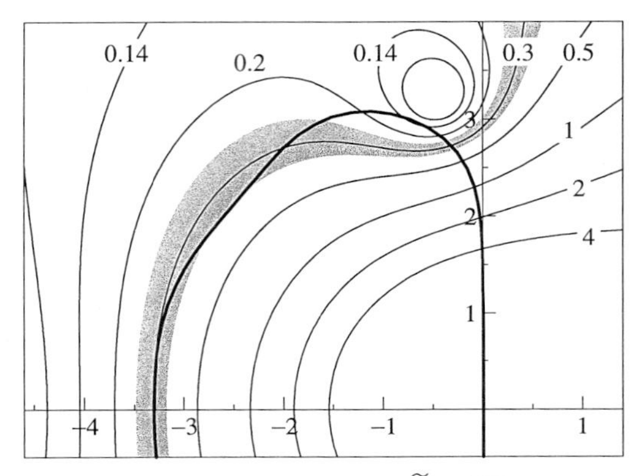

Fig. 2.4. Contour lines of $\widetilde{E}(z)/E(z)$

A second possibility for detecting stiffness is to estimate directly the dominant eigenvalue of the Jacobian of the problem. If v denotes an approximation to the corresponding eigenvector with $\|v\|$ sufficiently small then, by the mean value theorem,

$$|\lambda| \approx \frac{\|f(x, y+v) - f(x, y)\|}{\|v\|}$$

will be a good approximation to the leading eigenvalue. For the Dormand & Prince method (Table II.5.2) we have $c_6 = c_7 = 1$. Therefore, a natural choice is

$$\varrho = \frac{\|k_7 - k_6\|}{\|g_7 - g_6\|} \tag{2.26}$$

where $k_i = f(x_0 + c_i h, g_i)$ are the function values of the current step. Both values, $g_7 = y_1$ and g_6, approximate the exact solution $y(x_0 + h)$ and it can be shown by Taylor expansion that $g_7 - g_6 = \mathcal{O}(h^4)$. This difference is thus sufficiently small, in general. The same argument also shows that $g_7 - g_6 = \widetilde{E}(hJ)d_0$, where J is the Jacobian of the linearized differential equation and $\widetilde{E}(z)$ is a polynomial with subdegree 4. Hence, $g_7 - g_6$ is essentially the vector obtained by 4 iterations of the power method applied to the matrix hJ. It will be a good approximation to the

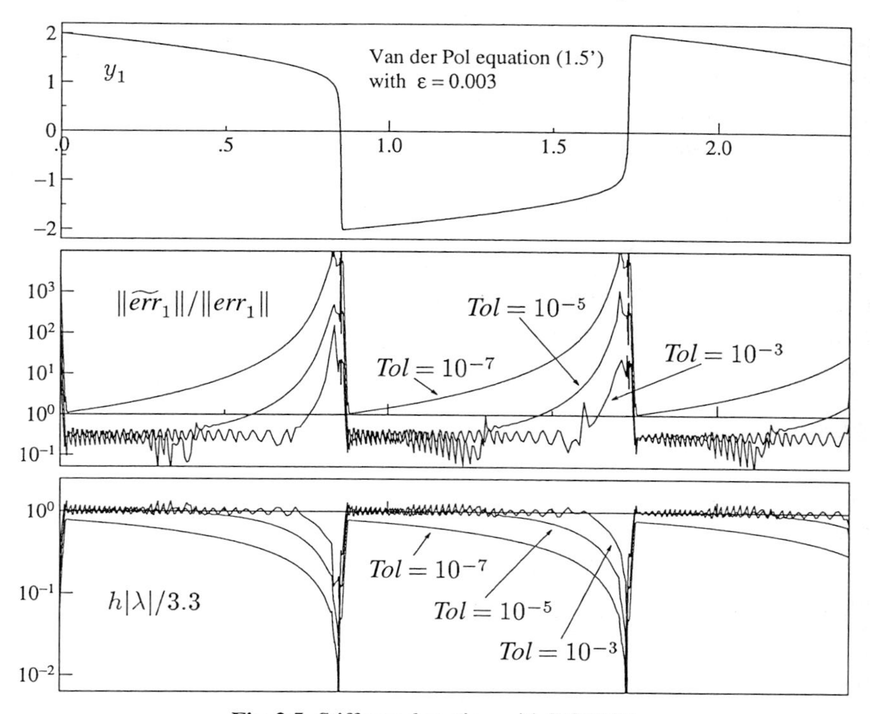

Fig. 2.5. Stiffness detection with DOPRI5

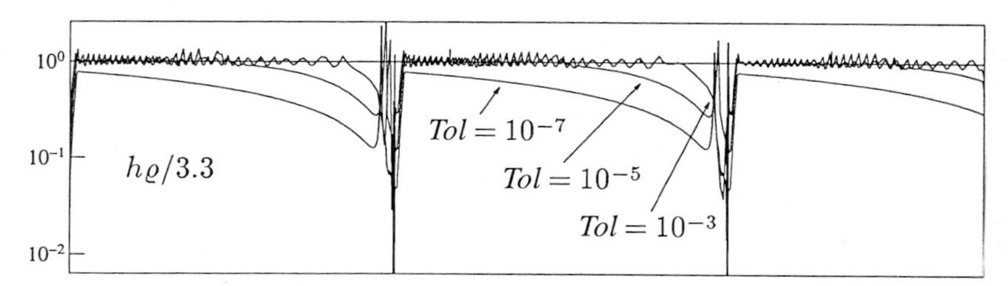

Fig. 2.6. Estimation of Lipschitz constant with DOPRI5

eigenvector corresponding to the leading eigenvalue. As in the above numerical experiment we applied the code DOPRI5 to the van der Pol equation (1.5') with $\varepsilon = 0.003$. Fig. 2.6 presents a plot of $h\varrho/3.3$ where h is the current step size and ϱ the estimate (2.26). This is in perfect agreement with the exact values $h|\lambda|/3.3$ (see third picture of Fig. 2.5).

Further numerical examples have shown that the estimate (2.26) also gives satisfactory approximations of $|\lambda|$ when the dominant eigenvalue λ is complex. However, if the argument of λ is needed too, one can extend the power method as proposed by Wilkinson (1965, page 579). This has been elaborated by Sottas (1984) and Robertson (1987).

The two techniques above allow us to detect the regions where the step size is restricted by stability. In order to decide whether a stiff integrator will be more efficient, one has to compare the expense of both methods. Studies on this question have been undertaken in Petzold (1983), Sottas (1984) and Butcher (1990).

Step-Control Stability

We now come to the explanation of another phenomenon encountered in Sect. IV.1, that of the ragged behaviour of the step size (e.g. Fig. 1.4 or 1.8), a research initiated by G. Hall (1985/86) and continued by G. Hall & D.J. Higham (1988). Do there exist methods or stiff equations for which the step sizes h_n behave smoothly and no frequent step rejections appear?

We make a numerical study on the equation

$$\begin{aligned} y_1' &= -2000\,(\cos x \cdot y_1 + \sin x \cdot y_2 + 1) \qquad & y_1(0) = 1 \\ y_2' &= -2000\,(-\sin x \cdot y_1 + \cos x \cdot y_2 + 1) \qquad & y_2(0) = 0 \end{aligned} \tag{2.27}$$

for $0 \le x \le 1.57$, whose eigenvalues move slowly on a large circle from -2000 to $\pm 2000i$. If we apply Fehlberg's method RKF5(4) (Table II.5.1) in local extrapolation mode (i.e., we continue the integration with the higher order result) and DOPRI5 to this equation (with Euclidean error norm without scaling), we obtain the step size behaviour presented in Fig. 2.7. There all rejected steps are crossed out (3 rejected steps for RKF5(4) and 104 for DOPRI5.

In order to explain this behaviour, we consider for $y' = \lambda y$ (of course!) the numerical process

$$\begin{aligned} y_{n+1} &= R(h_n\lambda) y_n \\ err_n &= E(h_n\lambda) y_n \\ h_{n+1} &= h_n \cdot \Big(\frac{Tol}{|err_n|}\Big)^{\alpha} \end{aligned} \tag{2.28}$$

(where err_n is the estimated error, $E(z) = \widehat{R}(z) - R(z)$, $\alpha = 1/(\widehat{p}+1)$ and $\widehat{p}$ is the order of $\widehat{R}$) as a dynamical system whose fixed points and stability we have to study. A possible safety factor ("*fac*" of formula (4.13) of Sect. II.4) can easily be incorporated into *Tol* and does not affect the theory. The analysis simplifies if we introduce logarithms

$$\eta_n = \log |y_n|, \qquad \chi_n = \log h_n \tag{2.29}$$

so that (2.28) becomes

$$\begin{aligned} \eta_{n+1} &= \log |R(e^{\chi_n}\lambda)| + \eta_n, \\ \chi_{n+1} &= \alpha\Big(\gamma - \log |E(e^{\chi_n}\lambda)| - \eta_n\Big) + \chi_n, \end{aligned} \tag{2.30}$$

where γ is a constant. This is now a map $\mathbb{R}^2 \to \mathbb{R}^2$. Its fixed point (η, χ) satisfies

$$|R(e^{\chi}\lambda)| = 1\,, \tag{2.31}$$

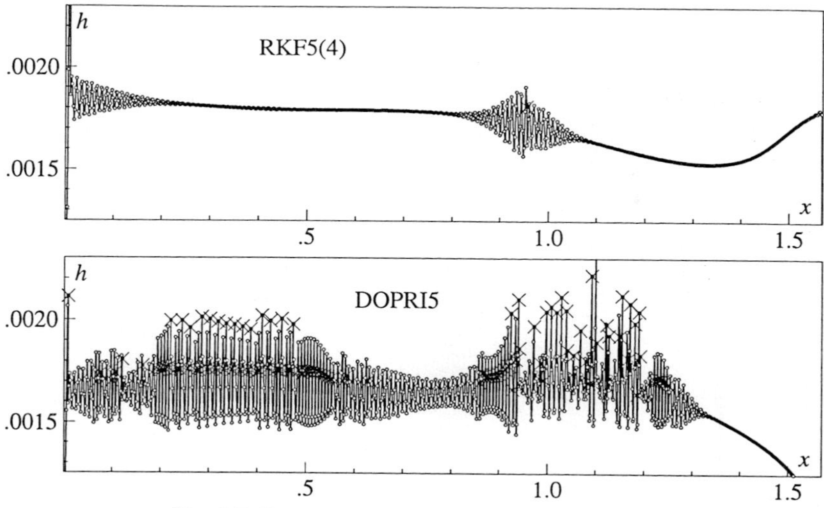

Fig. 2.7. Step sizes of RKF5(4) and DOPRI5 for (2.27)

which determines the step size e^{χ} so that the point $z = e^{\chi}\lambda$ must be on the border of the stability domain. Further

$$\eta = \gamma - \log |E(z)|$$

determines η. Now the Jacobian of the map (2.30) at this fixed point becomes

$$C = \frac{\partial(\eta_{n+1}, \chi_{n+1})}{\partial(\eta_n, \chi_n)} = \begin{pmatrix} 1 & u \\ -\alpha & 1-\alpha v \end{pmatrix} \qquad \begin{aligned} u &= \mathrm{Re}\left(\frac{R'(z)}{R(z)}\cdot z\right) \\ v &= \mathrm{Re}\left(\frac{E'(z)}{E(z)}\cdot z\right). \end{aligned} \tag{2.32}$$

Proposition 2.3. *The step-control mechanism is stable for $h\lambda = z$ on the boundary of the stability domain if and only if the spectral radius of C in (2.32) satisfies*

$$\varrho(C) < 1 .$$

We then call the method SC-stable at z. □

The matrix C is independent of the given differential equation and of the given tolerance. It is therefore a characteristic of the numerical method and the boundary of its stability domain. Let us study some methods of Sect. II.5.

a) RKF4(5) (Table 5.1), $\alpha = 1/5$:

$$R(z) = 1 + z + \frac{z^2}{2} + \frac{z^3}{6} + \frac{z^4}{24} + \frac{z^5}{104}, \quad E(z) = \frac{z^5}{780} - \frac{z^6}{2080} . \tag{2.33}$$

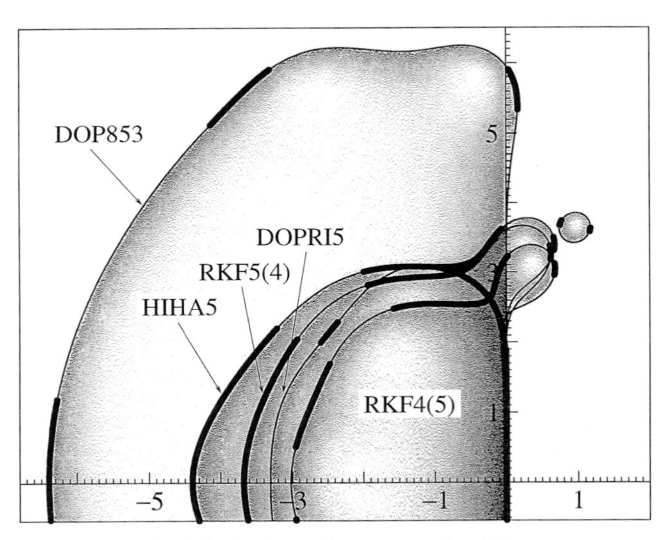

Fig. 2.8. Regions of step-control stability

b) DOPRI5 (Table 5.2), $\alpha = 1/5$:

$$R(z) = \text{ see (2.13)}, \quad E(z) = \frac{97}{120000}z^5 - \frac{13}{40000}z^6 + \frac{1}{24000}z^7 \tag{2.34}$$

c) RKF5(4) (Table 5.1, with local extrapolation), $\alpha = 1/5$:

$$R(z) = 1 + z + \frac{z^2}{2} + \frac{z^3}{6} + \frac{z^4}{24} + \frac{z^5}{120} + \frac{z^6}{2080}, \qquad E(z) \text{ same as (2.33).}$$

d) HIHA5 (Method of Higham & Hall, see Table 2.1 below), $\alpha = 1/5$:

$$R(z) = 1 + z + \frac{z^2}{2} + \frac{z^3}{6} + \frac{z^4}{24} + \frac{z^5}{120} + \frac{z^6}{1440}, \tag{2.35}$$

$$E(z) = -\frac{1}{1200}z^5 + \frac{1}{2400}z^6 + \frac{1}{14400}z^7 \tag{2.36}$$

The corresponding stability domains are represented in Fig. 2.8. There, the regions of the boundary, for which $\varrho(C) < 1$ is satisfied, are represented as **thick** lines. It can be observed that the phenomena of Fig. 2.7, as well as those of Sect. IV.1, are nicely verified.

DOP853. The step size control of the code DOP853 (Volume I) is slightly more complicated. It is based on a "stretched" error estimator (see Sect. II.10) and, for the test equation $y' = \lambda y$, it is equivalent to replacing $|E(z)|$ of (2.30) by

$$|E(z)| = \frac{|E_5(z)|^2}{\sqrt{|E_5(z)|^2 + 0.01 \cdot |E_3(z)|^2}}, \tag{2.37}$$

where $E_3(z) = \widehat{R}_3(z) - R(z)$, $E_5(z) = \widehat{R}_5(z) - R(z)$, and $\widehat{R}_3(z), \widehat{R}_5(z)$ are the stability functions of third and fifth order embedded methods, respectively. The above analysis is still valid if the expression v of (2.32) is replaced by the derivative

Table 2.1. Method HIHA5 of Higham and Hall

c_i							
0							
$\frac{2}{9}$	$\frac{2}{9}$						
$\frac{1}{3}$	$\frac{1}{12}$	$\frac{1}{4}$					
$\frac{1}{2}$	$\frac{1}{8}$	0	$\frac{3}{8}$				
$\frac{3}{5}$	$\frac{91}{500}$	$-\frac{27}{100}$	$\frac{78}{125}$	$\frac{8}{125}$			
1	$-\frac{11}{20}$	$\frac{27}{20}$	$\frac{12}{5}$	$-\frac{36}{5}$	5		
1	$\frac{1}{12}$	0	$\frac{27}{32}$	$-\frac{4}{3}$	$\frac{125}{96}$	$\frac{5}{48}$	
y_1	$\frac{1}{12}$	0	$\frac{27}{32}$	$-\frac{4}{3}$	$\frac{125}{96}$	$\frac{5}{48}$	0
$\widehat{y}_1$	$\frac{2}{15}$	0	$\frac{27}{80}$	$-\frac{2}{15}$	$\frac{25}{48}$	$\frac{1}{24}$	$\frac{1}{10}$

of $\log|E(e^{\chi}\lambda)|$ with respect to χ, which is

$$v = 2v_5 - \frac{v_5|E_5(z)|^2 + 0.01v_3|E_3(z)|^2}{|E_5(z)|^2 + 0.01|E_3(z)|^2}, \tag{2.38}$$

where $v_5 = \mathrm{Re}\,\big(zE_5'(z)/E_5(z)\big)$ and $v_3 = \mathrm{Re}\,\big(zE_3'(z)/E_3(z)\big)$. Since $|E(z)| = \mathcal{O}(|z|^8)$ for $|z| \to 0$, we have to use the value $\alpha = 1/8$ in (2.32). The regions of SC-stability are shown in Fig. 2.8.

SC-Stable Dormand and Prince Pairs of Order 5. We see from Fig. 2.8 that the method DOPRI5 is not SC-stable at the intersection of the real axis with the boundary of the stability region. We are therefore interested in finding 5(4)-th order explicit Runge-Kutta pairs from the family of Dormand & Prince (1980) with larger regions of SC-stability.

Requiring the simplifying assumption (II.5.15), Algorithm 5.2 of Sect. II.5 yields a class of Runge-Kutta methods with c_3, c_4, c_5 as free parameters. Higham & Hall (1990) have made an extensive computer search for good choices of these parameters in order to have a reasonable size of the stability domain, large parts of SC-stability and a small 6th order error constant. It turned out that the larger one wants the region of SC-stability, the larger the error constant becomes. A compromise choice between Scylla and Charybdis, which in addition yields nice rational coefficients, is given by $c_3 = 1/3$, $c_4 = 1/2$ and $c_5 = 3/5$. This then leads to the method of Table 2.1 which has satisfactory stability properties as can be seen from Fig. 2.8.

A PI Step Size Control

> We saw that it was an I-controler ... and a control-man knows that PI is always better than I ...
>
> (K. Gustafsson, June 1990)

In 1986/87 two students of control theory attended a course of numerical analysis at the University of Lund. The outcome of this contact was the idea to resolve the above instability phenomena in stiff computations by using the concept of "PID control" (Gustafsson, Lundh & Söderlind 1988). The motivation for PID control, a classic in control theory (Callender, Hartree & Porter 1936) is as follows:

Suppose we have a continuous-time control problem where $\theta(t)$ is the departure, at time t, of a quantity to be controlled from its normal value. Then one might suppose that

$$\dot{\theta}(t) = C(t) - m\theta(t) \tag{2.39}$$

where $C(t)$ denotes the effect of the control and the term $-m\theta(t)$ represents a self-regulating effect such as "a vessel in a constant temperature bath". The most simple assumption for the control would be

$$-\dot{C}(t) = n_1\theta(t) \tag{2.40}$$

which represents, say, a valve opened or closed in dependence of θ. The equations (2.39) and (2.40) together lead to

$$\ddot{\theta} + m\dot{\theta} + n_1\theta = 0 \tag{2.41}$$

which, for $n_1 > 0$, $m > 0$, is always stable. If, however, we assume (more realistically) that our system has some time-lag, we must replace (2.40) by

$$-\dot{C}(t) = n_1\theta(t-T) \tag{2.40'}$$

and the stability of the process may be destroyed. This is precisely the same effect as the instability of Equation (17.6) of Sect. II.17 and is discussed similarly. In order to preserve stability, one might replace (2.40') by

$$-\dot{C}(t) = n_1\theta(t-T) + n_2\dot{\theta}(t-T) \tag{2.40''}$$

or even by

$$-\dot{C}(t) = n_1\theta(t-T) + n_2\dot{\theta}(t-T) + n_3\ddot{\theta}(t-T). \tag{2.40'''}$$

Here, the first term on the right hand side represents the "Integral feedback" (I), the second term "Proportional feedback" (P) and the last term is the "Derivative feedback" (D). The P-term especially increases the constant m in (2.41), thus *adds extra friction* to the equation. It is thus natural to expect that the system becomes more stable. The precise tuning of the parameters n_1, n_2, n_3 is, however, a long task of analytic study and practical experience.

In order to adapt the continuous-time model (2.40'') to our situation, we replace

$$C(t) \longleftrightarrow \log h_n \qquad \text{(the "control variable")}$$
$$\theta(t) \longleftrightarrow \log|err_n| - \log Tol \qquad \text{(the "deviation")}$$

and replace derivatives in t by differences. Then the formula (see (2.28))

$$h_{n+1} = h_n \cdot \left(\frac{Tol}{|err_n|} \right)^{n_1},$$

which is

$$-(\log h_{n+1} - \log h_n) = n_1 (\log |err_n| - \log Tol),$$

corresponds to (2.40'). The PI-control (2.40") would read

$$\begin{aligned} -(\log h_{n+1} - \log h_n) &= n_1 (\log |err_n| - \log Tol) \\ &\quad + n_2 \big((\log |err_n| - \log Tol) - (\log |err_{n-1}| - \log Tol) \big), \end{aligned}$$

or when resolved,

$$h_{n+1} = h_n \cdot \left(\frac{Tol}{|err_n|} \right)^{n_1} \left(\frac{|err_{n-1}|}{|err_n|} \right)^{n_2}. \tag{2.42}$$

In order to perform a *theoretical analysis* of this new algorithm we again choose the problem $y' = \lambda y$ and have as in (2.28)

$$y_{n+1} = R(h_n \lambda) y_n \tag{2.43a}$$

$$err_n = E(h_n \lambda) y_n \tag{2.43b}$$

$$\begin{aligned} h_{n+1} &= h_n \cdot \left(\frac{Tol}{|err_n|} \right)^{n_1} \left(\frac{|err_{n-1}|}{|err_n|} \right)^{n_2} \\ &= h_n \left(\frac{Tol}{|err_n|} \right)^{\alpha} \left(\frac{|err_{n-1}|}{Tol} \right)^{\beta} \end{aligned} \tag{2.43c}$$

where $\alpha = n_1 + n_2$, $\beta = n_2$. With the notation (2.29) this process becomes

$$\begin{aligned} \eta_{n+1} &= \log |R(e^{\chi_n} \lambda)| + \eta_n \\ \chi_{n+1} &= \chi_n - \alpha \log |E(e^{\chi_n} \lambda)| - \alpha \eta_n + \beta \log |E(e^{\chi_{n-1}} \lambda)| + \beta \eta_{n-1} + \gamma \end{aligned} \tag{2.44}$$

with some constant γ. This can be considered as a map $(\eta_n, \chi_n, \eta_{n-1}, \chi_{n-1}) \to (\eta_{n+1}, \chi_{n+1}, \eta_n, \chi_n)$. At a fixed point (η, χ), which again satisfies (2.31), the Jacobian is given by

$$\widetilde{C} = \frac{\partial(\eta_{n+1}, \chi_{n+1}, \eta_n, \chi_n)}{\partial(\eta_n, \chi_n, \eta_{n-1}, \chi_{n-1})} = \begin{pmatrix} 1 & u & 0 & 0 \\ -\alpha & 1 - \alpha v & \beta & \beta v \\ 1 & 0 & 0 & 0 \\ 0 & 1 & 0 & 0 \end{pmatrix} \tag{2.45}$$

with u and v as in (2.32). A numerical study of the spectral radius $\varrho(\widetilde{C})$ with $\alpha = 1/p$ (where p is the exponent of h of the leading term in the error estimator), $\beta = 0.08$ along the boundary of the stability domains of the above RK-methods shows an impressive improvement (see Fig. 2.9) as compared to the standard algorithm of Fig. 2.8. The only exception is DOP853, which becomes unstable close to the real

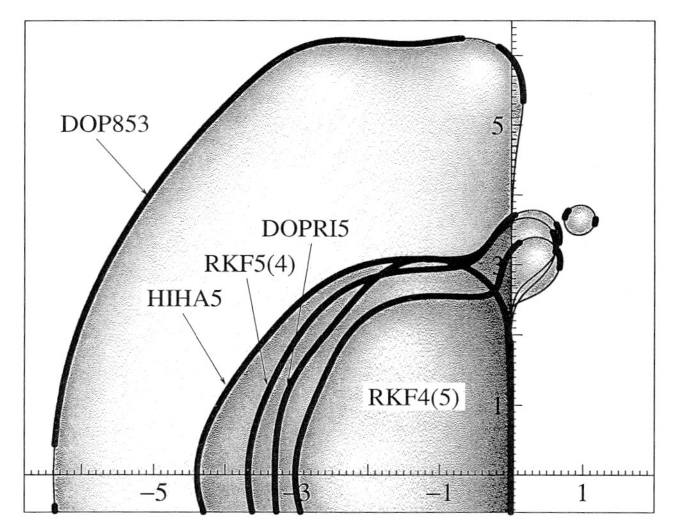

Fig. 2.9. Regions of step-control stability with stabilization factor $\beta = 0.08$

axis, whereas it was SC-stable for $\beta = 0$. For this method, the value $\beta = 0.04$ is more suitable.

The step size behaviour of DOPRI5 with the new strategy ($\beta = 0.13$) applied to the problem (1.6') is compared in Fig. 2.10 to the undamped step size control ($\beta = 0$). The improvement needs no comment. In order to make the difference clearly visible, we have chosen an extra-large tolerance $Atol = Rtol = 8 \cdot 10^{-2}$. With $\beta = 0.13$ the numerical solution becomes smooth in the time-direction. The zig-zag error in the x-direction represents the eigenvector corresponding to the largest eigenvalue of the Jacobian and its magnitude is below $Atol$.

> Man sieht dass selbst der frömmste Mann
> nicht allen Leuten gefallen kann.
> (W. Busch, Kritik des Herzens 1874)

Study for small h. For the non-stiff case the new step size strategy may be slightly less efficient. In order to understand this, we assume that $|err_n| \approx Ch_n^p$ so that (2.43c) becomes

$$h_{n+1} = h_n \left(\frac{Tol}{Ch_n^p} \right)^{\alpha} \left(\frac{Ch_{n-1}^p}{Tol} \right)^{\beta} \tag{2.46}$$

or, by taking logarithms,

$$\log h_{n+1} + (p\alpha - 1) \log h_n - p\beta \log h_{n-1} = (\alpha - \beta) \log \left(\frac{Tol}{C} \right).$$

This is a linear difference equation with characteristic equation

$$\lambda^2 + (p\alpha - 1)\lambda - p\beta = 0, \tag{2.47}$$

the roots of which govern the response of the system to variations in C. Obviously, the choice $\alpha = 1/p$ and $\beta = 0$ would be most perfect by making both roots equal to

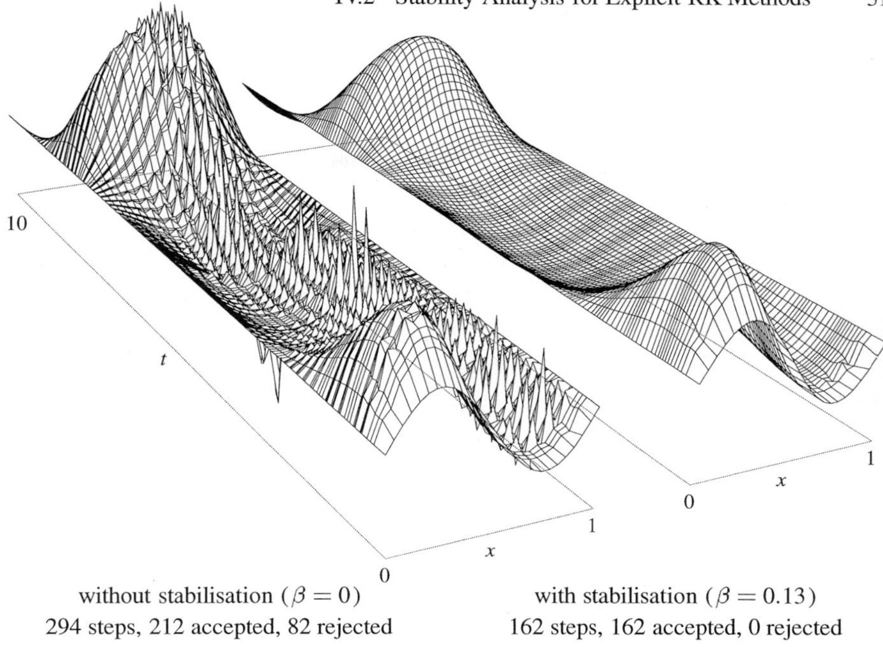

without stabilisation ($\beta = 0$) with stabilisation ($\beta = 0.13$)
294 steps, 212 accepted, 82 rejected 162 steps, 162 accepted, 0 rejected

Fig. 2.10. Numerical solution of (1.6') with $Tol = 8 \cdot 10^{-2}$

zero; but this is just the classical step size control. We therefore have to compromise by choosing α and β such that (2.45) remains stable for large parts of the stability boundary and at the same time keeping the roots of (2.47) significantly smaller than one. A fairly good choice, found by Gustafsson (1991) after some numerical computations, is

$$\alpha \approx 0.7/p, \qquad \beta \approx 0.4/p. \tag{2.48}$$

Stabilized Explicit Runge-Kutta Methods

For many problems, usually not very stiff, of large dimension, and with eigenvalues known to lie in a certain region, explicit methods with large stability domains can be very efficient. We consider here methods with extended stability domains along the negative real axis, which are, therefore, especially suited for the time integration of systems of parabolic PDEs. An excellent survey article with additional details and references is Verwer (1996).

Our problem is to find, for a given s, a polynomial of the form $R(z) = 1 + z + a_2 z^2 + \ldots + a_s z^s$ such that the corresponding stability domain is, in the direction of the negative axis, as large as possible. The main ingredient for these methods are the Chebyshev polynomials (Chebyshev 1854)

$$T_s(x) = \cos(s \arccos x) \tag{2.49}$$

or

$$T_s(x) = 2xT_{s-1}(x) - T_{s-2}(x), \qquad T_0(x) = 1, \quad T_1(x) = x \tag{2.49'}$$

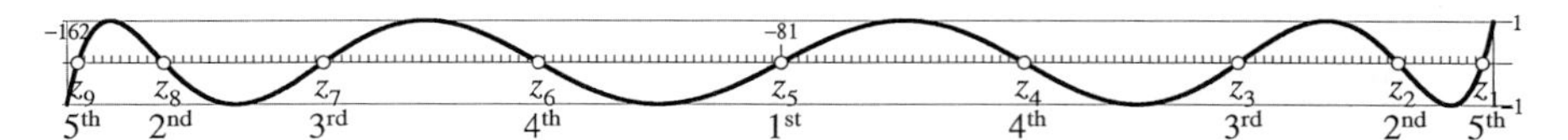

Fig. 2.11. Shifted Chebyshev polynomial $T_9(1+z/81)$ and its zeros

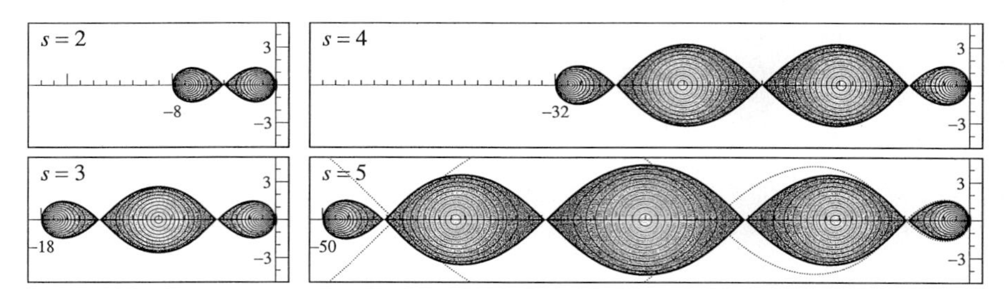

Fig. 2.12. Stability domains for shifted Chebyshev polynomials ($s=2,3,4,5$)
(dots represent limiting case $s \to \infty$, see Exercise 8 below)

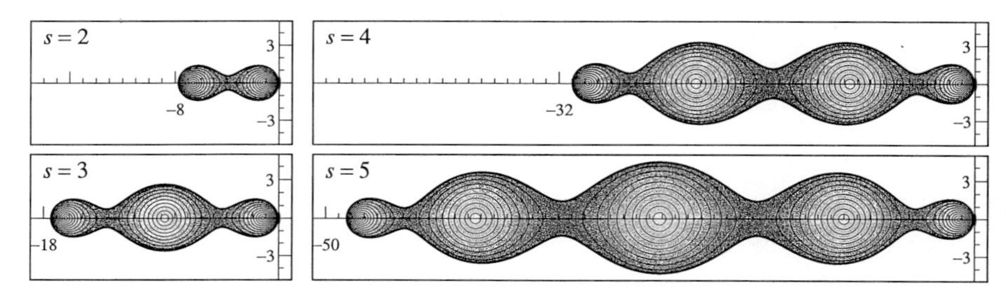

Fig. 2.13. Stability domains for *damped* Chebyshev stability functions, $\varepsilon = 0.05$

which remain for $-1 \le x \le 1$ between -1 and $+1$ and among these polynomials have the largest possible derivative $T_s'(1) = s^2$ (A.A. Markov 1890). Therefore, one must set (Saul'ev 1960, Saul'ev's postgraduate student Yuan Chzao Din 1958, Franklin 1959, Guillou & Lago 1961)

$$R_s(z) = T_s(1+z/s^2) \tag{2.50}$$

so that $R_s(0)=1$, $R_s'(0)=1$, and $|R_s(z)| \le 1$ for $-2s^2 \le z \le 0$ (see Fig. 2.11). In particular we have

$$\begin{aligned}
R_1(z) &= 1+z\\
R_2(z) &= 1+z+\tfrac{1}{8}z^2\\
R_3(z) &= 1+z+\tfrac{4}{27}z^2+\tfrac{4}{729}z^3\\
R_4(z) &= 1+z+\tfrac{5}{32}z^2+\tfrac{1}{128}z^3+\tfrac{1}{8192}z^4\\
R_5(z) &= 1+z+\tfrac{4}{25}z^2+\tfrac{28}{3125}z^3+\tfrac{16}{78125}z^4+\tfrac{16}{9765625}z^5 .
\end{aligned} \tag{2.50'}$$

whose stability domains are represented in Fig. 2.12.

Damping. In the points where $T_s(1+z/s^2)=\pm 1$, there is no damping at all of the higher frequencies and the stability domain has zero width. We therefore choose a small $\varepsilon>0$, say $\varepsilon=0.05$, and put (already suggested by Guillou & Lago 1961)

$$R_s(z)=\frac{1}{T_s(w_0)}T_s(w_0+w_1 z),\qquad w_0=1+\frac{\varepsilon}{s^2},\qquad w_1=\frac{T_s(w_0)}{T_s'(w_0)}. \tag{2.51}$$

These polynomials oscillate between approximately $1-\varepsilon$ and $-1+\varepsilon$ and again satisfy $R_s(z)=1+z+\mathcal{O}(z^2)$. The stability domains become a bit shorter (by $(4\varepsilon/3)s^2$), but the boundary is in a safe distance from the real axis (see Fig. 2.13).

Lebedev's Realization. Our next problem is to find Runge-Kutta methods which realize these stability polynomials. A first idea, mentioned by Saul'ev (1960) and Guillou & Lago (1961), is to write

$$R_s(z)=\prod_{i=1}^{s}(1+\delta_i z)\qquad\text{where}\quad \delta_i=-\frac{1}{z_i},\qquad z_i \text{ roots of } R(z) \tag{2.52}$$

and to represent the RK method as the *composition* of explicit Euler steps

$$g_0:=y_0,\qquad g_i:=g_{i-1}+h\delta_i f(g_{i-1}),\quad (i=1,2,\ldots,s),\qquad y_1:=g_s. \tag{2.53}$$

A disadvantage here is the fact that for the first of these roots, which in absolute value is much smaller than the others, we shall have a very large Euler step, which is surely not good. Lebedev's idea (Lebedev 1989, 1994) is therefore to group the roots symmetrically two-by-two together and to represent the corresponding quadratic factor

$$(1+\delta_i z)(1+\delta_i' z)=(1+2\alpha_i z+\beta_i z^2) \tag{2.54}$$

by a two-stage scheme

$$\begin{aligned} g_i &:= g_{i-1}+h\alpha_i f(g_{i-1})\\ g^{\star}_{i+1} &:= g_i+h\alpha_i f(g_i)\\ g_{i+1} &:= g^{\star}_{i+1}-h\alpha_i\gamma_i\big(f(g_i)-f(g_{i-1})\big)\\ &= g^{\star}_{i+1}-\gamma_i\big((g^{\star}_{i+1}-g_i)-(g_i-g_{i-1})\big) \end{aligned} \tag{2.55}$$

which produces (2.54) if $\beta_i=\alpha_i^2(1-\gamma_i)$. This halves nearly the largest Euler step size and allows also complex conjugate pairs of roots. The expression $(g^{\star}_{i+1}-g_i)-(g_i-g_{i-1})\approx h^2\alpha_i^2 y''$ can be used for error estimations and step size selections. For odd s, there remains one single root which gives rise to an Euler step (2.53).

Best Ordering. Some attention is now necessary for the decision in which order the roots shall be used (Lebedev & Finogenov 1976). This is done by two requirements: firstly, the quantities

$$S_j=\max_z|1+\delta_1 z|\prod_{i=1}^{j}|1+2\alpha_i z+\beta_i z^2|,$$

which express the stability of the internal stages, must be ≤ 1 (here, the $\max$ is taken over real z in the stability interval of the method). Secondly, the quantities

$$Q_j = \max_z \prod_{i=j+1}^{s} |1 + 2\alpha_i z + \beta_i z^2|,$$

which describe the propagation of rounding errors, must be as small as possible. These conditions, evaluated numerically for the case $s = 9$, lead to the ordering indicated in Fig. 2.11.

Second Order Methods. If the stability polynomial is a second order approximation to e^z, i.e., if

$$R_s(z) = 1 + z + \frac{z^2}{2} + a_3 z^3 + \ldots + a_s z^s \tag{2.56}$$

then it can be seen from (2.8) that any corresponding Runge-Kutta scheme is also of second order for nonlinear problems. Analytic expressions, in terms of an elliptic integral, for such optimal polynomials have been obtained by Lebedev & Medovikov (1994). Their stability region reaches to $-0.821842 \cdot s^2$ for $s \gg 1$. Their practical computation is usually done numerically (Remez 1957, Lebedev 1995). For example, in the case $s = 9$ and for a damping factor $\varepsilon = 0.015$, we obtain the roots

$$\begin{aligned} &z_9 = -64.64238389, \quad z_8 = -60.67479347, \quad z_7 = -53.21695488, \\ &z_6 = -43.16527010, \quad z_5 = -31.72471699, \quad z_4 = -20.25474163, \\ &z_3 = -10.05545938, \quad z_{2,1} = -1.30596166 \pm i \cdot 1.34047517 \end{aligned} \tag{2.57}$$

The corresponding stability polynomials, which are stable for $-65.15 \leq z \leq 0$, the stability domain, and the best ordering are shown in Fig. 2.14. We see that we now have a pair of complex roots.

Lebedev's computer code, called DUMKA, incorporates the formulas of the above type with automatic selection of h and s in a wide range.

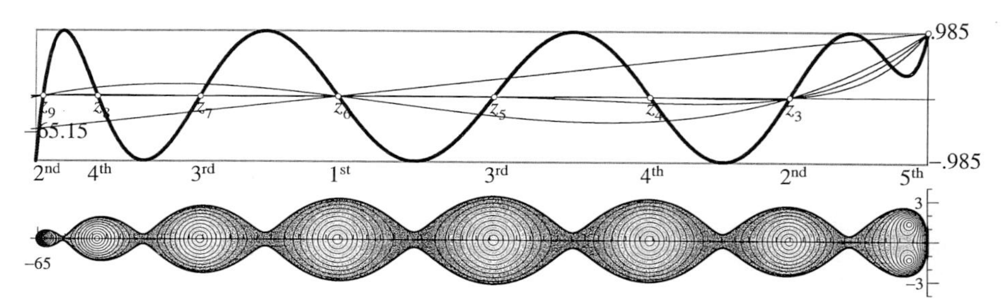

Fig. 2.14. Second order Zolotarev approximation with stability domain

Numerical Example. As an illustration, the method corresponding to (2.55) and (2.57) has been applied to problem (1.6'). Theory predicts stability for approximately $h \leq 65.15/135 = 0.4826$. The leftmost picture of Fig. 2.15 is computed

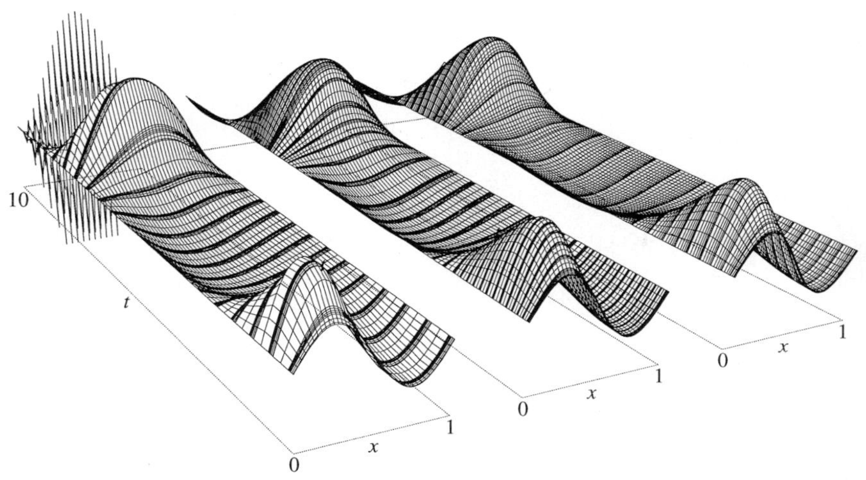

Fig. 2.15. Problem (1.6'): Lebedev9, $h = 0.48865$ (left), DUMKA (middle), RKC (right) (all internal stages drawn)

with $h = 0.48865$, which is a little too large and produces instability. The middle picture is produced by the code DUMKA with $Tol = 3 \cdot 10^{-3}$.

The Approach of van der Houwen & Sommeijer. An elegant idea for a second realization has been found by van der Houwen & Sommeijer (1980): apply scaled and shifted Chebyshev polynomials and use the three-term recusion formula (2.49') for defining the internal stages. We therefore, following Bakker (1973), set

$$R_s(z) = a_s + b_s T_s(w_0 + w_1 z) \qquad w_0 = 1 + \varepsilon/s^2, \ \ \varepsilon \approx 0.15. \tag{2.58}$$

The conditions for second order

$$R_s(0) = 1, \quad R_s'(0) = 1, \quad R_s''(0) = 1$$

lead to

$$w_1 = \frac{T_s'(w_0)}{T_s''(w_0)}, \qquad b_s = \frac{T_s''(w_0)}{(T_s'(w_0))^2}, \qquad a_s = 1 - b_s T_s(w_0), \tag{2.59}$$

with damping $a_s + b_s \approx 1 - \varepsilon/3$ (see Ex. 9). We now put for the internal stages

$$R_j(z) = a_j + b_j T_j(w_0 + w_1 z) \qquad j = 0, 1, \ldots, s-1. \tag{2.60}$$

It has been discovered by Sommeijer (see Sommeijer & Verwer 1980), that these $R_j(z)$ can, for $j \geq 2$, be approximations of second order at certain points $x_0 + c_j h$ if

$$R_j(0) = 1, \quad R_j'(0) = c_j, \quad R_j''(0) = c_j^2 \tag{2.61}$$

which gives

$$R_j(z) - 1 = b_j\big(T_j(w_0 + w_1 z) - T_j(w_0)\big), \qquad b_j = \frac{T_j''(w_0)}{(T_j'(w_0))^2}. \tag{2.62}$$

The three-term recurrence relation (2.49') now leads to

$$R_j(z)-1=\mu_j(R_{j-1}(z)-1)+\nu_j(R_{j-2}(z)-1)+\kappa_j\cdot z\cdot(R_{j-1}(z)-a_{j-1})$$

where

$$\mu_j=\frac{2b_j w_0}{b_{j-1}},\quad \nu_j=\frac{-b_j}{b_{j-2}},\quad \kappa_j=\frac{2b_j w_1}{b_{j-1}},\qquad j=2,3,\ldots,s. \tag{2.63}$$

This formula allows, in the case of a nonlinear differential system, to define the scheme

$$\begin{aligned} g_0-y_0&=0,\\ g_1-y_0&=\kappa_1 hf(g_0), \\ g_j-y_0&=\mu_j(g_{j-1}-y_0)+\nu_j(g_{j-2}-y_0)+\kappa_j hf(g_{j-1})-a_{j-1}\kappa_j hf(g_0), \end{aligned} \tag{2.64}$$

which, being of second order for $y'=\lambda y$, is of second order for nonlinear equations too (again because of (2.8)). For $j=1$ only first order is possible and κ_1 can be chosen freely. Sommeijer & Verwer (1980) suggest to put

$$b_0=b_2,\ \ b_1=b_2\quad \text{which gives}\quad \kappa_1=c_1=\frac{c_2}{T_2'(w_0)}\approx\frac{c_2}{4}.$$

Fig. 2.16 shows, for $s=9$ as usual, the functions $R_s(z)$ and $R_j(z)$, $j=2,\ldots,s-1$ together with the stability domain of $R_s(z)$ (the "Venus of Willendorf") in exactly the same frame as Lebedev's Zolotarev polynomial of Fig. 2.14. We see that the stability domain becomes a little shorter, but we have closed analytic expressions and a smoother behaviour of the c_i's (see Fig. 2.15, right). All internal stages satisfy $|R_j(z)|\le 1$, and the method can be seen to possess a satisfactory numerical stability (see Verwer, Hundsdorfer & Sommeijer 1990). The above formulas have been implemented in a research code RKC ("Runge-Kutta-Chebyshev") by Sommeijer (1991). As can be seen from Fig. 2.15, it performs well for equation (1.6'). More numerical results shall be reported in Sect. IV.10.

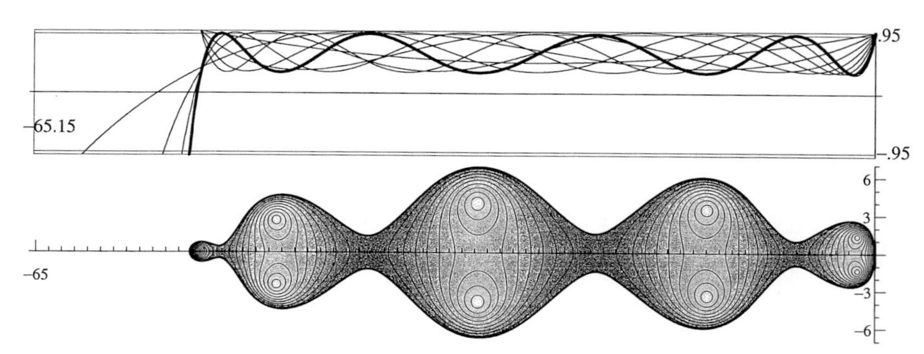

Fig. 2.16. Stability function and domain for RKC method, $s=9$, $\varepsilon=0.15$

Exercises

1. Prove that Runge-Kutta methods are invariant under linear transformations $y = Tz$ (i.e., if one applies the method to $y' = f(x,y)$ and to $z' = T^{-1}f(x,Tz)$ with initial values satisfying $y_0 = Tz_0$, then we have $y_1 = Tz_1$).

2. Consider the differential equation $y' = Ay$ and a numerical solution given by $y_{n+1} = R(hA)y_n$. Suppose that $R(z)$ is A-stable, i.e., it satisfies
$$|R(z)| \le 1 \qquad \text{for} \qquad \text{Re}\, z \le 0,$$
and show, by transforming A to Jordan canonical form, that
 a) if $y' = Ay$ is stable, then $\{y_n\}$ is bounded;
 b) if $y' = Ay$ is asymptotically stable, then $y_n \to 0$ for $n \to \infty$.

3. (Optimal stability for hyperbolic problems, van der Houwen (1968), (1977), p. 99): Given m, find a polynomial $R_m(z) = 1 + z + \ldots$ of degree $m+1$ such that $|R(iy)| \le 1$ for $-\beta \le y \le \beta$ with β as large as possible.

   Result. The solution (Sonneveld & van Leer 1985) is given by
$$R_m(z) = \frac{1}{2}V_{m-1}(\zeta) + V_m(\zeta) + \frac{1}{2}V_{m+1}(\zeta), \quad \zeta = \frac{z}{m} \tag{2.65}$$
   where $V_m(\zeta) = i^m T_m(\zeta/i)$ are the Chebyshev polynomials with positive coefficients. $R_m(iy)$ is stable for $-m \le y \le m$. The first R_m are (see Abramowitz & Stegun, p. 795)
$$\begin{aligned} R_1(z) &= 1 + \zeta + \zeta^2 \qquad\qquad \zeta = \frac{z}{m} \\ R_2(z) &= 1 + 2\zeta + 2\zeta^2 + 2\zeta^3 \\ R_3(z) &= 1 + 3\zeta + 5\zeta^2 + 4\zeta^3 + 4\zeta^4 \\ R_4(z) &= 1 + 4\zeta + 8\zeta^2 + 12\zeta^3 + 8\zeta^4 + 8\zeta^5 \end{aligned} \tag{2.66}$$
   Similar as for Chebyshev polynomials, they satisfy the recurrence relation $R_{m+1} = 2\zeta R_m + R_{m-1} (m \ge 2)$. Their stability domains are given in Fig. 2.17.

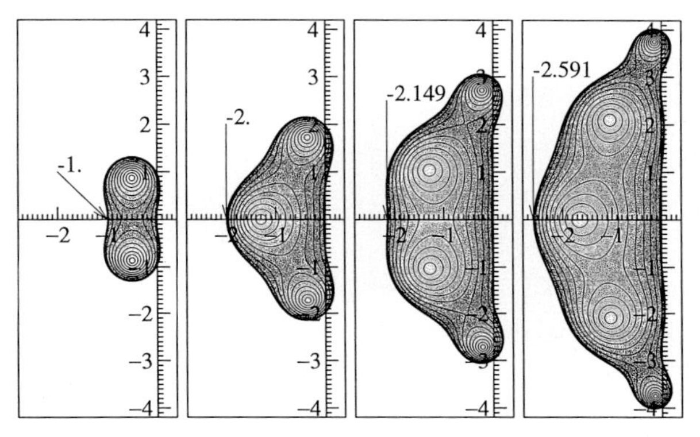

Fig. 2.17. Stability domains for hyperbolic approximations

4. Linearize the rope equation (1.24) in the neighbourhood of $\theta=\dot{\theta}=0$ and make a stability analysis. Re-obtain Lagrange's equation (I.6.2) from the linearized equation with the coordinate transformation

$$y=\begin{pmatrix}1 & & & \\ 1 & 1 & & \\ 1 & 1 & 1 & \\ \vdots & \vdots & \vdots & \ddots\end{pmatrix}\theta, \qquad \theta=\begin{pmatrix}1 & & & \\ -1 & 1 & & \\ & -1 & 1 & \\ & & \ddots & \ddots\end{pmatrix}y.$$

5. Fig. 2.18 shows the numerical results of the classical 4th order Runge-Kutta method with equidistant steps over $0 \leq t \leq 5$ for the beam problem (1.7)-(1.20) with $n=8$. Explain the result with the help of Fig. 2.1.

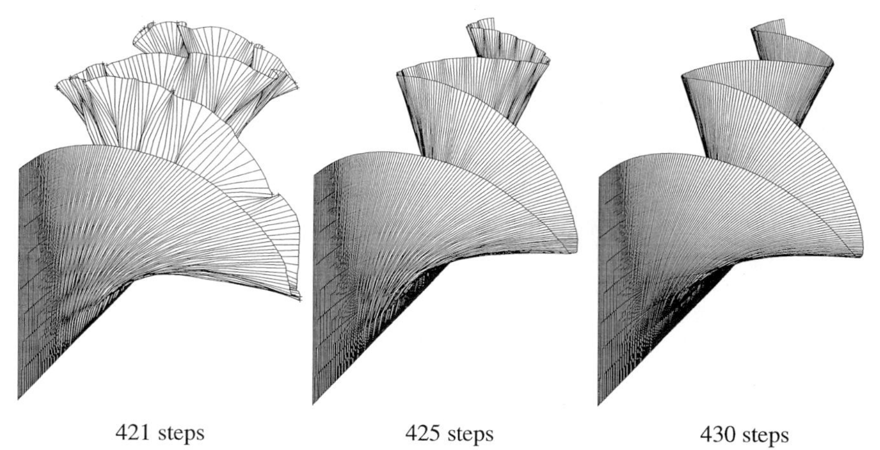

421 steps 425 steps 430 steps

Fig. 2.18. Classical Runge-Kutta method (constant step sizes) on the beam problem

6. For the example of Exercise 5, the explicit Euler method, although converging for $h \to 0$, is *never* stable (see Fig. 2.19). Why?

7. Let λ be an eigenvalue of the two-dimensional left upper submatrix of $\widetilde{C}$ in (2.45) (matrix C of (2.32)) and denote its analytic continuation as eigenvalue of $\widetilde{C}$ by $\lambda(\beta)$. Prove that

   a) If $\operatorname{Re}\lambda \neq 0$, then for some $y \in \mathbb{R}$

$$\lambda(\beta)=\lambda\cdot\Big(1-\frac{\beta}{\alpha}(1-\operatorname{Re}\lambda)+i\beta y+\mathcal{O}(\beta^2)\Big).$$

   This shows that $|\lambda(\beta)|<|\lambda|$ for small $\beta>0$ if $\operatorname{Re}\lambda<1$.

   b) If λ and μ are two distinct real eigenvalues of the above mentioned submatrix, then

$$\lambda(\beta)=\lambda\cdot\Big(1-\frac{\beta}{\alpha}\Big(1-\frac{1}{\lambda}\Big)^2\frac{1}{\lambda-\mu}+\mathcal{O}(\beta^2)\Big).$$

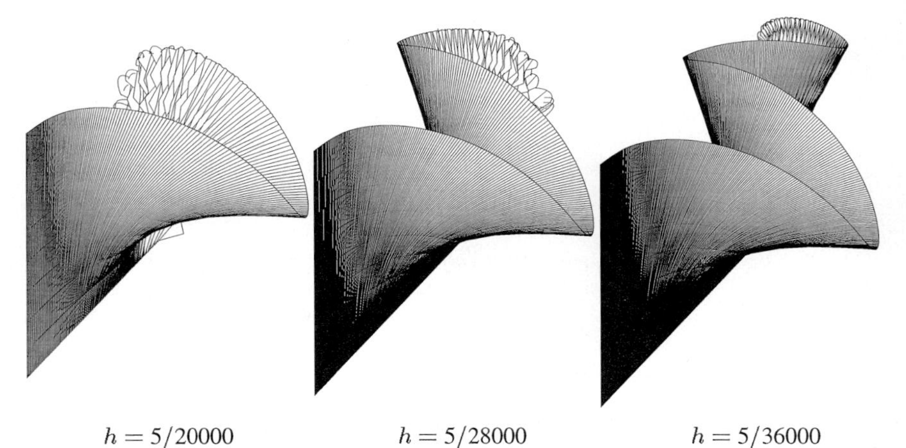

$h = 5/20000$ $\qquad$ $h = 5/28000$ $\qquad$ $h = 5/36000$

Fig. 2.19. Explicit Euler on the beam problem (every 50th step drawn)

Hint. Write the characteristic polynomial of $\widetilde{C}$ in the form

$$\det(\lambda I - \widetilde{C}) = \lambda\big(\lambda p(\lambda) + \beta q(\lambda)\big),$$

where $p(\lambda) = \det(\lambda I - C)$ is the characteristic polynomial of C, and differentiate with respect to β.

8. Show that for the Chebyshev stability functions (2.50) we have

$$\lim_{s\to\infty} R_s(z) = \cos(\sqrt{-2z}).$$

Hint. Insert $\arccos(1 - x^2/2) \approx x$ into (2.49) and (2.50). The corresponding stability domain is indicated by dotted lines in the last picture of Fig. 2.12.

9. Show (for example with the help of (2.49')) that for the Chebyshev polynomials

$$T_s'(1) = s^2, \qquad T_s''(1) = \frac{s^2(s^2-1)}{3}$$

and obtain asymptotic values (for $\varepsilon \to 0$) for w_1, b_s, a_s, the damping factor and the stability interval of the Bakker polynomials (2.58).

10. (Cross-shaped stability domains). For $-1 \le \varphi \le 1$ we put $z = -b \pm \sqrt{a(\varphi - 1) + b^2}$, so that z moves on a cross $-2b \le z \le 0$ and $z = -b \pm iy$. Thus (an idea of Lebedev)

$$R_{2s}(z) = T_s(\varphi(z))$$

is a stability function for eigenvalues on crosses (as, e.g., for the PLATE problem). Determine a in dependence of b from the condition $R'(0) = 1$ and find the maximal value for y.
Result. $R_{2s}(z) = T_s(1 + z/s^2 + z^2/(2bs^2))$; $y_{\max} = \sqrt{4bs^2 - b^2}$.

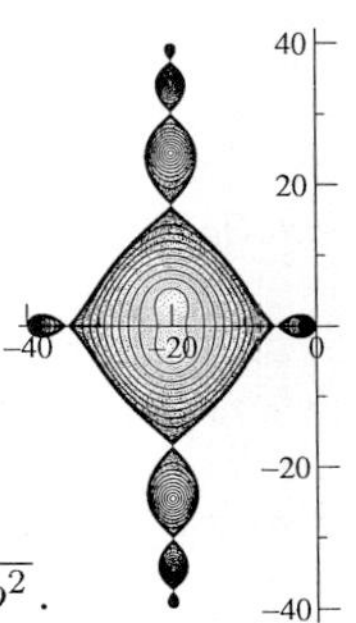

IV.3 Stability Function of Implicit RK-Methods

I didn't like all these "strong", "perfect", "absolute", "generalized", "super", "hyper", "complete" and so on in mathematical definitions, I wanted something neutral; and having been impressed by David Young's "property A", I chose the term "A-stable".

(G. Dahlquist, in 1979)

There are at least two ways to combat stiffness. One is to design a better computer, the other, to design a better algorithm.

(H. Lomax in Aiken 1985)

Methods are called A-stable if there are no stability restrictions for $y' = \lambda y$, $\operatorname{Re}\lambda < 0$ and $h > 0$. This concept was introduced by Dahlquist (1963) for linear multistep methods, but also applied to Runge-Kutta processes. Ehle (1968) and Axelsson (1969) then independently investigated the A-stability of implicit Runge-Kutta methods and proposed new classes of A-stable methods. A nice paper of Wright (1970) studied collocation methods.

The Stability Function

We start with the implicit Euler method. This method, $y_1 = y_0 + hf(x_1, y_1)$, applied to Dahlquist's equation $y' = \lambda y$ becomes $y_1 = y_0 + h\lambda y_1$ which, after solving for y_1, gives

$$y_1 = R(h\lambda)\, y_0 \qquad \text{with} \qquad R(z) = \frac{1}{1-z}.$$

This time, the stability domain is the *exterior* of the circle with radius 1 and centre $+1$. The stability domain thus covers the *entire* negative half-plane and a large part of the positive half-plane as well. The implicit Euler method is *very* stable.

Proposition 3.1. *The s-stage implicit Runge-Kutta method*

$$g_i = y_0 + h\sum_{j=1}^{s} a_{ij} f(x_0 + c_j h, g_j) \qquad i = 1,\ldots,s \tag{3.1a}$$

$$y_1 = y_0 + h\sum_{j=1}^{s} b_j f(x_0 + c_j h, g_j) \tag{3.1b}$$

applied to $y' = \lambda y$ *yields* $y_1 = R(h\lambda) y_0$ *with*

$$R(z) = 1 + z b^T (I - zA)^{-1} 1\!\!1, \tag{3.2}$$

where $\quad b^T = (b_1, \ldots, b_s), \quad A = (a_{ij})_{i,j=1}^{s}, \quad 1\!\!1 = (1,\ldots,1)^T.$

Remark. As in Definition 2.1, $R(z)$ is called the *stability function* of Method (3.1).

Proof. Equation (3.1a) with $f(x,y) = \lambda y$, $z = h\lambda$ becomes a linear system for the computation of $g_1, \ldots, g_s$. Solving this and inserting into (3.1b) leads to (3.2). □

Another useful formula for $R(z)$ is the following (Stetter 1973, Scherer 1979):

Proposition 3.2. *The stability function of (3.1) satisfies*

$$R(z) = \frac{\det(I - zA + z\mathbb{1}b^T)}{\det(I - zA)}. \tag{3.3}$$

Proof. Applying (3.1) to (2.9) yields the linear system

$$\begin{pmatrix} I - zA & 0 \\ -zb^T & 1 \end{pmatrix} \begin{pmatrix} g \\ y_1 \end{pmatrix} = y_0 \begin{pmatrix} \mathbb{1} \\ 1 \end{pmatrix}.$$

Cramer's rule (Cramer 1750) implies that the denominator of $R(z)$ is $\det(I - zA)$, and its numerator

$$\det \begin{pmatrix} I - zA & \mathbb{1} \\ -zb^T & 1 \end{pmatrix} = \det \begin{pmatrix} I - zA + z\mathbb{1}b^T & 0 \\ -zb^T & 1 \end{pmatrix} = \det(I - zA + z\mathbb{1}b^T). \qquad \square$$

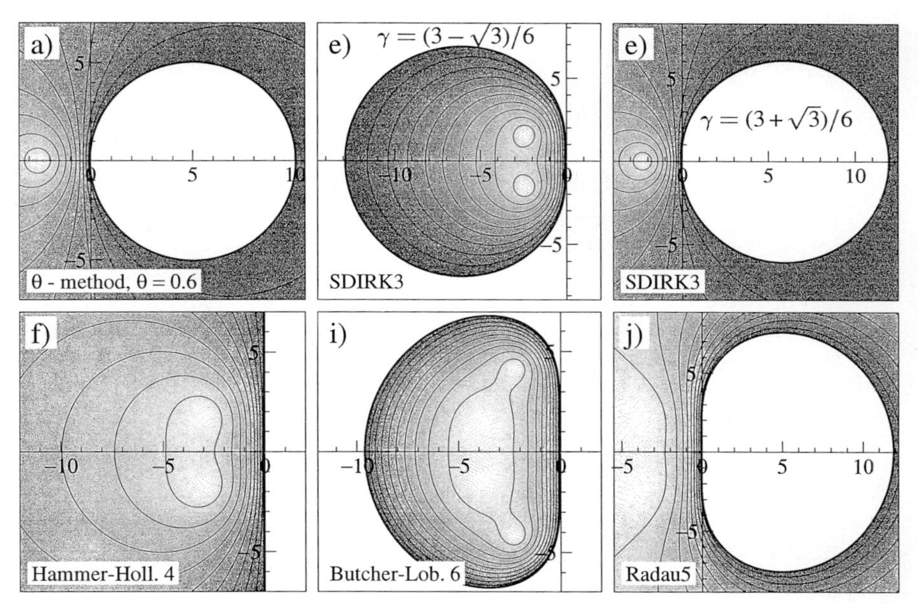

Fig. 3.1. Stability domains for implicit Runge-Kutta methods

The stability functions for the methods of Sect. II.7 are presented in Table 3.1. The corresponding stability domains are displayed in Fig. 3.1.

We see that for implicit methods $R(z)$ becomes a rational function with numerator and denominator of degree $\le s$. We write

$$R(z) = \frac{P(z)}{Q(z)}, \qquad \deg P = k, \quad \deg Q = j. \tag{3.4}$$

Table 3.1. Stability functions for implicit Runge-Kutta methods of Sect. II.7

	Method	$R(z)$
a)	θ-method (II.7.2)	$\frac{1+z(1-\theta)}{1-z\theta}$
b)	implicit Euler (II.7.3)	$\frac{1}{1-z}$
c)	implicit midpoint (II.7.4) } trapezoidal rule (II.7.5) }	$\frac{1+z/2}{1-z/2}$
d)	Hammer-Hollingsworth (II.7.6)	$\frac{1+4z/6+z^2/6}{1-z/3}$
e)	SDIRK order 3 (Table II.7.2)	$\frac{1+z(1-2\gamma)+z^2(1/2-2\gamma+\gamma^2)}{(1-\gamma z)^2}$
f)	Hammer-Hollingsw. 4 (Table II.7.3) } Lobatto IIIA, order 4 (Table II.7.7) }	$\frac{1+z/2+z^2/12}{1-z/2+z^2/12}$
g)	Kuntzm.-Butcher 6 (Table II.7.4)	$\frac{1+z/2+z^2/10+z^3/120}{1-z/2+z^2/10-z^3/120}$
h)	Butcher's Lobatto 4 (Table II.7.6)	$\frac{1+3z/4+z^2/4+z^3/24}{1-z/4}$
i)	Butcher's Lobatto 6 (Table II.7.6)	$\frac{1+2z/3+z^2/5+z^3/30+z^4/360}{1-z/3+z^2/30}$
j)	Radau IIA, order 5 (Table II.7.7)	$\frac{1+2z/5+z^2/20}{1-3z/5+3z^2/20-z^3/60}$

If the method is of order p, then

$$e^z - R(z) = Cz^{p+1} + \mathcal{O}(z^{p+2}) \qquad \text{for } z \to 0 \tag{3.5}$$

(see Theorem 2.2). The constant C is usually $\neq 0$. If not, we increase p in (3.5) until C becomes $\neq 0$. We then call $R(z)$ a *rational approximation to* e^z *of order* p and C its *error constant.*

A-Stability

We observe that some methods are stable on the entire left half-plane $\mathbb{C}^-$. This is precisely the set of eigenvalues, where the *exact* solution of (2.9) is stable too (Sect. I.13, Theorem 13.1). A desirable property for a numerical method is that it preserves this stability property.

Definition 3.3 (Dahlquist 1963). A method, whose stability domain satisfies

$$S \supset \mathbb{C}^- = \{z; \quad \operatorname{Re} z \leq 0\},$$

is called *A-stable.*

A Runge-Kutta method with (3.4) as stability function is A-stable if and only if

$$|R(iy)| \leq 1 \qquad \text{for all real } y \tag{3.6}$$

and

$$R(z) \quad \text{is analytic for } \operatorname{Re} z < 0. \tag{3.7}$$

This follows from the maximum principle applied to $\mathbb{C}^-$. By a slight abuse of language, we also call $R(z)$ A-stable in this case (or, as many authors say, "*A*-acceptable", Ehle 1968).

Condition (3.6) alone means stability on the imaginary axis and may be called *I-stability.* It is equivalent to the fact that the polynomial

$$E(y) = |Q(iy)|^2 - |P(iy)|^2 = Q(iy)Q(-iy) - P(iy)P(-iy) \tag{3.8}$$

satisfies

$$E(y) \geq 0 \qquad \text{for all } y \in \mathbb{R}. \tag{3.9}$$

Proposition 3.4. *$E(y)$, defined by (3.8), is an even polynomial of degree $\leq 2\max(\deg P, \deg Q)$. If $R(z)$ is an approximation of order p, then*

$$E(y) = \mathcal{O}(y^{p+1}) \qquad \textit{for } y \to 0.$$

Proof. Taking absolute values in (3.5) gives

$$|e^z| - \frac{|P(z)|}{|Q(z)|} = \mathcal{O}(z^{p+1}).$$

Putting $z = iy$ and using $|e^{iy}| = 1$ leads to

$$|Q(iy)| - |P(iy)| = \mathcal{O}(y^{p+1}).$$

The result now follows from

$$E(y) = (|Q(iy)| + |P(iy)|)(|Q(iy)| - |P(iy)|). \qquad \square$$

Examples 3.5. For the implicit midpoint rule, the trapezoidal rule, the Hammer & Hollingsworth, the Kuntzmann & Butcher and Lobatto IIIA methods (c, f, g of Table 3.1) we have $E(y) \equiv 0$ since $Q(z) = P(-z)$. This also follows from Proposition 3.4 because $p = 2j$. A straightforward computation shows that (3.7) is satisfied, hence these methods are A-stable.

For methods d, h, i of Table 3.1 we have $\deg P > \deg Q$ and the leading coefficient of E is negative. Therefore (3.9) cannot be true for $y \to \infty$ and these methods are not A-stable.

For the Radau IIA method of order 5 (case j) we obtain $E(y) = y^6/3600$ and by inspection of the zeros of $Q(z)$ the method is seen to be A-stable.

For the two-stage SDIRK method (case e) $E(y)$ becomes

$$E(y) = (\gamma - 1/2)^2(4\gamma - 1)y^4. \tag{3.10}$$

Thus the method is A-stable for $\gamma \geq 1/4$. The 3rd order method is A-stable for $\gamma = (3+\sqrt{3})/6$, but not for $\gamma = (3-\sqrt{3})/6$ (see Fig. 3.1).

The following general result explains the I-stability properties of the foregoing examples.

Proposition 3.6. *A rational function (3.4) of order* $p \geq 2j - 2$ *is* I*-stable if and only if* $|R(\infty)| \leq 1$.

Proof. $|R(\infty)| \leq 1$ implies $k \leq j$. By Proposition 3.4, $E(y)$ must be of the form $K \cdot y^{2j}$. By letting $y \to \infty$ in (3.6) and (3.9), we see that $|R(\infty)| \leq 1$ is equivalent to $K \geq 0$. □

L-Stability and $A(\alpha)$-Stability

> The trapezoidal rule for the numerical integration of first-order ordinary differential equations is shown to possess, for a certain type of problem, an undesirable property. (A.R. Gourlay 1970)

> A-stability is not the whole answer to the problem of stiff equations. (R. Alexander 1977)

Some of the above methods seem to be optimal in the sense that the stability region coincides *exactly* with the negative half-plane. This property is not as desirable as it may appear, since for a rational function

$$\lim_{z\to-\infty} R(z) = \lim_{z\to\infty} R(z) = \lim_{z=iy,\, y\to\infty} R(z).$$

The latter must then be 1 in modulus, since $|R(iy)| = 1$ for all real y. This means that for z close to the real axis with a very large negative real part, $|R(z)|$ is, although < 1, *very close* to one. As a consequence, stiff components in (2.6) are damped out *only very slowly.* We demonstrate this with the example

$$y' = -2000(y - \cos x), \qquad y(0) = 0, \qquad 0 \leq x \leq 1.5, \tag{3.11}$$

which is the same as (1.1), but with increased stiffness. The numerical results for the trapezoidal rule are compared to those of implicit Euler in Fig. 3.2. The implicit Euler damps out the transient phase much faster than the trapezoidal rule. It thus appears to be a desirable property of a method that $|R(z)|$ be much smaller than 1 for $z \to -\infty$.

Definition 3.7 (Ehle 1969). A method is called L*-stable* if it is A-stable and if in addition

$$\lim_{z\to\infty} R(z) = 0. \tag{3.12}$$

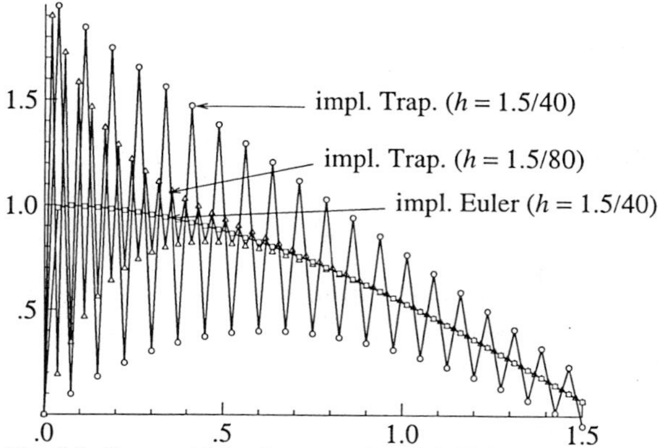

Fig. 3.2. Trapezoidal rule versus implicit Euler on (3.11)

Among the methods of Table 3.1, the implicit Euler, the SDIRK method (e) with $\gamma = (2 \pm \sqrt{2})/2$, as well as the Radau IIA formula (j) are L-stable.

Proposition 3.8. *If an implicit Runge-Kutta method with nonsingular A satisfies one of the following conditions:*

$$a_{sj} = b_j \quad j = 1, \ldots, s, \tag{3.13}$$

$$a_{i1} = b_1 \quad i = 1, \ldots, s, \tag{3.14}$$

then $R(\infty) = 0$. This makes A-stable methods L-stable.

Proof. By (3.2)

$$R(\infty) = 1 - b^T A^{-1} 1\!\!1 \tag{3.15}$$

and (3.13) means that $A^T e_s = b$ where $e_s = (0, \ldots, 0, 1)^T$. Therefore $R(\infty) = 1 - e_s^T 1\!\!1 = 1 - 1 = 0$. In the case of (3.14) use $Ae_1 = 1\!\!1 b_1$. □

Methods satisfying (3.13) are called *stiffly accurate* (Prothero & Robinson 1974). They are important for the solution of singularly perturbed problems and for differential-algebraic equations (see Chapters VI and VII).

The definition of A-stability is on the one hand too weak, as we have just seen, and on the other hand too strong in the sense that many methods which are not so bad at all are not A-stable. The following definition is a little weaker and will be specially useful in the chapter on multistep methods.

Definition 3.9 (Widlund 1967). A method is said to be *$A(\alpha)$-stable* if the sector

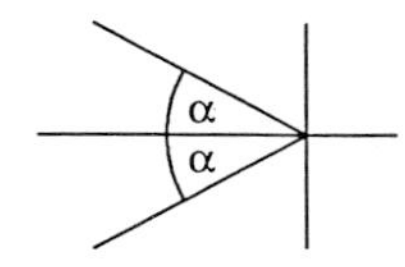

$$S_\alpha = \{z; \quad |\arg(-z)| < \alpha, \quad z \neq 0\}$$

is contained in the stability region.

For example, the Padé approximation $R_{03}(z) = \left(1 - z + \frac{z^2}{2!} - \frac{z^3}{3!}\right)^{-1}$ (see (3.29) below) is $A(\alpha)$-stable for $\alpha \le 88.23°$.

Numerical Results

To show the effects of good stability properties on the stiff examples of Sect. IV.1, we choose the 3-stage Radau IIA formula (Table 5.6 of Sect. IV.5) which, as we have seen, is *A*-stable, *L*-stable and of reasonably high order. It has been coded (Subroutine RADAU5 of the Appendix) and the details of this program will be discussed later (Sect. IV.8). This program integrates all the examples of Sect. IV.1 in a couple of steps and the plots of Fig. 1.3 and Fig. 1.5 show a clear difference.

The beam equation (1.10') with $n = 40$ is integrated, with *Rtol* = *Atol* = 10^{-3} (absolute) and smooth initial values, in 28 steps (Fig. 3.3).

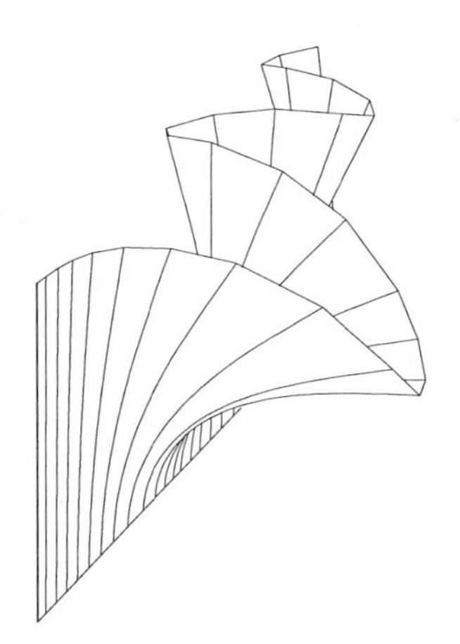

Fig. 3.3. RADAU5 on the beam (1.10'), every step drawn

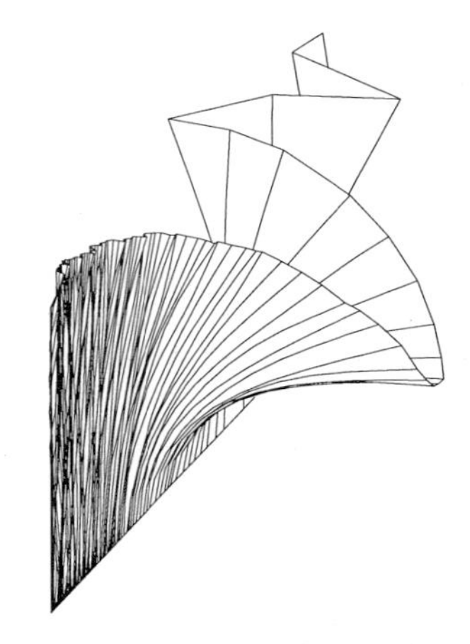

Fig. 3.4. RADAU5 on oscillatory beam with large *Tol* (107 steps, all drawn)

Since the Radau5 formula is L-stable, the stability domain also covers the imaginary axis and large parts of the right half-plane $\mathbb{C}^+$. This means that high oscillations of the true solution *may be damped* by the numerical method. This effect, sometimes judged undesirable (B. Lindberg (1974): "dangerous property ..."), may also be welcome to suppress uninteresting oscillations. This is demonstrated by applying RADAU5 with very large tolerance (*Rtol* = *Atol* = 1) to the beam equation (1.10') with $n = 10$ and the perturbed initial value $\theta_n(0) = 0.4$. Here, the high oscillations soon disappear and the numerical solution becomes perfectly smooth (Fig. 3.4). If, however, the tolerance requirement is increased, the program is forced to follow all the oscillations and the picture remains the same as in Fig. 1.11.

Stability Functions of Order $\geq s$

Consider rational functions $R(z)=P(z)/Q(z)$, where $Q(0)=1$, and both $P(z)$ and $Q(z)$ are polynomials of degree at most s. If $R(z)$ is an approximation of e^z of order $\geq s$, then is follows from (3.5) that

$$e^z Q(z) = P(z) + C_1 z^{s+1} + C_2 z^{s+2} + \ldots . \tag{3.16}$$

Consequently, the polynomial $P(z)$ and also the error constants $C_1, C_2, \ldots$ are uniquely determined in terms of the coefficients of $Q(z)$. For

$$Q(z) = q_0 + q_1 z + q_2 z^2 + \ldots + q_s z^s, \qquad q_0 = 1 \tag{3.17}$$

an expansion of $e^z Q(z)$ into powers of z yields

$$\begin{aligned} P(z) = q_0 &+ z\Big(\frac{q_0}{1!}+\frac{q_1}{0!}\Big) + z^2\Big(\frac{q_0}{2!}+\frac{q_1}{1!}+\frac{q_2}{0!}\Big) \\ &+ \ldots + z^s\Big(\frac{q_0}{s!}+\frac{q_1}{(s-1)!}+\ldots+\frac{q_s}{0!}\Big), \end{aligned} \tag{3.18}$$

and for the error constants

$$C_1 = \frac{q_0}{(s+1)!} + \frac{q_1}{s!} + \ldots + \frac{q_{s-1}}{2!} + \frac{q_s}{1!} \tag{3.19}$$

$$C_2 = \frac{q_0}{(s+2)!} + \frac{q_1}{(s+1)!} + \ldots + \frac{q_{s-1}}{3!} + \frac{q_s}{2!}. \tag{3.20}$$

The Polynomial $M(x)$. With help of the polynomial

$$M(x) = q_s + q_{s-1}\frac{x}{1!} + q_{s-2}\frac{x^2}{2!} + \ldots + q_0\frac{x^s}{s!} \tag{3.21}$$

the formulas for $Q(z)$ and $P(z)$ become more symmetric. We have

$$Q(z) = M^{(s)}(0) + M^{(s-1)}(0)z + \ldots + M(0)z^s \tag{3.22}$$

$$P(z) = M^{(s)}(1) + M^{(s-1)}(1)z + \ldots + M(1)z^s, \tag{3.23}$$

and the error constants are given by

$$C_1 = \int_0^1 M(x)\,dx, \qquad C_2 = \int_0^1 (1-x)M(x)\,dx. \tag{3.24}$$

For the stability function of collocation methods we have the following nice result.

Theorem 3.10 (K. Wright 1970, S.P. Nørsett 1975). *The stability function of the collocation method based on the points $c_1, c_2, \ldots, c_s$ is given by $R(z) = P(z)/Q(z)$, where $Q(z)$ and $P(z)$ are the polynomials of (3.22) and (3.23), respectively, with $M(x)$ given by*

$$M(x) = \frac{1}{s!}\prod_{i=1}^{s}(x-c_i). \tag{3.25}$$

Proof (Nørsett & Wanner 1979). We assume $x_0 = 0$, $h = 1$, $\lambda = z$, $y_0 = 1$ and let $u(x)$ be the collocation polynomial. Since $u'(x) - zu(x)$ is a polynomial of degree s which vanishes at the collocation points, there are constants K_0 and K such that

$$u'(x) - zu(x) = K_0 M(x) \qquad \text{or} \qquad \Big(1 - \frac{D}{z}\Big)u(x) = KM(x) \tag{3.26}$$

with the polynomial $M(x)$ of (3.25) (D denotes the differentiation operator). Expanding $(1 - D/z)^{-1}$ into a geometric series yields

$$u(x) = K\Big(1 + \frac{D}{z} + \frac{D^2}{z^2} + \ldots + \frac{D^s}{z^s}\Big)M(x), \tag{3.27}$$

because $M^{(j)}(x) \equiv 0$ for $j > s$. From $u(1) = R(z)u(0)$ we have the relation $R(z) = u(1)/u(0)$, which leads to (3.22) and (3.23). □

Padé Approximations to the Exponential Function

Comme cela est souvent le cas en ce qui concerne les découvertes scientifiques, leur inventeur n'est pas H. Padé.

(C. Brezinski 1984, Œuvres de H. Padé, p. 5)

Padé approximations (Padé 1892) are rational functions which, for a given degree of the numerator and the denominator, have highest order of approximation. Their origin lies in the theory of continued fractions and they played a fundamental role in Hermite's (1873) proof of the transcendency of e.

These optimal approximations can be obtained for the exponential function e^z from (3.22) and (3.23) by the following idea (Padé 1899): choose $M(x)$ such that in (3.22) and (3.23) as many terms as possible involving high powers of z become zero, i.e.,

$$M(x) = \frac{x^k(x-1)^j}{(k+j)!}\,; \tag{3.28}$$

then $M^{(i)}(0) = 0$ for $i = 0, \ldots, k-1$ and $M^{(i)}(1) = 0$ for $i = 0, \ldots, j-1$.

Theorem 3.11. *The* (k,j)*-Padé approximation to* e^z *is given by*

$$R_{kj}(z) = \frac{P_{kj}(z)}{Q_{kj}(z)} \tag{3.29}$$

where

$$P_{kj}(z) = 1 + \frac{k}{j+k}z + \frac{k(k-1)}{(j+k)(j+k-1)} \cdot \frac{z^2}{2!} + \ldots + \frac{k(k-1)\ldots 1}{(j+k)\ldots(j+1)} \cdot \frac{z^k}{k!}$$

$$\begin{aligned} Q_{kj}(z) &= 1 - \frac{j}{k+j}z + \frac{j(j-1)}{(k+j)(k+j-1)} \cdot \frac{z^2}{2!} - \ldots + (-1)^j \frac{j(j-1)\ldots 1}{(k+j)\ldots(k+1)} \cdot \frac{z^j}{j!} \\ &= P_{jk}(-z), \end{aligned}$$

with error

$$e^z - R_{kj}(z) = (-1)^j \frac{j!k!}{(j+k)!(j+k+1)!} z^{j+k+1} + \mathcal{O}(z^{j+k+2}). \tag{3.30}$$

It is the unique rational approximation to e^z *of order* $j+k$*, such that the degrees of numerator and denominator are* k *and* j*, respectively.*

Table 3.2. Padé approximations for e^z

$\frac{1}{1}$	$\frac{1+z}{1}$	$\frac{1+z+\frac{z^2}{2!}}{1}$
$\frac{1}{1-z}$	$\frac{1+\frac{1}{2}z}{1-\frac{1}{2}z}$	$\frac{1+\frac{2}{3}z+\frac{1}{3}\frac{z^2}{2!}}{1-\frac{1}{3}z}$
$\frac{1}{1-z+\frac{z^2}{2!}}$	$\frac{1+\frac{1}{3}z}{1-\frac{2}{3}z+\frac{1}{3}\frac{z^2}{2!}}$	$\frac{1+\frac{1}{2}z+\frac{1}{6}\frac{z^2}{2!}}{1-\frac{1}{2}z+\frac{1}{6}\frac{z^2}{2!}}$
$\frac{1}{1-z+\frac{z^2}{2!}-\frac{z^3}{3!}}$	$\frac{1+\frac{1}{4}z}{1-\frac{3}{4}z+\frac{1}{2}\frac{z^2}{2!}-\frac{1}{4}\frac{z^3}{3!}}$	$\frac{1+\frac{2}{5}z+\frac{1}{10}\frac{z^2}{2!}}{1-\frac{3}{5}z+\frac{3}{10}\frac{z^2}{2!}-\frac{1}{10}\frac{z^3}{3!}}$

Proof. Inserting (3.28) into (3.22) and (3.23) gives the formulas for $P_{kj}(z), Q_{kj}(z)$ and (3.30). The uniqueness is a consequence of the fact that the $(j+k)$-degree polynomial $M(x)$ of (3.21) must have a zero of multiplicity k at $x=0$, and one of multiplicity j at $x=1$. □

Table 3.2 shows the first Padé approximations to e^z. We observe that the stability function of many methods of Table 3.1 are Padé approximations. The *diagonal Padé approximations* are those with $k=j$.

Exercises

1. Let $R(z)$ be the stability function of (3.1) and $R^*(z)$ the stability function of its adjoint method (see Sect. II.8). Prove that

$$R^*(z) = \big(R(-z)\big)^{-1}.$$

2. Consider an implicit Runge-Kutta method of order $p \geq s$ with nonsingular A, distinct c_i and non-zero b_i. Show

   a) If $C(s)$ and $c_s = 1$ then (3.13);

   b) If $D(s)$ and $c_1 = 0$ then (3.14).

In both cases the stability function satisfies $R(\infty)=0$.

(For the definition of the assumptions $C(s)$ and $D(s)$ see Sect. IV.5).

3. Show that collocation methods can only be L-stable if $M(1)=0$, i.e., if one of the c's, usually c_s, equals 1.

4. (Padé (1899), see also Lagrange (1776)). Show that the continued fraction

$$e^x = 1 + \cfrac{x}{1-\cfrac{x}{2}+\cfrac{\frac{1}{1\cdot 3}\frac{x^2}{4}}{1+\cfrac{\frac{1}{3\cdot 5}\frac{x^2}{4}}{1+\cfrac{\frac{1}{5\cdot 7}\frac{x^2}{4}}{1+\cfrac{\frac{1}{7\cdot 9}\frac{x^2}{4}}{1+\ldots}}}}}$$

leads to the diagonal Padé approximations for e^x.

Hint. Compute the first partial fractions. If you don't succeed in finding a general proof, read Sect. IV.5.

5. The trapezoidal rule

$$\begin{array}{c|cc} 0 & 0 & 0 \\ 1 & 1/2 & 1/2 \\ \hline & 1/2 & 1/2 \end{array}$$

satisfies $a_{si}=b_i$, but not $R(\infty)=0$. Why doesn't this contradict Proposition 3.8?

6. Show that

$$y_1 = y_0 + hf(y_0+\theta(y_1-y_0))$$
$$y_1 = y_0 + h(1-\theta)f(y_0) + h\theta f(y_1)$$

are both nonlinear extensions of the θ-method. Find others.

7. The composition of a step of the θ-method with step-size αh, followed by a θ'-method with step-size $(1-2\alpha)h$ and again a θ-method with step-size αh leads to

$$R(z) = \Big(\frac{1+\alpha z(1-\theta)}{1-\alpha z\theta}\Big)^2 \Big(\frac{1+(1-2\alpha)z(1-\theta')}{1-(1-2\alpha)z\theta'}\Big)$$

Show that this method, for $\theta'=1-\theta$, is of order 2 if $\alpha = 1-\sqrt{2}/2$ and strongly A-stable (i.e., A-stable and $|R(\infty)|<1$) for $\theta > 1/2$. The authors Müller, Prohl, Rannacher & Turek (1994) call this method "fractional θ-method" and use it successfully for computations of the incompressible Navier-Stokes equations.

IV.4 Order Stars

> Mein hochgeehrter Lehrer, der vor wenigen Jahren verstorbene Geheime Hofrath *Gauss* in Göttingen, pflegte in vertraulichem Gespräche häufig zu äussern, die Mathematik sei weit mehr eine Wissenschaft für das Auge als eine für das Ohr. Was das Auge mit einem Blicke sogleich übersieht ...
> (J.F. Encke 1861, publ. in Kronecker's Werke, Vol. 5, page 391)

Order stars, discovered by searching for a better understanding of the stability properties of the Padé approximations to e^z (Wanner, Hairer & Nørsett 1978), offered nice and unexpected access to many other results: the "second barrier" of Dahlquist, the Daniel & Moore conjecture, highest possible order with real poles, comparison of stability domains (Jeltsch & Nevanlinna 1981, 1982), order bounds for hyperbolic or parabolic difference schemes (e.g., Iserles & Strang 1983, Iserles & Williamson 1983, Jeltsch 1988).

Introduction

> When I wrote my book in 1971 I wanted to draw "relative stability domains", but curious stars came out from the plotter. I thought of an error in the program and I threw them away ...
> (C.W. Gear, in 1979)

We present in Fig. 4.1 the stability domains for the Padé approximations R_{33}, R_{24}, R_{15}, R_{06} of Theorem 3.12, which are all 6th order approximations to $\exp(z)$. It can be observed that R_{33} and R_{24} are nicely A-stable. The other two are not, R_{15} violates (3.6) and R_{06} violates (3.7). After some meditation on these and similar figures, trying to obtain a better understanding of these phenomena, one is finally led to

Definition 4.1. The set

$$A = \Big\{ z \in \mathbb{C} \, ; \; |R(z)| > |e^z| \Big\} = \Big\{ z \in \mathbb{C} \, ; \; |q(z)| > 1 \Big\} \tag{4.1}$$

where $q(z) = R(z)/e^z$, is called the *order star* of R.

The order star does not compare $|R(z)|$ to 1, as does the stability domain, but to the exact solution $|e^z| = e^x$ and it is hoped that this might give more information. As we always assume that the coefficients of $R(z)$ are real, the order star is symmetric with respect to the real axis. Furthermore, since $|e^{iy}| = 1$, A is the complementary set of the stability domain S on the imaginary axis. Therefore we have from (3.6) and (3.7):

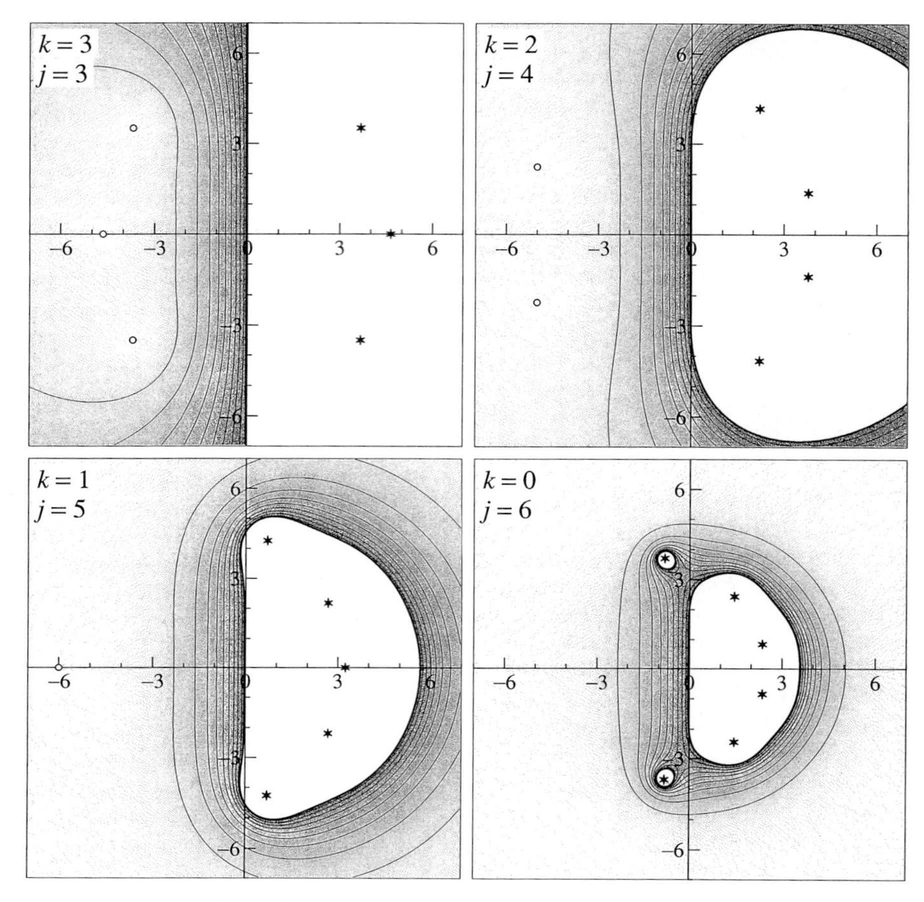

Fig. 4.1. Stability domains for Padé approximations

Lemma 4.2. *$R(z)$ is I-stable if and only if*

(i) $A \cap i\mathbb{R} = \emptyset$.

Further, $R(z)$ is A-stable if and only if (i) and

(ii) all poles of $R(z)$ (= poles of $q(z)$) lie in the positive half plane $\mathbb{C}^+$. □

Fig. 4.2 shows the order stars corresponding to the functions of Fig. 4.1. These order stars show a nice and regular behaviour: there are j black "fingers" to the right, each containing a pole of R_{kj}, and k white "fingers" to the left, each containing a zero. Exactly two boundary curves of A tend to infinity near to the imaginary axis. These properties are a consequence of the following three Lemmas.

Lemma 4.3. *If $R(z)$ is an approximation to e^z of order p, i.e., if*

$$e^z - R(z) = C z^{p+1} + \mathcal{O}(z^{p+2}) \tag{4.2}$$

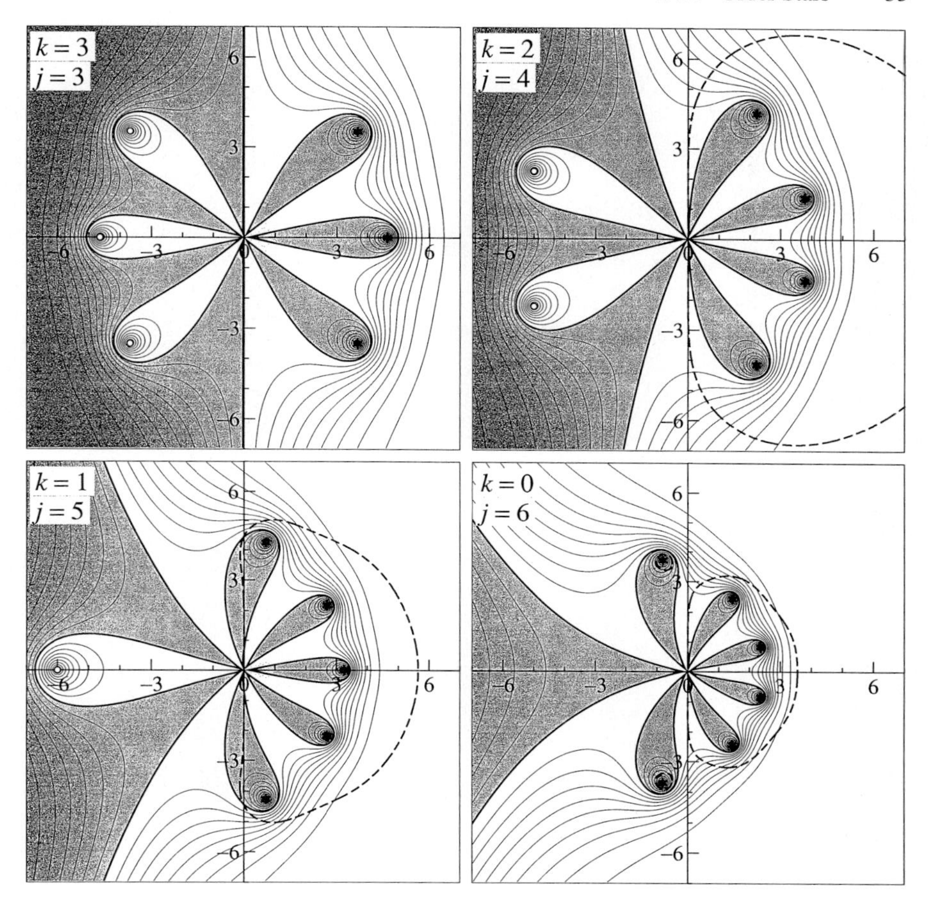

Fig. 4.2. Order stars for Padé approximations

with $C \neq 0$, then, for $z \to 0$, A behaves like a "star" with $p+1$ sectors of equal width $\pi/(p+1)$, separated by $p+1$ similar "white" sectors of the complementary set. The positive real axis is inside a black sector iff $C<0$ and inside a white sector iff $C>0$.

Proof. Dividing the error formula (4.2) by e^z gives

$$\frac{R(z)}{e^z} = 1 - Cz^{p+1} + \mathcal{O}(z^{p+2}).$$

Thus the value $R(z)/e^z$ surrounds the point 1 as often as z^{p+1} surrounds the origin, namely $p+1$ times. So, $R(z)/e^z$ is $p+1$ times alternatively inside or outside the unit circle. It lies inside for small positive real z whenever $C>0$. □

Lemma 4.4. *If $z = re^{i\theta}$ and $r \to \infty$, then $z \in A$ for $\pi/2 < \theta < 3\pi/2$ and $z \notin A$ for $-\pi/2 < \theta < \pi/2$. The border ∂A possesses only two branches which go to infinity. If*

$$R(z) = Kz^{\ell} + \mathcal{O}(z^{\ell-1}) \qquad \text{for} \quad z \to \infty, \tag{4.3}$$

these branches asymptotically approach

$$x = \log|K| + \ell \log|y| \tag{4.4}$$

Proof. The first assertion is the well-known fact that the exponential function, for $\mathrm{Re}\, z \to \pm\infty$ is much stronger than any polynomial or rational function. In order to show the uniqueness of the border lines, we consider for $r \to \infty$ the two functions

$$\varphi_1(\theta) = |e^z|^2 = e^{2r\cos\theta}$$
$$\varphi_2(\theta) = |R(z)|^2 = R(re^{i\theta})R(re^{-i\theta}).$$

Differentiation gives

$$\frac{\varphi_1'}{\varphi_1} = -2r\sin\theta, \qquad \frac{\varphi_2'}{\varphi_2} = 2r\mathrm{Re}\left(ie^{i\theta}\cdot\frac{R'(re^{i\theta})}{R(re^{i\theta})}\right). \tag{4.5}$$

Since $|R'/R| \to 0$ for $r \to \infty$, we have

$$\frac{d}{d\theta}\log\varphi_1(\theta) < \frac{d}{d\theta}\log\varphi_2(\theta) \qquad \text{for} \quad \theta \in [\varepsilon, \pi - \varepsilon].$$

Hence in this interval there can only be one value of θ with $\varphi_1(\theta) = \varphi_2(\theta)$. Formula (4.4) is obtained from (4.3) by

$$|K|(x^2+y^2)^{\ell/2} \approx e^x, \qquad \log|K| + \frac{\ell}{2}\log(x^2+y^2) \approx x$$

and by neglecting x^2, which is justified because $x/y \to 0$ whenever $x+iy$ tends to infinity on the border of A. □

It is clear from the maximum principle that each bounded "finger" of A in Fig. 4.2 must contain a pole of $q(z)$. A still stronger result is the following:

Lemma 4.5. *Each bounded subset $F \subset A$ with common boundary $\partial F \subset \partial A$ collecting m sectors at the origin must contain at least m poles of $q(z)$ (each counted according to its multiplicity). Analogously, each bounded "white" subset $F \subset \mathbb{C} \setminus A$ with m sectors at the origin must contain at least m zeros of $q(z)$.*

Proof. Suppose first that ∂F is represented by a parametrized positively oriented loop $c(t)$, $t_0 \le t \le t_1$. Let $\vec{a} = (c_1'(t), c_2'(t))$ be the tangent vector and $\vec{n} = (c_2'(t), -c_1'(t))$ an exterior normal vector. We write

$$q(z) = r(x,y)\cdot e^{i\varphi(x,y)}, \qquad z = x+iy$$

so that $\log q(z) = \log r(x,y) + i\varphi(x,y)$. Since the modulus increases inside F, we have

$$\frac{\partial(\log r)}{\partial \vec{n}} \le 0. \tag{4.6}$$

Now the Cauchy-Riemann differential equations for $\log q$ are

$$\frac{\partial(\log r)}{\partial x} = \frac{\partial\varphi}{\partial y}; \qquad \frac{\partial(\log r)}{\partial y} = -\frac{\partial\varphi}{\partial x}, \tag{4.7}$$

so that (4.6) becomes

$$\frac{\partial\varphi}{\partial \vec{a}} \le 0. \tag{4.8}$$

This inequality is strict except at a finite number of points, because $q'(c(t))\cdot c'(t) = i\cdot q(c(t))\cdot \partial\varphi/\partial\vec{a}$ and the number of zeros of $q'(z)$ is finite. Thus the *argument* of q decreases along c. If the contour curve $c(t)$ returns m times to the origin, where the argument is a multiple of 2π, the vector $q(z)$ must perform at least m complete revolutions in the negative sense (Fig. 4.3). Thus the argument principle (an idea which we have already encountered in Sect. I.13; see Volume I, pages 81 and 382), ensures the presence of at least m poles inside F (there are no zeros, because these are not in A).

If the boundary curve is represented by several curves, all rotation numbers are added up. For "white" subsets the proof is similar, just that $\partial(\log r)/\partial\vec{n} > 0$ and the argument rotates in the other sense. □

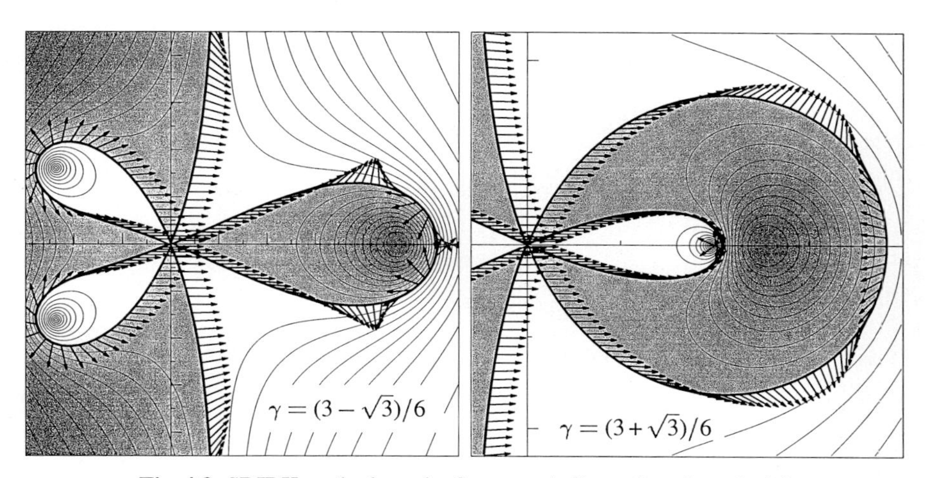

Fig. 4.3. SDIRK methods, order 3; arrows indicate direction of $q(z)$

Fig. 4.3 gives an illustration of two order stars for the SDIRK methods of order 3 (Table 3.1, case e). Here, $q(z)$ possesses a double pole at $z = 1/\gamma$. However, for $\gamma = (3-\sqrt{3})/6$, the bounded component F of A collects only *one* sector at the origin. Since the vector $q(z)$ performs two rotations, there is in addition to the origin a second point on ∂F for which $\arg(q) = 0$, i.e., $\arg(R(z)) = \arg(e^z)$. Thus, because $|R(z)| = |e^z|$ on ∂A, we have $R(z) = e^z$. These points are called *exponential fitting points.* Another version of Lemma 4.5 is thus (Iserles 1981):

Lemma 4.5'. *Each bounded subset $F \subset A$ with $\partial F \subset \partial A$ contains exactly as many poles as there are exponential fitting points on its boundary.* □

Order and Stability for Rational Approximations

In the sequel we suppose $R(z)$ to be an arbitrary rational approximation of order p with k zeros and j poles.

Theorem 4.6. *If $R(z)$ is A-stable, then $p \le 2k_1 + 2$, where k_1 is the number of different zeros of $R(z)$ in $\mathbb{C}^-$.*

Proof. At least $[(p+1)/2]$ sectors of A start in $\mathbb{C}^-$ (Lemma 4.3). By A-stability these have to be infinite and enclose at least $[(p+1)/2]-1$ bounded white fingers, each containing at least one zero by Lemma 4.5. Therefore $[(p+1)/2]-1 \le k_1$. □

Theorem 4.7. *If $R(z)$ is I-stable, then $p \le 2j_1$, where j_1 is the number of poles of $R(z)$ in $\mathbb{C}^+$.*

Proof. At least $[(p+1)/2]$ sectors of A start in $\mathbb{C}^+$. They cannot cross $i\mathbb{R}$ and must therefore be bounded (Lemma 4.4). Again by Lemma 4.5 we have $[(p+1)/2] \le j_1$. □

Theorem 4.8. *Suppose that $p \ge 2j-1$ and $|R(\infty)| \le 1$. Then, $R(z)$ is A-stable.*

Proof. By Proposition 3.6 the function $R(z)$ is I-stable. Applying Theorem 4.7 we get $j_1 \ge j$ so that I-stability implies A-stability. □

Theorem 4.9 (Crouzeix & Ruamps 1977). *Suppose $p \ge 2j-2$, $|R(\infty)| \le 1$, and the coefficients of the denominator $Q(z)$ have alternating signs. Then, $R(z)$ is A-stable.*

Proof. A similar argument as in the foregoing proof allows at most *one* pole in $\mathbb{C}^-$. It would then be real and its existence would contradict the hypothesis on signs of $Q(z)$. □

Theorem 4.10. *Suppose $p \geq 2j-3$, $R(z)$ is I-stable, and the coefficients of $Q(z)$ have alternating signs. Then, $R(z)$ is A-stable.*

Proof. For $p \geq 2j-3$ the argument of the foregoing proof is still valid. However Proposition 3.6 is no longer applicable and we need the hypothesis on I-stability. □

We see from Fig. 4.2 that all poles and all zeros for Padé approximations must be *simple.* Whenever two poles coalesce, the corresponding sectors create a bounded white finger between them with the need for an additional zero. Thus the presence of multiple zeros or poles will require an order reduction.

Theorem 4.11. *Let $R(z)$ possess k_0 distinct zeros and j_0 distinct poles. Then, $p \leq k_0 + j_0$.*

Fig. 4.4. Order star on Gaussian sphere

Proof. We identify the complex plane with the Gaussian sphere and the order star with a CW-complex decomposition of this sphere (Fig. 4.4). Let s_2 be the number of 2-cells f_i, s_1 the number of 1-cells l_i (paths), and s_0 the number of vertices. Then Euler's polyhedral formula ("Si enim numerus angulorum solidorum fuerit $=S$, numerus acierum $=A$ et numerus hedrarum $=H$, semper habetur $S+H=A+2$, hincque vel $S=A+2-H$ vel $H=A+2-S$ vel $A=S+H-2$, quae relationis simplicitas ob demonstrationis difficultatem ...", Euler (1752)), implies

$$s_0 - s_1 + s_2 = 2. \tag{4.9}$$

Modern versions are in any book on algebraic topology, for particularly easy reading see e.g. Massey (1980, p. 87, Corollary 4.4). Formula (4.9) is only true if all f_i are homeomorphic to disks. Otherwise, they have to be cut into disks by additional paths (dotted in Fig. 4.4). So, in general, we have

$$s_0 - s_1 + s_2 \geq 2. \tag{4.9'}$$

Since each vertex is reached by at least 2 paths, the origin by hypothesis by $2p+2$, and since every path has two extremities, we have

$$s_1 - s_0 \geq p. \tag{4.10}$$

By Lemma 4.5 each 2-cell, with the exception of two (the two "infinite" ones) must contain at least a pole or a zero, so we have

$$s_2 \leq k_0 + j_0 + 2. \tag{4.11}$$

These three inequalities give $p \leq k_0 + j_0$. □

Stability of Padé Approximations

... evidence is given to suggest that these are the only L-acceptable Padé approximations to the exponential.

(B.L. Ehle 1973)

Theorem 4.12. *A Padé approximation $R_{kj}(z)$, given in (3.30), is A-stable if and only if $k \leq j \leq k+2$. All zeros and all poles are simple.*

Proof. The "if"-part is a consequence of Theorem 4.9. The "only if"-part follows from Theorem 4.6 since $p = k + j$. For the same reason Theorem 4.11 shows that all poles and zeros are simple. □

Comparing Stability Domains

Da ist der allerärmste Mann
dem ander'n viel zu reich,
das Schicksal setzt den Hobel an
und hobelt beide gleich.

(F. Raimund, das Hobellied)

Jeltsch & Nevanlinna (1978) proved the following "disk theorem": *If S is the stability domain of an s-stage explicit Runge-Kutta method and D the disk with centre $-s$ and radius s* (i.e the stability domain of s explicit Euler steps with step size h/s), *then*

$$S \not\supset D \tag{4.12}$$

unless $S = D$ and the method in question is Euler's method. This curious result expresses the fact that Euler's method is "the most stable" of all methods with equal numerical work. After the discovery of order stars it became clear that the result is much more general and that *any* method has the same property (Jeltsch & Nevanlinna 1981). We shall also see in Chapter V that this result generalizes to many multistep methods. The main tool of this theory is

Definition 4.13. Let $R_1(z)$ and $R_2(z)$ be rational approximations to e^z, then their *relative order star* is defined as

$$B=\left\{z\in\mathbb{C}\,;\;\left|\frac{R_1(z)}{R_2(z)}\right|>1\right\}. \tag{4.13}$$

Here, the stability function for method 1 is compared to the stability function for method 2 instead of to the exact solution e^z. The following order relations

$$e^z-R_1(z)=C_1z^{p_1+1}+\ldots$$
$$e^z-R_2(z)=C_2z^{p_2+1}+\ldots$$

lead, by subtraction, to

$$\frac{R_1(z)}{R_2(z)}=1-Cz^{p+1}+\ldots \tag{4.14}$$

where $p=\min(p_1,p_2)$ and

$$C=\begin{cases}C_1-C_2 & \text{if } p_1=p_2\\ C_1 & \text{if } p_1>p_2\\ -C_2 & \text{if } p_1<p_2.\end{cases} \tag{4.15}$$

Remark 4.14. The statement of Lemma 4.3 remains unchanged for B, whenever $C\neq 0$. Since the fraction $R_1(z)/R_2(z)$ has no essential singularity at infinity, there is no analogue of Lemma 4.4. Further, the boundedness assumption on F can be omitted in Lemmas 4.5 and 4.5' (if ∞ is a pole of $R_1(z)/R_2(z)$, it has to be counted also). With the correspondences displayed in Table 4.1, the statements of Theorems 4.6 and 4.7 remain true for B.

Table 4.1. Correspondences between A and B

order star A (4.1)	$\longleftrightarrow$	relative order star B (4.13)
imaginary axis	$\longleftrightarrow$	∂S_2
$\mathbb{C}^-$	$\longleftrightarrow$	interior of S_2
$\mathbb{C}^+$	$\longleftrightarrow$	exterior of S_2
method A-stable	$\longleftrightarrow$	$S_1\supset S_2$
p	$\longleftrightarrow$	$\min(p_1,p_2)$

Theorem 4.15. *If $R_1(z)$ and $R_2(z)$ are polynomial stability functions of degree s and orders ≥ 1, then the corresponding stability domains satisfy*

$$S_1\not\supset S_2 \qquad \textit{and} \qquad S_1\not\subset S_2. \tag{4.16}$$

Proof. Suppose that $S_1\supset S_2$ (i.e., by Table 4.1, suppose "A-stability"). Then the analogue of Theorem 4.7 requires that $R_1(z)/R_2(z)$ have a pole *outside* S_2. Since

$R_1(z)$ and $R_2(z)$ have the same degree, $R_1(z)/R_2(z)$ has no pole at infinity. Therefore the only poles of $R_1(z)/R_2(z)$ are the zeros of $R_2(z)$ and these are *inside* S_2. This is a contradiction and proves the first part of (4.16). The second part is obtained by exchanging $R_1(z)$ and $R_2(z)$. □

In order to compare numerical methods with *different* numerical work, we consider scaled stability domains.

Definition 4.16. Let $R(z)$ be the stability function of degree s of an explicit Runge-Kutta method (usually with s stages), then

$$S^{scal} = \Big\{ z\,;\ |R(sz)| \le 1 \Big\} = \Big\{ z\,;\ s\cdot z \in S \Big\} = \frac{1}{s} S \tag{4.17}$$

will be called the *scaled stability domain* of the method.

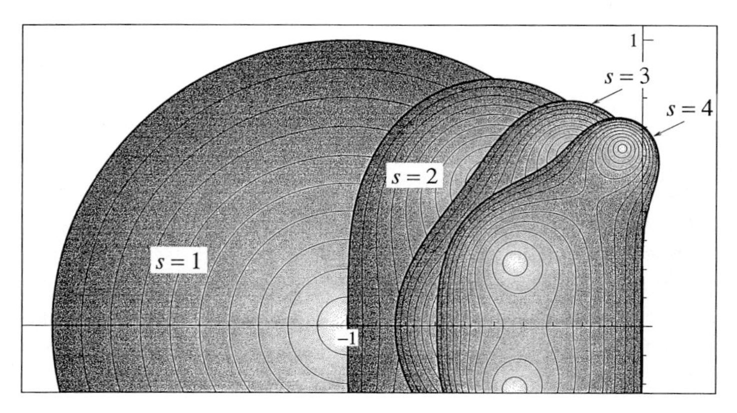

Fig. 4.5. Scaled stability domains for Taylor methods (2.12)

Theorem 4.17 (Jeltsch & Nevanlinna 1981). *If $R_1(z)$ and $R_2(z)$ are the stability functions of degrees s_1 resp. s_2 of two explicit Runge-Kutta methods of orders ≥ 1, then*

$$S_1^{scal} \not\supset S_2^{scal} \qquad \textit{and} \qquad S_1^{scal} \not\subset S_2^{scal}, \tag{4.18}$$

i.e., a scaled stability domain can never completely contain another.

The interesting interpretation of this result is that for any two methods, there exists a differential equation $y' = \lambda y$ such that one of them performs better than the other. No "miracle" method is possible.

Proof. We compare s_2 steps of method 1 with step size h/s_2 to s_1 steps of method 2 with step size h/s_1. Both procedures then have comparable numerical work for the same advance in step size. Applied to $y' = \lambda y$, this compares

$$\Big(R_1\big(\frac{z}{s_2}\big)\Big)^{s_2} \qquad \text{to} \qquad \Big(R_2\big(\frac{z}{s_1}\big)\Big)^{s_1}$$

of the same degree. Theorem 4.15 now gives

$$s_2 \cdot S_1 \not\supset s_1 \cdot S_2 \qquad \text{or} \qquad S_1^{scal} \not\supset S_2^{scal}. \qquad \square$$

As an illustration to this theorem, we present in Fig. 4.5 the *scaled* stability domains for the Taylor methods of orders 1, 2, 3, 4 (compare with Fig. 2.1). It can clearly be observed that none of them contains another.

Rational Approximations with Real Poles

The surprising result is that the maximum reachable order is $m+1$.
(Nørsett & Wolfbrandt 1977)

The stability functions of diagonally implicit Runge-Kutta methods (DIRK methods), i.e., methods with $a_{ij}=0$ for $i<j$, are

$$R(z)=\frac{P(z)}{(1-\gamma_1 z)(1-\gamma_2 z)\ldots(1-\gamma_s z)}, \tag{4.19}$$

where $\gamma_i = a_{ii}$ $(i=1,\ldots,s)$ and degree $P \le s$. This follows at once from Formula (3.3) of Proposition 3.2, since the determinant of a triangular matrix is the product of its diagonal elements. Thus $R(z)$ possesses *real poles* $1/\gamma_1, 1/\gamma_2, \ldots, 1/\gamma_s$. Such approximations to e^z will also appear in the next sections as stability functions of Rosenbrock methods and so-called singly-implicit Runge-Kutta methods. They thus merit a more thorough study. Research on these real-pole approximations was started by Nørsett (1974) and Wolfbrandt (1977). Many results are collected in their joint paper Nørsett & Wolfbrandt (1977).

If the method is of order at least s, $P(z)$ is given by (3.18). We shall here, and in the sequel, very often write the formulas for $s=3$ without always mentioning how trivial their extension to arbitrary s is. Hence for $s=3$

$$R(z)=\frac{1+z\Big(\frac{S_0}{1!}-\frac{S_1}{0!}\Big)+z^2\Big(\frac{S_0}{2!}-\frac{S_1}{1!}+\frac{S_2}{0!}\Big)+z^3\Big(\frac{S_0}{3!}-\frac{S_1}{2!}+\frac{S_2}{1!}-\frac{S_3}{0!}\Big)}{1-zS_1+z^2S_2-z^3S_3} \tag{4.20}$$

where

$$S_0=1, \qquad S_1=\gamma_1+\gamma_2+\gamma_3, \qquad S_2=\gamma_1\gamma_2+\gamma_1\gamma_3+\gamma_2\gamma_3, \qquad S_3=\gamma_1\gamma_2\gamma_3.$$

The error constant is for $p=s$

$$C=\frac{S_0}{4!}-\frac{S_1}{3!}+\frac{S_2}{2!}-\frac{S_3}{1!}. \tag{4.21}$$

Theorem 4.18. *Let $R(z)$ be an approximation to e^z of order p with real poles only and let k be the degree of its numerator. Then,*

$$p \le k+1.$$

Proof. If a sector of the order star A ends up with a pole on the real axis, then by symmetry the complex conjugate sector must join the first one. All white sectors enclosed by these two must therefore be finite (Fig. 4.6.). The same is true for sectors joining the infinite part of A. There is thus on each side of the real axis *at most one* white sector which can be infinite. Thus the remaining $p-1$ white sectors require together at least $p-1$ zeros by Lemma 4.5, i.e., we have $p-1 \leq k$. □

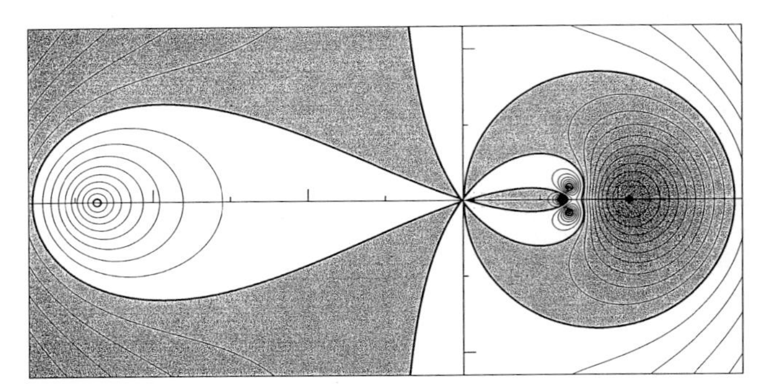

Fig. 4.6. An approximation with real poles, 3 zeros, order 4

Remark 4.19. If $p \geq k$, then at least one white sector must be unbounded. This is then either the first sector on the positive real axis, or, by symmetry, there is a pair of two sectors. By the proof of Theorem 4.18 the pair is unique and we shall call it *Cary Grant's part.*

Remark 4.20. If $p = k+1$, the optimal case, there are $k+2$ white sectors, two of them are infinite. Hence each of the remaining k sectors must then contain exactly one root of $P(z)$. As a consequence, $C < 0$ iff $P(z)$ has no positive real root between the origin and the first pole.

The Real-Pole Sandwich

We now analyze the approximations (4.19) with order $p \geq s$ in more detail (Nørsett & Wanner 1979). We are interested in two sets:

Definition 4.21. Let L be the set of $(\gamma_1, \ldots, \gamma_s)$ for which $\deg P(z)$ in (4.20) is $\leq s-1$, i.e., $R(\infty) = 0$ for $\gamma_i \neq 0$ $(i = 1, \ldots, s)$.

Definition 4.22. Denote by H the set of $(\gamma_1, \ldots, \gamma_s)$ for which the error constant (4.21) is zero, i.e., for which the approximation has highest possible order $p = s+1$.

A consequence of Theorem 4.18 is

$$L \cap H = \emptyset. \tag{4.22}$$

Written for the case $s = 3$ (generalizations to arbitrary s are straightforward) and using (4.20) and (4.21) the sets L and H become

$$\begin{aligned} L &= \Big\{(\gamma_1, \gamma_2, \gamma_3)\ ;\ \frac{1}{3!} - \frac{\gamma_1 + \gamma_2 + \gamma_3}{2!} + \frac{\gamma_1\gamma_2 + \gamma_1\gamma_3 + \gamma_2\gamma_3}{1!} - \frac{\gamma_1\gamma_2\gamma_3}{0!} = 0\Big\} \\ H &= \Big\{(\gamma_1, \gamma_2, \gamma_3)\ ;\ \frac{1}{4!} - \frac{\gamma_1 + \gamma_2 + \gamma_3}{3!} + \frac{\gamma_1\gamma_2 + \gamma_1\gamma_3 + \gamma_2\gamma_3}{2!} - \frac{\gamma_1\gamma_2\gamma_3}{1!} = 0\Big\}. \end{aligned} \tag{4.23}$$

Theorem 4.23 (Nørsett & Wanner 1979). *The surfaces H and L are each composed of s disjoint connected sheets*

$$L = L_1 \cup L_2 \cup \ldots \cup L_s, \qquad H = H_1 \cup H_2 \cup \ldots \cup H_s. \tag{4.24}$$

If a direction $\delta = (\delta_1, \ldots, \delta_s)$ is chosen with all $\delta_i \neq 0$ and if k of them are positive, then the ray

$$X = \Big\{(\gamma_1, \ldots, \gamma_s)\ ;\quad \gamma_i = t\delta_i,\ \ 0 \le t < \infty\Big\} \tag{4.25}$$

intersects the sheets $H_1, L_1, H_2, L_2, \ldots, H_k, L_k$ in this order and no others.

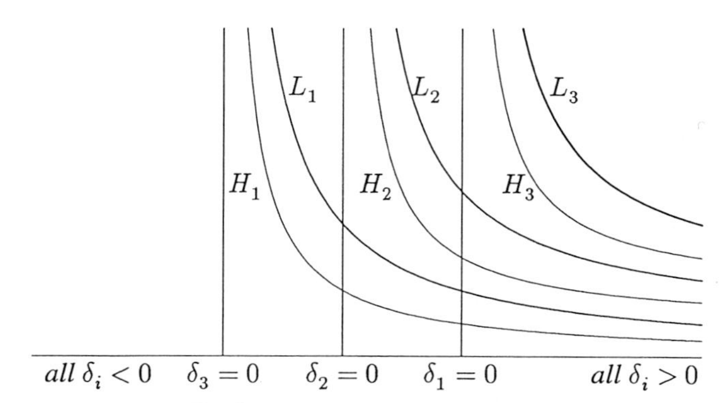

Fig. 4.7. Formation of the sandwich

Proof. When the δ_i have been chosen, inserting $\gamma_i = t\delta_i$ into (4.23) gives

$$\begin{aligned} &\frac{1}{3!} - t\frac{\delta_1 + \delta_2 + \delta_3}{2!} + t^2\frac{\delta_1\delta_2 + \delta_1\delta_3 + \delta_2\delta_3}{1!} - t^3\frac{\delta_1\delta_2\delta_3}{0!} = 0 \\ &\frac{1}{4!} - t\frac{\delta_1 + \delta_2 + \delta_3}{3!} + t^2\frac{\delta_1\delta_2 + \delta_1\delta_3 + \delta_2\delta_3}{2!} - t^3\frac{\delta_1\delta_2\delta_3}{1!} = 0 \end{aligned} \tag{4.26}$$

for L and H, respectively. These are third (in general sth) degree polynomials whose positive roots we have to study. We vary the δ's, and hence the ray X, starting with all δ's negative. The polynomials (4.26) then have all coefficients positive and obviously no positive real roots. When now *one* delta, say δ_3, changes

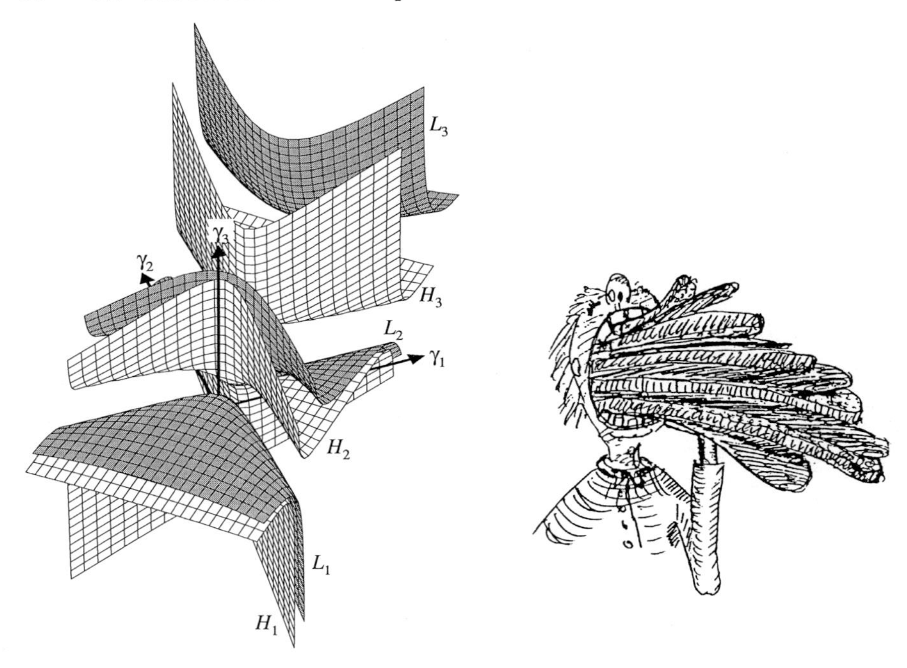

Fig. 4.8. The sandwich for $s=3\ldots$ and for $s=5$

sign, the leading coefficients of (4.26) become zero and *one* root becomes infinite for each equation and satisfies asymptotically

$$\begin{aligned} \frac{\delta_1\delta_2}{1!} - t\frac{\delta_1\delta_2\delta_3}{0!} &\approx 0 \qquad \Longrightarrow \qquad t \approx \frac{1}{\delta_3} \\ \frac{\delta_1\delta_2}{2!} - t\frac{\delta_1\delta_2\delta_3}{1!} &\approx 0 \qquad \Longrightarrow \qquad t \approx \frac{1}{2\delta_3} \end{aligned} \tag{4.27}$$

for L and H, respectively. Thus H comes below and L comes above. Because of $L \cap H = \emptyset$ (4.22) these two roots can never cross and must therefore remain in this configuration (see Fig. 4.7).

When then successively δ_2 and δ_1 change sign, the same scene repeats itself again and again, always two sheets of H and L descend from above in that order and are layed on the lower sheets like slices of bread and ham of a giant sandwich. Because $L \cap H = \emptyset$, these sheets can never cross, two roots for L or H can never come together and become complex. So all roots must remain real and the theorem must be true.

A three-dimensional view of these surfaces is given in Fig. 4.8. □

The following theorem describes the form of the corresponding order star in all these sheets.

Theorem 4.24. *Let $G_1, \ldots, G_s$ be the open connected components of $\mathbb{R}^s \setminus H$ such that L_i lies in G_i, and let G_0 be the component containing the origin. Then the order star of $R(z)$ given by (4.20) possesses exactly k bounded fingers to the right of Cary Grant's part if and only if*

$$(\gamma_1, \ldots, \gamma_s) \in G_k \cup H_k.$$

Proof. We prove this by a continuity argument letting the point $(\gamma_1, \ldots, \gamma_s)$ travel through the sandwich. Since Cary Grant's part is always present (Remark 4.19), the number of bounded sectors can change only where the error constant C (4.21) changes sign, i.e., on the surfaces $H_1, H_2, \ldots, H_s$. Fig. 4.9 gives some snap-shots from this voyage for $s = 3$ and $\gamma_1 = \gamma_2 = \gamma_3 = \gamma$. In this case the equations (4.23) become

$$\begin{aligned} \frac{1}{3!} - \frac{3\gamma}{2!} + \frac{3\gamma^2}{1!} - \frac{\gamma^3}{0!} &= 0 \\ \frac{1}{4!} - \frac{3\gamma}{3!} + \frac{3\gamma^2}{2!} - \frac{\gamma^3}{1!} &= 0 \end{aligned} \tag{4.28}$$

whose roots

$$\begin{aligned} &\lambda_1 = 0.158984, \qquad &&\lambda_2 = 0.435867, \qquad &&\lambda_3 = 2.40515 \\ &\chi_1 = 0.128886, &&\chi_2 = 0.302535, &&\chi_3 = 1.06858 \end{aligned} \tag{4.29}$$

do interlace nicely as required by Theorem 4.23. The affirmation of Theorem 4.24 for $s = 3$ can be clearly observed in Fig. 4.9.

For the proof of the general statement we also put $\gamma_1 = \ldots = \gamma_s = \gamma$ and investigate the two extreme cases:

1. $\gamma = 0$: Here $R(z)$ is the Taylor polynomial $1 + z + \ldots + z^s/s!$ whose order star has no bounded sector at all.

2. $\gamma \to \infty$: The numerator of $R(z)$ in (4.20) becomes for $s = 3$

$$P(z) = 1 + z\Big(\frac{1}{1!} - \frac{3\gamma}{0!}\Big) + z^2\Big(\frac{1}{2!} - \frac{3\gamma}{1!} + \frac{3\gamma^2}{0!}\Big) + z^3\Big(\frac{1}{3!} - \frac{3\gamma}{2!} + \frac{3\gamma^2}{1!} - \frac{\gamma^3}{0!}\Big). \tag{4.30}$$

If we let $\gamma \to \infty$, this becomes with $z\gamma = w$

$$1 - w\Big(3 + \mathcal{O}\big(\frac{1}{\gamma}\big)\Big) + w^2\Big(3 + \mathcal{O}\big(\frac{1}{\gamma}\big)\Big) - w^3\Big(1 + \mathcal{O}\big(\frac{1}{\gamma}\big)\Big).$$

Therefore all roots $w_i \to 1$, hence $z_i \to 1/\gamma$ (see the last picture of Fig. 4.9). Therefore *no* zero of $R(z)$ can remain left of Cary Grant's part and we have s bounded fingers.

Since between these extreme cases, there are at most s crossings of the surface H, Theorem 4.24 must be true. □

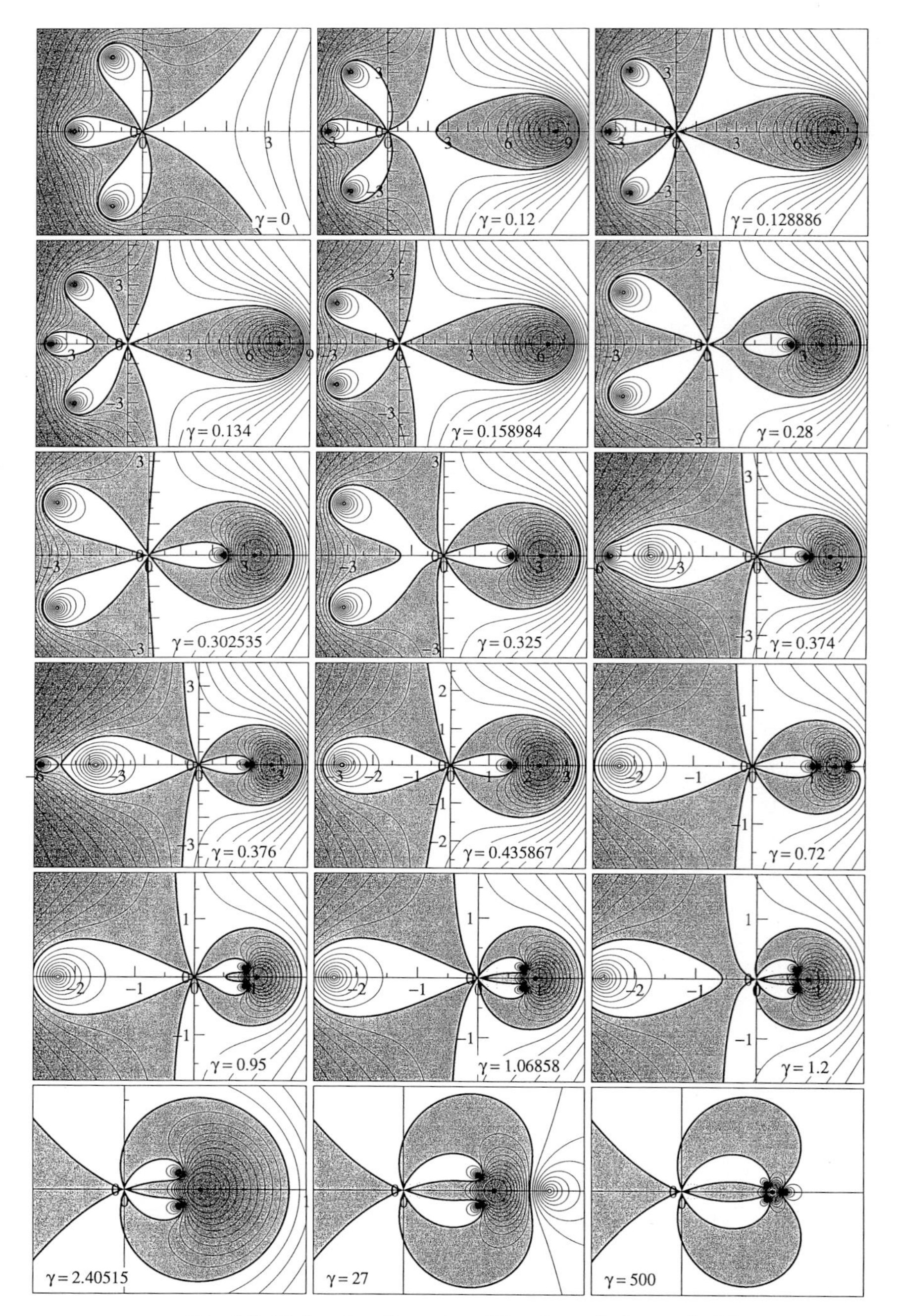

Fig. 4.9. Order stars for γ travelling through the sandwich

Theorem 4.25. *The function* $R(z)$ *defined by (4.20) can be I-stable only if*

$$(\gamma_1, \ldots, \gamma_s) \in H_q \cup G_q \cup H_{q+1} \qquad \text{if } s = 2q-1$$

and

$$(\gamma_1, \ldots, \gamma_s) \in G_q \cup H_{q+1} \cup G_{q+1} \qquad \text{if } s = 2q.$$

Proof. The reason for this result is similar to Theorem 4.12. For I-stability the imaginary axis cannot intersect the order star and must therefore reach the origin through Cary Grant's part. Thus I-stability (and hence A-stability) is only possible (roughly) *in the middle* of the sandwich. Since at most $[(p+2)/2]$ and at least $[(p+1)/2]$ of the $p+1$ sectors of A start in $\mathbb{C}^+$, the number k of bounded fingers satisfies

$$\left[\frac{p+2}{2}\right] \geq k \qquad \text{and} \qquad \left[\frac{p+1}{2}\right] \leq k.$$

Inserting $p = s+1$ on H and $p = s$ on G we get the above results. □

Multiple Real-Pole Approximations

> ... the next main result is obtained, saying that the least value of C is obtained when all the zeros of the denominator are equal ...
> (Nørsett & Wolfbrandt 1977)

Approximations for which all poles are equal, i.e., for which $\gamma_1 = \gamma_2 = \ldots = \gamma_s = \gamma$ are called *"multiple"* real-pole approximations (Nørsett 1974). We again consider only approximations for which the order is $\geq s$. These satisfy, for $s = 3$,

$$R(z) = \frac{P(z)}{(1-\gamma z)^3} \tag{4.31}$$

where $P(z)$ is given by (4.30), and their error constant is

$$C = \frac{1}{4!} - \frac{3\gamma}{3!} + \frac{3\gamma^2}{2!} - \frac{\gamma^3}{1!}. \tag{4.32}$$

Approximations with multiple poles have many computational advantages (the linear systems to be solved in Rosenbrock or DIRK methods have all the same matrix (see Sections IV.6 and IV.7)). We are now pleased to see that they also have the smallest error constants (Nørsett & Wolfbrandt 1977).

Theorem 4.26. *On each of the surfaces* L_i *and* H_i $(i=1, \ldots, s)$ *the error constant* C *of (4.20) is minimized (in absolute value) when* $\gamma_1 = \gamma_2 = \ldots = \gamma_s$.

Proof. Our proof uses relative order stars (similar to (4.13))

$$B = \Big\{ z \in \mathbb{C}\, ; \; |q(z)| > 1 \Big\}, \qquad q(z) = \frac{R_{new}(z)}{R_{old}(z)}, \tag{4.33}$$

where $R_{old}(z)$ is a real-pole approximation of order $p = s+1$ corresponding to $\gamma_1, \ldots, \gamma_s$ and $R_{new}(z)$ is obtained by an infinitely small change of the γ's. We assume that not all γ_i are identical and shall show that then the error constant can be decreased. After a permutation of the indices, we assume $\gamma_1 = \max(\gamma_i)$ (by Theorem 4.23 $\gamma_1 > 0$, so that $1/\gamma_1$ represents the pole on the positive real axis which is closest to the origin) and $\gamma_s < \gamma_1$. We don't allow arbitrary changes of the γ's but we *decrease* γ_1, keep $\gamma_2, \ldots, \gamma_{s-1}$ fixed and determine γ_s by the defining equations for H (see (4.23)). For example, for $s = 3$ we have

$$\gamma_3 = \frac{\dfrac{1}{4!} - \dfrac{\gamma_1+\gamma_2}{3!} + \dfrac{\gamma_1\gamma_2}{2!}}{\dfrac{1}{3!} - \dfrac{\gamma_1+\gamma_2}{2!} + \dfrac{\gamma_1\gamma_2}{1!}}. \tag{4.34}$$

Since the poles and zeros of $R_{old}(z)$ depend continuously on the γ_i, poles and zeros of $q(z)$ appear always in pairs (we call them dipoles). By the maximum principle or by Remark 4.14, each boundary curve of B leaving the origin must lead to at least one dipole before it rejoins the origin. Since there are $s+2 = p+1$ dipoles of $q(z)$ (identical poles for $R_{old}(z)$ and $R_{new}(z)$ don't give rise to a dipole of $q(z)$) and $p+1$ pairs of boundary curves of B leaving the origin (Remark 4.14), each such boundary curve passes through exactly one dipole before rejoining the origin. As a consequence no boundary curve of B can cross the real axis except at dipoles.

If the error constant of $R_{old}(z)$ satisfies $C_{old} < 0$, then, by Remark 4.20, $R_{old}(z)$ has no zero between $1/\gamma_1$ and the origin. Therefore also $q(z)$ possesses no dipole in this region. Since the pole of $R_{new}(z)$ is slightly larger than $1/\gamma_1$ (that of $R_{old}(z)$), the real axis between $1/\gamma_1$ and the origin must belong to the complement of B. Thus we have $C_{new} - C_{old} > 0$ by (4.14) and (4.15).

If $C_{old} > 0$ there is one additional dipole of $q(z)$ between $1/\gamma_1$ and the origin (see Remark 4.20). As above we conclude this time that $C_{new} - C_{old} < 0$.

In both cases $|C_{new}| < |C_{old}|$, since by continuity C_{new} is near to C_{old}. As a consequence no $(\gamma_1, \ldots, \gamma_s) \in H$ with at least two different γ_i can minimize the error constant. As it becomes large in modulus when at least one γ_i tends to ∞ (this follows from Theorem 4.18 and from the fact that in this case $R(z)$ tends to an approximation with s replaced by $s-1$) the minimal value of C must be attained when all poles are identical.

The proof for L is the same, there are only $s-1$ zeros of $R(z)$ and the order is $p = s$. □

An illustration of the order star B compared to A is given in Fig. 4.10. Another advantage of multiple real-pole approximations is exhibited by the following theorem:

Theorem 4.27 (Keeling 1989). *On each surface* $H_i \cap \{(\gamma_1, \ldots, \gamma_s);\ \gamma_j > 0\}$ *the value* $|\,R(\infty)\,|$ *of (4.20) is minimized when* $\gamma_1 = \gamma_2 = \ldots = \gamma_s$.

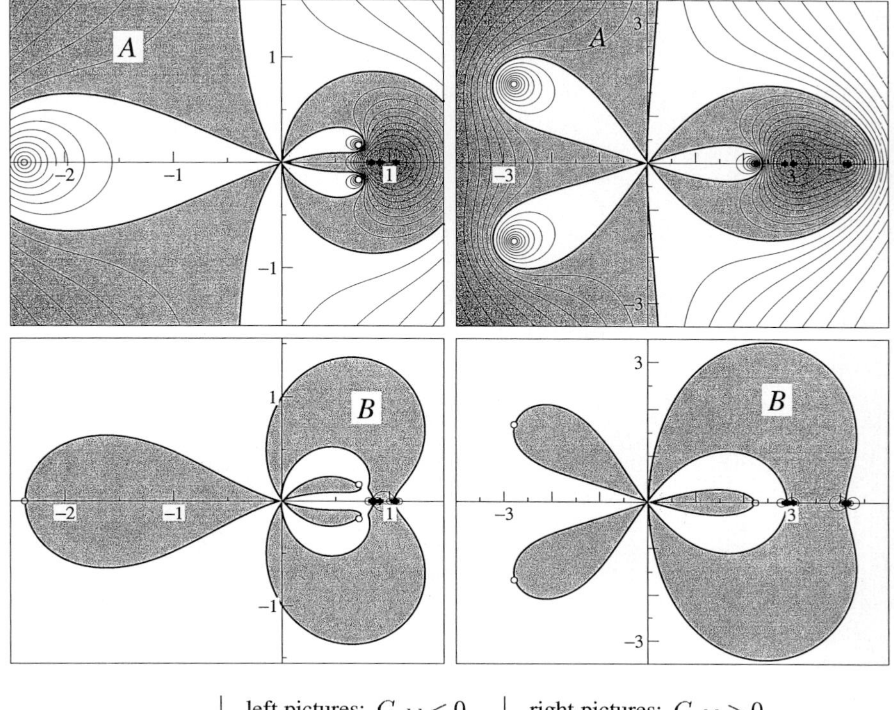

	left pictures: $C_{old} < 0$	right pictures: $C_{old} > 0$
R_{old}	$\gamma_1 = 1.2$	$\gamma_1 = 0.35$
	$\gamma_2 = 1.1$	$\gamma_2 = 0.33$
	$\gamma_3 = 0.9455446$	$\gamma_3 = 0.2406340$
R_{new}	$\gamma_1 = 1.17$	$\gamma_1 = 0.345$
	$\gamma_2 = 1.1$	$\gamma_2 = 0.33$
	$\gamma_3 = 0.9628661$	$\gamma_3 = 0.2440772$

Fig. 4.10. Order star A compared to B

Proof. The beginning of the proof is identical to that of Theorem 4.26. Besides $1/\gamma_1$ and $1/\gamma_s$ there is at best an even number of dipoles on the positive real axis to the right of $1/\gamma_1$. As in the proof above we conclude that a right-neighbourhood of $1/\gamma_1$ belongs to B so that ∞ must lie in its complement (cf. Fig. 4.10). This implies

$$| R_{new}(\infty) | < | R_{old}(\infty) |$$

As a consequence no element of $H \cap \{(\gamma_1, \dots, \gamma_s);\ \gamma_j > 0\}$ with at least two γ_j different can minimize $| R(\infty) |$. Also $| R(\infty) |$ increases if $\gamma_1 \to \infty$. The statement now follows from the fact that $| R(\infty) |$ tends to infinity when at least one γ_j approaches zero. $\square$

Exercises

1. (Ehle 1968). Compute the polynomial $E(y)$ for the third and fourth Padé subdiagonal $R_{k,k+3}(z)$ and $R_{k,k+4}(z)$ (which, by Proposition 3.4 consists of two terms only). Show that these approximations violate (3.6) and cannot be A-stable.

2. Prove the general formula

$$E(y)=\Big(\frac{k!}{(k+j)!}\Big)^2\sum_{r=\left[(k+j+2)/2\right]}^{j}\frac{(-1)^{j-r}}{(j-r)!}\Big(\prod_{q=1}^{j-r}(j-q+1)(k+q)(r-k-q)\Big)y^{2r}$$

for the Padé approximations $R_{kj}\ (j\geq k)$.

3. (For the fans of mathematical precision). Derive the following formulas for the roots λ_i and χ_i of (4.28)

$$\chi_1=\frac{1}{2}+\frac{1}{\sqrt{3}}\cos\frac{13\pi}{18},\qquad \lambda_1=1+\sqrt{2}\cos\Big(\frac{\theta+2\pi}{3}\Big),$$
$$\chi_2=\frac{1}{2}+\frac{1}{\sqrt{3}}\cos\frac{25\pi}{18},\qquad \lambda_2=1+\sqrt{2}\cos\Big(\frac{\theta+4\pi}{3}\Big),$$
$$\chi_3=\frac{1}{2}+\frac{1}{\sqrt{3}}\cos\frac{\pi}{18},\qquad \lambda_3=1+\sqrt{2}\cos\Big(\frac{\theta}{3}\Big),$$

where $\theta=\arctan(\sqrt{2}/4)$.

Hint. Use the Cardano-Viète formula (e.g., Hairer & Wanner (1995), page 66).

4. Prove that all zeros of

$$\frac{x^s}{s!}-S_1\frac{x^{s-1}}{(s-1)!}+S_2\frac{x^{s-2}}{(s-2)!}-\ldots\pm S_s$$

are real and distinct whenever all zeros of

$$Q(z)=1-zS_1+z^2S_2-\ldots\pm z^sS_s,\qquad S_s\neq 0$$

are real. Also, both polynomials have the same number of positive (and negative) zeros (Nørsett & Wanner 1979, Bales, Karakashian & Serbin 1988).

Hint. Apply Theorem 4.23. This furnishes a geometric proof of a classical result (see e.g., Pólya & Szegö (1925), Volume II, Part V, No.65) and allows us to interpret $R(z)$ as the stability function of a (real) collocation method.

5. Prove that $(\gamma,\ldots,\gamma)\in L$ (Definition 4.21) if and only if $L_s(1/\gamma)=0$, where $L_s(x)$ denotes the *Laguerre polynomial of degree* s (see Abramowitz & Stegun (1964), Formula 22.3.9 or Formula (6.11) below).

IV.5 Construction of Implicit Runge-Kutta Methods

Although most of these methods appear at the moment to be largely of theoretical interest ... (B.L. Ehle 1968)

In Sect. II.7 the first implicit Runge-Kutta methods were introduced. As we saw in Sect. IV.3, not all of them are suitable for the solution of stiff differential equations. This section is devoted to the collection of several classes of fully implicit Runge-Kutta methods possessing good stability properties.

The construction of such methods relies heavily on the simplifying assumptions

$$B(p): \quad \sum_{i=1}^{s} b_i c_i^{q-1} = \frac{1}{q} \qquad q=1,\dots,p;$$

$$C(\eta): \quad \sum_{j=1}^{s} a_{ij} c_j^{q-1} = \frac{c_i^q}{q} \qquad i=1,\dots,s, \quad q=1,\dots,\eta;$$

$$D(\zeta): \quad \sum_{i=1}^{s} b_i c_i^{q-1} a_{ij} = \frac{b_j}{q}(1-c_j^q) \quad j=1,\dots,s, \quad q=1,\dots,\zeta.$$

Condition $B(p)$ simply means that the quadrature formula (b_i, c_i) is of order p. The importance of the other two conditions is seen from the following fundamental theorem, which was derived in Sect. II.7.

Theorem 5.1 (Butcher 1964). *If the coefficients* b_i, c_i, a_{ij} *of a Runge-Kutta method satisfy* $B(p)$, $C(\eta)$, $D(\zeta)$ *with* $p \le \eta + \zeta + 1$ *and* $p \le 2\eta + 2$, *then the method is of order* p. □

Gauss Methods

These processes, named "Kuntzmann-Butcher methods" in Sect. II.7, are collocation methods based on the Gaussian quadrature formulas, i.e., $c_1, \dots, c_s$ are the zeros of the shifted Legendre polynomial of degree s,

$$\frac{d^s}{dx^s}\Big(x^s(x-1)^s\Big).$$

For the sake of completeness we present the first of these in Tables 5.1 and 5.2.

Table 5.1. Gauss methods of order 2 and 4

c	A
$\frac{1}{2}$	$\frac{1}{2}$
	1

c	A	
$\frac{1}{2}-\frac{\sqrt{3}}{6}$	$\frac{1}{4}$	$\frac{1}{4}-\frac{\sqrt{3}}{6}$
$\frac{1}{2}+\frac{\sqrt{3}}{6}$	$\frac{1}{4}+\frac{\sqrt{3}}{6}$	$\frac{1}{4}$
	$\frac{1}{2}$	$\frac{1}{2}$

Table 5.2. Gauss method of order 6

c	A		
$\frac{1}{2}-\frac{\sqrt{15}}{10}$	$\frac{5}{36}$	$\frac{2}{9}-\frac{\sqrt{15}}{15}$	$\frac{5}{36}-\frac{\sqrt{15}}{30}$
$\frac{1}{2}$	$\frac{5}{36}+\frac{\sqrt{15}}{24}$	$\frac{2}{9}$	$\frac{5}{36}-\frac{\sqrt{15}}{24}$
$\frac{1}{2}+\frac{\sqrt{15}}{10}$	$\frac{5}{36}+\frac{\sqrt{15}}{30}$	$\frac{2}{9}+\frac{\sqrt{15}}{15}$	$\frac{5}{36}$
	$\frac{5}{18}$	$\frac{4}{9}$	$\frac{5}{18}$

Theorem 5.2 (Butcher 1964, Ehle 1968). *The s-stage Gauss method is of order $2s$. Its stability function is the (s,s)-Padé approximation and the method is A-stable.*

Proof. The order result has already been proved in Sect. II.7. Since the degrees of the numerator and the denominator are not larger than s for any s-stage Runge-Kutta method, the stability function of this $2s$-order method is the (s,s)-Padé approximation by Theorem 3.11. A-stability thus follows from Theorem 4.12. □

Radau IA and Radau IIA Methods

Butcher (1964) introduced Runge-Kutta methods based on the Radau and Lobatto quadrature formulas. He called them processes of type I, II or III according to whether $c_1,\dots,c_s$ are the zeros of

$$\text{I:}\quad \frac{d^{s-1}}{dx^{s-1}}\Big(x^s(x-1)^{s-1}\Big), \qquad \text{(Radau left)} \tag{5.1}$$

$$\text{II:}\quad \frac{d^{s-1}}{dx^{s-1}}\Big(x^{s-1}(x-1)^s\Big), \qquad \text{(Radau right)} \tag{5.2}$$

$$\text{III:}\quad \frac{d^{s-2}}{dx^{s-2}}\Big(x^{s-1}(x-1)^{s-1}\Big). \qquad \text{(Lobatto)} \tag{5.3}$$

The weights $b_1,\dots,b_s$ are chosen such that the quadrature formula satisfies $B(s)$, which implies $B(2s-1)$ in the Radau case and $B(2s-2)$ in the Lobatto case (see Lemma 5.15 below). Unfortunately, none of these methods of Butcher turned out to be A-stable (see e.g. Table 3.1). Ehle (1969) took up the ideas of Butcher and constructed methods of type I, II and III with excellent stability properties. Independently, Axelsson (1969) found the Radau IIA methods together with an elegant proof of their A-stability.

The s-stage Radau IA method is of type I, where the coefficients a_{ij}, $(i,j=1,\dots,s)$ are defined by condition $D(s)$. This is uniquely possible since the c_i are distinct and the b_i not zero. Tables 5.3 and 5.4 present the first of these methods.

Table 5.3. Radau IA methods of orders 1 and 3

$$\begin{array}{c|c} 0 & 1 \\ \hline & 1 \end{array} \qquad \begin{array}{c|cc} 0 & \frac{1}{4} & -\frac{1}{4} \\ \frac{2}{3} & \frac{1}{4} & \frac{5}{12} \\ \hline & \frac{1}{4} & \frac{3}{4} \end{array}$$

Table 5.4. Radau IA method of order 5

$$\begin{array}{c|ccc} 0 & \frac{1}{9} & \frac{-1-\sqrt{6}}{18} & \frac{-1+\sqrt{6}}{18} \\ \frac{6-\sqrt{6}}{10} & \frac{1}{9} & \frac{88+7\sqrt{6}}{360} & \frac{88-43\sqrt{6}}{360} \\ \frac{6+\sqrt{6}}{10} & \frac{1}{9} & \frac{88+43\sqrt{6}}{360} & \frac{88-7\sqrt{6}}{360} \\ \hline & \frac{1}{9} & \frac{16+\sqrt{6}}{36} & \frac{16-\sqrt{6}}{36} \end{array}$$

Ehle's type II processes are obtained by imposing condition $C(s)$. By Theorem II.7.7 this results in the collocation methods based on the zeros of (5.2). They are called Radau IIA methods. Examples are given in Tables 5.5 and 5.6. For $s=1$ we obtain the implicit Euler method.

Theorem 5.3. *The s-stage Radau IA method and the s-stage Radau IIA method are of order $2s-1$. Their stability function is the $(s-1,s)$ subdiagonal Padé approximation. Both methods are A-stable.*

Proof. The stated orders follow from Theorem 5.1 and Lemma 5.4 below. Since $c_1=0$ for the Radau IA method, $D(s)$ with $j=1$ and $B(2s-1)$ imply (3.14). Similarly, for the Radau IIA method, $c_s=1$ and $C(s)$ imply (3.13). Therefore, in both cases, the numerator of the stability function is of degree $\le s-1$ by Proposition 3.8. The statement now follows from Theorem 3.11 and Theorem 4.12. □

Table 5.5. Radau IIA methods of orders 1 and 3

$$\begin{array}{c|c} 1 & 1 \\ \hline & 1 \end{array} \qquad \begin{array}{c|cc} \frac{1}{3} & \frac{5}{12} & -\frac{1}{12} \\ 1 & \frac{3}{4} & \frac{1}{4} \\ \hline & \frac{3}{4} & \frac{1}{4} \end{array}$$

Table 5.6. Radau IIA method of order 5

$$\begin{array}{c|ccc} \frac{4-\sqrt{6}}{10} & \frac{88-7\sqrt{6}}{360} & \frac{296-169\sqrt{6}}{1800} & \frac{-2+3\sqrt{6}}{225} \\ \frac{4+\sqrt{6}}{10} & \frac{296+169\sqrt{6}}{1800} & \frac{88+7\sqrt{6}}{360} & \frac{-2-3\sqrt{6}}{225} \\ 1 & \frac{16-\sqrt{6}}{36} & \frac{16+\sqrt{6}}{36} & \frac{1}{9} \\ \hline & \frac{16-\sqrt{6}}{36} & \frac{16+\sqrt{6}}{36} & \frac{1}{9} \end{array}$$

Lemma 5.4. *Let an s-stage Runge-Kutta method have distinct $c_1, \ldots, c_s$ and non-zero weights $b_1, \ldots, b_s$. Then we have*

a) $C(s)$ and $B(s+\nu)$ imply $D(\nu)$;

b) $D(s)$ and $B(s+\nu)$ imply $C(\nu)$.

Proof. Put

$$d_j^{(q)} := \sum_{i=1}^{s} b_i c_i^{q-1} a_{ij} - \frac{b_j}{q}(1 - c_j^q). \tag{5.4}$$

Conditions $C(s)$ and $B(s+\nu)$ imply

$$\sum_{j=1}^{s} d_j^{(q)} c_j^{k-1} = 0 \qquad \text{for} \quad k = 1, \ldots, s \text{ and } q = 1, \ldots, \nu.$$

The vector $(d_1^{(q)}, \ldots, d_s^{(q)})$ must vanish, because it is the solution of a homogeneous linear system with a non singular matrix of Vandermonde type. This proves $D(\nu)$.

For part (b) one defines

$$e_i^{(q)} := \sum_{j=1}^{s} a_{ij} c_j^{q-1} - \frac{c_i^q}{q}$$

and applies a similar argument to

$$\sum_{i=1}^{s} b_i c_i^{k-1} e_i^{(q)} = 0, \qquad k = 1, \ldots, s, \qquad q = 1, \ldots, \nu. \qquad \square$$

Lobatto IIIA, IIIB and IIIC Methods

For all type III processes the c_i are the zeros of the polynomial (5.3) and the weights b_i are such that $B(2s-2)$ is satisfied.

The coefficients a_{ij} are defined by $C(s)$ for the Lobatto IIIA methods. It is therefore a collocation method. For the Lobatto IIIB methods we impose $D(s)$ and, finally, for the Lobatto IIIC methods we put

$$a_{i1} = b_1 \quad \text{for} \quad i = 1, \ldots, s \tag{5.5}$$

and determine the remaining a_{ij} by $C(s-1)$. Ehle (1969) introduced the first two classes, and presented the IIIC methods for $s \leq 3$. The general definition of the IIIC methods is due to Chipman (1971); see also Axelsson (1972). Examples are given in Tables 5.7-5.12.

Table 5.7. Lobatto IIIA methods of orders 2 and 4

c_i	a_{i1}	a_{i2}
0	0	0
1	$\frac{1}{2}$	$\frac{1}{2}$
	$\frac{1}{2}$	$\frac{1}{2}$

c_i	a_{i1}	a_{i2}	a_{i3}
0	0	0	0
$\frac{1}{2}$	$\frac{5}{24}$	$\frac{1}{3}$	$-\frac{1}{24}$
1	$\frac{1}{6}$	$\frac{2}{3}$	$\frac{1}{6}$
	$\frac{1}{6}$	$\frac{2}{3}$	$\frac{1}{6}$

Table 5.8. Lobatto IIIA method of order 6

c_i	a_{i1}	a_{i2}	a_{i3}	a_{i4}
0	0	0	0	0
$\frac{5-\sqrt{5}}{10}$	$\frac{11+\sqrt{5}}{120}$	$\frac{25-\sqrt{5}}{120}$	$\frac{25-13\sqrt{5}}{120}$	$\frac{-1+\sqrt{5}}{120}$
$\frac{5+\sqrt{5}}{10}$	$\frac{11-\sqrt{5}}{120}$	$\frac{25+13\sqrt{5}}{120}$	$\frac{25+\sqrt{5}}{120}$	$\frac{-1-\sqrt{5}}{120}$
1	$\frac{1}{12}$	$\frac{5}{12}$	$\frac{5}{12}$	$\frac{1}{12}$
	$\frac{1}{12}$	$\frac{5}{12}$	$\frac{5}{12}$	$\frac{1}{12}$

Theorem 5.5. *The s-stage Lobatto IIIA, IIIB and IIIC methods are of order $2s-2$. The stability function for the Lobatto IIIA and IIIB methods is the diagonal $(s-1, s-1)$-Padé approximation. For the Lobatto IIIC method it is the $(s-2, s)$-Padé approximation. All these methods are A-stable.*

Proof. We first prove that the IIIC methods satisfy $D(s-1)$. Condition (5.5) implies $d_1^{(q)} = 0$ $(q = 1, \ldots, s-1)$ for $d_1^{(q)}$ given by (5.4). Conditions $C(s-1)$

Table 5.9. Lobatto IIIB methods of orders 2 and 4

c		
0	$\frac{1}{2}$	0
1	$\frac{1}{2}$	0
	$\frac{1}{2}$	$\frac{1}{2}$

c			
0	$\frac{1}{6}$	$-\frac{1}{6}$	0
$\frac{1}{2}$	$\frac{1}{6}$	$\frac{1}{3}$	0
1	$\frac{1}{6}$	$\frac{5}{6}$	0
	$\frac{1}{6}$	$\frac{2}{3}$	$\frac{1}{6}$

Table 5.10. Lobatto IIIB method of order 6

c				
0	$\frac{1}{12}$	$\frac{-1-\sqrt{5}}{24}$	$\frac{-1+\sqrt{5}}{24}$	0
$\frac{5-\sqrt{5}}{10}$	$\frac{1}{12}$	$\frac{25+\sqrt{5}}{120}$	$\frac{25-13\sqrt{5}}{120}$	0
$\frac{5+\sqrt{5}}{10}$	$\frac{1}{12}$	$\frac{25+13\sqrt{5}}{120}$	$\frac{25-\sqrt{5}}{120}$	0
1	$\frac{1}{12}$	$\frac{11-\sqrt{5}}{24}$	$\frac{11+\sqrt{5}}{24}$	0
	$\frac{1}{12}$	$\frac{5}{12}$	$\frac{5}{12}$	$\frac{1}{12}$

Table 5.11. Lobatto IIIC methods of orders 2 and 4

c		
0	$\frac{1}{2}$	$-\frac{1}{2}$
1	$\frac{1}{2}$	$\frac{1}{2}$
	$\frac{1}{2}$	$\frac{1}{2}$

c			
0	$\frac{1}{6}$	$-\frac{1}{3}$	$\frac{1}{6}$
$\frac{1}{2}$	$\frac{1}{6}$	$\frac{5}{12}$	$-\frac{1}{12}$
1	$\frac{1}{6}$	$\frac{2}{3}$	$\frac{1}{6}$
	$\frac{1}{6}$	$\frac{2}{3}$	$\frac{1}{6}$

Table 5.12. Lobatto IIIC method of order 6

c				
0	$\frac{1}{12}$	$\frac{-\sqrt{5}}{12}$	$\frac{\sqrt{5}}{12}$	$\frac{-1}{12}$
$\frac{5-\sqrt{5}}{10}$	$\frac{1}{12}$	$\frac{1}{4}$	$\frac{10-7\sqrt{5}}{60}$	$\frac{\sqrt{5}}{60}$
$\frac{5+\sqrt{5}}{10}$	$\frac{1}{12}$	$\frac{10+7\sqrt{5}}{60}$	$\frac{1}{4}$	$\frac{-\sqrt{5}}{60}$
1	$\frac{1}{12}$	$\frac{5}{12}$	$\frac{5}{12}$	$\frac{1}{12}$
	$\frac{1}{12}$	$\frac{5}{12}$	$\frac{5}{12}$	$\frac{1}{12}$

and $B(2s-2)$ then yield

$$\sum_{j=2}^{s} d_j^{(q)} c_j^{k-1} = 0 \qquad \text{for} \quad k=1,\ldots,s-1 \text{ and } q=1,\ldots,s-1.$$

As in the proof of Lemma 5.4 we deduce $D(s-1)$. All order statements now follow from Lemma 5.4 and Theorem 5.1.

By definition, the first row of the Runge-Kutta matrix A vanishes for the IIIA methods, and its last column vanishes for the IIIB methods. The denominator of the stability function is therefore of degree $\le s-1$. Similarly, the last row of $A - 1\!\!1 b^T$ vanishes for IIIA, and the first column of $A - 1\!\!1 b^T$ for IIIB. Therefore, the numerator of the stability function is also of degree $\le s-1$ by Formula (3.3). It now follows from Theorem 3.11 that both methods have the $(s-1, s-1)$-Padé approximation as stability function.

For the IIIC process the first column as well as the last row of $A - 1\!\!1 b^T$ vanish. Thus the degree of the numerator of the stability function is at most $s-2$ by Formula (3.3). Again, Theorem 3.11 and Theorem 4.12 imply the statement. □

For a summary of these statements see Table 5.13.

Table 5.13. Fully implicit Runge-Kutta methods

method	simplifying assumptions			order	stability function
Gauss	$B(2s)$	$C(s)$	$D(s)$	$2s$	(s,s)-Padé
Radau IA	$B(2s-1)$	$C(s-1)$	$D(s)$	$2s-1$	$(s-1,s)$-Padé
Radau IIA	$B(2s-1)$	$C(s)$	$D(s-1)$	$2s-1$	$(s-1,s)$-Padé
Lobatto IIIA	$B(2s-2)$	$C(s)$	$D(s-2)$	$2s-2$	$(s-1,s-1)$-Padé
Lobatto IIIB	$B(2s-2)$	$C(s-2)$	$D(s)$	$2s-2$	$(s-1,s-1)$-Padé
Lobatto IIIC	$B(2s-2)$	$C(s-1)$	$D(s-1)$	$2s-2$	$(s-2,s)$-Padé

The W-Transformation

We now attack the explicit construction of all Runge-Kutta methods covered by Theorem 5.1. The first observation is (Chipman 1971, Burrage 1978) that $C(\eta)$ can be written as

$$\begin{pmatrix} a_{11} & \ldots & a_{1s} \\ \vdots & & \vdots \\ \vdots & & \vdots \\ a_{s1} & \ldots & a_{ss} \end{pmatrix} \begin{pmatrix} 1 & c_1 & \ldots & c_1^{\eta-1} \\ \vdots & \vdots & & \vdots \\ \vdots & \vdots & & \vdots \\ 1 & c_s & \ldots & c_s^{\eta-1} \end{pmatrix} = \begin{pmatrix} 1 & c_1 & \ldots & c_1^{\eta} \\ \vdots & \vdots & & \vdots \\ \vdots & \vdots & & \vdots \\ 1 & c_s & \ldots & c_s^{\eta} \end{pmatrix} \begin{pmatrix} 0 & 0 & \ldots & 0 \\ 1 & 0 & \ldots & 0 \\ 0 & \frac{1}{2} & \ldots & 0 \\ \vdots & \vdots & \ddots & \vdots \\ 0 & 0 & \ldots & \frac{1}{\eta} \end{pmatrix}. \tag{5.6}$$

Hence, if V is the Vandermonde matrix

$$V = \begin{pmatrix} 1 & c_1 & \dots & c_1^{s-1} \\ \vdots & \vdots & & \vdots \\ 1 & c_s & \dots & c_s^{s-1} \end{pmatrix},$$

then the first η (for $\eta \le s-1$) columns of $V^{-1}AV$ must have the special structure (with many zeros) of the rightmost matrix in (5.6). This "V-transformation" already considerably simplifies the discussion of order and stability of methods governed by $C(\eta)$ with η close to s (Burrage 1978). Thus, *collocation methods* $(\eta = s)$ are characterized by

$$V^{-1}AV = \begin{pmatrix} 0 & & & & & -\varrho_0/s \\ 1 & 0 & & & & -\varrho_1/s \\ & 1/2 & 0 & & & -\varrho_2/s \\ & & \ddots & \ddots & & \vdots \\ & & & \ddots & 0 & -\varrho_{s-2}/s \\ & & & & 1/(s-1) & -\varrho_{s-1}/s \end{pmatrix} \tag{5.7}$$

where the ϱ's are the coefficients of $M(t) = \prod_{i=1}^{s}(t-c_i)$ and appear when the c_i^s in (5.6) are replaced by lower powers. Whenever some of the columns of $V^{-1}AV$ are *not* as in (5.7), a nice idea of Nørsett allows one to interpret the method as a *perturbed collocation* method (see Nørsett & Wanner (1981) for more details).

However, the V-transformation has some drawbacks: it does not allow a similar characterization of $D(\zeta)$, and the discussions of A- and B-stability remain fairly complicated (see e.g. the above cited papers). It was then discovered (Hairer & Wanner 1981, 1982) that nicer results are obtained, if the Vandermonde matrix V is replaced by a matrix W whose elements are *orthogonal polynomials* evaluated at c_i. We therefore use the (non standard) notation

$$P_k(x) = \frac{\sqrt{2k+1}}{k!}\frac{d^k}{dx^k}\Big(x^k(x-1)^k\Big) = \sqrt{2k+1}\sum_{j=0}^{k}(-1)^{j+k}\binom{k}{j}\binom{j+k}{j}x^j \tag{5.8}$$

for the *shifted Legendre polynomials* normalized so that

$$\int_0^1 P_k^2(x)dx = 1. \tag{5.9}$$

These polynomials satisfy the integration formulas

$$\begin{aligned} \int_0^x P_0(t)dt &= \xi_1 P_1(x) + \frac{1}{2}P_0(x) \\ \int_0^x P_k(t)dt &= \xi_{k+1}P_{k+1}(x) - \xi_k P_{k-1}(x) \qquad k = 1,2,\dots \end{aligned} \tag{5.10}$$

with

$$\xi_k = \frac{1}{2\sqrt{4k^2-1}} \tag{5.11}$$

(Exercise 1). Instead of (5.7) we now have the following result.

Theorem 5.6. *Let* W *be defined by*

$$w_{ij} = P_{j-1}(c_i), \qquad i = 1, \dots, s, \quad j = 1, \dots, s, \tag{5.12}$$

and let A *be the coefficient matrix for the Gauss method of order* $2s$. *Then,*

$$W^{-1}AW = \begin{pmatrix} 1/2 & -\xi_1 & & & \\ \xi_1 & 0 & -\xi_2 & & \\ & \xi_2 & \ddots & \ddots & \\ & & \ddots & 0 & -\xi_{s-1} \\ & & & \xi_{s-1} & 0 \end{pmatrix} =: X_G. \tag{5.13}$$

Proof. We first write $C(\eta)$ in the form

$$\sum_{j=1}^{s} a_{ij}\, p(c_j) = \int_0^{c_i} p(x)\, dx \qquad \text{if} \quad \deg(p) \le \eta - 1, \tag{5.14}$$

which, by (5.10), is equivalent to

$$\begin{aligned} &\sum_{j=1}^{s} a_{ij} P_0(c_j) = \xi_1 P_1(c_i) + \frac{1}{2} P_0(c_i) \\ &\sum_{j=1}^{s} a_{ij} P_k(c_j) = \xi_{k+1} P_{k+1}(c_i) - \xi_k P_{k-1}(c_i) \qquad k = 1, \dots, \eta - 1. \end{aligned} \tag{5.15}$$

For $\eta = s$, inserting (5.12), and using matrix notation, this becomes

$$\begin{pmatrix} a_{11} & \dots & a_{1s} \\ \vdots & & \vdots \\ \vdots & & \vdots \\ a_{s1} & \dots & a_{ss} \end{pmatrix} \begin{pmatrix} w_{11} & \dots & w_{1s} \\ \vdots & & \vdots \\ \vdots & & \vdots \\ w_{s1} & \dots & w_{ss} \end{pmatrix} = \tag{5.16}$$

$$\begin{pmatrix} w_{11} & \dots & w_{1s} & P_s(c_1) \\ \vdots & & \vdots & \vdots \\ \vdots & & \vdots & \vdots \\ w_{s1} & \dots & w_{ss} & P_s(c_s) \end{pmatrix} \begin{pmatrix} 1/2 & -\xi_1 & & & \\ \xi_1 & 0 & -\xi_2 & & \\ & \xi_2 & \ddots & \ddots & \\ & & \ddots & 0 & -\xi_{s-1} \\ & & & \xi_{s-1} & 0 \\ & & & & \xi_s \end{pmatrix}.$$

Since for the Gauss processes we have $P_s(c_1) = \ldots = P_s(c_s) = 0$, the last column respectively row of the right hand matrices can be dropped and we obtain (5.13). □

In what follows we shall study similar results for other implicit Runge-Kutta methods. We first formulate the following lemma, which is an immediate consequence of (5.15) and (5.16).

Lemma 5.7. *Let A be the coefficient matrix of an implicit Runge-Kutta method and let W be a nonsingular matrix with*

$$w_{ij} = P_{j-1}(c_i) \qquad \textit{for} \quad i=1,\dots,s, \quad j=1,\dots,\eta+1.$$

Then $C(\eta)$ (with $\eta \le s-1$) is equivalent to the fact that the first η **columns** *of $W^{-1}AW$ are equal to those of X_G in (5.13).* □

The second type of simplifying assumption, $D(\zeta)$, is now written in the form

$$\sum_{i=1}^{s} b_i p(c_i) a_{ij} = b_j \int_{c_j}^{1} p(x)dx \qquad \text{if} \quad \deg(p) \le \zeta - 1. \tag{5.17}$$

The integration formulas (5.10) together with orthogonality relations

$$\int_0^1 P_0(x)dx = 1, \quad \int_0^1 P_k(x)\,dx = \int_0^1 P_0(x)P_k(x)\,dx = 0 \quad \text{for} \quad k=1,2,\dots$$

show that $D(\zeta)$ (i.e., (5.17)) is equivalent to

$$\sum_{i=1}^{s} P_0(c_i) b_i a_{ij} = \Big(\frac{1}{2}P_0(c_j) - \xi_1 P_1(c_j)\Big) b_j \tag{5.18}$$

$$\sum_{i=1}^{s} P_k(c_i) b_i a_{ij} = \Big(\xi_k P_{k-1}(c_j) - \xi_{k+1} P_{k+1}(c_j)\Big) b_j \qquad k=1,\dots,\zeta-1.$$

This can be stated as

Lemma 5.8. *As in the preceding lemma, let W be a nonsingular matrix with*

$$w_{ij} = P_{j-1}(c_i) \qquad \textit{for} \quad i=1,\dots,s, \quad j=1,\dots,\zeta+1,$$

and let $B = \mathrm{diag}(b_1,\dots,b_s)$ with $b_i \ne 0$. Then $D(\zeta)$ (with $\zeta \le s-1$) is equivalent to the condition that the first ζ **rows** *of the matrix $(W^TB)A(W^TB)^{-1}$ are equal to those of X_G in (5.13) (if B is singular, we still have (5.19) below).*

Proof. Formulas (5.18), written in matrix form, give

$$W^TBA = \begin{pmatrix} 1/2 & -\xi_1 & & & & \\ \xi_1 & 0 & \ddots & & & \\ & \ddots & \ddots & -\xi_{\zeta-1} & & \\ & & \xi_{\zeta-1} & 0 & -\xi_\zeta & \\ * & * & \cdot\cdot & \cdot\cdot & \cdot\cdot & * \\ * & * & \cdot\cdot & \cdot\cdot & \cdot\cdot & * \end{pmatrix} W^TB. \tag{5.19}$$

□

It is now a natural and interesting question, whether both transformation matrices of the foregoing lemmas can be made equal, i.e., whether

$$W^TB = W^{-1} \qquad \text{or} \qquad W^TBW = I. \tag{5.20}$$

A first result is:

Lemma 5.9. *For any quadrature formula of order* $\geq 2s-1$ *the matrix*

$$W = \Big(P_{j-1}(c_i)\Big)_{i,j=1,\ldots,s} \tag{5.21}$$

satisfies (5.20).

Proof. If the quadrature formula is of sufficiently high order, the polynomials $P_k(x)P_l(x)$ $(k+l \leq 2s-2)$ are integrated exactly, i.e.,

$$\sum_{i=1}^{s} b_i P_k(c_i) P_l(c_i) = \int_0^1 P_k(x)P_l(x)dx = \delta_{kl}\ ; \tag{5.22}$$

this, however, is simply $W^T B W = I$. □

Unfortunately, Condition (5.20) is too restrictive for many methods. We therefore relax our requirements as follows:

Definition 5.10. Let η, ζ be given integers between 0 and $s-1$. We say that an $s\times s$-matrix W satisfies $T(\eta,\zeta)$ for the quadrature formula $(b_i,c_i)_{i=1}^s$ if

$$\left.\begin{array}{ll} \text{a)} & W \text{ is nonsingular} \\ \text{b)} & w_{ij} = P_{j-1}(c_i) \quad i=1,\ldots,s, \quad j=1,\ldots,\max(\eta,\zeta)+1 \\ \text{c)} & W^T B W = \begin{pmatrix} I & 0 \\ 0 & R \end{pmatrix} \end{array}\right\} \qquad T(\eta,\zeta)$$

where I is the $(\zeta+1)\times(\zeta+1)$ identity matrix; R is an arbitrary $(s-\zeta-1)\times(s-\zeta-1)$ matrix.

The main result is given in the following theorem. Together with Theorem 5.1 it is very helpful for the construction of high order methods (see Examples 5.16 and 5.24, and Theorem 13.15).

Theorem 5.11. *Let* W *satisfy* $T(\eta,\zeta)$ *for the quadrature formula* $(b_i,c_i)_{i=1}^s$. *Then for a Runge-Kutta method based on* (b_i,c_i) *we have, for the matrix* $X = W^{-1}AW$,

a) the first η *columns of* X *are those of* $X_G \iff C(\eta)$,

b) the first ζ *rows of* X *are those of* $X_G \iff D(\zeta)$.

Proof. The equivalence of (a) with $C(\eta)$ follows from Lemma 5.7. For the proof of (b) we multiply (5.19) from the right by W and obtain

$$W^T B W \cdot X = \widetilde{X} \cdot W^T B W$$

where $\widetilde{X}$ is the large matrix of (5.19). Because of Condition (c) of $T(\eta,\zeta)$ the first ζ rows of $\widetilde{X}$ and X must be the same (write them as block matrices). The statement now follows from Lemma 5.8. □

We have still left open the question of the existence of W satisfying $T(\eta,\zeta)$. The following two lemmas and Theorem 5.14 give an answer.

Lemma 5.12. *If the quadrature formula has distinct nodes c_i and all weights positive ($b_i > 0$) and if it is of order p with $p \geq 2\eta + 1$ and $p \geq 2\zeta + 1$, then the matrix*

$$W = \Big(p_{j-1}(c_i)\Big)_{i,j=1,\dots,s} \tag{5.23}$$

possesses property $T(\eta,\zeta)$ and satisfies (5.20). Here $p_j(x)$ is the polynomial of degree j orthonormalized for the scalar product

$$\langle p, r\rangle = \sum_{i=1}^{s} b_i p(c_i) r(c_i). \tag{5.24}$$

Proof. The positivity of the b's makes (5.24) a scalar product on the space of polynomials of degree $\leq s-1$. Because of the order property (compare with (5.22)), the orthonormalized $p_j(x)$ must coincide for $j \leq \max(\eta,\zeta)$ with the Legendre polynomials $P_j(x)$. Orthonormality with respect to (5.24) means that $W^T B W = I$. □

Lemma 5.13. *If the quadrature formula has distinct nodes c_i and is of order $p \geq s+\zeta$, then W defined by (5.21) has property $T(\eta,\zeta)$.*

Proof. Because of $p \geq s+\zeta$, (5.22) holds for $k = 0,\dots,s-1$ and $l = 0,\dots,\zeta$. This ensures (c) of Definition 5.10. □

Theorem 5.14. *Let the quadrature formula be of order p. Then there exists a transformation with property $T(\eta,\zeta)$ if and only if*

$$p \geq \eta + \zeta + 1 \qquad \text{and} \qquad p \geq 2\zeta + 1, \tag{5.25}$$

and at least $\max(\eta,\zeta)+1$ numbers among $c_1,\dots,c_s$ are distinct.

Proof. Set $\nu = \max(\eta,\zeta)$ and denote the columns of the transformation W by $w_1,\dots,w_s$. In virtue of (b) of $T(\eta,\zeta)$ we have

$$w_j = \Big(P_{j-1}(c_1),\dots,P_{j-1}(c_s)\Big)^T \qquad \text{for} \quad j = 1,\dots,\nu+1.$$

These $\nu+1$ columns are linearly independent only if at least $\nu+1$ among $c_1,\dots,$ c_s are distinct. Now condition (c) of $T(\eta,\zeta)$ means that $w_1,\dots,w_{\zeta+1}$ are orthonormal to $w_1,\dots,w_s$ for the bilinear form $u^T B v$. In particular, the orthonormality of $w_1,\dots,w_{\zeta+1}$ to $w_1,\dots,w_{\nu+1}$ (compare with (5.22)) means that the quadrature formula is exact for all polynomials of degree $\nu+\zeta$. Therefore, $p \geq \nu+\zeta+1$ (which is the same as (5.25)) is a necessary condition for $T(\eta,\zeta)$.

To show its sufficiency, we complete $w_1, \ldots, w_{\nu+1}$ to a basis of $\mathbb{R}^s$. The new basis vectors $\widehat{w}_{\nu+2}, \ldots, \widehat{w}_s$ are then projected into the orthogonal complement of $\mathrm{span}\langle w_1, \ldots, w_{\zeta+1}\rangle$ with respect to $u^T B v$ by a Gram-Schmidt type orthogonalization. This yields

$$w_j = \widehat{w}_j - \sum_{k=1}^{\zeta+1} (w_k^T B \widehat{w}_j) w_k \qquad \text{for } j = \nu+2, \ldots, s. \qquad \square$$

Construction of Implicit Runge-Kutta Methods

For the construction of implicit Runge-Kutta methods satisfying $B(p)$, $C(\eta)$ and $D(\zeta)$ with the help of Theorem 5.11, we first have to choose a quadrature formula of order p. The following lemma is the basic result for Gaussian integration.

Lemma 5.15. *Let $c_1, \ldots, c_s$ be real and distinct and let $b_1, \ldots, b_s$ be determined by condition $B(s)$ (i.e., the formula is "interpolatory"). Then this quadrature formula is of order $2s-k$ if and only if the polynomial $M(x) = (x-c_1)(x-c_2)\ldots(x-c_s)$ is orthogonal to all polynomials of degree $\leq s-k-1$, i.e., if and only if*

$$M(x) = C\Big(P_s(x) + \alpha_1 P_{s-1}(x) + \ldots + \alpha_k P_{s-k}(x)\Big). \tag{5.26}$$

For a *proof* see Exercise 2. $\square$

We see from (5.26) that all quadrature formulas of order $2s-k$ can be specified in terms of k parameters $\alpha_1, \alpha_2, \ldots, \alpha_k$.

Next, if the integers η and ζ satisfy $\eta+\zeta+1 \leq 2s-k$ and $2\zeta+1 \leq 2s-k$ (cf. (5.25)), we can compute a matrix W satisfying $T(\eta,\zeta)$ from Theorem 5.14 (or one of Lemmas 5.12 and 5.13). Finally a matrix X is chosen which satisfies (a) and (b) of Theorem 5.11. Then the Runge-Kutta method with coefficients $A = WXW^{-1}$ is of order at least $\min(\eta+\zeta+1, 2\eta+2)$ by Theorem 5.1.

Example 5.16. We search for all implicit Runge-Kutta methods satisfying $B(2s-2)$, $C(s-1)$ and $D(s-2)$, i.e., methods which are of order at least $2s-2$ by Theorem 5.1. As in (5.26), we put

$$M(x) = C\Big(P_s(x) + \alpha_1 P_{s-1}(x) + \alpha_2 P_{s-2}(x)\Big). \tag{5.27}$$

If α_2 satisfies

$$\alpha_2 < \frac{s-1}{s}\frac{\sqrt{2s+1}}{\sqrt{2s-3}},$$

then the roots of M are real and distinct (see Exercise 7). The matrix W given in (5.21) has Property $T(s-1,s-2)$ by Lemma 5.13. Finally we put

$$X = \begin{pmatrix} 1/2 & -\xi_1 & & & \\ \xi_1 & 0 & \ddots & & \\ & \ddots & \ddots & -\xi_{s-2} & \\ & & \xi_{s-2} & 0 & \beta_{s-1} \\ & & & \xi_{s-1} & \beta_s \end{pmatrix} \tag{5.28}$$

(see Theorem 5.11), and obtain with $A = WXW^{-1}$ a family of implicit Runge-Kutta methods of order $2s-2$ with the four parameters $\alpha_1, \alpha_2, \beta_s, \beta_{s-1}$.

All methods of Table 5.13 (with the exception of Lobatto IIIB) must be special cases. The corresponding parameter values are indicated in Table 5.14 (for their computation see Exercise 3). If we put $\alpha_1 = 0$ and $\alpha_2 = -\sqrt{2s+1}/\sqrt{2s-3}$ (Lobatto quadrature), we obtain the two-parameter family of Chipman (1976).

Table 5.14. Special cases of method (5.27, 5.28)

Method	α_1	α_2	β_s	β_{s-1}
Gauss	0	0	0	$-\xi_{s-1}$
Radau IA	$\sqrt{2s+1}/\sqrt{2s-1}$	0	$1/(4s-2)$	$-\xi_{s-1}$
Radau IIA	$-\sqrt{2s+1}/\sqrt{2s-1}$	0	$1/(4s-2)$	$-\xi_{s-1}$
Lobatto IIIA	0	$-\sqrt{2s+1}/\sqrt{2s-3}$	0	0
Lobatto IIIC	0	$-\sqrt{2s+1}/\sqrt{2s-3}$	$1/(2s-2)$	$-\xi_{s-1}(2s-1)/(s-1)$

Stability Function

We try to express the stability function of an implicit Runge-Kutta method in terms of the transformed Runge-Kutta matrix $X = W^{-1}AW$. From (b) and (c) of Property $T(\eta,\zeta)$ it follows that

$$We_1 = \mathbb{1}, \qquad W^T B\mathbb{1} = e_1, \qquad e_1 = (1,0,\ldots,0)^T. \tag{5.29}$$

Hence Formulas (3.2) and (3.3) become

$$R(z) = 1 + z e_1^T (I - zX)^{-1} e_1, \tag{5.30}$$

$$R(z) = \frac{\det(I - zX + z e_1 e_1^T)}{\det(I - zX)}. \tag{5.31}$$

The stability function depends only on X and not on the underlying quadrature formula. Hence, the stability function of the method of Example 5.16 depends on β_s and β_{s-1} only. Formula (5.31) becomes more symmetric (Hairer & Türke 1984) if we introduce the arithmetic mean of the matrices X and $X - e_1 e_1^T$ and define

$$Y = X - \frac{1}{2} e_1 e_1^T, \tag{5.32}$$

which is just the matrix X without the 1/2 in the $(1,1)$-position.

Proposition 5.17. *For a Runge-Kutta method (3.1) let W satisfy $T(\eta,\zeta)$ for some $\eta,\ \zeta\geq 0$, and let Y be given by (5.32) where $X=W^{-1}AW$. The stability function then satisfies*

$$R(z)=\frac{1+\frac{1}{2}\Psi(z)}{1-\frac{1}{2}\Psi(z)} \tag{5.33}$$

with

$$\Psi(z)=ze_1^T(I-zY)^{-1}e_1. \tag{5.34}$$

Proof. Applying the Runge-Kutta method to the test equation (2.9) yields

$$g=\mathbb{1}y_0+zAg,\qquad y_1=y_0+zb^Tg.$$

With $W^{-1}g=\widehat{g}=(\widehat{g}_1,\ldots,\widehat{g}_s)^T$ this becomes

$$(I-zY)\widehat{g}=e_1(y_0+\frac{z}{2}\widehat{g}_1),\qquad y_1=y_0+z\widehat{g}_1, \tag{5.35}$$

where we have used (5.29). Computing $\widehat{g}_1$ from the first equation of (5.35) and inserting this into the second one gives the result. □

If the Runge-Kutta method satisfies $B(2\nu+1)$, $C(\nu)$ and $D(\nu)$ for some integer ν, then Y is given by (see Theorem 5.11)

$$Y=\begin{pmatrix} 0 & -\xi_1 & & & \\ \xi_1 & \ddots & \ddots & & \\ & \ddots & 0 & -\xi_\nu & \\ & & \xi_\nu & \multicolumn{2}{|c}{Y_\nu} \end{pmatrix}. \tag{5.36}$$

In this case the computation of (5.34) for the (s,s)-matrix Y can be reduced to that of the smaller $(s-\nu,s-\nu)$-matrix Y_ν as follows:

Theorem 5.18. *If Y is given by (5.36), the function $\Psi(z)$ of (5.34) has the continued fraction representation*

$$\Psi(z)=\frac{z|}{|1}+\frac{\xi_1^2z^2|}{|\ 1}+\ldots+\frac{\xi_{\nu-1}^2z^2|}{|\ 1}+\xi_\nu^2z\Psi_\nu(z) \tag{5.37}$$

where $\quad\Psi_\nu(z)=ze_1^T(I-zY_\nu)^{-1}e_1.$

Proof. Let Y_j (for $0\leq j\leq\nu+1$) denote the $(s-j,s-j)$ principal minors of Y, where the first j rows and columns are suppressed. Expanding the determinant of $I-zY_{j-1}$ with respect to the first row (and then the first column) gives for $j=1,\ldots,\nu$

$$\det(I-zY_{j-1})=\det(I-zY_j)+\xi_j^2z^2\det(I-zY_{j+1}). \tag{5.38}$$

By Cramer's rule, the functions $\Psi_j(z)$ can also be written as

$$\Psi_j(z) = z e_1^T (I - zY_j)^{-1} e_1 = z \frac{\det(I - zY_{j+1})}{\det(I - zY_j)} . \tag{5.39}$$

Dividing (5.38) by $\det(I - zY_j)$ yields

$$\Psi_{j-1}(z) = \frac{z}{1 + \xi_j^2 z \Psi_j(z)} . \tag{5.40}$$

A repeated use of (5.40) gives (5.37) since $\Psi(z) = \Psi_0(z)$. □

We are thus naturally led to continued fraction expansions, a technique which was historically the earliest one. Birkhoff & Varga (1965) used it in their proof of the A-stability of the diagonal Padé approximations. Later, Ehle (1969, 1973) tried to extend "Varga's proof" to verify the A-stability of the first and second subdiagonals of the Padé table ("This was unsuccessful because the resulting continued fraction expansions were not easily related to one another."). Therefore, Ehle (1973), Ehle & Picel (1975), proved A-stability results for the first and second subdiagonal and some generalizations by a completely different method. The following study of A-stability (see Butcher 1977, Hairer 1982, Hairer & Türke 1984) combines the above continued fraction expansion with properties of positive functions.

Positive Functions

> Many stability conditions for numerical methods can be expressed in the form that some associated function is positive.
>
> (G. Dahlquist 1978)

A-stability of an implicit Runge-Kutta method is defined by the property

$$|R(z)| < 1 \qquad \text{for} \quad \operatorname{Re} z < 0. \tag{5.41}$$

Since the transformation $(1+\zeta)/(1-\zeta)$ occurring in (5.33) maps the negative half-plane onto the open unit disc, (5.41) is equivalent to

$$\operatorname{Re} \Psi(z) < 0 \qquad \text{for} \quad \operatorname{Re} z < 0. \tag{5.42}$$

This condition means that $-\Psi(-z)$ is a positive function; for rational functions the concept of positivity can be defined as follows:

Definition 5.19. A rational function $f(z)$ is called *positive* if

$$\operatorname{Re} f(z) > 0 \qquad \text{for} \quad \operatorname{Re} z > 0.$$

A nice survey on the relevance of positive functions to numerical analysis is given by Dahlquist (1978). The following lemmas collect some properties of positive functions.

Lemma 5.20. *Let $f(z)$ and $g(z)$ be positive functions. Then we have*

a) $\alpha f(z)+\beta g(z)$ *is positive, if* $\alpha > 0$ *and* $\beta \geq 0$;

b) $1/f(z)$ *is positive;*

c) $f(g(z))$ *is positive.* □

Observe that the poles of a positive function cannot lie in the positive half-plane, but poles on the imaginary axis are possible, e.g., the function $1/z$ is positive.

Lemma 5.21. *Suppose that*

$$f(z)=\frac{c}{z}+g(z) \quad \textit{with} \quad g(z)=\mathcal{O}(1) \quad \text{for } z\to 0\,,$$

and $g(z)\not\equiv 0$. *Then* $f(z)$ *is positive if and only if* $c\geq 0$ *and* $g(z)$ *is positive.*

Proof. The "if-part" follows from Lemma 5.20. Suppose now that $f(z)$ is positive. The constant c has to be non-negative, since for small positive values of z we have $\operatorname{Re} f(z)>0$. On the imaginary axis we have (apart from poles) $\operatorname{Re} g(iy)= \operatorname{Re} f(iy)\geq 0$ or more precisely

$$\liminf_{z\to iy,\ \operatorname{Re} z>0} \operatorname{Re} g(z)\geq 0 \qquad \text{for} \quad y\in\mathbb{R}.$$

The maximum principle for harmonic functions then implies that either $g(z)\equiv 0$ or $g(z)$ is positive. □

A consequence of this lemma is the following characterization of A-stability.

Theorem 5.22. *Consider a Runge-Kutta method whose stability function is given by (5.33) with Y as in (5.36). It is A-stable if and only if*

$$\operatorname{Re}\Psi_\nu(z)<0 \qquad \textit{for} \quad \operatorname{Re} z<0 \tag{5.43}$$

where $\Psi_\nu(z)=ze_1^T(I-zY_\nu)^{-1}e_1$ *as in (5.37).*

Proof. We consider the submatrices Y_j of Y and the functions $\Psi_j(z)$ of (5.39). As we prefer to work with positive functions we put

$$\chi_j(z)=-\Psi_j(-z)=ze_1^T(I+zY_\nu)^{-1}e_1. \tag{5.44}$$

By (5.42), A-stability is equivalent to the positivity of $\chi_0(z)$ and condition (5.43) means that $\chi_\nu(z)$ is a positive function. Relation (5.40) becomes

$$\big(\chi_{j-1}(z)\big)^{-1}=\frac{1}{z}+\xi_j^2\chi_j(z).$$

Since all $\chi_j(z)$ are bounded near the origin and do not vanish identically (see (5.44)), it follows from Lemma 5.21 that $\chi_j(z)$ is a positive function iff $\chi_{j-1}(z)$ is positive. This proves the theorem. □

Example 5.23. For the Runge-Kutta method of Example 5.16 with X given by (5.28) we have

$$\Psi_{s-2}(z) = \frac{z(1-\beta_s z)}{1-\beta_s z - \xi_{s-1}\beta_{s-1}z^2}.$$

Since

$$\Big(\Psi_{s-2}(z)\Big)^{-1} = \frac{1}{z} - \xi_{s-1}\beta_{s-1}\frac{z}{1-\beta_s z}$$

it follows from Lemma 5.21 and Theorem 5.22 that the method is A-stable iff

$$\beta_{s-1} = 0 \quad \text{or} \quad (\beta_{s-1} < 0 \text{ and } \beta_s \geq 0). \tag{5.45}$$

Comparing this result with Tables 5.14 and 5.13 leads to a second proof for the A-stability of the diagonal and the first two subdiagonal Padé approximations for e^z (see Theorem 4.12).

Example 5.24 (Construction of all A-stable Runge-Kutta methods satisfying $B(2s-4)$, $C(s-2)$ and $D(s-3)$). We take a quadrature formula of order $2s-4$ and construct, by Theorem 5.14, a matrix W satisfying Property $T(s-2, s-3)$. The Runge-Kutta matrix A is then of the form

$$A = W(Y + \frac{1}{2}e_1 e_1^T)W^{-1}$$

with Y given by (5.36), $\nu = s-3$ and

$$Y_{s-3} = \begin{pmatrix} 0 & \gamma_{s-2} & \beta_{s-2} \\ \xi_{s-2} & \gamma_{s-1} & \beta_{s-1} \\ 0 & \gamma_s & \beta_s \end{pmatrix}.$$

For the study of A-stability we have to compute $\Psi_{s-3}(z)$ from (5.39). Expanding $\det(I - zY_{s-3})$ with respect to its first column we obtain

$$\Big(\Psi_{s-3}(z)\Big)^{-1} = \frac{1}{z} + \frac{z\xi_{s-2}(g_0 - g_1 z)}{1 - f_1 z + f_2 z^2}$$

where

$$\begin{aligned} f_1 &= \beta_s + \gamma_{s-1}, \qquad & f_2 &= \beta_s\gamma_{s-1} - \beta_{s-1}\gamma_s, \\ g_0 &= -\gamma_{s-2}, \qquad & g_1 &= -\beta_s\gamma_{s-2} + \beta_{s-2}\gamma_s. \end{aligned} \tag{5.46}$$

By Lemma 5.21 and Theorem 5.22 we have A-stability iff either $g_0 = g_1 = 0$ or

$$\frac{z(g_0 + g_1 z)}{1 + f_1 z + f_2 z^2} \tag{5.47}$$

is a positive function, which is equivalent to (see Exercise 4b)

$$g_0 > 0, \quad g_1 \geq 0, \quad f_2 \geq 0, \quad g_0 f_1 - g_1 \geq 0. \tag{5.48}$$

A similar characterization of A-stable Runge-Kutta methods of order $2s-4$ is given in Wanner (1980).

Exercises

1. Verify the integration formulas (5.10) for the shifted Legendre polynomials.

   Hint. By orthogonality $\int_0^x P_k(t)dt$ must be a linear combination of P_{k+1}, P_k and P_{k-1} only. The coefficient of P_k vanishes by symmetry. For the rest just look at the coefficients of x^{k+1} and x^{k-1}.

2. Give a proof of Lemma 5.15.

   Hint (Jacobi 1826). If $f(x)$ is a polynomial of degree $2s-k-1$, and $r(x)$ the interpolation polynomial of degree $s-1$, then $f(x)=q(x)M(x)+r(x)$, where $\deg q(x)\le s-k-1$.

3. Let $R(z)$ be the stability function of the Runge-Kutta method of Example 5.16.

   a) The degree of its denominator is $\le s-1$ iff $\beta_s=\beta_{s-1}\xi_{s-1}2(2s-3)$.

   Hint. Use Formula (5.31) and the fact that $\det(I-zX_G)$ is the denominator of the diagonal Padé approximation.

   b) The degree of the numerator of $R(z)$ is $\le s-1$ iff

   $$\beta_s=-\beta_{s-1}\xi_{s-1}2(2s-3). \tag{5.49}$$

   c) The degree of the numerator of $R(z)$ is $\le s-2$ iff in addition to (5.49) it holds $\beta_s=1/(2s-2)$.

   d) Verify the entries of Table 5.14.

4. a) The function

   $$s(z)=\frac{\alpha+\beta z}{\gamma+\delta z}$$

   with $\gamma>0$ satisfies $\operatorname{Re} s(z)\ge 0$ for $\operatorname{Re} z>0$ iff $\alpha\ge 0$, $\beta\ge 0$ and $\delta\ge 0$.

   b) Use the identity (for $g_0>0$)

   $$\frac{1+f_1z+f_2z^2}{z(g_0+g_1z)}-\frac{1}{zg_0}=\frac{(f_1-g_1/g_0)+f_2z}{g_0+g_1z}$$

   to verify that the function given in (5.47) is positive iff (5.48) holds.

5. Suppose that

   $$f(z)=cz+g(z)\qquad \text{with}\qquad g(z)=\mathcal{O}(1)\qquad \text{for } z\to\infty$$

   and $g(z)\not\equiv 0$. Using the transformation $z\to 1/z$ in Lemma 5.21, show that $f(z)$ is a positive function, if and only if $c\ge 0$ and $g(z)$ is positive.

6. Give an alternative proof of the Routh criterion (Theorem 13.4 of Chapter I): *All zeros of the real polynomial*

   $$p(z)=a_0z^n+a_1z^{n-1}+\ldots+a_n\quad (a_0>0)$$

lie in the negative half-plane $\mathrm{Re}\, z < 0$ *if and only if*

$$c_{i0} > 0 \qquad \text{for } i = 0, 1, \ldots, n.$$

The c_{ij} *are the coefficients of the polynomials*

$$p_i(z) = c_{i0} z^{n-i} + c_{i1} z^{n-i-2} + c_{i2} z^{n-i-4} + \ldots$$

where

$$p_0(z) = a_0 z^n + a_2 z^{n-2} + \ldots, \qquad \textit{i.e.,}\ c_{0j} = a_{2j}$$

$$p_1(z) = a_1 z^{n-1} + a_3 z^{n-3} + \ldots, \qquad \textit{i.e.,}\ c_{1j} = a_{2j+1}.$$

and

$$p_{i+1}(z) = c_{i0} p_{i-1}(z) - c_{i-1,0} z p_i(z), \qquad i = 1, \ldots, n-1. \tag{5.50}$$

Hint. By the maximum principle for harmonic functions the condition "$p(z) \neq 0$ for $\mathrm{Re}\, z \geq 0$" is equivalent to "$|p(-z)/p(z)| < 1$ for $\mathrm{Re}\, z > 0$" and the condition that $p_0(z)$ and $p_1(z)$ are irreducible. Using the transformation (5.33) this becomes equivalent to the positivity of $p_0(z)/p_1(z)$. Now divide (5.50) by $c_{i-1,0} p_i(z)$ and use Exercise 5 recursively.

7. Show that

$$\alpha_2 < \frac{s-1}{s} \frac{\sqrt{2s+1}}{\sqrt{2s-3}} \tag{5.51}$$

is a sufficient condition for $M(x) = P_s(x) + \alpha_1 P_{s-1}(x) + \alpha_2 P_{s-2}(x)$ to have real and pairwise distinct roots.

Hint. (See "Lemma 18" of Nørsett & Wanner 1981). Consider the set D of all pairs (α_1, α_2) for which the roots c_i of $M(x)$ are real and distinct, and the corresponding interpolatory quadrature formula has positive b_i. Verify that $(0,0) \in D$, and show that for $(\alpha_1, \alpha_2) \in \partial D$ either one b_i becomes zero or two c_i coalesce but the quadrature formula remains of order $2s-2$. Therefore it must be the Gaussian formula with $s-1$ nodes of order $2s-2$ and we must have

$$P_s(x) + \alpha_1 P_{s-1}(x) + \alpha_2 P_{s-2}(x) = c(x-\beta) P_{s-1}(x). \tag{5.52}$$

Now use the three-term recursion formula

$$s\xi_s P_s(x) = (x - 1/2) P_{s-1}(x) - (s-1)\xi_{s-1} P_{s-2}(x) \tag{5.53}$$

(Abramowitz & Stegun p. 782, modified) to eliminate xP_{s-1} on the right of (5.52). Then obtain by comparing the coefficients of P_s, P_{s-1} and P_{s-2}

$$c = \frac{1}{s\xi_s} \qquad \alpha_1 = \frac{1}{s\xi_s}\Big(\frac{1}{2} - \beta\Big), \qquad \alpha_2 = \frac{s-1}{s} \frac{\sqrt{2s+1}}{\sqrt{2s-3}}. \tag{5.54}$$

If β is one of the roots of P_{s-1}, then (5.52) has a double root and the estimate (5.51) for α_2 is optimal.

IV.6 Diagonally Implicit RK Methods

> ... they called their methods "diagonally implicit", a term which is reserved here for the special case where all diagonal entries are equal ... (R. Alexander 1977)

We continue to quote from this nice paper: "To integrate a system of n differential equations, an implicit method with a full $s \times s$ matrix requires the solution of ns simultaneous implicit (in general nonlinear) equations in each time step (...) One way to circumvent this difficulty is to use a lower triangular matrix (a_{ij}) (i.e., a matrix with $a_{ij} = 0$ for $i < j$); the equations may then be solved in s successive stages with only an n-dimensional system to be solved at each stage". In accordance with many authors, and in disaccordance with others (see above), we call such a method *diagonally implicit* (DIRK).

"In solving the n-dimensional systems by Newton-type iterations one solves linear systems at each stage with a coefficient matrix of the form $I - ha_{ii}\partial f/\partial y$. If all a_{ii} are equal one may hope to use repeatedly the stored LU-factorization of a single such matrix". When we want to emphasize this additional property for a DIRK method, we shall call it a *singly diagonally implicit* (SDIRK) method.

It is a curious coincidence that in the early seventies at least four theses dedicated a large part of their research to DIRK and SDIRK methods, very often having in mind their usefulness for the treatment of partial differential equations (R. Alt 1971, M. Crouzeix 1975, A. Kurdi 1974, S.P. Nørsett 1974). The classical paper on the subject is Alexander (1977).

Order Conditions

> The traditional problem of choosing the coefficients leads to a nonlinear algebraic jungle, to which civilization and order were brought in the pioneering work of J.C. Butcher, further refined in the Thesis of M. Crouzeix. (R. Alexander 1977)

We want to make the "jungle" still a little more civilized by the following idea: consider a SDIRK scheme

$$
\begin{array}{c|ccccc}
c_1 & \gamma & & & \\
c_2 & a_{21} & \gamma & & \\
\vdots & \vdots & \vdots & \ddots & \\
c_s & a_{s1} & a_{s2} & \cdots & \gamma \\
\hline
 & b_1 & b_2 & \cdots & b_s
\end{array}
$$

with s stages. The order conditions (see Vol. I, Sect. II.2) consist of sums such as

$$\sum_{j,k,l} b_j a_{jk} a_{kl} = \frac{1}{6}. \tag{6.1}$$

Because there are now more non-zero entries in the matrix A than for explicit methods, this sum contains far more terms as it did before. The trick is to transfer all expressions containing a γ to the right-hand side of (6.1). The resulting sum, denoted by $\sum{}'$, is then only built upon the subdiagonal entries as for explicit Runge-Kutta methods. The right-hand side becomes (for this example)

$$\sum_{j,k,l}{}' b_j a_{jk} a_{kl} = \sum_{j,k,l} b_j (a_{jk} - \gamma\delta_{jk})(a_{kl} - \gamma\delta_{kl}) \tag{6.1'}$$

where δ_{jk} denotes the Kronecker delta. Multiplying out we obtain

$$\sum_{j,k,l}{}' b_j a_{jk} a_{kl} = \sum_{j,k,l} b_j a_{jk} a_{kl} - \gamma\Big(\sum_{j,l} b_j a_{jl} + \sum_{j,k} b_j a_{jk}\Big) + \gamma^2 \sum_j b_j .$$

For all sums on the right we insert order conditions (e.g. from Theorem 2.1 of Sect. II.2) and obtain

$$\sum_{j,k,l}{}' b_j a_{jk} a_{kl} = \frac{1}{6} - \gamma + \gamma^2. \tag{6.1''}$$

The general rule is that there appears an alternating polynomial in γ whose coefficients are sums of $1/\gamma(u)$, where u runs through all trees which are obtained by "short-circuiting" one, two, three, etc. vertices of t (with exception of the root). The conditions for order 4 obtained in this way are summarized in Table 6.1. For $s=2$, $p=3$ and $s=3$, $p=4$ these simplified conditions have only very few non-zero terms and the equations become especially simple to solve (see Exercise 1).

Stiffly Accurate SDIRK Methods

Our main interest here lies in methods satisfying

$$a_{sj} = b_j \quad \text{for} \quad j = 1, \ldots, s, \tag{6.2}$$

i.e., in methods for which the numerical solution y_1 is identical to the last internal stage. A first consequence of this property is that $R(\infty) = 0$ (see Proposition 3.8). The order conditions for such methods can, instead of (6.1"), be simplified still further: consider again the example (6.1), which can now be written as

$$\sum_{j,k,l} a_{sj} a_{jk} a_{kl} = \frac{1}{6}.$$

Table 6.1. Order conditions for SDIRK methods

t	$\varrho(t)$	previous conditions	simplified conditions
j	1	$\sum b_j = 1$	$\sum b_j = 1$
k, j	2	$\sum b_j a_{jk} = \frac{1}{2}$	$\sum' b_j a_{jk} = \frac{1}{2} - \gamma$
k, l, j	3	$\sum b_j a_{jk} a_{jl} = \frac{1}{3}$	$\sum' b_j a_{jk} a_{jl} = \frac{1}{3} - \gamma + \gamma^2$
l, k, j	3	$\sum b_j a_{jk} a_{kl} = \frac{1}{6}$	$\sum' b_j a_{jk} a_{kl} = \frac{1}{6} - \gamma + \gamma^2$
k, l, m, j	4	$\sum b_j a_{jk} a_{jl} a_{jm} = \frac{1}{4}$	$\sum' b_j a_{jk} a_{jl} a_{jm} = \frac{1}{4} - \gamma + \frac{3}{2}\gamma^2 - \gamma^3$
l, m, k, j	4	$\sum b_j a_{jk} a_{kl} a_{jm} = \frac{1}{8}$	$\sum' b_j a_{jk} a_{kl} a_{jm} = \frac{1}{8} - \frac{5}{6}\gamma + \frac{3}{2}\gamma^2 - \gamma^3$
l, m, k, j	4	$\sum b_j a_{jk} a_{kl} a_{km} = \frac{1}{12}$	$\sum' b_j a_{jk} a_{kl} a_{km} = \frac{1}{12} - \frac{2}{3}\gamma + \frac{3}{2}\gamma^2 - \gamma^3$
m, l, k, j	4	$\sum b_j a_{jk} a_{kl} a_{lm} = \frac{1}{24}$	$\sum' b_j a_{jk} a_{kl} a_{lm} = \frac{1}{24} - \frac{1}{2}\gamma + \frac{3}{2}\gamma^2 - \gamma^3$

This time we have, instead of (6.1')

$$\begin{aligned}\sum_{j,k,l}{}' a_{sj} a_{jk} a_{kl} &= \sum_{j,k,l} (a_{sj} - \gamma\delta_{sj})(a_{jk} - \gamma\delta_{jk})(a_{kl} - \gamma\delta_{kl}) \\ &= \sum_{j,k,l} a_{sj} a_{jk} a_{kl} - \gamma\Big(\sum_{j,k} a_{sj} a_{jk} + \sum_{j,l} a_{sj} a_{jl} + \sum_{k,l} a_{sk} a_{kl}\Big) \\ &\quad + \gamma^2\Big(\sum_j a_{sj} + \sum_k a_{sk} + \sum_l a_{sl}\Big) - \gamma^3 \cdot 1.\end{aligned}$$

Again inserting known order conditions, we now obtain

$$\sum_{j,k,l}{}' a_{sj} a_{jk} a_{kl} = \frac{1}{6} - \frac{3}{2}\gamma + 3\gamma^2 - \gamma^3. \qquad (6.1''')$$

The general rule is similar to the one above: the difference is that *all* vertices (including the root) are now available for being short-circuited. Another example, for the tree t_{42}, is sketched in Fig. 6.1 and leads to the following right-hand side:

$$\begin{aligned}&\frac{1}{8} - \gamma\Big(\frac{1}{3} + \frac{1}{3} + 1\cdot\frac{1}{2} + \frac{1}{6}\Big) + \gamma^2\Big(\frac{1}{2} + 1\cdot 1 + 1\cdot 1 + \frac{1}{2} + \frac{1}{2} + \frac{1}{2}\Big) \\ &\quad - \gamma^3(1+1+1+1) + \gamma^4 = \frac{1}{8} - \frac{4}{3}\gamma + 4\gamma^2 - 4\gamma^3 + \gamma^4.\end{aligned}$$

The order conditions obtained in this manner are displayed in Table 6.2 for all trees of order ≤ 4. The expressions $\sum'$ are written explicitly for the SDIRK method

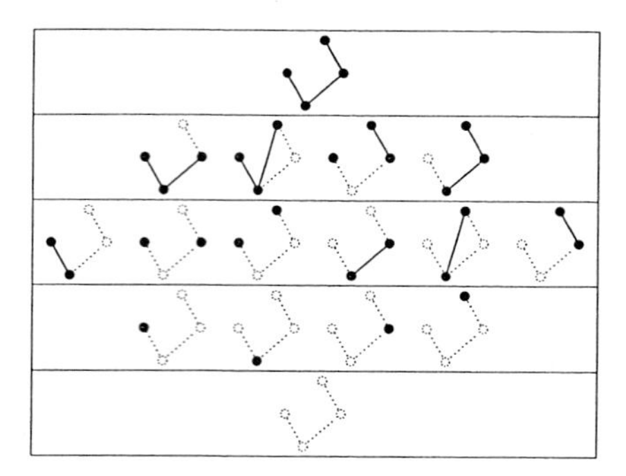

Fig. 6.1. Short-circuiting tree t_{42}

(6.3) with $s=5$ satisfying condition (6.2)

$$
\begin{array}{c|ccccc}
 & \gamma & & & & \\
 & a_{21} & \gamma & & & \\
 & a_{31} & a_{32} & \gamma & & \\
 & a_{41} & a_{42} & a_{43} & \gamma & \\
 & b_1 & b_2 & b_3 & b_4 & \gamma \\
\hline
 & b_1 & b_2 & b_3 & b_4 & \gamma
\end{array}
\qquad
\begin{aligned}
c_2' &= a_{21} \\
c_3' &= a_{31} + a_{32} \\
c_4' &= a_{41} + a_{42} + a_{43}
\end{aligned}
\tag{6.3}
$$

Observe that they become very similar to those of Formulas (1.11) in Sect. II.1.

Table 6.2. Order conditions for method (6.3)

Condition	
$\sum' a_{sj} = b_1 + b_2 + b_3 + b_4 = p_1$	(6.4;1)
$\sum' a_{sj}a_{jk} = b_2c_2' + b_3c_3' + b_4c_4' = p_2$	(6.4;2)
$\sum' a_{sj}a_{jk}a_{jl} = b_2c_2'^2 + b_3c_3'^2 + b_4c_4'^2 = p_3$	(6.4;3)
$\sum' a_{sj}a_{jk}a_{kl} = b_3a_{32}c_2' + b_4(a_{42}c_2' + a_{43}c_3') = p_4$	(6.4;4)
$\sum' a_{sj}a_{jk}a_{jl}a_{jm} = b_2c_2'^3 + b_3c_3'^3 + b_4c_4'^3 = p_5$	(6.4;5)
$\sum' a_{sj}a_{jk}a_{jl}a_{lm} = b_3c_3'a_{32}c_2' + b_4c_4'(a_{42}c_2' + a_{43}c_3') = p_6$	(6.4;6)
$\sum' a_{sj}a_{jk}a_{kl}a_{km} = b_3a_{32}c_2'^2 + b_4(a_{42}c_2'^2 + a_{43}c_3'^2) = p_7$	(6.4;7)
$\sum' a_{sj}a_{jk}a_{kl}a_{lm} = b_4a_{43}a_{32}c_2' = p_8$	(6.4;8)

$$
\begin{aligned}
p_1 &= 1-\gamma & p_5 &= \frac{1}{4} - 2\gamma + \frac{9}{2}\gamma^2 - 4\gamma^3 + \gamma^4 \\
p_2 &= \frac{1}{2} - 2\gamma + \gamma^2 & p_6 &= \frac{1}{8} - \frac{4}{3}\gamma + 4\gamma^2 - 4\gamma^3 + \gamma^4 \\
p_3 &= \frac{1}{3} - 2\gamma + 3\gamma^2 - \gamma^3 & p_7 &= \frac{1}{12} - \gamma + \frac{7}{2}\gamma^2 - 4\gamma^3 + \gamma^4 \\
p_4 &= \frac{1}{6} - \frac{3}{2}\gamma + 3\gamma^2 - \gamma^3 & p_8 &= \frac{1}{24} - \frac{2}{3}\gamma + 3\gamma^2 - 4\gamma^3 + \gamma^4
\end{aligned}
$$

Solution of Equations (6.4). By clever elimination from equations (6.4;4) and (6.4;6) as well as (6.4;4) and (6.4;7) we obtain

$$\begin{aligned} b_3 a_{32} c_2' (c_4' - c_3') &= c_4' p_4 - p_6 \\ b_4 c_3' a_{43} (c_2' - c_3') &= c_2' p_4 - p_7. \end{aligned} \tag{6.5}$$

Multiplying these two equations and using (6.4;8) gives

$$p_8 b_3 (c_4' - c_3')(c_2' - c_3') c_3' = (c_4' p_4 - p_6)(c_2' p_4 - p_7).$$

We now compute b_2, b_3, b_4 from (6.4;2), (6.4;3), (6.4;5). This gives

$$b_3 = \big(-p_2 c_2' c_4' + p_3 (c_4' + c_2') - p_5\big) / \big(c_3' (c_3' - c_2')(c_4' - c_3')\big) \tag{6.6}$$

and b_2 as well as b_4 by cyclic permutation. Comparing the last two equations leads to

$$c_4' = \frac{p_8 p_3 c_2' - p_8 p_5 - c_2' p_6 p_4 + p_6 p_7}{p_8 p_2 c_2' - p_8 p_3 - c_2' p_4 p_4 + p_4 p_7}. \tag{6.7}$$

We now choose γ, c_2' and c_3' as free parameters. Then c_4' is obtained from (6.7); b_2, b_3, b_4 from (6.6), b_1 from (6.4;1), a_{32} and a_{43} from (6.5), a_{42} from (6.4;4), and finally a_{21}, a_{31}, a_{41} from (6.3).

Embedded 3rd order formula: As proposed by Cash (1979), we can append to the above formula a third order expression

$$\widehat{y}_1 = y_0 + h \sum_{i=1}^{4} \widehat{b}_i k_i$$

(thus by omitting the term $b_5 = \gamma$) for the sake of step size control. The coefficients $\widehat{b}_1, \ldots, \widehat{b}_4$ are simply obtained by solving the first 4 equations of Table 6.1 (linear system). *Continuous* embedded 3rd order formulas can be obtained in this way too (see Theorem 6.1 of Sect. II.6)

$$y(x_0 + \theta h) \approx y_0 + h \sum_{i=1}^{4} b_i(\theta) k_i.$$

The coefficients $b_1(\theta), \ldots, b_4(\theta)$ are obtained by solving the first 4 (simplified) conditions of Table 6.1, with the right-hand sides replaced by

$$\theta, \quad \frac{\theta^2}{2} - \gamma\theta, \quad \frac{\theta^3}{3} - \gamma\theta^2 + \gamma^2\theta, \quad \frac{\theta^3}{6} - \gamma\theta^2 + \gamma^2\theta,$$

respectively. The continuous solution obtained in this way becomes $\widehat{y}_1$ for $\theta = 1$ instead of the 4-th order solution y_1. The global continuous solution would therefore be discontinuous. In order to avoid this discontinuity, we add $b_5(\theta)$ and include the fifth equation from Table 6.1 with right-hand side

$$\frac{\theta^4}{4} - \gamma\theta^3 + \frac{3\gamma^2\theta^2}{2} - \gamma^3\theta.$$

The Stability Function

By Formula (3.3), the stability function $R(z)$ for a DIRK method is of the form

$$R(z) = \frac{P(z)}{(1-a_{11}z)(1-a_{22}z)\ldots(1-a_{ss}z)}, \tag{6.8}$$

because the determinant of a triangular matrix is the product of its diagonal entries. The numerator $P(z)$ is a polynomial of degree s at most. If the method is of order $p \geq s$, this polynomial is uniquely determined by Formula (3.18). It is simply obtained from the first terms of the power series for $(1-a_{11}z)\ldots(1-a_{ss}z)\cdot e^z$.

For SDIRK methods, with $a_{11} = \ldots = a_{ss} = \gamma$, we obtain (see also Formula (3.18) with $q_j = (-\gamma)^j \binom{s}{j}$)

$$R(z) = \frac{P(z)}{(1-\gamma z)^s}, \qquad P(z) = (-1)^s \sum_{j=0}^{s} L_s^{(s-j)}\Big(\frac{1}{\gamma}\Big)(\gamma z)^j \tag{6.9}$$

with error constant

$$C = \frac{\gamma^s(-1)^{s+1}}{s+1} L_{s+1}^{(1)}\Big(\frac{1}{\gamma}\Big) \tag{6.10}$$

where

$$L_s(x) = \sum_{j=0}^{s} (-1)^j \binom{s}{j} \frac{x^j}{j!} \tag{6.11}$$

is the s-degree Laguerre polynomial. $L_s^{(k)}(x)$ denotes its k-th derivative. Since the function (6.9) is analytic in $\mathbb{C}^-$ for $\gamma > 0$, A-stability is equivalent to

$$E(y) = Q(iy)Q(-iy) - P(iy)P(-iy) \geq 0 \qquad \text{for all } y \tag{6.12}$$

(see (3.8)). This is an even polynomial of degree $2s$ (in general) and subdegree $2j$ where $j = [(p+2)/2]$ (see Proposition 3.4). We therefore define the polynomial $F(x)$ by

$$F(y^2) = E(y)/y^{2j} \quad j = [(p+2)/2].$$

and check the condition $F(x) \geq 0$ for $x \geq 0$ using Sturm sequences. We display the results obtained (similar to Burrage 1978) in Table 6.3.

For completeness, we give the following explicit formulas for $E(y)$.

$s = 1;\ p = 1:$

$$E = y^2(2\gamma - 1)$$

$s = 2;\ p = 2:$

$$E = y^4\Big(-\tfrac{1}{4} + 2\gamma - 5\gamma^2 + 4\gamma^3\Big) = y^4(2\gamma-1)^2\Big(\gamma - \tfrac{1}{4}\Big)$$

$s = 3;\ p = 3:$

$$E = y^4\Big(\tfrac{1}{12} - \gamma + 3\gamma^2 - 2\gamma^3\Big) + y^6\Big(-\tfrac{1}{36} + \tfrac{\gamma}{2} - \tfrac{13\gamma^2}{4} + \tfrac{28\gamma^3}{3} - 12\gamma^4 + 6\gamma^5\Big)$$

Table 6.3. A-stability of (6.9), order $p \geq s$

s	A-stability	A-stability and $p = s+1$
1	$1/2 \leq \gamma < \infty$	$1/2$
2	$1/4 \leq \gamma < \infty$	$(3+\sqrt{3})/6$
3	$1/3 \leq \gamma \leq 1.06857902$	1.06857902
4	$0.39433757 \leq \gamma \leq 1.28057976$	—
5	$\begin{cases} 0.24650519 \leq \gamma \leq 0.36180340 \\ 0.42078251 \leq \gamma \leq 0.47326839 \end{cases}$	0.47326839
6	$0.28406464 \leq \gamma \leq 0.54090688$	—
7	—	—
8	$0.21704974 \leq \gamma \leq 0.26471425$	—

$s = 4;\ p = 4:$

$$E = y^6\Big(\tfrac{1}{72} - \tfrac{\gamma}{3} + \tfrac{17\gamma^2}{6} - \tfrac{32\gamma^3}{3} + 17\gamma^4 - 8\gamma^5\Big)$$

$$+ y^8\Big(-\tfrac{1}{576} + \tfrac{\gamma}{18} - \tfrac{25\gamma^2}{36} + \tfrac{13\gamma^3}{3} - \tfrac{173\gamma^4}{12} + \tfrac{76\gamma^5}{3} - 22\gamma^6 + 8\gamma^7\Big).$$

A-stability means here that all coefficients must be non-negative. A general formula is as follows.

Lemma 6.1. *The E-polynomial for (6.8) with $a_{11} = \ldots = a_{ss} = \gamma$ and $p \geq s$ satisfies*

$$E(y) = \Big(1 - L_s\Big(\frac{1}{\gamma}\Big)^2\Big)(\gamma y)^{2s}$$
$$-2 \sum_{j=[(p+2)/2]}^{s-1} (-1)^{s+j}(\gamma y)^{2j} \int_0^{1/\gamma} L_s(x) L_s^{(2s+1-2j)}(x)\,dx. \tag{6.13}$$

Proof. Inserting Formula (6.9) into the definition of $E(y)$

$$E(y) = (1+\gamma^2 y^2)^s - P(iy)P(-iy)$$
$$= (1+\gamma^2 y^2)^s - \sum_k \sum_l L_s^{(s-k)}\Big(\frac{1}{\gamma}\Big) L_s^{(s-l)}\Big(\frac{1}{\gamma}\Big)(\gamma i y)^{k+l}(-1)^l$$

and using integration by parts for the verification of

$$2\int_0^{\alpha} L_s(x) L_s^{(2s+1-2j)}(x)\,dx = (-1)^s \sum_{k+l=2j} (-1)^l L_s^{(s-k)}(x) L_s^{(s-l)}(x)\Big|_0^{\alpha}$$

one obtains the result, since

$$\sum_{k+l=2j} (-1)^l L_s^{(s-k)}(0) L_s^{(s-l)}(0) = (-1)^j \binom{s}{j}. \qquad \square$$

Multiple Real-Pole Approximations with $R(\infty)=0$

For methods satisfying (6.2) we have $R(\infty)=0$. Therefore the highest coefficient of $P(z)$ in (6.9) is zero. If the order of the method is known to be $p \geq s-1$, the remaining coefficients of $P(z)$ are still uniquely determined by γ and we have

$$P(z) = (-1)^s \sum_{j=0}^{s-1} L_s^{(s-j)}\Big(\frac{1}{\gamma}\Big)(\gamma z)^j \tag{6.14}$$

with error constant

$$C = (-1)^s L_s\Big(\frac{1}{\gamma}\Big)\gamma^s . \tag{6.15}$$

The first polynomials $E(y)$ of (6.12) are now:

$s=2,\ p=1:$

$$E = y^2(-1+4\gamma-2\gamma^2) + y^4\gamma^4$$

$s=3,\ p=2:$

$$E = y^4\Big(-\tfrac{1}{4} + 3\gamma - 12\gamma^2 + 18\gamma^3 - 6\gamma^4\Big) + y^6\gamma^6$$

$s=4,\ p=3:$

$$\begin{aligned} E &= y^4\Big(\tfrac{1}{12} - \tfrac{4\gamma}{3} + 6\gamma^2 - 8\gamma^3 + 2\gamma^4\Big) \\ &\quad + y^6\Big(-\tfrac{1}{36} + \tfrac{2\gamma}{3} - 6\gamma^2 + \tfrac{76\gamma^3}{3} - 52\gamma^4 + 48\gamma^5 - 12\gamma^6\Big) + y^8\gamma^8. \end{aligned}$$

The regions of γ for A-(and hence L-)stability are displayed in Table 6.4.

Table 6.4. L-stability of $R(z)$ with P from (6.14), order $p \geq s-1$

s	L-stability	L-stab. and $p=s$
2	$(2-\sqrt{2})/2 \leq \gamma \leq (2+\sqrt{2})/2$	$\gamma = (2 \pm \sqrt{2})/2$
3	$0.18042531 \leq \gamma \leq 2.18560010$	$\gamma = 0.43586652$
4	$0.22364780 \leq \gamma \leq 0.57281606$	$\gamma = 0.57281606$
5	$0.24799464 \leq \gamma \leq 0.67604239$	$\gamma = 0.27805384$
6	$0.18391465 \leq \gamma \leq 0.33414237$	$\gamma = 0.33414237$
7	$0.20408345 \leq \gamma \leq 0.37886489$	—
8	$0.15665860 \leq \gamma \leq 0.23437316$	$\gamma = 0.23437316$

Choice of Method

We now determine the free parameters for method (6.3) with $s = 5$ and order 4. For a good choice of γ, we have displayed in Fig. 6.2 the error constant C as well as the regions for A- and $A(0)$-stability.

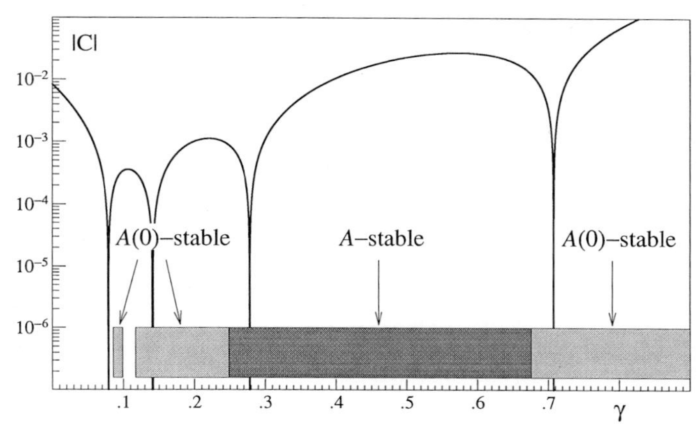

Fig. 6.2 Error constant and A-stability domain for $s = 5$, $p = 4$

This suggests that γ between 0.25 and 0.29 is a good choice. The method is then L-stable and the error constant is small. For various values of γ in this range, we determined (by a nonlinear Gauss-Newton code) c_2' and c_3' in order to minimize the fifth-order error terms. It turned out that

$$c_2' = 0.5, \qquad c_3' = 0.3$$

is close to optimal. With this we coded two different choices of γ: $\gamma = 4/15 = 0.2666\ldots$, which was numerically the better choice and $\gamma = 1/4$, which gave, via Formulas (6.4), (6.5), (6.6) and (6.7), especially nice rational coefficients. These latter are displayed in Table 6.5. We have included a continuous solution to this method

$$y(x_0 + \theta h) \approx y_0 + h \sum_{j=1}^{5} b_j(\theta) k_j,$$

which is third order for $0 < \theta < 1$ and updates to the fourth order approximation y_1 for $\theta = 1$.

Table 6.5. L-stable SDIRK method of order 4

$$
\begin{array}{c|ccccc}
\frac{1}{4} & \frac{1}{4} \\
\frac{3}{4} & \frac{1}{2} & \frac{1}{4} \\
\frac{11}{20} & \frac{17}{50} & -\frac{1}{25} & \frac{1}{4} \\
\frac{1}{2} & \frac{371}{1360} & -\frac{137}{2720} & \frac{15}{544} & \frac{1}{4} \\
1 & \frac{25}{24} & -\frac{49}{48} & \frac{125}{16} & -\frac{85}{12} & \frac{1}{4} \\
\hline
y_1 = & \frac{25}{24} & -\frac{49}{48} & \frac{125}{16} & -\frac{85}{12} & \frac{1}{4} \\
\widehat{y}_1 = & \frac{59}{48} & -\frac{17}{96} & \frac{225}{32} & -\frac{85}{12} & 0 \\
err = & -\frac{3}{16} & -\frac{27}{32} & \frac{25}{32} & 0 & \frac{1}{4}
\end{array}
\tag{6.16}
$$

$$
\begin{aligned}
b_1(\theta) &= \frac{11}{3}\theta - \frac{463}{72}\theta^2 + \frac{217}{36}\theta^3 - \frac{20}{9}\theta^4 \\
b_2(\theta) &= \frac{11}{2}\theta - \frac{385}{16}\theta^2 + \frac{661}{24}\theta^3 - 10\theta^4 \\
b_3(\theta) &= -\frac{125}{18}\theta + \frac{20125}{432}\theta^2 - \frac{8875}{216}\theta^3 + \frac{250}{27}\theta^4 \\
b_4(\theta) &= -\frac{85}{4}\theta^2 + \frac{85}{6}\theta^3 \\
b_5(\theta) &= -\frac{11}{9}\theta + \frac{557}{108}\theta^2 - \frac{359}{54}\theta^3 + \frac{80}{27}\theta^4 .
\end{aligned}
\tag{6.17}
$$

Exercises

1. (Crouzeix & Raviart 1980). Compute the SDIRK methods (Table 6.1) for $s = 3,\ p = 4$. Obtain also (for $s = 2, p = 3$) once again the method of Table 7.2, Sect. II.7.

   Result. The last order condition is in both cases just a polynomial in γ. Among the different solutions, the following presents an A-stable scheme:

$$
\begin{array}{c|ccc}
\gamma & \gamma \\
\frac{1}{2} & \frac{1}{2}-\gamma & \gamma \\
1-\gamma & 2\gamma & 1-4\gamma & \gamma \\
\hline
 & \delta & 1-2\delta & \delta
\end{array}
\qquad
\begin{aligned}
\gamma &= \frac{1}{\sqrt{3}}\cos\left(\frac{\pi}{18}\right) + \frac{1}{2} \\
\delta &= \frac{1}{6(2\gamma-1)^2}.
\end{aligned}
\tag{6.18}
$$

2. Verify all details of Tables 6.1 and 6.2.

3. The four cases of A-stable SDIRK methods of order $p = s + 1$ indicated in Table 6.3 (right) are the *only* ones existing. This fact has not yet been *rigorously* proved, because the "proof" given in Wanner, Hairer & Nørsett (1978) uses an asymptotic formula without error estimation. Do better.

4. Cooper & Sayfy (1979) have derived many DIRK (which they call "semi-explicit") methods of high order. Their main aim was to *minimize* the number of implicit stages and *not* to maximize stability. One of their methods is

$$
\begin{array}{c|cccccc}
\frac{6-\sqrt{6}}{10} & \frac{6-\sqrt{6}}{10} & & & & & \\
\frac{6+9\sqrt{6}}{35} & \frac{-6+5\sqrt{6}}{14} & \frac{6-\sqrt{6}}{10} & & & & \\
1 & \frac{888+607\sqrt{6}}{2850} & \frac{126-161\sqrt{6}}{1425} & \frac{6-\sqrt{6}}{10} & & & \\
\frac{4-\sqrt{6}}{10} & \frac{3153-3082\sqrt{6}}{14250} & \frac{3213+1148\sqrt{6}}{28500} & \frac{-267+88\sqrt{6}}{500} & \frac{6-\sqrt{6}}{10} & & \\
\frac{4+\sqrt{6}}{10} & \frac{-32583+14638\sqrt{6}}{71250} & \frac{-17199+364\sqrt{6}}{142500} & \frac{1329-544\sqrt{6}}{2500} & \frac{-96+131\sqrt{6}}{625} & \frac{6-\sqrt{6}}{10} & \\
\hline
1 & 0 & 0 & \frac{1}{9} & \frac{16-\sqrt{6}}{36} & \frac{16+\sqrt{6}}{36} & 0
\end{array}
$$

Show that it is of order 5 and A-stable, but not L-stable.

5. It can be seen in Table 6.4 that for $s = 2, 4, 6$, and 8 the L-stability superconvergence point coincides with the right end of the A-stability interval. Explain this with the help of order star theory (Fig. 6.3.a).

   Further, for $s = 7$, a superconvergence point is given by $\gamma = 0.20406693$, which misses the A-stability interval given there by less than $2 \cdot 10^{-5}$. Should the above argument also apply here and must there be a computation error somewhere? Study the corresponding order star to show that this is not the case (Fig. 6.3.b).

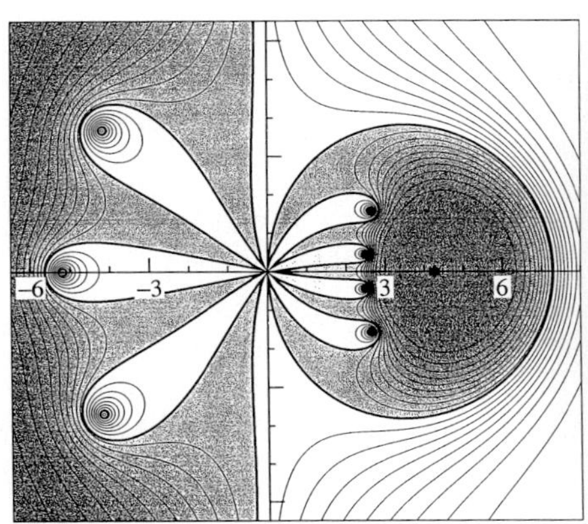

Fig. 6.3.a.
Multiple pole order star
$s = 8,\ \gamma = 0.23437316$

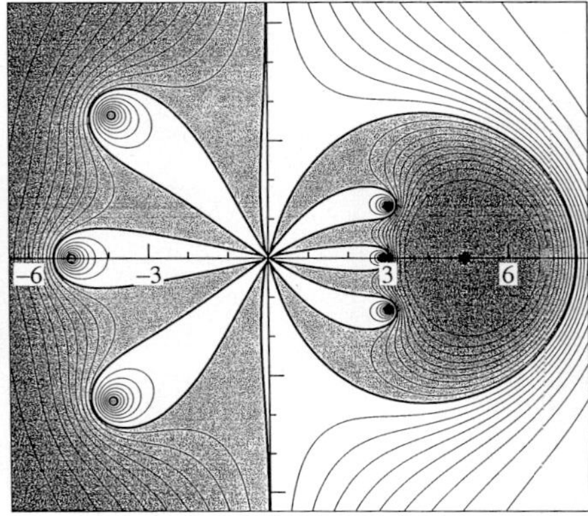

Fig. 6.3.b.
Multiple pole order star
$s = 7,\ \gamma = 0.20406693$

IV.7 Rosenbrock-Type Methods

> When the functions φ are non-linear, implicit equations can in general be solved only by iteration. This is a severe drawback, as it adds to the problem of stability, that of convergence of the iterative process. An alternative, which avoids this difficulty, is ...
>
> (H.H. Rosenbrock 1962/63)

... is discussed in this section. Among the methods which already give satisfactory results for stiff equations, Rosenbrock methods are the easiest to program. We shall describe their theory in this section, which will lead us to our first "stiff" code. Rosenbrock methods belong to a large class of methods which try to avoid nonlinear systems and replace them by a sequence of linear systems. We therefore call these methods *linearly implicit Runge-Kutta methods.* In the literature such methods are often called "semi-implicit" (or was it "semi-explicit"?), or "generalized" or "modified" or "adaptive" or "additive" Runge-Kutta methods.

Derivation of the Method

We start, say, with a diagonally implicit Runge-Kutta method

$$\begin{aligned} k_i &= hf\Big(y_0 + \sum_{j=1}^{i-1} a_{ij} k_j + a_{ii} k_i \Big) \qquad i = 1, \ldots, s \\ y_1 &= y_0 + \sum_{i=1}^{s} b_i k_i \end{aligned} \tag{7.1}$$

applied to the autonomous differential equation

$$y' = f(y). \tag{7.2}$$

The main idea is to linearize Formula (7.1). This yields

$$\begin{aligned} k_i &= hf(g_i) + hf'(g_i) a_{ii} k_i \\ g_i &= y_0 + \sum_{j=1}^{i-1} a_{ij} k_j, \end{aligned} \tag{7.3}$$

and can be interpreted as the application of *one* Newton iteration to each stage in (7.1) with starting values $k_i^{(0)} = 0$. Instead of continuing the iterations until convergence, we consider (7.3) as a new class of methods and investigate anew its order and stability properties.

Important computational advantage is obtained by replacing the Jacobians $f'(g_i)$ by $J = f'(y_0)$, so that the method requires its calculation only once (Calahan 1968). Many methods of this type and much numerical experience with them have been obtained by van der Houwen (1973), Cash (1976) and Nørsett (1975).

We gain further freedom by introducing additional linear combinations of the terms Jk_j into (7.3) (Nørsett & Wolfbrandt 1979, Kaps & Rentrop 1979). We then arrive at the following class of methods:

Definition 7.1. An s-stage *Rosenbrock method* is given by the formulas

$$\begin{aligned} k_i &= hf\Big(y_0 + \sum_{j=1}^{i-1} \alpha_{ij} k_j\Big) + hJ \sum_{j=1}^{i} \gamma_{ij} k_j, \qquad i = 1, \ldots, s \\ y_1 &= y_0 + \sum_{j=1}^{s} b_j k_j \end{aligned} \tag{7.4}$$

where $\alpha_{ij}, \gamma_{ij}, b_i$ are the determining coefficients and $J = f'(y_0)$.

Each stage of this method consists of a system of linear equations with unknowns k_i and with matrix $I - h\gamma_{ii} J$. Of special interest are methods for which $\gamma_{11} = \ldots = \gamma_{ss} = \gamma$, so that we need only one LU-decomposition per step.

Non-autonomous problems. The equation

$$y' = f(x, y) \tag{7.2a}$$

can be converted to autonomous form by adding $x' = 1$. If method (7.4) is applied to the augmented system, the components corresponding to the x-variable can be computed explicitly and we arrive at

$$\begin{aligned} k_i &= hf\Big(x_0 + \alpha_i h, y_0 + \sum_{j=1}^{i-1} \alpha_{ij} k_j\Big) + \gamma_i h^2 \frac{\partial f}{\partial x}(x_0, y_0) + h \frac{\partial f}{\partial y}(x_0, y_0) \sum_{j=1}^{i} \gamma_{ij} k_j \\ y_1 &= y_0 + \sum_{j=1}^{s} b_j k_j, \end{aligned} \tag{7.4a}$$

where the additional coefficients are given by

$$\alpha_i = \sum_{j=1}^{i-1} \alpha_{ij}, \qquad \gamma_i = \sum_{j=1}^{i} \gamma_{ij}. \tag{7.5}$$

Implicit differential equations. Suppose the problem is of the form

$$My' = f(x, y) \tag{7.2b}$$

where M is a constant matrix (nonsingular for the moment). If we formally multiply (7.2b) with M^{-1}, apply method (7.4a), and then multiply the resulting formula with M, we obtain

$$Mk_i = hf\Big(x_0 + \alpha_i h, y_0 + \sum_{j=1}^{i-1} \alpha_{ij} k_j\Big) + \gamma_i h^2 \frac{\partial f}{\partial x}(x_0, y_0) + h\frac{\partial f}{\partial y}(x_0, y_0) \sum_{j=1}^{i} \gamma_{ij} k_j$$

$$y_1 = y_0 + \sum_{j=1}^{s} b_j k_j. \tag{7.4b}$$

An advantage of this formulation is that the inversion of M is avoided and that possible band-structures of the matrices M and $\partial f/\partial y$ are preserved.

Order Conditions

Conditions on the free parameters which ensure that the method is of order p, i.e., the local error satisfies

$$y(x_0 + h) - y_1 = \mathcal{O}(h^{p+1}),$$

can be obtained either by straightforward differentiation or by the use of the theorems on B-series (Sect. II.12). We follow here the first approach, since it requires only the knowledge of Sect. II.2. The second possibility is sketched in Exercise 2.

As in Sect. II.2, we write the system (7.2) in tensor notation and Method (7.4) as [1]

$$\begin{aligned} k_j^J &= hf^J(g_j) + h\sum_K f_K^J(y_0)\sum_k \gamma_{jk} k_k^K \\ g_i^J &= y_0^J + \sum_j \alpha_{ij} k_j^J, \\ y_1^J &= y_0^J + \sum_j b_j k_j^J. \end{aligned} \tag{7.4'}$$

Again, we use Leibniz's rule (cf. (II.2.4))

$$(k_j^J)^{(q)}\big|_{h=0} = q\big(f^J(g_j)\big)^{(q-1)}\big|_{h=0} + q\sum_K f_K^J(y_0)\sum_k \gamma_{jk}(k_k^K)^{(q-1)}\big|_{h=0} \tag{7.6}$$

and have from the chain rule (cf. Sect. II.2, (2.6;1), (2.6;2))

$$\begin{aligned} \big(f^J(g_j)\big)' &= \sum_K f_K^J(g_j)\cdot(g_j^K)' \\ \big(f^J(g_j)\big)'' &= \sum_{K,L} f_{KL}^J(g_j)\cdot(g_j^K)'\cdot(g_j^L)' + \sum_K f_K^J(g_j)\cdot(g_j^K)'' \end{aligned}$$

---

[1] In the sequel, the reader will find many k's of different meaning; on the one hand the "k" in Formula (7.1) which goes back to Runge and Kutta, on the other hand "k" as summation index as since ever in numerical analysis. Although this looks somewhat strange in certain formulas, we prefer to retain the notation of previous sections.

etc. Inserting this into (7.6) we obtain recursively

$$(k_j^J)^{(0)}\big|_{h=0} = 0 \tag{7.7;0}$$

$$(k_j^J)^{(1)}\big|_{h=0} = f^J \tag{7.7;1}$$

$$\begin{aligned}(k_j^J)^{(2)}\big|_{h=0} &= 2\sum_K f_K^J f^K \sum_k \alpha_{jk} + 2\sum_K f_K^J f^K \sum_k \gamma_{jk} \\ &= 2\sum_K f_K^J f^K \sum_k (\alpha_{jk}+\gamma_{jk})\end{aligned} \tag{7.7;2}$$

$$\begin{aligned}(k_j^J)^{(3)}\big|_{h=0} &= 3\sum_{K,L} f_{KL}^J f^K f^L \sum_{k,l} \alpha_{jk}\alpha_{jl} \\ &+ 3\cdot 2\sum_{K,L} f_K^J f_L^K f^L \sum_{k,l} (\alpha_{jk}+\gamma_{jk})(\alpha_{kl}+\gamma_{kl})\end{aligned} \tag{7.7;3}$$

etc. All elementary differentials are evaluated at y_0. Comparing the derivatives of the numerical solution $(q \geq 1)$

$$(y_1^J)^{(q)}\big|_{h=0} = \sum_j b_j (k_j^J)^{(q)}\big|_{h=0} \tag{7.8}$$

with those of the true solution (Sect. II.2, Formula (2.7;1), (2.7;2), (2.7;3)), we arrive at the following conditions for order three:

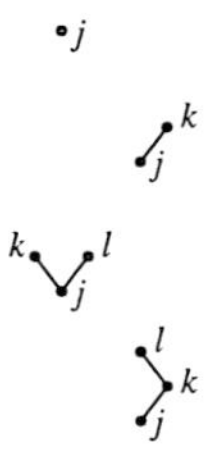

$$\sum b_j = 1$$

$$\sum b_j(\alpha_{jk}+\gamma_{jk}) = \frac{1}{2}$$

$$\sum b_j \alpha_{jk}\alpha_{jl} = \frac{1}{3}$$

$$\sum b_j(\alpha_{jk}+\gamma_{jk})(\alpha_{kl}+\gamma_{kl}) = \frac{1}{6}.$$

The only difference with the order conditions for Runge-Kutta methods is that at singly-branched vertices of the corresponding trees α_{jk} is replaced by $\alpha_{jk}+\gamma_{jk}$. In order to arrive at a general result, the formulas obtained motivate the following definition:

Definition 7.2. Let t be a labelled tree of order q with root j; we denote by

$$\Phi_j(t) = \sum_{k,l,\ldots} \varphi_{j,k,l,\ldots}$$

the sum over the remaining $q-1$ indices $k,l,\ldots$ etc. The summand $\varphi_{j,k,l,\ldots}$ is a product of $q-1$ factors, which are

$\alpha_{kl}+\gamma_{kl}$ if l is the only son of k;

α_{kl} if l is a son of k and k has at least two sons.

Using the recursive representation of trees (Def. II.2.12) we have $\Phi_j(\tau)=1$ for the only tree of order 1 and, as in (II.2.19),

$$\Phi_j(t)=\begin{cases}\sum\limits_{k_1,\ldots,k_m}\alpha_{jk_1}\ldots\alpha_{jk_m}\Phi_{k_1}(t_1)\ldots\Phi_{k_m}(t_m) & \text{if } t=[t_1,\ldots,t_m],\\ & m\geq 2\\ \sum\limits_{k}(\alpha_{jk}+\gamma_{jk})\Phi_k(t_1) & \text{if } t=[t_1]\,.\end{cases}\tag{7.9}$$

Theorem 7.3. *The derivatives of* k_j^J, *given by (7.4'), satisfy*

$$(k_j^J)^{(q)}\big|_{h=0}=\sum_{t\in LT_q}\gamma(t)\Phi_j(t)F^J(t)(y_0)\tag{7.7;q}$$

and the numerical solution y_1^J *satisfies*

$$(y_1^J)^{(q)}\big|_{h=0}=\sum_{t\in LT_q}\gamma(t)\sum_j b_j\Phi_j(t)F^J(t)(y_0),\tag{7.10}$$

where $F^J(t)$ *are the elementary differentials (Definition II.2.3).*

Proof. Because of (7.8) we only have to prove the first formula. This is done by induction on q and follows exactly the lines of the proof of Theorem II.2.11. We use (7.6), replace the expression $f^J(g_j)^{(q-1)}$ by Faà di Bruno's formula (Lemma II.2.8), use

$$(g_j^K)^{(\delta)}=\sum_k\alpha_{jk}(k_k^K)^{(\delta)}$$

for the derivatives of g_j and insert the induction hypothesis (7.7;δ) with $\delta\leq q-1$. This gives

$$\begin{aligned}(k_j^J)^{(q)}\big|_{h=0}=q&\sum_{u\in LS_q}\sum_{t_1\in LT_{\delta_1}}\cdots\sum_{t_m\in LT_{\delta_m}}\gamma(t_1)\ldots\gamma(t_m)\\ &\cdot\sum_{k_1}\alpha_{jk_1}\Phi_{k_1}(t_1)\ldots\sum_{k_m}\alpha_{jk_m}\Phi_{k_m}(t_m)\\ &\cdot\sum_{K_1,\ldots,K_m}f^J_{K_1\ldots K_m}(y_0)F^{K_1}(t_1)(y_0)\ldots F^{K_m}(t_m)(y_0)\\ &+q\sum_{t_1\in LT_{q-1}}\gamma(t_1)\sum_k\gamma_{jk}\Phi_k(t_1)\sum_K f^J_K(y_0)F^K(t_1)(y_0).\end{aligned}$$

The one-to-one correspondence between the summation set $\{(u,t_1,\ldots,t_m)|u\in LS_q,\ t_j\in LT_{\delta_j}\}$ and LT_q together with the recursion formulas (7.9), (II.2.17), (II.2.18) now yields the result. □

Comparing Theorems 7.3 and II.2.6 we obtain:

Table 7.1. Trees and order conditions up to order 5

$\varrho(t)$	t	graph	$\gamma(t)$	$\Phi_j(t)$	$p_t(\gamma)$
1	τ		1	1	1
2	t_{21}		2	$\sum_k \beta_{jk}$	$1/2-\gamma$
3	t_{31}		3	$\sum_{k,l} \alpha_{jk}\alpha_{jl}$	$1/3$
	t_{32}		6	$\sum_{k,l} \beta_{jk}\beta_{kl}$	$1/6-\gamma+\gamma^2$
4	t_{41}		4	$\sum_{k,l,m} \alpha_{jk}\alpha_{jl}\alpha_{jm}$	$1/4$
	t_{42}		8	$\sum_{k,l,m} \alpha_{jk}\beta_{kl}\alpha_{jm}$	$1/8-\gamma/3$
	t_{43}	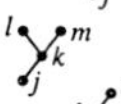	12	$\sum_{k,l,m} \beta_{jk}\alpha_{kl}\alpha_{km}$	$1/12-\gamma/3$
	t_{44}	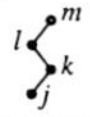	24	$\sum_{k,l,m} \beta_{jk}\beta_{kl}\beta_{lm}$	$1/24-\gamma/2+3\gamma^2/2-\gamma^3$
5	t_{51}	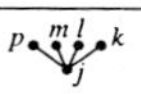	5	$\sum \alpha_{jk}\alpha_{jl}\alpha_{jm}\alpha_{jp}$	$1/5$
	t_{52}	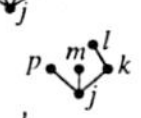	10	$\sum \alpha_{jk}\beta_{kl}\alpha_{jm}\alpha_{jp}$	$1/10-\gamma/4$
	t_{53}	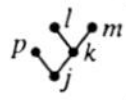	15	$\sum \alpha_{jk}\alpha_{kl}\alpha_{km}\alpha_{jp}$	$1/15$
	t_{54}	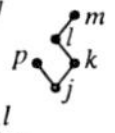	30	$\sum \alpha_{jk}\beta_{kl}\beta_{lm}\alpha_{jp}$	$1/30-\gamma/4+\gamma^2/3$
	t_{55}	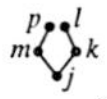	20	$\sum \alpha_{jk}\beta_{kl}\alpha_{jm}\beta_{mp}$	$1/20-\gamma/4+\gamma^2/3$
	t_{56}	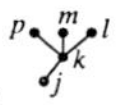	20	$\sum \beta_{jk}\alpha_{kl}\alpha_{km}\alpha_{kp}$	$1/20-\gamma/4$
	t_{57}	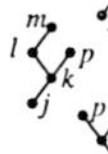	40	$\sum \beta_{jk}\alpha_{kl}\beta_{lm}\alpha_{kp}$	$1/40-5\gamma/24+\gamma^2/3$
	t_{58}	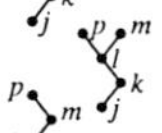	60	$\sum \beta_{jk}\beta_{kl}\alpha_{lm}\alpha_{lp}$	$1/60-\gamma/6+\gamma^2/3$
	t_{59}		120	$\sum \beta_{jk}\beta_{kl}\beta_{lm}\beta_{mp}$	$1/120-\gamma/6+\gamma^2-2\gamma^3+\gamma^4$

Theorem 7.4. *A Rosenbrock method (7.4) with $J=f'(y_0)$ is of order p iff*

$$\sum_j b_j\Phi_j(t)=\frac{1}{\gamma(t)} \qquad \textit{for} \quad \varrho(t)\le p. \tag{7.11}$$

□

The expressions $\Phi_j(t)$ simplify, if we introduce the abbreviation

$$\beta_{ij}=\alpha_{ij}+\gamma_{ij}. \tag{7.12}$$

The order conditions (7.11) for all trees up to order 5 are given in Table 7.1.

A further simplification of the order conditions (7.11) is possible if

$$\gamma_{ii}=\gamma \qquad \text{for all } i \tag{7.13}$$

(It is unfortunate that in the current literature the letter γ is used for the parameter in (7.4) as well as for $\gamma(t)$ in (7.11) and we hope that no confusion will arise). In the same way as for DIRK methods, the summations in the expressions for $\Phi_j(t)$

in the 5th column of Table 7.1 again contain *more* terms than the corresponding expressions for explicit Runge-Kutta methods, since the matrix γ_{ij} (and hence β_{ij}) contains non-zero elements in the diagonal. The difference is that here these diagonal γ appear only for singly-branched vertices (see Definition 7.2). Therefore the procedure explained in Sect. IV.6 (see Formulas (6.1') and (6.1")) must be slightly modified and leads to order conditions of the form

$$\sum_j{}' b_j \Phi_j(t) = p_t(\gamma) \tag{7.11'}$$

where the polynomials $p_t(\gamma)$ are listed in the last column of Table 7.1.

The Stability Function

If we apply Method (7.4) to the test equation $y' = \lambda y$ and if we assume $J = f'(y_0) = \lambda$ then the numerical solution becomes $y_1 = R(h\lambda)y_0$ with

$$R(z) = 1 + zb^T(I - zB)^{-1}1\!\!1 \tag{7.14}$$

where we have used the notation $b^T = (b_1, \ldots, b_s)$ and $B = (\beta_{ij})_{i,j=1}^{s}$. Since B is a lower triangular matrix, the stability function (7.14) is equal to that of a DIRK-method with RK-matrix B. Properties of such stability functions have already been investigated in Sect. IV.6.

Construction of Methods of Order 4

In order to construct 4-stage Rosenbrock methods of order 4 we list, for convenience, the whole set of order conditions (c.f. Table 7.1.).

$$b_1 + b_2 + b_3 + b_4 = 1 \tag{7.15a}$$

$$b_2\beta_2' + b_3\beta_3' + b_4\beta_4' = \frac{1}{2} - \gamma = p_{21}(\gamma) \tag{7.15b}$$

$$b_2\alpha_2^2 + b_3\alpha_3^2 + b_4\alpha_4^2 = \frac{1}{3} \tag{7.15c}$$

$$b_3\beta_{32}\beta_2' + b_4(\beta_{42}\beta_2' + \beta_{43}\beta_3') = \frac{1}{6} - \gamma + \gamma^2 = p_{32}(\gamma) \tag{7.15d}$$

$$b_2\alpha_2^3 + b_3\alpha_3^3 + b_4\alpha_4^3 = \frac{1}{4} \tag{7.15e}$$

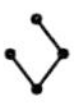

$$b_3\alpha_3\alpha_{32}\beta_2' + b_4\alpha_4(\alpha_{42}\beta_2' + \alpha_{43}\beta_3') = \frac{1}{8} - \frac{\gamma}{3} = p_{42}(\gamma) \tag{7.15f}$$

$$b_3\beta_{32}\alpha_2^2 + b_4(\beta_{42}\alpha_2^2 + \beta_{43}\alpha_3^2) = \frac{1}{12} - \frac{\gamma}{3} = p_{43}(\gamma) \tag{7.15g}$$

$$b_4\beta_{43}\beta_{32}\beta_2' = \frac{1}{24} - \frac{\gamma}{2} + \frac{3}{2}\gamma^2 - \gamma^3 = p_{44}(\gamma) \tag{7.15h}$$

Here we have used the abbreviations

$$\alpha_i = \sum_{j=1}^{i-1} \alpha_{ij}, \qquad \beta_i' = \sum_{j=1}^{i-1} \beta_{ij}. \tag{7.16}$$

For the sake of step size control we also look for an embedded formula (Wolfbrandt 1977, Kaps & Rentrop 1979)

$$\widehat{y}_1 = y_0 + \sum_{j=1}^{s} \widehat{b}_j k_j \tag{7.17}$$

which uses the same k_j-values as (7.4), but has different weights. This method should have order 3, i.e., the four conditions (7.15a)-(7.15d) should be satisfied also for the $\widehat{b}_i$. These equations constitute the linear system

$$\begin{pmatrix} 1 & 1 & 1 & 1 \\ 0 & \beta_2' & \beta_3' & \beta_4' \\ 0 & \alpha_2^2 & \alpha_3^2 & \alpha_4^2 \\ 0 & 0 & \beta_{32}\beta_2' & \sum' \beta_{4j}\beta_j' \end{pmatrix} \begin{pmatrix} \widehat{b}_1 \\ \widehat{b}_2 \\ \widehat{b}_3 \\ \widehat{b}_4 \end{pmatrix} = \begin{pmatrix} 1 \\ 1/2 - \gamma \\ 1/3 \\ 1/6 - \gamma + \gamma^2 \end{pmatrix}. \tag{7.18}$$

Whenever the matrix in (7.18) is regular, uniqueness of the solutions of the linear system implies $\widehat{b}_i = b_i \;\; (i = 1, \ldots, 4)$ and the approximation $\widehat{y}_1$ cannot be used for step size control. We therefore have to require that the matrix (7.18) be singular, i.e.,

$$(\beta_2'\alpha_4^2 - \beta_4'\alpha_2^2)\beta_{32}\beta_2' = (\beta_2'\alpha_3^2 - \beta_3'\alpha_2^2) \sum_{j=2}^{3} \beta_{4j}\beta_j'. \tag{7.19}$$

This condition guarantees the existence of a 3rd order embedded method (7.17), whenever (7.15) possesses a solution. The computation of the coefficients α_{ij}, β_{ij}, γ, b_i satisfying (7.15), (7.16) and (7.19) is now done in the following steps:

Step 1. Choose $\gamma > 0$ such that the stability function (7.14) has desirable stability properties (c.f. Table 6.3).

Step 2. Choose $\alpha_2, \alpha_3, \alpha_4$ and b_1, b_2, b_3, b_4 in such a way that the three conditions (7.15a), (7.15c), (7.15e) are fulfilled. One obviously has four degrees of freedom in this choice. Observe that the (b_i, α_i) need not be the coefficients of a standard quadrature formula, since $\sum b_i \alpha_i = 1/2$ need not be satisfied.

Step 3. Take β_{43} as a free parameter and compute $\beta_{32}\beta_2'$ from (7.15h), then $(\beta_{42}\beta_2' + \beta_{43}\beta_3')$ from (7.15d). These expressions, inserted into (7.19) yield a second relation between $\beta_2', \beta_3', \beta_4'$ (the first one is (7.15b)). Eliminating $(b_4\beta_{42} + b_3\beta_{32})$ from (7.15d) and (7.15g) gives

$$b_4\beta_{43}(\beta_2'\alpha_3^2 - \beta_3'\alpha_2^2) = \beta_2' p_{43}(\gamma) - \alpha_2^2 p_{32}(\gamma),$$

a third linear relation for $\beta_2', \beta_3', \beta_4'$. The resulting linear system is regular iff $b_4\beta_{43}\alpha_2\gamma(3\gamma - 1) \neq 0$.

Step 4. Once the β_i' are known we can find β_{32} and β_{42} from the values of $\beta_{32}\beta_2'$, $(\beta_{42}\beta_2' + \beta_{43}\beta_3')$ obtained in Step 3.
Step 5. Choose $\alpha_{32}, \alpha_{42}, \alpha_{43}$ according to (7.15f). One has two degrees of freedom to do this. Finally, the values α_i, β_i' yield α_{i1}, β_{i1} via condition (7.16).

Table 7.2 Rosenbrock methods of order 4

method	γ	parameter choices	$A(\alpha)$-stable	$\lvert R(\infty)\rvert$
GRK4A (Kaps-Rentrop 79)	0.395	$\alpha_2 = 0.438,\ \alpha_3 = 0.87$ $b_4 = 0.25$	$\pi/2$	0.995
GRK4T (Kaps-Rentrop 79)	0.231	$\alpha_2 = 2\gamma$, (7.22), $b_3 = 0$	89.3°	0.454
Shampine (1982)	0.5	$\alpha_2 = 2\gamma$, (7.22), $b_3 = 0$	$\pi/2$	1/3
Veldhuizen (1984)	0.225708	$\alpha_2 = 2\gamma$, (7.22), $b_3 = 0$	89.5°	0.24
Veldhuizen (1984)	0.5	$\alpha_2 = 2\gamma,\ \alpha_3 = 0.5,\ b_3 = 0$	$\pi/2$	1/3
L-stable method	0.572816	$\alpha_2 = 2\gamma$, (7.22), $b_3 = 0$	$\pi/2$	0

Most of the popular Rosenbrock methods are special cases of this construction (see Table 7.2). Usually the remaining free parameters are chosen as follows: if we require

$$\alpha_{43} = 0, \qquad \alpha_{42} = \alpha_{32} \quad \text{and} \quad \alpha_{41} = \alpha_{31} \tag{7.20}$$

then the argument of f in (7.4) is the same for $i = 3$ and $i = 4$. Hence, the number of function evaluations is reduced by one. Further free parameters can be determined so that several order conditions of order five are satisfied. Multiplying the condition (7.15g) with α_2 and subtracting it from the order condition for the tree t_{56} yields

$$b_4\beta_{43}\alpha_3^2(\alpha_3 - \alpha_2) = p_{56}(\gamma) - \alpha_2 p_{43}(\gamma). \tag{7.21}$$

This determines β_{43}. The order condition for t_{51} can also easily be fulfilled in Step 2. If $\alpha_3 = \alpha_4$ (see (7.20)) this leads to the restriction

$$\alpha_3 = \frac{1/5 - \alpha_2/4}{1/4 - \alpha_2/3}. \tag{7.22}$$

In Table 7.2 we collect some well-known methods. All of them satisfy (7.20) and (7.21) (Only exception: the second method of van Veldhuizen for $\gamma = 0.5$ has $\beta_{43} = 0$ instead of (7.21)). The definition of the remaining free parameters is given in the first two columns. The last columns indicate some properties of the stability function.

Higher Order Methods

As for explicit Runge-Kutta methods the construction of higher order methods is facilitated by the use of *simplifying assumptions*. First, the condition

$$\sum_{i=j}^{s} b_i \beta_{ij} = b_j(1-\alpha_j), \qquad j=1,\ldots,s \tag{7.23}$$

plays a role similar to that of (II.1.12) for explicit Runge-Kutta methods. It implies that the order condition of the left-hand tree in Fig. 7.1 is a consequence of the two on the right-hand side. A difference to Runge-Kutta methods is that here the vertex directly above the root has to be multiply-branched.

The second type of simplifying asumption is (with $\beta_k = \sum_{l=1}^{k} \beta_{kl}$)

$$\sum_{k=1}^{j-1} \alpha_{jk}\beta_k = \frac{\alpha_j^2}{2}, \qquad j=2,\ldots,s. \tag{7.24}$$

It has an effect similar to that of (II.5.7). As a consequence of (7.24) the order conditions of the two trees in Fig. 7.2 are equivalent. Again the vertex marked by an arrow has to be multiply-branched.

The use of the above simplifying assumptions has been exploited by Kaps & Wanner (1981) for their construction of methods up to order 6. Still higher order methods would need generalizations of the above simplifying assumptions (in analogy to $C(\eta)$ and $D(\zeta)$ of Sect. II.7).

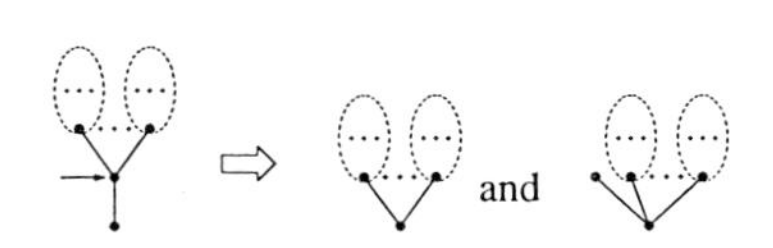

Fig. 7.1. Reduction with (7.23)

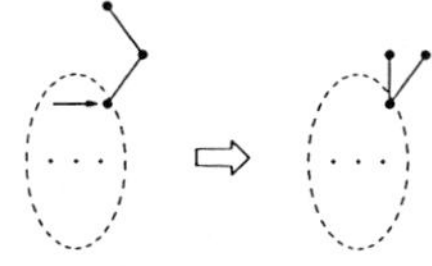

Fig. 7.2. Reduction with (7.24)

Implementation of Rosenbrock-Type Methods

A direct implementation of (7.4) requires, at each stage, the solution of a linear system with the matrix $I - h\gamma_{ii}J$ and also the matrix-vector multiplication $J \cdot \sum \gamma_{ij}k_j$. The latter can be avoided by the introduction of the new variables

$$u_i = \sum_{j=1}^{i} \gamma_{ij}k_j, \qquad i=1,\ldots,s.$$

If $\gamma_{ii} \neq 0$ for all i, the matrix $\Gamma = (\gamma_{ij})$ is invertible and the k_i can be recovered from the u_i:

$$k_i = \frac{1}{\gamma_{ii}} u_i - \sum_{j=1}^{i-1} c_{ij}u_j, \qquad C = \operatorname{diag}(\gamma_{11}^{-1},\ldots,\gamma_{ss}^{-1}) - \Gamma^{-1}.$$

Inserting this formula into (7.4) and dividing by h yields

$$\begin{aligned}\Big(\frac{1}{h\gamma_{ii}}I - J\Big)u_i &= f\Big(y_0 + \sum_{j=1}^{i-1} a_{ij}u_j\Big) + \sum_{j=1}^{i-1}\Big(\frac{c_{ij}}{h}\Big)u_j, \qquad i = 1,\ldots,s \\ y_1 &= y_0 + \sum_{j=1}^{s} m_j u_j,\end{aligned} \tag{7.25}$$

where

$$(a_{ij}) = (\alpha_{ij})\Gamma^{-1}, \qquad (m_1,\ldots,m_s) = (b_1,\ldots,b_s)\Gamma^{-1}.$$

Compared to (7.4) the formulation (7.25) of a Rosenbrock method avoids not only the above mentioned matrix-vector multiplication, but also the n^2 multiplications for $(\gamma_{ii}h)J$. Similar transformations were first proposed by Wolfbrandt (1977), Kaps & Wanner (1981) and Shampine (1982). The formulation (7.25) can be found in Kaps, Poon & Bui (1985).

For *non-autonomous* problems this transformation yields

$$\begin{aligned}\Big(\frac{1}{h\gamma_{ii}}I - \frac{\partial f}{\partial y}(x_0,y_0)\Big)u_i =& f\Big(x_0 + \alpha_i h,\ y_0 + \sum_{j=1}^{i-1} a_{ij}u_j\Big) \\ &+ \sum_{j=1}^{i-1}\Big(\frac{c_{ij}}{h}\Big)u_j + \gamma_i h\frac{\partial f}{\partial x}(x_0,y_0)\end{aligned} \tag{7.26}$$

with α_i and γ_i given by (7.5).

For *implicit differential equations* of the form (7.2b) the transformed Rosenbrock method becomes

$$\begin{aligned}\Big(\frac{1}{h\gamma_{ii}}M - \frac{\partial f}{\partial y}(x_0,y_0)\Big)u_i =& f\Big(x_0 + \alpha_i h,\ y_0 + \sum_{j=1}^{i-1} a_{ij}u_j\Big) \\ &+ M\sum_{j=1}^{i-1}\Big(\frac{c_{ij}}{h}\Big)u_j + \gamma_i h\frac{\partial f}{\partial x}(x_0,y_0).\end{aligned} \tag{7.27}$$

Coding. Rosenbrock methods are nearly as simple to implement as explicit Runge-Kutta methods. The only difference is that at each step the Jacobian $\partial f/\partial y$ has to be evaluated and s linear systems have to be solved. Thus, one can take an explicit RK code (say DOPRI5), add four lines which compute $\partial f/\partial y$ by finite differences (or call a user-supplied subroutine JAC which furnishes it analytically); add further a call to a Gaussian DECOmposition routine, and add to each evaluation-stage a call to a linear SOLver. Since the method is of order 4(3), the step size prediction formula

$$h_{new} = h\cdot\min\Big\{6.,\ \max\Big(0.2,\ 0.9\cdot(Tol/err)^{1/4}\Big)\Big\} \tag{7.28}$$

seems appropriate.

However, we want the code to work economically for non-autonomous problems as well as for implicit equations. Further, if the dimension of the system is large, it becomes crucial that the linear algebra be done, whenever possible, in banded form. All these possibilities, autonomous or not, implicit or explicit, $\partial f/\partial y$ banded or not, B banded or not, $\partial f/\partial y$ analytic or not, ("... that is the question") lead to 2^5 different cases, for each of which the code contains special parts for high efficiency. Needless to say, it works well on all stiff problems of Sect. IV.1. A more thorough comparison and testing will be given in Sect. IV.10.

The "Hump"

On some very stiff equations, however, the code shows a curious behaviour: consider the van der Pol equation in singular perturbation form (1.5') with

$$\varepsilon = 10^{-6}, \quad y_1(0) = 2, \quad y_2(0) = -0.66. \tag{7.29}$$

We further select method GRK4T (Table 7.2; each other method there behaves similarly) and $Tol = 7 \cdot 10^{-5}$. Fig. 7.3 shows the numerical solution y_1 as well as the step sizes chosen by the code. There all rejected steps are indicated by an $\times$.

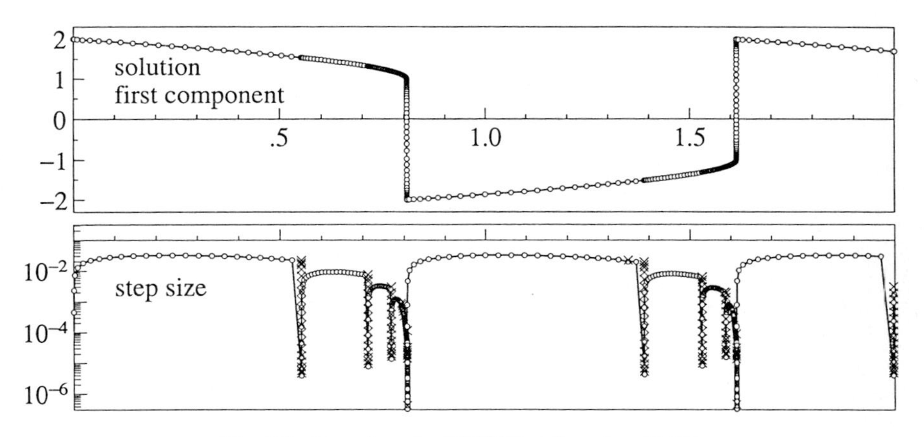

Fig. 7.3. Step sizes for GRK4T at Equation (1.5')

Curious step size drops (by a factor of about 10^{-3}) occur without any apparent exterior reason. Further, these drops are accompanied by a huge number of step rejections (up to 20). In order to understand this phenomenon, we present in the left picture of Fig. 7.4 the *exact local error* as well as the *estimated local error* $\|y_1 - \widehat{y}_1\|$ at $x = 0.55139$ as a function of the step size h (both in logarithmic scale). The current step size is marked by large symbols. The error behaves like $C \cdot h^5$ only for very small h $(\leq 10^{-6} = \varepsilon)$. Between $h = 10^{-5}$ and the step size actually used $(\approx 10^{-2})$ the error is more or less constant. Whenever this constant is larger than *Tol* (horizontal broken line), the code is forced to decrease the step

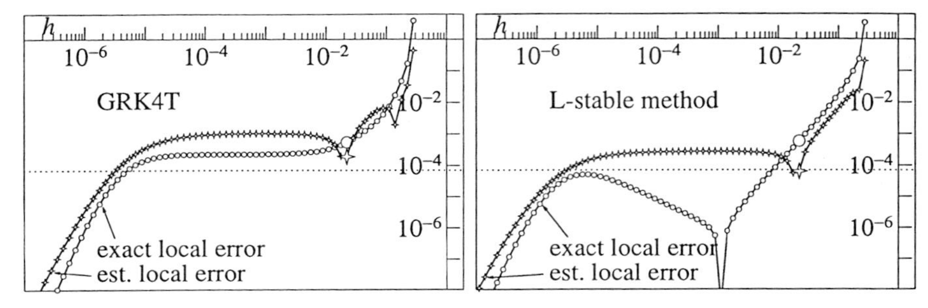

Fig. 7.4. Study of local error for (1.5') at $x = 0.55139$

size until $h \approx \varepsilon$. As a first remedy, we accelerate this lengthy process, as Shampine (1982) also did, by more drastical step size reductions ($h_{new} = h/10$) after each second consecutive step rejection. It also turns out (see right picture of Fig. 7.4) that the effect disappears in the neighbourhood of the actual step size for the L-stable method (where $R(\infty) = 0$). Methods with $R(\infty) = 0$ and also $\widehat{R}(\infty) = 0$ have been derived by Kaps & Ostermann (1990).

A more thorough understanding of these phenomena is possible by the consideration of singular perturbation problems (Chapter VI).

Methods with Inexact Jacobian (W-Methods)

> The relevant question is now, what is the cheapest type of implicitness we have to require. (Steihaug & Wolfbrandt 1979)

All the above theory is built on the assumption that J is the exact Jacobian $\partial f/\partial y$. This implies that the matrix must be evaluated at every step, which can make the computations costly. The following attempt, due to Steihaug & Wolfbrandt (1979), searches for order conditions which assure classical order for *all* approximations A of $\partial f/\partial y$. The latter is then maintained over several steps and is just used to assure stability. The derivation of the order conditions must now be done somewhat differently: if J is replaced by an arbitrary matrix A, Formula (7.6) becomes

$$(k_j^J)^{(q)}\big|_{h=0} = q(f^J(g_j))^{(q-1)}\big|_{h=0} + q\sum_K A_K^J \sum_k \gamma_{jk}(k_k^K)^{(q-1)}\big|_{h=0} \tag{7.30}$$

where $A = (A_K^J)_{J,K=1}^n$, and we obtain

$$(k_j^J)^{(2)}\big|_{h=0} = 2\sum_K f_K^J f^K \sum_k \alpha_{jk} + 2\sum_K A_K^J f^K \sum_k \gamma_{jk}. \tag{7.31;2}$$

Inserted into (7.8), the first term must equal the derivative of the exact solution and the second must be zero. Similarly, we obtain instead of (7.7;3)

$$
\begin{aligned}
(k_j^J)^{(3)}\big|_{h=0} = 3\sum_{K,L} f^J_{KL} f^K f^L \sum_{k,l} \alpha_{jk}\alpha_{jl} \\
+3\cdot 2\sum_{K,L} f^J_K f^K_L f^L \sum_{k,l} \alpha_{jk}\alpha_{kl} + 3\cdot 2\sum_{K,L} f^J_K A^K_L f^L \sum_{k,l} \alpha_{jk}\gamma_{kl} \\
+3\cdot 2\sum_{K,L} A^J_K f^K_L f^L \sum_{k,l} \gamma_{jk}\alpha_{kl} + 3\cdot 2\sum_{K,L} A^J_K A^K_L f^L \sum_{k,l} \gamma_{jk}\gamma_{kl}
\end{aligned}
\tag{7.31;3}
$$

and the order conditions for order three become

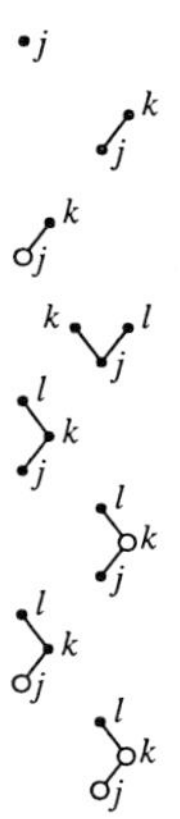

$$
\begin{aligned}
&\sum b_j = 1 \\
&\sum b_j\alpha_{jk} = 1/2 \\
&\sum b_j\gamma_{jk} = 0 \\
&\sum b_j\alpha_{jk}\alpha_{jl} = 1/3 \\
&\sum b_j\alpha_{jk}\alpha_{kl} = 1/6 \\
&\sum b_j\alpha_{jk}\gamma_{kl} = 0 \\
&\sum b_j\gamma_{jk}\alpha_{kl} = 0 \\
&\sum b_j\gamma_{jk}\gamma_{kl} = 0.
\end{aligned}
\tag{7.32}
$$

For a graphical representation of the elementary differentials in (7.31;q) and of the order conditions (7.32) we need trees with two different kinds of vertices (one representing f and the other A). As in Sect. II.15 we use "meagre" and "fat" vertices (see Definitions II.15.1 to II.15.4). Not all trees with meagre and fat vertices (P-trees) have to be considered. From the above derivation we see that fat vertices have to be singly-branched (derivatives of the constant matrix A are zero) and that they cannot be at the end of a branch. We therefore use the notation

$$
TW = \{\ P\text{-trees}\ ;\ \text{end-vertices are meagre and fat vertices are singly-branched}\ \}
\tag{7.33}
$$

and if the vertices are labelled monotonically, we write LTW.

Definition 7.5. The *elementary differentials* for trees $t \in TW$ are defined recursively by $F^J(\tau)(y) = f^J(y)$ and

$$
F^J(t)(y) = \begin{cases} \displaystyle\sum_{K_1,\ldots,K_m} f^J_{K_1,\ldots,K_m}(y)\cdot\Big(F^{K_1}(t_1)(y),\ldots,F^{K_m}(t_m)(y)\Big) & \text{if } t = {}_a[t_1,\ldots,t_m] \quad \text{(meagre root)} \\[2ex] \displaystyle\sum_K A^J_K \cdot F^K(t_1)(y) \quad \text{if } t = {}_b[t_1] & \text{(fat root).} \end{cases}
$$

Definition 7.6. For $t \in TW$ we let $\Phi_j(\tau) = 1$ and

$$\Phi_j(t) = \begin{cases} \sum\limits_{k_1,\ldots,k_m} \alpha_{jk_1} \ldots \alpha_{jk_m} \Phi_{k_1}(t_1) \ldots \Phi_{k_m}(t_m) & \text{if } t = {}_a[t_1,\ldots,t_m] \\ \sum\limits_{k} \gamma_{jk} \Phi_k(t_1) & \text{if } t = {}_b[t_1] \,. \end{cases}$$

We remark that T (the set of trees as considered for Runge-Kutta methods) is a subset of *TW* and that the above definitions coincide with Definitions II.2.3 and II.2.9 (c.f. also Formulas (II.2.18) and (II.2.19)). The general result is now the following

Theorem 7.7. *A W-method (7.4) with $J = A$ arbitrary is of order p iff*

$$\sum_j b_j \Phi_j(t) = \frac{1}{\gamma(t)} \qquad \textit{for } t \in T \textit{ with } \varrho(t) \le p, \textit{ and}$$

$$\sum_j b_j \Phi_j(t) = 0 \qquad \textit{for } t \in TW \setminus T \textit{ with } \varrho(t) \le p\,.$$

The *proof* is essentially the same as for Theorems 7.3 and 7.4. □

Table 7.3. Number of order conditions for W-methods

order p	1	2	3	4	5	6	7	8
no. of conditions	1	3	8	21	58	166	498	1540

The number of order conditions for W-methods is rather large (see Table 7.3), since each tree of T with κ singly-branched vertices gives rise to 2^κ order conditions (in the case of symmetry some may be identical). Therefore, W-methods of higher order are best obtained by extrapolation (see Sect. IV.9).

The *stability* investigation for linearly implicit methods with $A \neq f'(y_0)$ is very complicated. If we linearize the differential equation (as in the beginning of Sect. IV.2) and assume the Jacobian to be constant, we arrive at a recursion of the form

$$y_1 = R(hf'(y_0), hA)y_0.$$

Since, in general, the matrices $f'(y_0)$ and A cannot be diagonalized simultaneously, the consideration of scalar test equations is not justified. Stability investigations for the case when $\|f'(y_0) - A\|$ is small will be considered in Sect. IV.11.

Exercises

1. (Kaps 1977). There exists no Rosenbrock method (7.4) with $s=4$ and $p=5$. Prove this.

2. (Nørsett & Wolfbrandt 1979). Generalize the derivation of order conditions for Runge-Kutta methods with the help of B-series (Sect. II.11, page 247) to Rosenbrock methods.

   Hint. Prove that, for a B-series $B(\mathbf{a}, y_0)$ with $\mathbf{a}: T \longrightarrow \mathbb{R}$ satisfying $\mathbf{a}(\emptyset)=0$,

$$hf'(y_0)B(\mathbf{a}, y_0) = B(\widehat{\mathbf{a}},\, y_0)$$

   is again a B-series with coefficients

$$\widehat{\mathbf{a}}(t) = \begin{cases} \varrho(t)\mathbf{a}(t_1) & \text{if } t=[t_1] \\ 0 & \text{else .} \end{cases}$$

3. Cooper & Sayfy (1983) consider *additive* Runge-Kutta methods

$$\begin{aligned} g_i &= y_0 + h\sum_{j=1}^{i-1} \alpha_{ij} f(x_0+c_j h,\, g_j) + hJ\sum_{j=1}^{i} \eta_{ij} g_j \qquad i=1,\ldots,s+1 \\ y_1 &= g_{s+1} \end{aligned} \tag{7.34}$$

   whose coefficients satisfy $\sum_{j=1}^{i-1}\alpha_{ij}=c_i$, $\sum_{j=1}^{i}\eta_{ij}=0$.

   a) Prove that (7.34) is equivalent to (7.4) whenever $\alpha_{s+1,i}=b_i$ and

$$(\eta_{ij})(\alpha_{ij}) = (\alpha_{ij})(\gamma_{ij}). \tag{7.35}$$

   Here all matrices are of dimension $(s+1)\times(s+1)$. The last line of (γ_{ij}) need not be specified since the last column of (α_{ij}) is zero.

   b) If the coefficients of (7.34) satisfy $\alpha_{i,i-1}\neq 0$ for all i, then we can always find an equivalent method of type (7.4).

4. (Verwer 1980, Verwer & Scholz 1983). Derive order conditions for Rosenbrock methods "with time-lagged Jacobian", i.e., methods of type (7.4) where J is assumed to be $f'(y(x_0-\omega h))$. If ω is the step ratio h_{old}/h, this allows re-use of the Jacobian of the previous step.

5. (Kaps & Ostermann 1989). Show that some order conditions of (7.32) can be shifted to higher orders if it is assumed that

$$f'(y_0) - J = \mathcal{O}(h).$$

   This makes the conditions of Exercise 4 independent of ω.

   Result. The number of order-shifts is equal to the number of fat nodes.

IV.8 Implementation of Implicit Runge-Kutta Methods

> These have not been used to any great extent ...
> (S.P. Nørsett 1976)

> However, the implementation difficulties of these methods have precluded their general use; ...
> (J.M. Varah 1979)

> Although Runge-Kutta methods present an attractive alternative, especially for stiff problems, ... it is generally believed that they will never be competitive with multistep methods.
> (K. Burrage, J.C. Butcher & F.H. Chipman 1980)

> Runge-Kutta methods for stiff problems, we are just beginning to explore them ...
> (L. Shampine in Aiken 1985)

If the dimension of the differential equation $y' = f(x,y)$ is n, then the s-stage fully implicit Runge-Kutta method (3.1) involves a $n \cdot s$-dimensional nonlinear system for the unknowns $g_1, \ldots, g_s$. An efficient solution of this system is the main problem in the implementation of an implicit Runge-Kutta method.

Among the methods discussed in Sect. IV.5, the processes Radau IIA of Ehle, which are L-stable and of high order, seem to be particularly promising. Most of the questions arising (starting values and stopping criteria for the simplified Newton iterations, efficient solution of the linear systems, and the selection of the step sizes) are discussed here for the particular Ehle method with $s = 3$ and $p = 5$. This then constitutes a description of the code RADAU5 of the appendix. An adaptation of the described techniques to other fully implicit Runge-Kutta methods is more or less straight-forward, if the Runge-Kutta matrix has at least one real eigenvalue. We also describe briefly our implementation of the diagonal implicit method SDIRK4 (Formula (6.16)).

Reformulation of the Nonlinear System

In order to reduce the influence of round-off errors we prefer to work with the smaller quantities

$$z_i = g_i - y_0. \tag{8.1}$$

Then (3.1a) becomes

$$z_i = h \sum_{j=1}^{s} a_{ij} f(x_0 + c_j h, y_0 + z_j) \qquad i = 1, \ldots, s. \tag{8.2a}$$

Whenever the solution $z_1, \ldots, z_s$ of the system (8.2a) is known, then (3.1b) is an explicit formula for y_1. A direct application of this requires s additional function evaluations. These can be avoided, if the matrix $A = (a_{ij})$ of the Runge-Kutta coefficients is nonsingular. Indeed, (8.2a) can be written as

$$\begin{pmatrix} z_1 \\ \vdots \\ z_s \end{pmatrix} = A \begin{pmatrix} hf(x_0 + c_1 h, y_0 + z_1) \\ \vdots \\ hf(x_0 + c_s h, y_0 + z_s) \end{pmatrix},$$

so that (3.1b) is seen to be equivalent to

$$y_1 = y_0 + \sum_{i=1}^{s} d_i z_i \tag{8.2b}$$

where

$$(d_1, \ldots, d_s) = (b_1, \ldots, b_s) A^{-1}. \tag{8.3}$$

For the 3-stage Radau IIA method (Table 5.6) the vector d is simply $(0,0,1)$, since $b_i = a_{si}$ for all i.

Another advantage of Formula (8.2b) is the following: the quantities $z_1, \ldots, z_s$ are computed iteratively and are therefore affected by iteration errors. The evaluation of $f(x_0 + c_i h, y_0 + z_i)$ in Eq. (3.1b) would then, due to the large Lipschitz constant of f, amplify these errors, which then "can be disastrously inaccurate for a stiff problem" (L.F. Shampine 1980).

Simplified Newton Iterations

For a general nonlinear differential equation the system (8.2a) has to be solved iteratively. In the stone-age of stiff computation (i.e., before 1967) people were usually thinking of simple fixed-point iteration. But this transforms the algorithm into an explicit method and destroys the good stability properties. The paper of Liniger & Willoughby (1970) then showed the advantages of using Newton's method for this purpose. Newton's method applied to system (8.2a) needs for each iteration the solution of a linear system with matrix

$$\begin{pmatrix} I - ha_{11}\frac{\partial f}{\partial y}(x_0 + c_1 h, y_0 + z_1) & \ldots & -ha_{1s}\frac{\partial f}{\partial y}(x_0 + c_s h, y_0 + z_s) \\ \vdots & & \vdots \\ -ha_{s1}\frac{\partial f}{\partial y}(x_0 + c_1 h, y_0 + z_1) & \ldots & I - ha_{ss}\frac{\partial f}{\partial y}(x_0 + c_s h, y_0 + z_s) \end{pmatrix}.$$

In order to simplify this, we replace all Jacobians $\frac{\partial f}{\partial y}(x_0 + c_i h, y_0 + z_i)$ by an approximation

$$J \approx \frac{\partial f}{\partial y}(x_0, y_0).$$

Then the simplified Newton iterations for (8.2a) become

$$\begin{aligned} (I - hA \otimes J)\Delta Z^k &= -Z^k + h(A \otimes I)F(Z^k) \\ Z^{k+1} &= Z^k + \Delta Z^k. \end{aligned} \tag{8.4}$$

Here $Z^k = (z_1^k, \ldots, z_s^k)^T$ is the k-th approximation to the solution, and $\Delta Z^k = (\Delta z_1^k, \ldots, \Delta z_s^k)^T$ are the increments. $F(Z^k)$ is an abbreviation for

$$F(Z^k) = \big(f(x_0 + c_1 h, y_0 + z_1^k), \ldots, f(x_0 + c_s h, y_0 + z_s^k)\big)^T.$$

Each iteration requires s evaluations of f and the solution of a $n \cdot s$-dimensional linear system. The matrix $(I - hA \otimes J)$ is the same for all iterations. Its LU-decomposition is done only once and is usually very costly.

Starting Values for the Newton Iteration. A natural and simple choice for the starting values in the iteration (8.4) (or equivalently (8.13) below), since the exact solution of (8.2a) satisfies $z_i = \mathcal{O}(h)$, would be

$$z_i^0 = 0, \qquad i = 1, \ldots, s. \tag{8.5}$$

However, better choices are possible in general. If the implicit Runge-Kutta method satisfies the condition $C(\eta)$ (see Sections IV.5 and II.7) for some $\eta \le s$, then

$$z_i = y(x_0 + c_i h) - y_0 + \mathcal{O}(h^{\eta+1}). \tag{8.6}$$

Suppose now that $c_i \neq 0$ $(i = 1, \ldots, s)$ and consider the interpolation polynomial of degree s, defined by

$$q(0) = 0, \qquad q(c_i) = z_i \qquad i = 1, \ldots, s.$$

Since the interpolation error is of size $\mathcal{O}(h^{s+1})$ we obtain together with (8.6)

$$y(x_0 + th) - y_0 - q(t) = \mathcal{O}(h^{\eta+1})$$

(cf. Theorem 7.10 of Chapter II for collocation methods). We use the values of $q(t)$ also beyond the interval $[0, 1]$ and take

$$z_i^0 = q(1 + wc_i) + y_0 - y_1, \quad i = 1, \ldots, s, \quad w = h_{new}/h_{old} \tag{8.5'}$$

as starting values for the Newton iteration in the subsequent step. Numerical experiments with the 3-stage Radau IIA method have shown that (8.5') usually leads to a faster convergence than (8.5).

Stopping Criterion. This question is closely related to an estimation of the iteration error. Since convergence is linear, we have

$$\|\Delta Z^{k+1}\| \le \Theta \|\Delta Z^k\|, \qquad \text{hopefully with} \quad \Theta < 1. \tag{8.7}$$

Applying the triangle inequality to

$$Z^{k+1} - Z^* = (Z^{k+1} - Z^{k+2}) + (Z^{k+2} - Z^{k+3}) + \ldots$$

(where Z^* is the exact solution of (8.2a)) yields the estimate

$$\|Z^{k+1} - Z^*\| \le \frac{\Theta}{1 - \Theta} \|\Delta Z^k\|. \tag{8.8}$$

The convergence rate Θ can be estimated by the computed quantities

$$\Theta_k = \|\Delta Z^k\| / \|\Delta Z^{k-1}\|, \qquad k \ge 1. \tag{8.9}$$

It is clear that the iteration error should not be larger than the local discretization error, which is usually kept close to *Tol*. We therefore stop the iteration when

$$\eta_k \|\Delta Z^k\| \le \kappa \cdot Tol \quad \text{with} \quad \eta_k = \frac{\Theta_k}{1-\Theta_k} \tag{8.10}$$

and accept Z^{k+1} as approximation to Z^*. This strategy can only be applied after at least two iterations. In order to be able to stop the computations after the first iteration already (which is especially advantageous for linear systems) we take for $k=0$ the quantity

$$\eta_0 = (\max(\eta_{old}, Uround))^{0.8}$$

where η_{old} is the last η_k of the preceding step. It remains to make a good choice for the parameter κ in (8.10). To this end we applied the code RADAU5 for many different values of κ between 10 and 10^{-4} and with some different tolerances *Tol* to several differential equations. The observation was that the code works most efficiently for values of κ around 10^{-1} or 10^{-2}.

It is our experience that the code becomes more efficient when we allow a relatively high number of iterations (e.g., $k_{max} = 7$ or 10). During these k_{max} iterations, the computations are interrupted and restarted with a smaller stepsize (for example with $h := h/2$) if one of the following situations occurs
a) there is a k with $\Theta_k \ge 1$ (the iteration "diverges");
b) for some k,

$$\frac{\Theta_k^{k_{max}-k}}{1-\Theta_k} \|\Delta Z^k\| > \kappa \cdot Tol. \tag{8.11}$$

The left-hand expression in (8.11) is a rough estimate of the iteration error to be expected after $k_{max}-1$ iterations. The norm, used in all these formulas, should be the same as the one used for the local error estimator.

If only one Newton iteration was necessary to satisfy (8.10) or if the last Θ_k was very small, say $\le 10^{-3}$, then we don't recompute the Jacobian in the next step. As a consequence, the Jacobian is computed only once for linear problems with constant coefficients (as long as no step rejection occurs).

The Linear System

An essential gain of numerical work for the solution of the linear system (8.4) is obtained by the following method, introduced independently by Butcher (1976) and Bickart (1977), which exploits with much profit the special structure of the matrix $I - hA \otimes J$ in (8.4).

The idea is to premultiply (8.4) by $(hA)^{-1} \otimes I$ (we suppose here that A is invertible) and to transform A^{-1} to a simple matrix (diagonal, block diagonal, triangular or Jordan canonical form)

$$T^{-1}A^{-1}T = \Lambda. \tag{8.12}$$

With the transformed variables $W^k = (T^{-1} \otimes I)Z^k$, the iteration (8.4) becomes equivalent to

$$\begin{aligned}(h^{-1}\Lambda \otimes I - I \otimes J)\Delta W^k &= -h^{-1}(\Lambda \otimes I)W^k + (T^{-1} \otimes I)F((T \otimes I)W^k) \\ W^{k+1} &= W^k + \Delta W^k. \end{aligned} \tag{8.13}$$

We also replace Z^k and ΔZ^k by W^k and ΔW^k in the formulas (8.7)–(8.11) (and thereby again save some work).

For the sequel, we suppose that the matrix A^{-1} has one real eigenvalue $\widehat{\gamma}$ and one complex conjugate eigenvalue pair $\widehat{\alpha} \pm i\widehat{\beta}$. This is a typical situation for 3-stage implicit Runge-Kutta methods such as Radau IIA. With $\gamma = h^{-1}\widehat{\gamma}, \alpha = h^{-1}\widehat{\alpha}, \beta = h^{-1}\widehat{\beta}$ the matrix in (8.13) becomes

$$\begin{pmatrix} \gamma I - J & 0 & 0 \\ 0 & \alpha I - J & -\beta I \\ 0 & \beta I & \alpha I - J \end{pmatrix} \tag{8.14}$$

so that (8.13) splits into two linear systems of dimension n and $2n$, respectively. Several ideas are possible to exploit the special structure of the $2n \times 2n$-submatrix. The easiest and numerically most stable way has turned out to be the following: transform the real subsystem of dimension $2n$ into an n-dimensional, complex system

$$\big((\alpha + i\beta)I - J\big)(u + iv) = a + ib \tag{8.14'}$$

and apply simple Gaussian elimination. For machines without complex arithmetic, one just has to modify the linear algebra routines. Then a complex multiplication consists of 4 real multiplications and the amount of work for the solution of (8.14') becomes approximately $4n^3/3$ operations. Thus the total work for system (8.14) is about $5n^3/3$ operations. Compared to $(3n)^3/3$, which would be the number of operations necessary for decomposing the untransformed matrix $I - hA \otimes J$ in (8.4), we gain a factor of about 5 in arithmetical operations. Observe that the transformations, such as $Z^k = (T \otimes I)W^k$, need only $\mathcal{O}(n)$ additions and multiplications. The gain is still more drastic for methods with more than 3 stages.

Transformation to Hessenberg Form. For large systems with a full Jacobian J a further gain is possible by transforming J to Hessenberg form

$$S^{-1}JS = H = \begin{pmatrix} * & \dots & * & * \\ * & & & * \\ & \ddots & & \vdots \\ & & * & * \end{pmatrix}. \tag{8.15}$$

This procedure was originally proposed for multistep methods by Enright (1978) and extended to the Runge-Kutta case by Varah (1979). With the code ELMHES, taken from LINPACK, this is performed with $2n^3/3$ operations. Because the multiplication of S with a vector needs only $n^2/2$ operations (observe that S is triangular) the solution of (8.13) is found in $\mathcal{O}(n^2)$ operations, if the Hessenberg matrix H is known. This transformation is especially advantageous, if the Jacobian J is not changed during several steps.

Step Size Selection

One possibility to select the step sizes is Richardson extrapolation (cf. Sect. II.4). We describe here the use of an embedded pair of methods which is easier to program and which makes the code more flexible. The following formulas are for the special case of the 3-stage Radau IIA methods; the same ideas are applicable to all implicit Runge-Kutta methods, whose Runge-Kutta matrix has at least one real eigenvalue.

Embedded Formula. Since our method is of optimal order, it is impossible to embed it efficiently into one of still higher order. Therefore we search for a lower order method of the form

$$\widehat{y}_1 = y_0 + h\Big(\widehat{b}_0 f(x_0, y_0) + \sum_{i=1}^{3} \widehat{b}_i f(x_0 + c_i h, g_i)\Big) \tag{8.16}$$

where g_1, g_2, g_3 are the values obtained from the Radau IIA method and $\widehat{b}_0 \neq 0$ (the choice $\widehat{b}_0 = \gamma_0 = \widehat{\gamma}^{-1}$, where $\widehat{\gamma}$ is the real eigenvalue of the matrix A^{-1}, again saves some multiplications). The difference

$$\widehat{y}_1 - y_1 = \gamma_0 h f(x_0, y_0) + \sum_{i=1}^{3} (\widehat{b}_i - b_i) h f(x_0 + c_i h, g_i),$$

which can also be written in the form

$$\widehat{y}_1 - y_1 = \gamma_0 h f(x_0, y_0) + e_1 z_1 + e_2 z_2 + e_3 z_3, \tag{8.17}$$

then serves for error estimation. In order that $\widehat{y}_1 - y_1 = \mathcal{O}(h^4)$ the coefficients have to satisfy

$$(e_1, e_2, e_3) = \frac{\gamma_0}{3}(-13 - 7\sqrt{6}, -13 + 7\sqrt{6}, -1). \tag{8.18}$$

Unfortunately, for $y' = \lambda y$ and $h\lambda \to \infty$ the difference (8.17) behaves like $\widehat{y}_1 - y_1 \approx \gamma_0 h \lambda y_0$, which is unbounded and therefore not suitable for stiff equations. We propose (an idea of Shampine) to use instead

$$err = (I - h\gamma_0 J)^{-1}(\widehat{y}_1 - y_1). \tag{8.19}$$

The LU-decomposition of $((h\gamma_0)^{-1} I - J)$ is available anyway from the previous work, so that the computation of (8.19) is cheap. For $h \to 0$ we still have $err = \mathcal{O}(h^4)$, and for $h\lambda \to \infty$ (if $y' = \lambda y$ and $J = \lambda$) we obtain $err \to -1$.

This behaviour (for $h\lambda \to \infty$) is already much better than that for $\widehat{y}_1 - y_1$, but it is not good enough in order to avoid the "hump" phenomenon, described in Sect. IV.7. In the first step and after every rejected step for which $\|err\| > 1$, we therefore use instead of (8.19) the expression

$$\widetilde{err} = (I - h\gamma_0 J)^{-1}\big(\gamma_0 h f(x_0, y_0 + err) + e_1 z_1 + e_2 z_2 + e_3 z_3\big) \tag{8.20}$$

for step size prediction. This requires one additional function evaluation, but satisfies $\widetilde{err} \to 0$ for $h\lambda \to \infty$, as does the error of the numerical solution.

Standard Step Size Controller. Since the expressions (8.19) and (8.20) behave like $\mathcal{O}(h^4)$ for $h \to 0$, the standard step size prediction leads to

$$h_{\text{new}} = fac \cdot h_{\text{old}} \cdot \|err\|^{-1/4}. \tag{8.21}$$

where

$$\|err\| = \sqrt{\frac{1}{n}\sum_{i=1}^{n}\Big(\frac{err_i}{sc_i}\Big)^2},$$

and $sc_i = Atol_i + \max(|y_{0i}|, |y_{1i}|) \cdot Rtol_i$ as in (4.11) of Sect. II.4. Here, the safety factor *fac* is proposed to depend on *Newt*, the number of Newton iterations of the current step and on the maximal number of Newton iterations $k_{\max}$, say, as: $fac = 0.9 \times (2k_{\max} + 1)/(2k_{\max} + Newt)$.

In order to save LU-decompositions of the matrix (8.14), we also include the following strategy: if no Jacobian is recomputed and if the step size h_{new}, defined by (8.21), satisfies

$$c_1 h_{\text{old}} \le h_{\text{new}} \le c_2 h_{\text{old}} \tag{8.22}$$

with, say $c_1 = 1.0$ and $c_2 = 1.2$, then we retain h_{old} for the following step.

Predictive Controller. The step size prediction by formula (8.21) has the disadvantage that step size reductions by more than the factor *fac* are not possible without step rejections (observe that $h_{\text{new}} < fac \cdot h_{\text{old}}$ implies $\|err\| > 1$). For stiff differential equations, however, a rapid decrease of the step size is often required (see for example the situation of Fig. 8.1, where the step size drops from 10^{-2} to 10^{-7} within a very small time interval). Denoting by err_{n+1} the error expression (8.19) (or (8.20)), computed in the nth step with step size h_n, step size predictions are typically derived from the asymptotic formula

$$\|err_{n+1}\| = C_n h_n^4. \tag{8.23}$$

The strategy (8.21) is based on the additional assumption $C_{n+1} \approx C_n$, which, as we have seen, is not always very realistic.

A careful control-theoretic study of step size strategies has been undertaken by Gustafsson (1994). He came to the conclusion that a better model is to assume that $\log C_n$ is a linear function of n. This means that $\log C_{n+1} - \log C_n$ is constant or, equivalently,

$$C_{n+1}/C_n \approx C_n/C_{n-1}. \tag{8.24}$$

Inserting C_n and C_{n-1} from (8.23) and C_{n+1} from $1 = C_{n+1}h_{\text{new}}^4$ into (8.24) yields

$$h_{\text{new}} = fac \cdot h_n \Big(\frac{1}{\|err_{n+1}\|}\Big)^{1/4} \cdot \frac{h_n}{h_{n-1}}\Big(\frac{\|err_n\|}{\|err_{n+1}\|}\Big)^{1/4}. \tag{8.25}$$

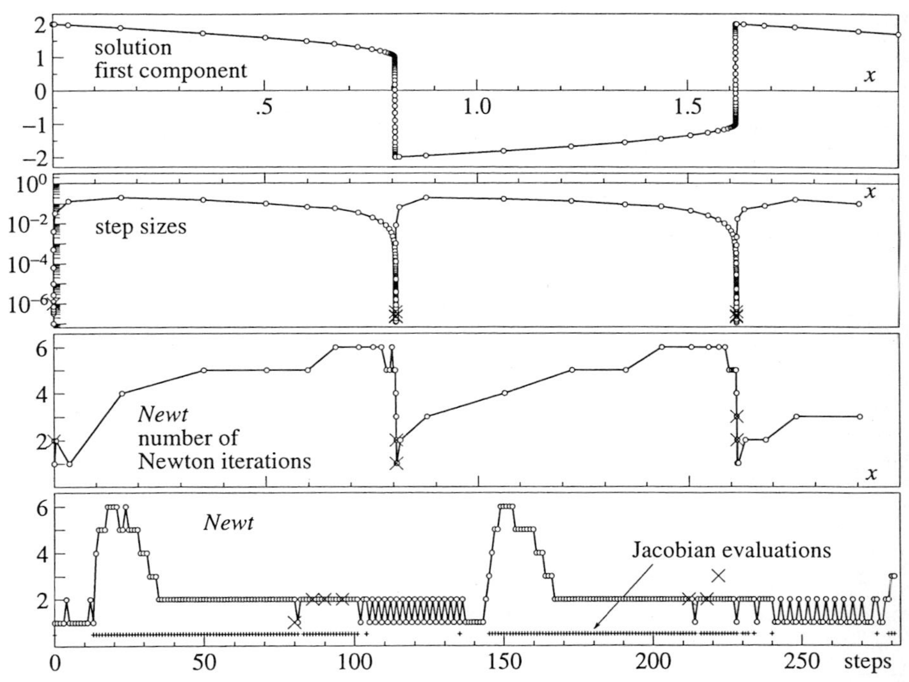

Fig. 8.1. Solution, step sizes and Newton iterations for RADAU5

In our code RADAU5 we take the minimum of the two step sizes (8.21) and (8.25). For the problem considered in Fig. 8.1, this new strategy reduces the number of rejected steps from 27 to 7.

Numerical Study of the Step-Control Mechanism. As a representative example we choose the van der Pol equation (1.5') with $\varepsilon = 10^{-6}$, initial values $y_1(0) = 2$, $y_2(0) = -0.6$ and integration interval $0 \le x \le 2$. Fig. 8.1 shows four pictures. The first one presents the solution $y_1(x)$ with all accepted integration steps for $Atol = Rtol = 10^{-4}$. Below this, the step sizes obtained by RADAU5 are plotted as function of x. The solid line represents the accepted steps. The rejected steps are indicated by $\times$'s. Observe the very small step sizes which are required in the rapid transients between the smooth parts of the solution. The lowest two pictures give the number of Newton iterations needed for solving the nonlinear system (8.2a), once as function of x, and once as function of the step-number. The last picture also indicates the steps where the Jacobian has been recomputed.

Another numerical experiment (Fig. 8.2) illustrates the quality of the error estimates. We applied the code RADAU5 with $Atol = Rtol = 10^{-4}$ and initial step size $h = 10^{-4}$ to the above problem and plotted at several chosen points of the numerical solution

a) the exact local error (marked by small circles)

b) the estimates (8.19) and (8.20) (marked by ✧ and ⌘ respectively)

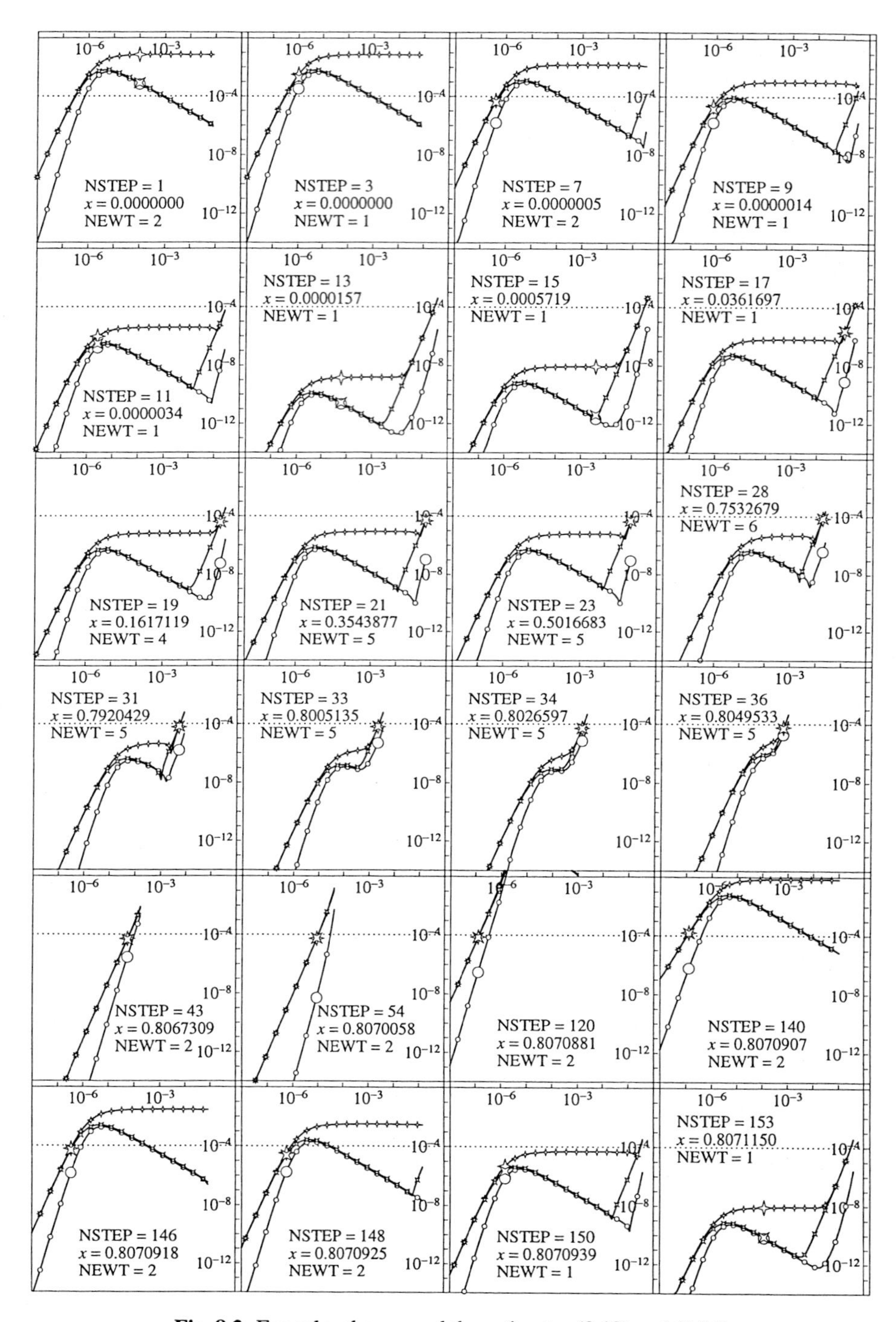

Fig. 8.2. Exact local error and the estimates (8.19) and (8.20)

as functions of h. The large symbols indicate the position of the actually used step size. *Newt* is the number of required Newton iterations.

It is interesting to note that the local error behaves like $\mathcal{O}(h^6)$ (straight line of slope 6) only for $h \le \varepsilon$ and for large h. Between these regions, the local error *grows* like $\mathcal{O}(h^{-1})$ with decreasing h. This is the only region where the error estimate (8.20) is significantly better than (8.19). Therefore, we use the more expensive estimator (8.20) only in the first and after each rejected step. In any way, *both* error estimators are always above the actual local error, so that the code usually produces very precise results.

Implicit Differential Equations

Many applications (such as space discretizations of parabolic differential equations) often lead to systems of the form

$$My' = f(x,y), \qquad y(x_0) = y_0 \tag{8.26}$$

with a constant matrix M. For such problems we formally replace all f's by $M^{-1}f$ and multiply the resulting equations by M. Formulas (8.13) and (8.19) then have to be replaced by

$$(h^{-1}\Lambda \otimes M - I \otimes J)\,\Delta W^k = -h^{-1}\,(\Lambda \otimes M)W^k + (T^{-1} \otimes I)F((T \otimes I)W^k) \tag{8.13a}$$

$$err = ((h\gamma_0)^{-1}M - J)^{-1}\,\big(f(x_0,y_0) + (h\gamma_0)^{-1}M(e_1 z_1 + e_2 z_2 + e_3 z_3)\big)\,. \tag{8.19a}$$

Here the matrix J is again an approximation to $\partial f/\partial y$. These formulas may even be applied to certain problems (8.26) with singular M (for more details see Chapters VI and VII).

Solving the linear system (8.13a) is done by a decomposition of the matrix (see (8.14), (8.14'))

$$\begin{pmatrix} \gamma M - J & 0 \\ 0 & (\alpha + i\beta)M - J \end{pmatrix}. \tag{8.27}$$

If M and J are banded or sparse, the matrices $\gamma M - J$ and $(\alpha + i\beta)M - J$ remain banded or sparse, respectively. The code RADAU5 of the appendix has options for banded structures.

An SDIRK-Code

We have also coded, using many of the above ideas, the SDIRK formula (6.16) together with the global solution (6.17). For this method also, it was again very important to replace the error estimator $y_1 - \widehat{y}_1$ by (8.19).

Here, in contrast to fully implicit Runge-Kutta methods, one can treat the stages one after the other. Such a serial computation has the advantage that the information of the already computed stages can be used for a good choice of the starting values for the Newton iterations in the subsequent stages. For example, suppose that

$$
\begin{aligned}
z_1 &= \gamma h f(x_0 + \gamma h, y_0 + z_1)\\
z_2 &= \gamma h f(x_0 + c_2 h, y_0 + z_2) + a_{21} h f(x_0 + \gamma h, y_0 + z_1)
\end{aligned}
$$

are already available. Since for all i

$$
z_i = c_i h f(x_0, y_0) + \big(\sum_j a_{ij} c_j\big) h^2 (f_x + f_y f)(x_0, y_0) + \mathcal{O}(h^3),
$$

by solving

$$
\begin{pmatrix} c_1 & c_2 \\ \sum_j a_{1j} c_j & \sum_j a_{2j} c_j \end{pmatrix} \begin{pmatrix} \alpha_1 \\ \alpha_2 \end{pmatrix} = \begin{pmatrix} c_3 \\ \sum_j a_{3j} c_j \end{pmatrix}
$$

one finds α_1, α_2 such that

$$
\alpha_1 z_1 + \alpha_2 z_2 = z_3 + \mathcal{O}(h^3).
$$

The expression $z_3^{(0)} = \alpha_1 z_1 + \alpha_2 z_2$ then serves as starting value for the computation of z_3. In the last stage one can take $\widehat{y}_1$, which is then available, for starting the Newton iterations for $g_s = y_1$. The computation of z_3, z_4, y_1, done in this way, needs few Newton iterations and a failure of convergence is usually already detected in the first stage.

However, when *parallel* processors are available, the exploitation of the triangular structure of the Runge-Kutta matrix may be less desirable. Whereas in the iteration (8.13) all s function evaluations and much of the linear algebra can be done in parallel, this is no longer possible for DIRK-methods, when $z_1, \ldots, z_k$ is used in the computations of z_{k+1}.

SIRK-Methods

> The fact that singly-implicit methods have a coefficient matrix with a one-point spectrum is the key to reducing the operation count for these methods to the level which prevails in linear multistep methods.
>
> (J.C. Butcher, K. Burrage & F.H. Chipman 1980)

In order to avoid the difficulties (in writing a Runge-Kutta code) caused by the complex eigenvalues of the Runge-Kutta matrix A, one may look for methods with real eigenvalues, especially with a single s-fold real eigenvalue. Such methods were introduced by Nørsett (1976). Burrage (1978) provided them with error estimators, and codes in ALGOL and FORTRAN are presented in Butcher, Burrage & Chipman (1980). The basic methods for their code STRIDE are given by the following lemma.

Lemma 8.1. *For collocation methods (i.e., for Runge-Kutta methods satisfying condition $C(s)$ of Sect. IV.5), we have*

$$\det(I - zA) = (1-\gamma z)^s, \tag{8.28}$$

if and only if

$$c_i = \gamma x_i, \qquad i=1,\ldots,s, \tag{8.29}$$

where $x_1,\ldots,x_s$ are the zeros of the Laguerre polynomial $L_s(x)$ (c.f. Formula (6.11)).

Proof. The polynomial $\det(I-zA)$ is the denominator of the stability function (Formula (3.3)), so that by Theorem 3.10

$$M^{(s)}(0) + M^{(s-1)}(0)z + \ldots + M(0)z^s = (1-\gamma z)^s \tag{8.30}$$

with $M(x)$ given by (3.25). Computing $M^{(j)}(0)$ from (8.30) we obtain

$$\frac{1}{s!}\prod_{i=1}^{s}(x-c_i) = M(x) = \sum_{j=0}^{s}\binom{s}{j}(-\gamma)^{s-j}\frac{x^j}{j!} = (-\gamma)^s L_s\Big(\frac{x}{\gamma}\Big)$$

which leads to (8.29). □

The stability function of the method of Lemma 8.1 has been studied in Sections IV.4 (multiple real-pole approximations) and IV.6. We have further seen (Proposition 3.8) that $R(\infty)=0$ when x_0+h is a collocation point. This means that $c_q=1$ or $\gamma=1/x_q$ for $q\in\{1,\ldots,s\}$ where $0<x_1<\ldots<x_s$ are the zeros of $L_s(x)$. However, if we want A-stable methods, Theorem 4.25 restricts this point to be *in the middle* (more precisely: $q=s/2$ or $s/2+1$ for s even, $q=(s+1)/2$ for s odd). An apparently undesirable consequence of this is that many of the collocation points lie *outside* the integration interval (for example, for $s=5$ and $q=3$ we have $c_1=0.073$, $c_2=0.393$, $c_3=1$, $c_4=1.970$, $c_5=3.515$).

Since these methods with $\gamma=1/x_q$ are of order $p=s$ only, it is easy to embed them into a method of higher order. Burrage (1978) added a further stage

$$g_{s+1} = y_0 + h\sum_{j=1}^{s+1} a_{s+1,j}f(x_0+c_jh, g_j)$$

where c_{s+1} and $a_{s+1,s+1}$ are arbitrary and the other $a_{s+1,j}$ are determined so that the $(s+1)$-stage method satisfies $C(s)$ too. In order to avoid a new LU-decomposition we choose $a_{s+1,s+1}=\gamma$. The coefficient c_{s+1} is fixed arbitrarily as $c_{s+1}=0$. We then find a unique method

$$\widehat{y}_1 = y_0 + h\sum_{j=1}^{s+1}\widehat{b}_j f(x_0+c_jh, g_j)$$

of order $s+1$ by computing the coefficients of the interpolatory quadrature rule. An explicit formula for the matrix T which transforms the Runge-Kutta matrix A to Jordan canonical form and A^{-1} to a very simple lower triangular matrix Λ is given in Exercise 1. It can be used for economically solving the linear system (8.13).

Exercises

1. (Butcher 1979). For the collocation method with $c_1,\ldots,c_s$ given by (8.29) prove that (e.g. for $s=4$)

$$T^{-1}AT=\gamma\begin{pmatrix}1&&&\\-1&1&&\\&-1&1&\\&&-1&1\end{pmatrix},\qquad T^{-1}A^{-1}T=\frac{1}{\gamma}\begin{pmatrix}1&&&\\1&1&&\\1&1&1&\\1&1&1&1\end{pmatrix}$$

where the transformation T satisfies

$$T=\big(L_{j-1}(x_i)\big)_{i,j=1}^{s},\qquad T^{-1}=\left(\frac{x_jL_{i-1}(x_j)}{s^2L_{s-1}(x_j)^2}\right)_{i,j=1}^{s}$$

and $L_{j-1}(x)$ are the Laguerre polynomials.

Hint. Use the identities

$$L_n'(x)=L_{n-1}'(x)-L_{n-1}(x),\qquad L_n(x)=L_{n-1}(x)+\frac{x}{n}L_n'(x)$$

and the Christoffel-Darboux formula

$$\sum_{j=0}^{n}L_j(x)L_j(y)=\frac{n+1}{y-x}\Big(L_{n+1}(x)L_n(y)-L_{n+1}(y)L_n(x)\Big)$$

which, in the limit $y\to x$, becomes

$$\sum_{j=0}^{n}(L_j(x))^2=(n+1)\Big(L_{n+1}(x)L_n'(x)-L_{n+1}'(x)L_n(x)\Big).$$

IV.9 Extrapolation Methods

> It seems that a suitable version of an IEM (implicit extrapolation method) which takes care of these difficulties may become a very strong competitor to any of the general discretization methods for stiff systems presently known.
>
> (the very last sentence of Stetter's book, 1973)

Extrapolation of explicit methods is an interesting approach to solving nonstiff differential equations (see Sect. II.9). Here we show to what extent the idea of extrapolation can also be used for stiff problems. We shall use the results of Sect. II.8 for the existence of asymptotic expansions and apply them to the study of those implicit and linearly implicit methods, which seem to be most suitable for the computation of stiff differential equations. Our theory here is restricted to classical $h \to 0$ order, the study of stability domains and A-stability.

A big difficulty, however, is the fact that the coefficients and remainders of the asymptotic expansion can explode with increasing stiffness and the h-interval, for which the expansion is meaningful, may tend to zero. Bounds on the remainder which hold uniformly for a class of arbitrarily stiff problems, will be discussed later in Sect. VI.5.

Extrapolation of Symmetric Methods

It is most natural to look first for symmetric one-step methods as the basic integration scheme. Promising candidates are the trapezoidal rule

$$y_{i+1} = y_i + \frac{h}{2}\Big(f(x_i, y_i) + f(x_{i+1}, y_{i+1})\Big) \tag{9.1}$$

and the implicit mid-point rule

$$y_{i+1} = y_i + hf\Big(x_i + \frac{h}{2}, \frac{1}{2}\left(y_{i+1} + y_i\right)\Big). \tag{9.2}$$

We take some step-number sequence $n_1 < n_2 < n_3 < \ldots$, set $h_j = H/n_j$ and define

$$T_{j1} = y_{h_j}(x_0 + H), \tag{9.3}$$

the numerical solution obtained by performing n_j steps with step size h_j. As described in Sect. II.9 we extrapolate these values according to

$$T_{j,k+1} = T_{j,k} + \frac{T_{j,k} - T_{j-1,k}}{(n_j/n_{j-k})^2 - 1}. \tag{9.4}$$

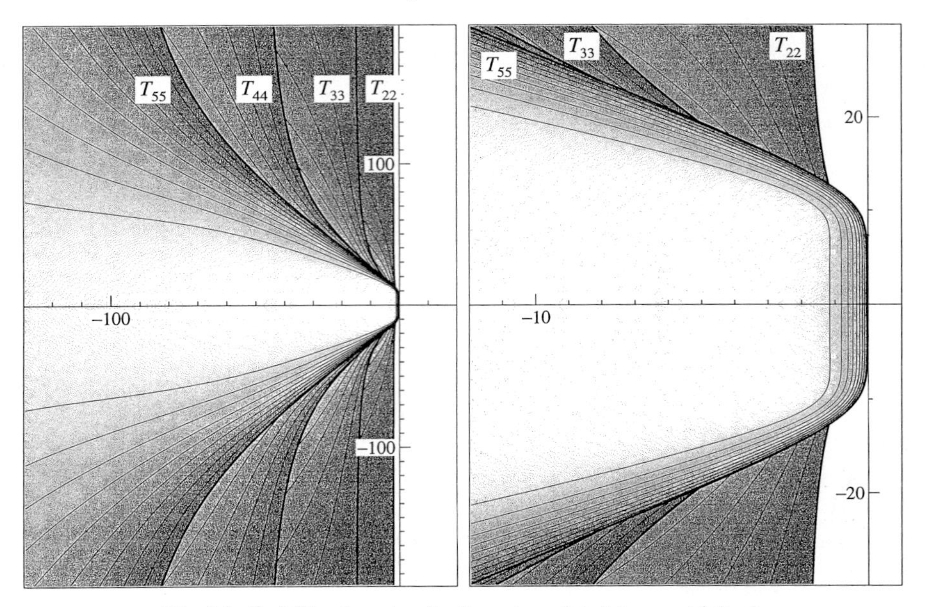

Fig. 9.1. Stability domains for the extrapolated trapezoidal rule

This provides an extrapolation tableau

$$\begin{array}{llll} T_{11} & & & \\ T_{21} & T_{22} & & \\ T_{31} & T_{32} & T_{33} & \\ \vdots & \vdots & \vdots & \ddots, \end{array} \tag{9.5}$$

all entries of which represent diagonally implicit Runge-Kutta methods (see Exercise 1). Due to the symmetry of the basic schemes (9.1) and (9.2), T_{jk} is a DIRK-method of order $2k$. In order to study the stability properties of these methods, we apply them to the test equation $y' = \lambda y$. For both methods, (9.1) and (9.2), we obtain

$$y_{i+1} = \frac{1 + \frac{h\lambda}{2}}{1 - \frac{h\lambda}{2}} y_i$$

so that the stability function $R_{jk}(z)$ of the method T_{jk} is given recursively by $(z = H\lambda)$

$$R_{j1}(z) = \left(\frac{1 + \frac{z}{2n_j}}{1 - \frac{z}{2n_j}} \right)^{n_j}, \tag{9.6a}$$

$$R_{j,k+1}(z) = R_{j,k}(z) + \frac{R_{j,k}(z) - R_{j-1,k}(z)}{(n_j/n_{j-k})^2 - 1}. \tag{9.6b}$$

Already Dahlquist (1963) noticed that for $n_1 = 1$ and $n_2 = 2$ we have

$$R_{22}(z) = \frac{1}{3}\left(4\left(\frac{1+\frac{z}{4}}{1-\frac{z}{4}}\right)^2 - \left(\frac{1+\frac{z}{2}}{1-\frac{z}{2}}\right)\right) \to \frac{5}{3} > 1 \quad \text{for } z \to \infty, \tag{9.7}$$

an undesirable property when solving stiff problems. Stetter (1973) proposed taking only even or only odd numbers in the step-number sequence $\{n_j\}$. Then, all stability functions of the extrapolation tableau tend for $z \to \infty$ to 1 or -1, respectively. But even in this situation extrapolation immediately destroys the A-stability of the underlying scheme (Exercise 2). Fig. 9.1 shows the stability domains $\{z\ ;\ |R_{kk}(z)| \le 1\}$ for the sequence $\{1, 3, 5, 7, 9, \ldots\}$.

Smoothing

> Some numerical examples reveal the power of the smoothing combined with extrapolation. (B. Lindberg 1971)

Another possibility to overcome the difficulty encountered in (9.7) is smoothing (Lindberg 1971). The idea is to replace the definition (9.3) by Gragg's smoothing step

$$\widehat{T}_{j1} = S_{h_j}(x_0 + H), \tag{9.8}$$

$$S_h(x) = \frac{1}{4}\big(y_h(x-h) + 2y_h(x) + y_h(x+h)\big). \tag{9.9}$$

With $y_h(x)$, $S_h(x)$ also possesses an asymptotic expansion in even powers of h. Therefore, extrapolation according to (9.4) is justified. For the stability function of $\widehat{T}_{j1}$ we now obtain

$$\widehat{R}_{j1}(z) = \frac{1}{4}\left\{\left(\frac{1+\frac{z}{2n_j}}{1-\frac{z}{2n_j}}\right)^{n_j-1} + 2\left(\frac{1+\frac{z}{2n_j}}{1-\frac{z}{2n_j}}\right)^{n_j} + \left(\frac{1+\frac{z}{2n_j}}{1-\frac{z}{2n_j}}\right)^{n_j+1}\right\}$$

$$= \frac{1}{\left(1-\frac{z}{2n_j}\right)^2}\left(\frac{1+\frac{z}{2n_j}}{1-\frac{z}{2n_j}}\right)^{n_j-1} \tag{9.10}$$

which is an L-stable approximation to the exponential function. The stability functions $\widehat{R}_{jk}(z)$ (obtained from (9.6b)) all satisfy $\widehat{R}_{jk}(z) = \mathcal{O}(z^{-2})$ for $z \to \infty$. For the step-number sequence

$$\{n_j\} = \{1, 2, 3, 4, 5, 6, 7, \ldots\} \tag{9.11}$$

the stability domains of $\widehat{R}_{kk}(z)$ are plotted in Fig. 9.2.

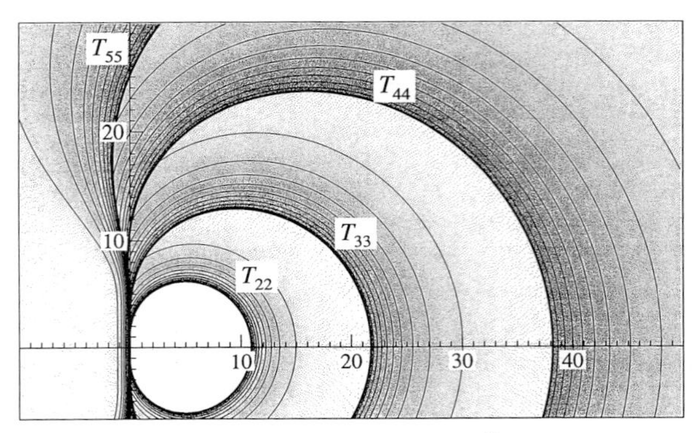

Fig. 9.2. Stability domains of $\widehat{R}_{kk}(z)$

The Linearly Implicit Mid-Point Rule

Extrapolation codes based on fully implicit methods are difficult to implement efficiently. After extensive numerical computations, G. Bader and P. Deuflhard (1983) found that a linearly implicit (Rosenbrock-type) extension of the GBS method of Sect. II.9 gave promising results for stiff equations. This method is based on a two-step algorithm, since one-step Rosenbrock methods (7.4) cannot be symmetric for nonlinear differential equations.

The motivation for the Bader & Deuflhard method is based on Lawson's transformation (Lawson 1967)

$$y(x) = e^{Jx} \cdot c(x), \tag{9.12}$$

where it is hoped that the matrix $J \approx f'(y)$ will neutralize the stiffness. Differentiation gives

$$c' = e^{-Jx} \cdot g(x, e^{Jx}c) \quad \text{with} \quad g(x,y) = f(x,y) - Jy. \tag{9.13}$$

We now solve (9.13) by the Gragg algorithm (II.9.13b)

$$c_{i+1} = c_{i-1} + 2he^{-Jx_i} \cdot g(x_i, e^{Jx_i}c_i)$$

and obtain by back-substitution of (9.12)

$$e^{-hJ}y_{i+1} = e^{hJ}y_{i-1} + 2hg(x_i, y_i). \tag{9.14}$$

For evident reasons of computational ease we now replace $e^{\pm hJ}$ by the approximations $I \pm hJ$ and obtain, adding an appropriate starting and final smoothing step,

$$(I - hJ)y_1 = y_0 + hg(x_0, y_0) \tag{9.15a}$$

$$(I - hJ)y_{i+1} = (I + hJ)y_{i-1} + 2hg(x_i, y_i) \tag{9.15b}$$

$$S_h(x) = \frac{1}{2}(y_{2m-1} + y_{2m+1}) \quad \text{where } x = x_0 + 2mh. \tag{9.15c}$$

Substituting finally g from (9.13), we arrive at (with $x = x_0 + 2mh$, $x_i = x_0 + ih$)

$$(I - hJ)(y_1 - y_0) = hf(x_0, y_0) \tag{9.16a}$$

$$(I - hJ)(y_{i+1} - y_i) = -(I + hJ)(y_i - y_{i-1}) + 2hf(x_i, y_i) \tag{9.16b}$$

$$S_h(x) = \frac{1}{2}(y_{2m-1} + y_{2m+1}) \tag{9.16c}$$

where J stands for some approximation to the Jacobian $\frac{\partial f}{\partial y}(x_0, y_0)$. Putting $J = 0$, Formulas (9.16a) and (9.16b) become equivalent to those of the GBS method. The scheme (9.16b) is the linearly implicit (or semi-implicit) mid-point rule, Formula (9.16a) the linearly implicit Euler method.

Theorem 9.1 (Bader & Deuflhard 1983). *Let $f(x, y)$ be sufficiently often differentiable and let J be an arbitrary matrix; then the numerical solution defined by (9.16a,b,c) possesses an asymptotic expansion of the form*

$$y(x) - S_h(x) = \sum_{j=1}^{l} e_j(x) h^{2j} + h^{2l+2} C(x, h) \tag{9.17}$$

where $C(x, h)$ is bounded for $x_0 \le x \le \overline{x}$ and $0 \le h \le h_0$. For $J \ne 0$ we have in general $e_j(x_0) \ne 0$.

Proof. As in Stetter's proof for the GBS algorithm we introduce the variables

$$\begin{aligned} h^* &= 2h, \quad x_k^* = x_0 + kh^*, \quad u_0 = v_0 = y_0, \quad u_k = y_{2k}, \\ v_k &= (I - hJ) y_{2k+1} + hJy_{2k} - hf(x_{2k}, y_{2k}) \\ &= (I + hJ) y_{2k-1} - hJy_{2k} + hf(x_{2k}, y_{2k}). \end{aligned} \tag{9.18}$$

Method (9.16a,b) can then be rewritten as

$$\begin{pmatrix} u_{k+1} \\ v_{k+1} \end{pmatrix} = \begin{pmatrix} u_k \\ v_k \end{pmatrix} + h^* \begin{pmatrix} f(x_k^* + \frac{h^*}{2}, y_{2k+1}) - Jy_{2k+1} + J\left(\frac{u_{k+1} + u_k}{2}\right) \\ \frac{1}{2}\Big(f(x_k^* + h^*, u_{k+1}) + f(x_k^*, u_k)\Big) + Jy_{2k+1} - J\left(\frac{u_{k+1} + u_k}{2}\right) \end{pmatrix} \tag{9.19}$$

where, from (9.18), we obtain the symmetric representation

$$y_{2k+1} = \frac{v_{k+1} + v_k}{2} + h^* J\left(\frac{u_{k+1} - u_k}{4}\right) - \frac{h^*}{4}\Big(f(x_{k+1}^*, u_{k+1}) - f(x_k^*, u_k)\Big).$$

The symmetry of (9.19) is illustrated in Fig. 9.3 and can be checked analytically by exchanging $u_{k+1} \leftrightarrow u_k$, $v_{k+1} \leftrightarrow v_k$, $h^* \leftrightarrow -h^*$, and $x_k^* \leftrightarrow x_k^* + h^*$. Method (9.19) is consistent with the differential equation

$$\begin{aligned} u' &= f(x, v) - J(v - u), \qquad & u(x_0) &= y_0 \\ v' &= f(x, u) + J(v - u), \qquad & v(x_0) &= y_0 \end{aligned}$$

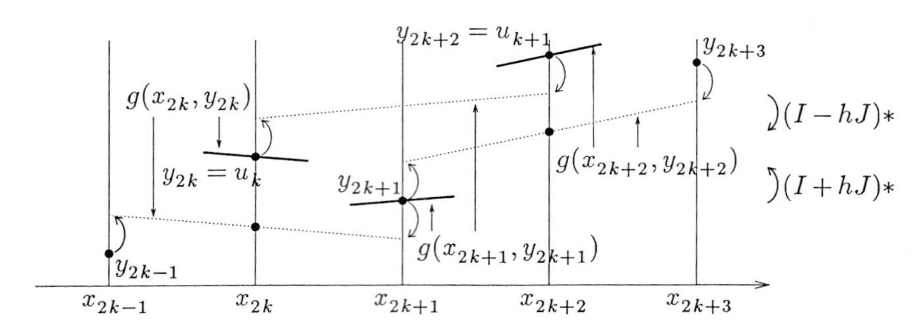

Fig. 9.3. Symmetry of Method (9.19) (see (9.16b))

whose exact solution is $u(x) = v(x) = y(x)$, where $y(x)$ is the solution of the original equation $y' = f(x, y)$. Applying Theorem II.8.10 we obtain

$$
\begin{aligned}
y(x) - u_{h^*}(x) &= \sum_{j=1}^{l} a_j(x) h^{2j} + h^{2l+2} A(x, h) \\
y(x) - v_{h^*}(x) &= \sum_{j=1}^{l} b_j(x) h^{2j} + h^{2l+2} B(x, h)
\end{aligned}
\tag{9.20}
$$

with $a_j(x_0) = b_j(x_0) = 0$. With the help of Formulas (9.18) we can express the numerical solution (9.16c) in terms of u_m and v_m as follows:

$$
\frac{1}{2}(y_{2m+1} + y_{2m-1}) = (I - h^2 J^2)^{-1} \Big(v_m + h^2 J \big(f(x_{2m}, u_m) - J u_m \big) \Big),
$$

and we obtain for $x = x_0 + 2mh$,

$$
\begin{aligned}
y(x) - S_h(x) = (I - h^2 J^2)^{-1} \Big(& y(x) - v_{h^*}(x) \\
& - h^2 J \Big(f(x, u_{h^*}(x)) + J \big(y(x) - u_{h^*}(x) \big) \Big) \Big).
\end{aligned}
$$

Inserting the expansions (9.20) we find (9.17). □

As an application of this theorem we obtain an interesting theoretical result on the existence of W-methods (7.4) (with inexact Jacobian). We saw in Volume I (Exercise 1 of Sect. II.9 and Theorem II.9.4) that the $T_{j,k}$ of the extrapolated GBS method represent explicit Runge-Kutta methods. By analogy, it is not difficult to guess that the $T_{j,k}$ for the above linearly implicit midpoint rule represent W-*methods* (more details in Exercise 3) and we have the following existence result for such methods.

Theorem 9.2. *For p even, there exists a W-method (7.4) of order p with $s = p(p+2)/4$ stages.*

Proof. It follows from (9.20) that for $x = x_0 + 2mh$ the numerical solution $y_h(x) = y_{2m}$ possesses an h^2-expansion of the form (9.17) with $e_j(x_0) = 0$. Therefore, extrapolation yields W-methods of order $2k$ (in the k-th column). The result follows by taking $\{n_j\} = \{2, 4, 6, 8, 10, 12, \ldots\}$ and counting the number of necessary function evaluations. □

Table 9.1. $A(\alpha)$-stability of extrapolated linearly implicit mid-point rule

90°						
90°	90°					
90°	90°	90°				
90°	89.34°	87.55°	87.34°			
90°	88.80°	86.87°	86.10°	86.02°		
90°	88.49°	87.30°	86.61°	86.36°	86.33°	
90°	88.43°	87.42°	87.00°	86.78°	86.70°	86.69°

For a stability analysis we apply the method (9.16) with $J = \lambda$ to the test equation $y' = \lambda y$. In this case Formula (9.16b) reduces to

$$y_{i+1} = \frac{1 + h\lambda}{1 - h\lambda} \, y_{i-1}$$

and the numerical result is given by

$$S_h(x_0 + 2mh) = \frac{1}{(1 - h\lambda)^2} \left(\frac{1 + h\lambda}{1 - h\lambda} \right)^{m-1} y_0, \tag{9.21}$$

exactly the same as that obtained from the trapezoidal rule with smoothing (see Formula (9.10)). We next have to choose a step-number sequence $\{n_j\}$. Clearly, $n_j = 2m_j$ must be even. Bader & Deuflhard (1983) proposed taking only odd numbers m_j, since then $S_h(x_0 + 2m_j h)$ in (9.21) has the same sign as the exact solution $e^{\lambda 2 m_j h} y_0$ for all real $h\lambda \le 0$. Consequently they were led to

$$\{n_j\} = \{2, 6, 10, 14, 22, 34, 50, \ldots\}. \tag{9.22}$$

Putting $T_{j1} = S_{h_j}(x_0 + H)$ with $h_j = H/n_j$ and defining T_{jk} by (9.4) we obtain a tableau of W-methods (7.4) (Exercise 3). By Theorem 9.1 the k-th column of this tableau represents methods of order $2k - 1$ independent of the choice of J (the methods are not of order $2k$, since $e_l(x_0) \neq 0$ in (9.17)). The stability function of T_{j1} is given by

$$R_{j1}(z) = \frac{1}{(1 - \frac{z}{n_j})^2} \left(\frac{1 + \frac{z}{n_j}}{1 - \frac{z}{n_j}} \right)^{n_j/2 - 1} \tag{9.23}$$

and those of T_{jk} can be computed with the recursion (9.6b). An investigation of the E-polynomial (3.8) for these rational functions shows that not only T_{j1}, but

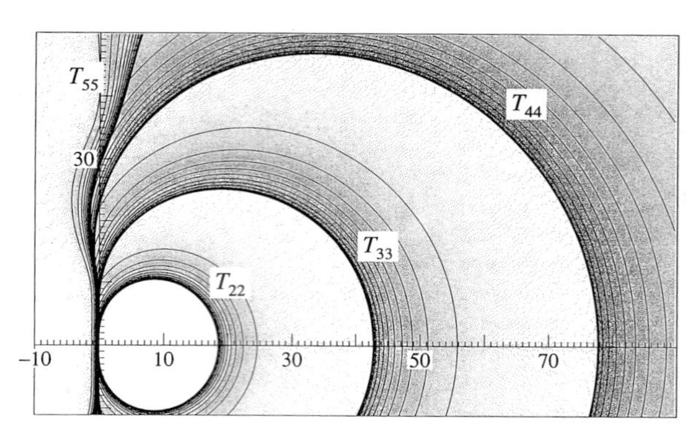

Fig. 9.4. Stability domains of extrapolated linearly implicit mid-point rule

also T_{22}, T_{32} and T_{33} are A-stable (Hairer, Bader & Lubich 1982). The angle of $A(\alpha)$-stability for some further elements in the extrapolation tableau are listed in Table 9.1. Stability domains of T_{kk} for $k = 2, 3, 4, 5, 6$ are plotted in Fig. 9.4.

Implicit and Linearly Implicit Euler Method

Why not consider also non-symmetric methods as basic integration schemes? Deuflhard (1985) reports on experiments with extrapolation of the implicit Euler method

$$y_{i+1} = y_i + hf(x_{i+1}, y_{i+1}) \tag{9.24}$$

and of the linearly implicit Euler method

$$(I - hJ)(y_{i+1} - y_i) = hf(x_i, y_i), \tag{9.25}$$

where, again, J is an approximation to $\frac{\partial f}{\partial y}(x_0, y_0)$. These methods are not symmetric and have only a h-expansion of their global error. We therefore have to extrapolate the numerical solutions at $x_0 + H$ according to

$$T_{j,k+1} = T_{j,k} + \frac{T_{j,k} - T_{j-1,k}}{(n_j/n_{j-k}) - 1}, \tag{9.26}$$

so that T_{jk} represents a method of order k.

For both basic methods, (9.24) and (9.25), the stability function of T_{jk} is the same and defined recursively by

$$R_{j1}(z) = \Big(1 - \frac{z}{n_j}\Big)^{-n_j} \tag{9.27a}$$

$$R_{j,k+1}(z) = R_{j,k}(z) + \frac{R_{j,k}(z) - R_{j-1,k}(z)}{(n_j/n_{j-k}) - 1}. \tag{9.27b}$$

Taking the step-number sequence

$$\{n_j\} = \{1, 2, 3, 4, 5, 6, 7, \ldots\} \tag{9.28}$$

we have plotted in Fig. 9.5 the stability domains of $R_{kk}(z)$ (left picture) and $R_{k,k-1}(z)$ (right picture). All these methods are seen to be $A(\alpha)$-stable with α close to 90°. The values of α (computed numerically) for $R_{jk}(z)$ with $j \leq 8$ are given in Table 9.2.

We shall see in the chapter on differential algebraic systems that it is preferable to use the first subdiagonal of the extrapolation tableau resulting from (9.28). This is equivalent to the use of the step number sequence $\{n_i\} = \{2, 3, 4, 5, \ldots\}$. Also an effective construction of a *dense output* can best be motivated in the setting of differential-algebraic equations (Sect. VI.5).

Table 9.2. $A(\alpha)$-stabiliy of extrapolated Euler

90°							
90°	90°						
90°	90°	89.85°					
90°	90°	89.90°	89.77°				
90°	90°	89.93°	89.84°	89.77°			
90°	90°	89.95°	89.88°	89.82°	89.78°		
90°	90°	89.96°	89.91°	89.86°	89.82°	89.80°	
90°	90°	89.97°	89.93°	89.89°	89.85°	89.83°	89.81°

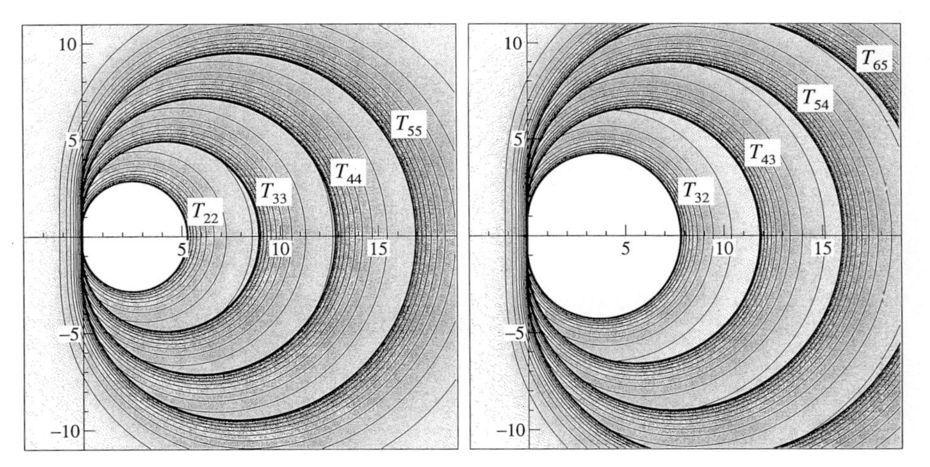

Fig. 9.5. Stability domains of extrapolated Euler

Implementation

Extrapolation methods based on implicit discretizations are in general less efficient than those based on linearly implicit discretizations. The reason is that the arising nonlinear systems have to be solved very accurately, so that the asymptotic expansion of the error is not destroyed. The first successful extrapolation code for stiff differential equations is METAN1 of Bader & Deuflhard (1983), which implements the linearly implicit mid-point rule (9.16). In fact, Formula (9.16b) is replaced by the equivalent formulation

$$\Delta y_i = \Delta y_{i-1} + 2(I - hJ)^{-1}\Big(hf(x_i, y_i) - \Delta y_{i-1}\Big), \qquad \Delta y_i = y_{i+1} - y_i \quad (9.29)$$

which avoids a matrix-vector multiplication. The step size and order selection of this code is described in Deuflhard (1983). Modifications in the control of step size and order are proposed by Shampine (1987). We have implemented the following two extrapolation codes (see Appendix):

SODEX is based on the linearly implicit mid-point rule (9.16), uses the step-number sequence (9.22) and is mathematically equivalent to METAN1. The step size and order selection in SODEX is with some minor changes that of the non-stiff code ODEX of Sect. II.9. We just mention that in the formula for the work per unit step (II.9.26) the number A_k is augmented by the dimension of the differential equation in order to take into account the Jacobian evaluation.

SEULEX is an implementation of the linearly implicit Euler method (9.25) using the step-number sequence $\{2, 3, 4, 5, 6, 7, \ldots\}$ (other sequences can be chosen as internal options). The step size and order selection is that of SODEX. The original code (EULSIM, first discussed by Deuflhard 1985) uses the same numerical method, but a different implementation.

> Neither code can solve the van der Pol equation problem in a straightforward way because of overflow ...
>
> (L.F. Shampine 1987)

A big difficulty in the implementation of extrapolation methods is the use of "large" step sizes. During the computation of T_{j1} one may easily get into trouble with exponential overflow when evaluating the right-hand side of the differential equation. As a remedy we propose the following strategies:

a) In establishing the extrapolation tableau we compare the estimated error $err_j = \|T_{j,j-1} - T_{jj}\|$ with the preceding one. Whenever $err_j \geq err_{j-1}$ for some $j \geq 3$ we restart the computation of the step with a smaller H, say, $H = 0.5 \cdot H$.

b) In order to be able to interrupt the computations already after the first f-evaluations, we require that the step sizes $h = H/n_i$ (for $i = 1$ and $i = 2$) be small enough so that a simplified Newton iteration applied to the implicit Euler method $y = y_0 + hf(x, y)$, $x = x_0 + h$ would converge ("stability check", an idea of Deuflhard). The first two iterations read

$$\begin{aligned} (I - hJ)\Delta_0 &= hf(x_0, y_0), \qquad y^{(1)} = y_0 + \Delta_0 \\ (I - hJ)\Delta_1 &= hf(x_0 + h, y^{(1)}) - \Delta_0. \end{aligned} \quad (9.30)$$

The computations for the step are restarted with a smaller H, if $\|\Delta_1\| \geq \|\Delta_0\|$

(divergence of the iteration). Observe that for both methods, (9.16) and (9.25), no additional function evaluations are necessary. For the linearly implicit mid-point rule we have the simple relations $\Delta_0 = \Delta y_0$, $\Delta_1 = \frac{1}{2}(\Delta y_1 - \Delta y_0)$ (see (9.29)).

Non-Autonomous Differential Equations. Given a non-autonomous differential equation $y' = f(x, y)$, one has several possibilities to apply the above extrapolation algorithms:

i) apply the Formula (9.16) or (9.25) directly (this is justified, since all asymptotic expansions hold for general non-autonomous problems);

ii) transform the differential equation into an autonomous system by adding $x' = 1$ and then apply the algorithm. This yields

$$(I - hJ)(y_{i+1} - y_i) = hf(x_i, y_i) + h^2 \frac{\partial f}{\partial x}(x_0, y_0) \tag{9.31}$$

for the linearly implicit Euler method (the derivative $\frac{\partial f}{\partial x}(x_0, y_0)$ can also be replaced by some approximation). For the linearly implicit mid-point rule, (9.16a) has to be replaced by (9.31) with $i = 0$, the remaining two formulas (9.16b) and (9.16c) are not changed.

iii) apply one simplified Newton iteration to the implicit Euler discretization (9.24). This gives

$$(I - hJ)(y_{i+1} - y_i) = hf(x_{i+1}, y_i). \tag{9.32}$$

The use of this formula avoids the computation of the derivative $\partial f / \partial x$, but requires one additional function evaluation for each T_{j1}. In the case of the linearly implicit mid-point rule the replacement of (9.16a) by (9.32) would destroy symmetry and the expansions in h^2.

A theoretical study of the three different approaches for the linearly implicit Euler method applied to the Prothero-Robinson equation (see Exercise 4 below) indicates that the third approach is preferable. More theoretical insight into this question will be obtained from the study of singular perturbation problems (Chapter VI).

Implicit Differential Equations. Our codes in the appendix are written for problems of the form

$$My' = f(x, y) \tag{9.33}$$

where M is a constant square matrix. The necessary modifications in the basic formulas are obtained, as usual, by replacing all $f's$ and $J's$ by $M^{-1}f$ and $M^{-1}J$, and premultiplying by M. The linearly implicit Euler method then reads

$$(M - hJ)(y_{i+1} - y_i) = hf(x_i, y_i) \tag{9.34}$$

and the linearly implicit mid-point rule becomes, with $\Delta y_i = y_{i+1} - y_i$,

$$\Delta y_i = \Delta y_{i-1} + 2(M - hJ)^{-1}\Big(hf(x_i, y_i) - M\Delta y_{i-1}\Big). \tag{9.35}$$

Exercises

1. Consider the implicit mid-point rule (9.2) as basic integration scheme and define T_{jk} by (9.3) and (9.4).

   a) Prove that T_{jk} represents a DIRK-method of order $p=2k$ with $s=n_1+n_2+\ldots+n_j$ stages.

   b) $\widehat{T}_{jk}$, defined by (9.8) and (9.4), is equivalent to a DIRK-method of order $p=2k-1$ only.

2. Let $R_{jk}(z)$ be given by (9.6) and assume that the step-number sequence consists of even numbers only. Prove that $R_{j2}(z)$ cannot be A-stable. More precisely, show that at most a finite number of points of the imaginary axis can lie in the stability domain of $R_{j2}(z)$ (interpret Fig. 9.6).

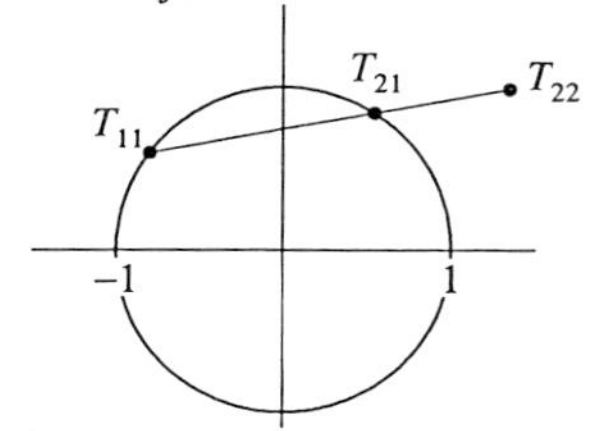

Fig. 9.6. How extrapolation destroys A-stability

3. Prove that $S_h(x)$, defined by (9.16), is the numerical result of the $(2n+1)$-stage W-method (7.4) with the following coefficients $(n=2m)$:

$$\alpha_{ij}=\begin{cases}1/n & \text{if } j=1 \text{ and } i \text{ even,}\\ 2/n & \text{if } 1<j<i \text{ and } i-j \text{ odd,}\\ 0 & \text{else.}\end{cases}$$

$$\gamma_{ij}=\begin{cases}(-1)^{i-j}/n & \text{if } j=1 \text{ or } j=i,\\ 2(-1)^{i-j}/n & \text{if } 1<j<i.\end{cases}$$

$$b_i=\alpha_{n+1,i}+\gamma_{n+1,i} \qquad \text{for all } i.$$

4. Apply the three different versions of the linearly implicit Euler method (9.25), (9.31) and (9.32) to the problem $y'=\lambda(y-\varphi(x))+\varphi'(x)$. Prove that the errors $e_i=y_i-\varphi(x_i)$ satisfy $e_{i+1}=(1-h\lambda)^{-1}e_i+\delta_h(x_i)$, where for $h\to 0$ and $h\lambda\to\infty$,

$$\delta_h(x)=-h\varphi'(x)+\mathcal{O}(h^2)+\mathcal{O}(\lambda^{-1}),$$

$$\delta_h(x)=-\frac{h^2}{2}\varphi''(x)+(1-h\lambda)^{-1}h^2\lambda(\varphi'(x)-\varphi'(x_0))+\mathcal{O}(h^3)+\mathcal{O}(h\lambda^{-1}),$$

$$\delta_h(x)=(1-h\lambda)^{-1}\Big(\frac{h^2}{2}\varphi''(x)+\mathcal{O}(h^3)\Big),$$

respectively.

IV.10 Numerical Experiments

> Theory without practice cannot survive and dies as quickly as it lives. (Leonardo da Vinci 1452-1519, cited from M. Kline, Math. Thought 1972, p. 224)

> Sine experientia nihil sufficienter scrire potest (Without experience it is not possible to know anything adequately).
> (Inscription overlooking Botanic Garden, Oxford; found in *The Latin Citation Calendar*, Oxford 1996)

After having seen so many different methods and ideas in the foregoing sections, it is legitimate to study how all these theoretical properties pay off in numerical efficiency.

The Codes Used

We compared the following codes, some of which are described in the Appendix:

RADAU5 and SDIRK4 are implicit Runge-Kutta codes. The first one is based on the Radau IIA method with $s=3$ of order 5 (Table 5.6), whereas the second one is based on the SDIRK method (6.16) of order 4. Both methods are L-stable. Details of their implementation are given in Sect. IV.8.

RODAS and ROS4 are Rosenbrock codes of order 4 with an embedded 3rd order error estimator. ROS4 implements the methods of Table 7.2. A switch allows one to choose between the different coefficient sets. The underlying method of RODAS satisfies additional order conditions for differential-algebraic equations (see Sect. VI.4 below), but requires a little more work per step. RODAS5 is an extension of RODAS to order 5. Its coefficients are constructed by Di Marzo (1992).

SEULEX and SODEX are extrapolation codes. They implement the (Stiff) linearly implicit EULer EXtrapolation method (9.32) and the extrapolation algorithm based on the linearly implicit mid-point rule (method (9.16) of Bader & Deuflhard 1983), respectively. Both methods are discussed in Sect. IV.9.

In the numerical experiments of this section we have also included the results of LSODE (a BDF code of Hindmarsh 1980). It is a representative of the class of multistep methods to be described in Chapter V.

Many of the treated examples are very stiff and *explicit* methods would require hours to compute the solution. On some examples, however, it was also interesting to see their performance, especially for the methods with extended region of stability (e.g., the Runge-Kutta-Chebyshev code RKC of Sommeijer (1991), explained in Sect. IV.2), as well as for a standard explicit Runge-Kutta code, such as DOPRI5 of Volume I.

Twelve Test Problems

> Man hüte sich, auf Grund einzelner Beispiele allgemeine Schlüsse über den Wert oder Unwert einer Methode zu ziehen. Dazu gehört sehr viel Erfahrung. (L. Collatz 1950)

The first extensive numerical comparisons for stiff equations were made by Enright, Hull & Lindberg (1975). Their STIFF-DETEST set of problems has become a veritable "must" for generations of software writers (see also the critical remarks of Shampine 1981). Several additional test problems, usually from chemical kinetics, have been proposed by Enright & Hull (1976). An interesting review article containing also problems of large dimension is due to Byrne & Hindmarsh (1987).

The problems chosen here for our tests are the following:

VDPOL — the van der Pol oscillator (see (1.5') and Fig. 8.1)

$$\begin{aligned} y_1' &= y_2 \\ y_2' &= \big((1-y_1^2)y_2 - y_1\big)/\varepsilon, \qquad \varepsilon = 10^{-6} \\ y_1(0) = 2, \quad y_2(0) &= 0; \quad x_{\text{out}} = 1,2,3,4,\ldots,11. \end{aligned} \tag{10.1}$$

ROBER — the reaction of Robertson (1966) (see (1.3) and (1.4))

$$\begin{aligned} y_1' &= -0.04y_1 + 10^4 y_2 y_3 & y_1(0) &= 1 \\ y_2' &= 0.04y_1 - 10^4 y_2 y_3 - 3\cdot 10^7 y_2^2 & y_2(0) &= 0 \\ y_3' &= 3\cdot 10^7 y_2^2 & y_3(0) &= 0, \end{aligned} \tag{10.2}$$

one of the most prominent examples of the "stiff" literature. It was usually treated on the interval $0 \le x \le 40$, until Hindmarsh discovered that many codes fail if x becomes very large (10^{11} say). The reason is that whenever the numerical solution of y_2 accidentally becomes negative, it then tends to $-\infty$ and the run ends by overflow. We have therefore chosen $x_{\text{out}} = 1, 10, 10^2, 10^3, \ldots, 10^{11}$.

OREGO — the Oregonator, the famous model with a periodic solution describing the Belusov-Zhabotinskii reaction (Field & Noyes 1974, see also Enright & Hull 1976)

$$\begin{aligned} y_1' &= 77.27\Big(y_2 + y_1(1 - 8.375\cdot 10^{-6} y_1 - y_2)\Big) \\ y_2' &= \frac{1}{77.27}\big(y_3 - (1+y_1)y_2\big) \\ y_3' &= 0.161(y_1 - y_3) \end{aligned} \tag{10.3}$$

$$y_1(0) = 1, \quad y_2(0) = 2, \quad y_3(0) = 3, \qquad x_{\text{out}} = 30, 60, 90, \ldots, 360.$$

For pictures see Volume I, p. 119.

HIRES — this chemical reaction involving eight reactants was proposed by Schäfer (1975) to explain "the growth and differentiation of plant tissue independent of photosynthesis at high levels of irradiance by light". It has been promoted as a test example by Gottwald (1977). The corresponding equations are

$$\begin{aligned}
y_1' &= -1.71 \cdot y_1 + 0.43 \cdot y_2 + 8.32 \cdot y_3 + 0.0007 \\
y_2' &= 1.71 \cdot y_1 - 8.75 \cdot y_2 \\
y_3' &= -10.03 \cdot y_3 + 0.43 \cdot y_4 + 0.035 \cdot y_5 \\
y_4' &= 8.32 \cdot y_2 + 1.71 \cdot y_3 - 1.12 \cdot y_4 \\
y_5' &= -1.745 \cdot y_5 + 0.43 \cdot y_6 + 0.43 \cdot y_7 \\
y_6' &= -280 \cdot y_6 y_8 + 0.69 \cdot y_4 + 1.71 \cdot y_5 - 0.43 \cdot y_6 + 0.69 \cdot y_7 \\
y_7' &= 280 \cdot y_6 y_8 - 1.81 \cdot y_7 \\
y_8' &= -y_7'
\end{aligned} \tag{10.4}$$

$$y_1(0) = 1, \quad y_2(0) = y_3(0) = \ldots = y_7(0) = 0, \quad y_8(0) = 0.0057$$

and chosen output values are $x_{\text{out}} = 321.8122$ and 421.8122.

E5 — is another chemical recation problem, called "E5" in the collection by Enright, Hull & Lindberg (1975). It is given by

$$\begin{aligned}
y_1' &= -Ay_1 - By_1y_3 & & & y_1(0) &= 1.76 \cdot 10^{-3} \\
y_2' &= Ay_1 & - MCy_2y_3 & & y_2(0) &= 0 \\
y_3' &= Ay_1 - By_1y_3 & - MCy_2y_3 & + Cy_4 & y_3(0) &= 0 \\
y_4' &= By_1y_3 & & - Cy_4 & y_4(0) &= 0,
\end{aligned} \tag{10.5}$$

where $A = 7.89 \cdot 10^{-10}$, $B = 1.1 \cdot 10^{7}$, $C = 1.13 \cdot 10^{3}$, and $M = 10^{6}$. As we can see from Fig. 10.1 the variables are badly scaled ($y_1 \approx 10^{-3}$ at the beginning, all other components do not exceed the value $1.46 \cdot 10^{-10}$), and "... a scalar absolute error tolerance is quite unsuitable" (Shampine 1981). The differential equation possesses the invariant $y_2 - y_3 - y_4 = 0$, and it is recommended to use the relation $y_3' = y_2' - y_4'$ in the function subroutine (because of eventual cancellation of digits).

Originally the problem was posed on the interval $0 \le x \le 1000$, but Alexander (1997) discovered that the solutions possess interesting properties on a much longer interval. We follow this suggestion and consider output values at $x_{\text{out}} = 10, 10^3, 10^5, 10^7, \ldots, 10^{13}$.

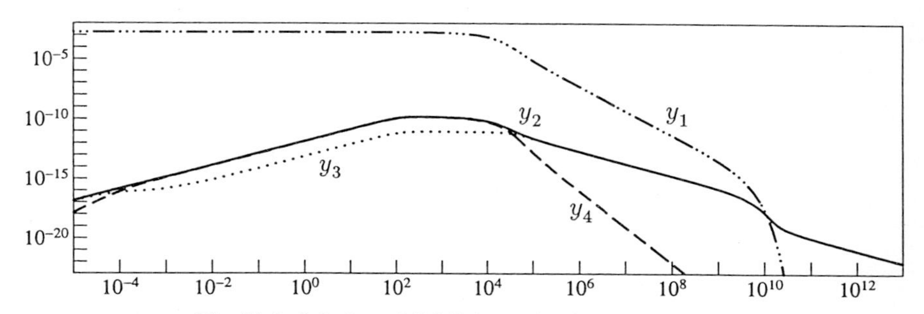

Fig. 10.1. Solution of (10.5) in double logarithmic scale

PLATE — this is a linear and non-autonomous example of medium stiffness and medium size. It describes the movement of a rectangular plate under the load of a car passing across it:

$$\frac{\partial^2 u}{\partial t^2} + \omega \frac{\partial u}{\partial t} + \sigma \Delta\Delta u = f(x,y,t). \tag{10.6}$$

The plate $\Omega = \{(x,y)\ ;\ 0 \le x \le 2,\ 0 \le y \le 4/3\}$ is discretized on a grid of 8×5 interior points $x_i = ih$, $y_j = jh$, $h = 2/9$ with initial and boundary conditions

$$u|_{\partial\Omega} = 0, \quad \Delta u|_{\partial\Omega} = 0, \quad u(x,y,0) = 0, \quad \frac{\partial u}{\partial t}(x,y,0) = 0. \tag{10.7}$$

The integration interval is $0 \le t \le 7$. The load $f(x,y,t)$ is idealized by the sum of two Gaussian curves which move in the x-direction and which reside on "four wheels"

$$f(x,y,t) = \begin{cases} 200\left(e^{-5(t-x-2)^2} + e^{-5(t-x-5)^2}\right) & \text{if } y = y_2 \text{ or } y_4 \\ 0 & \text{for all other } y. \end{cases}$$

The plate operator $\Delta\Delta$ is discretized via the standard "computational molecule"

$$\begin{array}{ccccc} & & 1 & & \\ & 2 & -8 & 2 & \\ 1 & -8 & 20 & -8 & 1 \\ & 2 & -8 & 2 & \\ & & 1 & & \end{array}$$

and the friction and stiffness parameters are chosen as $\omega = 1000$ and $\sigma = 100$. The resulting system is then of dimension 80 with negative real as well as complex eigenvalues ranging between $-500 \le \operatorname{Re}\lambda < 0$ with maximal angle $\alpha \approx 71°$ (see Definition 3.9).

BEAM — the elastic beam (1.10) of Sect. IV.1. We choose $n = 40$ in (1.10') so that the differential system is of dimension 80, and $0 \le t \le 5$ as integration interval. The eigenvalues of the Jacobian are purely imaginary and vary between $-6400i$ and $+6400i$ (see Eq. (2.23)). The initial conditions (1.18) and (1.19) are chosen such that the solution nevertheless appears to be smooth. However, a detailed numerical study shows that the exact solution possesses high oscillations with period $\approx 2\pi/6400$ and amplitude $\approx 10^{-6}$ (see Fig. 10.2).

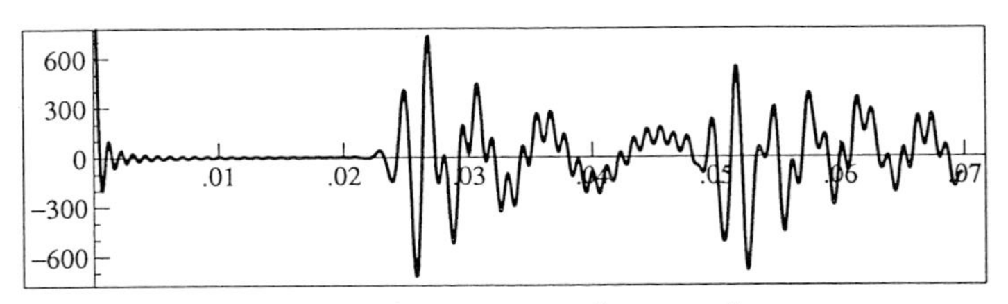

Fig. 10.2. Third finite differences $\Delta^3 y_{80}/\Delta x^3$ of solutions of the beam equation (1.10') with $n = 40$ for $0 \le x \le 0.07$

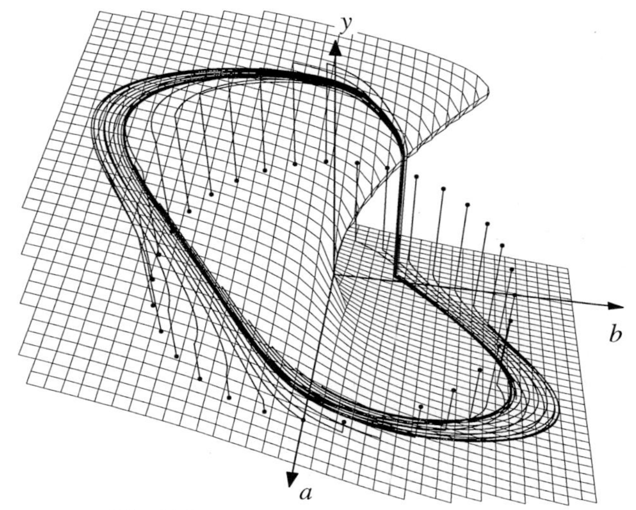

Fig. 10.3. The cusp catastrophe with $N = 32$.

CUSP — this is a combination of Zeeman's "cusp catastrophe" model ($-\varepsilon\dot{y} = y^3 + ay + b$) for the nerve impulse mechanism (Zeeman 1972) combined with the van der Pol oscillator (see Fig. 10.3)

$$\begin{aligned}\frac{\partial y}{\partial t} &= -\frac{1}{\varepsilon}(y^3 + ay + b) + \sigma\frac{\partial^2 y}{\partial x^2}\\ \frac{\partial a}{\partial t} &= b + 0.07v + \sigma\frac{\partial^2 a}{\partial x^2}\\ \frac{\partial b}{\partial t} &= (1-a^2)b - a - 0.4y + 0.035v + \sigma\frac{\partial^2 b}{\partial x^2}\end{aligned} \tag{10.8}$$

where

$$v = \frac{u}{u+0.1}, \qquad u = (y-0.7)(y-1.3).$$

We put $\sigma = 1/144$ and make the problem stiff by choosing $\varepsilon = 10^{-4}$. We discretize the diffusion terms by the method of lines

$$\begin{aligned}\dot{y}_i &= -10^4(y_i^3 + a_i y_i + b_i) + D(y_{i-1} - 2y_i + y_{i+1})\\ \dot{a}_i &= b_i + 0.07v_i + D(a_{i-1} - 2a_i + a_{i+1}) \qquad\qquad i = 1,\ldots,N\\ \dot{b}_i &= (1-a_i^2)b_i - a_i - 0.4y_i + 0.035v_i + D(b_{i-1} - 2b_i + b_{i+1})\end{aligned} \tag{10.8'}$$

where

$$N = 32, \quad v_i = \frac{u_i}{u_i + 0.1}, \quad u_i = (y_i - 0.7)(y_i - 1.3), \quad D = \sigma N^2 = \frac{N^2}{144},$$

with periodic boundary conditions

$$\begin{aligned}y_0 &:= y_N, & a_0 &:= a_N, & b_0 &:= b_N,\\ y_{N+1} &:= y_1, & a_{N+1} &:= a_1, & b_{N+1} &:= b_1,\end{aligned}$$

and obtain a system of dimension $3 \cdot N = 96$. We take the initial values

$$y_i(0) = 0, \quad a_i(0) = -2\cos\Big(\frac{2i\pi}{N}\Big), \quad b_i(0) = 2\sin\Big(\frac{2i\pi}{N}\Big) \qquad i = 1,\dots,N.$$

and $t_{\text{out}} = 1.1$.

BRUSS — this is the equation (1.6') with $\alpha = 1/50$, the same initial conditions as in Sect. IV.1, and integration interval $0 \le t \le 10$. But we now let $N = 500$ so that (1.6') becomes a system of 1000 differential equations with largest eigenvalue close to -20000. The equations therefore become considerably stiff. The Jacobian of this system is banded with upper and lower bandwidth 2 (if the solution components are ordered as $u_1, v_1, u_2, v_2, u_3, v_3$, etc.).

KS — is the one-dimensional Kuramoto-Sivashinsky equation

$$\frac{\partial U}{\partial t} = -\frac{\partial^2 U}{\partial x^2} - \frac{\partial^4 U}{\partial x^4} - \frac{1}{2}\frac{\partial U^2}{\partial x} \tag{10.9}$$

with periodic boundary conditions $u(x+L,t) = u(x,t)$, taken from Collet, Eckmann, Epstein & Stubbe (1993). We choose $L = 2\pi/q$, $q = 0.025$, and take as initial condition

$$U(x,0) = 16 \cdot \max(0, \eta_1, \eta_2, \eta_3, \eta_4), \qquad \begin{aligned} \eta_1 &= \min(x/L, 0.1 - x/L), \\ \eta_2 &= 20(x/L - 0.2)(0.3 - x/L), \\ \eta_3 &= \min(x/L - 0.6, 0.7 - x/L), \\ \eta_4 &= \min(x/L - 0.9, 1 - x/L), \end{aligned}$$

The inverse heat equation term $-\partial^2 U/\partial x^2$ creates instability, which is stabilized for the higher oscillations by the beam equation term $-\partial^4 U/\partial x^4$. The nonlinear transport term $\partial U^2/\partial x$ couples the modes and ensures that the solution remains bounded. All this creates wonderful chaos (see Fig. 10.4).

We solve Eq. (10.9) using the pseudo-spectral method, i.e., we consider the Fourier coefficients

$$\widehat{U}_j(t) = \frac{1}{L}\int_0^L U(x,t)e^{-iqjx}\,dx, \qquad U(x,t) = \sum_{j\in\mathbb{Z}} \widehat{U}_j(t)e^{iqjx}, \tag{10.10}$$

so that (10.9) takes the form of an infinite dimensional ordinary differential equation

$$\widehat{U}_j' = \big((qj)^2 - (qj)^4\big)\widehat{U}_j - \frac{iqj}{2}(\widehat{U \cdot U})_j.$$

We truncate this system as follows: for a fixed N, say $N = 1024$, we consider the N-periodic sequence $u(t) = \{u_j(t)\}$ solving the ordinary differential equation

$$u' = (d^2 - d^4)u - \frac{id}{2}\mathcal{F}_N(\mathcal{F}_N^{-1}u \cdot \mathcal{F}_N^{-1}u), \tag{10.11}$$

where d denotes the N-periodic sequence given by $d_j = qj$ for $|j| < N/2$ and $d_{N/2} = 0$, and the product of sequences in (10.11) is componentwise. The discrete

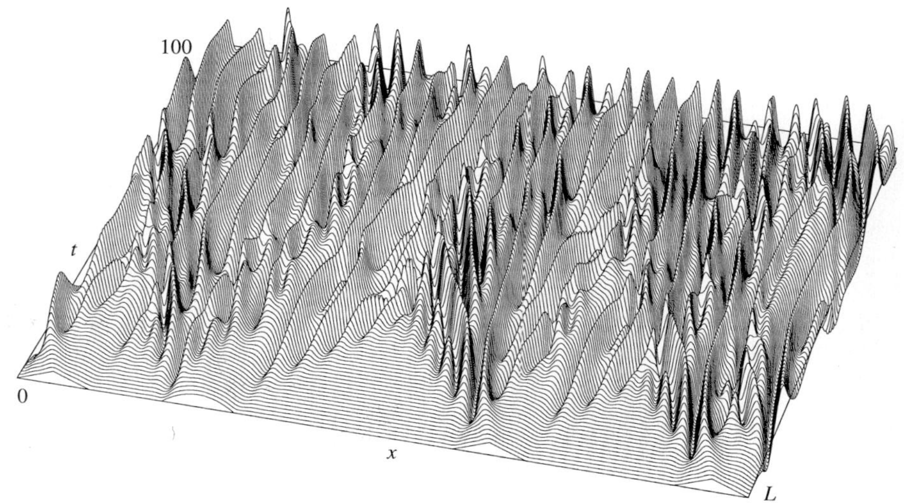

Fig. 10.4. Solution of Kuramoto-Sivashinsky equation

Fourier transform $\mathcal{F}_N$ can be computed by FFT. From the fact that $U(x,t)$ is real it follows that the sequence u is hermitian, i.e., $u_{-j} = \overline{u}_j$. Hence, the routine REALFT from Press, Flannery, Teukolsky & Vetterling (1986,1989), Chapter 12, is best suited for computing the right-hand side of (10.11). Since $d_0 = d_{N/2} = 0$, the components $u_0(t)$ and $u_{N/2}(t)$ are constant and need not be integrated. We thus are concerned with an ordinary differential equation of real dimension $N-2 = 1022$. As initial values we take the discrete Fourier transform of $\{U(jL/N,0)\}$ with the $(N/2)$th component put to zero. In our tests we solve the differential equation (10.11) on the interval $0 \le t \le 100$ (see Fig. 10.4).

It can be seen from Fig. 10.5 that the Fourier modes tend to zero for $j \to \infty$, behave chaotically, and, by computing their mean values over a long period, that the modes for $qj \approx \sqrt{2}/2$ are dominant.

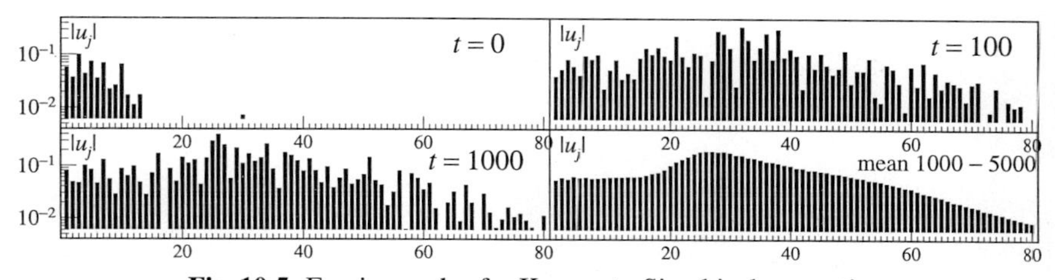

Fig. 10.5. Fourier modes for Kuramoto-Sivashinsky equation

BECKDO — the Becker-Döring model describes the dynamics of a system with a large number of identical particles which can coagulate to form clusters. We let y_k denote the expected number of k-particle clusters per unit volume. Assuming that clusters can gain or loose only single particles, we are led to the system

$$\begin{aligned} y_1' &= -J_1 - \sum_{k=1}^{N-1} J_k, \qquad y_N' = J_{N-1} \\ y_k' &= J_{k-1} - J_k, \qquad k = 2,3,\ldots,N-1, \end{aligned} \tag{10.12}$$

where $J_k = y_1 y_k - b_{k+1} y_{k+1}$ and $b_{k+1} = \exp\big(k^{2/3} - (k-1)^{2/3}\big)$. For a detailed description of this system we refer to the article by Carr, Duncan & Walshaw (1995). This equation is especially interesting because of its *metastability* (extremely slow variations in the solution over very long time intervals; see Fig. 10.6).

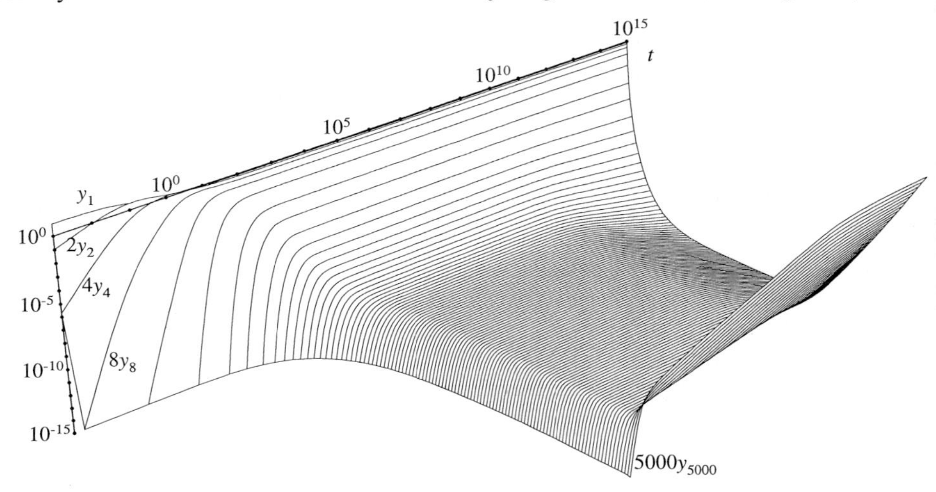

Fig. 10.6. Solutions of Becker-Döring equation (10.12)

As initial condition we take

$$y_1(0) = \varrho, \qquad y_k(0) = 0 \qquad \text{for} \quad k = 2,\ldots,N \tag{10.13}$$

(no clusters at the beginning). It can be seen by differentiation that the density (total number of particles per unit volume)

$$\sum_{k=1}^{N} k y_k \quad (= \varrho) \tag{10.14}$$

is an invariant of the system (10.12). Most numerical schemes (in particular Runge-Kutta methods and multistep methods) preserve automatically such linear invariants in the absence of round-off errors. Whenever the relation (10.14) is not satisfactorily preserved, there is the possibility to re-establish it during the computations by projections (see "differential equations with invariants", Sect. VII.2). This precautionary measure was not used in the subsequent numerical tests.

In order to be able to observe the metastable states of the system, the dimension N has to be sufficiently large. Following the experiments of Carr, Duncan &

Walshaw (1995) we take $N = 5000$ and $\varrho = 7.5$, and consider the solution on the interval $0 \le t \le 10^{15}$. We compare the errors at $x_{\text{out}} = 1, 10, 10^2, 10^3, \ldots, 10^{15}$.

The Jacobian of this system is tri-diagonal with an additional non-zero first row and a non-zero first column. A Gershgorin test reveals that its eigenvalues can not go, except for the initial phase, beyond -10. Stiffness, in this example, is therefore not created by large eigenvalues of J, but by the extremely long integration interval.

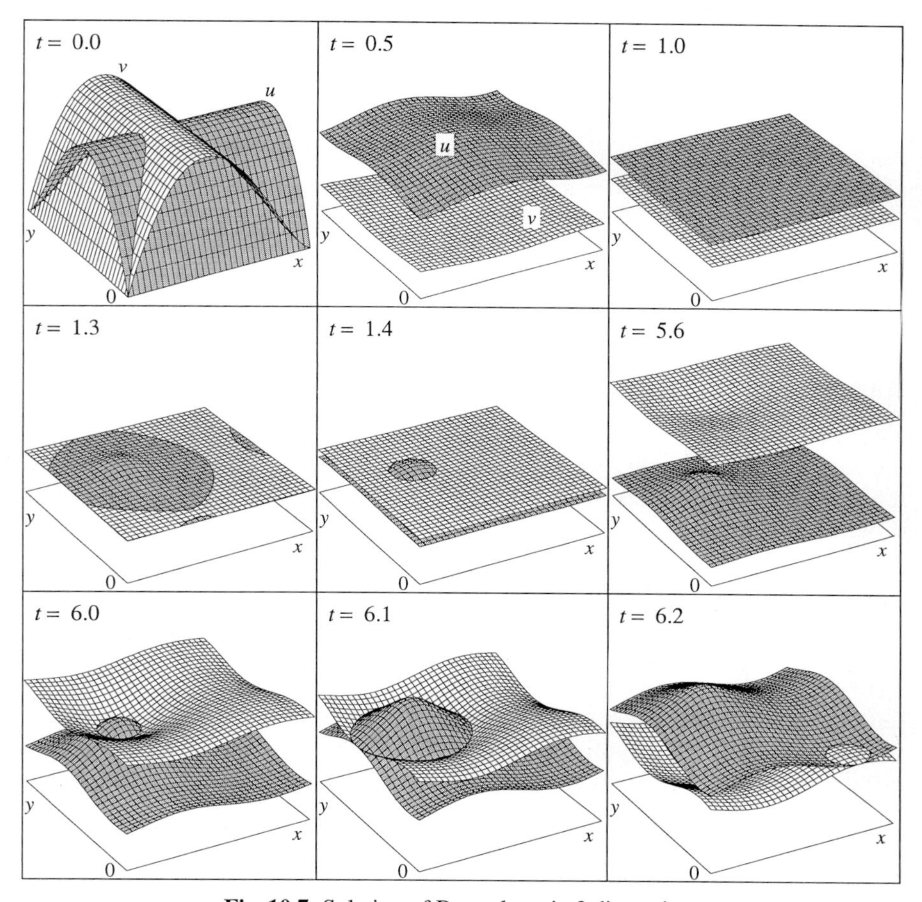

Fig. 10.7. Solution of Brusselator in 2 dimensions

BRUSS-2D — the two-dimensional Brusselator reaction-diffusion problem of Sect. II.10

$$\begin{aligned}
\frac{\partial u}{\partial t} &= 1 + u^2 v - 4.4u + \alpha\Big(\frac{\partial^2 u}{\partial x^2} + \frac{\partial^2 u}{\partial y^2}\Big) + f(x,y,t) \\
\frac{\partial v}{\partial t} &= 3.4u - u^2 v + \alpha\Big(\frac{\partial^2 v}{\partial x^2} + \frac{\partial^2 v}{\partial y^2}\Big)
\end{aligned} \tag{10.15}$$

in its discretized form (II.10.14), but this time we make the problem stiff by increasing the coefficient α (which was 0.002) to $\alpha = 0.1$ and by increasing the number of grid points to $N = 128$. This gives an ordinary differential equation of dimension $2N^2 = 32768$. The initial conditions, chosen here as

$$u(x,y,0) = 22 \cdot y(1-y)^{3/2}, \qquad v(x,y,0) = 27 \cdot x(1-x)^{3/2}, \tag{10.16}$$

are quickly wiped out by the strong diffusion (see Fig. 10.7 for $t = 1$), we therefore suppose that the inhomogeneity $f(x,y,t)$ defined by

$$f(x,y,t) = \begin{cases} 5 & \text{if } (x-0.3)^2 + (y-0.6)^2 \le 0.1^2 \text{ and } t \ge 1.1 \\ 0 & \text{else} \end{cases}$$

models an extra addition of substance u in a small disc. In order to be able to solve the linear algebra comfortably by a double FFT routine we replace the Neumann conditions of Sect. II.10 by periodic boundary conditions

$$u(x+1,y,t) = u(x,y,t), \qquad u(x,y+1,t) = u(x,y,t).$$

As output points we choose $x_{\text{out}} = 1.5$ and 11.5.

Results and Discussion

For each of these examples we have computed very carefully the exact solution at the specified output points. Then, the above codes have been applied with many different tolerances

$$Tol = 10^{-2-m/4}, \qquad m = 0, 1, 2, \ldots, 32.$$

More precisely, we set the relative error tolerance to be $Rtol = Tol$ and the absolute error tolerance $Atol = 10^{-6} \cdot Tol$ for the problems OREGO and ROBER, $Atol = 10^{-4} \cdot Tol$ for HIRES, $Atol = 10^{-3} \cdot Tol$ for PLATE and BECKDO, $Atol = 1.7 \cdot 10^{-24}$ for E5, and $Atol = Tol$ for all other problems. Several codes returned numerical results which were considerably less precise than the required precision, while other methods turned out to be more reliable. As a reasonable measure of efficiency we have therefore chosen to compare

- the actual *error* (a norm taken over all components and all output points)
- the *computing time* (of a SUN Sparc 20 Workstation) in seconds.

The obtained data are then displayed as a polygonal line in a "precision-work diagram" in double logarithmic scales. The integer-exponent tolerances 10^{-2}, 10^{-3}, $10^{-4}, \ldots$ are displayed as enlarged symbols. The symbol for $Tol = 10^{-5}$ is specially distinguished by its gray colour. The more this line is to the right, the higher was the obtained precision; the higher this line is to the top, the slower was the code. The "slope" of the curve expresses the (effective) order of the formula: lower order methods are steeper than higher order methods. The results of the above codes on the 12 test examples are displayed in Figs. 10.8 and 10.9.

VDPOL, ROBER, OREGO — are very stiff problems of small dimension. We see from Fig. 10.8 that the Rosenbrock code RODAS is best for low tolerances (10^{-2} to 10^{-5}), whereas the extrapolation code SEULEX is superiour for stringent tolerances. Due to the cheapness of the function evaluations the multistep code LSODE requires in general slightly more computing time than the one-step codes. We also remark that for a given tolerance (the position of the gray symbol for $Tol = 10^{-5}$) the code RADAU5 gives the precisest result, followed by RODAS, SEULEX, and LSODE.

HIRES — this problem is less stiff and can also be solved by explicit methods. The computing times for the explicit code DOPRI5 are initially perfectly horizontal. This is, of course, no surprise, because the step size is restricted by stability. The (explicit, but stabilized) Runge-Kutta-Chebyshev code RKC shows a considerable improvement over DOPRI5 for low tolerances. The stiff codes are still more efficient.

E5 — is a stiff and badly scaled problem, which is integrated over a very long time. Codes cannot work correctly, if the absolute tolerance *Atol* is too large. The codes RODAS (for low tolerances) and RADAU5 (for $Tol \le 10^{-4}$) give the best results. LSODE works safely only for $Tol \le 10^{-5}$, whereas SEULEX has problems with round-off errors at high precision.

PLATE and BEAM — are both problems of the type $y'' = f(x, y, y')$, implemented as the first order system $y' = v$, $v' = f(x, y, v)$. For stiff codes the linear systems to be solved have a matrix of the form

$$\begin{pmatrix} \alpha I & I \\ B & C \end{pmatrix} \tag{10.17}$$

(where I is the identity matrix). Using the option `IWORK(9)=N/2` (where N is the dimension of the first order system) our codes do the first $N/2$ elimination sweeps analytically and the dimension of the linear system is halved. Without this option, the computing times for the codes RADAU5, RODAS, and SEULEX would be larger by a factor of about 3.0, 1.7, and 2.6, respectively (these numbers are for the BEAM problem at $Tol = 10^{-5}$). We did not include here the results of LSODE, for which we did not have an easy possibility for such a reduction. For the PLATE problem we also exploited the banded structure of $\partial f/\partial y$ and $\partial f/\partial v$ by putting `MLJAC=16` and `MUJAC=16`.

For both problems the explicit code DOPRI5 was applicable too. A curious phenomenon arose for DOPRI5 at the PLATE problem: as expected, for low tolerance requirements ($Tol \ge 10^{-5}$), the code appeared to be restricted by stability, gave computing times independent of *Tol* and issued the message "the problem seems to be stiff". But for more stringent tolerances the code was restricted by precision, with computing times unexpectedly high above those of the implicit code RADAU5. The analysis of Sect. IV.15 for the Prothero & Robinson problem (15.1) gives an

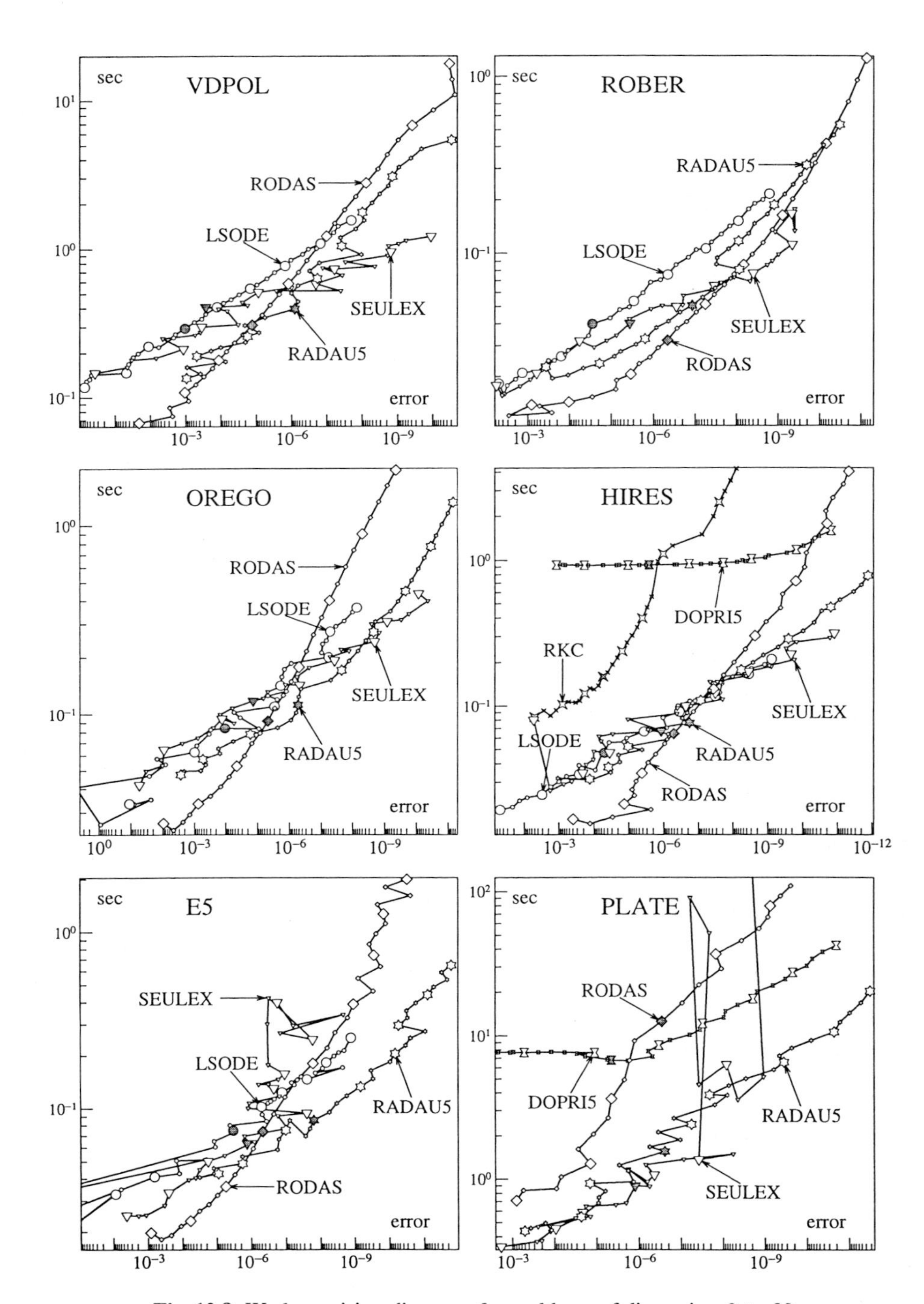

Fig. 10.8. Work-precision diagrams for problems of dimension 2 to 80

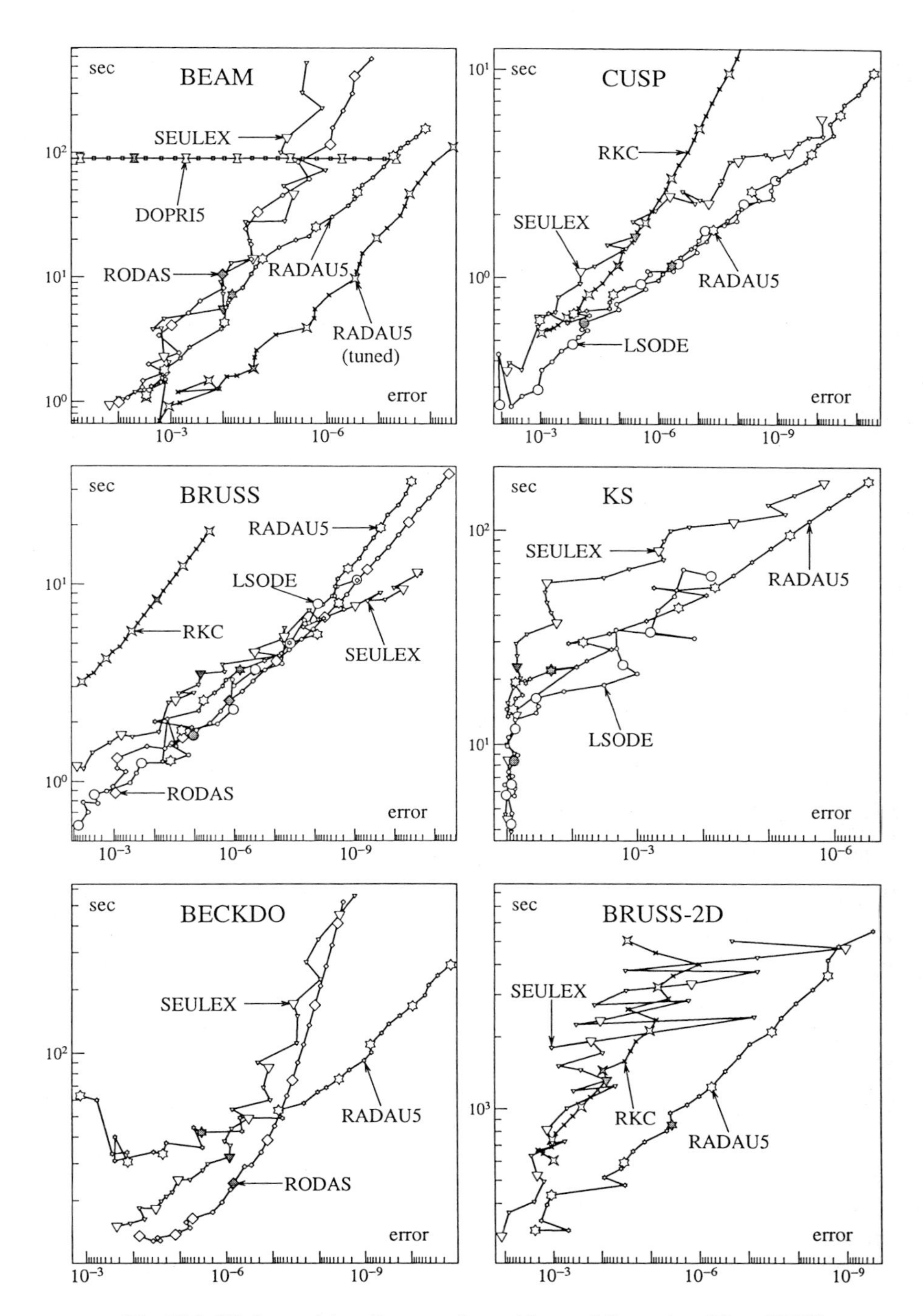

Fig. 10.9. Work-precision diagrams for problems of dimension 80 to 32768

explanation for this fact. We see that stiff problems not only create loss of stability, but also loss of precision for explicit integrators.

Especially for the BEAM problem, a problem with expensive linear algebra, the efficiency of the codes can be considerably increased by *tuning the parameters*. If, for the integration with RADAU5, we put

```
WORK(3)=0.1      (Jacobian less often recomputed)
WORK(4)=0.3      (Newton iterations stopped earlier)
WORK(5)=0.99 }   (Step size changed less often,
WORK(6)=2.   }   decreasing number of LU-decompositions)
```

then the computing time decreases by a factor between 2 and 5. Fig. 10.9 shows the spectacular improvement of this "tuned" run.

CUSP — the Jacobian of this problem is of the form

$$J = \begin{pmatrix} A_1 & B_1 & & D_1 \\ C_2 & A_2 & \ddots & \\ & \ddots & \ddots & B_{N-1} \\ D_N & & C_N & A_N \end{pmatrix} \tag{10.18}$$

where A_i, B_i, C_i, D_i are 3×3 matrices, and an efficient solution of the linear system needs a special treatment (see Exercise 1). However the considered methods, with the exception of the Rosenbrock methods, do not require an exact Jacobian. Therefore, an easy possibility for a considerable reduction of computing time is simply to use the codes in the banded version by putting `ML=MU=3`. The D_1 and D_N are neglected and we obtain the computing times displayed in Fig. 10.9. If the Jacobian were treated as a full matrix, the computing times would increase by a factor of 8.3, 6.6, and 4.8 for the codes RADAU5, SEULEX, and LSODE, respectively (these numbers are for $Tol = 10^{-5}$). The explicit code RKC gives excellent results for low precision, whereas the results of DOPRI5 (more than 30 seconds) are outside of the picture.

BRUSS — for this one-dimensional reaction-diffusion problem the linear algebra is done in the "banded" version with "analytical Jacobian". The problem is very stiff (large diffusion constant and small Δx) and an explicit method, such as DOPRI5, would require close to 60000 steps of integration. The code RKC works well, although less efficiently than the stiff integrators.

KS — the solution of this problem is sensitive with respect to changes in the initial values, a phenomenon already encountered in the LRNZ problem of Sect. II.10. Similarly as there, the precision increases only for *Tol* beyond a certain threshold. The Jacobian of this problem is full. Numerical experiments revealed that the codes worked best when the Jacobian is replaced by a diagonal matrix with $(qj)^2 - (qj)^4$ in its jth entry. Rosenbrock methods, which require an exact Jacobian, are not efficient here. The explicit codes RKC and DOPRI5 need too much computing time.

BECKDO — for this problem, the stiff codes (the only ones which work) require the solution of linear systems of the form

$$\begin{pmatrix} u & v^T \\ w & T \end{pmatrix} \begin{pmatrix} x \\ y \end{pmatrix} = \begin{pmatrix} a \\ b \end{pmatrix}, \tag{10.19}$$

where v, w, b are $(n-1)$-dimensional vectors and T is a tri-diagonal matrix. Since the linear algebra routines are completely separated from the codes RADAU5, RODAS and SEULEX, it is easy to replace these routines by a special program which solves (10.19) efficiently as follows

$$\begin{aligned} x &= \big(a - v^T T^{-1} b\big) / \big(u - v^T T^{-1} w\big) \\ y &= T^{-1} b - x T^{-1} w. \end{aligned} \tag{10.20}$$

It is not necessary to alter the stiff integrator itself.

Fig. 10.9 shows that, as usual, RODAS is best for low tolerances and RADAU5 is preferable for high precision. *Not* as usual is the fact that RODAS performs very badly for stringent tolerances. We explain this by the fact that the linear system (10.19) is sensitive to round-off errors, or, as Wilkinson would turn it, delivers a solution for a *wrong* Jacobian. Thus, the order of the Rosenbrock method drops to 1.

BRUSS-2D — due to its large dimension ($n = 2 \cdot 128^2 = 32768$), this problem makes no sense in full or even banded linear algebra. We therefore solved the linear equations (in the codes with separated linear algebra, see the corresponding remarks in the BECKDO problem) by FFT methods, taking into account only the (stiff) diffusion terms and neglecting the (in this problem non-stiff) reaction terms. The FFT codes used were those of Press, Flannery, Teukolsky & Vetterling (1986,1989) in the chapter on partial differential equations. A special advantage of the Radau method is here that the complex algebra, which is anyway used in FFT, crunches the complex eigenvalues of the Runge-Kutta matrix without further harm.

For this problem, which is a typical parabolic partial differential equation with non-stiff nonlinearities, we have made a detailed comparison of the performances of the implicit code RADAU5, the "stabilized" explicit code RKC, and the explicit code DOPRI5, in dependence of the discretization parameter $\Delta x = \Delta y = 1/N$ and the diffusion parameter α (see Eqs. (10.15) and (II.10.14)). The results (number of function calls and computing times) are displayed in Table 10.1, where the best performances are displayed in boldface characters. We can see how the olympic fire goes over from DOPRI5, which is best for low stiffness ($\alpha N^2 \le 1$), by increasing the stiffness first to RKC, and then (for $\alpha N^2 \ge 1000$) to the implicit RADAU5 code. We also observe that the number of function evaluations is nearly independent of the stiffness for RADAU5, behaves like $Const \cdot \sqrt{\alpha} \cdot N$ for RKC, and like $Const \cdot \alpha \cdot N^2$ for DOPRI5.

Comparisons Between Codes of the Same Type. Figs. 10.8 and 10.9, which are a sort of "Final Competition of Wimbledon", contain only one code from each class of integration methods (Radau methods, Implicit Runge-Kutta, Rosenbrock,

Table 10.1. Function evaluations / computing times at $Tol = 10^{-5}$

RADAU5	$N=16$	$N=32$	$N=64$	$N=128$	$N=256$
$\alpha=10^{-3}$	3372/19.8	3233/84.9	3271/413.5	3290/2215.6	3261/14902.1
$\alpha=10^{-2}$	1286/7.7	1322/36.2	1295/167.4	1381/868.8	1380/6459.3
$\alpha=10^{-1}$	1150/6.8	1131/30.9	1227/**172.3**	1173/**854.9**	1204/**5664.9**
$\alpha=1$	1195/7.8	1199/**33.0**	1247/**177.3**	1242/**945.9**	1258/**5961.2**
RKC	$N=16$	$N=32$	$N=64$	$N=128$	$N=256$
$\alpha=10^{-3}$	2367/4.7	2277/18.6	2249/76.3	2311/**352.5**	2911/**1912.0**
$\alpha=10^{-2}$	1661/3.2	1674/**13.8**	2078/**70.4**	3379/**511.5**	6259/**4086.9**
$\alpha=10^{-1}$	1899/**3.6**	2823/**22.5**	5047/**176.8**	9666/1446.2	18911/12312.2
$\alpha=1$	4013/**7.2**	7565/58.9	14631/503.4	29022/4328.8	
DOPRI5	$N=16$	$N=32$	$N=64$	$N=128$	$N=256$
$\alpha=10^{-3}$	976/**2.0**	1030/**8.5**	1408/**48.5**	3286/509.4	11464/7704.2
$\alpha=10^{-2}$	784/**1.6**	1894/15.4	6976/240.6	27478/4369.6	
$\alpha=10^{-1}$	4366/9.0	17176/145.5	68446/2419.7	273568/43982.2	
$\alpha=1$	42832/90.6	171010/1505.8	683836/24362.7		

and extrapolation methods). Following are some comparisons within each of these classes.

Radau Methods. For a comparison of Radau methods of various orders (see also the results of Reymond (1989) in the first edition), we have written a code RADAUP, which allows to choose with the help of a method flag `IWORK(11)=3,5,7` to choose between $s = 3, 5$, or 7 (i.e., between orders $p = 5, 9$, or 13). The code is for $s = 3$ mathematically equivalent to RADAU5, but, due to a different coding, slightly slower. We can see in Fig. 10.10 how the higher order pays off for higher precision, but for lower precision arise problems due to large step sizes and bad convergence of the Newton iterations.

Implicit Runge-Kutta Methods. It has for a long time been taken for granted that only DIRK and SDIRK methods could be implemented efficiently. Our experience shows that the diagonally implicit method SDIRK4, constructed in Section IV.6, gives rather disappointing results (see Fig. 10.11). An exception is the BEAM problem with its, microscopically, highly oscillatory solutions. Since the code SDIRK4 has not the option for "second order" linear algebra, we have also applied RADAU5 without this option. The computing times for RADAU5 are therefore not the same as in Fig. 10.9.

Rosenbrock Methods. There is usually not much difference between the performance of the different Rosenbrock methods (see Fig. 10.12). In spite of their larger number of stages, the codes RODAS5 (order 5) and RODAS (order 4) give often

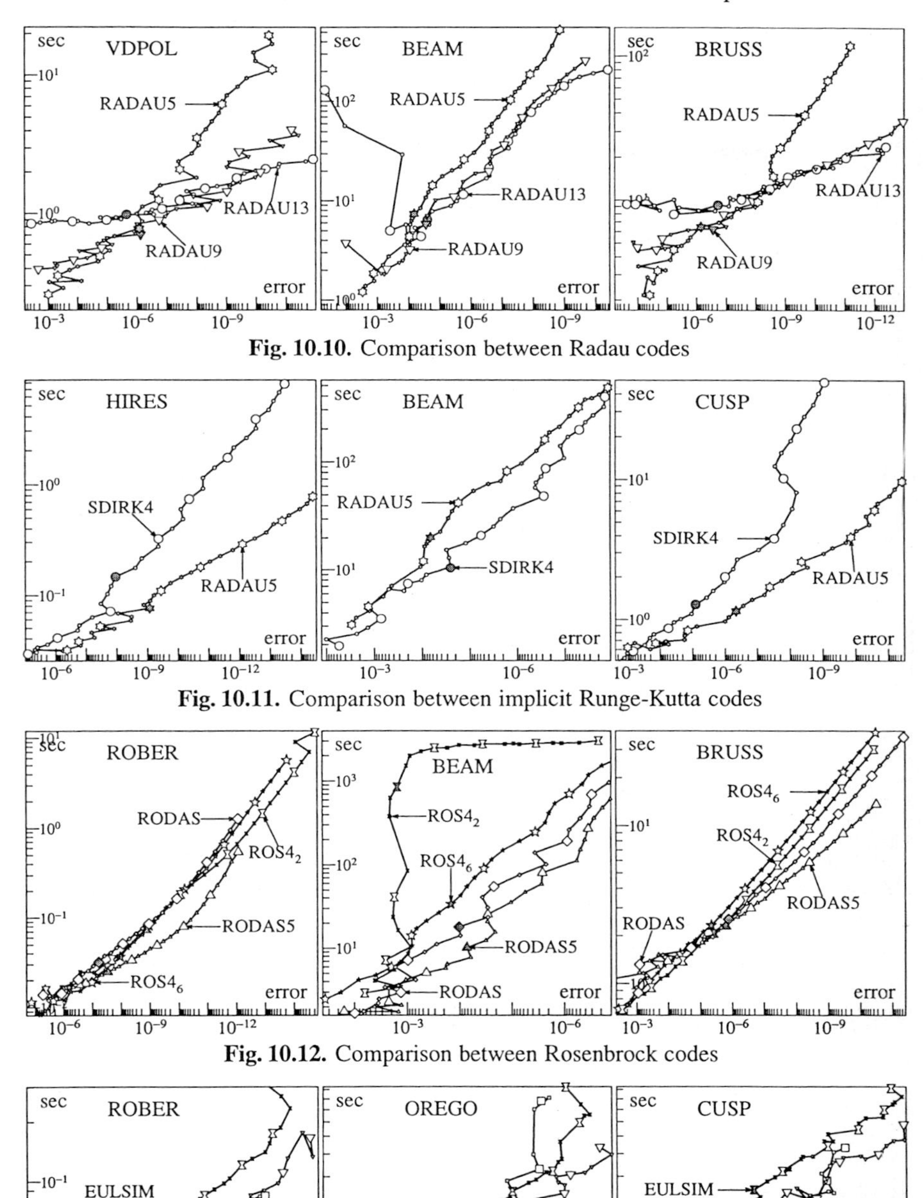

Fig. 10.10. Comparison between Radau codes

Fig. 10.11. Comparison between implicit Runge-Kutta codes

Fig. 10.12. Comparison between Rosenbrock codes

Fig. 10.13. Comparison between extrapolation codes

the best results. Among the 4th order "classical" Rosenbrok methods of Table 7.2 the best is in general "method 2" with its small error constant; it fails completely, however, on the Beam problem due to lack of A-stability. "Method 6" corresponds to the choice of coefficients which give an L-stable method.

Extrapolation Methods. The code SODEX, which is based on an h^2-extrapolation of the semi-implicit midpoint rule, is clearly superiour to SEULEX for low precision (see Fig. 10.13). The opposite situation appears for more stringent tolerances; here we observe an order reduction phenomenon, which is explained in Sect. VI.5 below. We have also included in these tests the results of the code EULSIM by Deuflhard, Novak & Poehle (poehle@sc.zib-berlin.de) which is another implementation of the extrapolated semi-implicit Euler method, with a different stepsize sequence.

Chebyshev Methods. During the final realization of these experiments we have received a code DUMKA3 (written by A. Medovikov, nucrect@inm.ras.ru) which implements an extension of the optimal Chebyshev methods of Lebedev (see Sect. IV.2) to third order. This code is still in a very experimental stage, but the results, presented in Fig. 10.14, are very promising.

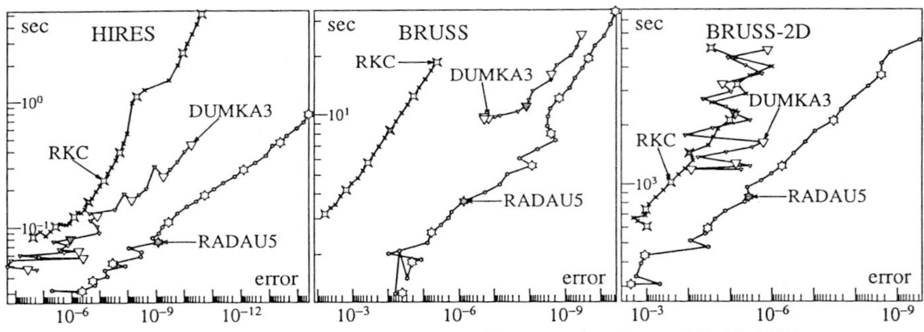

Fig. 10.14. Comparison between Chebyshev codes and RADAU5

Partitioning and Projection Methods

> Most codes for solving stiff systems ... spend most of their time solving systems of linear equations ...
>
> (Watkins & HansonSmith 1983)

Further spectacular reductions of the work for the linear algebra are often possible. One of the oldest ideas is to *partition* a stiff system into a (hopefully) small stiff system and a large nonstiff part,

$$\begin{aligned} y_a' &= f_a(y_a, y_b) \qquad \text{(stiff)} \\ y_b' &= f_b(y_a, y_b) \qquad \text{(nonstiff),} \end{aligned} \tag{10.21}$$

so that the two systems can be treated by two different methods, one implicit and the other explicit (e.g. Hofer 1976). The theory of P-series in Sect. II.14 had its origin in the study of the order properties of such methods. A difficulty of this approach is, of course, to decide *which* equations should be the stiff ones. Further, stiffness may affect subspaces which are *not* parallel to the coordinate axes. We shall therefore turn our attention to procedures which do not adapt the underlying *numerical method* to the partitioning, but the *linear algebra* only. An excellent survey of the older literature on these methods is given by Söderlind (1981). The following definition describes an especially promising class of problems:

Definition 10.1 (Björck 1983, 1984). The system $y' = f(x,y)$ is called *separably stiff* at a position x_0, y_0 if the Jacobian $J = \frac{\partial f}{\partial y}(x_0, y_0)$ possesses $k < n$ eigenvalues $\lambda_1, \ldots, \lambda_k$ such that

$$\min_{1\le i\le k} |\lambda_i| >> \max_{k+1\le i\le n} |\lambda_i|.$$

The eigenvalues $\lambda_1, \ldots, \lambda_k$ are called the *stiff eigenvalues* and

$$\mu = \min_{1\le i\le k} |\lambda_i| \Big/ \max_{k+1\le i\le n} |\lambda_i| \tag{10.22}$$

the *relative separation*. The space D spanned by the *stiff eigenvectors* is called the *dominant invariant subspace*.

For example, the Robertson problem (10.2) possesses only *one* stiff eigenvalue (close to -2000), and is therefore separably stiff with $k=1$. The CUSP problem (10.8') of dimension 96 has 32 large eigenvalues which range, except for transient phases, between -20000 and -60000. All other eigenvalues satisfy approximately $|\lambda| < 30$. This problem is, in fact, a singular perturbation problem (see Sect. VI.1), and such problems are all separably stiff. The other large problems of this section have eigenvalues scattered all around. A.R. Curtis' study (1983) points out that in *practical* problems separably stiff problems are rather seldom.

The Method of Gear and Saad. Implicit methods such as (transformed) Runge-Kutta or multistep formulas require the solution of a linear system (where we denote, as usual in linear algebra, the unknown vector by x)

$$Ax = b \qquad \text{where} \qquad A = \frac{1}{h\gamma} I - J \tag{10.23}$$

with *residual* $r = b - Ax$. We choose k (usually) orthogonal vectors $q_1, \ldots, q_k$ in such a way that the span $\{q_1, \ldots, q_k\} = \widetilde{D}$ is an *approximation* to the dominant subspace D, and denote by Q the $k \times n$-matrix formed by the columns q_j,

$$Q = (q_1, \ldots, q_k). \tag{10.24}$$

There are now several possibilities for replacing the solution x of (10.23) by an approximate solution $\widetilde{x} \in \widetilde{D}$. One of the most natural is to require (Saad 1981, Gear & Saad 1983; in fact, Galerkin 1915) that the residual of $\widetilde{x}$,

$$\widetilde{r} = b - A\widetilde{x} = A(x - \widetilde{x}), \tag{10.25}$$

be *orthogonal* to $\widetilde{D}$, i.e., that $Q^T(b - A\widetilde{x}) = 0$. If we write $\widetilde{x}$ in the basis of (10.24) as $\widetilde{x} = Q\widetilde{y}$, this yields

$$H\widetilde{y} = Q^T b, \tag{10.26}$$

where

$$H = Q^T AQ \qquad \text{or} \qquad QH = AQ, \tag{10.27}$$

which means that we have to solve a linear system of dimension k with matrix H. A particularly good choice for $\widetilde{D}$ is a *Krylov subspace* spanned by an arbitrary vector r_0 (usually the residual of a well chosen initial approximation x_0),

$$\widetilde{D} = \text{span}\,\{r_0, Ar_0, A^2 r_0, \ldots, A^{k-1} r_0\}. \tag{10.28}$$

The vectors (10.28) constitute the sequence created by the well-known power method. Therefore, in the case of a separably stiff system, as analyzed by D.J. Higham (1989), the space $\widetilde{D}$ approaches the space D extremely well as soon as its dimension is sufficiently high. In the *Arnoldi process* (Arnoldi 1951) the vectors of (10.28) are successively orthonormalized (Gram-Schmidt) as

$$\begin{aligned} &q_1 = r_0/\|r_0\| \\ &\widehat{q}_2 = Aq_1 - h_{11}q_1, \qquad q_2 = \widehat{q}_2/h_{21} \quad \text{with} \quad h_{21} = \|\widehat{q}_2\| \end{aligned}$$

and so on, and we see that

$$\begin{aligned} Aq_1 &= h_{21}q_2 + h_{11}q_1 \\ Aq_2 &= h_{32}q_3 + h_{22}q_2 + h_{12}q_1 \\ &\ldots \end{aligned} \tag{10.29}$$

which, compared to (10.28), shows that H is *Hessenberg*. For A symmetric, H is also symmetric, hence tridiagonal, so that the method is equivalent to the conjugate gradient method.

Two features are important for this method: Firstly, the matrix A need never be computed nor stored. All that is needed are the matrix-vector multiplications in (10.29), which can be obtained from the "directional derivative"

$$Jv \approx [f(x, y + \delta v) - f(x, y)]/\delta. \tag{10.30}$$

Several people therefore call such methods "matrix-free". Secondly, the dimension k does not have to be known: one simply computes one column of H after the other and periodically estimates the residual. As soon as this estimate is small enough (or k becomes too large) the algorithm stops. We also mention two variants of the method:

1. (Gear & Saad 1983, p. 595). Before starting the computation of the Krylov subspace, perform some initial iteration of the power method on the initial vector r_0, using either the matrix A or the matrix J. Lopez & Trigiante (1989) report excellent numerical results for this procedure.

2. *Incomplete Orthogonalization* (Saad 1982). The new vector Aq_j is only orthogonalized against the previous p vectors, where p is some small integer. This

makes H a banded matrix and saves computing time and memory. For symmetric matrices, the ideal choice is of course $p=2$, for matrices more and more unsymmetric p usually is increased to 10 or 15.

The EKBWH-Method (this tongue-twister stands for Enright, Kamel, Björck, Watkins and HansonSmith). Here, the matrices A (and J) in (10.23) are replaced by approximations

$$\widetilde{A} = \frac{1}{h\gamma} I - \widetilde{J} \tag{10.31}$$

where $\widetilde{J}$ should approach J sufficiently well and the matrix $\widetilde{A}$ be relatively easy to invert. $\widetilde{J}$ is determined as follows: Complete (theoretically) the vectors (10.24) to an orthogonal basis $(Q,\widehat{Q})$ of $\mathbb{R}^n$. In the new basis J becomes

$$\begin{pmatrix} Q^T \\ \widehat{Q}^T \end{pmatrix} J\,(Q,\widehat{Q}) = \begin{pmatrix} T_{11} & T_{12} \\ T_{21} & T_{22} \end{pmatrix} \tag{10.32}$$

and we have

$$Q^T J\, Q = T_{11}. \tag{10.33}$$

If span $Q=\widetilde{D}$ approaches D, then T_{11} will contain the stiff eigenvalues and T_{21} will tend to zero. If $\widetilde{D}=D$ exactly, then $T_{21}=0$ and (10.32) is a block-Schur decomposition of J. For separably stiff systems $\|T_{22}\|$ will become small compared to $(h\gamma)^{-1}$ and we define

$$\widetilde{J} = (Q,\widehat{Q}) \begin{pmatrix} T_{11} & T_{12} \\ 0 & 0 \end{pmatrix} \begin{pmatrix} Q^T \\ \widehat{Q}^T \end{pmatrix} = Q(T_{11}Q^T + T_{12}\widehat{Q}^T) \overset{(10.32)}{=} QQ^T J.$$

This shows $\widetilde{J}$ to be the orthogonal projection of J onto $\widetilde{D}$. The inverse of $\widetilde{A}$ is computed by developing $(I-B)^{-1} = I + B + B^2 + \ldots$ as a geometric series

$$\begin{aligned} \widetilde{A}^{-1} &= h\gamma\big(I - h\gamma QQ^T J\big)^{-1} \\ &= h\gamma\big(I + h\gamma QQ^T J + h^2\gamma^2 Q \underbrace{Q^T J Q}_{T_{11}} Q^T J + \ldots\big) \\ &= h\gamma\big(I + Q(h\gamma I + h^2\gamma^2 T_{11} + h^3\gamma^3 T_{11}^2 + \ldots)Q^T J\big) \\ &= h\gamma\big(I + Q\big(\frac{1}{h\gamma} I - T_{11}\big)^{-1} Q^T J\big) \end{aligned} \tag{10.34}$$

which only requires the solution of the "small" system with matrix $(I/h\gamma - T_{11})$ (the last expression is called the Sherman-Morrison-Woodbury formula).

Choice of Q:

— Björck (1983) computes the precise span of D, by Householder transforms followed by block-QR iterations. For separably stiff systems the block T_{21} converges to zero linearly with ratio μ^{-1} so that usually 2 or 3 iterations are sufficient. A disadvantage of the method is that an estimate for the dimension k of D must be known in advance.

— Enright & Kamel (1979) transform J to Hessenberg form and stop the transformations when $\|T_{21}\|+\|T_{22}\|$ become sufficiently small (remark that T_{21} is non zero in its last column only). Thus the dimension k can be discovered dynamically. Enright & Kamel combine the Householder reflexions with a pivoting strategy and repeated row & column permutations in order to make T_{22} small as fast as possible. It was first observed numerically (by Carlsson) and then shown theoretically (Söderlind 1981) that this pivoting strategy "needs some comments": if we start from (10.32), by knowing that

$$\begin{pmatrix} T_{11} & T_{12} \\ T_{21} & T_{22} \end{pmatrix}$$

is Hessenberg in its first k columns, (with $h_{21}\neq 0,\, h_{32}\neq 0,\ldots$) and do the analysis of formulas (10.29) backwards, we see that the space $\widetilde{D}$ for the Enright & Kamel method is a Krylov subspace created by q_1 (D.J. Higham 1989). Thus only the first permutation influences the result.

— Watkins & HansonSmith (1983) start from an arbitrary $Q^{(0)}$ followed by several steps of the block power method

$$JQ^{(i)} = Q^{(i+1)}R^{(i+1)} \tag{10.35}$$

where $R^{(i+1)}$ re-orthogonalizes the vectors of the product $JQ^{(i)}$. A great advantage of this procedure is that no large matrix needs to be computed nor stored. The formulas (10.35) as well as (10.34) only contain matrix-vector products which are computed by (10.30). The disadvantage is that the dimension of the space must be known.

Stopping Criteria. The above methods need a criterion on the goodness of the approximation $\widetilde{J}$ to decide whether the dimension k is sufficient. Suppose that we solve the linear equation (10.23) by a modified Newton correction which uses $\widetilde{A}$ as "approximate Jacobian"

$$\widetilde{x} = x_0 + \widetilde{A}^{-1}(b - Ax_0),$$

then the convergence of this iteration is governed by the condition

$$\varrho(I-\widetilde{A}^{-1}A) = \varrho(\widetilde{A}^{-1}(\widetilde{A}-A)) = \varrho(\widetilde{A}^{-1}(J-\widetilde{J})) < 1. \tag{10.36}$$

A reasonable condition is therefore that the spectral radius ϱ of $\widetilde{A}^{-1}(J-\widetilde{J})$ is plainly smaller than 1. Let us compute this value for the Björk method ($T_{21}=0$): since the eigenvalues of a matrix C are invariant under the similarity transformation $T^{-1}CT$, we have

$$\begin{aligned}\varrho(\widetilde{A}^{-1}(J-\widetilde{J})) &= \varrho\left(\left(\frac{1}{h\gamma}I-\begin{pmatrix} T_{11} & T_{12} \\ 0 & 0 \end{pmatrix}\right)^{-1}\begin{pmatrix} 0 & 0 \\ 0 & T_{22} \end{pmatrix}\right) \\ &= \varrho\left(\begin{pmatrix} (\frac{1}{h\gamma}I-T_{11})^{-1} & \times\times\times \\ 0 & h\gamma I \end{pmatrix}\begin{pmatrix} 0 & 0 \\ 0 & T_{22} \end{pmatrix}\right) \\ &= \varrho\left(\begin{pmatrix} 0 & \times\times\times \\ 0 & h\gamma T_{22} \end{pmatrix}\right) = \varrho(h\gamma T_{22}).\end{aligned}$$

In practice, a condition of the form

$$\|h\gamma T_{22}\| < 1, \tag{10.37}$$

where $\|\cdot\|$ is usually the Frobenius norm $\sqrt{\sum_{i,j} a_{ij}^2}$, ensures a reasonable rate of convergence. For an analogous condition in the Enright-Kamel case see Exercise 3 below.

Exercises

1. (The red-black reduction). The Jacobian matrix of the (periodic) cusp catastrophe model (10.8') is of the form

$$\begin{pmatrix} A_1 & B_1 & & & C_1 \\ C_2 & A_2 & B_2 & & \\ & \ddots & \ddots & \ddots & \\ & & C_{2m-1} & A_{2m-1} & B_{2m-1} \\ B_{2m} & & & C_{2m} & A_{2m} \end{pmatrix} \tag{10.38}$$

where A_i, B_i, C_i are (3×3)-matrices. Write a solver which solves linear equations with matrix (10.38) using the "red-black ordering reduction". This means that $A_1, A_3, A_5, \ldots$ are used as (matricial) pivots to eliminate $C_2, C_4, \ldots, B_2, B_4, \ldots$ above and below by Gaussian block-elimination. Then the resulting system is again of the same structure as (10.38) with *halved* dimension. If the original system's dimension contains 2^k as prime factor, this process can be iterated k times. Study the increase of performance which this algorithm allows for the RADAU5 and Rosenbrock codes on model (10.8'). The algorithm is also highly parallelizable.

2. Show by numerical experiments that the circular nerve (10.8') loses its limit cycle when the diffusion coefficient D becomes either too small (the message does not go across the water fall) or too large (the limit cycle then melts down across the origin).

3. (Stopping criterion for Enright & Kamel method; D.J. Higham 1989). Suppose that the matrix J has been transformed to partial Hessenberg form (see

(10.32))

$$\begin{pmatrix} Q^T \\ \widehat{Q}^T \end{pmatrix} J\,(Q,\widehat{Q}) = \begin{matrix} & k & n-k \\ k & H & T_{12} \\ n-k & (0\;b) & T_{22} \end{matrix}$$

where H is upper Hessenberg and b a column vector. Show that the criterion (10.36) then becomes

$$\varrho(h\gamma B) < 1$$

where

$$B = \begin{matrix} & k-1 & 1+n-k \\ k & 0 & -h\gamma\overline{H}^{-1}T_{12}(b\,T_{22}) \\ n-k & 0 & (b\,T_{22}) \end{matrix}$$

with $\overline{H} = (I - h\gamma H)$. Since $\varrho(B)$ is the same as the spectral radius of its lower $1+n-k$ by $1+n-k$ principal submatrix, a sufficient condition for convergence is

$$|h\gamma|\sqrt{\|T_{22}\|^2 + \|b\|^2 + \|y\|^2} < 1$$

where y^T is the k-th row of the matrix $-h\gamma\overline{H}^{-1}T_{12}(b\,T_{22})$.

IV.11 Contractivity for Linear Problems

> He who loves practice without theory is like the sailor who boards ship without a rudder and compass and never knows where he may be cast. (Leonardo da Vinci 1452-1519, cited from M. Kline, Mathematical Thought ... 1972, p. 224)

The stability analysis of the preceeding sections is based on the transformation of the Jacobian $J \approx \partial f / \partial y$ to diagonal form (see Formulas (2.5), (2.6) of Sect. IV.2). Especially for large-dimensional problems, however, the matrix which performs this transformation may be badly conditioned and destroy all the nice estimations which have been obtained.

Example 11.1. The discretization of the hyperbolic problem

$$\frac{\partial u}{\partial t} = \frac{\partial u}{\partial x} \tag{11.1}$$

by the method of lines leads to

$$y' = Ay, \qquad A = \lambda \begin{pmatrix} -1 & 1 & & \\ & -1 & \ddots & \\ & & \ddots & 1 \\ & & & -1 \end{pmatrix}, \qquad \lambda = \frac{1}{\Delta x} > 0. \tag{11.2}$$

This matrix has all eigenvalues at $-\lambda$ and the above spectral stability analysis would indicate fast asymptotic convergence to zero. But neither the solution of (11.1), which just represents a travelling wave, nor the solution of (11.2), if the dimension becomes large, have this property. So our interest in this section is to obtain rigorous bounds for the numerical solution (see (2.3))

$$y_{m+1} = R(hA)y_m \tag{11.3}$$

in different norms of $\mathbb{R}^n$ or $\mathbb{C}^n$. Here $R(z)$ represents the stability function of the method employed. We have from (11.3)

$$\|y_{m+1}\| \le \|R(hA)\| \cdot \|y_m\| \tag{11.4}$$

(see Volume I, Sect. I.9, Formula (9.10)), and contractivity is assured if

$$\|R(hA)\| \le 1.$$

Euclidean Norms (Theorem of von Neumann)

> People in mathematics and science should be reminded that many of the things we take for granted today owe their birth to perhaps one of the most brilliant people of the twentieth century — John von Neumann.
>
> (John Impagliazzo, quoted from SIAM News September 1988)

Let the considered norm be Euclidean with the corresponding scalar product denoted by $\langle\cdot,\cdot\rangle$. Then, for the solution of $y' = Ay$ we have

$$\frac{d}{dx}\|y\|^2 = \frac{d}{dx}\langle y, y\rangle = 2\mathrm{Re}\,\langle y, y'\rangle = 2\mathrm{Re}\,\langle y, Ay\rangle, \tag{11.5}$$

hence the solutions are decaying in this norm if

$$\mathrm{Re}\,\langle y, Ay\rangle \le 0 \quad \text{for all} \quad y \in \mathbb{C}^n. \tag{11.6}$$

This result is related to Theorem 10.6 of Sect. I.10, because

$$\mathrm{Re}\,\langle y, Ay\rangle \le \mu_2(A)\,\|y\|^2, \tag{11.7}$$

where $\mu_2(A)$ is the logarithmic norm of A (Eq. (10.20) of Sect. I.10).

Theorem 11.2. *Let the rational function $R(z)$ be bounded for $\mathrm{Re}\, z \le 0$ and assume that the matrix A satisfies (11.6). Then, in the matrix norm corresponding to the scalar product we have*

$$\|R(A)\| \le \sup_{\mathrm{Re}\, z \le 0} |R(z)|. \tag{11.8}$$

Remark. This is a finite-dimensional version of a result of J. von Neumann (1951). A short proof is given in Hairer, Bader & Lubich (1982). The idea of the following proof is due to M. Crouzeix (unpublished).

Proof. a) Normal matrices can be transformed to diagonal form by a unitary matrix Q (see Exercise 3 of Section I.12). Hence, $A = QDQ^*$, where $D = \mathrm{diag}\{\lambda_1, \ldots, \lambda_n\}$. In this case we have

$$\|R(A)\| = \|QR(D)Q^*\| = \|R(D)\| = \max_{i=1,\ldots,n} |R(\lambda_i)|,$$

and (11.8) follows from (11.6), because the eigenvalues of A satisfy $\mathrm{Re}\,\lambda_i \le 0$.

b) For a general A we consider the matrix function

$$A(\omega) = \frac{\omega}{2}(A + A^*) + \frac{1}{2}(A - A^*).$$

We see from the identity

$$\langle v, A(\omega)v\rangle = \omega \mathrm{Re}\,\langle v, Av\rangle + i\,\mathrm{Im}\,\langle v, Av\rangle$$

that $A(\omega)$ satisfies (11.6) for all ω with $\mathrm{Re}\,\omega \ge 0$, so that also the eigenvalues of $A(\omega)$ satisfy $\mathrm{Re}\,\lambda(\omega) \le 0$ for $\mathrm{Re}\,\omega \ge 0$. Therefore, the rational function

$$\varphi(\omega) = \langle u, R(A(\omega))v\rangle$$

(u, v fixed) has no poles in $\operatorname{Re}\omega \geq 0$. Using $A(1) = A$ we obtain from the maximum principle that

$$\begin{aligned}\langle u, R(A)v\rangle = \varphi(1) &\leq \sup_{y\in\mathbb{R}} \varphi(iy) \leq \sup_{y\in\mathbb{R}} \|R(A(iy))\| \|u\| \|v\| \\ &\leq \sup_{\operatorname{Re} z\leq 0} |R(z)| \|u\| \|v\|.\end{aligned} \tag{11.9}$$

The last inequality of (11.9) follows from part (a), because $A(iy)$ is a normal matrix (i.e., $A(iy)A(iy)^* = A(iy)^* A(iy)$). Formula (11.8) is now an immediate consequence of (11.9) and of the fact that $\|C\| = \sup_{\|u\|\leq 1, \|v\|\leq 1} \langle u, Cv\rangle$. □

Corollary 11.3. *If the rational function $R(z)$ is A-stable, then the numerical solution $y_{n+1} = R(hA)y_n$ is contractive in the Euclidean norm (i.e., $\|y_{n+1}\| \leq \|y_n\|$), whenever (11.6) is satisfied.*

Proof. A-stability implies that $\max_{\operatorname{Re} z\leq 0} |R(z)| \leq 1$. □

Corollary 11.4. *If a matrix A satisfies $\operatorname{Re}\langle v, Av\rangle \leq \nu\|v\|^2$ for all $v \in \mathbb{C}^n$, then*

$$\|R(A)\| \leq \sup_{\operatorname{Re} z\leq \nu} |R(z)|. \tag{11.10}$$

Proof. Apply Theorem 11.2 to $\widetilde{R}(z) = R(z+\nu)$ and $\widetilde{A} = A - \nu I$. □

Error Growth Function for Linear Problems

Guided by the above estimate, we define

$$\varphi_R(x) := \sup_{\operatorname{Re} z\leq x} |R(z)|. \tag{11.11}$$

This function is called *error growth function* (for linear problems). It is continuous and monotonically increasing. If $R(z)$ is analytic in the half-plane $\operatorname{Re} z < x$, the maximum principle implies that

$$\varphi_R(x) = \sup_{y\in\mathbb{R}} |R(x+iy)|.$$

Examples.

1. Implicit Euler method:

$$R(z) = \frac{1}{1-z} \qquad \varphi_R(x) = \begin{cases} R(x) & \text{if } -\infty < x < 1 \\ \infty & \text{if } 1 \leq x\,. \end{cases} \tag{11.12}$$

2. The stability function of the θ-method (or of a one-stage Rosenbrock method):

$$R(z)=\frac{1+(1-\theta)z}{1-\theta z}\qquad \varphi_R(x)=\begin{cases}|R(\infty)| & \text{if } x\le\xi_0\\ R(x) & \text{if } \xi_0\le x<1/\theta\\ \infty & \text{if } 1/\theta\le x\,,\end{cases}\tag{11.13}$$

where $\xi_0=(1-2\theta)/(2\theta(1-\theta))$ for $0<\theta<1$ and $\xi_0=-\infty$ for $\theta\ge 1$.

3. The (0,2)-Padé approximation:

$$R(z)=\frac{1}{1-z+z^2/2}\qquad \varphi_R(x)=\begin{cases}R(x) & \text{if } -\infty<x\le 0\\ \dfrac{1}{1-x} & \text{if } 0\le x<1\\ \infty & \text{if } 1\le x\,.\end{cases}\tag{11.14}$$

4. The (1,2)-Padé approximation $R(z)=\dfrac{1+z/3}{1-2z/3+z^2/6}$:

$$\varphi_R(x)=\begin{cases}|R(x)| & \text{if } -\infty<x\le\xi_0\\ \dfrac{\sqrt{3\sqrt{12x^2+12x+9}+10x+7}}{2(2-x)} & \text{if } \xi_0\le x<2\\ \infty & \text{if } 2\le x\,,\end{cases}\tag{11.15}$$

where $\xi_0=-6-3\sqrt{10}$.

5. The (2,2)-Padé approximation $R(z)=\dfrac{1+z/2+z^2/12}{1-z/2+z^2/12}$:

$$\varphi_R(x)=\begin{cases}1 & \text{if } -\infty<x\le 0\\ \dfrac{2x+\sqrt{9+3x^2}}{3-x} & \text{if } 0\le x<3\\ \infty & \text{if } 3\le x\,.\end{cases}\tag{11.16}$$

The next two theorems give some general results on the shape of $\varphi_R(x)$.

Theorem 11.5. *Let $R(z)$ be an A-stable approximation to e^z of exact order p, i.e., $R(z)=e^z-Cz^{p+1}+\mathcal{O}(z^{p+2})$ with $C\ne 0$. If additionally $|R(iy)|<1$ for $y\ne 0$ and $|R(\infty)|<1$, then we have*

a) if p is odd

$$\varphi_R(x)=e^x+\mathcal{O}(x^{p+1})\quad \textit{for}\quad x\to 0.\tag{11.17}$$

b) if p is even we have (11.17) only for $(-1)^{p/2}Cx>0$, otherwise

$$\varphi_R(x)=e^x+\mathcal{O}(x^{r+1})\quad \textit{for}\quad x\to 0\tag{11.18}$$

for some positive rational number $r\le p/2$.

Proof. The assumptions imply that for $x \to 0$ the maximum of $\{|R(x+iy)|; y \in \mathbb{R}\}$ must be located near the origin. We further observe that it must lie within the order star $A = \{z \in \mathbb{C};\ |R(z)| > |e^z|\}$. If p is odd, the order star consists of $p+1$ sectors near the origin (Lemma 4.3) and, asymptotically for $z \to \infty$, all elements of A satisfy $|z| \le D|x|$, $D < \infty$. Therefore

$$|R(z)| = e^x + \mathcal{O}(|z|^{p+1}) = e^x + \mathcal{O}(x^{p+1}) \quad \text{for} \quad x \to 0.$$

The same argument applies if p is even and $(-1)^{p/2}Cx > 0$. In the remaining case (p even and $(-1)^{p/2}Cx < 0$) the maximum of $\{|R(x+iy)|; y \in \mathbb{R}\}$ is attained near the imaginary axis and a more detailed analysis is necessary (Hairer, Bader & Lubich 1982). □

Theorem 11.6 (Hairer & Zennaro 1996). *For an A-stable approximation to e^z the function $\varphi_R(x)$ is* superexponential, *i.e., it satisfies $\varphi_R(0) = 1$ and*

$$\varphi_R(x_1)\,\varphi_R(x_2) \le \varphi_R(x_1 + x_2) \tag{11.19}$$

for all x_1, x_2 having the same sign.

Proof. A-stability is equivalent to $\varphi_R(0) = 1$. It therefore remains to verify (11.19). Let x_1 and x_2 be fixed (both ≤ 0 or both ≥ 0) and assume $\varphi_R(x_1 + x_2) < \infty$. The idea is to consider the rational function

$$S(z) = R(a - z)R(z)$$

where $a \in \mathbb{C}$ is a parameter satisfying $\operatorname{Re} a \le x_1 + x_2$. Due to A-stability and $\varphi_R(x_1 + x_2) < \infty$, $S(z)$ is analytic on the stripe $0 \le \operatorname{Re} z \le x_1 + x_2$ (or $x_1 + x_2 \le \operatorname{Re} z \le 0$), and its modulus is bounded by $\varphi_R(x_1 + x_2)$ on the border. By the maximum principle we therefore have for all z in the considered stripe

$$|R(a - z)R(z)| \le \varphi_R(x_1 + x_2).$$

We now choose z on the line $\operatorname{Re} z = x_2$ in such a way that $|R(z)|$ becomes maximal; then, we choose a on the line $\operatorname{Re} a = x_1 + x_2$ (i.e., $\operatorname{Re}(a - z) = x_1$) such that $|R(a - z)|$ becomes maximal (eventually one has to consider limits). This proves (11.19). □

Property (11.19) has an interesting practical interpretation. Consider a numerical solution y_n obtained with variable step sizes. Repeated application of (11.4) and Corollary 11.4 implies

$$\|y_m\| \le \Big(\prod_{k=0}^{m-1} \varphi_R(h_k \mu) \Big) \cdot \|y_0\|, \tag{11.20}$$

if the problem $y' = Ay$ satisfies (11.7) with $\mu = \mu_2(A)$. For $\mu < 0$ and for an A-stable method all factors $\varphi_R(h_k \mu)$ are smaller than one. If in addition $|R(\infty)| < 1$, these factors are close to one only for $h_k \to 0$. The inequality (11.19), written as

$$\varphi_R(h_k\mu)\,\varphi_R(h_{k+1}\mu) \le \varphi((h_k + h_{k+1})\mu),$$

means that replacing two consecutive steps by one large step of size $h_k + h_{k+1}$ increases the upper bound (11.20). Therefore, after combining several consecutive steps (if necessary), we may assume $h_k \ge h > 0$ for all k. This implies that $\|y_m\| \le \varrho^m \|y_0\|$ with $\varrho = \varphi_R(h\mu) < 1$. Hence, for any mesh $x_0, x_1, \ldots$ with $x_m \to \infty$, we have asymptotic stability, i.e., $\|y_m\| \to 0$ for $m \to \infty$. Under additional restrictions on the step size, sharper bounds on $\|y_m\|$ can be obtained (Exercise 3).

Small Nonlinear Perturbations

The above estimates, valid only for linear autonomous equations $y' = Jy$, can be extended to problems with small nonlinear perturbations, so-called *semi-linear* problems

$$y' = Jy + g(x, y) \tag{11.21}$$

where

$$\langle y, Jy\rangle \le \mu \|y\|^2 \tag{11.22a}$$

$$\|g(x,y) - g(x,z)\| \le L\|y - z\| \tag{11.22b}$$

with L assumed to be small.

Here, in the presence of nonlinearities, stability properties are obtained by estimating the *distance* of two neighbouring solutions $y(x)$ and $\widehat{y}(x)$. Instead of (11.5) we therefore have

$$\frac{d}{dx}\|y(x) - \widehat{y}(x)\|^2 = 2\langle y' - \widehat{y}\,', y - \widehat{y}\rangle$$

which gives, after inserting (11.21) for y' and $\widehat{y}\,'$, using the Cauchy-Schwarz inequality and the estimates (11.22)

$$\frac{d}{dx}\|y(x) - \widehat{y}(x)\|^2 \le 2\big(\mu + L\big)\ \|y(x) - \widehat{y}(x)\|^2. \tag{11.23}$$

We thus have contractivity whenever $\mu + L \le 0$.

We now want to establish the same property for the *numerical* solutions. In principle, these estimates can be carried out for all methods of this chapter; however, since the subsequent sections will deal with so many nice properties of implicit Runge-Kutta methods, we shall concentrate here on Rosenbrock methods.

Example 11.7. Consider the 1-stage Rosenbrock method

$$\begin{aligned}(I - \gamma hJ)k_1 &= hf(x_0, y_0)\\ y_1 &= y_0 + k_1\end{aligned} \tag{11.24}$$

with $\gamma > 0$ as a free parameter. Its stability function is

$$R(z) = \frac{1+(1-\gamma)z}{1-\gamma z}$$

and we have A-stability for $\gamma \geq 1/2$. Application of (11.24) to (11.21) yields

$$y_1 = R(hJ)y_0 + (I-\gamma hJ)^{-1}hg(x_0, y_0). \tag{11.25}$$

From von Neumann's theorem (Corollary 11.4) we obtain $\|(I-\gamma hJ)^{-1}\| \leq (1-\gamma h\mu)^{-1}$ and $\|R(hJ)\| \leq \varphi_R(h\mu)$ with φ_R given in (11.13). If we take a second numerical solution $\widehat{y}_1$, also defined by (11.25), its difference to y_1 can be estimated by

$$\|y_1 - \widehat{y}_1\| \leq \Big(R(h\mu) + \frac{hL}{1-\gamma h\mu}\Big)\|y_0 - \widehat{y}_0\| = \Big(1 + \frac{h(\mu+L)}{1-\gamma h\mu}\Big)\|y_0 - \widehat{y}_0\|$$

whenever $\xi_0 < h\mu < 1/\gamma$ with ξ_0 given in (11.13). Therefore contractivity occurs for $\mu + L \leq 0$, as desired.

For the general Rosenbrock method (7.4) applied to problem (11.21)

$$k_i = hg(x_0 + c_i h, u_i) + hJy_0 + hJ\sum_{j=1}^{i}(\alpha_{ij} + \gamma_{ij})k_j$$

$$u_i = y_0 + \sum_{j=1}^{i-1}\alpha_{ij}k_j, \qquad y_1 = y_0 + \sum_{i=1}^{s} b_i k_i$$

we easily find the following analogue of the variation of constants formula.

Theorem 11.8. *The numerical solution of a Rosenbrock method applied to (11.21) can be written as*

$$\begin{aligned} y_1 &= R(hJ)y_0 + h\sum_{i=1}^{s} b_i(hJ)g(x_0 + c_i h, u_i) \\ u_i &= R_i(hJ)y_0 + h\sum_{j=1}^{i-1} a_{ij}(hJ)g(x_0 + c_j h, u_j), \qquad i = 1,\ldots,s. \end{aligned} \tag{11.26}$$

Here $R(z)$ is the stability function, $R_i(z)$ are the so-called internal stability functions and $b_i(z)$, $a_{ij}(z)$ are rational functions whose only pole is $1/\gamma$ and which satisfy $b_i(\infty) = 0$, $a_{ij}(\infty) = 0$. □

Remark. For many classes of linearly implicit methods (e.g., the methods of van der Houwen (1977), Friedli (1978), Strehmel & Weiner (1982), etc.), the numerical solution can be expressed by (11.26) with certain rational functions. Thus the following analysis can be applied to these methods as well.

We now take a second numerical solution $\widehat{y}_0, \widehat{u}_i, \widehat{y}_1$ (again defined by (11.26)), take the difference to y_1 and apply the triangle inequality. Using von Neumann's theorem (Corollary 11.4) the assumptions (11.22) then imply

$$\begin{aligned} \|\widehat{y}_1 - y_1\| &\le \varphi_R(h\mu)\|\widehat{y}_0 - y_0\| + hL\sum_{i=1}^{s}\varphi_{b_i}(h\mu)\|\widehat{u}_i - u_i\| \\ \|\widehat{u}_i - u_i\| &\le \varphi_{R_i}(h\mu)\|\widehat{y}_0 - y_0\| + hL\sum_{j=1}^{i-1}\varphi_{a_{ij}}(h\mu)\|e\widehat{u}_j - u_j\|. \end{aligned} \tag{11.27}$$

Inserting the second inequality of (11.27) repeatedly into the first one yields

Theorem 11.9. *Under the assumption (11.22) the difference of two numerical solutions of (7.4) can be estimated by*

$$\|\widehat{y}_1 - y_1\| \le (\varphi_R(h\mu) + chL)\|\widehat{y}_0 - y_0\| \tag{11.28}$$

where $\varphi_R(x)$ is given by (11.11) ($R(z)$ is the stability function of (7.4)) and c is a constant depending smoothly on hL and $h\mu$ but not on $\|J\|$ (which represents the stiffness of the problem). . □

This estimate shows numerical contractivity whenever $\varphi_R(h\mu) + hL^* \le 0$. In Theorem 11.5 we have shown under certain assumptions that $\varphi_R(x) = 1 + x + o(x)$, so contractivity holds essentially for $\mu + L^* \le 0$. In any case we have that A-stability implies

$$\|\widehat{y}_1 - y_1\| \le (1 + hC^*)\,\|\widehat{y}_0 - y_0\|$$

for $h\mu \le Const$. Here, C^* is a constant independent of the stiffness of (11.21).

Remark. Since the rational functions b_i and a_{ij} in (11.26) vanish at infinity, also $(1-\gamma hJ)b_i(hJ)$ and $(1-\gamma hJ)a_{ij}(hJ)$ are uniformly bounded for J satisfying (11.22) and for $h\mu \le C < \gamma^{-1}$. Instead of the second condition of (11.22) we may therefore require that

$$\|(I - \gamma hJ)^{-1}h(g(x,y) - g(x,z))\| \le \ell\|y - z\|, \tag{11.29}$$

and the statement of Theorem 11.9 holds with hL replaced by ℓ. Observe that the assumption (11.22) implies (11.29) with $\ell = hL/(1-\gamma h\mu)$. However, in some special situations the number ℓ may be significantly smaller than hL. Related techniques are used by Hundsdorfer (1985) and Strehmel & Weiner (1987) to prove contractivity and convergence for linearly implicit methods. Ostermann (1988) applies these ideas to nonlinear singular perturbation problems, where $hL = \mathcal{O}(h\varepsilon^{-1})$ with some very small ε $(\varepsilon \ll h)$, but ℓ can be bounded independently of ε^{-1}.

Contractivity in $\|\cdot\|_\infty$ and $\|\cdot\|_1$

The study of contractivity in general norms has been carried out mainly by Spijker (1983, 1985) and his collaborators. Similar techniques of proof can be found in Bolley & Crouzeix (1978), where a related problem (monotonicity) is treated.

The following theorem gives a condition which is *necessary* for contractivity just for the special equation (11.2) and for one of the two norms $\|\cdot\|_\infty$ or $\|\cdot\|_1$. Later, the same condition will also turn out to be *sufficient* for general problems and *all* norms.

Theorem 11.10. *Let A be the n-dimensional matrix of (11.2) with fixed $\lambda \geq 0$. For a rational function $R(z)$ satisfying $R(0)=1$ we have*

$$\|R(hA)\|_\infty \leq 1 \quad \text{in all dimensions} \quad n=1,2,\ldots \tag{11.30}$$

only if

$$R^{(j)}(x) \geq 0 \quad \text{for} \quad x \in [-\lambda h, 0] \quad \text{and} \quad j=0,1,2,\ldots \tag{11.31}$$

(The same statement is true, if $\|\cdot\|_\infty$ in (11.30) is replaced by $\|\cdot\|_1$).

Proof. We put $h=1$ and write $A=-\lambda I+\lambda N$, where N is a nilpotent matrix. In a suitable norm, $\|N\|$ is arbitrarily small and therefore we have by Taylor expansion and $N^n=0$

$$R(A)=\sum_{j=0}^{n-1} R^{(j)}(-\lambda)\frac{(\lambda N)^j}{j!}.$$

This means (e.g. for $n=4$)

$$R(A)=\begin{pmatrix} R(-\lambda) & \lambda R'(-\lambda) & \frac{\lambda^2}{2!}R''(-\lambda) & \frac{\lambda^3}{3!}R'''(-\lambda) \\ & R(-\lambda) & \lambda R'(-\lambda) & \frac{\lambda^2}{2!}R''(-\lambda) \\ & & R(-\lambda) & \lambda R'(-\lambda) \\ & & & R(-\lambda) \end{pmatrix}.$$

Application of Formula (I.9.11') shows that $\|R(A)\|_\infty \leq 1$ (or $\|R(A)\|_1 \leq 1$) is equivalent to

$$\sum_{j=0}^{n-1} |R^{(j)}(-\lambda)|\frac{\lambda^j}{j!} \leq 1. \tag{11.32}$$

If (11.32) is valid for all $n \geq 1$, the series

$$\sum_{j\geq 0} R^{(j)}(-\lambda)\frac{\lambda^j}{j!} \tag{11.33}$$

is absolutely convergent, and therefore we have

$$1=R(0)=\sum_{j\geq 0} R^{(j)}(-\lambda)\frac{\lambda^j}{j!} \leq \sum_{j\geq 0} |R^{(j)}(-\lambda)|\frac{\lambda^j}{j!} \leq 1$$

implying $R^{(j)}(-\lambda) \geq 0$ for all $j \geq 0$. Since the Taylor expansion

$$R^{(j)}(x) = \sum_{k \geq j} R^{(k)}(-\lambda) \frac{(x+\lambda)^{k-j}}{(k-j)!}$$

consists for $x \geq -\lambda$ only of non-negative terms, we have (11.31). □

The next theorem shows that condition (11.31) is sufficient for contractivity in arbitrary norms. It can readily be applied to the system (11.2), since its matrix satisfies $\|A+\lambda I\|_\infty = \lambda$.

Theorem 11.11. *Consider an arbitrary norm and let* A *be such that for some* $\lambda \geq 0$,

$$\|A+\lambda I\| \leq \lambda. \tag{11.34}$$

If the stability function of a method satisfies $R(0) = 1$ *and*

$$R^{(j)}(x) \geq 0 \quad \text{for} \quad x \in [-\varrho, 0] \quad \text{and} \quad j = 0, 1, 2, \ldots \tag{11.35}$$

then we have numerical contractivity $\|R(hA)\| \leq 1$, *whenever* $h\lambda \leq \varrho$.

Proof. We again put $h = 1$. Since for $0 \leq \lambda \leq \varrho$ we have $R^{(j)}(-\lambda) \geq 0$ for all j, the function

$$R(z) = \sum_{j \geq 0} R^{(j)}(-\lambda) \frac{(z+\lambda)^j}{j!} \tag{11.36}$$

satisfies $|R(z)| \leq R(-\lambda + r)$ for all complex z in the disk $|z+\lambda| \leq r$. This property and (11.35) imply that no pole of $R(z)$ can lie in $|z+\lambda| \leq \lambda$, so that the radius of convergence of (11.36) is strictly larger than λ. Consequently we have from (11.34)

$$R(A) = \sum_{j \geq 0} R^{(j)}(-\lambda) \frac{(A+\lambda I)^j}{j!}. \tag{11.37}$$

The triangle inequality applied to (11.37) yields the conclusion. □

Study of the Threshold Factor

Definition 11.12. The largest ϱ satisfying (11.35) is called the *threshold-factor* of $R(z)$.

Example 11.13. The implicit Euler method, for which

$$R^{(j)}(x) = \frac{j!}{(1-x)^{j+1}}, \qquad j = 0, 1, 2, \ldots,$$

satisfies (11.35) for all $\varrho > 0$. It possesses a threshold-factor $\varrho = \infty$.

Example 11.14 (Threshold-factor for Padé-approximations). The derivatives of the polynomials

$$R_{k0}(z) = 1 + z + \frac{z^2}{2!} + \ldots + \frac{z^k}{k!}$$

are easily calculated; the most dangerous one is $1 + z$, therefore $\varrho = 1$ for all k.

The Padé approximations $R_{k1}(z)$ possess one simple pole only, so they can be written in the form

$$R_{k1}(z) = \frac{a}{1 - bz} + \text{polynomial in } z,$$

which has only a finite number of derivatives which can change sign (see Example 11.13). The numerical values obtained are shown in Table 11.1.

The functions $R_{k2}(z)$ possess no real pole (see Sect. IV.4). But the property $|R(z)| \le R(-\varrho + r)$ for $|z + \varrho| \le r$ (see proof of Theorem 11.10) means that the maximum of $|R(z)|$ on the circle with center $-\varrho$ and radius r is assumed to the right on the real axis. For increasing r, the first pole met by this circle must therefore be real and to the right of $-\varrho$. This is not possible here and therefore the approximations $R_{k2}(z)$ *never* satisfy property (11.35). This is indicated by an asterisk ($*$) in Table 11.1.

All further values of Table 11.1 were computed using the decomposition of $R(z)$ into partial fractions and are cited from Kraaijevanger (1986) and van de Griend & Kraaijevanger (1986).

Table 11.1. Threshold-factors of Padé approximations

k	0	1	2	3	4	5	6
$j = 0$	—	1	1	1	1	1	1
$j = 1$	∞	2	2.196	2.350	2.477	2.586	2.682
$j = 2$	$*$	$*$	$*$	$*$	$*$	$*$	$*$
$j = 3$	0.584	1.195	1.703	2.208	2.710	3.212	3.713
$j = 4$	$*$	$*$	$*$	$*$	$*$	$*$	$*$
$j = 5$	0.353	0.770	1.081	1.424	1.794	2.185	2.590

It is curious to observe that in this table the methods with the largest threshold-factors are precisely those which are not A-stable. An exception is the implicit Euler method ($k = 0, j = 1$) for which $\varrho = \infty$.

Absolutely Monotonic Functions

> ... on peut définir la fonction e^x comme la seule fonction absolument monotone sur tout le demi-axe négatif qui prend à l'origine, ainsi que sa dérivée première [*sic*] la valeur un.
>
> (S. Bernstein 1928)

A thorough study of real functions satisfying (11.35) was begun by S. Bernstein (1914) and continued by F. Hausdorff (1921). Such functions are called *absolutely monotonic* in $[-\varrho, 0]$. Later, S. Bernstein (1928) gave the following characterization of functions which are absolutely monotonic in $(-\infty, 0]$ (see also D.V. Widder 1946).

Theorem 11.15 (Bernstein 1928). *A necessary and sufficient condition that $R(x)$ be absolutely monotonic in $(-\infty, 0]$ is that*

$$R(x) = \int_0^\infty e^{xt} d\alpha(t), \tag{11.38}$$

where $\alpha(t)$ is bounded and non-decreasing and the integral converges for $-\infty < x \leq 0$.

This is a hard result and the main key for the next two theorems. It does not seem to permit an elementary and easy proof. We therefore refer to the original literature, S. Bernstein (1928). For a more recent description see e.g. Widder (1946), p. 160. From this result we immediately get the "limit case $\lambda \to \infty$" of Theorem 11.11, which also holds for an arbitrary norm.

Theorem 11.16. *Let $R(x)$ be absolutely monotonic in $(-\infty, 0]$, $R(0) = 1$ and A a matrix with non-positive logarithmic norm $\mu(A) \leq 0$, then*

$$\|R(A)\| \leq 1.$$

Proof. By Theorem I.10.6 we have for the solution $y(x) = e^{Ax} y_0$ of $y' = Ay$ that $\|y(x)\| \leq \|y_0\|$, hence also $\|e^{Ax}\| \leq 1$ for $x \geq 0$. The statement now follows from

$$\|R(A)\| = \| \int_0^\infty e^{At} d\alpha(t) \| \leq \int_0^\infty \|e^{At}\| d\alpha(t) \leq \int_0^\infty d\alpha(t) = R(0) = 1$$

since $\alpha(t)$ is non-decreasing. □

The following result proves that no Runge-Kutta method of order $p > 1$ can have a stability function which is absolutely monotonic in $(-\infty, 0]$.

Theorem 11.17. *If $R(x)$ is absolutely monotonic in $(-\infty, 0]$ and*

$$R(x) = 1 + x + x^2/2 + \mathcal{O}(x^3) \qquad \textit{for} \quad x \to 0,$$

then $R(x) = e^x$.

Proof (Bolley & Crouzeix 1978). It follows from (11.38) that

$$R^{(j)}(0) = \int_0^\infty t^j \, d\alpha(t).$$

Since $R(0) = R'(0) = R''(0) = 1$, this yields

$$\int_0^\infty (1-t)^2 \, d\alpha(t) = 0.$$

Consequently, $\alpha(t)$ must be the Heaviside function ($\alpha(t) = 0$ for $t \le 1$ and $\alpha(t) = 1$ for $t > 1$). Inserted into (11.38) this gives $R(x) = e^x$. □

Exercises

1. Prove Formula (11.14). For given x, study the set of y-values for which $|R(x+iy)|$ attains its maximum.

2. Show that the error growth function (11.11) for an A-stable $R(z)$ of order $p \ge 1$ satisfies

$$\varphi_R(x) > e^x \quad \text{for all} \quad x \ne 0.$$

   Hint. You can study the order star on parallel lines $\{x+iy, y \in \mathbb{R}\}$ (Hairer, Bader & Lubich 1982), or you can use the fact that $\varphi_R(x)$ is superexponential.

3. (Hairer & Zennaro 1996). Let $|R(\infty)| < 1$ and consider a mesh $x_0, x_1, \ldots$ with step sizes $h_k = x_{k+1} - x_k$ satisfying $h_{k+1} \le c h_k$ $(c > 1)$. Prove the existence of constants $C > 0$ and $\alpha > 0$ such that

$$\|y_m\| \le C(x_m - x_0)^{-\alpha} \|y_0\| \qquad \text{for} \quad m = 1, 2, \ldots .$$

4. (Kraaijevanger 1986). Let $R(z)$ be a polynomial of degree s satisfying $R(z) = e^z + \mathcal{O}(z^{p+1})$. Then the threshold factor ϱ (Definition 11.11) is restricted by

$$\varrho \le s - p + 1.$$

   Hint. Justify the formula

$$R^{(p-1)}(z) = \sum_{j=0}^{s-p+1} \alpha_j \Big(1 + \frac{z}{\varrho}\Big)^j, \quad \alpha_j \ge 0$$

   and deduce the result from $R^{(p-1)}(0) = R^{(p)}(0) = 1$.

5. Let ϱ be the threshold factor of the rational function $R(z)$. Show that its stability domain contains the disc $|z + \varrho| \le \varrho$.

IV.12 B-Stability and Contractivity

> Next we need a generalization of the notion of A-stability. The most natural generalization would be to consider the case that $x(t)$ is a uniform-asymptotically stable solution ... in the sense of Liapunov theory ... but this case seems to be a little too wide.
>
> (G. Dahlquist 1963)

> The theoretical analysis of the application of numerical methods on stiff nonlinear problems is still fairly incomplete.
>
> (G. Dahlquist 1975)

Here we enter a new era, the study of stability and convergence for general *non-linear* systems. All the "crimes" and diverse omissions of which we have been guilty in earlier sections, especially in Sect. IV.2, shall now be repaired.

Large parts of Dahlquist's (1963) paper deal with a generalization of A-stability to nonlinear problems. His search for a sufficiently general class of nonlinear systems was finally successful 12 years later. In his talk at the Dundee conference of July 1975 he proposed to consider differential equations satisfying a one-sided Lipschitz condition, and he presented some first results for multistep methods. J.C. Butcher (1975) then extended (on the flight back from the conference) the ideas to implicit Runge-Kutta methods and the concept of B-stability was born.

One-Sided Lipschitz Condition

We consider the nonlinear differential equation

$$y' = f(x, y) \tag{12.1}$$

such that for the Euclidean norm the *one-sided Lipschitz condition*

$$\langle f(x,y) - f(x,z), y - z \rangle \le \nu \, \|y - z\|^2 \tag{12.2}$$

holds. The number ν is the *one-sided Lipschitz constant* of f. This definition is motivated by the following result.

Lemma 12.1. *Let $f(x,y)$ be continuous and satisfy (12.2). Then, for any two solutions $y(x)$ and $z(x)$ of (12.1) we have*

$$\|y(x) - z(x)\| \le \|y(x_0) - z(x_0)\| \cdot e^{\nu(x-x_0)} \qquad \textit{for} \quad x \ge x_0 .$$

Proof. Differentiation of $m(x) = \|y(x) - z(x)\|^2$ yields

$$m'(x) = 2\big\langle f(x,y(x)) - f(x,z(x)),\ y(x) - z(x) \big\rangle \le 2\nu \, m(x).$$

This differential inequality can be solved to give (see Theorem I.10.3)

$$m(x) \le m(x_0) e^{2\nu(x-x_0)} \qquad \text{for } x \ge x_0,$$

which is equivalent to the statement. □

Remarks. a) In an open convex set, condition (12.2) is equivalent to $\mu(\frac{\partial f}{\partial y}) \leq \nu$ (see Sect. I.10, Exercise 6), if f is continuously differentiable. Lemma 12.1 then becomes a special case of Theorem I.10.6.

b) For complex-valued y and f condition (12.2) has to be replaced by

$$\text{Re}\,\langle f(x,y) - f(x,z), y - z\rangle \leq \nu\,\|y-z\|^2, \qquad y,z \in \mathbb{C}^n, \tag{12.2'}$$

and Lemma 12.1 remains valid.

B-Stability and Algebraic Stability

Whenever $\nu \leq 0$ in (12.2) the distance between any two solutions of (12.1) is a non-increasing function of x. The same property is then also desirable for the numerical solutions. We consider here implicit Runge-Kutta methods

$$y_1 = y_0 + h\sum_{i=1}^{s} b_i f(x_0 + c_i h, g_i), \tag{12.3a}$$

$$g_i = y_0 + h\sum_{j=1}^{s} a_{ij} f(x_0 + c_j h, g_j), \qquad i = 1,\ldots,s. \tag{12.3b}$$

Definition 12.2 (Butcher 1975). A Runge-Kutta method is called B-*stable*, if the contractivity condition

$$\langle f(x,y) - f(x,z), y - z\rangle \leq 0 \tag{12.2''}$$

implies for all $h \geq 0$

$$\|y_1 - \widehat{y}_1\| \leq \|y_0 - \widehat{y}_0\|.$$

Here, y_1 and $\widehat{y}_1$ are the numerical approximations after one step starting with initial values y_0 and $\widehat{y}_0$, respectively.

Clearly, B-stability implies A-stability. This is seen by applying the above definition to $y' = \lambda y, \lambda \in \mathbb{C}$ or, more precisely, to

$$\begin{pmatrix} y_1' \\ y_2' \end{pmatrix} = \begin{pmatrix} \alpha & -\beta \\ \beta & \alpha \end{pmatrix}\begin{pmatrix} y_1 \\ y_2 \end{pmatrix}. \tag{12.4}$$

Example 12.3. For the collocation methods based on Gaussian quadrature a simple proof of B-stability is possible (Wanner 1976). We denote by $u(x)$ and $\widehat{u}(x)$ the collocation polynomials (see Definition II.7.6) for the initial values y_0 and $\widehat{y}_0$ and differentiate the function $m(x) = \|u(x) - \widehat{u}(x)\|^2$. At the collocation points $\xi_i = x_0 + c_i h$ we obtain

$$m'(\xi_i) = 2\big\langle f(\xi_i, u(\xi_i)) - f(\xi_i, \widehat{u}(\xi_i)),\ u(\xi_i) - \widehat{u}(\xi_i)\big\rangle \leq 0.$$

The result then follows from the fact that Gaussian quadrature integrates the polynomial $m'(x)$ (which is of degree $2s-1$) exactly and that the weights b_i are positive:

$$\|y_1-\widehat{y}_1\|^2 = m(x_0+h) = m(x_0)+\int_{x_0}^{x_0+h} m'(x)\,dx$$
$$= m(x_0)+h\sum_{i=1}^{s} b_i m'(x_0+c_i h) \le m(x_0) = \|y_0-\widehat{y}_0\|^2.$$

An *algebraic criterion* for B-stability was found independently by Burrage & Butcher (1979) and Crouzeix (1979). The result is

Theorem 12.4. *If the coefficients of a Runge-Kutta method (12.3) satisfy*

i) $b_i \ge 0$ *for* $i=1,\dots,s$,
ii) $M=(m_{ij})=(b_i a_{ij}+b_j a_{ji}-b_i b_j)_{i,j=1}^{s}$ *is non-negative definite,*

then the method is B*-stable.*

Definition 12.5. A Runge-Kutta method, satisfying (i) and (ii) of Theorem 12.4, is called *algebraically stable.*

Proof of Theorem 12.4. We introduce the differences

$$\Delta y_0 = y_0-\widehat{y}_0, \qquad \Delta y_1 = y_1-\widehat{y}_1, \qquad \Delta g_i = g_i-\widehat{g}_i,$$
$$\Delta f_i = h\big(f(x_0+c_i h, g_i)-f(x_0+c_i h,\widehat{g}_i)\big),$$

and subtract the Runge-Kutta formulas (12.3) for y and $\widehat{y}$

$$\Delta y_1 = \Delta y_0+\sum_{i=1}^{s} b_i\,\Delta f_i, \tag{12.5a}$$

$$\Delta g_i = \Delta y_0+\sum_{j=1}^{s} a_{ij}\,\Delta f_j. \tag{12.5b}$$

Next we take the square of Formula (12.5a)

$$\|\Delta y_1\|^2 = \|\Delta y_0\|^2+2\sum_{i=1}^{s} b_i\langle \Delta f_i,\Delta y_0\rangle+\sum_{i=1}^{s}\sum_{j=1}^{s} b_i b_j\langle \Delta f_i,\Delta f_j\rangle. \tag{12.6}$$

The main idea of the proof is now to compute Δy_0 from (12.5b) and insert this into (12.6). This gives

$$\|\Delta y_1\|^2 = \|\Delta y_0\|^2+2\sum_{i=1}^{s} b_i\langle \Delta f_i,\Delta g_i\rangle-\sum_{i=1}^{s}\sum_{j=1}^{s} m_{ij}\langle \Delta f_i,\Delta f_j\rangle. \tag{12.7}$$

The statement now follows from the fact that $\langle \Delta f_i,\Delta g_i\rangle \le 0$ by (12.2") and that $\sum_{i,j=1}^{s} m_{ij}\langle \Delta f_i,\Delta f_j\rangle \ge 0$ (see Exercise 2). □

Example 12.6. For the SDIRK method of Table 7.2 (Chapter II) the weights b_i are seen to be positive and the matrix M becomes

$$M = (\gamma - 1/4) \cdot \begin{pmatrix} 1 & -1 \\ -1 & 1 \end{pmatrix}.$$

For $\gamma \geq 1/4$ this matrix is non-negative definite and we have B-stability. Exactly the same condition was obtained by studying its A-stability (c.f. (3.10)).

Some Algebraically Stable IRK Methods

La première de ces propriétés consiste en ce que tous les A_k sont positifs. (T.-J. Stieltjes 1884)

The general study of algebraic stability falls naturally into two steps: the positivity of the quadrature weights and the nonnegative-definitness of the matrix M.

Theorem 12.7. *Consider a quadrature formula* $(c_i, b_i)_{i=1}^s$ *of order* p.

a) If $p \geq 2s-1$ *then* $b_i > 0$ *for all* i.

b) If c_i *are the zeros of (5.3) (Lobatto quadrature) then* $b_i > 0$ *for all* i.

Proof (Stieltjes 1884). The first statement follows from the fact that for $p \geq 2s-1$ polynomials of degree $2s-2$ are integrated exactly, hence

$$b_i = \int_0^1 \prod_{j \neq i} \Big(\frac{x - c_j}{c_i - c_j} \Big)^2 dx > 0. \tag{12.8}$$

In the case of the Lobatto quadrature ($c_1 = 0$, $c_s = 1$ and $p = 2s-2$) the factors for the indices $j = 1$ and $j = s$ are taken without squaring and the same argument applies. □

In order to verify condition (ii) of Theorem 12.4 we find it convenient to use the W-transformation of Sect. IV.5 and to consider $W^T M W$ instead of M. In vector notation ($b = (b_1, \ldots, b_s)^T$, $B = \operatorname{diag}(b_1, \ldots, b_s)$, $A = (a_{ij})$) we have

$$M = BA + A^T B - bb^T. \tag{12.9}$$

If we choose W according to Lemma 5.12, then $W^T B W = I$ and, since $W^T b = e_1 = (1, 0, \ldots, 0)^T$, condition (ii) becomes equivalent to

$$W^T M W = X + X^T - e_1 e_1^T \quad \text{is non-negative definite} \tag{12.10}$$

where $X = W^{-1} A W = W^T B A W$ as in Theorem 5.11.

Theorem 12.8. *Suppose that a Runge-Kutta method with distinct* c_i *and positive* b_i *satisfies the simplifying assumptions* $B(2s-2), C(s-1), D(s-1)$ *(see beginning of Sect. IV.5). Then the method is algebraically stable if and only if* $|R(\infty)| \leq 1$ *(where* $R(z)$ *denotes the stability function).*

Proof. Since the order of the quadrature formula is $p \geq 2s-2$, the matrix W of Lemma 5.12 is

$$W = W_G D, \qquad D = \mathrm{diag}(1,\ldots,1,\alpha^{-1}), \tag{12.11}$$

where $W_G = (P_{j-1}(c_i))_{i,j=1}^s$ is as in (5.21), and $\alpha^2 = \sum_{i=1}^s b_i P_{s-1}^2(c_i) \neq 0$. Using the relation (observe that $W^T B W = I$)

$$X = W^{-1}AW = D^{-1}W_G^{-1}AW_G D = DW_G^T BA(W_G^T B)^{-1}D^{-1}$$

and applying Lemma 5.7 with $\eta = s-1$ and Lemma 5.8 with $\xi = s-1$, we obtain

$$X = \begin{pmatrix} 1/2 & -\xi_1 & & & \\ \xi_1 & 0 & \ddots & & \\ & \ddots & \ddots & -\xi_{s-2} & \\ & & \xi_{s-2} & 0 & -\alpha\xi_{s-1} \\ & & & \alpha\xi_{s-1} & \beta \end{pmatrix}.$$

If this matrix is inserted into (12.10) then, marvellous surprise, everything cancels with the exception of β. Therefore, condition (ii) of Theorem 12.4 is equivalent to $\beta \geq 0$.

Using the representation (5.31) of the stability function we obtain by developing the determinants

$$|R(\infty)| = \left|\frac{\det(X - e_1 e_1^T)}{\det X}\right| = \left|\frac{\beta d_{s-1} - \alpha^2\xi_{s-1}^2 d_{s-2}}{\beta d_{s-1} + \alpha^2\xi_{s-1}^2 d_{s-2}}\right|, \tag{12.12}$$

where $d_k = k!/(2k)!$ is the determinant of the k-dimensional matrix X_G of (5.13). Since $\alpha^2\xi_{s-1}^2 d_{s-2} > 0$, the expression (12.12) is bounded by 1 iff $\beta \geq 0$. This proves the statement. □

Comparing these theorems with Table 5.13 yields

Theorem 12.9. *The methods Gauss, Radau IA, Radau IIA and Lobatto IIIC are algebraically stable and therefore also B-stable.* □

AN-Stability

A-stability theory is based on the autonomous linear equation $y' = \lambda y$, whereas B-stability is based on general nonlinear systems $y' = f(x,y)$. The question arises whether there is a reasonable stability theory *between* these two extremes. A natural approach would be to study the scalar, linear, nonautonomous equation

$$y' = \lambda(x)y, \qquad \mathrm{Re}\,\lambda(x) \leq 0, \tag{12.13}$$

where $\lambda(x)$ is an arbitrarily varying complex-valued function (Burrage & Butcher 1979, Scherer 1979). The somewhat surprising result of this subsection will be that stability for (12.13) will, for most RK-methods, be equivalent to B-stability.

For the problem (12.13) the Runge-Kutta method (12.3) becomes (in vector notation $g=(g_1,\ldots,g_s)^T$, $1\!\!1=(1,\ldots,1)^T$)

$$g = 1\!\!1 y_0 + AZg, \quad Z=\operatorname{diag}(z_1,\ldots,z_s), \quad z_j = h\lambda(x_0+c_jh). \tag{12.14}$$

Computing g from (12.14) and inserting into (12.3a) gives

$$y_1 = K(Z)y_0, \quad K(Z) = 1 + b^T Z(I-AZ)^{-1}1\!\!1. \tag{12.15}$$

Definition 12.10. A Runge-Kutta method is called *AN-stable*, if

$$|K(Z)| \le 1 \quad \begin{cases} \text{for all } Z=\operatorname{diag}(z_1,\ldots,z_s) \text{ satisfying } \operatorname{Re} z_j \le 0 \\ \text{and } z_j = z_k \text{ whenever } c_j = c_k \ \ (j,k=1,\ldots,s). \end{cases}$$

Comparing (12.15) with (3.2) we find that

$$K\big(\operatorname{diag}(z,z,\ldots,z)\big) = R(z), \tag{12.16}$$

the usual stability function. Further, arguing as with (12.4), B-stability implies AN-stability. Therefore we have:

Theorem 12.11. *For Runge-Kutta methods it holds*

$$B\text{-stable} \quad \Rightarrow \quad AN\text{-stable} \quad \Rightarrow \quad A\text{-stable}. \qquad \square$$

For the trapezoidal rule $y_1 = y_0 + \frac{h}{2}\big(f(x_0,y_0)+f(x_1,y_1)\big)$ the function $K(Z)$ of (12.15) is given by

$$K(Z) = \frac{1+z_1/2}{1-z_2/2}. \tag{12.17}$$

Putting $z_2=0$ and $z_1\to-\infty$ we see that this method is not AN-stable. More generally we have the following result.

Theorem 12.12 (Scherer 1979). *The Lobatto IIIA and Lobatto IIIB methods are not AN-stable and therefore not B-stable.*

Proof. As in Proposition 3.2 we find that

$$K(Z) = \frac{\det(I-(A-1\!\!1 b^T)Z)}{\det(I-AZ)}. \tag{12.18}$$

By definition, the first row of A and the last row of $A-1\!\!1 b^T$ vanish for the Lobatto IIIA methods (compare also the proof of Theorem 5.5). Therefore the denominator of $K(Z)$ does not depend on z_1 and the numerator not on z_s. If we put for example $z_2=\ldots=z_s=0$, the function $K(Z)$ is unbounded for $z_1\to-\infty$. This contradicts AN-stability.

For the Lobatto IIIB methods, one uses in a similar way that the last column of A and the first column of $A-1\!\!1 b^T$ vanish. $\square$

The following result shows, as mentioned above, that AN-stability is much closer to B-stability than to A-stability.

Theorem 12.13 (Burrage & Butcher 1979). *Suppose that*

$$|K(Z)| \leq 1 \qquad \begin{cases} \textit{for all } Z = \operatorname{diag}(z_1, \ldots, z_s) \textit{ with } \operatorname{Re} z_j \leq 0 \\ \textit{and } |z_j| \leq \varepsilon \textit{ for some } \varepsilon > 0, \end{cases} \tag{12.19}$$

then the method is algebraically stable (and hence also B-stable).

Proof. For $\Delta f_i := z_i \Delta g_i$ and $\Delta y_0 = 1$ the result of (12.5) is $\Delta y_1 = K(Z)$. Taking care of the fact that z_i need not be real, the computation of the proof of Theorem 12.4 shows that

$$|K(Z)|^2 - 1 = 2\sum_{i=1}^{s} b_i \operatorname{Re} z_i |g_i|^2 - \sum_{i,j=1}^{s} m_{ij} \bar{z}_i \bar{g}_i z_j g_j. \tag{12.20}$$

Here $g = (g_1, \ldots, g_s)^T$ is a solution of (12.14) with $y_0 = 1$.

To prove that $b_i \geq 0$, choose $z_i = -\varepsilon < 0$ and $z_j = 0$ for $j \neq i$. Assumption (12.19) together with (12.20) implies

$$-2\varepsilon b_i |g_i|^2 - m_{ii}\varepsilon^2 |g_i|^2 \leq 0. \tag{12.21}$$

For sufficiently small ε, g_i is close to 1 and the second term in (12.21) is negligible for $b_i \neq 0$. Therefore, b_i must be non-negative.

To verify the second condition of algebraic stability we choose the purely imaginary numbers $z_j = i\varepsilon\xi_j$ ($\xi_j \in \mathbb{R}$). Since again $g_i = 1 + \mathcal{O}(\varepsilon)$ for $\varepsilon \to 0$, we have from (12.20) that

$$-\varepsilon^2 \sum_{i,j=1}^{s} m_{ij} \xi_i \xi_j + \mathcal{O}(\varepsilon^3) \leq 0.$$

Therefore, $M = (m_{ij})$ has to be non-negative definite. □

Combining this result with those of Theorems 12.4 and 12.11 we obtain

Corollary 12.14. *For non-confluent Runge-Kutta methods (i.e., methods with all c_j distinct) the concepts of AN-stability, B-stability and algebraic stability are equivalent.* □

An equivalence result (between B- and algebraic stability) for *confluent* Runge-Kutta methods is much more difficult to prove (see Theorem 12.18 below) and will be our next goal. To this end we first have to discuss *reducible* methods.

Reducible Runge-Kutta Methods

For an RK-method (12.3) it may happen that for all differential equations (12.1)

i) some stages don't influence the numerical solution;

ii) several g_i are identical.

In both situations the Runge-Kutta method can be simplified to an "equivalent" one with fewer stages.

For an illustration of situation (i) consider the method of Table 12.1. Its numerical solution is independent of g_2 and equivalent to the implicit Euler solution. For the method of Table 12.2 one easily verifies that $g_1 = g_2$, whenever the system (12.3b) possesses a unique solution. The method is thus equivalent to the implicit mid-point rule.

Table 12.1.
DJ-reducible method

1	1	0
1/2	1/4	1/4
	1	0

Table 12.2.
S-reducible method

1/2	1/2	0
1/2	1/4	1/4
	1/2	1/2

The situation (i) above can be made more precise as follows:

Definition 12.15 (Dahlquist & Jeltsch 1979). A Runge-Kutta method is called *DJ-reducible*, if for some non-empty index set $T \subset \{1,\ldots,s\}$,

$$b_j = 0 \quad \text{for} \quad j \in T \qquad \text{and} \qquad a_{ij} = 0 \quad \text{for} \quad i \notin T, j \in T. \tag{12.22}$$

Otherwise it is called *DJ-irreducible.*

Condition (12.22) implies that the stages $j \in T$ don't influence the numerical solution. This is best seen by permuting the stages so that the elements of T are the last ones (Cooper 1985). Then the Runge-Kutta tableau becomes that of Table 12.3.

Table 12.3. DJ-reducibility

$$\begin{array}{c|cc} c_1 & A_{11} & 0 \\ c_2 & A_{21} & A_{22} \\ \hline & b_1^T & 0 \end{array} \qquad \Rightarrow \qquad \begin{array}{c|c} c_1 & A_{11} \\ \hline & b_1^T \end{array}$$

An interesting property of DJ-irreducible and algebraically stable Runge-Kutta methods was discovered by Dahlquist & Jeltsch (1979).

Theorem 12.16. *A DJ-irreducible, algebraically stable Runge-Kutta method satisfies*

$$b_i > 0 \quad \textit{for} \quad i = 1,\ldots,s.$$

Proof. Suppose $b_j = 0$ for some index j. Then $m_{jj} = 0$ by definition of M. Since M is non-negative definite, all elements in the jth column of M must vanish (Exercise 11) so that $b_i a_{ij} = 0$ for all i. This implies (12.22) for the set $T = \{j | b_j = 0\}$, a contradiction to DJ-irreducibility. □

An algebraic criterion for the situation (ii) was given for the first time (but incompletely) by Stetter (1973, p. 127) and finally by Hundsdorfer & Spijker (1981), see also Butcher (1987), p. 319, and Dekker & Verwer (1984), p. 108.

Definition 12.17. A Runge-Kutta method is *S-reducible*, if for some partition $(S_1, \ldots, S_r)$ of $\{1, \ldots, s\}$ with $r < s$ we have for all l and m

$$\sum_{k \in S_m} a_{ik} = \sum_{k \in S_m} a_{jk} \quad \text{if} \quad i, j \in S_l. \tag{12.23}$$

Otherwise it is called *S-irreducible*. Methods which are neither DJ-reducible nor S-reducible are called *irreducible*.

In order to understand condition (12.23) we assume that, after a certain permutation of the stages, $l \in S_l$ for $l = 1, \ldots, r$. We then consider the r-stage method with coefficients

$$c_i^* = c_i, \qquad a_{ij}^* = \sum_{k \in S_j} a_{ik}, \qquad b_j^* = \sum_{k \in S_j} b_k. \tag{12.24}$$

Application of this new method to (12.1) yields $g_1^*, \ldots, g_r^*, y_1^*$ and one easily verifies that g_i and y_1 defined by

$$g_i = g_l^* \quad \text{if} \quad i \in S_l, \qquad y_1 = y_1^*,$$

are a solution of the original method (12.3). For the method of Table 12.2 we have $S_1 = \{1, 2\}$. A further example of an S-reducible method is given in Table 12.4 of Sect. II.12 ($S_1 = \{1, 2, 3\}$ and $S_2 = \{4\}$).

The Equivalence Theorem for S-Irreducible Methods

Theorem 12.18 (Hundsdorfer & Spijker 1981). *For S-irreducible Runge-Kutta methods,*

$$B\text{-stable} \quad \Longleftrightarrow \quad \text{algebraically stable.}$$

Proof. Because of Corollary 12.14, which covers nearly all cases of practical importance — and which was much easier to prove — this theorem seems to be of little practical interest. However, it is an interesting result which had been conjectured by many people for many years, so we reproduce its proof, which also includes the three Lemmas 12.19-12.21. The counter example of Exercise 6 below shows that S-irreducibility is a necessary hypethesis.

By Theorem 12.4 it is sufficient to prove that B-stability and S-irreducibility imply algebraic stability. For this we take s complex numbers $z_1, \dots, z_s$ which satisfy $\operatorname{Re} z_j < 0$ and $|z_j| \le \varepsilon$ for some sufficiently small $\varepsilon > 0$. We show that there exists a continuous function $f : \mathbb{C} \to \mathbb{C}$ satisfying

$$\operatorname{Re} \langle f(u) - f(v),\; u - v \rangle \le 0 \quad \text{for all } u, v \in \mathbb{C}\,, \tag{12.25}$$

such that the Runge-Kutta solutions y_1, g_i and $\widehat{y}_1, \widehat{g}_i$ corresponding to $y_0 = 0$, $\widehat{y}_0 = 1$, $h = 1$ satisfy

$$f(\widehat{g}_i) - f(g_i) = z_i(\widehat{g}_i - g_i). \tag{12.26}$$

This yields $\widehat{y}_1 - y_1 = K(Z)$ with $K(Z)$ given by (12.15). B-stability then implies $|K(Z)| \le 1$. By continuity of $K(Z)$ near the origin we then have $|K(Z)| \le 1$ for all z_j which satisfy $\operatorname{Re} z_j \le 0$ and $|z_j| \le \varepsilon$, so that Theorem 12.13 proves the statement.

Construction of the function f: we denote by Δg_i the solution of

$$\Delta g_i = 1 + \sum_{j=1}^{s} a_{ij} z_j \Delta g_j$$

(the solution exists uniquely if $|z_j| \le \varepsilon$ and ε is sufficiently small). With ξ, η given by Lemma 12.19 (below) we define

$$\begin{aligned} g_i &= t\eta_i, & f(g_i) &= t\xi_i \\ \widehat{g}_i &= g_i + \Delta g_i, & f(\widehat{g}_i) &= f(g_i) + z_i \Delta g_i. \end{aligned} \tag{12.27}$$

This is well-defined for sufficiently large t (to be fixed later), because the η_i are distinct. Clearly, g_i and $\widehat{g}_i$ represent a Runge-Kutta solution for $y_0 = 0$ and $\widehat{y}_0 = 1$, and (12.26) is satisfied by definition.

We next show that

$$\operatorname{Re} \langle f(u) - f(v),\; u - v \rangle < 0 \qquad \text{if} \qquad u \ne v \tag{12.28}$$

is satisfied for $u, v \in D = \{g_1, \dots, g_s, \widehat{g}_1, \dots, \widehat{g}_s\}$. This follows from the construction of ξ, η, if $u, v \in \{g_1, \dots, g_s\}$. If $u = g_i$ and $v = \widehat{g}_i$ this is a consequence of (12.26). For the remaining case $u = \widehat{g}_i, v \in D \setminus \{g_i, \widehat{g}_i\}$ we have

$$\langle f(u) - f(v), u - v \rangle = t^2 (\xi_i - \xi_j)(\eta_i - \eta_j) + \mathcal{O}(t) \quad \text{for} \quad t \to \infty,$$

so that (12.28) is satisfied, if t is sufficiently large. Applying Lemma 12.20 below we find a continuous function $f : \mathbb{C} \to \mathbb{C}$ that extends (12.27) and satisfies (12.25). □

To complete the above proof we still need the following three lemmas:

Lemma 12.19. *Let A be the coefficient matrix of an S-irreducible Runge-Kutta method. Then there exist vectors $\xi \in \mathbb{R}^s$ and $\eta = A\xi$ such that*

$$(\xi_i - \xi_j)(\eta_i - \eta_j) < 0 \quad \textit{for} \quad i \ne j. \tag{12.29}$$

Proof (see Butcher 1982). The first idea is to put

$$\xi = 1\!\!1 - \varepsilon A 1\!\!1 \quad \text{with} \quad 1\!\!1 = (1, 1, \ldots, 1)^T,$$

so that η becomes

$$\eta = A\xi = A 1\!\!1 - \varepsilon A^2 1\!\!1.$$

If $c_i \neq c_j$ for all i, j, then $\xi_i - \xi_j \neq 0$ and for ε sufficiently small we have $\eta_i - \eta_j$ of opposite sign, thus (12.29) is true.

For a proof of the remaining cases, we shall construct recursively vectors $v_0, v_1, v_2, \ldots$ and denote by P_k the partition of $\{1, \ldots, s\}$ defined by the equivalence relation

$$i \sim j \quad \Longleftrightarrow \quad (v_q)_i = (v_q)_j \quad \text{for} \quad q = 0, 1, \ldots, k. \tag{12.30}$$

For a given partition P of $\{1, 2, \ldots, s\}$ we introduce the space

$$X(P) = \{v \in \mathbb{R}^s;\ (v)_i = (v)_j \quad \text{if} \quad i \sim j \quad \text{with respect to } P\}.$$

With this notation, the method is S-irreducible if and only if

$$AX(P) \not\subset X(P) \tag{12.31}$$

for every partition other than $\{\{1\}, \{2\}, \ldots, \{s\}\}$.

We start with $v_0 = 1\!\!1$ and $P_0 = \{\{1, \ldots, s\}\}$ and define

$$v_{k+1} = \begin{cases} Av_k & \text{if } Av_k \notin X(P_k) \\ \omega & \text{if } Av_k \in X(P_k) \end{cases}$$

where ω is an arbitrary vector of $X(P_k)$ satisfying $A\omega \notin X(P_k)$. Such a choice is possible by (12.31). After a finite number of steps, say m, we arrive at $P_m = \{\{1\}, \{2\}, \ldots, \{s\}\}$, because the number of components of P_k is increasing, and strictly increasing after every second step. Therefore all elements of the vector

$$\xi = v_0 - \varepsilon v_1 + \varepsilon^2 v_2 - \ldots + (-\varepsilon)^m v_m$$

are distinct (for sufficiently small $\varepsilon > 0$) and (12.29) is satisfied. □

Lemma 12.20 (Minty 1962). *Let $u_1, \ldots, u_k$ and $f(u_1), \ldots, f(u_k)$ be elements of $\mathbb{R}^n$ with*

$$\langle f(u_i) - f(u_j), u_i - u_j \rangle < 0 \qquad \textit{for} \quad i \neq j.$$

Then there exists a continuous extension $f : \mathbb{R}^n \to \mathbb{R}^n$ satisfying

$$\langle f(u) - f(v), u - v \rangle \leq 0 \qquad \textit{for all} \quad u, v \in \mathbb{R}^n.$$

Proof (Wakker 1985). Define

$$\gamma = \max_{i \neq j} \frac{\langle f(u_i) - f(u_j), u_i - u_j \rangle}{\|f(u_i) - f(u_j)\|^2} < 0$$

and put $g(u_i) = 2\gamma f(u_i) - u_i$, so that $\|g(u_i) - g(u_j)\| \le \|u_i - u_j\|$. An application of Lemma 12.21 shows that there exists a continuous extension $g : \mathbb{R}^n \to \mathbb{R}^n$ satisfying $\|g(u) - g(v)\| \le \|u - v\|$ (i.e., g is non-expansive). The function

$$f(u) = \frac{1}{2\gamma}\big(g(u) + u\big)$$

then satisfies the requirements. □

Lemma 12.21 (Kirszbraun 1934). *Let* $u_1, \ldots, u_k$ *and* $g(u_1), \ldots, g(u_k) \in \mathbb{R}^n$ *be such that*

$$\|g(u_i) - g(u_j)\| \le \|u_i - u_j\| \qquad \textit{for } i, j = 1, \ldots, k. \tag{12.32}$$

Then there exists a continuous extension $g : \mathbb{R}^n \to \mathbb{R}^n$ *such that*

$$\|g(u) - g(v)\| \le \|u - v\| \qquad \textit{for all} \quad u, v \in \mathbb{R}^n. \tag{12.33}$$

Proof. This was once a difficult result in set-theory. A particularly nice proof, of which we give here a "dynamic" version, has been found by I.J. Schoenberg (1953).

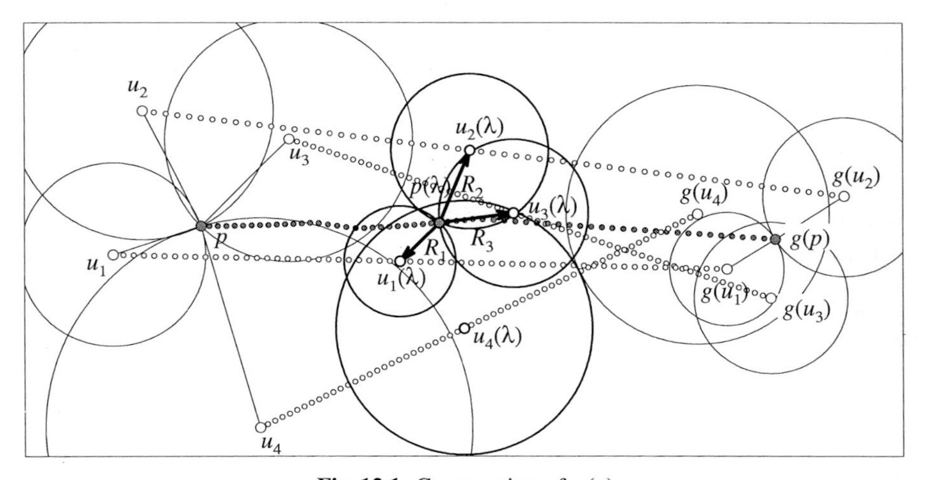

Fig. 12.1. Construction of $g(p)$

a) The main problem is to construct for *one* given point p the extension $g(p)$ such that (12.33) remains satisfied. We move the points u_i into their images $g(u_i)$ by an affine map

$$u_i(\lambda) = u_i + \lambda(g(u_i) - u_i), \qquad 0 \le \lambda \le 1, \;\; i = 1, \ldots, k. \tag{12.34}$$

We define $r_i = \|u_i - p\|$ and shrink, for each λ, the balls with center $u_i(\lambda)$ and radius $r_i\mu$ until their intersection consists of one point only

$$\mu(\lambda) := \min\Big\{\mu \; ; \; \bigcap_{i=1}^{k} \{u \; ; \; \|u_i(\lambda) - u\| \le r_i\mu\} \ne \emptyset\Big\}. \tag{12.35}$$

This intersection point, denoted by $p(\lambda)$ (see Fig. 12.1), depends continuously (except for a possible sudden decrease of μ if $\lambda=0$) and piecewise differentiably on λ. We shall show that $\mu(\lambda)$ is non-increasing, which means that $g(p):=p(1)$ is a point satisfying (12.33).

We denote the vectors

$$R_i := u_i(\lambda) - p(\lambda), \tag{12.36}$$

and have from the hypothesis (12.32) that $\|R_i - R_j\|^2$ is non-increasing, hence that $\langle R_i - R_j, dR_i - dR_j\rangle \le 0$ or

$$\langle R_i, dR_j\rangle + \langle R_j, dR_i\rangle \ge \langle R_i, dR_i\rangle + \langle R_j, dR_j\rangle. \tag{12.37}$$

As can be seen in Fig. 12.1, not all points $u_i(\lambda)$ are always "active" in (12.35), i.e., $p(\lambda)$ lies on the boundary of the shrinked ball centered in $u_i(\lambda)$. While for $\lambda=0$ (for which $\|R_i\| = r_i\mu$) all four are active, at $\lambda = 1/2$ the active points are $u_1(\lambda)$, $u_2(\lambda)$, $u_3(\lambda)$, and finally for $\lambda=1$ we only have $u_1(\lambda)$ and $u_2(\lambda)$ active. We suppose, for a given λ, that $u_1(\lambda),\ldots,u_m(\lambda)$ $(m\le k)$ are the active points, which may sometimes require a proper renumbering. The crucial idea of Schoenberg is the fact that $p(\lambda)$ lies in the convex hull of $u_1(\lambda),\ldots,u_m(\lambda)$, i.e., there are positive values $c_1(\lambda),\ldots,c_m(\lambda)$ with $\sum_{i=1}^m c_iR_i = 0$. This means that

$$\langle \textstyle\sum_i c_iR_i, \sum_j c_j\, dR_j\rangle = 0.$$

We here apply (12.37) pairwise to i,j and j,i, and obtain

$$0 = \langle \textstyle\sum_i c_iR_i, \sum_j c_j\, dR_j\rangle \ge \sum_i \langle R_i, dR_i\rangle \big(c_i \sum_j c_j\big).$$

Since by construction (see (12.36)) all $\|R_i\|$ decrease or increase simultaneously with μ, and since all $c_i>0$, we see that $d\mu\le 0$, i.e., μ is non-increasing.

b) The rest is now standard (Kirszbraun): we choose a countable dense sequence of points $p_1,p_2,p_3,\ldots$ in $\mathbb{R}^n$ and extend g gradually to these points, so that (12.33) is always satisfied. By continuity (see (12.33)), our function is then defined everywhere. This completes the proof of Lemma 12.21 and with it the proof of Theorem 12.18. □

> Nous ne connaissons pas d'exemples de méthodes qui soient B-stables au sens de Butcher et qui ne soient pas B-stables suivant notre définition. (M. Crouzeix 1979)

Remark. Burrage & Butcher (1979) distinguish between BN-stability (based on non-autonomous systems) and B-stability (based on autonomous systems). Since the differential equation constructed in the above proof (see (12.25)) is *autonomous*, both concepts are equivalent for irreducible methods.

Error Growth Function

All the above theory deals only with contractivity when the one-sided Lipschitz constant ν in (12.2) is zero (see Definition 12.2). The question arises whether we can sharpen the estimate when it is known that $\nu < 0$, and whether we can obtain estimates also in the case when (12.2) holds only for some $\nu > 0$.

Definition 12.22 (Burrage & Butcher 1979). Let ν be given and set $x = h\nu$, where h is the step size. We then denote by $\varphi_B(x)$ the smallest number for which the estimate

$$\|y_1 - \widehat{y}_1\| \le \varphi_B(x)\, \|y_0 - \widehat{y}_0\| \tag{12.38}$$

holds for all problems satisfying

$$\mathrm{Re}\,\big\langle f(x,y) - f(x,z),\, y - z\big\rangle \le \nu\, \|y - z\|^2. \tag{12.39}$$

We call $\varphi_B(x)$ the *error growth function* of the method.

We consider here complex-valued functions $f : \mathbb{R} \times \mathbb{C}^n \to \mathbb{C}^n$. This is not more general (any such system can be written in real form by considering real and imaginary parts, see Eq. (12.4)), but it is more convenient when working with problems $y' = \lambda(x)y$, where $\lambda(x)$ is complex-valued.

In the case of a linear nonautonomous problem $y' = A(x)y$, condition (12.39) becomes $\mu\big(A(x)\big) \le \nu$ (where $\mu(\cdot)$ denotes the logarithmic norm; see Sect. I.10). Putting $Z_i := hA(x_0 + c_i h)$, the difference of two numerical solutions becomes

$$y_1 - \widehat{y}_1 = K(Z_1, \ldots, Z_s)(y_0 - \widehat{y}_0),$$

where

$$K(Z_1, \ldots, Z_s) = I + (b^T \otimes I)Z\big(I \otimes I - (A \otimes I)Z\big)^{-1}(1\!\!1 \otimes I), \tag{12.40}$$

and Z is the block diagonal matrix with $Z_1, \ldots, Z_s$ as entries in the diagonal.

Theorem 12.23. *The error growth function of an implicit Runge-Kutta method satisfies*

$$\varphi_B(x) = \sup_{\mu(Z_1)\le x, \ldots, \mu(Z_s)\le x} \|K(Z_1, \ldots, Z_s)\|. \tag{12.41}$$

Proof. Upper Bound. The difference $\Delta y_1 = y_1 - \widehat{y}_1$ of two Runge-Kutta solutions satisfies (12.5). The assumption (12.39) implies that $\mathrm{Re}\,\langle \Delta f_i, \Delta g_i\rangle \le x\|\Delta g_i\|^2$. We shall prove that there exist matrices Z_i $(i = 1, \ldots, s)$ with $\mu(Z_i) \le x$ such that $\Delta f_i = Z_i \Delta g_i$. This implies $\Delta y_1 = K(Z_1, \ldots, Z_s)\Delta y_0$ and, as a consequence, that the right-hand expression of Eq. (12.41) is an upper bound of $\varphi_B(x)$.

If $\Delta g_i = 0$ then $\Delta f_i = 0$ and we can take an arbitrary matrix satisfying $\mu(Z_i) \le x$. Therefore, let us consider vectors f, g (with $g \ne 0$) in $\mathbb{C}^n$ satisfying $\mathrm{Re}\,\langle f, g\rangle \le x\|g\|^2$. We put $u_1 := g/\|g\|$, and complete it to an orthonormal basis $u_1, \ldots, u_n$

of $\mathbb{C}^n$. Then we define the matrix Z by

$$Zu_1 := f/\|g\|, \qquad Zu_i := xu_i - \langle u_i, f\rangle u_1/\|g\|, \qquad i = 2,\ldots,n.$$

We have $Zg = f$, and one readily verifies that $\mathrm{Re}\,\langle Zv, v\rangle \le x\|v\|^2$ for all $v = \sum_{i=1}^n \alpha_i u_i$.

Lower Bound. We first consider nonconfluent Runge-Kutta methods. For given $Z_1,\ldots,Z_s$ with $\mu(Z_i)\le x$ let $A(x)$ be a continuous function satisfying $hA(x_0 + c_i h) = Z_i$ and $\mu(A(x)) \le x$ for all x ($A(x)$ is, for example, obtained by linear interpolation). Then we have $\Delta y_1 = K(Z_1,\ldots,Z_s)\Delta y_0$ and, consequently, also $\varphi_B(x) \ge \|K(Z_1,\ldots,Z_s)\|$ for all $Z_1,\ldots,Z_s$ with $\mu(Z_i)\le x$.

For confluent methods the proof is more complicated. Without loss of generality we can assume that the method is irreducible, because neither the value $\varphi_B(x)$ nor the right-hand expression of Eq. (12.41) change, when the method is replaced by an equivalent one. The main observation is now that the Lemmata 12.20 and 12.21 are valid in arbitrary dimensions. Consider $Z_1,\ldots,Z_s$ with $\mu(Z_i)\le x$, such that the linear system $\Delta g_i = \Delta y_0 + \sum_{j=1}^s a_{ij} Z_j \Delta g_j$ has a solution. Exactly as in the proof of Theorem 12.18 we can construct a continuous function $f:\mathbb{C}^n \to \mathbb{C}^n$, which satisfies (12.39) with $\nu = x$ (we put $h=1$) and $f(g_i) - f(\widehat{g}_i) = Z_i(g_i - \widehat{g}_i)$. This completes the proof of the theorem. □

For 1-stage methods ($s=1$) the Theorem of von Neumann (Corollary 11.4) implies that it is sufficient to consider scalar, complex-valued z_1 in Eq. (12.41). Since $K(z) = R(z)$ in this case, we have

$$\varphi_B(x) = \varphi_R(x) \qquad \text{for all 1-stage methods.} \tag{12.42}$$

For the moment it is not clear, whether one can restrict the supremum in Eq. (12.41) to scalar, complex-valued z_i also for $s \ge 2$. This would require a generalization of the Theorem of von Neumann to functions of more than one variables (Hairer & Wanner 1996). We shall come back to this question later in this section.

Theorem 12.24 (Hairer & Zennaro 1996). *For B-stable Runge-Kutta methods the error growth function is superexponential, i.e., $\varphi_B(0) = 1$ and*

$$\varphi_B(x_1)\,\varphi_B(x_2) \le \varphi_B(x_1 + x_2) \qquad \textit{for } x_1, x_2 \textit{ having the same sign.}$$

Proof. The property $\varphi_B(0)=1$ follows from Definition 12.5. For the proof of the inequality we consider the rational function

$$S(z) = u_A^* K(A_1 - zI,\ldots,A_s - zI) v_A u_B^* K(B_1 + zI,\ldots,B_s + zI) v_B,$$

where the matrices A_j, B_j satisfy $\mu(A_j) \le x_1 + x_2$ and $\mu(B_j) \le 0$, and u_A, v_A, u_B, v_B are arbitrary vectors of $\mathbb{C}^n$. Using the property $\mu(A_j - zI) = \mu(A_j) - \mathrm{Re}\, z$ and the fact that $\|C\| = \sup_{\|u\|=1,\|v\|=1} |u^* C v|$, the inequality is obtained exactly as in the proof of Theorem 11.6. □

The fact that $\varphi_B(x)$ is superexponential together with $\varphi_B(-\infty) = |R(\infty)|$ (see Exercise 8) allows us to draw the same conclusions on asymptotic stability of numerical solutions as in Sect. IV.11.

Computation of $\varphi_B(x)$

The idea is to search for the maximum of $\|\Delta y_1\|$ under the restriction (12.39). More precisely, we consider the following inequality constrained optimization problem:

$$\begin{aligned} &\|\Delta y_1\|^2 \to \max, \\ &\mathrm{Re}\,\langle \Delta f_i, \Delta g_i\rangle \le x\|\Delta g_i\|^2, \quad i=1,\dots,s. \end{aligned} \tag{12.43}$$

Here $\Delta f_1,\dots,\Delta f_s$ are regarded as independent variables in $\mathbb{C}^n$, Δy_1 and Δg_i are defined by (12.5), and Δy_0 is considered as a parameter. A classical approach for solving the optimization problem (12.43) is to introduce Lagrange multipliers $d_1,\dots,d_s$, and to consider the Lagrangian

$$\begin{aligned} \mathcal{L}(\Delta f, D) &= \frac{1}{2}\|\Delta y_1\|^2 - \sum_{i=1}^{s} d_i\Big(\mathrm{Re}\,\langle \Delta f_i, \Delta g_i\rangle - x\|\Delta g_i\|^2\Big) \\ &= -\frac{1}{2}\,(\Delta y_0^*, \Delta f^*)\left(\begin{pmatrix} \alpha & u^T \\ u & W \end{pmatrix}\otimes I\right)\begin{pmatrix} \Delta y_0 \\ \Delta f \end{pmatrix}, \end{aligned} \tag{12.44}$$

where $\Delta f = (\Delta f_1,\dots,\Delta f_s)^T$, $D = \mathrm{diag}\,(d_1,\dots,d_s)$, and

$$\alpha = -1 - 2x\,\mathbb{1}^T D\mathbb{1}, \tag{12.45a}$$
$$u = D\mathbb{1} - b - 2xA^T D\mathbb{1}, \tag{12.45b}$$
$$W = DA + A^T D - bb^T - 2xA^T DA. \tag{12.45c}$$

Theorem 12.25 (Burrage & Butcher 1980). *If the matrix*

$$\begin{pmatrix} \alpha + \varphi^2 & u^T \\ u & W \end{pmatrix} \quad \textit{is positive semi-definite} \tag{12.46}$$

for some $d_1 \ge 0,\dots,d_s \ge 0$, then it holds $\|\Delta y_1\| \le \varphi\|\Delta y_0\|$ for all problems satisfying (12.39) with $h\nu \le x$. Consequently, we have $\varphi_B(x) \le \varphi$.

Proof. Substracting $\varphi^2\|\Delta y_0\|^2/2$ from both sides of (12.44) yields

$$\frac{1}{2}\Big(\|\Delta y_1\|^2 - \varphi^2\|\Delta y_0\|^2\Big) - \sum_{i=1}^{s} d_i\Big(\mathrm{Re}\,\langle \Delta f_i, \Delta g_i\rangle - x\|\Delta g_i\|^2\Big) \le 0.$$

The statement then follows from $d_i \ge 0$ and $\mathrm{Re}\,\langle \Delta f_i, \Delta g_i\rangle \le x\|\Delta g_i\|^2$. □

With the help of this theorem, Burrage & Butcher (1980) computed an upper bound of $\varphi_B(x)$ for many 2-stage methods. It turned out that for all these 2-stage methods $\varphi_B(x) = \varphi_K(x)$, where

$$\varphi_K(x) = \sup_{\mathrm{Re}\, z_1 \le x, \ldots, \mathrm{Re}\, z_s \le x} |K(z_1, \ldots, z_s)|. \tag{12.47}$$

There naturally arises the question: *is it true that $\varphi_B(x) = \varphi_K(x)$ for all Runge-Kutta methods?* If we want to check the validity of $\varphi_B(x) = \varphi_K(x)$ for a given Runge-Kutta method, we have to find non-negative Lagrange multipliers d_i, such that (12.46) is satisfied. The following lemmas will be useful for this purpose.

We denote by $z_1^0, \ldots, z_s^0$ the values, for which the supremum in Eq. (12.47) is attained. By the maximum principle we have $z_j^0 = x + iy_j^0$ ($y_j^0 = \infty$ is admitted). We further put $z^0 = (z_1^0, \ldots, z_s^0)$ and let $\partial_j K(z^0)$ be the derivative of $K(z_1, \ldots, z_s)$ with respect to the jth argument, evaluated at z^0.

Lemma 12.26. *Let x be fixed with $\varphi_K(x) < \infty$. The condition (12.46) with $\varphi = \varphi_K(x)$ then uniquely determines the Lagrange multipliers $d_1, \ldots, d_s$ (see Eq. (12.53) below). They are real and positive.*

Proof. Consider the identity (12.44) for the special case, where Δf_j is scalar, $\Delta f_j = z_j \Delta g_j$, and hence $\Delta y_1 = K(z_1, \ldots, z_s)$. For $\mathrm{Re}\, z_j = x$ this identity becomes

$$|K(z_1, \ldots, z_s)|^2 - \varphi^2 = -(1, \Delta f^*) \begin{pmatrix} \alpha + \varphi^2 & u^T \\ u & W \end{pmatrix} \begin{pmatrix} 1 \\ \Delta f \end{pmatrix}. \tag{12.48}$$

Putting $\varphi := \varphi_K(x)$ and $z_j := z_j^0$ (eventually one has to consider limits) the left-hand expression of Eq. (12.48) vanishes. This together with assumption (12.46) implies that $u + W\Delta f = 0$, i.e.,

$$D1\!\!1 - b - 2xA^T D1\!\!1 + \left(DA + A^T D - bb^T - 2xA^T DA\right) \Delta f = 0.$$

Collecting suitable terms, and using $\Delta f = Z_0 \Delta g$ and $\Delta g = 1\!\!1 + A\Delta f$, where $Z_0 = \mathrm{diag}\,(z_1^0, \ldots, z_s^0)$, this relation becomes

$$D\,\Delta g = (I - A^T Z_0^*)^{-1} b \cdot K(z^0). \tag{12.49}$$

We shall show that all components of $\Delta g = (I - AZ_0)^{-1} 1\!\!1$ are different from zero, so that (12.49) uniquely determines the Lagrange multipliers $d_1, \ldots, d_s$.

Expanding $K(z_1, \ldots, z_s)$ into a Taylor series with respect to z_j, we obtain

$$K(z_1^0, \ldots, z_j, \ldots, z_s^0) = K(z^0)\Big(1 + c(z_j - z_j^0) + \mathcal{O}\big((z_j - z_j^0)^2\big)\Big),$$

where $c = \partial_j K(z^0)/K(z^0)$. Since $|K(z_1^0, \ldots, z_j, \ldots, z_s^0)| \le |K(z^0)|$ for $\mathrm{Re}\, z_j \le \mathrm{Re}\, z_j^0$, we have $c > 0$, and consequently also

$$\partial_j K(z^0) \ne 0, \qquad 0 < \partial_j K(z^0)/K(z^0) < \infty. \tag{12.50}$$

Differentiating $K(z_1,\ldots,z_s)=1+b^TZ(I-AZ)^{-1}\mathbb{1}$ with respect to z_j yields

$$\partial_j K(z^0)=b^T(I-Z_0A)^{-1}e_je_j^T(I-AZ_0)^{-1}\mathbb{1}, \tag{12.51}$$

and we obtain from (12.50) that

$$b^T(I-Z_0A)^{-1}e_j\neq 0, \qquad \Delta g_j=e_j^T(I-AZ_0)^{-1}\mathbb{1}\neq 0, \tag{12.52}$$

so that $d_1,\ldots,d_s$ are uniquely determined by (12.49). Dividing the j th component of (12.49) by Δg_j, it follows from (12.51) that

$$d_j=|b^T(I-Z_0A)^{-1}e_j|^2\cdot\frac{K(z^0)}{\partial_j K(z^0)}, \tag{12.53}$$

which is a strictly positive real number by (12.50) and (12.52).

In this proof we have implicitly assumed that all z_j^0 are finite. If $z_j^0=x+i\infty$ for some j, one has to apply the standard transformation $\omega_j=x+1/(z_j-x)$, which maps the half-plane $\mathrm{Re}\,z_j\le x$ onto $\mathrm{Re}\,\omega_j\le x$, and ∞ into 0. □

Lemma 12.27. *If the matrix W of Eq. (12.45c), with $d_1,\ldots,d_s$ given by Lemma 12.26, is positive semi-definite, then we have $\varphi_B(x)=\varphi_K(x)$.*

Proof. It follows from

$$\begin{pmatrix}\alpha+\varphi_K^2(x) & u^T\\ u & W\end{pmatrix}\begin{pmatrix}1\\ \Delta f\end{pmatrix}=0 \tag{12.54}$$

(see Eq. (12.48)) and from $v^TWv\ge 0$ for all $v\in\mathbb{R}^s$ that the matrix in (12.54) is positive semi-definite. The statement then follows from Theorem 12.25. □

With the above results it is possible to check for a given Runge-Kutta method, whether $\varphi_B(x)=\varphi_K(x)$ is satisfied. This can be done by the following algorithm:

- compute $\varphi=\varphi_K(x)$ of Eq. (12.47) either numerically or with help of a formula manipulation program;
- compute the Lagrange multipliers $d_1,\ldots,d_s$ from Lemma 12.26;
- check, whether the matrix W of Eq. (12.45c) is positive semi-definite. If this is the case, it holds $\varphi_B(x)=\varphi_K(x)$ by Lemma 12.27.

Example 12.28. For the two-stage Radau IIA method (see Table 5.5) the function $K(z_1,z_2)$ is given by

$$K(z_1,z_2)=\frac{1+z_1/3}{1-5z_1/12-z_2/4+z_1z_2/6}.$$

The maximum of $|K(z_1,z_2)|$ on the set $\operatorname{Re} z_i \le x$ is attained at

$$z_1^0 = \begin{cases} x + i\infty & \text{for } x \le \xi \\ x + ix\sqrt{\dfrac{45-42x+8x^2}{9+18x-8x^2}} & \text{for } \xi \le x < 3/2 \end{cases}$$

$$z_2^0 = \begin{cases} x & \text{for } x \le \xi \\ x + i\dfrac{x\sqrt{(45-42x+8x^2)(9+18x-8x^2)}}{8x^2-6x-9} & \text{for } \xi \le x < 3/2 \end{cases}$$

(the value $\xi = (9-3\sqrt{17})/8$ is a root of $9+18x-8x^2 = 0$) and it is given by

$$\varphi_K(x) = \begin{cases} \dfrac{4}{5-2x} & \text{if } \quad x \le \xi \\ \dfrac{3+4x}{\sqrt{(3-2x)(3+4x-2x^2)}} & \text{if } \quad \xi \le x < 3/2. \end{cases}$$

The function $K(z_1,z_2)$ is not bounded on $\operatorname{Re} z_i \le x$ for $x \ge 3/2$. From the proof of Lemma 12.26 we compute d_1 and d_2, and obtain

$$d_1 = \begin{cases} \dfrac{9}{(3-x)(5-2x)} & \text{for } x \le \xi \\ \dfrac{(3+4x)^2}{4(3+4x-2x^2)} & \text{for } \xi \le x \end{cases} \qquad d_2 = \begin{cases} \dfrac{2}{5-2x} & \text{for } x \le \xi \\ \dfrac{3+4x}{4(3+4x-2x^2)} & \text{for } \xi \le x. \end{cases}$$

With these values one checks straight-forwardly that the matrix W of Eq. (12.45c) is semi-definite positive, so that $\varphi_B(x) = \varphi_K(x)$; see also Burrage & Butcher (1980). Actually, the matrix W is non-singular for $x < \xi$, and of rank one for $\xi \le x < 3/2$.

A comparison with Eq. (11.15) shows that we do not obtain the same estimate as for linear autonomous problems.

The above algorithm can easily be applied to other two-stage methods. We thus obtain for the two-stage Gauss method

$$\varphi_B(x) = \begin{cases} 1 & \text{if } -\infty < x \le 0 \\ \dfrac{2x+\sqrt{9+3x^2}}{3-x} & \text{if } 0 \le x < 3\,, \end{cases}$$

and for the two-stage Lobatto IIIC method

$$\varphi_B(x) = \begin{cases} \dfrac{1}{1-x+x^2} & \text{if } -\infty < x \le 0 \\ \dfrac{1}{1-x} & \text{if } 0 \le x < 1\,. \end{cases}$$

For methods with more than two stages, explicit formulas are difficult to obtain, and one has to apply numerical methods for the computation of z_j^0 (supremum in Eq. (12.47)).

Exercises

1. Prove, directly from Def. 12.2, that the implicit Euler method is B-stable.

2. Let M be a symmetric $s \times s$-matrix and $\langle \cdot, \cdot \rangle$ the scalar product of $\mathbb{R}^n$. Then M is non-negative definite, if and only if

$$\sum_{i=1}^{s} \sum_{j=1}^{s} m_{ij} \langle u_i, u_j \rangle \geq 0 \quad \text{for all} \quad u_i \in \mathbb{R}^n.$$

Hint. Use $M = Q^T D Q$ where D is diagonal.

3. Give a simple proof for the B-stability of the Radau IIA methods by extending the ideas of Example 12.3.

Hint. For the quadrature, based on the zeros of (5.2), we have

$$\int_0^1 \varphi(x)dx = \sum_{i=1}^{s} b_i \varphi(c_i) + C\varphi^{(2s-1)}(\xi), \quad 0 < \xi < 1.$$

with $C < 0$ (see e.g. Abramowitz & Stegun (1964, Formula 25.4.31)).

4. (Dahlquist & Jeltsch 1987). Prove that Method I of Table 12.4 is S-reducible with respect to the partition ($\{1\}, \{2,3\}$). The reduced method II itself is DJ-reducible and reduces to Method III.

For the initial value problem $y' = f(y)$, $y(0) = 1$, where $f(y) = y^2$ for $y \geq 0$ and $f(y) = 0$ for $y < 0$, and for $h = 2$, Methods I and III have unique solutions which are different. Explain this apparent contradiction.

Table 12.4. Reduction of RK-methods

Method I

c			
0	0	0	0
1/2	0	1	-1/2
1/2	0	1/2	0
	1	b	$-b$

Method II

c		
0	0	0
1/2	0	1/2
	1	0

Method III

c	
0	0
	1

5. Give a counterexample of an irreducible AN-stable but not algebraically stable, and hence not B-stable method.

Hint. Start with any algebraically stable method with, say, two stages and modify it as indicated in Table 12.5. Find conditions on the free parameters d, e, α such that the two methods are identical for equations $y' = \lambda(x)y$. This ensures AN-stability of the second method. Then play with the parameters to destroy algebraic stability.

6. Show that the method of Table 12.1 is DJ-reducible, but not S-reducible; show that it is algebraically stable together with the reduced method.

Table 12.5. Construction of AN-stable but not B-stable method

$$\begin{array}{c|cc} c_1 & a_{11} & a_{12} \\ c_2 & a_{21} & a_{22} \\ \hline & b_1 & b_2 \end{array} \quad \Rightarrow \quad \begin{array}{c|ccc} c_1 & a_{11} & a_{12}\alpha & a_{12}(1-\alpha) \\ c_2 & c_2-d & d\alpha & d(1-\alpha) \\ c_2 & c_2-e & e\alpha & e(1-\alpha) \\ \hline & b_1 & b_2\alpha & b_2(1-\alpha) \end{array}$$

Show that the method of Table 12.2 is S-reducible, but not DJ-reducible; show that it is not algebraically stable, but that the reduced method is.

7. (Sandberg & Shichman 1968, Vanselow 1979, Hundsdorfer 1985). Prove that Rosenbrock methods are not B-stable in the sense of Definition 11.2.

   Hint. Apply the method to the scalar problem $y' = f(y)$, $y_0 = 1$ where $f(y)$ is a non-increasing function satisfying (for a small ε)

$$f(y) = \begin{cases} -y & \text{if } \ |y-1| \geq 2\varepsilon \\ -1 & \text{if } \ |y-1| \leq \varepsilon \,. \end{cases}$$

8. (Hairer & Zennaro 1996). For irreducible, algebraically stable Runge-Kutta methods the error growth function satisfies

$$\varphi_B(x) \leq \frac{\sqrt{1-2x\gamma(1-\varrho^2)} - 2x\gamma\varrho}{1-2x\gamma} \qquad \text{for} \quad x \leq 0,$$

   where $\varrho = |R(\infty)|$ ($R(z)$ is the stability function), $\gamma = \Big(\sum_{j=1}^{s} b_j^{-1} v_j^2\Big)^{-1}$, and $(v_1,\ldots,v_s)^T = \lim_{\varepsilon\to 0} b^T(A+\varepsilon I)^{-1}$.

   Hint. From (12.7) we have $\|\Delta y_1\|^2 \leq \|\Delta y_0\|^2 + 2x\sum_i b_i\|\Delta g_i\|^2$. Then, compute Δf_i from (12.5b) (if A is invertible), insert it into (12.5a) and conclude $\Delta y_1 = R(\infty)\Delta y_0 + \sum_j(\sum_i b_i\omega_{ij})\Delta g_j$, where $(\omega_{ij}) = A^{-1}$. The Cauchy-Schwarz inequality yields $\sum_i b_i\|\Delta g_i\|^2 \geq \gamma\big(\|\Delta y_1\| - \varrho\|\Delta y_0\|\big)^2$ which, inserted into the first estimate, gives a second degree inequality for Δy_1.

9. Prove that for the 3-stage Gauss method we have for $x \geq 0$

$$\varphi_B(x) \geq (1+x/2)/(1-x/2).$$

   Hint. Using (12.18), compute $K(Z)$ for $z_1 \to -\infty$, $z_2 = x$, $z_3 \to -\infty$.

10. If the matrix W of Eq. (12.45c), with $d_1,\ldots,d_s$ given by Lemma 12.26, is either non-singular or of rank ≤ 1, then it holds $\varphi_B(x) = \varphi_K(x)$.

    Hint. Exploit the fact that the expression in Eq. (12.48) with $\varphi = \varphi_K(x)$ is non-positive for all z_j with $\operatorname{Re} z_j \leq x$.

11. Show that for a non-negative definite symmetric matrix $M = (m_{ij})$ one has

$$|m_{ij}| \leq \sqrt{m_{ii}m_{jj}}.$$

IV.13 Positive Quadrature Formulas and B-Stable RK-Methods

Bien que le problème (des quadratures) ait une durée de deux cents ans à peu près, bien qu'il était l'objet de nombreuses recherches de plusieurs géomètres: Newton, Cotes, Gauss, Jacobi, Hermite, Tchébychef, Christoffel, Heine, Radeau [*sic*], A. Markov, T. Stitjes [*sic*], C. Possé, C. Andréev, N. Sonin et d'autres, il ne peut être considéré, cependant, comme suffisamment épuisé.
(V. Steklov 1917)

We shall give a constructive characterization of all irreducible B-stable Runge-Kutta methods (Theorem 13.15). Because of Theorem 12.16 we first have to study quadrature formulas with positive weights.

Quadrature Formulas and Related Continued Fractions

Steklov (1916) proved that a family of interpolatory quadrature formulas converges for all Riemann integrable functions, if all weights of the formulas are positive ("Il faut remarquer cependant que de tels théorèmes généraux ne peuvent avoir aucune valeur pratique ..."). This theorem, rediscovered around 1922 by Fejér, initiated an extensive search for quadrature formulas with positive weights. Fejér (1933, "weiter habe ich noch auf sehr kurzem Wege das folgende Resultat erhalten ...") found the result:

"If $P_s(z)$ are the Legendre polynomials normalized as in (13.4) and $c_1, \ldots, c_s$ are the zeros of $M(z) = P_s(z) + \alpha_1 P_{s-1}(z) + \alpha_2 P_{s-2}(z)$ with $\alpha_2 \leq 0$, then the weights b_i are all positive".

The theory of B-stable methods renewed the interest in positive quadrature formulas and Burrage (1978) obtained the sharp bound

$$\alpha_2 < \frac{(s-1)^2}{4(2s-1)(2s-3)} \tag{13.1}$$

for the positivity of the b_i in the above case. This is the same as condition (5.51) in a different normalization. A short proof of this result (see "Lemma 18" of Nørsett & Wanner 1981) then led to a complete characterization of positive quadrature formulas by Sottas & Wanner (1982). An independent proof of an equivalent result was found by Peherstorfer (1981). In what follows, we give a new approach using continued fractions.

Consider a quadrature formula

$$\sum_{j=1}^{s} b_j f(c_j) \approx \int_0^1 f(x)dx$$

with distinct nodes c_i and non-zero weights b_i. The main idea is to consider the rational function

$$Q(z)=\sum_{j=1}^{s} b_j \frac{1}{z-c_j}=\frac{N(z)}{M(z)} \tag{13.2}$$

where, as usual, $M(z)=(z-c_1)\cdot\ldots\cdot(z-c_s)$. We first express the order of the quadrature formula in terms of the function $Q(z)$.

Lemma 13.1. *A quadrature formula is of order* p *if and only if* $Q(z)$*, defined by (13.2), satisfies*

$$Q(z)=-\log\Big(1-\frac{1}{z}\Big)+\mathcal{O}\Big(\frac{1}{z^{p+1}}\Big) \qquad \textit{for} \quad z\to\infty. \tag{13.3}$$

Proof. Inserting the geometric series for $(1-c_j/z)^{-1}$ into (13.2) we obtain

$$Q(z)=\sum_{k\geq 1}\Big(\sum_{j=1}^{s} b_j c_j^{k-1}\Big)\frac{1}{z^k}.$$

Therefore (13.3) is equivalent to

$$\sum_{j=1}^{s} b_j c_j^{k-1}=\frac{1}{k} \qquad \text{for} \quad k=1,\ldots,p. \qquad \square$$

We now study the case of the *Gaussian quadrature formulas*, where the function (13.2) will be denoted by $Q_s^G(z)=N_s^G(z)/M_s^G(z)$; here the c_i are the zeros of the s-degree shifted Legendre polynomial

$$P_s(z)=\frac{s!}{(2s)!}\frac{d^s}{dz^s}\big(z^s(z-1)^s\big), \tag{13.4}$$

which is normalized so that the coefficient of z^s is 1. The polynomials (13.4) satisfy the recurrence relation (see Eq. (5.53) or Abramowitz & Stegun, p. 782)

$$P_{s+1}(z)=\Big(z-\frac{1}{2}\Big)P_s(z)-\tau_s P_{s-1}(z), \qquad \tau_s=\frac{s^2}{4(4s^2-1)} \tag{13.5}$$

and $P_0(z)=1$, $P_{-1}(z)=0$. Since this quadrature formula is of optimal order $2s$, it follows from (13.3) that

$$N_s^G(z)=-M_s^G(z)\log\Big(1-\frac{1}{z}\Big)+\mathcal{O}\Big(\frac{1}{z^{s+1}}\Big). \tag{13.6}$$

We now insert $M_s^G(z)=P_s(z)$ (see (13.2)) into (13.5) and multiply by $\log(1-1/z)$ (which is $\mathcal{O}(1/z)$ for $z\to\infty$). A comparison with (13.6) shows that the polynomials $N_s^G(z)$ must also satisfy the recurrence formula (13.5) (with $N_0^G(z)=0$, $N_1^G(z)=1$). It thus follows from elementary properties of continued fractions

(Exercise 1 or Perron (1913), page 4) that

$$Q_s^G(z) = \frac{1 \mid}{\mid z - \frac{1}{2}} - \frac{\tau_1 \mid}{\mid z - \frac{1}{2}} - \ldots - \frac{\tau_{s-1} \mid}{\mid z - \frac{1}{2}}. \tag{13.7}$$

For an arbitrary quadrature formula we have

Lemma 13.2. *An irreducible rational function* $Q(z) = N(z)/M(z)$ *(with* $\deg M = s$, $\deg N = s-1$*) satisfies (13.3) with* $p \geq 2(s-k)$*, if and only if*

$$Q(z) = \frac{1 \mid}{\mid z - \frac{1}{2}} - \frac{\tau_1 \mid}{\mid z - \frac{1}{2}} - \ldots - \frac{\tau_{s-k-1} \mid}{\mid z - \frac{1}{2}} - \frac{g(z) \mid}{\mid f(z)} \tag{13.7'}$$

with $\deg f = k$ *and* $\deg g \leq k-1$.

Proof. From Lemma 13.1 we know that $Q(z) = Q_s^G(z) + \mathcal{O}(1/z^{2(s-k)+1})$. Therefore the first $2(s-k)$ coefficients in the continued fraction expansions for $Q(z)$ and $Q_s^G(z)$ must be the same. □

> Endlich sei noch die folgende Formel wegen ihrer häufigen Anwendungen ausdrücklich hervorgehoben:
>
> (O. Perron 1913, page 5)

Lemma 13.3. *The functions* $M(z)$ *and* $N(z)$ *of Lemma 13.2 are related to* $f(z)$ *and* $g(z)$ *of (13.7') as follows:*

$$\begin{aligned} M(z) &= P_{s-k}(z)f(z) - P_{s-k-1}(z)g(z), \\ N(z) &= N_{s-k}^G(z)f(z) - N_{s-k-1}^G(z)g(z). \end{aligned} \tag{13.8}$$

Proof. This follows from the recursion (13.30) and Exercise 1 below, if we put there $b_0 = 0$, $b_1 = \ldots = b_{s-k} = z - 1/2$, $b_{s-k+1} = f(z)$ and $a_1 = 1$, $a_j = -\tau_{j-1} (j = 2, \ldots, s-k)$, $a_{s-k+1} = -g(z)$. □

Solving the linear system (13.8) for $f(z)$ and $g(z)$ gives, with the use of Exercise 2,

$$\begin{aligned} f(z) \cdot \tau_1 \cdots \tau_{s-k-1} &= N(z)P_{s-k-1}(z) - M(z)N_{s-k-1}^G(z) \\ g(z) \cdot \tau_1 \cdots \tau_{s-k-1} &= N(z)P_{s-k}(z) - M(z)N_{s-k}^G(z). \end{aligned} \tag{13.9}$$

Number of Positive Weights

For a given rational function (13.2), the weights are determined by

$$b_i = \frac{N(c_i)}{M'(c_i)}. \tag{13.10}$$

But we want our theory to work also for *confluent* nodes for which $M'(c_i) = 0$.

Therefore we suppose that $c_1, \ldots, c_m$ $(m \le s)$ are the *real and distinct* zeros of $M(z)$ *of multiplicities* $l_1, \ldots, l_m$. Then we let

$$b_i = \frac{N(c_i)}{M^{(l_i)}(c_i)} \qquad i = 1, \ldots, m. \tag{13.10'}$$

For $l_i = 1$ this is just (13.10); otherwise we are considering the weights for the highest derivative of a Hermitian quadrature formula (see Exercise 3).

The main idea (following Sottas & Wanner 1982) is now to consider the path $\gamma(t) = (f(t), g(t))$ in the plane $\mathbb{R}^2$, where f and g are the polynomials of (13.7'). For $t \to \pm\infty$ this path tends to infinity with horizontal limiting directions, since the degree of f is higher than that of g. Equation (13.8) tells us that for an irreducible $Q(z)$ this path does not pass through the origin.

Definition 13.4. The *rotation number* r of γ is the integer for which $r\pi$ is the total angle of rotation around the origin for the path $\gamma(t)$ $(-\infty < t < \infty)$ measured in the negative (clockwise) sense. Counter-clockwise rotations are negative.

An algebraic definition of r is possible as

$$r = \sum_i \text{sign}\,\big(f^{(l_i)}(t_i) g(t_i)\big),$$

where the summation is over all real zeros of $f(t)$ with *odd* multiplicity l_i.

Theorem 13.5 (Sottas & Wanner 1982). *Let* $Q(z) = N(z)/M(z)$ *be an irreducible rational function as in Lemma 13.2. Suppose that* $c_1, \ldots, c_m$ *are the (distinct) real zeros of* $M(z)$ *with odd multiplicity and denote by* n_+ *(respectively* n_-*) the number of positive (respectively negative)* b_i. *Further, let* r *be the rotation number of* $\gamma = (f, g)$ *(Definition 13.4). Then*

$$n_+ - n_- = s - k + r. \tag{13.11}$$

Proof. The proof is by counting the number of crossings of the vectors $\gamma(t) = (f(t), g(t))$ and $\beta(t) = (P_{s-k-1}(t), P_{s-k}(t))$, like the crossings of hands on a Swiss cuckoo clock.

From (13.9) we see that when t equals a zero c_i of M, these two vectors are parallel in the same sense $(N(c_i) > 0)$ or in the opposite sense $(N(c_i) < 0)$. From (13.8) we observe that $M(t)$ is just the exterior product $\gamma(t) \times \beta(t)$. By elementary geometry, and taking into account Formula (13.10'), we see that at every zero c_i with odd multiplicity we have

i) $b_i > 0$, if the crossing of $\gamma(t)$ with $\beta(t)$ is clockwise;

ii) $b_i < 0$, if this crossing is counter-clockwise.

Zeros of $M(t)$ with even multiplicity don't give rise to crossings.

Since the zeros of P_{s-k} and P_{s-k-1} interlace (see e.g. Theorem 3.3.2 of Szegö 1939), the vector $\beta(t)$ turns counter-clockwise with a total angle of $-(s-$

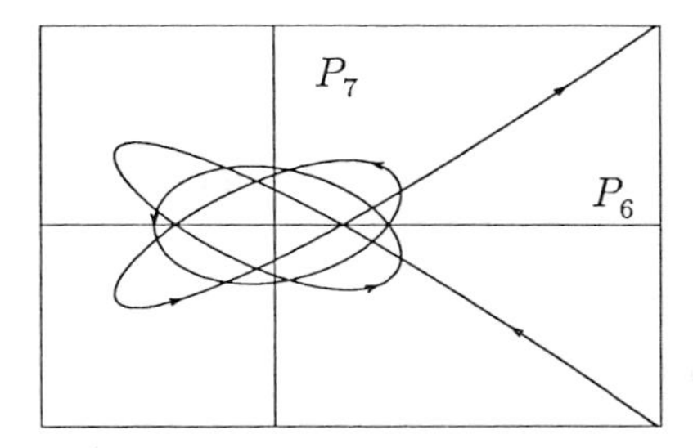

Fig. 13.1. The path $(P_{s-k-1}(t), P_{s-k}(t))$ for $s-k=7$

$k)\pi$ (see Fig. 13.1). The vector $\gamma(t)$ turns with a total angle $r\pi$ measured clockwise (Definition 13.4). Since the limiting directions of $\gamma(t)$ and $\beta(t)$ are different (horizontal for $\gamma(t)$ and vertical for $\beta(t)$), $\gamma(t)$ must cross $\beta(t)$, as t increases from $-\infty$ to $+\infty$, exactly $s-k+r$ times more often clockwise than counterclockwise. This gives Formula (13.11). □

Corollary 13.6. *Under the assumptions of Theorem 13.5, all zeros of $M(z)$ are real and simple, and the b_i are positive if and only if*

$$r=k.$$

Proof. $r=k$ means by (13.11) that $n_+ - n_- = s$. Because of $n_- \geq 0$ and $n_+ \leq s$, this is equivalent to $n_+ = s$ and $n_- = 0$. □

Characterization of Positive Quadrature Formulas

The following theorem gives a constructive characterization of all quadrature formulas with positive weights.

Theorem 13.7. *Let*

$$\sigma_1 < \varrho_1 < \sigma_2 < \varrho_2 < \ldots < \varrho_{k-1} < \sigma_k$$

be arbitrary real numbers and C a positive constant. Then, putting

$$f(z) = (z-\sigma_1)\ldots(z-\sigma_k), \quad g(z) = C(z-\varrho_1)\ldots(z-\varrho_{k-1}), \tag{13.12}$$

computing $M(z)$, $N(z)$ from (13.8), taking $c_1,\ldots,c_s$ as the zeros of $M(z)$ and b_i from (13.10), one obtains all quadrature formulas with positive weights of order $p \geq 2(s-k)$. If $C=\tau_{s-k}$ the order is $p \geq 2(s-k)+1$.

Proof. The functions $f(z)$ and $g(z)$ are irreducible, so that also the fraction $N(z)/M(z)$ is irreducible by (13.9). The statement now follows from Corollary 13.6, since the polynomials (13.12) are all possible polynomials for which $r=k$. The stated order properties follow from Lemma 13.2. □

Example 13.8. Let $c_1, \ldots, c_s$ be the zeros of

$$M(z) = P_s(z) + \alpha_1 P_{s-1}(z) + \alpha_2 P_{s-2}(z). \tag{13.13}$$

In order to study when the corresponding quadrature formula has positive weights, we use (13.5) to write (13.13) as

$$M(z) = P_{s-1}(z)\Big(z - \frac{1}{2} + \alpha_1\Big) - P_{s-2}(z)(\tau_{s-1} - \alpha_2).$$

Consequently $f(z) = z - 1/2 + \alpha_1$, $g(z) = \tau_{s-1} - \alpha_2$ and Theorem 13.7 implies that the zeros of $M(z)$ are real and the weights positive, if and only if $\alpha_2 < \tau_{s-1}$, hence (13.1) is proved.

For $k > 1$ the rotation number r of $(f(t), g(t))$ can be computed with Sturm's algorithm (Lemma 13.3 of Sect. I.13). Consider, for example,

$$\begin{aligned} M(z) &= P_s(z) + \alpha_1 P_{s-1}(z) + \alpha_2 P_{s-2}(z) + \alpha_3 P_{s-3}(z) \\ &= P_{s-2}(z)\big[(z - \tfrac{1}{2})(z - \tfrac{1}{2} + \alpha_1) + \alpha_2 - \tau_{s-1}\big] \\ &\qquad\qquad - P_{s-3}(z)\big[\tau_{s-2}(z - \tfrac{1}{2} + \alpha_1) - \alpha_3\big]. \end{aligned}$$

Application of Lemma I.13.3 to the polynomials $f(z) = (z - \frac{1}{2})(z - \frac{1}{2} + \alpha_1) + \alpha_2 - \tau_{s-1}$ and $g(z) = \tau_{s-2}(z - \frac{1}{2} + \alpha_1) - \alpha_3$ shows that the corresponding quadrature formula has positive weights iff

$$\frac{\alpha_3}{\tau_{s-2}}\Big(\alpha_1 - \frac{\alpha_3}{\tau_{s-2}}\Big) - \alpha_2 + \tau_{s-1} > 0, \tag{13.14}$$

a result first found by Burrage (1978).

Necessary Conditions for Algebraic Stability

We now turn our attention to algebraic stability. We again use the notation $B(p)$, $C(\eta)$, $D(\xi)$ of Sect. IV.5.

Lemma 13.9 (Burrage 1982). *Consider Runge-Kutta methods, which satisfy $B(2)$ and the second condition for algebraic stability (i.e. M non-negative). Then,*

a) $C(k)$ implies $B(2k-1)$;

b) $D(k)$ implies $B(2k-1)$.

Proof. Instead of considering M, we work with the transformed matrix $\widehat{M} = V^T M V$ where $V = (c_i^{j-1})_{i,j=1}^s$ is the Vandermonde matrix. The elements of $\widehat{M}$ are given by

$$\widehat{m}_{qr} = \sum_{i=1}^s b_i c_i^{q-1} \sum_{j=1}^s a_{ij} c_j^{r-1} + \sum_{j=1}^s b_j c_j^{r-1} \sum_{i=1}^s a_{ji} c_i^{q-1} - \sum_{i=1}^s b_i c_i^{q-1} \sum_{j=1}^s b_j c_j^{r-1}. \tag{13.15}$$

We further introduce

$$g_r = r\sum_{j=1}^{s} b_j c_j^{r-1} - 1$$

so that $B(\nu)$ is equivalent to $g_r = 0$ $(r = 1,\dots,\nu)$. Then $C(k)$ simplifies (13.15) to

$$\widehat{m}_{qr} = \frac{1}{q\cdot r}\left(g_{q+r} + 1 - (g_q+1)(g_r+1)\right) \qquad q\le k,\ r\le k.$$

Similarly, $D(k)$ implies

$$\widehat{m}_{qr} = -\frac{1}{q\cdot r}\left(g_{q+r} + g_q\cdot g_r\right) \qquad q\le k,\ r\le k.$$

We now start with the hypothesis $B(2)$, i.e., $B(2l)$ for $l=1$. This means that $g_1 = \dots = g_{2l} = 0$, so that, in both cases, $\widehat{m}_{ll} = 0$. But if for a non-negative definite matrix a diagonal element is zero, the whole corresponding column must also be zero (see Exercise 11 of Sect. IV.12). This leads to $g_{l+q} = 0$ for $q = 1,\dots,k$; so we have $B(k+l)$. We then repeat the argument inductively until we arrive at $B(2k-1)$. □

Since s-stage collocation methods satisfy $B(s)$ and $C(s)$ (see Theorem 7.8 of Chapter II) we have

Corollary 13.10 (Burrage 1978). *An s-stage algebraically stable collocation method must be of order at least $2s-1$.* □

Because *symmetric* methods have even order this gives:

Corollary 13.11 (Ascher & Bader 1986). *A symmetric algebraically stable collocation scheme has to be at Gaussian points.* □

The next result states the necessity of the simplifying assumption $C(k)$. Observe that by Theorem 12.16 the weights b_i of DJ-irreducible, algebraically stable methods have to be positive.

Lemma 13.12. *If a Runge-Kutta method of order $p\ge 2k+1$ satisfies $b_i>0$ for $i=1,\dots,s$, then the condition $C(k)$ holds.*

Proof (Dahlquist & Jeltsch (1979) attribute this idea to Butcher). The order conditions (see Sect. II.2)

$$\sum_{i=1}^{s} b_i c_i^{2q} = \frac{1}{2q+1}$$

$$\sum_{i,j=1}^{s} b_i c_i^q a_{ij} c_j^{q-1} = \frac{1}{(2q+1)q}$$

$$\sum_{i,j,m=1}^{s} b_i a_{ij} c_j^{q-1} a_{im} c_m^{q-1} = \frac{1}{(2q+1)q^2}$$

imply that

$$\sum_{i=1}^{s} b_i \Big(\sum_{j=1}^{s} a_{ij} c_j^{q-1} - \frac{c_i^q}{q} \Big)^2 = 0$$

for $2q+1 \le p$. Since the b_i are positive, the individual terms of this sum must be zero for $q \le k$. □

A simple consequence of this lemma are the following *order barriers* for diagonally implicit DIRK ($a_{ij} = 0$ for $i < j$) and singly diagonally implicit SDIRK ($a_{ij} = 0$ for $i < j$ and $a_{ii} = \gamma$ for all i) methods.

Theorem 13.13 (Hairer 1980).
a) A DIRK method with all b_i positive has order at most 6;
b) An SDIRK method with all b_i positive has order at most 4;
c) An algebraically stable DIRK method has order at most 4.

Proof. a) Suppose the order is greater than 6 and let i be the smallest index such that $c_i \neq 0$. Then by Lemma 13.12

$$a_{ii} c_i = \frac{c_i^2}{2}, \qquad a_{ii} c_i^2 = \frac{c_i^3}{3},$$

contradicting $c_i \neq 0$.

b) As above, we arrive for order greater than 4 at

$$a_{ii} c_i = \frac{c_i^2}{2} \quad \text{or} \quad a_{ii} = \frac{c_i}{2} \ (\neq 0).$$

Since for SDIRK methods we have $a_{ii} = a_{11}$, this leads to $c_1 = a_{11} \neq 0$, hence $i = 1$. Now $a_{11} = c_1/2$ contradicts $a_{11} = c_1$.

c) It is sufficient to consider DJ-irreducible methods, since the reduction process (see Table 12.3) leaves the class of DIRK methods invariant. From Theorem 12.16 and Lemma 13.12 we obtain that algebraic stability and order greater than 4 imply

$$a_{11} = c_1, \qquad a_{11} c_1 = \frac{c_1^2}{2},$$

and hence $a_{11} = 0$. Inserted into m_{11} this yields $m_{11} = -b_1^2 < 0$, contradicting the non-negativity of the matrix M. □

Similarly to Lemma 13.12 we have the following result for the second type of simplifying assumptions.

Lemma 13.14. *If a Runge-Kutta method of order $p \geq 2k+1$ is algebraically stable and satisfies $b_i > 0$ for all i, then the condition $D(k)$ holds.*

Proof. The main idea is to use the W-transformation of Sect. IV.5 and to consider $W^T MW$ instead of M (see also the proof of Theorem 12.8). By Theorem 5.14 there exists a matrix W satisfying $T(k,k)$ (see Definition 5.10). With the help of Lemma 13.12 and Theorem 5.11a we obtain that the first k diagonal elements of

$$W^T MW = (W^T BW)X + X^T (W^T BW)^T - e_1 e_1^T \tag{13.16}$$

are zero. Since M and hence also $W^T MW$ is non-negative definite, the first k columns and rows of $W^T MW$ have to vanish. Thus the matrix $(W^T BW)X$ must be skew-symmetric in these regions (with exception of the first element). Because of $C(k)$ the first k columns and rows of $(W^T BW)X$ and X are identical. Thus the result follows from Theorem 5.11. □

Characterization of Algebraically Stable Methods

Theorem 12.16, Lemma 13.12 and Lemma 13.14 imply that DJ-irreducible and algebraically stable RK-methods of order $p \geq 2k+1$ satisfy $b_i > 0$ for all i, and the simplifying assumptions $C(k)$ and $D(k)$. These properties allow the following constructive characterization of all irreducible B-stable RK-methods.

Theorem 13.15 (Hairer & Wanner 1981). *Consider a pth order quadrature formula $(b_i, c_i)_{i=1}^s$ with positive weights and let W satisfy Property $T(k,k)$ of Definition 5.10 with $k = [(p-1)/2]$. Then all pth order algebraically stable Runge-Kutta methods corresponding to this quadrature formula are given by*

$$A = WXW^{-1} \tag{13.17}$$

where

$$(W^T BW)X = \frac{1}{2} e_1 e_1^T + \begin{pmatrix} 0 & -\xi_1 & & & \\ \xi_1 & \ddots & \ddots & & \\ & \ddots & 0 & -\xi_k & \\ & & \xi_k & \multicolumn{2}{|c}{} \\ & & & \multicolumn{2}{|c}{Q} \end{pmatrix} \tag{13.18}$$

and Q is an arbitrary matrix of dimension $s-k$ for which $Q+Q^T$ is non-negative definite. For p even we have to require that $q_{11} = 0$.

Proof. Algebraic stability and the positivity of the weights b_i imply $C(k)$ and $D(k)$ with $k = [(p-1)/2]$. The matrix A of such a method can be written as

(13.17) with X given by (13.18). This follows from Theorem 5.11 and the fact that multiplication with W^TBW does not change the first k columns and rows of X. This method is algebraically stable iff M (or W^TMW) is non-negative definite. By (13.16) this means that $Q+Q^T$ is non-negative definite.

Conversely, any Runge-Kutta method given by (13.17), (13.18) with $Q+Q^T$ non-negative definite is algebraically stable and satisfies $C(k)$ and $D(k)$. Therefore it follows from Theorem 5.1 in the case of odd $p=2k+1$ that the Runge-Kutta method is of order p.

If p is even, say $p=2k+2$, the situation is slightly more complicated. Because of

$$q_{11} = \sum_{i,j=1}^{s} b_i P_k(c_i) a_{ij} P_k(c_j)$$

it follows from $B(2k+2)$, $C(k)$, $D(k)$ that the order condition (13.19) below (with $\xi=\eta=k$) is equivalent to $q_{11}=0$. The stated order p of the RK-method now follows from Lemma 13.16. □

In the above proof we used the following modification of Theorem 5.1.

Lemma 13.16. *If the coefficients b_i, c_i, a_{ij} of an RK-method satisfy*

$$\sum_{i,j=1}^{s} b_i c_i^{\xi} a_{ij} c_j^{\eta} = \frac{1}{(\eta+\xi+2)(\eta+1)} \tag{13.19}$$

and $B(p)$, $C(\eta)$, $D(\xi)$ with $p\le\eta+\xi+2$ and $p\le 2\eta+2$, then the method is of order p.

Proof. The reduction process with the help of $C(\eta)$ and $D(\xi)$ as described in Sect. II.7 (Volume I) reduces all trees to the bushy trees covered by $B(p)$. The only exception is the tree corresponding to order condition (13.19). □

Example 13.17 (Three-stage B-stable SIRK methods). Choose a third order quadrature formula with positive weights and let W satisfy $W^TBW=I$. Then (13.18) becomes

$$X = \begin{pmatrix} \frac{1}{2} & -\xi_1 & 0 \\ \xi_1 & a & b \\ 0 & c & d \end{pmatrix}, \qquad \xi_1 = \frac{1}{2\sqrt{3}}.$$

The method is B-stable if $X^T+X-e_1e_1^T$ is non-negative, i.e. if

$$a\ge 0, \quad d\ge 0, \quad 4ad\ge (c+b)^2. \tag{13.20}$$

If we want this method to be singly-implicit, we must have for the characteristic polynomial of A

$$\chi_A(z) = (1-\gamma z)^3 = 1-3\gamma z+3\gamma^2 z^2-\gamma^3 z^3.$$

This means that (see (13.17))

$$\begin{aligned} \frac{1}{2}+a+d &= 3\gamma \\ \frac{a}{2}+\frac{1}{12}+\frac{d}{2}+ad-cb &= 3\gamma^2 \\ \frac{ad-cb}{2}+\frac{1}{12}d &= \gamma^3. \end{aligned}$$

Some elementary algebra shows that these equations can be solved and the inequalities (13.20) satisfied if $1/3 \le \gamma \le 1.06857902$, i.e., *exactly if* the corresponding rational approximation is A-stable (cf. Table 6.3; see also Hairer & Wanner (1981), where the analogous case with $s=p=5$ is treated).

The "Equivalence" of A- and B-Stability

Many A-stable RK-methods are not B-stable (e.g., the trapezoidal rule, the Lobatto IIIA and Lobatto IIIB methods; see Theorem 12.12). On the other hand there is the famous result of Dahlquist (1978), saying that *every* A-stable *one-leg-method* is B-stable, which we shall prove in Sect. V.6. We have further seen in Example 13.17 that for a certain class of A-stable methods there is always a B-stable method with the same stability function. The general truth of this result was conjectured for many years and is as follows:

Theorem 13.18 (Hairer & Türke 1984, Hairer 1986). *Let $R(z)=P(z)/Q(z)$ ($P(0)=Q(0)=1$, $\deg P \le s$, $\deg Q = s$) be an irreducible, A-stable function satisfying $R(z)-e^z=\mathcal{O}(z^{p+1})$ for some $p\ge 1$. Then there exists an s-stage B-stable Runge-Kutta method of order p with $R(z)$ as stability function.*

Proof. Since $R(z)$ is an approximation to e^z of order p, it can be written in the form

$$R(z)=\frac{1+\frac{1}{2}\Psi(z)}{1-\frac{1}{2}\Psi(z)}, \qquad \Psi(z)=\frac{z|}{|1}+\frac{\xi_1^2 z^2|}{|\ 1}+\ldots+\frac{\xi_{k-1}^2 z^2|}{|\ 1}+\xi_k^2 z\Psi_k(z) \tag{13.21}$$

where $k=[(p-1)/2]$, $\xi_j^2=1/(4(4j^2-1))$ and $\Psi_k(z)=zg(z)/f(z)$ with $g(0)=f(0)=1$, $\deg f\le s-k$, $\deg g \le s-k-1$ (for p even we have in addition $g'(0)=f'(0)$). For the diagonal Padé-approximation $R^G(z)$ of order $2s$ this follows from Theorem 5.18 with $\nu=s-1$ and $\Psi_\nu=z$:

$$R^G(z)=\frac{1+\frac{1}{2}\Psi^G(z)}{1-\frac{1}{2}\Psi^G(z)}, \qquad \Psi^G(z)=\frac{z|}{|1}+\frac{\xi_1^2 z^2|}{|\ 1}+\ldots+\frac{\xi_{s-1}^2 z^2|}{|\ 1}. \tag{13.22}$$

For an arbitrary $R(z)$ (satisfying the assumptions of the theorem) this is then a consequence of $R(z)=R^G(z)+\mathcal{O}(z^{p+1})$, or equivalently $\Psi(z)=\Psi^G(z)+\mathcal{O}(z^{p+1})$.

The function $R(z)$ of (13.21) is A-stable iff (Theorem 5.22)

$$\operatorname{Re}\Psi_k(z) < 0 \quad \text{for} \quad \operatorname{Re} z < 0.$$

Therefore, the function $\chi(z) = -\Psi_k(-1/z)$ is positive (c.f. Definition 5.19) and by Lemma 13.19 below there exists an $(s-k)$-dimensional matrix Q such that

$$\chi(z) = e_1^T(Q+zI)^{-1}e_1 \quad \text{and} \quad Q+Q^T \quad \text{non-negative definite.}$$

We now fix an arbitrary quadrature formula of order p with positive weights b_i and (for the sake of simplicity) distinct nodes c_i. We let W be a matrix satisfying $W^TBW = I$ and Property $T(k,k)$ with $k = [(p-1)/2]$ (c.f. Lemma 5.12), and define the Runge-Kutta coefficients (a_{ij}) by (13.17) and (13.18). This Runge-Kutta method is algebraically stable, because $Q+Q^T$ is non-negative definite and of order p (observe that $g'(0) = f'(0)$ implies that the upper left element of Q vanishes). Finally, it follows from Theorem 5.18 and $\Psi_k(z) = -\chi(-1/z) = ze_1^T(I-zQ)^{-1}e_1$ that its stability function is $R(z)$. □

It remains to prove the following lemma.

Lemma 13.19. *Let $\chi(z) = \alpha(z)/\beta(z)$ be an irreducible rational function with real polynomials*

$$\alpha(z) = z^{n-1} + \alpha_1 z^{n-2} + \ldots, \qquad \beta(z) = z^n + \beta_1 z^{n-1} + \ldots. \tag{13.23}$$

Then $\chi(z)$ is a positive function iff there exists an n-dimensional real matrix Q, such that

$$\chi(z) = e_1^T(Q+zI)^{-1}e_1 \quad \text{and} \quad Q+Q^T \quad \text{non-negative definite.} \tag{13.24}$$

Proof. a) The *sufficiency* follows from

$$\operatorname{Re}\chi(z) = q(z)^*\{\operatorname{Re} z\cdot I + \tfrac{1}{2}(Q+Q^T)\}q(z)$$

with $q(z) = (Q+zI)^{-1}e_1$, since $Q+Q^T$ is non-negative definite.

b) For the proof of *necessity*, the hard part, we use Lemma 6.8 of Sect. V.6 below. This lemma is the essential ingredient for Dahlquist's equivalence result and will be proved in the chapter on multistep methods. It states that the positivity of $\chi(z)$ is equivalent to the existence of real, symmetric and non-negative definite matrices A and B, such that for arbitrary $z, w \in \mathbb{C}$ ($\vec{z} = (z^{n-1},\ldots,z,1)^T$, $\vec{w} = (w^{n-1},\ldots,w,1)^T$),

$$\alpha(z)\beta(w) + \alpha(w)\beta(z) = (z+w)\vec{z}^{\,T}A\vec{w} + \vec{z}^{\,T}B\vec{w}. \tag{13.25}$$

The matrix A is positive definite, if $\alpha(z)$ and $\beta(z)$ are relatively prime.

Comparing the coefficients of w^n in (13.25) we get

$$\alpha(z) = \vec{z}^{\,T}Ae_1 \tag{13.26}$$

and observe that the first column of A consists of the coefficients of $\alpha(z)$. For the Cholesky decomposition of A, $A = U^TU$ (U is an upper triangular matrix) we thus have $Ue_1 = e_1$. We next consider the possible computation of the matrix Q from the relation

$$(Q + zI)U\vec{z} = \beta(z) \cdot e_1 \tag{13.27}$$

or equivalently

$$QU\vec{z} = \beta(z) \cdot e_1 - zU\vec{z}. \tag{13.28}$$

The right-hand side of (13.28) is a known polynomial of degree $n-1$, since $Ue_1 = e_1$. Therefore, a comparison of the coefficients in (13.28) yields the matrix QU and hence also Q. It remains to prove that this matrix Q satisfies (13.24).

Using (13.27), the formula $Ae_1 = U^T U e_1 = U^T e_1$ and (13.26) we obtain

$$e_1^T (Q + zI)^{-1} e_1 \cdot \beta(z) = e_1^T U\vec{z} = e_1^T A^T \vec{z} = \alpha(z), \tag{13.29}$$

which verifies the first relation of (13.24). Further, from (13.27) and $\alpha(z) = e_1^T U\vec{z}$ we get

$$\vec{z}^T U^T (Q + wI) U\vec{w} = \alpha(z)\beta(w).$$

Inserting this formula and the analogous one (with z and w exchanged) into (13.25) yields $0 = \vec{z}^T (B - U^T(Q + Q^T)U)\vec{w}$, so that $B = U^T(Q + Q^T)U$. This verifies the second relation of (13.24), since B is symmetric and non-negative definite. □

Exercises

1. (Perron (1913) attributes this result to Wallis, Arithmetica infinitorum 1655 and Euler 1737). Let the sequences $\{A_k\}$ and $\{B_k\}$ be given by

$$\begin{aligned} A_k &= b_k A_{k-1} + a_k A_{k-2}, \qquad & A_{-1} &= 1, \quad & A_0 &= b_0 \\ B_k &= b_k B_{k-1} + a_k B_{k-2}, \qquad & B_{-1} &= 0, \quad & B_0 &= 1 \end{aligned} \tag{13.30}$$

then

$$\frac{A_n}{B_n} = b_0 + \frac{a_1 |}{|b_1} + \ldots + \frac{a_n |}{|b_n}. \tag{13.31}$$

Hint. Let $x = (x_0, x_1, \ldots, x_{n+1})^T$ be the solution of $Mx = (0, \ldots, 0, 1)^T$, where

$$M = \begin{pmatrix} 1 & -b_0 & -a_1 & & & \\ & 1 & -b_1 & -a_2 & & \\ & & \ddots & \ddots & \ddots & \\ & & & 1 & -b_{n-1} & -a_n \\ & & & & 1 & -b_n \\ & & & & & 1 \end{pmatrix}.$$

One easily finds

$$\frac{x_0}{x_1} = b_0 + \frac{a_1 \quad |}{|x_1/x_2} = b_0 + \frac{a_1 |}{|b_1} + \frac{a_2 \quad |}{|x_2/x_3} = \ldots$$

so that x_0/x_1 is equal to the right hand side of (13.31). The statement now follows from the fact that

$$(A_{-1}, A_0, \ldots, A_n)M = (1, 0, \ldots, 0)$$
$$(B_{-1}, B_0, \ldots, B_n)M = (0, 1, 0, \ldots, 0).$$

implying $x_0 = A_n$ and $x_1 = B_n$.

2. Let $P_s(z)$ be the Legendre polynomial (13.4) and $N_s^G(z)$ defined by the recursion (13.5) with $N_0^G(z) = 0$, $N_1^G(z) = 1$. Prove that

$$N_{s-k}^G(z)P_{s-k-1}(z) - N_{s-k-1}^G(z)P_{s-k}(z) = \tau_1 \cdot \tau_2 \cdot \ldots \cdot \tau_{s-k-1}.$$

Hint. Use the relation

$$\begin{pmatrix} N_m^G(z) & P_m(z) \\ N_{m-1}^G(z) & P_{m-1}(z) \end{pmatrix} = \begin{pmatrix} z - \frac{1}{2} & -\tau_{m-1} \\ 1 & 0 \end{pmatrix} \begin{pmatrix} N_{m-1}^G(z) & P_{m-1}(z) \\ N_{m-2}^G(z) & P_{m-2}(z) \end{pmatrix}.$$

3. Consider the Hermitian quadrature formula

$$\int_0^1 f(x)dx = b_1 f(c_1) + \alpha f(c_2) + \beta \frac{f'(c_2)}{1!} + \gamma \frac{f''(c_2)}{2!}. \tag{13.32}$$

Replace $f'(c_2)$ and $f''(c_2)$ by finite divided differences based on $f(c_2 - \varepsilon)$, $f(c_2)$, $f(c_2 + \varepsilon)$ to obtain a quadrature formula

$$\int_0^1 f(x)dx = \overline{b}_1 f(c_1) + \overline{b}_2 f(c_2 - \varepsilon) + \overline{b}_3 f(c_2) + \overline{b}_4 f(c_2 + \varepsilon). \tag{13.33}$$

a) Compute $Q(z)$ for Formula (13.33) and obtain, by letting $\varepsilon \to 0$, an expression which generalizes (13.2) to Hermitian quadrature formulas.

b) Compute the values of b_1 and b_2 $(l_1 = 1, l_2 = 3)$ of (13.10').

c) Show that $n_+ - n_-$ (see Theorem 13.5) is the same for (13.32) and (13.33) with ε sufficiently small.

Results. a) $$Q(z) = \frac{b_1}{z - c_1} + \frac{\alpha}{z - c_2} + \frac{\beta}{(z - c_2)^2} + \frac{\gamma}{(z - c_2)^3}$$

b) $b_1 = b_1$ (sic!), $b_2 = \gamma/3!$.

4. The rational function $\chi(z) = \alpha(z)/\beta(z)$ with $\alpha(z) = z + \alpha_1$, $\beta(z) = z^2 + \beta_1 z + \beta_2$ is positive, iff $\alpha_1 \geq 0$, $\beta_2 \geq 0$, $\beta_1 - \alpha_1 \geq 0$ (compare (5.48))

a) Find real, symmetric and non-negative definite matrices A and B such that (13.25) holds.

b) Show that these matrices are, in general, not unique.

c) As in the proof of Lemma 13.19, compute the matrix Q such that (13.24) holds.
Hint. Begin with the construction of B by putting $w = -z$ in (13.25).

IV.14 Existence and Uniqueness of IRK Solutions

Jusqu'à présent, nous avons supposé que le schéma admettait une solution. Pour en démontrer l'existence ...

(Crouzeix & Raviart 1980)

Since contractivity without feasibility makes little sense ...

(M.N. Spijker 1985)

Since the Runge-Kutta methods studied in the foregoing sections are all implicit, we have to ensure that the numerical solutions, for which we have derived so many nice results, also really exist. The existence theory for implicit Runge-Kutta methods, presented in Volume I (Theorem II.7.2), is for the non-stiff case only, where hL is small (L the Lipschitz constant). This is not a reasonable assumption for the stiff case.

We shall study the existence of a Runge-Kutta solution, defined implicitly by

$$g_i = y_0 + h\sum_{j=1}^{s} a_{ij} f(x_0 + c_j h, g_j), \quad i = 1, \ldots, s \tag{14.1a}$$

$$y_1 = y_0 + h\sum_{j=1}^{s} b_j f(x_0 + c_j h, g_j), \tag{14.1b}$$

for differential equations which satisfy the one-sided Lipschitz condition

$$\langle f(x,y) - f(x,z), y - z\rangle \le \nu \|y - z\|^2. \tag{14.2}$$

Existence

It was first pointed out by Crouzeix & Raviart (1980) that the coercivity of the Runge-Kutta matrix A (or of its inverse) plays an important role for the proof of existence.

Definition 14.1. We consider the inner product $\langle u, v\rangle_D = u^T D v$, where $D = \operatorname{diag}(d_1, \ldots, d_s)$ with $d_i > 0$. We then denote by $\alpha_D(A^{-1})$ the largest number α such that

$$\langle u, A^{-1}u\rangle_D \ge \alpha \langle u, u\rangle_D \qquad \text{for all} \quad u \in \mathbb{R}^s. \tag{14.3}$$

We also set

$$\alpha_0(A^{-1}) = \sup_{D>0} \alpha_D(A^{-1}). \tag{14.4}$$

The first existence results for the above problem were given by Crouzeix & Raviart (1980), Dekker (1982) and Crouzeix, Hundsdorfer & Spijker (1983). Their results can be summarized as follows:

Theorem 14.2. *Let f be continuously differentiable and satisfy (14.2). If the Runge-Kutta matrix A is invertible and*

$$h\nu < \alpha_0(A^{-1}) \tag{14.5}$$

then the nonlinear system (14.1a) possesses a solution $(g_1,\ldots,g_s)$.

Proof. The original proofs are based on the "uniform monotonicity theorem" or on similar results. We present here a more elementary version which, however, has the disadvantage of requiring the differentiability hypothesis for f. The idea is to consider the homotopy

$$g_i = y_0 + h\sum_{j=1}^{s} a_{ij} f(x_0 + c_j h, g_j) + (\tau - 1)h\sum_{j=1}^{s} a_{ij} f(x_0 + c_j h, y_0), \tag{14.6}$$

which is constructed in such a way that for $\tau = 0$ the system (14.6) has the solution $g_i = y_0$, and for $\tau = 1$ it is equivalent to (14.1a). We consider g_i as functions of τ and differentiate (14.6) with respect to this parameter. This gives

$$\dot g_i = h\sum_{j=1}^{s} a_{ij}\frac{\partial f}{\partial y}(x_0 + c_j h, g_j)\cdot \dot g_j + h\sum_{j=1}^{s} a_{ij} f(x_0 + c_j h, y_0)$$

or equivalently

$$\big(I - h(A\otimes I)\{f_y\}\big)\,\dot g = h(A\otimes I) f_0 \tag{14.7}$$

where we have used the notations

$$\dot g = (\dot g_1,\ldots,\dot g_s)^T, \qquad f_0 = \big(f(x_0 + c_1 h, y_0),\ldots, f(x_0 + c_s h, y_0)\big)^T$$

(more precisely, $\dot g$ should be written as $(\dot g_1^T,\ldots,\dot g_s^T)^T$) and

$$\{f_y\} = \mathrm{blockdiag}\Big(\frac{\partial f}{\partial y}(x_0 + c_1 h, g_1),\ldots,\frac{\partial f}{\partial y}(x_0 + c_s h, g_s)\Big).$$

In order to show that $\dot g$ can be expressed as $\dot g = G(g)$ with a globally bounded $G(g)$, we take a D satisfying $h\nu < \alpha_D(A^{-1})$, multiply (14.7) by $\dot g^T(DA^{-1}\otimes I)$ and so obtain

$$\dot g^T(DA^{-1}\otimes I)\dot g - h\dot g^T(D\otimes I)\{f_y\}\dot g = h\dot g^T(D\otimes I) f_0. \tag{14.8}$$

We now estimate the three individual terms of this equation.

The estimate

$$\dot g^T(DA^{-1}\otimes I)\dot g \ge \alpha_D(A^{-1})\ |||\dot g|||_D^2, \tag{14.9}$$

where we have introduced the notation $|||\dot g|||_D^2 = \dot g^T(D\otimes I)\dot g$, is (14.3) in the case of *scalar* differential equations (absence of "$\otimes I$"). In the general case we must apply the ideas of Exercise 1 of Sect. IV.12 to the matrix $\frac{1}{2}(DA^{-1} + (DA^{-1})^T) - \alpha_D(A^{-1})D$, which is non-negative definite by Definition 14.1. It follows from (14.2) with $y = z + \varepsilon u$ that

$$\Big\langle \varepsilon\frac{\partial f}{\partial y}(x,z)u + o(\varepsilon), \varepsilon u\Big\rangle \le \nu\varepsilon^2\|u\|^2.$$

Dividing by ε^2 and taking the limit $\varepsilon \to 0$ we obtain $\langle u, \frac{\partial f}{\partial y}(x,z)u\rangle \le \nu\|u\|^2$ for all (x,z) and all u. Consequently we also have

$$\dot{g}^T(D\otimes I)\{f_y\}\dot{g} \le \nu |||\dot{g}|||_D^2. \tag{14.10}$$

The right-hand term of (14.8) is bounded by $h|||\dot{g}|||_D \cdot |||f_0|||_D$ by the Cauchy-Schwarz-Bunjakowski inequality.

Inserting these three estimates into (14.8) yields

$$\big(\alpha_D(A^{-1}) - h\nu\big)\ |||\dot{g}|||_D^2 \le h\ |||\dot{g}|||_D \cdot |||f_0|||_D.$$

This proves that $\dot{g}$ can be written as $\dot{g} = G(g)$ with

$$|||G(g)|||_D \le \frac{h|||f_0|||_D}{\alpha_D(A^{-1}) - h\nu}.$$

It now follows from Theorem 7.4 (Sect. I.7) that this differential equation with initial values $g_i(0) = y_0$ possesses a solution for all τ, in particular also for $\tau = 1$. This proves the existence of a solution of (14.1a). □

Remark. It has recently been shown by Kraaijevanger & Schneid (1991, Theorem 2.12) that Condition (14.5) is "essentially optimal".

A Counterexample

After our discussion that Monday afternoon (October 1980) I went for a walk and I got the idea for the counterexample.
(M.N. Spijker)

The inequality in (14.5) is *strict*, therefore Theorem 14.2 (together with Exercise 1 below) does not yet answer the simple question: "does a B-stable method on a contractive problem ($\nu = 0$) always admit a solution". A first counterexample to this statement has been given by Crouzeix, Hundsdorfer & Spijker (1983). An easy idea for constructing another counterexample is to use the W-transformation (see Sections IV.5 and IV.13) as follows:

We put $s = 4$ and take a quadrature formula with positive weights, say,

$$(c_i) = (0, 1/3, 2/3, 1), \quad (b_i) = (1/8, 3/8, 3/8, 1/8).$$

We then construct a matrix W satisfying property $T(1,1)$ according to Lemma 5.12. This yields for the above quadrature formula

$$W = \begin{pmatrix} 1 & -\sqrt{3} & \sqrt{3} & -1 \\ 1 & -\sqrt{3}/3 & -\sqrt{3}/3 & 1 \\ 1 & \sqrt{3}/3 & -\sqrt{3}/3 & -1 \\ 1 & \sqrt{3} & \sqrt{3} & 1 \end{pmatrix}.$$

Finally, we put (with $\xi_1 = 1/(2\sqrt{3})$)

$$A = WXW^{-1} \quad \text{with} \quad X = \begin{pmatrix} 1/2 & -\xi_1 & 0 & 0 \\ \xi_1 & 0 & 0 & 0 \\ 0 & 0 & 0 & -\beta \\ 0 & 0 & \beta & 0 \end{pmatrix}.$$

For $\beta = 1/(4\sqrt{3})$ this gives nice rational coefficients for the RK-matrix, namely

$$A = \frac{1}{48}\begin{pmatrix} 3 & 0 & 3 & -6 \\ 6 & 9 & 0 & 1 \\ 5 & 18 & 9 & 0 \\ 12 & 15 & 18 & 3 \end{pmatrix}.$$

It follows from Theorem 13.15 that this method is algebraically stable and of order 4. However, $\pm i\beta$ is an eigenvalue pair of X and hence also of A.

We thus choose the differential equation

$$y' = Jy + f(x) \quad \text{with} \quad J = \begin{pmatrix} 0 & -1/\beta \\ 1/\beta & 0 \end{pmatrix},$$

which satisfies (14.2) with $\nu = 0$ independent of the choice of $f(x)$. If we apply the above method with $h = 1$ to this problem and initial values $x_0 = 0$, $y_0 = (0,0)^T$, Eq. (14.1a) becomes equivalent to the linear system

$$(I - A \otimes J)g = (A \otimes I)f_0,$$

where $g = (g_1, \ldots, g_4)^T$ and $f_0 = (f(c_1), \ldots, f(c_4))^T$. The matrix $(I - A \otimes J)$ is singular because the eigenvalues of $I - A \otimes J$ are just $1 - \lambda\mu$ where λ and μ are the eigenvalues of A and J, respectively. However, A is regular, therefore it is possible to choose $f(x)$ in such a way that this equation does not have a solution.

Influence of Perturbations and Uniqueness

Our next problem is the question, how *perturbations* in the Runge-Kutta equations influence the numerical solution. Research into this problem was initiated independently by Frank, Schneid & Ueberhuber (preprint 1981, published 1985) and Dekker (1982).

As above, we use the notations

$$\|u\|_D = \sqrt{u^T D u} = \sqrt{\langle u, u \rangle_D} \qquad u \in \mathbb{R}^s$$

$$|||g|||_D = \sqrt{g^T (D \otimes I) g} \qquad g \in \mathbb{R}^{sn}$$

and $\|A\|_D$ for the corresponding matrix norm.

Theorem 14.3 (Dekker 1984). *Let g_i and y_1 be given by (14.1) and consider perturbed values $\widehat{g}_i$ and $\widehat{y}_1$ satisfying*

$$\widehat{g}_i = y_0 + h\sum_{j=1}^{s} a_{ij} f(x_0 + c_j h, \widehat{g}_j) + \delta_i \tag{14.11a}$$

$$\widehat{y}_1 = y_0 + h\sum_{j=1}^{s} b_j f(x_0 + c_j h, \widehat{g}_j). \tag{14.11b}$$

If the Runge-Kutta matrix A is invertible, if the one-sided Lipschitz condition (14.2) is satisfied, and $h\nu < \alpha_D(A^{-1})$ for some positive diagonal matrix D, then we have the estimates

$$|||\widehat{g} - g|||_D \le \frac{\|A^{-1}\|_D}{\alpha_D(A^{-1}) - h\nu}\, |||\delta|||_D \tag{14.12}$$

$$\|\widehat{y}_1 - y_1\| \le \|b^T A^{-1}\|_D \left(1 + \frac{\|A^{-1}\|_D}{\alpha_D(A^{-1}) - h\nu}\right) |||\delta|||_D, \tag{14.13}$$

where $g = (g_1, \ldots, g_s)^T$, $\widehat{g} = (\widehat{g}_1, \ldots, \widehat{g}_s)^T$, and $\delta = (\delta_1, \ldots, \delta_s)^T$.

Proof. With the notation $\Delta g = \widehat{g} - g$ and

$$\Delta f = \Big(f(x_0 + c_1 h, \widehat{g}_1) - f(x_0 + c_1 h, g_1), \ldots, f(x_0 + c_s h, \widehat{g}_s) - f(x_0 + c_s h, g_s)\Big)^T$$

the difference of (14.11a) and (14.1a) can be written as

$$\Delta g = h(A \otimes I)\Delta f + \delta.$$

As in the proof of Theorem 14.2 we multiply this equation by $\Delta g^T(DA^{-1} \otimes I)$ and obtain

$$\Delta g^T(DA^{-1} \otimes I)\Delta g - h\Delta g^T(D \otimes I)\Delta f = \Delta g^T(DA^{-1} \otimes I)\delta. \tag{14.14}$$

This equation is very similar to Eq. (14.8) and we estimate it in the same way: since D is a diagonal matrix with positive entries, it follows from (14.2) that

$$\Delta g^T(D \otimes I)\Delta f \le \nu\, |||\Delta g|||_D^2. \tag{14.15}$$

Inserting (14.15) and (14.9) (with $\dot{g}$ replaced by Δg) into (14.14) we get

$$(\alpha_D(A^{-1}) - h\nu)\, |||\Delta g|||_D^2 \le |||\Delta g|||_D\, |||(A^{-1} \otimes I)\delta|||_D$$

which implies (14.12). The estimate (14.13) then follows immediately from

$$\widehat{y}_1 - y_1 = h(b^T \otimes I)\Delta f = (b^T A^{-1} \otimes I)(\Delta g - \delta). \qquad \square$$

Putting $\delta = 0$ in Theorem 14.3 we get the following uniqueness result.

Theorem 14.4. *Consider a differential equation satisfying (14.2). If the Runge-Kutta matrix A is invertible and $h\nu < \alpha_0(A^{-1})$, then the system (14.1a) possesses at most one solution.* $\square$

Computation of $\alpha_0(A^{-1})$

... the determination of a suitable matrix D ... This task does not seem easy at first glance ... (K. Dekker 1984)

The value $\alpha_D(A^{-1})$ of Definition 14.1 is the smallest eigenvalue of the symmetric matrix $\big(D^{1/2}A^{-1}D^{-1/2}+(D^{1/2}A^{-1}D^{-1/2})^T\big)/2$. The computation of $\alpha_0(A^{-1})$ is more difficult, because the optimal D is not known in general.

An upper bound for $\alpha_0(A^{-1})$ is

$$\alpha_0(A^{-1})\le \min_{i=1,\ldots,s}\omega_{ii} \tag{14.16}$$

where ω_{ij} are the entries of A^{-1}. This follows from (14.3) by putting $u=e_i$, the ith unit vector.

Lower bounds for $\alpha_0(A^{-1})$ were first given by Frank, Schneid & Ueberhuber in 1981. Following are the exact values due to Dekker (1984), Dekker & Verwer (1984, p. 55-164), and Dekker & Hairer (1985) (see also Liu & Kraaijevanger 1988 and Kraaijevanger & Schneid 1991).

Theorem 14.5. *For the methods of Sect. IV.5 we have:*

Gauss $\quad \alpha_0(A^{-1})=\min_{i=1,..,s}\dfrac{1}{2c_i(1-c_i)},$

Radau IA $\quad \alpha_0(A^{-1})=\begin{cases}1 & \text{if } s=1,\\ \dfrac{1}{2(1-c_2)} & \text{if } s>1,\end{cases}$

Radau IIA $\quad \alpha_0(A^{-1})=\begin{cases}1 & \text{if } s=1,\\ \dfrac{1}{2c_{s-1}} & \text{if } s>1,\end{cases}$

Lobatto IIIC $\quad \alpha_0(A^{-1})=\begin{cases}1 & \text{if } s=2,\\ 0 & \text{if } s>2.\end{cases}$

Proof. a) Gauss methods: written out in "symmetricized form", estimate (14.3) reads

$$\frac{1}{2}u^T\big(DA^{-1}+(DA^{-1})^T\big)u\ge \alpha u^T Du.$$

Evidently the sharpest estimates come out if D is such that the left-hand matrix is as "close to diagonal as possible". After many numerical computations, Dekker had the nice surprise that with the choice $D=B(C^{-1}-I)$, where $B=\operatorname{diag}(b_1,\ldots,b_s)$ and $C=\operatorname{diag}(c_1,\ldots,c_s)$, the matrix

$$DA^{-1}+(DA^{-1})^T=BC^{-2} \tag{14.17}$$

becomes completely diagonal. Then the optimal α is simply obtained by testing the unit vectors $u=e_k$, which gives

$$\alpha_0(A^{-1})=\min_i\frac{b_i}{2c_i^2d_i}=\min_i\frac{b_i}{2c_i^2b_i(1/c_i-1)}=\min_i\frac{1}{2c_i(1-c_i)}.$$

It remains to prove (14.17): we verify the equivalent formula

$$V^T(A^T D + DA - A^T BC^{-2}A)V = 0 \tag{14.18}$$

where $V = (c_i^{j-1})$ is the Vandermonde matrix. The (l,m)-element of the matrix (14.18) is

$$\begin{aligned}\sum_{i,j} b_j\Big(\frac{1}{c_j}-1\Big)a_{ji}c_i^{l-1}c_j^{m-1} + \sum_{i,j} b_i\Big(\frac{1}{c_i}-1\Big)a_{ij}c_i^{l-1}c_j^{m-1}\\ - \sum_{i,j,k} b_i\frac{1}{c_i^2}a_{ik}c_k^{l-1}a_{ij}c_j^{m-1}.\end{aligned} \tag{14.19}$$

With the help of the simplifying assumptions $C(s)$ and $B(2s)$ the expression (14.19) can be seen to be zero.

b) For the Radau IA methods we take $D = B(I-C)$ and show that

$$DA^{-1} + (DA^{-1})^T = B + e_1 e_1^T. \tag{14.20}$$

The stated formula for $\alpha_0(A^{-1})$ then follows from $0 = c_1 < c_2 < \ldots < c_s$ and from

$$\frac{b_1+1}{b_1} \ge \frac{1}{1-c_2},$$

which is a simple consequence of $b_1 = 1/s^2$ (see Abramowitz & Stegun (1964), Formula 25.4.31). For the verification of (14.20) one shows that $V^T(DA^{-1} + (DA^{-1})^T - B - e_1 e_1^T)V = 0$. Helpful formulas for this verification are $A^{-1}Ve_1 = b_1^{-1}e_1$, $V^T e_1 = e_1$ and $A^{-1}Ve_j = (j-1)Ve_{j-1}$ for $j \ge 2$.

c) Similarly, the statement for the Radau IIA methods follows with $D = BC^{-1}$ from the identity

$$DA^{-1} + (DA^{-1})^T = BC^{-2} + e_s e_s^T.$$

d) As in part (b) one proves for the Lobatto IIIC methods that

$$BA^{-1} + (BA^{-1})^T = e_1 e_1^T + e_s e_s^T. \tag{14.21}$$

Since this matrix is diagonal, we obtain $\alpha_0(A^{-1}) = 1$ for $s = 2$ and $\alpha_0(A^{-1}) = 0$ for $s > 2$. □

For diagonally implicit Runge-Kutta methods we have the following result.

Theorem 14.6 (Montijano 1983). *For a DIRK-method with positive a_{ii} we have*

$$\alpha_0(A^{-1}) = \min_{i=1,\ldots,s} \frac{1}{a_{ii}}. \tag{14.22}$$

Proof. With $D = \operatorname{diag}(1, \varepsilon^2, \varepsilon^4, \ldots, \varepsilon^{2s-2})$ we obtain

$$D^{1/2}A^{-1}D^{-1/2} + (D^{1/2}A^{-1}D^{-1/2})^T = \operatorname{diag}(a_{11}^{-1}, \ldots, a_{ss}^{-1}) + \mathcal{O}(\varepsilon),$$

so that $\alpha_0(A^{-1}) \ge \min_i a_{ii}^{-1} + \mathcal{O}(\varepsilon)$. This inequality for $\varepsilon \to 0$ and (14.16) prove the statement. □

Methods with Singular A

For the Lobatto IIIA methods the first stage is explicit (the first row of A vanishes) and for the Lobatto IIIB methods the last stage is explicit (the last column of A vanishes). For these methods the Runge-Kutta matrix is of the form

$$A=\begin{pmatrix} 0 & 0 \\ a & \widetilde{A} \end{pmatrix} \qquad \text{or} \qquad A=\begin{pmatrix} \widetilde{A} & 0 \\ a^T & 0 \end{pmatrix} \tag{14.23}$$

and we have the following variant of Theorem 14.2.

Theorem 14.7. *Let f be continuously differentiable and satisfy (14.2). If the Runge-Kutta matrix is given by one of the matrices in (14.23) with invertible $\widetilde{A}$, then the assumption*

$$h\nu < \alpha_0(\widetilde{A}^{-1})$$

implies that the nonlinear system (14.1a) has a solution.

Proof. The explicit stage poses no problem for the existence of a solution. To obtain the result we repeat the proof of Theorem 14.2 for the $s-1$ implicit stages (i.e., A is replaced by $\widetilde{A}$ and the inhomogenity in (14.6) may be different). □

An explicit formula for $\alpha_0(\widetilde{A}^{-1})$ for the Lobatto IIIB methods has been given by Dekker & Verwer (1984), and for the Lobatto IIIA methods by Liu, Dekker & Spijker (1987). The result is

Theorem 14.8. *We have for*

$$\text{Lobatto IIIA} \qquad \alpha_0(\widetilde{A}^{-1})=\begin{cases} 2 & \text{if } s=2, \\ c_{s-1}^{-1} & \text{if } s>2, \end{cases}$$

$$\text{Lobatto IIIB} \qquad \alpha_0(\widetilde{A}^{-1})=\begin{cases} 2 & \text{if } s=2, \\ (1-c_2)^{-1} & \text{if } s>2. \end{cases}$$

Proof. For the Lobatto IIIA methods we put $D=\widetilde{B}\widetilde{C}^{-2}$ with the diagonal matrices $\widetilde{B}=\operatorname{diag}(b_2,\ldots,b_s)$ and $\widetilde{C}=\operatorname{diag}(c_2,\ldots,c_s)$. As in part (a) of the proof of Theorem 14.5 we get

$$D\widetilde{A}^{-1}+(D\widetilde{A}^{-1})^T=e_{s-1}e_{s-1}^T+2\widetilde{B}\widetilde{C}^{-3}$$

which implies the formula for $\alpha_0(\widetilde{A}^{-1})$, because $b_s=(s(s-1))^{-1}$ and $(1+2b_s)\geq b_s/c_{s-1}$ for $s>2$.

For the Lobatto IIIB methods the choice $D=\widetilde{B}(I-\widetilde{C})^2$ (with the matrices $\widetilde{B}=\operatorname{diag}(b_1,\ldots,b_{s-1})$, $\widetilde{C}=\operatorname{diag}(c_1,\ldots,c_{s-1})$) leads to

$$D\widetilde{A}^{-1}+(D\widetilde{A}^{-1})^T=e_1e_1^T+2\widetilde{B}(I-\widetilde{C}).$$

This proves the second statement. □

Methods with explicit stages (such as Lobatto IIIA and IIIB) don't allow estimates of the numerical solution in the presence of arbitrary perturbations. They are usually not AN-stable and $K(Z)$ is not bounded (see Theorem 12.12). Nevertheless we have the following uniqueness result.

Theorem 14.9. *Consider a differential equation satisfying (14.2). If the Runge-Kutta matrix is of the form (14.23) with invertible* $\widetilde{A}$ *and if* $h\nu < \alpha_0(\widetilde{A}^{-1})$, *then the nonlinear system (14.1a) has at most one solution.*

Proof. Suppose, there exists a second solution $\widehat{g}_i$ satisfying (14.11a) with $\delta_i = 0$.

a) If the first stage is explicit we have $\widehat{g}_1 = g_1$. The difference of the two Runge-Kutta formulas then yields

$$\Delta g = h(\widetilde{A} \otimes I)\Delta f$$

with $\Delta g = (\widehat{g}_i - g_i)_{i=2}^{s}$ and $\Delta f = (f(x_0 + c_i h, \widehat{g}_i) - f(x_0 + c_i h, g_i))_{i=2}^{s}$. As in the proof of Theorem 14.3 we then conclude that $\Delta g = 0$.

b) In the second case we can apply Theorem 14.3 to the first $s-1$ stages, which yields uniqueness of $g_1, \ldots, g_{s-1}$. Clearly, g_s also is unique, because the last stage is explicit. □

Lobatto IIIC Methods

For the Lobatto IIIC methods with $s \geq 3$ we have $\alpha_0(A^{-1}) = 0$ (see Theorem 14.5). Since these methods are algebraically stable it is natural to ask whether the nonlinear system (14.1a) also has a solution for differential equations satisfying (14.2) with $\nu = 0$. A positive answer to this question has been given by Hundsdorfer & Spijker (1987) for the case $s = 3$, and by Liu & Kraaijevanger (1988) for the general case $s \geq 3$ (see Exercise 6 below; see also Kraaijevanger & Schneid 1991).

Exercises

1. Prove that $\alpha_0(A) \geq 0$ for algebraically stable Runge-Kutta methods. Also, $\alpha_0(A^{-1}) \geq 0$ if in addition the matrix A is invertible.

2. Let A be a real matrix. Show that $\alpha_0(A) \leq \operatorname{Re}\lambda$, where λ is an eigenvalue of A.

3. (Hundsdorfer 1985, Cooper 1986). Prove that Theorem 14.2 remains valid for singular A, if $h\nu < \alpha$ with α satisfying

$$\langle u, Au\rangle_D \geq \alpha \langle Au, Au\rangle_D \qquad \text{for all} \quad u \in \mathbb{R}^s.$$

Hint. Use the transformation $g = 1\!\!1 \otimes y_0 + (A \otimes I)k$ and apply the ideas of the proof of Theorem 14.2 to the homotopy

$$k_i = f(x_0 + c_i h, y_0 + h\sum_{j=1}^{s} a_{ij}k_j) + (\tau - 1)f(x_0 + c_i h, y_0).$$

4. (Barker, Berman & Plemmons 1978, Montijano 1983). Prove that for any two-stage method the condition

$$a_{11} > 0, \qquad a_{22} > 0, \qquad \det(A) > 0 \tag{14.24}$$

is equivalent to $\alpha_0(A^{-1}) > 0$.

Remark. For a generalization of this result to three-stage methods see Kraaijevanger (1991).

5. For the two-stage Radau IIA method we have $\alpha_0(A^{-1}) = 3/2$. Construct a differential equation $y' = \lambda(x)y$ with $\operatorname{Re}\lambda(x) = 3/2 + \varepsilon$ ($\varepsilon > 0$ arbitrarily small) such that the Runge-Kutta equations do not admit a unique solution for all $h > 0$.

6. Prove that for the Lobatto IIIC methods (with $s \geq 3$) the matrix

$$I - (A \otimes I)J \quad \text{with} \quad J = \operatorname{blockdiag}(J_1, \ldots, J_s)$$

is non-singular, if $\mu_2(J_k) \leq 0$. This implies that the Runge-Kutta equations (14.1a) have a unique solution for all problems $y' = A(x)y + f(x)$ with $\mu_2(A(x)) \leq 0$.

Hint (Liu & Kraaijevanger 1988, Liu, Dekker & Spijker 1987). Let $v = (v_1, \ldots, v_s)^T$ be a solution of $(I - (A \otimes I)J)v = 0$. With the help of (14.21) show first that $v_1 = v_s = 0$. Then consider the $(s-2)$-dimensional submatrix $\widetilde{A} = (a_{ij})_{i,j=2}^{s-1}$ and prove $\alpha_0(\widetilde{A}^{-1}) > 0$ by considering the diagonal matrix $\widetilde{D} = \operatorname{diag}(b_i(c_i^{-1} - 1)^2)_{i=2}^{s-1}$.

7. Consider an algebraically stable Runge-Kutta method with invertible A and apply it to the differential equation $y' = (J(x) - \varepsilon I)y + f(x)$ where $\mu(J(x)) \leq 0$ and $\varepsilon > 0$. Prove that the numerical solution $y_1(\varepsilon)$ converges to a limit for $\varepsilon \to 0$, whereas the internal stages $g_i(\varepsilon)$ need not converge.

Hint. Expand the $g_i(\varepsilon)$ in a series $g_i(\varepsilon) = \varepsilon^{-1}g_i^{(-1)} + g_i^{(0)} + \varepsilon g_i^{(1)} + \ldots$ and prove the implication

$$g = (A \otimes I)Jg \quad \Longrightarrow \quad (b^T \otimes I)Jg = 0$$

where $J = \operatorname{blockdiag}(J(x_0 + c_1 h), \ldots, J(x_0 + c_s h))$.

IV.15 B-Convergence

> In using A-stable one-step methods to solve large systems of stiff nonlinear differential equations, we have found that
> — (a) some A-stable methods give highly unstable solutions, and
> — (b) the accuracy of the solutions obtained when the equations are stiff often appears to be unrelated to the order of the method used.
> This has caused us to re-examine the form of stability required when stiff systems of equations are solved, and to question the relevance of the concept of (nonstiff) order of accuracy for stiff problems. (A. Prothero & A. Robinson 1974)

Prothero & Robinson (1974) were the first to discover the order reduction of implicit Runge-Kutta methods when applied to stiff differential equations. Frank, Schneid & Ueberhuber (1981) then introduced the "concept of B-convergence", which furnishes global error estimates independent of the stiffness.

The Order Reduction Phenomenon

For the study of the accuracy of Runge-Kutta methods applied to stiff differential equations, Prothero & Robinson (1974) proposed considering the problem

$$y' = \lambda\big(y - \varphi(x)\big) + \varphi'(x), \qquad y(x_0) = \varphi(x_0), \qquad \operatorname{Re}\lambda \le 0. \tag{15.1}$$

This allows explicit formulas for the local and global errors and provides much new insight.

Applying a Runge-Kutta method to (15.1) yields

$$\begin{aligned} g_i &= y_0 + h\sum_{j=1}^{s} a_{ij}\Big(\lambda\big(g_j - \varphi(x_0+c_jh)\big) + \varphi'(x_0+c_jh)\Big) \\ y_1 &= y_0 + h\sum_{j=1}^{s} b_j\Big(\lambda\big(g_j - \varphi(x_0+c_jh)\big) + \varphi'(x_0+c_jh)\Big). \end{aligned} \tag{15.2}$$

If we replace here the g_i, y_0 and y_1 by the exact solution values $\varphi(x_0+c_ih)$, $\varphi(x_0)$ and $\varphi(x_0+h)$, respectively, we obtain a defect which is given by

$$\begin{aligned} \varphi(x_0+c_ih) &= \varphi(x_0) + h\sum_{j=1}^{s} a_{ij}\varphi'(x_0+c_jh) + \Delta_{i,h}(x_0) \\ \varphi(x_0+h) &= \varphi(x_0) + h\sum_{j=1}^{s} b_j\varphi'(x_0+c_jh) + \Delta_{0,h}(x_0). \end{aligned} \tag{15.3}$$

Taylor series expansion of the functions in (15.3) shows that

$$\Delta_{0,h}(x_0) = \mathcal{O}(h^{p+1}), \qquad \Delta_{i,h}(x_0) = \mathcal{O}(h^{q+1}), \tag{15.4}$$

where p is the order of the quadrature formula (b_i, c_i) and q is the largest number such that the condition $C(q)$ (see Sect. IV.5), i.e.,

$$\sum_{j=1}^{s} a_{ij} c_j^{k-1} = \frac{c_i^k}{k} \qquad \text{for} \quad k = 1, \ldots, q \quad \text{and all} \quad i, \tag{15.5}$$

holds. The minimum of q and p is often called the *stage order* of the Runge-Kutta method. Subtracting (15.3) from (15.2) and eliminating the internal stages we get

$$y_1 - \varphi(x_0 + h) = R(z)(y_0 - \varphi(x_0)) - zb^T(I - zA)^{-1}\Delta_h(x_0) - \Delta_{0,h}(x_0) \tag{15.6}$$

where we have used the notation $z = \lambda h$, $R(z) = 1 + zb^T(I - zA)^{-1}\mathbb{1}$ for the stability function and $\Delta_h(x) = (\Delta_{1,h}(x), \ldots, \Delta_{s,h}(x))^T$. We also denote the *local error*, which we get from (15.6) on putting $y_0 = \varphi(x_0)$, by

$$\delta_h(x) = -zb^T(I - zA)^{-1}\Delta_h(x) - \Delta_{0,h}(x). \tag{15.7}$$

If we repeat the above calculation with x_n instead of x_0 we obtain the recursion

$$y_{n+1} - \varphi(x_{n+1}) = R(z)(y_n - \varphi(x_n)) + \delta_h(x_n) \tag{15.8}$$

which leads to the following formula for the *global error*

$$y_{n+1} - \varphi(x_{n+1}) = R(z)^{n+1}(y_0 - \varphi(x_0)) + \sum_{j=0}^{n} R(z)^{n-j}\delta_h(x_j). \tag{15.9}$$

The classical (non-stiff) theory treats the case where $z = \mathcal{O}(h)$ and in this situation the global error behaves like $\mathcal{O}(h^p)$. When solving stiff differential equations one is interested in step sizes h which are much larger than $|\lambda|^{-1}$. We therefore study the global error (15.9) under the assumption that simultaneously $h \to 0$ and $z = \lambda h \to \infty$. In Table 15.1 we collect the results for the Runge-Kutta methods of Sect. IV.5. There in the last column (variable h) the symbols h and z have to be interpreted as $\max h_i$ and $z = \lambda \min h_i$. We remark that Formulas (15.7) and (15.8) (but not (15.9)) remain valid for variable h, if z is replaced by $z_n = h_n\lambda$.

Table 15.1. Error for (15.1) when $h \to 0$ and $z = h\lambda \to \infty$

Method		local error	global error	
			constant h	variable h
Gauss	s odd	h^{s+1}	h^{s+1}	h^s
	s even		h^s	
Radau IA		h^s	h^s	h^s
Radau IIA		$z^{-1}h^{s+1}$	$z^{-1}h^{s+1}$	$z^{-1}h^{s+1}$
Lobatto IIIA	s odd	$z^{-1}h^{s+1}$	$z^{-1}h^s$	$z^{-1}h^s$
	s even		$z^{-1}h^{s+1}$	
Lobatto IIIB	s odd	zh^{s-1}	zh^{s-2}	zh^{s-2}
	s even		zh^{s-1}	
Lobatto IIIC		$z^{-1}h^s$	$z^{-1}h^s$	$z^{-1}h^s$

Verification of Table 15.1.

Gauss. Since the Runge-Kutta matrix A is invertible, we have $-zb^T(I-zA)^{-1} = b^TA^{-1}+\mathcal{O}(z^{-1})$ and (15.4) inserted into (15.7) gives $\delta_h(x)=\mathcal{O}(h^{s+1})$ (observe that $q=s$). It then follows from (15.8) (for constant and variable h) that the global error behaves like $\mathcal{O}(h^s)$ because $|R(z)|\le 1$. For odd s we have $R(\infty)=-1$ and the global error estimate can be improved in the case of constant step sizes. This follows from partial summation

$$\sum_{j=0}^{n} \varrho^{n-j}\delta(x_j) = \frac{1-\varrho^{n+1}}{1-\varrho}\delta(x_0) + \sum_{j=1}^{n}\frac{1-\varrho^{n+1-j}}{1-\varrho}\big(\delta(x_j)-\delta(x_{j-1})\big) \quad (15.10)$$

of the sum in (15.9) and from the fact that $\delta_h(x_j)-\delta_h(x_{j-1})=\mathcal{O}(h^{q+2})$.

Radau IA. The local error estimate follows in the same way as for the Gauss methods. Since $R(z)=\mathcal{O}(z^{-1})$ the error propagation in (15.8) is negligible and the local and global errors have the same asymptotic behaviour.

Radau IIA and **Lobatto IIIC**. These methods have $a_{si}=b_i$ for all i. Therefore the last internal stage is identical to the numerical solution and the local error can be written as

$$\delta_h(x) = -e_s^T(I-zA)^{-1}\Delta_h(x).$$

Since A is invertible this formula shows the presence of z^{-1} in the local error. Again we have $R(\infty)=0$, so that the global error is essentially equal to the local error.

Lobatto IIIA. The first stage is explicit, $g_1=y_0$, and is done without introducing an error. Therefore $\Delta_{1,h}(x)=0$ and (because of $a_{si}=b_i$) the local error has the form

$$\delta_h(x) = -e_{s-1}^T(I-z\widetilde{A})^{-1}\widetilde{\Delta}_h(x)$$

where $\widetilde{A}=(a_{ij})_{i,j=2}^{s}$ and $\widetilde{\Delta}_h=(\Delta_{2,h},\ldots,\Delta_{s,h})^T$. The statements of Table 15.1 now follow as for the Gauss methods.

Lobatto IIIB. The matrix A is singular (its last column vanishes), therefore the two "z" in (15.7) do not simply cancel for $z\to\infty$. A more detailed analysis (see Exercise 5 below) shows that the local error is not bounded if $z\to\infty$. Although A-stable, these methods are not suited for the solution of stiff problems. □

We observe from Table 15.1 that the order of convergence for problem (15.1) with large λ is considerably smaller than the classical order. Further we see that methods satisfying $a_{si}=b_i$ (Radau IIA, Lobatto IIIA and Lobatto IIIC) give an asymptotically exact result for $z\to\infty$. Prothero & Robinson (1974) call such methods *stiffly accurate*. The importance of this condition will appear again when we treat singularly perturbed and differential-algebraic problems (Chapter VI).

The Local Error

Das besondere Schmerzenskind sind die Fehlerabschätzungen.
(L. Collatz 1950)

Our next aim is to extend the above results to general nonlinear differential equations $y' = f(x,y)$ satisfying a one-sided Lipschitz condition

$$\langle f(x,y) - f(x,z), y - z \rangle \le \nu \|y - z\|^2. \tag{15.11}$$

The following analysis, begun by Frank, Schneid & Ueberhuber (1981), was elaborated by Frank, Schneid & Ueberhuber (1985) and Dekker & Verwer (1984). We again denote the local error by

$$\delta_h(x) = y_1 - y(x+h),$$

where y_1 is the numerical solution with initial value $y_0 = y(x)$ on the exact solution.

Proposition 15.1. *Consider a differential equation which satisfies (15.11). Assume that the Runge-Kutta matrix A is invertible, $\alpha_0(A^{-1}) \ge 0$ (see Definition 14.1), and that the stage order is q.*
a) If $\alpha_0(A^{-1}) > 0$ then

$$\|\delta_h(x)\| \le C\, h^{q+1} \max_{\xi \in [x,x+h]} \|y^{(q+1)}(\xi)\| \quad \textit{for} \quad h\nu \le \alpha < \alpha_0(A^{-1}).$$

b) If $\alpha_D(A^{-1}) = 0$ for some positive diagonal matrix D and $\nu < 0$ then

$$\|\delta_h(x)\| \le C\Big(h + \frac{1}{|\nu|}\Big)\, h^q \max_{\xi \in [x,x+h]} \|y^{(q+1)}(\xi)\|.$$

In both cases the constant C depends only on the coefficients of the Runge-Kutta matrix and on α (for case (a)).

Remarks. a) The crucial fact in these estimates is that the right-hand side depends only on derivatives of the exact solution and not on the stiffness of the problem. These estimates are useful when a "smooth" solution of a stiff problem has to be approximated.

b) The hypothesis $\alpha_D(A^{-1}) = 0$ (see case (b)) is stronger than $\alpha_0(A^{-1}) = 0$ (see Exercise 4 below). For the Lobatto IIIC methods, for which $\alpha_0(A^{-1})=0$ ($s > 2$), we have $\alpha_D(A^{-1}) = 0$ with $D = B$ (see (14.21)). For stiffly accurate methods the estimate of part (b) can be improved by using (14.12) instead of (14.13).

c) In the estimates of the above proposition the maximum is taken over $\xi \in [x, x+h]$. In the case where $0 \le c_i \le 1$ is not satisfied, this interval must of course be correspondingly enlarged.

Proof. We put $\widehat{g}_i = y(x_0 + c_i h)$, so that the relation (14.11a) is satisfied with

$$\delta_i = y(x_0 + c_i h) - y(x_0) - h \sum_{j=1}^{s} a_{ij} y'(x_0 + c_j h).$$

Taylor expansion shows that

$$\|\delta_i\| \le C_i h^{q+1} \max_{x \in [x_0, x_1]} \|y^{(q+1)}(x)\|$$

where $C_i = (|c_i|^{q+1} + (q+1) \sum_{j=1}^{s} |a_{ij}| \cdot |c_j|^q)/(q+1)!$ is a method-dependent constant. Similarly, the value $\widehat{y}_1$ of (14.11b) satisfies

$$y(x_0 + h) - \widehat{y}_1 = y(x_0 + h) - y(x_0) - h \sum_{j=1}^{s} b_j y'(x_0 + c_j h) = \mathcal{O}(h^{q+1}), \quad (15.12)$$

because the order of the quadrature formula (b_i, c_i) is $\ge q$. Since

$$\|\delta_h(x)\| \le \|y_1 - \widehat{y}_1\| + \|\widehat{y}_1 - y(x_0 + h)\|$$

the desired estimates follow from (14.13) of Theorem 14.3. □

Error Propagation

At the end of Sect. IV.12 we derived for some particular Runge-Kutta methods sharp estimates of the form

$$\|\widehat{y}_1 - y_1\| \le \varphi_B(h\nu)\, \|\widehat{y}_0 - y_0\|, \quad (15.13)$$

where $\widehat{y}_1, y_1$ are the numerical solutions corresponding to $\widehat{y}_0, y_0$, respectively, and where the differential equation satisfies (15.11). We give here a simple proof of a crude estimate of $\varphi_B(h\nu)$ which, however, will be sufficient to derive interesting convergence results.

Proposition 15.2 (Dekker & Verwer 1984). *Suppose that the differential equation satisfies (15.11) and apply an algebraically stable Runge-Kutta method with invertible A and $\alpha_0(A^{-1}) > 0$. Then for any α with $0 < \alpha < \alpha_0(A^{-1})$ there exists a constant $C > 0$ such that*

$$\|\widehat{y}_1 - y_1\| \le (1 + Ch\nu)\|\widehat{y}_0 - y_0\| \qquad \textit{for} \quad 0 \le h\nu \le \alpha.$$

Proof. From (12.7) we have (using the notation of the proof of Theorem 12.4)

$$\|\Delta y_1\|^2 = \|\Delta y_0\|^2 + 2\sum_{i=1}^{s} b_i \langle \Delta f_i, \Delta g_i \rangle - \sum_{i=1}^{s} \sum_{j=1}^{s} m_{ij} \langle \Delta f_i, \Delta f_j \rangle. \quad (15.14)$$

By algebraic stability the last term in (15.14) is non-positive and can be neglected. Using (15.11) and the estimate (14.12) with $\delta_i = \widehat{y}_0 - y_0$ we obtain

$$2\sum_{i=1}^{s} b_i\langle \Delta f_i, \Delta g_i\rangle \le 2h\nu \sum_{i=1}^{s} b_i\,\|\Delta g_i\|^2$$

$$\le 2h\nu\, C_1\, |||\Delta g|||_D^2 \le \frac{2h\nu C_2}{(\alpha_D(A^{-1}) - h\nu)^2}\,\|\Delta y_0\|^2.$$

Inserting this into (15.14) yields

$$\|\Delta y_1\| \le \Big(1 + \frac{h\nu C_2}{(\alpha_D(A^{-1}) - h\nu)^2}\Big)\|\Delta y_0\|$$

which proves the desired estimate. □

B-Convergence for Variable Step Sizes

We are now in a position to present the main result of this section.

Theorem 15.3. *Consider an algebraically stable Runge-Kutta method with invertible A and stage order $q \le p$ and suppose that (15.11) holds.*
a) If $0 < \alpha < \alpha_0(A^{-1})$ and $\nu > 0$ then the global error satisfies

$$\|y_n - y(x_n)\| \le h^q \frac{(e^{C_1\nu(x_n - x_0)} - 1)}{C_1\nu} C_2 \max_{x\in[x_0,x_n]} \|y^{(q+1)}(x)\| \quad \textit{for} \quad h\nu \le \alpha.$$

b) If $\alpha_0(A^{-1}) > 0$ and $\nu \le 0$ then

$$\|y_n - y(x_n)\| \le h^q (x_n - x_0) C_2 \max_{x\in[x_0,x_n]} \|y^{(q+1)}(x)\| \quad \textit{for all} \quad h > 0.$$

c) If $\alpha_D(A^{-1}) = 0$ for some positive diagonal matrix D and $\nu < 0$ then

$$\|y_n - y(x_n)\| \le h^{q-1} C\Big(h + \frac{1}{|\nu|}\Big)(x_n - x_0) \max_{x\in[x_0,x_n]} \|y^{(q+1)}(x)\|.$$

The constants C_1, C_2, C depend only on the coefficients of the Runge-Kutta matrix. In the case of variable step sizes, h has to be interpreted as $h = \max h_i$.

Proof. This convergence result is obtained in exactly the same way as that for nonstiff problems (Theorem II.3.6). For the transported errors E_j (see Fig. II.3.2) we have the estimate (for $\nu \ge 0$)

$$\|E_j\| \le e^{C\nu(x_n - x_j)}\,\|\delta_{h_{j-1}}(x_{j-1})\| \tag{15.15}$$

by Proposition 15.2, because $1 + Ch\nu \le e^{C\nu h}$. We next insert the local error estimate of Proposition 15.1 into (15.15) and sum up the transported errors E_j. This yields the desired estimate for $\nu \geq 0$ because

$$\sum_{j=1}^{n} h_{j-1} e^{C\nu(x_n - x_j)} \leq \int_{x_0}^{x_n} e^{C\nu(x_n - x)} dx$$

$$= \begin{cases} (e^{C\nu(x_n - x_0)} - 1)/(C\nu) & \text{for } \nu > 0 \\ x_n - x_0 & \text{for } \nu = 0. \end{cases}$$

If $\nu < 0$ we have $\|E_j\| \leq \|\delta_{h_{j-1}}(x_{j-1})\|$ by algebraic stability and the same arguments apply. □

Motivated by this result we define the order of B-convergence as follows:

Definition 15.4 (Frank, Schneid & Ueberhuber 1981). A Runge-Kutta method is called *B-convergent of order r* for problems $y' = f(x,y)$ satisfying (15.11), if the global error admits an estimate

$$\|y_n - y(x_n)\| \leq h^r \gamma(x_n - x_0, \nu) \max_{j=1,\ldots,l} \max_{x \in [x_0, x_n]} \|y^{(j)}(x)\| \quad \text{for} \quad h\nu \leq \alpha, \tag{15.16}$$

where $h = \max h_i$. Here γ is a method-dependent function and α also depends only on the coefficients of the method.

As an application of the above theorem we have

Theorem 15.5. *The Gauss and Radau IIA methods are B-convergent of order s (number of stages). The Radau IA methods are B-convergent of order $s-1$. The 2-stage Lobatto IIIC method is B-convergent of order 1.* □

For the Lobatto IIIC methods with $s \geq 3$ stages $(\alpha_0(A^{-1}) = 0$ and $q = s-1)$ Theorem 15.3 shows B-convergence of order $s-2$ if $\nu < 0$. This is not an optimal result. Spijker (1986) proved B-convergence of order $s - 3/2$ for $\nu < 0$ and constant step sizes. Schneid (1987) improved this result to $s-1$. Recently, Dekker, Kraaijevanger & Schneid (1991) showed that these methods are B-convergent of order $s-1$ for general step size sequences, if one allows the function γ in Definition 15.4 to depend also on the ratio $\max h_i / \min h_i$.

The Lobatto IIIA and IIIB methods cannot be B-convergent since they are not algebraically stable. This will be the content of the next subsection.

B-Convergence Implies Algebraic Stability

In order to find necessary conditions for B-convergence we consider the problem

$$y' = \lambda(x)\big(y - \varphi(x)\big) + \varphi'(x), \qquad \operatorname{Re}\lambda(x) \le \nu \tag{15.17}$$

with exact solution $\varphi(x) = x^{q+1}$. We apply a Runge-Kutta method with stage order q and obtain for the global error $\varepsilon_n = y_n - \varphi(x_n)$ the simple recursion

$$\varepsilon_{n+1} = K(Z_n)\varepsilon_n - L(Z_n)h^{q+1} \tag{15.18}$$

(cf. Eq. (15.8) of the beginning of this section, where the case $\lambda(x) = \lambda$ was treated). Here $Z_n = \operatorname{diag}\big(h\lambda(x_n + c_1 h), \ldots, h\lambda(x_n + c_s h)\big)$ and

$$K(Z) = 1 + b^T Z(I - AZ)^{-1}1\!\!1, \quad L(Z) = d_0 + b^T Z(I - AZ)^{-1} d. \tag{15.19}$$

The function $K(Z)$ was already encountered in Definition 12.10, when treating AN-stability. The vector $d = (d_1, \ldots, d_s)^T$ and d_0 in $L(Z)$ characterize the local error and are given by

$$d_0 = 1 - (q+1)\sum_{j=1}^{s} b_j c_j^q, \qquad d_i = c_i^{q+1} - (q+1)\sum_{j=1}^{s} a_{ij} c_j^q. \tag{15.20}$$

Observe that by definition of the stage order we have either $d_0 \ne 0$ or $d \ne 0$ (or both). We are now in the position to prove

Theorem 15.6 (Dekker, Kraaijevanger & Schneid 1991). *Consider a DJ-irreducible Runge-Kutta method which satisfies $0 \le c_1 < c_2 < \ldots < c_s \le 1$. If, for some r, l and $\nu < 0$, the global error satisfies the B-convergence estimate (15.16), then the method is algebraically stable.*

Proof. Suppose that the method is not algebraically stable. Then, by Theorem 12.13 and Lemma 15.17 below, there exists $Z = \operatorname{diag}(z_1, \ldots, z_s)$ with $\operatorname{Re} z_j < 0$ such that $(I - AZ)^{-1}$ exists and

$$|K(Z)| > 1, \qquad L(Z) \ne 0. \tag{15.21}$$

We consider the interval $[0, (1+\theta)/2]$ and for even N the step size sequence $(h_n)_{n=0}^{N-1}$ given by

$$h_n = 1/N \quad (\text{for } n \text{ even}), \quad h_n = \theta/N \quad (\text{for } n \text{ odd}).$$

If N is sufficiently large it is possible to define a function $\lambda(x)$ which satisfies $\operatorname{Re}\lambda(x) \le \nu$ and

$$\lambda(x_n + c_i h_n) = \begin{cases} N z_i & \text{for } n \text{ even} \\ N z_{s+1-i} & \text{for } n \text{ odd}\,. \end{cases}$$

Because of (15.18) the global error $\varepsilon_n = y_n - \varphi(x_n)$ for the problem (15.17) satisfies (with $h = 1/N$)

$$\varepsilon_{n+1} = K(Z)\varepsilon_n - h^{q+1}L(Z) \quad \text{for } n \text{ even}$$
$$\varepsilon_{n+1} = K(\widetilde{Z})\varepsilon_n - h^{q+1}L(\widetilde{Z}) \quad \text{for } n \text{ odd}$$

where $\widetilde{Z} = \operatorname{diag}(\theta z_s, \ldots, \theta z_1)$. Consequently we have

$$\varepsilon_{2m+2} = K(\widetilde{Z})K(Z)\varepsilon_{2m} - h^{q+1}(K(\widetilde{Z})L(Z) + \theta^{q+1}L(\widetilde{Z}))$$

and the error at $X = (1+\theta)/2$ is given by

$$\varepsilon_N = -\frac{1}{N^{q+1}}\,(K(\widetilde{Z})L(Z) + \theta^{q+1}L(\widetilde{Z}))\,\frac{(K(\widetilde{Z})K(Z))^{N/2} - 1}{K(\widetilde{Z})K(Z) - 1}. \tag{15.22}$$

If θ is sufficiently small, $K(\widetilde{Z}) \to 1$ and $L(\widetilde{Z}) \to d_0$, so that by (15.21)

$$|K(\widetilde{Z})K(Z)| > 1 \quad \text{and} \quad K(\widetilde{Z})L(Z) + \theta^{q+1}L(\widetilde{Z}) \neq 0.$$

Therefore $|\varepsilon_N| \to \infty$ as $N \to \infty$ (N even), which contradicts the estimate (15.16) of B-convergence. □

To complete the above proof we give the following lemma:

Lemma 15.7 (Dekker, Kraaijevanger & Schneid 1990). *Consider a DJ-irreducible Runge-Kutta method and suppose*

$$b^T Z(I - AZ)^{-1} d = 0 \tag{15.23}$$

for all $Z = \operatorname{diag}(z_1, \ldots, z_s)$ with $I - AZ$ invertible; then $d = 0$.

Proof. We define

$$T = \{j \mid b_{i_1} a_{i_1 i_2} a_{i_2 i_3} \cdots a_{i_{k-1} i_k} = 0 \quad \text{for all } k \text{ and } i_l \text{ with } i_k = j\}.$$

Putting $k = 1$ we obtain $b_j = 0$ for $j \in T$. Further, if $i \notin T$ and $j \in T$ there exists $(i_1, \ldots, i_k)$ with $i_k = i$ such that

$$b_{i_1} a_{i_1 i_2} \cdots a_{i_{k-1} i_k} \neq 0, \quad b_{i_1} a_{i_1 i_2} \cdots a_{i_{k-1} i_k} a_{ij} = 0$$

implying $a_{ij} = 0$. Therefore the method is DJ-reducible if $T \neq \emptyset$. For the proof of the statement it thus suffices to show that $d \neq 0$ implies $T \neq \emptyset$.

Replacing $(I - AZ)^{-1}$ by its geometric series, assumption (15.23) becomes equivalent to

$$b^T Z(AZ)^{k-1} d = 0 \quad \text{for all } k \text{ and} \quad Z = \operatorname{diag}(z_1, \ldots, z_s). \tag{15.24}$$

Comparing the coefficient of $z_{i_1} \cdots z_{i_k}$ gives

$$\sum b_{j_1} a_{j_1 j_2} \cdots a_{j_{k-1} j_k} d_{j_k} = 0, \tag{15.25}$$

where the summation is over all permutations $(j_1,\ldots,j_k)$ of $(i_1,\ldots,i_k)$. Suppose now that $d_j \neq 0$ for some index j. We shall prove by induction on k that

$$b_{i_1} a_{i_1 i_2} \ldots a_{i_{k-1} i_k} = 0 \quad \text{for all} \quad i_\ell \,(\ell = 1,\ldots,k) \quad \text{with} \quad i_k = j, \tag{15.26}$$

so that $j \in T$ and consequently $T \neq \emptyset$.

For $k = 1$ this follows immediately from (15.25). In order to prove (15.26) for $k+1$ we suppose, by contradiction, that $(i_1,\ldots,i_{k+1})$ with $i_{k+1} = j$ exists such that $b_{i_1} a_{i_1 i_2} \ldots a_{i_k i_{k+1}} \neq 0$. The relation (15.25) then implies the existence of a permutation $(j_1,\ldots,j_{k+1})$ of $(i_1,\ldots,i_{k+1})$ such that $b_{j_1} a_{j_1 j_2} \ldots a_{j_k j_{k+1}} \neq 0$, too. We now denote by q the smallest index for which $i_q \neq j_q$. Then $i_q = j_r$ for some $r > q$ and

$$b_{i_1} a_{i_1 i_2} \cdots a_{i_{q-1} i_q} a_{j_r j_{r+1}} \cdots a_{j_k j_{k+1}} \neq 0 \tag{15.27}$$

contradicts the induction hypothesis, because the expression in (15.27) contains at most k factors. □

The Trapezoidal Rule

The trapezoidal rule

$$y_{k+1} = y_k + \frac{h_k}{2}\Big(f(x_k, y_k) + f(x_{k+1}, y_{k+1})\Big) \tag{15.28}$$

is not algebraically stable. Therefore (Theorem 15.6) it cannot be B-convergent in the sense of Definition 15.4. Nevertheless it is possible to derive estimates (15.16), if we restrict ourselves to special step size sequences (constant, monotonic, ...). This was first proved by Stetter (unpublished) and investigated in detail by Kraaijevanger (1985). The result is

Theorem 15.8 (Kraaijevanger 1985). *If the differential equation satisfies (15.11), then the global error of the trapezoidal rule permits for $h_j \nu \leq \alpha < 2$ the estimate*

$$\|y_n - y(x_n)\| \leq C \max_{x \in [x_0, x_n]} \|y^{(3)}(x)\| \sum_{k=0}^{n-1} \Big(\prod_{j=k+1}^{n-1} \max\,(1, h_j / h_{j-1}) \Big) h_k^3.$$

Proof. We denote by $\widehat{y}_k = y(x_k)$ the exact solution at the grid points. From the Taylor expansion we then get

$$\widehat{y}_{k+1} = \widehat{y}_k + \frac{h_k}{2}\Big(f(x_k, \widehat{y}_k) + f(x_{k+1}, \widehat{y}_{k+1})\Big) + \delta_k \tag{15.29}$$

where

$$\|\delta_k\| \leq \frac{1}{12} h_k^3 \max_{x \in [x_k, x_{k+1}]} \|y^{(3)}(x)\|. \tag{15.30}$$

The main idea is now to introduce the intermediate values

$$\begin{aligned} y_{k+1/2} &= y_k + \frac{h_k}{2} f(x_k, y_k) = y_{k+1} - \frac{h_k}{2} f(x_{k+1}, y_{k+1}) \\ \widehat{y}_{k+1/2} &= \widehat{y}_k + \frac{h_k}{2} f(x_k, \widehat{y}_k) + \delta_k = \widehat{y}_{k+1} - \frac{h_k}{2} f(x_{k+1}, \widehat{y}_{k+1}). \end{aligned} \tag{15.31}$$

The transition $y_{k-1/2} \to y_{k+1/2}$

$$y_{k+1/2} = y_{k-1/2} + \frac{1}{2}(h_{k-1} + h_k) f(x_k, y_k)$$

can then be interpreted as one step of the θ-method

$$y_{m+1} = y_m + hf\big(x_m + \theta h,\ y_m + \theta(y_{m+1} - y_m)\big)$$

with $\theta = h_{k-1}/(h_{k-1} + h_k)$ and step size $h = (h_{k-1} + h_k)/2$. A similar calculation shows that the same θ-method maps $\widehat{y}_{k-1/2}$ to $\widehat{y}_{k+1/2} - \delta_k$. Therefore we have

$$\|\widehat{y}_{k+1/2} - y_{k+1/2} - \delta_k\| \le \varphi_B(h\nu)\ \|\widehat{y}_{k-1/2} - y_{k-1/2}\|,$$

where the growth function $\varphi_B(h\nu)$ is given by (see Eqs. (12.42) and (11.13))

$$\begin{aligned} \varphi_B(h\nu) &= \max\{(1-\theta)/\theta,\ (1 + (1-\theta)h\nu)/(1 - \theta h\nu)\} \\ &= \max\{h_k/h_{k-1},\ (1 + \frac{1}{2}h_k\nu)/(1 - \frac{1}{2}h_{k-1}\nu)\} =: \varphi_k. \end{aligned} \tag{15.32}$$

By the triangle inequality we also get

$$\|\widehat{y}_{k+1/2} - y_{k+1/2}\| \le \varphi_k \|\widehat{y}_{k-1/2} - y_{k-1/2}\| + \|\delta_k\|. \tag{15.33}$$

Further it follows from (15.31) with $k = 0$ and from $\widehat{y}_0 = y_0$ that

$$\|\widehat{y}_{1/2} - y_{1/2}\| = \|\delta_0\|, \tag{15.34}$$

whereas the backward Euler steps $y_{n-1/2} \to y_n$ and $\widehat{y}_{n-1/2} \to \widehat{y}_n$ (see (15.31)) imply

$$\|\widehat{y}_n - y_n\| \le \frac{1}{(1 - \frac{1}{2}h_{n-1}\nu)} \|\widehat{y}_{n-1/2} - y_{n-1/2}\| \tag{15.35}$$

again by Example 12.24 with $\theta = 1$. A combination of (15.33), (15.34) and (15.35) yields

$$\|\widehat{y}_n - y_n\| \le \frac{1}{(1 - \frac{1}{2}h_{n-1}\nu)} \sum_{k=0}^{n-1} \Big(\prod_{j=k+1}^{n-1} \varphi_j \Big) \|\delta_k\|. \tag{15.36}$$

For $\nu \le 0$ we have $\varphi_k \le \max(1, h_k/h_{k-1})$ and the statement follows if we insert (15.30) into (15.36). For $\nu \ge 0$ we use the estimate $(h_{k-1}\nu \le 1)$

$$\frac{1 + \frac{1}{2}h_k\nu}{1 - \frac{1}{2}h_{k-1}\nu} = \frac{1 + \frac{1}{2}h_{k-1}\nu}{1 - \frac{1}{2}h_{k-1}\nu} \cdot \frac{1 + \frac{1}{2}h_k\nu}{1 + \frac{1}{2}h_{k-1}\nu} \le e^{2h_{k-1}\nu} \cdot \max\Big(1, \frac{h_k}{h_{k-1}}\Big)$$

so that the statement holds with $C = e^{2\nu(x_n - x_0)}/12$. □

Corollary 15.9. If the step size sequence $(h_k)_{k=0}^{N-1}$ is constant or monotonic, then for $h = \max h_i$

$$\|y_n - y(x_n)\| \le C \max_{x\in[x_0,x_n]} \|y^{(3)}(x)\| \cdot h^2. \qquad \square$$

Order Reduction for Rosenbrock Methods

Obviously, Rosenbrock methods (Definition 7.1) cannot be B-convergent in the sense of Definition 15.4 (see also Exercise 7 of Sect. IV.12). Nevertheless it is interesting to study their behaviour on stiff problems such as the Prothero & Robinson model (15.1). Since this equation is non-autonomous we have to use the formulation (7.4a). A straightforward calculation shows that the global error $\varepsilon_n = y_n - \varphi(x_n)$ satisfies the recursion

$$\varepsilon_{n+1} = R(z)\varepsilon_n + \delta_h(x_n) \tag{15.37}$$

where $R(z)$ is the stability function (7.14) and the local error is given by

$$\delta_h(x) = \varphi(x) - \varphi(x+h) + b^T(I - zB)^{-1}\Delta \tag{15.38}$$

with $B = (\alpha_{ij} + \gamma_{ij})$, $b = (b_1, \dots, b_s)^T$, $\Delta = (\Delta_1, \dots, \Delta_s)^T$ and

$$\Delta_i = z(\varphi(x) - \varphi(x + \alpha_i h) - \gamma_i h\varphi'(x)) + h\varphi'(x + \alpha_i h) + \gamma_i h^2\varphi''(x).$$

Taylor expansion gives the following result.

Lemma 15.10. *The local error $\delta_h(x)$ of a Rosenbrock method applied to (15.1) satisfies for $h \to 0$ and $z = \lambda h \to \infty$*

$$\delta_h(x) = \Big(\sum_{i,j} b_i\omega_{ij}\alpha_j^2 - 1\Big)\frac{h^2}{2}\varphi''(x) + \mathcal{O}(h^3) + \mathcal{O}\Big(\frac{h^2}{z}\Big),$$

where ω_{ij} are the entries of B^{-1}. $\square$

Remarks. a) Unless the Rosenbrock method satisfies the new order condition

$$\sum_{i,j=1}^{s} b_i\omega_{ij}\alpha_j^2 = 1, \tag{15.39}$$

the local error and the global error (if $|R(\infty)| < 1$) are of size $\mathcal{O}(h^2)$. Since none of the classical Rosenbrock methods of Sect. IV.7 satisfies (15.39), their order of convergence is only 2 for the problem (15.1) if λ is very large.

b) A convenient way to satisfy (15.39) is to require

$$\alpha_{si} + \gamma_{si} = b_i \ (i = 1, \dots, s) \quad \text{and} \quad \alpha_s = 1. \tag{15.40}$$

This is the analogue of the condition $a_{si} = b_i$ for Runge-Kutta methods. It implies not only (15.39) but even

$$\delta_h(x) = \mathcal{O}\Big(\frac{h^2}{z}\Big),$$

so that such methods yield asymptotically exact results for $z \to \infty$.

c) A deeper understanding of Eq. (15.39) will be possible when studying the error of Rosenbrock methods for singular perturbation and differential-algebraic problems (Chapter VI). We shall construct there methods satisfying (15.40).

d) Scholz (1989) writes the local error $\delta_h(x)$ in the form

$$\delta_h(x) = \sum_{j\geq 2} C_j(z) h^j \varphi^{(j)}(x) \tag{15.41}$$

and investigates the possibility of having $C_j(z) \equiv 0$ for $j = 2$ (and also $j > 2$). Hundsdorfer (1986) and Strehmel & Weiner (1987) extend the above analysis to semi-linear problems (11.21) which satisfy (11.22). Their results are rather technical but allow the construction of "B-convergent" methods of order $p > 1$.

Exercises

1. Prove that the stage order of an SDIRK method is at most 1, that of a DIRK method at most 2.

2. Consider a Runge-Kutta method with $0 \leq c_1 < \ldots < c_s \leq 1$ which has stage order q. Prove that the method cannot be B-convergent (for variable step sizes) of order $q+1$.

   Hint. Use Formula (15.22) and prove that

$$\frac{K(\widetilde{Z})L(Z) + \theta^{q+1} L(\widetilde{Z})}{K(\widetilde{Z})K(Z) - 1} \tag{15.42}$$

   cannot be uniformly bounded for

$$Z = \operatorname{diag}(z_1, \ldots, z_s), \qquad \widetilde{Z} = \operatorname{diag}(\widetilde{z}_1, \ldots, \widetilde{z}_s)$$

   with $\operatorname{Re} z_i \leq 0$, $\operatorname{Re} \widetilde{z}_i \leq 0$ (in the case $c_1 = 0$ and $c_s = 1$ one has to prove this under the restriction $\widetilde{z}_1 = \theta z_s$, $\widetilde{z}_s = \theta z_1$). For this consider values z_j, $\widetilde{z}_j$ close to the origin.

3. (Burrage & Hundsdorfer 1987). Assume $c_i - c_j$ is not an integer for $1 \leq i < j \leq s$, and the order of B-convergence (for constant step sizes) of a Runge-Kutta method is $q+1$ (q denotes the stage order). Then $d_0 = 0$ and all components of $d = (d_1, \ldots, d_s)^T$ are equal (see (15.20) for the definition of d_j).

   Hint. Study the uniform boundedness of the function $L(Z)/(K(Z) - 1)$.

4. (Kraaijevanger). Show that for

$$A^{-1} = \begin{pmatrix} 0 & 1 & 0 \\ -1 & 0 & 0 \\ 1 & 1 & 1 \end{pmatrix} \tag{15.43}$$

we have $\alpha_0(A^{-1}) = 0$, but there exists no positive diagonal matrix D such that $\alpha_D(A^{-1}) = 0$. For more insight see "Corollary 2.15" of Kraaijevanger & Schneid (1991).

5. Prove that for the Lobatto IIIB methods, with

$$A = \begin{pmatrix} \widetilde{A} & 0 \\ a^T & 0 \end{pmatrix}$$

the dominant term of the local error (15.7) is (for $h \to 0$ and $z = h\lambda \to \infty$)

$$zb_s(a^T\widetilde{A}^{-1}c^{q+1} - 1)\frac{h^{q+1}}{(q+1)!}\varphi^{(q+1)}(x).$$

Here $q = s-2$ is the stage order and $c = (c_1, \dots, c_{s-1})^T$. Show further that

$$a^T\widetilde{A}^{-1}c^k = 1 \qquad \text{for} \quad k = 1, 2, \dots, q \tag{15.44}$$

$$a^T\widetilde{A}^{-1}c^k \neq 1 \qquad \text{for} \quad k = q+1. \tag{15.45}$$

Hint. Equation (15.44) follows from $C(q)$. Show (15.45) by supposing that $a^T\widetilde{A}^{-1}c^{q+1} = 1$ which together with (15.44) implies that

$$\sum_{i=1}^{s-1} d_i p(c_i) = p(1) \qquad \text{where} \qquad d^T = a^T\widetilde{A}^{-1}$$

for every polynomial of $\deg p \le q+1 = s-1$ satisfying $p(0) = 0$. Arrive at a contradiction with

$$p(x) = (x - c_1)(x - c_2) \cdot \ldots \cdot (x - c_{s-1}).$$

Chapter V. Multistep Methods for Stiff Problems

Multistep methods (BDF) were the first numerical methods to be proposed for stiff differential equations (Curtiss & Hirschfelder 1952) and since Gear's book (1971) computer codes based on these methods have been the most prominent and most widely used for all stiff computations.

This chapter introduces the linear stability theory for multistep methods in Sect. V.1, and arrives at the famous theorem of Dahlquist which says that A-stable multistep methods cannot have high order. Attempts to circumvent this barrier proceed mainly in two directions: either study methods with slightly weaker stability requirements (Sect. V.2) or introduce new classes of methods (Sect. V.3). Order star theory on Riemann surfaces (Sect. V.4) then helps to extend Dahlquist's barrier to generalized methods and to explain various properties of stability domains. Section V.5 presents numerical experiments with several codes based on the methods introduced.

Since all the foregoing stability theory is based uniquely on linear autonomous problems $y' = Ay$, the question arises of their validity for general nonlinear problems. This leads to the concepts of G-stability for multistep methods (Sect. V.6) and algebraic stability for general linear methods (Sect. V.9).

Another important subject is convergence estimates for $h \to 0$ which are independent of the stiffness (the analogue of B-convergence in Sect. IV.15). We describe various techniques for obtaining such estimates in Sections V.7 (for linear problems) as well as V.6 and V.8 (for nonlinear problems). These techniques are: use of G-stability, the Kreiss matrix theorem, the multiplier technique and, last but not least, a discrete variation of constants formula.

V.1 Stability of Multistep Methods

A general k-step multistep method is of the form

$$\alpha_k y_{m+k} + \alpha_{k-1} y_{m+k-1} + \ldots + \alpha_0 y_m = h(\beta_k f_{m+k} + \ldots + \beta_0 f_m). \tag{1.1}$$

For this method, we can do the same stability analysis as in Sect. IV.2 for Euler's method. This means that we apply it to the linearized and autonomous system

$$y' = Jy \tag{1.2}$$

(see (IV.2.2')); this gives

$$\alpha_k y_{m+k} + \ldots + \alpha_0 y_m = hJ(\beta_k y_{m+k} + \ldots + \beta_0 y_m). \tag{1.3}$$

We again introduce a new basis for the vectors y_{m+i} consisting of the eigenvectors of J. Then for the *coefficients* of y_{m+i}, with respect to an eigenvector v of J, we obtain exactly the same reccurrence equation as (1.3), with J replaced by the corresponding eigenvalue λ. This gives [1]

$$(\alpha_k - \mu\beta_k) y_{m+k} + \ldots + (\alpha_0 - \mu\beta_0) y_m = 0, \qquad \mu = h\lambda \tag{1.4}$$

and is the same as Method (1.1) applied to Dahlquist's test equation

$$y' = \lambda y. \tag{1.5}$$

The Stability Region

The difference equation (1.4) is solved using Lagrange's method (see Volume I, Sect. III.3): we set $y_j = \zeta^j$, divide by ζ^m and obtain the characteristic equation

$$(\alpha_k - \mu\beta_k)\zeta^k + \ldots + (\alpha_0 - \mu\beta_0) = \varrho(\zeta) - \mu\sigma(\zeta) = 0 \tag{1.6}$$

which depends on the complex parameter μ. The polynomials $\varrho(\zeta)$ and $\sigma(\zeta)$ are our old friends from (III.2.4). The difference equation (1.4) has stable solutions (for arbitrary starting values) iff all roots of (1.6) are ≤ 1 in modulus. In addition, *multiple* roots must be *strictly* smaller than 1 (see Volume I, Sect. III.3, Exercise 1).

[1] In contrast to Chapter IV, where the product $h\lambda$ was denoted throughout by z, we write $h\lambda = \mu$ here, since in multistep theory (Sect. III.3) z denotes the Cayley transform of ζ.

Definition 1.1. The set

$$S = \left\{ \mu \in \mathbb{C}\,;\ \begin{array}{l} \text{all roots } \zeta_j(\mu) \text{ of (1.6) satisfy } |\zeta_j(\mu)| \le 1, \\ \text{multiple roots satisfy } |\zeta_j(\mu)| < 1 \end{array} \right\} \tag{1.7}$$

is called the *stability domain* or *stability region* or *region of absolute stability* of Method (1.1). We have A*-stability* if $S \supset \mathbb{C}^-$.

It is sometimes desirable to consider S as a subset of the compactified complex plane $\overline{\mathbb{C}}$. In this case, for $\mu \to \infty$, the roots of Eq. (1.6) tend to those of $\sigma(\zeta) = 0$.

For $\mu = 0$, Eq. (1.6) becomes $\varrho(\zeta) = 0$. Thus the usual stability (in the sense of Definition III.3.2) is equivalent to $0 \in S$.

Theorem 1.2. *All numerical solutions of Method (1.1) are bounded for the linearized equation (1.2) with a diagonalizable matrix J iff $h\lambda \in S$ for all eigenvalues λ of J.* □

We explain the computation of the stability domain at a particular example, the explicit Adams method of order 4 (see Sect. III.1, Eq. (1.5)),

$$y_{m+4} = y_{m+3} + h\Big(\frac{55}{24}f_{m+3} - \frac{59}{24}f_{m+2} + \frac{37}{24}f_{m+1} - \frac{9}{24}f_m\Big),$$

for which Eq. (1.6) becomes

$$\zeta^4 - \Big(1 + \frac{55}{24}\mu\Big)\zeta^3 + \frac{59}{24}\mu\zeta^2 - \frac{37}{24}\mu\zeta + \frac{9}{24}\mu = 0. \tag{1.8}$$

In Fig. 1.1 we display the complicated behavior of the roots of this equation. We choose the μ values as the dots surrounding the white horse, and plot the corresponding 4 roots $\zeta_1, \zeta_2, \zeta_3, \zeta_4$ in the ζ-plane, which can be observed to emerge from the roots $1, 0, 0, 0$ of the ϱ-polynomial.

Complex mappings are *conformal*, i.e., preserve angles and orientation. The angle of rotation and the magnification of a complex map is (locally) determined by its *derivative*. Differentiating (1.8) with respect to μ and putting $\mu = 0$, $\zeta = 1$ gives

$$\varrho'(1) \cdot \zeta_1'(0) - \sigma(1) = 0,$$

hence $\zeta_1'(0) = 1$ (because of the consistency conditions $\varrho'(1) \neq 0$, $\sigma(1) = \varrho'(1)$, see Volume I, Eq. (III.2.6)). This explains the fact that the map $\mu \mapsto \zeta_1$ is close to $1 + \mu$ in the neighbourhood of $\mu = 0$, and $\zeta_1(\mu)$ moves *inside* the unit disc when μ moves *inside* $\mathbb{C}^-$.

The Root Locus Curve. The key for computing S is the fact that the *inverse* map $\zeta \mapsto \mu$, since (1.8) is linear in μ, can easily be computed and is one-valued

$$\mu = \frac{\varrho(\zeta)}{\sigma(\zeta)} = \frac{\zeta^4 - \zeta^3}{\frac{55}{24}\zeta^3 - \frac{59}{24}\zeta^2 + \frac{37}{24}\zeta - \frac{9}{24}}. \tag{1.9}$$

The *outside* of the unit circle in the ζ-plane mapped back into the μ-plane by this formula (see the zodiac of gray horses in Fig. 1.1) produces the forbidden μ-values, for which at least one root $\zeta_i(\mu)$ generates instability. The image of the boundary curve of the unit circle $\zeta = e^{i\theta}$, $0 \leq \theta \leq 2\pi$, is called the *root locus curve*. It must be considered as an oriented curve and the stability region, whenever it is not empty, must lie *to the left* of it.

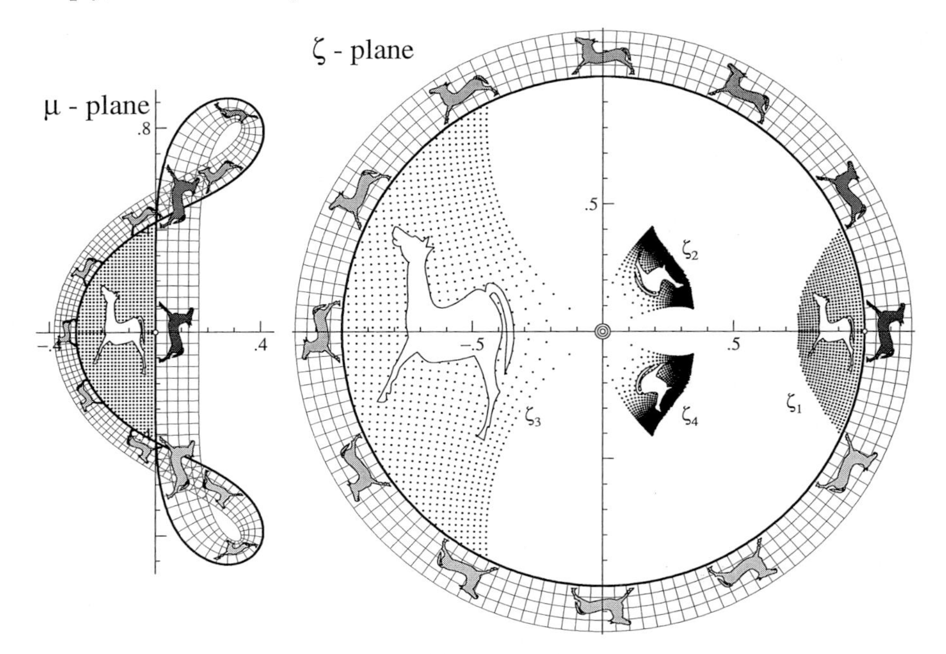

Fig. 1.1. Plot of the stability function (1.8) with root locus curve

We conclude that the stability domain of Adams4 is precisely the small diamond shaped region surrounded by the root locus curve in the positive direction located between the origin and the point $\mu = 2 \cdot 24/(-55-59-37-9) = -0.3$.

Adams Methods

The *explicit Adams methods* (III.1.5) applied to $y' = \lambda y$ give

$$y_{n+1} = y_n + \mu \sum_{j=0}^{k-1} \gamma_j \nabla^j y_n, \qquad \gamma_0 = 1,\ \gamma_1 = \frac{1}{2},\ \gamma_2 = \frac{5}{12},\ \gamma_3 = \frac{3}{8}, \ldots \tag{1.10}$$

or, after putting $y_n = \zeta^n$ and dividing by ζ^n,

$$\zeta - 1 = \mu\Big(\gamma_0 + \gamma_1\Big(1 - \frac{1}{\zeta}\Big) + \gamma_2\Big(1 - \frac{2}{\zeta} + \frac{1}{\zeta^2}\Big) + \ldots\Big).$$

Hence the root locus curve becomes

$$\mu = \frac{\zeta - 1}{\sum_{j=0}^{k-1} \gamma_j (1 - 1/\zeta)^j}, \qquad \zeta = e^{i\theta}. \tag{1.10'}$$

For $k = 1$ we again obtain the circle of Euler's method, centred at -1. These curves are plotted in Fig. 1.2 for $k = 2, 3, \dots, 6$ and show stability domains of rapidly decreasing sizes. These methods are thus surely not appropriate for stiff problems.

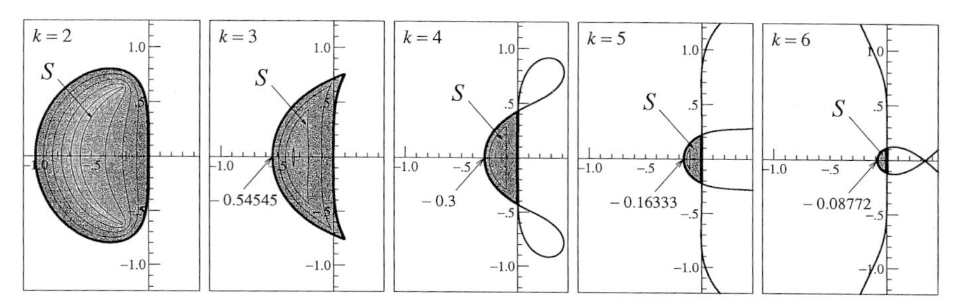

Fig. 1.2. Stability domains for explicit Adams methods

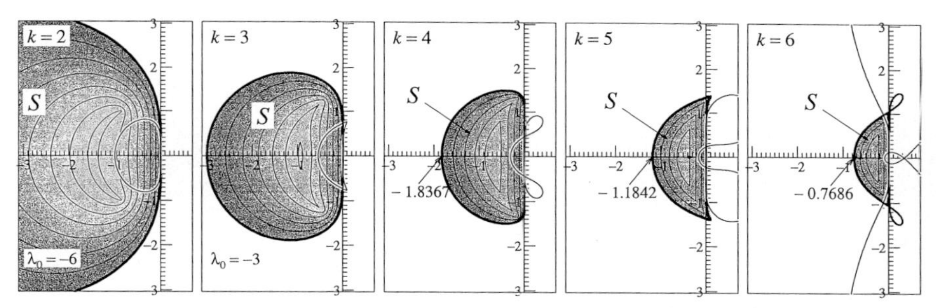

Fig. 1.3. Stability domains of implicit Adams methods, compared to those of the explicit ones

The *implicit Adams methods* (III.1.8) lead to

$$y_{n+1} = y_n + \mu \sum_{j=0}^{k} \gamma_j^* \nabla^j y_{n+1}, \qquad \gamma_0^* = 1,\ \gamma_1^* = -\frac{1}{2},\ \gamma_2^* = -\frac{1}{12}, \dots \tag{1.11}$$

Here we put $y_n = \zeta^n$ and divide by ζ^{n+1}. This gives

$$\mu = \frac{1 - 1/\zeta}{\sum_{j=0}^{k} \gamma_j^* (1 - 1/\zeta)^j}, \qquad \zeta = e^{i\theta}. \tag{1.11'}$$

For $k = 1$ this is the implicit trapezoidal rule and is A-stable. For $k = 2, 3, \dots, 6$ the stability domains, though much larger than those of the explicit methods, do not cover $\mathbb{C}^-$ (see Fig. 1.3). Hence these methods are *not* A-stable.

Predictor-Corrector Schemes

> The inadequacy of the theory incorporating the effect of the corrector equation only for predictor-corrector methods was first discovered through experimental computations on the prototype linear equation
>
> $$y' = f(x,y) = -100y + 100, \qquad y(0) = 0,$$
>
> (...) Very poor correlation of actual errors with the errors expected on the basis of the properties of the corrector equation alone was obtained. This motivated the development of the theory ...
>
> (P.E. Chase 1962)

As we have seen in Sect. III.1, the classical way of computing y_{n+1} from the implicit equations (III.1.8) is to use the result y^*_{n+1} of the explicit Adams method as a *predictor* in $\beta_k f(x_{n+1}, y_{n+1})$. This destroys a good deal of the stability properties of the method (Chase 1962). The stability analysis changes as follows: the predictor formula

$$y^*_{n+1} = y_n + \mu\big(\gamma_0 y_n + \gamma_1(y_n - y_{n-1}) + \gamma_2(y_n - 2y_{n-1} + y_{n-2}) + \ldots\big) \tag{1.12}$$

must be inserted into the corrector formula

$$\begin{aligned} y_{n+1} = y_n + \mu\Big(&\gamma_0^* y^*_{n+1} + \\ &\gamma_1^*(y^*_{n+1} - y_n) + \\ &\gamma_2^*(y^*_{n+1} - 2y_n + y_{n-1}) + \\ &\gamma_3^*(y^*_{n+1} - 3y_n + 3y_{n-1} - y_{n-2}) + \ldots\Big). \end{aligned} \tag{1.13}$$

Since there is a μ in (1.12) and in (1.13), we obtain this time, by putting $y_n = \zeta^n$ and dividing by ζ^n, a *quadratic* equation for μ,

$$A\mu^2 + B\mu + C = 0, \tag{1.14}$$

$$A = \Big(\sum_{j=0}^{k} \gamma_j^*\Big)\Big(\sum_{j=0}^{k-1} \gamma_j\big(1 - \frac{1}{\zeta}\big)^j\Big),$$

$$B = (1-\zeta)\sum_{j=0}^{k} \gamma_j^* + \zeta \sum_{j=0}^{k} \gamma_j^* \big(1 - \frac{1}{\zeta}\big)^j,$$

$$C = 1 - \zeta.$$

For each $\zeta = e^{i\theta}$, Eq. (1.14) has two roots. These give rise to two root locus curves which determine the stability domain. These curves are represented in Fig. 1.4 and compared to those of the original implicit methods. It can be seen that we loose a lot of stability. In particular, for $k = 1$ the trapezoidal rule becomes an explicit second order Runge Kutta method and the A-stability is destroyed.

While Chase (1962) studied real eigenvalues only, the general complex case has been stated by Crane & Klopfenstein (1965) and, with beautiful figures, by

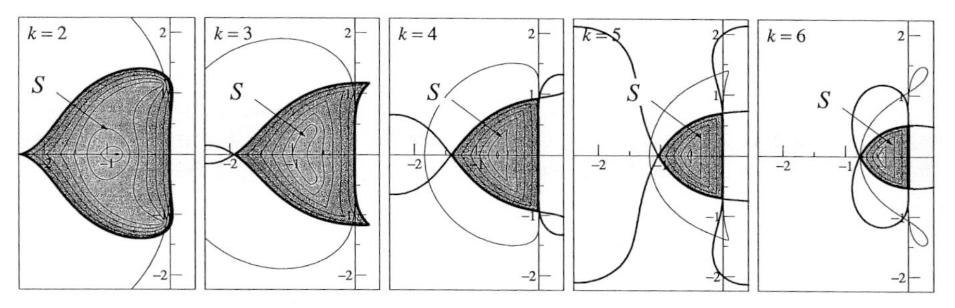

Fig. 1.4. Stability domains for PECE compared to original implicit methods

Krogh (1966). All three papers also searched for procedures with increased stability domains. This research was brought to perfection by Stetter (1968).

Nyström Methods

> Thus we see that Milne's method will not handle so simple an equation as $y' = -y,\ y(0) = 1 \ldots$ (R.W. Hamming 1959)

> ... Milne's method has a number of virtues not possessed by its principal rival, the Runge-Kutta method, which are especially important when the order of the system of equations is fairly high (N=10 to 30 or more) ... (R.W. Hamming 1959)

The *explicit Nyström method* (III.1.13) for $k = 1$ and 2 is the "explicit midpoint rule"

$$y_{n+1} = y_{n-1} + 2hf_n \tag{1.15}$$

and leads to the root locus curve

$$\mu = \frac{e^{i\theta} - e^{-i\theta}}{2} = i\sin\theta. \tag{1.15'}$$

This curve moves up and down the imaginary axis between $\pm i$ and leaves as stability domain just the interval $(-i, +i)$ (see Fig. 1.5). All eigenvalues in the interior of the negative half plane lead to instabilities. This is caused by the second root -1 of $\varrho(\zeta)$ which moves out of the unit circle when μ goes West. This famous phenomenon is called the "weak instability" of the midpoint rule and was the "entry point" of Dahlquist's stability-career (Dahlquist 1951). The graphs of Fig. III.9.2 nicely show the (weak) instability of the numerical solution.

The *implicit Milne-Simpson method* (III.1.15) for $k = 2$ and 3 is

$$y_{n+1} = y_{n-1} + h\Big(\frac{1}{3}f_{n+1} + \frac{4}{3}f_n + \frac{1}{3}f_{n-1}\Big) \tag{1.16}$$

and has the root locus curve

$$\mu = \frac{e^{i\theta} - e^{-i\theta}}{\frac{1}{3}e^{i\theta} + \frac{4}{3} + \frac{1}{3}e^{-i\theta}} = 3i\,\frac{\sin\theta}{\cos\theta + 2}, \tag{1.16'}$$

which moves up and down the imaginary axis between $\pm i\sqrt{3}$. Thus its behaviour is similar to the explicit Nyström method with a slightly larger stability interval.

The higher order Nyström and Milne-Simpson methods have root locus curves which are oriented the wrong way round (see Fig. 1.5). Their stability domains therefore reduce to the smallest possible set (for stable methods): *just the origin.*

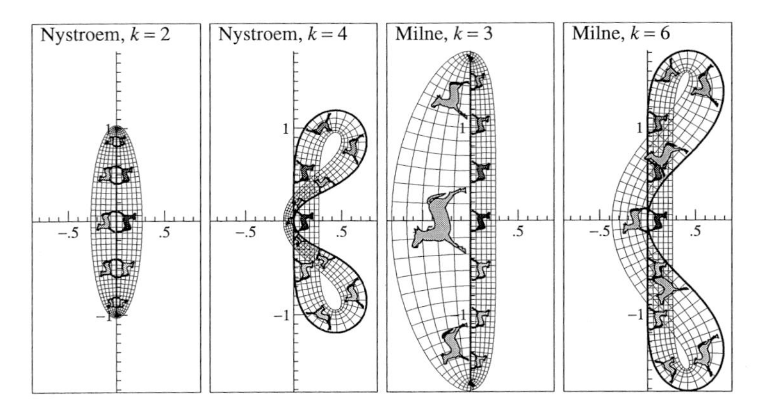

Fig. 1.5. Root locus curves for Nyström and Milne methods

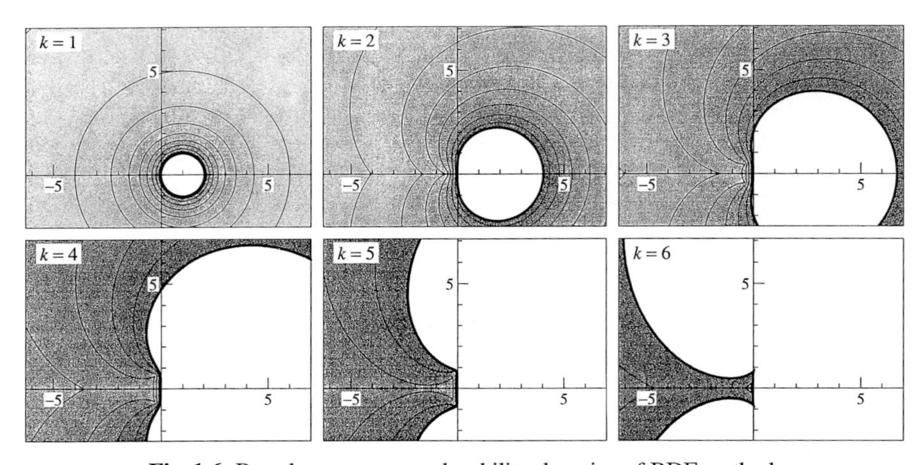

Fig. 1.6. Root locus curves and stability domains of BDF methods

BDF

The backward differentiation formulas $\sum_{j=1}^{k} \frac{1}{j}\nabla^j y_{n+1} = hf_{n+1}$ (see Equation (III.1.22')) have the root locus curves given by

$$\mu = \sum_{j=1}^{k} \frac{1}{j}\Big(1-\frac{1}{\zeta}\Big)^j = \sum_{j=1}^{k} \frac{1}{j}\big(1-e^{-i\theta}\big)^j. \tag{1.17}$$

For $k=1$ we have the implicit Euler method with stability domain $S=\{\mu\,;\,|\mu-1|\ge 1\}$. For $k=2$ the root locus curve (see Fig. 1.6) has $\mathrm{Re}\,(\mu)=\frac{3}{2}-2\cos\theta+\frac{1}{2}\cos 2\theta$ which is ≥ 0 for all θ. Therefore the method is A-stable and of order 2. However, for $k=3,4,5$ and 6, we see that the methods loose more and more stability on a part of the imaginary axis. For $k\ge 7$, as we know, the formulas are unstable anyway, even at the origin.

The Second Dahlquist Barrier

I searched for a long time, finally Professor Lax showed me the Riesz-Herglotz theorem and I knew that I had my theorem.

(G. Dahlquist 1979)

Theorem 1.3. *If the multistep method (1.1) is A-stable, then*

$$\mathrm{Re}\left(\frac{\varrho(\zeta)}{\sigma(\zeta)}\right)>0 \qquad \text{for} \qquad |\zeta|>1. \tag{1.18}$$

For irreducible methods the converse is also true: (1.18) implies A-stability.

Proof. If the method is A-stable then all roots of (1.6) must satisfy $|\zeta|\le 1$ whenever $\mathrm{Re}\,\mu\le 0$. The logically equivalent statement ($\mathrm{Re}\,\mu>0$ whenever $|\zeta|>1$) yields (1.18) since by (1.6) $\mu=\varrho(\zeta)/\sigma(\zeta)$.

Suppose now that (1.18) holds and that the method is irreducible. Fix a μ_0 with $\mathrm{Re}\,\mu_0\le 0$ and let ζ_0 be a root of (1.6). We then have $\sigma(\zeta_0)\ne 0$ (otherwise the method would be reducible). Hence $\mu_0=\varrho(\zeta_0)/\sigma(\zeta_0)$ and it follows from (1.18) that $|\zeta_0|\le 1$. We still have to show that ζ_0 is a simple root if $|\zeta_0|=1$. By a continuity argument it follows from (1.18) that $|\zeta_0|=1$ and $\mathrm{Re}\,\mu_0<0$ are contradictory. Therefore, it remains to prove that for $\mathrm{Re}\,\mu_0=0$ a root satisfying $|\zeta_0|=1$ must be simple. In a neighbourbood of such a root we have

$$\frac{\varrho(\zeta)}{\sigma(\zeta)}-\mu_0=C_1(\zeta-\zeta_0)+C_2(\zeta-\zeta_0)^2+\ldots$$

and (1.18) implies that $C_1\ne 0$. This, however, is only possible if ζ_0 is a simple root of (1.6). □

In all the above examples we have not yet seen an A-stable multistep formula of order $p\ge 3$. The following famous theorem explains this observation.

Theorem 1.4 (Dahlquist 1963). *An A-stable multistep method must be of order $p\le 2$. If the order is 2, then the error constant satisfies*

$$C\le -\frac{1}{12}. \tag{1.19}$$

The trapezoidal rule is the only A-stable method of order 2 with $C=-1/12$.

Proof. Dahlquist's first proof of this theorem is difficult. More elementary versions emerged in Widlund (1967), in lecture notes of W. Liniger (Univ. of Neuchâtel 1971) and in the book of Grigorieff (1977, vol.2, p. 218).

We start by recalling some formulas from Volume I: Eq. (ii) of Theorem III.2.4 and Eq. (III.2.7) yield

$$\varrho(e^h) - h\sigma(e^h) = C_{p+1}h^{p+1} + \ldots \qquad \text{for} \quad h \to 0. \tag{1.20}$$

From the consistency conditions (III.2.6) we have

$$\varrho(e^h) = \varrho(1+h+\ldots) = \varrho(1) + \varrho'(1)h + \ldots = \sigma(1)h + \ldots .$$

We divide (1.20) by $h\varrho(e^h)$ and obtain

$$\frac{1}{h} - \frac{\sigma(e^h)}{\varrho(e^h)} = Ch^{p-1} + \ldots \qquad \text{for} \quad h \to 0 \tag{1.21}$$

where C is the *error constant* (III.2.13). With $\zeta = e^h$ this becomes

$$\frac{1}{\log \zeta} - \frac{\sigma(\zeta)}{\varrho(\zeta)} = C(\zeta - 1)^{p-1} + \ldots \qquad \text{for} \quad \zeta \to 1. \tag{1.22}$$

In this formula we put $p = 2$. Whenever the method is of higher order, we have $C = 0$. When the order of the method is one, we have nothing to prove. The same formula for the trapezoidal rule for which $\varrho_T(\zeta) = \zeta - 1$, $\sigma_T(\zeta) = \frac{1}{2}(\zeta + 1)$, becomes by series expansion (or by using Table III.2.1)

$$\frac{1}{\log \zeta} - \frac{\sigma_T(\zeta)}{\varrho_T(\zeta)} = -\frac{1}{12}(\zeta - 1) + \ldots \qquad \text{for} \quad \zeta \to 1. \tag{1.23}$$

The idea is now to subtract the two formulas and obtain

$$d(\zeta) := \frac{\sigma(\zeta)}{\varrho(\zeta)} - \frac{\sigma_T(\zeta)}{\varrho_T(\zeta)} = \Big(-C - \frac{1}{12}\Big)(\zeta - 1) + \ldots \qquad \text{for} \quad \zeta \to 1. \tag{1.24}$$

From (1.18) we have that

$$\mathrm{Re}\left(\frac{\varrho(\zeta)}{\sigma(\zeta)}\right) > 0 \quad \text{or equivalently} \quad \mathrm{Re}\left(\frac{\sigma(\zeta)}{\varrho(\zeta)}\right) > 0 \quad \text{for } |\zeta| > 1. \tag{1.25}$$

The point here is that for the trapezoidal rule this $\mathrm{Re}\,(\ldots)$ is *zero* for $|\zeta| = 1$ since this method has precisely $\mathbb{C}^-$ as stability domain. Hence from (1.24) we obtain

$$\lim_{\substack{\zeta \to \zeta_0 \\ |\zeta| > 1}} \mathrm{Re}\, d(\zeta) \geq 0 \qquad \text{for} \quad |\zeta_0| = 1. \tag{1.26}$$

The poles of $d(\zeta)$ are the roots of $\varrho(\zeta)$, which, by stability, are not allowed outside the unit circle. Thus, by the maximum principle, (1.26) remains true everywhere outside the unit circle. Choosing then $\zeta = 1 + \varepsilon$ with $\mathrm{Re}\, \varepsilon > 0$ and $|\varepsilon|$ small, we see from (1.24) that either $-C - \frac{1}{12} > 0$ or $d(\zeta) \equiv 0$. This concludes the proof. □

Exercises

1. The Milne-Simpson methods for $k=4$ and 5 satisfy $\mathrm{Re}\,(\varrho(\zeta)/\sigma(\zeta))\geq 0$ for $|\zeta|=1$. Since their order is higher than 2, this seems to be in contradiction with the above proof of Theorem 1.4. Explain.

2. For the explicit midpoint rule (1.15), do the endpoints $\pm i$ of the stability region belong to S ? Study the (possible) stability of this method applied with $h=1$ to $u'=v$, $v'=-u$.

3. Compute for the explicit and implicit Adams methods the largest $\lambda_0\in\mathbb{R}$ such that the real interval $[-\lambda_0,0]$ lies in S. Show that for the k-step explicit Adams methods we have $\lambda_0=2/u_k$ with $u_k=\sum_{j=0}^{k-1}2^j\gamma_j$ ($u_1=1$, $u_2=2$, $u_3=11/3$, $u_4=20/3$, $u_5=551/45,\ldots$). The use of generating functions (see Sect. III.1) allow us to show that

$$\sum_{j=1}^{\infty}u_kt^k=\Big(-1+\frac{2}{1-t}-\frac{1}{1-2t}\Big)\log(1-2t),$$

a series with convergence radius $1/2$. This explains why these stability domains decrease so rapidly.

Hint. Just set $\theta=\pi$ in the root locus curve.

4. Prove that the stability region of the k-step, implicit Adams methods is of finite size for every $k\geq 2$.

Hint. Show that $(-1)^k\sigma(-1)<0$, so that σ has a real negative root, smaller than -1.

5. a) Show that all 2-step methods of order 2 are given by

$$\begin{aligned}\varrho(\zeta)&=(\zeta-1)(\alpha\zeta+1-\alpha)\\ \sigma(\zeta)&=(\zeta-1)^2\beta+(\zeta-1)\alpha+(\zeta+1)/2\end{aligned}$$

(which are irreducible for $\alpha\neq 2\beta$).

b) The method is stable at 0 iff $\alpha\geq 1/2$.

c) The method is stable at ∞ iff

$$\alpha\geq 1/2 \qquad \text{and} \qquad \beta>\alpha/2. \tag{1.27}$$

Apply the Schur-Cohn criterion (Sect. III.3, Exercise 4).

d) The method is A-stable iff (1.27) holds.

Hint.

$$\frac{\sigma(\zeta)}{\varrho(\zeta)}=\frac{1}{2}\cdot\frac{\zeta+1}{\zeta-1}+\Big(\beta-\frac{\alpha}{2}\Big)\cdot\frac{\zeta-1}{\alpha\zeta+1-\alpha}.$$

V.2 "Nearly" A-Stable Multistep Methods

> We are not attempting to disprove Dahlquist's theorems but are trying to get round the conditions they impose ...
>
> (J. Cash 1979)

Dahlquist's condition $p \le 2$ for the order of an A-stable linear multistep method is a severe restriction for efficient practical calculations of high precision. There are only two ways of "breaking" this barrier:

- either weaken the condition;
- or strengthen the method.

These two points will occupy our attention in this and in the following section.

$A(\alpha)$-Stability and Stiff Stability

> It is the purpose of this note to show that a slightly different stability requirement permits methods of higher accuracy.
>
> (O. Widlund 1967)

> The angle α is only one of a number of parameters which have been proposed for measuring the extent of the stability region. But it is probably the best such measure ...
>
> (Skeel & Kong 1977)

Many important classes of practical problems do not require stability on the entire left half-plane $\mathbb{C}^-$. Further, for eigenvalues on the imaginary axis, the solutions are often highly oscillatory and one is then forced anyhow to restrict the step size "to the highest frequency present in order to represent the signal" (Gear 1971, p. 214).

Definition 2.1 (Widlund 1967). A convergent linear multistep method is $A(\alpha)$-*stable*, $0 < \alpha < \pi/2$, if

$$S \supset S_\alpha = \{\mu \,;\, |\arg(-\mu)| < \alpha,\ \mu \neq 0\}. \tag{2.1}$$

A method is $A(0)$-stable if it is $A(\alpha)$-stable for some (sufficiently small) $\alpha > 0$.

Similarly, Gear (1971) required in his famous concept of *"stiff stability"* that

$$S \supset \{\mu \,;\, \operatorname{Re}\mu < -D\} \tag{2.2}$$

for some $D > 0$ and that the method be "accurate" in a rectangle $-D \le \operatorname{Re}\mu \le a$, $-\theta \le \operatorname{Im}\mu \le \theta$ for some $a > 0$ and θ about $\pi/5$. Many subsequent writers didn't like the inaccurate meaning of "accurate" in this definition and replaced it by something else. For example Jeltsch (1976) required that in addition to (2.2),

$$|\zeta_1(\mu)| > |\zeta_i(\mu)|, \quad i = 2, \ldots, k \qquad \text{in} \quad |\mathrm{Re}\,\mu| \le a, \; |\mathrm{Im}\,\mu| \le \theta, \tag{2.3}$$

where $\zeta_1(\mu)$ is the analytic continuation of the principal root $\zeta_1(0) = 1$ of (1.6). Also, the rectangle given by

$$|\mathrm{Im}\,\mu| \le \theta, \qquad -D \le \mathrm{Re}\,\mu \le -a$$

should belong to S.

Other concepts are A_0*-stable* (Cryer 1973) if

$$|\zeta_i(x)| < 1, \quad i = 1, \ldots, k \qquad \text{for} \quad -\infty < x < 0 \tag{2.4}$$

and $\mathring{A}$*-stable* (a joke of O. Nevanlinna 1979) if

$$(-\infty, 0] \subset S. \tag{2.5}$$

Of course, we have

$$A(0)\text{-stable} \Longrightarrow A_0\text{-stable} \Longrightarrow \mathring{A}\text{-stable} \tag{2.6}$$

but neither implication is reversible (Exercise 3; see also "Theorem 1" of Jeltsch 1976).

The BDF methods (1.18) satisfy (2.1) for $A(\alpha)$-stability and (2.2) for stiff stability with the values

k	1	2	3	4	5	6
α	90°	90°	86.03°	73.35°	51.84°	17.84°
D	0	0	0.083	0.667	2.327	6.075

(2.7)

High Order $A(\alpha)$-Stable Methods

> Dill and Gear ... and Jain and Srivastava ... have used computers to construct stiffly stable methods of orders eight and eleven, respectively, but were unable to construct higher order stiffly stable methods. Even though we have shown here that A_0-stable methods of arbitrarily high order exist, we conjecture that $A(0)$-stable linear multistep methods of higher order, of order greater than 20 say, do not exist. (Cryer 1973)

Widlund (1967) showed that for every $\alpha < \pi/2$, α arbitrarily close to $\pi/2$, there exist $A(\alpha)$-stable multistep methods of order $p = k$ for $p = 3$ and $p = 4$. It is now an interesting question whether such methods also exist for higher orders. Well, the answer consists of good news and bad news.

First the good news. The conjecture of Cryer (see quotation) was quickly disproved by combining Cryer's A_0-stable methods with the result of Jeltsch (1976) which says that certain A_0-stable methods are also $A(\alpha)$-stable. The following theorem shows that α can even be chosen arbitrarily close to $\pi/2$:

Theorem 2.2 (Grigorieff & Schroll 1978). *Let $\alpha < \pi/2$ be given. Then for every $k \in \mathbb{N}$ there exists an $A(\alpha)$-stable linear k-step method of order $p = k$.*

Proof. For $p = k = 2$ the two-step BDF method which is A-stable, and hence $A(\alpha_2)$-stable for every $\alpha_2 \leq \pi/2$, does the job. For k arbitrary, we intercalate $k-2$ values between α and $\pi/2$,

$$\alpha < \alpha_{k-1} < \alpha_{k-2} < \ldots < \alpha_3 < \alpha_2 \leq \frac{\pi}{2}, \tag{2.8}$$

and extend the method step by step with the help of Lemma 2.3. □

Lemma 2.3. *Suppose an $A(\alpha)$-stable k-step method of order p is given with*

$$\varrho(\zeta) \neq 0 \qquad \textit{if} \;\; |\zeta| = 1, \;\; \zeta \neq 1 \tag{2.9a}$$

$$\sigma(\zeta) \neq 0 \qquad \textit{if} \;\; |\zeta| = 1. \tag{2.9b}$$

Then for every $\widetilde{\alpha} < \alpha$ there exists an $A(\widetilde{\alpha})$-stable $(k+1)$-step method of order $p+1$ which also satisfies (2.9).

The *proof* follows very closely the ideas of Jeltsch & Nevanlinna (1982): Let $\varrho(\zeta)$ and $\sigma(\zeta)$ represent the given k-step method with order condition

$$\frac{\varrho(\zeta)}{\log \zeta} - \sigma(\zeta) = C_{p+1}(\zeta-1)^p + \mathcal{O}\big((\zeta-1)^{p+1}\big). \tag{2.10}$$

If we multiply ϱ and σ by $(\zeta - 1)$ we formally increase the order by 1 and at the same time leave the root locus curve unchanged. Everything seems to be proved. However, the new ϱ-polynomial would have a double root at $\zeta = 1$ and would thus lead to an unstable method. We therefore choose $\varepsilon > 0$ and multiply (2.10) by $(\zeta - 1 + \varepsilon)$, which moves the root slightly inside the unit circle. We then obtain a new method of order $p+1$ if we put

$$\begin{aligned} \widetilde{\varrho}(\zeta) &= \varrho(\zeta)\,(\zeta - 1 + \varepsilon) \\ \widetilde{\sigma}(\zeta) &= \sigma(\zeta)\,(\zeta - 1 + \varepsilon) + \varepsilon\, C_{p+1}(\zeta-1)^p. \end{aligned} \tag{2.11}$$

Since $p = k+2$ is excluded (by Theorem III.3.9 methods with $p = k+2$ are symmetric and violate Hypothesis (2.9a)), both polynomials $\widetilde{\varrho}$ and $\widetilde{\sigma}$ are of degree $\leq k+1$. Now the formula

$$\frac{\widetilde{\sigma}(\zeta)}{\widetilde{\varrho}(\zeta)} - \frac{\sigma(\zeta)}{\varrho(\zeta)} = \frac{\varepsilon C_{p+1}(\zeta-1)^p}{\varrho(\zeta)(\zeta - 1 + \varepsilon)} \tag{2.12}$$

allows us to compare, for ε small, the root-locus curves of the two methods. The fact that we are working with $\sigma(e^{i\theta})/\varrho(e^{i\theta}) = 1/\mu$ instead of $\mu = \varrho(e^{i\theta})/\sigma(e^{i\theta})$

does not matter, because the transformation $\mu \mapsto 1/\mu$ maps the sector of Definition 2.1 onto itself. Because of Hypothesis (2.9a), 1 is the only (simple) root of $\varrho(\zeta)$ on the unit circle, therefore

$$\left|\frac{\widetilde{\sigma}(\zeta)}{\widetilde{\varrho}(\zeta)} - \frac{\sigma(\zeta)}{\varrho(\zeta)}\right| \le C\cdot\varepsilon\frac{|\zeta-1|^{p-1}}{|\zeta-1+\varepsilon|} \qquad \text{for} \quad \zeta=e^{i\theta}. \tag{2.13}$$

A small obstacle still separates us from "endless pleasure, endless love, Semele enjoys above": the denominator $|\zeta-1+\varepsilon|$, which becomes small for $\varepsilon\to 0$ and $\theta\to 0$. For $p>1$, this "small" denominator is simply balanced by one of the factors $|\zeta-1|$ from the numerator and we have

$$\left|\frac{\widetilde{\sigma}(\zeta)}{\widetilde{\varrho}(\zeta)} - \frac{\sigma(\zeta)}{\varrho(\zeta)}\right| \le \widehat{C}\cdot\varepsilon \tag{2.14}$$

which means uniform pointwise convergence of $\widetilde{\sigma}(\zeta)/\widetilde{\varrho}(\zeta)$ to $\sigma(\zeta)/\varrho(\zeta)$ if $\varepsilon\to 0$. Since $\sigma(\zeta)/\varrho(\zeta)$ is bounded away from the origin (Hypothesis (2.9b)), this also means uniform convergence of the angles.

This is already sufficient to prove Theorem 2.2, where we always have $p\ge 2$. However, Lemma 2.3 remains valid for $p=1$ too: the critical region is when $\theta\to 0$, in which case $|\sigma(e^{i\theta})/\varrho(e^{i\theta})|$ and $|\widetilde{\sigma}(e^{i\theta})/\widetilde{\varrho}(e^{i\theta})|$ tend to infinity like $Const/\theta$. Instead of (2.14) we have for $p=1$

$$\left|\frac{\widetilde{\sigma}(\zeta)}{\widetilde{\varrho}(\zeta)} - \frac{\sigma(\zeta)}{\varrho(\zeta)}\right| \le \frac{C\varepsilon}{|\zeta-1+\varepsilon|} = \mathcal{O}\Big(\frac{\varepsilon}{\theta}\Big).$$

Thus the *angle* (seen from the origin) between $\widetilde{\sigma}(\zeta)/\widetilde{\varrho}(\zeta)$ and $\sigma(\zeta)/\varrho(\zeta)$ is $\mathcal{O}(\varepsilon)$. □

Approximating Low Order Methods with High Order Ones

The above proof of Lemma 2.3 actually shows more than angle-boundedness of the root locus curve, namely uniform convergence of the root locus curve of a high order method to that of a lower order one. This leads to the following theorem of Jeltsch & Nevanlinna (1982):

Theorem 2.4. *Let a linear stable k-step method of order p and stability domain S be given which satisfies (2.9a). Then to any closed set $\Omega\subset \operatorname{Int} S\subset\overline{\mathbb{C}}$ and any $K\in\mathbb{N}$ there exists a linear $k+K$-step method of order $p+K$ whose stability domain $\widetilde{S}$ satisfies*

$$\widetilde{S}\supset\Omega.$$

Moreover if the first method is explicit, the higher-order method is also explicit.

Proof. The proof is similar to that of Lemma 2.3. Instead of the sequence (2.8) we use a sequence of embedded closed and open subsets between Ω and S (Urysohn's Lemma). Hypothesis (2.9b) is ruled out by passing to the compactified topology of $\overline{\mathbb{C}}=\mathbb{C}\cup\{\infty\}$. □

Remark. No method with non-empty $\operatorname{Int} S$ of practical interest violates Hypothesis (2.9a). Nevertheless, Theorem 2.4 remains valid *without* this hypothesis, but the proof becomes more complicated (see "Lemma 3.6" of Jeltsch & Nevanlinna 1982).

A Disc Theorem

Another weakening of A-stability is to require stability for

$$D_r = \{\mu\,;\ |\mu + r| \le r\}, \tag{2.15}$$

which is a disc of radius r in $\mathbb{C}^-$ tangent to the imaginary axis at the origin. Theorems about stability in D_r are stronger than theorems about $A(\alpha)$-stability for eigenvalues close to the origin. The following result is, again, due to Jeltsch & Nevanlinna (1982):

Theorem 2.5. *Let a linear k-step method of order p be given with $S \supset D_r$. Then for any $\widetilde{r} < r$ and any $K \in \mathbb{N}$ there exists a linear $k+K$-step method of order $p+K$ whose stability domain $\widetilde{S}$ satisfies $\widetilde{S} \supset D_{\widetilde{r}}$.*

Proof. The map $\mu \mapsto 1/\mu$ used in the proof of Lemma 2.3 maps the exterior of D_r onto the half-plane

$$\Big\{\mu \in \mathbb{C}\,;\ \operatorname{Re}\mu > -\frac{1}{2r}\Big\}. \tag{2.16}$$

Therefore the uniform convergence established in (2.14) also covers the new situation if $p > 1$. The case $p = 1$, however, needs a more careful study and we refer to the original paper of Jeltsch & Nevanlinna (1982, pp. 277-279). □

Accuracy Barriers for Linear Multistep Methods

Now here is the "bad news": high order $A(\alpha)$-stable methods, for α close to $\pi/2$, cannot be of practical use, or in other words: "the second Dahlquist barrier cannot be broken". The reason is simply that high order alone is not sufficient for high accuracy, because the methods then have enormous error constants. Jeltsch & Nevanlinna (1982) give an impressive staccato (from "Theorem 4.1" to "Lemma 4.15") of lower bounds for error constants and Peano kernels of methods having large stability domains. The Peano kernels, the most serious measures for the error, are defined by the formulas (see (III.2.15) and (III.2.3) of Volume I)

$$L(x) = h^{q+1} \int_{-\infty}^{\infty} \check{K}_q(-s)\, y^{(q+1)}(x + sh)\, ds \tag{2.17}$$

$$= \sum_{j=0}^{k} \big(\alpha_j y(x + jh) - h\beta_j y'(x + jh)\big). \tag{2.18}$$

The kernels $\check{K}_q(-s) = K_q(s)$ are zero outside the interval $0 \le s \le k$ and are piecewise polynomials given by complicated formulas (see (III.2.16)) which appear not very attractive to work with.

However, the formulas simplify if we use the *Fourier transform* which, for a function $f(x)$, is defined by

$$\widehat{f}(\xi) = \int_{-\infty}^{\infty} e^{-ix\xi} f(x)dx. \tag{2.19}$$

We obtain $\widehat{L}$ from (2.17) by insertion of the definitions, several integrations by parts and transformations of double integrals:

$$\widehat{L}(\xi) = h^{q+1}\widehat{\check{K}_q}(h\xi)\cdot \widehat{y^{(q+1)}}(\xi) \tag{2.20}$$

$$= \widehat{\check{K}_q}(h\xi)(ih\xi)^{q+1}\widehat{y}(\xi), \tag{2.21}$$

and from (2.18)

$$\widehat{L}(\xi) = \big(\varrho(e^{ih\xi}) - ih\xi\,\sigma(e^{ih\xi})\big)\cdot\widehat{y}(\xi). \tag{2.22}$$

Thus (2.20) and (2.22) give

$$\widehat{K_q}(-\xi) = \widehat{\check{K}_q}(\xi) = \big(\varrho(e^{i\xi}) - i\xi\,\sigma(e^{i\xi})\big)(i\xi)^{-(q+1)}, \tag{2.23}$$

a nice formula, involving the polynomials ϱ and σ, with which we are better acquainted.

What about the usefulness of $\widehat{K_q}$ for error estimates? Well, it is the *Parseval identity* (Exercise 4)

$$\|f\|_{L^2(-\infty,\infty)} = \frac{1}{\sqrt{2\pi}}\,\|\widehat{f}\,\|_{L^2(-\infty,\infty)} \tag{2.24}$$

which allows us to obtain the L^2-estimate for the error

$$\|L\|_{L^2(-\infty,\infty)} \le h^{q+1}\|\widehat{K_q}\|_{L^\infty}\cdot\|y^{(q+1)}\|_{L^2}, \tag{2.25}$$

as follows:

$$\begin{aligned}
\|L\|^2_{L^2(-\infty,\infty)} &= \frac{1}{2\pi}\,\|\widehat{L}\|^2_{L^2(-\infty,\infty)} && \text{(from (2.24))}\\
&= \frac{h^{2q+2}}{2\pi}\int_{-\infty}^{\infty} \big|\widehat{\check{K}_q}(\xi)\big|^2 \big|\widehat{y^{(q+1)}}(\xi)\big|^2 d\xi && \text{(from (2.20))}\\
&\le \frac{h^{2q+2}}{2\pi}\max|\widehat{\check{K}_q}(\xi)|^2\cdot\int_{-\infty}^{\infty}\big|\widehat{y^{(q+1)}}(\xi)\big|^2 d\xi && \text{(estimation)}\\
&= \frac{h^{2q+2}}{2\pi}\,\|\widehat{\check{K}_q}\|^2_{L^\infty}\cdot\|\widehat{y^{(q+1)}}\|^2_{L^2} && \text{(definitions)}\\
&= h^{2q+2}\|\widehat{K_q}\,\|^2_{L^\infty}\cdot\|y^{(q+1)}\|^2_{L^2}. && \text{(from (2.23), (2.24))}
\end{aligned}$$

In order that the obtained estimates (2.25) for L express the *actual errors* of the numerical solution, we adopt throughout this section the normalization $\sigma(1) = 1$ (cf. Eq. (III.2.13)).

And here is the theorem which tells us that linear multistep methods of order $p>2$ and "large" stability domain cannot be precise:

Theorem 2.6 (Jeltsch & Nevanlinna 1982). *Consider k-step methods of order $p>2$, normalized by $\sigma(1)=1$, for which the disc D_r of (2.15) is in the stability domain S. Then there exists a constant $C>0$ (depending on k,p,q; but independent of r) such that the Fourier transform of the Peano kernel K_q $(q\le p)$ satisfies*

$$\|\widehat{K_q}\|_\infty \ge C\Big(\frac{r}{3}\Big)^{p-2}. \tag{2.26}$$

The *proof* of Jeltsch & Nevanlinna is in two steps:

a) The stability requirement forces some coefficients a_j of $R(z)$ to be large (Lemma 2.7 below), where as in (III.3.17)

$$R(z)=\Big(\frac{z-1}{2}\Big)^k\varrho\Big(\frac{z+1}{z-1}\Big)=\sum_{j=0}^k a_jz^j \tag{2.27}$$

$$S(z)=\Big(\frac{z-1}{2}\Big)^k\sigma\Big(\frac{z+1}{z-1}\Big)=\sum_{j=0}^k b_jz^j. \tag{2.28}$$

b) $\|\widehat{K_q}\|_{L^\infty}$ can be bounded from below by $\max_j a_j$ (Lemma 2.8).

Lemma 2.7. *If $D_r\subset S$ and $p>2$ then*

$$a_{k-j}\ge\Big(\frac{r}{3}\Big)^{j-1}\cdot a_{k-1}=\Big(\frac{r}{3}\Big)^{j-1}\cdot 2^{1-k}\qquad \textit{for}\quad j=2,\ldots,p-1. \tag{2.29}$$

Proof. Stability in D_r means that for $\mu\in D_r$ all roots of $\varrho(\zeta)-\mu\sigma(\zeta)=0$ lie in $|\zeta|\le 1$. Hence

$$\varrho(\zeta)/\sigma(\zeta)\notin D_r\qquad\text{for}\quad |\zeta|>1. \tag{2.30}$$

Applying the Graeco-Roman transformation $\zeta=(z+1)/(z-1)$ and using (2.16) this means that

$$\mathrm{Re}\,\frac{S(z)}{R(z)}>-\frac{1}{2r}\qquad\text{for}\quad \mathrm{Re}\,z>0 \tag{2.31}$$

or

$$\mathrm{Re}\,\frac{2rS(z)+R(z)}{R(z)}>0\qquad\text{for}\quad \mathrm{Re}\,z>0. \tag{2.32}$$

Next, we must consider the order conditions (Lemma III.3.7 and Exercise 9 of Sect. III.3)

$$R(z)\Big(\frac{z}{2}-\frac{1}{6z}-\frac{2}{45z^3}-\ldots\Big)-S(z)=\mathcal{O}\Big(\Big(\frac{1}{z}\Big)^{p-k}\Big),\qquad z\to\infty. \tag{2.33}$$

This shows that $R(z)=\mathcal{O}(z^{k-1})$, $S(z)=\mathcal{O}(z^k)$, but $2S(z)-zR(z)=\mathcal{O}(z^{k-1})$. Thus we subtract rz from (2.32) in order to lower the degree of the numerator. The resulting function again satisfies

$$\mathrm{Re}\ \frac{r(2S(z)-zR(z))+R(z)}{R(z)}>0 \qquad \text{for} \quad \mathrm{Re}\, z>0 \tag{2.34}$$

because of $\mathrm{Re}\,(rz)=0$ on $z=iy$ and the maximum principle (an idea similar to that of Lemma IV.5.21). The function (2.34) can therefore have no zeros in $\mathbb{C}^+$ (since by Taylor expansion all arguments of a function appear in a complex neighbourhood of a zero). Therefore the *numerator* of (2.34) must have non-negative coefficients (cf. the proof of Lemma III.3.6). Multiplying out (2.33) and (2.34) we obtain for the coefficient of z^{k-j} $(j\le p-1)$:

$$0\le r\Big(-\frac{1}{3}\,a_{k-j+1}-\frac{4}{45}\,a_{k-j+3}-\ldots\Big)+a_{k-j}$$

or by simplifying (cf. Lemmas III.3.8 and III.3.6)

$$\frac{r}{3}\,a_{k-j+1}\ \le\ a_{k-j}.$$

Using $a_{k-1}=2^{1-k}\ \varrho'(1)=2^{1-k}$ (see Lemma III.3.6), this leads to (2.29). □

Lemma 2.8. *There exists $C>0$ (depending on k,p and q with $q=0,1,\ldots,p$) with the following property: if $0\in S$, then*

$$\|\widehat{K_q}\|_{L^\infty}\ \ge\ C\cdot\max_j a_j. \tag{2.35}$$

Proof. We set $e^{i\xi}=\zeta,\ \xi=-i\log\zeta$ in Eq. (2.23) so that the maximum must be taken over the set $|\zeta|=1$. Then we introduce $\zeta=(z+1)/(z-1)$ and take the maximum over the imaginary axis. This gives with (2.27) and (2.28)

$$\|\widehat{K_q}\|_{L^\infty}=\sup_t\underbrace{\left|\frac{1}{(it)^k}\left(\frac{R(it)}{\log\frac{it+1}{it-1}}-S(it)\right)\right|}_{\Phi(t)}\cdot\underbrace{\left|\left(\frac{2it}{it-1}\right)^k\right|\cdot\left|\log\left(\frac{it+1}{it-1}\right)\right|^{-q}}_{\Psi(t)}. \tag{2.36}$$

We now insert, for $|t|>1$, Eqs. (III.3.19), (III.3.21) and (III.3.22) to obtain

$$|\Phi(t)|=\Big|P_k\Big(\frac{1}{it}\Big)+\frac{d_1}{(it)^{k+1}}+\frac{d_2}{(it)^{k+2}}+\ldots\Big| \tag{2.37}$$

where P_k is a polynomial of degree k and subdegree p (see Lemma III.3.7), determined by the method. Since we want our estimates to be true for *all* methods, we treat P_k as an *arbitrary* polynomial. Separating real and imaginary parts and substituting $1/t=s$ gives

$$\begin{aligned}|\Phi(t)|^2=&\big|Q_{k-1}(s)+d_1s^{k+1}-d_3s^{k+3}+-\ldots\big|^2\\&+\big|Q_k(s)+d_2s^{k+2}-d_4s^{k+4}+-\ldots\big|^2=|\Phi_1(t)|^2+|\Phi_2(t)|^2\end{aligned} \tag{2.38}$$

where $Q_{k-1}(s)$ and $Q_k(s)$ are arbitrary (even or odd) polynomials of subdegree p and degree $k-1$ and k, respectively. Both terms are minorized separately, e.g. for the first we write

$$|\Phi_1(t)| \geq |Q_{k-1}(s) + d_1 s^{k+1}| - |d_3 s^{k+3} - d_5 s^{k+5} + - \ldots|. \tag{2.39}$$

Since $\mu_1 < \mu_3 < \mu_5 < \ldots < 0$ (Exercise 6 below) and $a_i \geq 0$ we have from (III.3.22)

$$d_1 \leq d_3 \leq d_5 \leq \ldots \leq 0 \quad \text{and} \quad d_2 \leq d_4 \leq d_6 \leq \ldots \leq 0. \tag{2.40}$$

Therefore, the second term in (2.39) is majorized by the alternating series argument for $0 < s < 1$ as

$$|d_3 s^{k+3} - d_5 s^{k+5} + - \ldots| \leq |d_3| s^{k+3} \leq |d_1| s^{k+3}.$$

Since $Q_{k-1}(s)$ is an arbitrary polynomial, we can replace it by $|d_1| Q_{k-1}(s)$ so that $|d_1|$ becomes a common factor of the whole expression

$$|\Phi_1(t)| \geq |d_1| \Big(|Q_{k-1}(s) + s^{k+1}| - s^{k+3} \Big). \tag{2.41}$$

This suggests that we define the constants

$$D_1 = \inf_{Q_{k-1}} \left\{ \sup_{0<s<1} \left[\Big(|Q_{k-1}(s) + s^{k+1}| - s^{k+3} \Big) \left(\frac{2}{\sqrt{1+s^2}} \right)^k \left(\frac{1}{2\arctan s} \right)^q \right] \right\}$$

$$D_2 = \inf_{Q_k} \left\{ \sup_{0<s<1} \left[\Big(|Q_k(s) + s^{k+2}| - s^{k+4} \Big) \left(\frac{2}{\sqrt{1+s^2}} \right)^k \left(\frac{1}{2\arctan s} \right)^q \right] \right\} \tag{2.42}$$

where the inf is taken over all polynomials $Q_{k-1}(s) = c_{k-1}s^{k-1} + c_{k-3}s^{k-3} + c_{k-5}s^{k-5} + \ldots$ respectively $Q_k(s) = c_k s^k + c_{k-2}s^{k-2} + c_{k-4}s^{k-4} + \ldots$ of subdegree p. The last two factors represent $\Psi(t)$ of (2.36). Since s^{k+1} dominates s^{k+3} for small s, D_1 and D_2 are *positive* constants (see Exercise 8). We then have from (2.38) and (2.36)

$$\|\widehat{K}_q\|_{L^\infty} \geq \sqrt{d_1^2 D_1^2 + d_2^2 D_2^2} \tag{2.43}$$

Since both d_1 and d_2 are sums of a_j with negative coefficients (see (III.3.22) and Lemma III.3.8), $\|\widehat{K}_q\|_\infty$ must be large if one of the coefficient a_j is large. □

This concludes the proof of Theorem 2.6 which, by the way, also proves Theorem 1.4 again. □

Exercises

1. Show that no explicit method can be $A(0)$-stable.

2. Show that $\beta_k/\alpha_k > 0$ is a necessary condition for an $A(\alpha)$-stable linear k-step method.

3. a) Show that the method

$$y_{n+2} - y_{n+1} = \frac{h}{4}(f_{n+2} + 2f_{n+1} + f_n)$$

has a stability domain bounded by a parabola. It is therefore A_0-stable, but not $A(0)$-stable (Cryer 1973).

b) Find a "deformation" of the 5th order BDF scheme

$$\sum_{j=1}^{5} \frac{1}{j}\nabla^j y_{n+1} + \beta\nabla^6 y_{n+1} = hf_{n+1}$$

with $\beta \approx 0.232\ldots$ which is $\mathring{A}$-stable, but not A_0-stable.

c) Find a method which is A_0-stable, but not stable at infinity.

Hint for (c). If you "lift up your heads, o ye gates" (just a few lines, not to heaven), the answer is easy to find.

4. (Parseval 1799). Prove the identity (2.24).

Hint. Insert the definitions into

$$\|\widehat{f}\|_{L^2}^2 = \int_{-\infty}^{\infty} \widehat{f}(\xi)\overline{\widehat{f}}(\xi)d\xi$$

to get a triple integral. Two of these integrals then disappear with the Fourier inversion formula.

Remark. You may be astonished to see that Parseval's identity is older than Fourier series and Fourier transforms. Well, Parseval's identity was originally a formula between an infinite sum and an integral, which was later re-interpreted and generalized to become what it is today.

5. Substitute $\xi = \pi$ in Formula (2.23) to obtain an easy minorization for $\|\widehat{K_q}\|_{L^\infty}$. Then compute for the methods defined in the proof of Lemma 2.3 (normalized by $\sigma(1) = 1$) the value $\sigma(-1)$ for ε small. This then shows that $\widehat{K_q}$ becomes very large.

6. Use the formula (see the proof of Lemma III.3.8)

$$\mu_{2j+1} = \int_{-1}^{+1} x^{2j}\left(\left(\log\frac{1+x}{1-x}\right)^2 + \pi^2\right)^{-1} dx$$

to show that $\mu_1 > \mu_3 > \mu_5 > \ldots > 0$.

7. Show that for $q = p$ Eq. (2.23) becomes, by substituting $i\xi = h$ and letting $h \to 0$ in Eq. (1.20), $\widehat{K_p}(0) = C_{p+1}$, where C_{p+1} is, for $\sigma(1) = 1$, the *error constant*.

   Formula (2.36) then provides, for $p = k$ and $t \to \infty$, lower bounds for the error constant (see "Theorem 4.5" of Jeltsch & Nevanlinna 1982).

8. For $p = k + 1$, the polynomials Q_{k-1} and Q_k in (2.42) vanish identically, because the subdegree must be p. Compute in this case the constants D_1 and D_2. It is also easy to compute them for $p = k - 1$. In the general case the optimal solution satisfies a sort of "Tchebysheff alternative".

   Results.

   Case $p = k + 1$ $(Q = 0)$:

D_1	$p=3$, $k=2$	$p=4$, $k=3$	$p=5$, $k=4$	$p=6$, $k=5$
$q=0$	0.4742	0.5695	0.7020	0.8813
$q=1$	0.3876	0.4435	0.5298	0.6505
$q=2$	0.3524	0.3659	0.4152	0.4933
$q=3$	0.5000	0.3381	0.3459	0.3891
$q=4$		0.5000	0.3251	0.3275
$q=5$			0.5000	0.3131
$q=6$				0.5000

D_2	$p=3$, $k=2$	$p=4$, $k=3$	$p=5$, $k=4$	$p=6$, $k=5$
$q=0$	0.3607	0.4501	0.5706	0.7319
$q=1$	0.2754	0.3347	0.4163	0.5263
$q=2$	0.2205	0.2570	0.3108	0.3852
$q=3$	0.1935	0.2075	0.2400	0.2888
$q=4$		0.1849	0.1956	0.2244
$q=5$			0.1770	0.1845
$q=6$				0.1698

   Case $p = k - 1$ (one free constant in Q):

D_1	$p=3$, $k=4$	$p=4$, $k=5$	$p=5$, $k=6$	$p=6$, $k=7$
$q=0$	0.0511	0.0362	0.0262	0.0193
$q=1$	0.0727	0.0499	0.0353	0.0256
$q=2$	0.1100	0.0709	0.0486	0.0344
$q=3$	0.2031	0.1070	0.0691	0.0474
$q=4$		0.1962	0.1041	0.0673
$q=5$			0.1894	0.1012
$q=6$				0.1828

D_2	$p=3$, $k=4$	$p=4$, $k=5$	$p=5$, $k=6$	$p=6$, $k=7$
$q=0$	0.0195	0.0142	0.0104	0.0077
$q=1$	0.0269	0.0191	0.0138	0.0101
$q=2$	0.0384	0.0263	0.0186	0.0135
$q=3$	0.0583	0.0374	0.0256	0.0181
$q=4$		0.0567	0.0365	0.0250
$q=5$			0.0552	0.0356
$q=6$				0.0537

   Case $p = k - 3$ (two free constants in Q):

D_1	$p=3$, $k=6$	$p=4$, $k=7$	$p=5$, $k=8$	$p=6$, $k=9$
$q=0$	0.0030	0.0014	0.0007	0.0003
$q=1$	0.0066	0.0029	0.0014	0.0007
$q=2$	0.0160	0.0066	0.0029	0.0014
$q=3$	0.0457	0.0158	0.0065	0.0029
$q=4$		0.0448	0.0156	0.0064
$q=5$			0.0439	0.0154
$q=6$				0.0431

D_2	$p=3$, $k=6$	$p=4$, $k=7$	$p=5$, $k=8$	$p=6$, $k=9$
$q=0$	0.0007	0.0004	0.0002	0.0001
$q=1$	0.0015	0.0007	0.0003	0.0002
$q=2$	0.0034	0.0015	0.0007	0.0003
$q=3$	0.0082	0.0034	0.0015	0.0007
$q=4$		0.0081	0.0033	0.0015
$q=5$			0.0080	0.0033
$q=6$				0.0079

V.3 Generalized Multistep Methods

> The Dahlquist bound of two on the order of A-stable multistep methods was the imperative to propound ... weaker stability properties, ... An alternative approach for circumventing Dahlquist's bound is to modify the class of methods, rather than the property.
> (T.A. Bickart & W.B. Rubin 1974)

The search for higher order A-stable multistep methods is carried out in two main directions:

- Use higher derivatives of the solutions;
- Throw in additional stages, off-step points, super-future points and the like, which leads into the large field of general linear methods.

Second Derivative Multistep Methods of Enright

> Hermite's formulas are rediscovered and republished every four years.
> (P.J. Davis 1963)

Differentiation of a differential equation

$$y' = f(x, y) \tag{3.1}$$

with respect to x gives the second derivative of the solution

$$y'' = f_x + f_y \cdot f =: g(x, y), \tag{3.2}$$

which we shall denote by g. Now a straightforward generalization of both multistep formulas (1.1) and, say, the Taylor series method (see I.8.13)

$$y_{n+1} = y_n + hf_n + \frac{h^2}{2!} g_n$$

can be written in the form

$$\sum_{i=0}^{k} \alpha_i y_{n+i} = h \sum_{i=0}^{k} \beta_i f_{n+i} + h^2 \sum_{i=0}^{k} \gamma_i g_{n+i} \tag{3.3}$$

where the α_i, β_i, γ_i are parameters which must be chosen appropriately. Most of the theory of linear multistep methods (Sect. III.2) generalizes without difficulty. Taylor expansion similar to (III.2.5) shows that method (3.3) is of *order* p if and only if

$$\sum_{i=0}^{k} \alpha_i i^q = q \sum_{i=0}^{k} \beta_i i^{q-1} + q(q-1) \sum_{i=0}^{k} \gamma_i i^{q-2} \tag{3.4}$$

for $0 \le q \le p$. The first two of these formulas are identical to (III.2.6), i.e., to

$$\varrho(1)=0, \qquad \varrho'(1)=\sigma(1). \tag{3.5}$$

The *error constant* (see Eq. (III.2.13) and Exercise 2 of Sect. III.4) is given by

$$C=\frac{1}{\sigma(1)(p+1)!}\Big(\sum_{i=0}^{k}\alpha_i\, i^{p+1}-(p+1)\sum_{i=0}^{k}\beta_i\, i^{p}-(p+1)p\sum_{i=0}^{k}\gamma_i\, i^{p-1}\Big). \tag{3.6}$$

A search for a good choice of the free parameters α_i, β_i, γ_i was undertaken by Enright (1974) with the following ideas:

(i) Set $\alpha_k=1$, $\alpha_{k-1}=-1$, $\alpha_{k-2}=\ldots=\alpha_0=0$ to ensure reasonable stability in a neighbourhood of the origin as in the standard Adams formulas;

(ii) Set $\gamma_k\neq 0$, $\gamma_{k-1}=\ldots=\gamma_0=0$ to ensure stability at infinity as in the BDF formulas;

(iii) Determine the remaining $k+2$ coefficients γ_k, β_k, $\beta_{k-1},\ldots,\beta_0$ from Equations (3.4) for $q=1,2,\ldots,k+2$ ($q=0$ is satisfied with (i)) to ensure a reasonably high order.

The result is a class of k-step formulas of order $k+2$, which are of the form

$$y_{n+1}=y_n+h\sum_{i=0}^{k}\beta_i f_{n+i-k+1}+h^2\gamma_k g_{n+1}. \tag{3.7}$$

The first few of these methods are

$$k=1:\; y_{n+1}=y_n+h\Big(\frac{2}{3}f_{n+1}+\frac{1}{3}f_n\Big)-\frac{1}{6}h^2 g_{n+1}$$

$$k=2:\; y_{n+1}=y_n+h\Big(\frac{29}{48}f_{n+1}+\frac{5}{12}f_n-\frac{1}{48}f_{n-1}\Big)-\frac{1}{8}h^2 g_{n+1}$$

$$k=3:\; y_{n+1}=y_n+h\Big(\frac{307}{540}f_{n+1}+\frac{19}{40}f_n-\frac{1}{20}f_{n-1}+\frac{7}{1080}f_{n-2}\Big)-\frac{19}{180}h^2 g_{n+1}$$

$$k=4:\; y_{n+1}=y_n+h\Big(\frac{3133}{5760}f_{n+1}+\frac{47}{90}f_n-\frac{41}{480}f_{n-1}+\frac{1}{45}f_{n-2}-\frac{17}{5760}f_{n-3}\Big)-\frac{3}{32}h^2 g_{n+1}$$

For a general expression, see Eq. (3.12) below and Exercise 1.

The stability analysis for second derivative methods is again done by linearizing and leads to

$$y'=\lambda y \qquad \text{for which} \qquad y''=\lambda^2 y. \tag{3.8}$$

This, inserted into (3.3), gives as the characteristic equation

$$\sum_{i=0}^{k}(\alpha_i-\mu\beta_i-\mu^2\gamma_i)\zeta^i=0, \qquad \mu=h\lambda \tag{3.9}$$

instead of (1.6). Equation (3.9) is, for $\zeta = e^{i\theta}$, a quadratic equation which gives rise to *two* root locus curves which, together, describe the stability domain. The Enright methods (3.7) turn out to be A-stable for $k = 1$ and 2 (hence for $p = 3$ and 4) and are stiffly stable for $k = 3, 4, 5, 6$ and 7. The corresponding values α (for $A(\alpha)$-stability), D and the error constants C are given in Table 3.1. Pictures are shown in Fig. 3.1.

Table 3.1. Stability characteristics and error constants for Enright methods

k	1	2	3	4	5	6	7
p	3	4	5	6	7	8	9
α	90°	90°	87.88°	82.03°	73.10°	59.95°	37.61°
D	0.	0.	0.103	0.526	1.339	2.728	5.182
C	0.01389	0.00486	0.00236	0.00136	0.00086	0.00059	0.00042

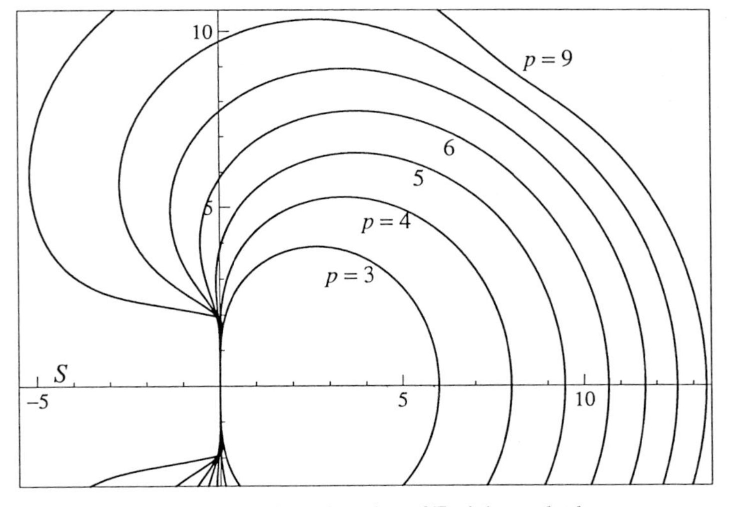

Fig. 3.1. Stability domains of Enright methods

Dense Output for Enright Methods. We have seen in Sect. III.1 that Newton's interpolation formula, based on the data $x_{n+1}, x_n, \ldots, x_{n-k+1}$,

- when integrated from x_n to x_{n+1}, leads to the implicit Adams methods;
- when differentiated at x_{n+1}, leads to the BDF methods.

It is natural to apply the same idea to *Hermite* interpolation (Addison 1979): guided by much previous experience (see above) we choose the data points

$$x_{n+1} \text{ (double node)},\ x_n,\ x_{n-1}, \ldots, x_{n-k+1} \text{ (simple nodes)}. \tag{3.10}$$

This gives the following scheme of divided differences

$$\begin{array}{lllll}
s=1 & f_1 & & & \\
 & & hf_1' & & \\
s=1 & f_1 & & hf_1'-\nabla f_1 & \dfrac{hf_1'-\nabla f_1-\frac{1}{2}\nabla^2 f_1}{2!} \\
 & & \nabla f_1 & & \\
s=0 & f_0 & & \dfrac{\nabla^2 f_1}{2!} & \\
 & & \nabla f_0 & & \\
s=-1 & f_{-1} & & &
\end{array}$$

where $x = x_n + sh$. For these "confluent" data, Newton's interpolation formula becomes

$$\begin{aligned}
f(x_n+sh) = f_1 &+ (s-1)hf_1' + (s-1)^2(hf_1' - \nabla f_1) \\
&+ (s-1)^2 s\, \frac{hf_1' - \nabla f_1 - \frac{1}{2}\nabla^2 f_1}{2!} \\
&+ (s-1)^2 s(s+1)\, \frac{hf_1' - \nabla f_1 - \frac{1}{2}\nabla^2 f_1 - \frac{1}{3}\nabla^3 f_1}{3!} + \ldots
\end{aligned} \tag{3.11}$$

We now interpret f as the derivative $f(x, y(x))$ of the solution, so that f' becomes the second derivative. Integrating Formula (3.11) from x_n to x_{n+1} we obtain

$$y_{n+1} = y_n + hf_{n+1} - h\sum_{j=1}^{k} \frac{\nabla^j f_{n+1}}{j}\Big(\sum_{i=j}^{k} \nu_i\Big) + h^2 g_{n+1} \cdot \Big(\sum_{i=0}^{k} \nu_i\Big) \tag{3.12}$$

where

$$\nu_i = \int_0^1 \frac{(s-1)^2 s(s+1)\ldots(s+i-2)}{i!}\, ds = (-1)^i \int_0^1 (s-1)\binom{1-s}{i} ds. \tag{3.13}$$

Table 3.2. Coefficients for Enright methods

j	0	1	2	3	4	5	6	7
ν_j	$-\frac{1}{2}$	$\frac{1}{3}$	$\frac{1}{24}$	$\frac{7}{360}$	$\frac{17}{1440}$	$\frac{41}{5040}$	$\frac{731}{120960}$	$\frac{8563}{1814400}$

The first few values of ν_i are given in Table 3.2 and Eq. (3.12) is seen to be identical with (3.7). Dense output, of course, is obtained by integrating (3.11) from x_n to $x_n + \theta h$:

$$y(x_n+\theta h) \approx y_n + \theta h f_{n+1} - h\sum_{j=1}^{k} \frac{\nabla^j f_{n+1}}{j}\Big(\sum_{i=j}^{k} \nu_i(\theta)\Big) + h^2 g_{n+1} \cdot \Big(\sum_{i=0}^{k} \nu_i(\theta)\Big)$$

where

$$\nu_i(\theta) = (-1)^i \int_0^\theta (s-1)\binom{1-s}{i} ds.$$

Second Derivative BDF Methods

If we are interested in a "second derivative" analogue of the BDF methods, we replace all f's by y's in (3.11) and differentiate twice at x_{n+1}. This, on setting $y''(x_{n+1}) = g_{n+1}$, results in the methods

$$\frac{h^2}{2} g_{n+1} = \Big(\sum_{i=1}^{k} \frac{1}{i}\Big) h f_{n+1} - \sum_{j=1}^{k} \Big(\sum_{i=j}^{k} \frac{1}{i}\Big) \frac{\nabla^j y_{n+1}}{j} \tag{3.14}$$

which we call *"Second derivative BDF methods"* (SDBDF, the reader is cautioned against confusion: Cash (1981) uses this expression for the class of "Enright methods"). Analyzing the stability of these methods leads to the parameters of Table 3.3. The root locus curves are drawn in Fig. 3.2.

In complete analogy to the behaviour of implicit Adams compared to BDF methods, the second derivative BDF methods have larger error constants than the Enright methods, but allow stiffly stable methods of higher order.

Table 3.3. Stability characteristics and error constants for SDBDF methods

k	1	2	3	4	5	6	7	8	9	10
p	2	3	4	5	6	7	8	9	10	11
α	90°	90°	90°	89.36°	86.35°	80.82°	72.53°	60.71°	43.39°	12.34°
D	0.	0.	0.	0.015	0.128	0.401	0.886	1.646	2.770	4.373
C	.1667	.0556	.0273	.0160	.0104	.0073	.0054	.0041	.0032	.0026

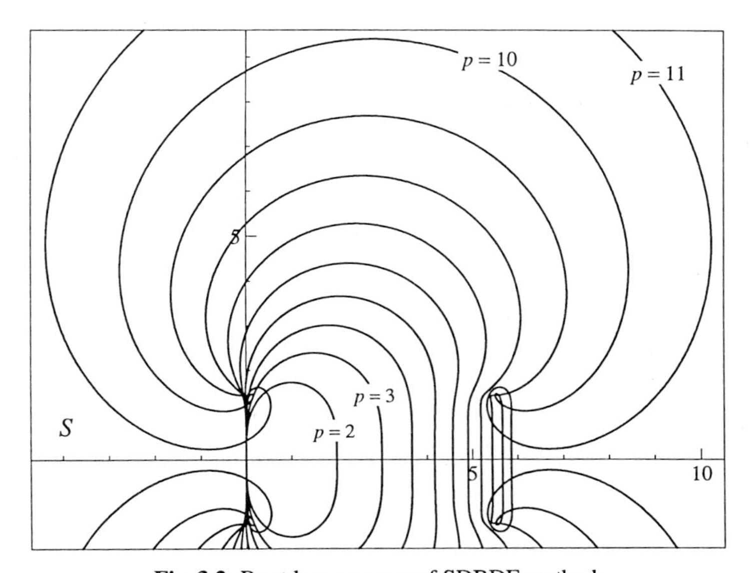

Fig. 3.2. Root locus curves of SDBDF methods

Blended Multistep Methods

The original motivation for *blended methods* goes as follows (Skeel & Kong 1977): We know that Adams methods

$$-y_{n+1}+y_n+h\big(\beta_k f_{n+1}+\beta_{k-1}f_n+\ldots+\beta_0 f_{n-k+1}\big)=0 \qquad (AMF^{(k+1)})$$

are a very good choice for nonstiff problems, and that BDF methods

$$-\big(\alpha_k y_{n+1}+\alpha_{k-1}y_n+\ldots+\alpha_0 y_{n-k+1}\big)+hf_{n+1}=0 \qquad (BDF^{(k)})$$

are a very good choice for stiff problems. Nonstiff problems are characterized by the fact that $-h\,\partial f/\partial y$ is *small*, while stiff problems are characterized by *large* $-h\,\partial f/\partial y$ (at first this makes sense only for scalar equations; but it works as well for systems of equations if we descend into the eigenspaces of the Jacobian matrix $\partial f/\partial y=J$). The idea is now to use a weighted mean ("blend", a term suggested by C.W. Gear) of the two methods such as

$$\{AMF^{(k+1)}\}-\gamma^{(k)}hJ\{BDF^{(k)}\}=0 \qquad (3.15)$$

where $\gamma^{(k)}$ is a free parameter. The factor $-hJ$, when small or large, just puts the weight at the right place, as required by the above motivation. Taylor expansion shows that Eq. (3.15) is for all $\gamma^{(k)}$ of order $p=k+1$ (the factor "h" in the second term saves one order), even if J differs from $\partial f/\partial y$. This method is thus a multistep analogue to the W-methods discussed in Sect. IV.7.

Example. We put $k=2$ in (3.15) and insert the values from Sect. III.1, Formulas (III.1.8") and (III.1.22"):

$$\begin{aligned} y_{n+1}=y_n&+h\Big(\frac{5}{12}f_{n+1}+\frac{8}{12}f_n-\frac{1}{12}f_{n-1}\Big)\\ &-\gamma^{(2)}hJ\Big(-\frac{3}{2}y_{n+1}+2y_n-\frac{1}{2}y_{n-1}+hf_{n+1}\Big). \end{aligned} \qquad (3.16)$$

If we now suppose that our differential equation is linear and autonomous $y'=Jy$, then $Jy_{n+i}=f_{n+i}$ and the equation simplifies. Two special choices for $\gamma^{(2)}$ are then interesting:

a) $\gamma^{(2)}=1/6$: In this case the f_{n-1} cancels with Jy_{n-1} and Eq. (3.16) becomes the $(k-1)$-step Enright formula of order $k+1$;

b) $\gamma^{(2)}=1/8$: This is a "superconvergence point" for linear equations and we obtain the k-step Enright formula of order $k+2$.

Both properties generalize to arbitrary k; in the first case we have to put $\gamma^{(k)}=-k\gamma_k^*$, where the γ_k^* are the values of Table III.1.2, and in the second case we use $\gamma^{(k)}=-\sum_{i=0}^k \nu_i$ as in (3.12). Blended methods therefore share the excellent stability properties of the Enright methods and seem, at the same time, easier to implement. A third possibility is to choose $\gamma^{(k)}$ in order to maximize the angle α

Table 3.4. Values for $\gamma^{(k)}$ and corresponding angles for blended methods

k	p	$-k\gamma_k^*$	α for $\gamma^{(k)} = -k\gamma_k^*$	$\gamma_{opt}^{(k)}$	α for $\gamma^{(k)} = \gamma_{opt}^{(k)}$
1	2	.5	90°	$[0, +\infty)$	90°
2	3	.1666667	90°	$[.125, +\infty)$	90°
3	4	.125	90°	$[.12189, .68379]$	90°
4	5	.1055556	87.88°	.1284997	89.42°
5	6	.09375	82.03°	.1087264	86.97°
6	7	.08561508	73.10°	.0962596	82.94°
7	8	.07957176	59.95°	.08754864	77.43°
8	9	.07485229	37.61°	.08105624	70.22°
9	10	.07103299	−−	.07599875	60.68°
10	11	.06785850	−−	.07192937	47.63°
11	12	.06516462	−−	.06857226	28.68°

for $A(\alpha)$-stability. The root-locus-curve equation for general $\gamma^{(k)}$ becomes

$$\mu^2 \cdot \gamma^{(k)} + \mu\Big(-\sum_{j=0}^{k} \gamma_j^*(1-e^{-i\theta})^j - \gamma^{(k)} \sum_{j=1}^{k} \frac{1}{j}(1-e^{-i\theta})^j\Big) + (1-e^{-i\theta}) = 0.$$

Skeel & Kong (1977) have carefully computed the optimal $\gamma^{(k)}$ (see Table 3.4, the imprecise values for the "Enright column" have been corrected) and arrived thereby at stiffly stable methods up to order 12.

Extended Multistep Methods of Cash

The second possibility for circumventing Dahlquist's barrier, instead of adding higher derivatives, is to add further stages, additional nodes, or off-step points. This leads into the huge desert ("A fable of K. Burrage") of general linear methods which have been discussed in Sect. III.8. Pioneering results for stiff differential equations are the "composite multistep methods" of Sloate & Bickart (1973), Bickart & Rubin (1974), the "hybrid" methods of England (1982), and the "extended" BDF methods of Cash (1980). We shall present the basic ideas for the latter in some detail. In order to increase stability of the BDF methods, we extend them by adding a "super-future" point at x_{n+k+1}

$$\sum_{j=0}^{k} \alpha_j y_{n+j} = h\beta_k f_{n+k} + h\beta_{k+1} f_{n+k+1}, \tag{3.17}$$

where the coefficients are obtained by solving $\sum_j \alpha_j j^q = q \sum_j \beta_j j^{q-1}$ for $q = 0, 1, \ldots, k+1$ with the normalization $\alpha_k = 1$. Formula (3.17) is then used as follows (see Fig. 3.3):

(i) Suppose that the solution values $y_n, y_{n+1}, \ldots, y_{n+k-1}$ are available. Compute $\overline{y}_{n+k}$ as the solution of the conventional BDF formula

$$\sum_{j=0}^{k} \widehat{\alpha}_j\, y_{n+j} = h\widehat{\beta}_k\, f_{n+k}, \qquad \widehat{\alpha}_k = 1; \tag{3.17i}$$

(ii) Compute $\overline{y}_{n+k+1}$ as the solution of the same BDF formula advanced by one step (using $\overline{y}_{n+k}$ for y_{n+k})

$$\sum_{j=0}^{k} \widehat{\alpha}_j\, y_{n+j+1} = h\widehat{\beta}_k\, f_{n+k+1} \qquad (y_{n+k} := \overline{y}_{n+k}) \tag{3.17ii}$$

and set $\overline{f}_{n+k+1} = f(x_{n+k+1}, \overline{y}_{n+k+1})$;

(iii) Discard $\overline{y}_{n+k}$, insert $\overline{f}_{n+k+1}$ into (3.17) and solve for a new y_{n+k} which serves as the final numerical solution of the method.

The advance of the numerical solution by *one step* thus requires the solution of *three* nonlinear systems of dimension n. In stage (i) and stage (iii) we have excellent initial approximations: the super future point of the previous step and the value $\overline{y}_{n+k}$, respectively.

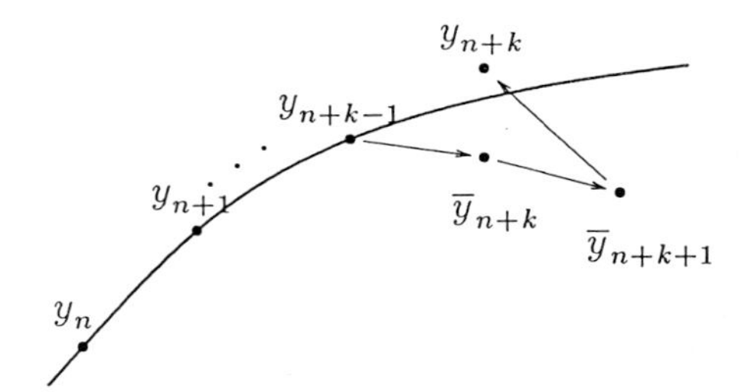

Fig. 3.3. Errors of Cash's algorithm

Lemma 3.1 (Cash 1980). *If Formula (3.17) is of order $k+1$ and the BDF methods used in (3.17i) and (3.17ii) are of order k, then the whole predictor-corrector algorithm (i)–(iii) is of order $k+1$.*

Proof. Suppose that $y_n, \ldots, y_{n+k-1}$ are on the exact solution (Fig. 3.3). Then a simple calculation (as in the proof of Lemma III.2.2, see also Eq. (III.2.7)) shows that

$$y(x_{n+k}) - \overline{y}_{n+k} = C_1 h^{k+1} y^{(k+1)}(x_{n+k}) + \mathcal{O}(h^{k+2}) \tag{3.18}$$

$$y(x_{n+k+1}) - \overline{y}_{n+k+1} = C_1 \Big(1 - \frac{\widehat{\alpha}_{k-1}}{\widehat{\alpha}_k}\Big) h^{k+1} y^{(k+1)}(x_{n+k}) + \mathcal{O}(h^{k+2}) \tag{3.19}$$

where C_1 depends on the BDF method used. If now $C_2 h^{k+2} y^{(k+2)}(\xi)$ is the defect of Eq. (3.17) (for the exact solution), replacing $hf(x_{n+k+1}, y(x_{n+k+1}))$ by

$hf(x_{n+k+1}, \overline{y}_{n+k+1})$ adds the expression obtained in (3.19) to this error and we obtain

$$y(x_{n+k}) - y_{n+k} = h^{k+2}\Big(C_2 y^{(k+2)} + \beta_{k+1} C_1 \big(1 - \frac{\widehat{\alpha}_{k-1}}{\alpha_k}\big) \frac{\partial f}{\partial y} \cdot y^{(k+1)}\Big)(x_{n+k}) + \mathcal{O}(h^{k+3}). \tag{3.20}$$

The method is thus of order $k+1$. Like Runge-Kutta methods, but unlike linear multistep methods, the principal error term is composed of several "elementary differentials". □

Modified EBDF Methods. A disadvantage of the above algorithm is that stages (i) and (ii) represent nonlinear systems with the same Jacobian $I - h\widehat{\beta}_k J$, but stage (iii) has a different Jacobian $I - h\beta_k J$. This requires an extra LU-decomposition. The idea is to modify Eq. (3.17) for stage (iii) as follows (Cash 1983):

$$\sum_{j=0}^{k} \alpha_j y_{n+j} = h\widehat{\beta}_k f_{n+k} + h(\beta_k - \widehat{\beta}_k)\overline{f}_{n+k} + h\beta_{k+1}\overline{f}_{n+k+1}. \tag{3.17.mod}$$

This just adds an extra h^{k+2}-term to the above proof and does not alter the order of the method. It allows the same Jacobian to be used in the Newton iteration for all three stages, and, possibly, to preserve it over several steps as well.

Stability Analysis. We insert $hf_j = \mu y_j$ in (3.17.mod), (3.17i) and (3.17ii), set $y_n = 1$, $y_{n+1} = \zeta, \ldots, y_{n+k-1} = \zeta^{k-1}$ and compute, following the algorithm (i), (ii), (iii), the solution $y_{n+k} =: \zeta^k$. This gives the characteristic equation

$$A\mu^3 + B\mu^2 + C\mu + D = 0 \tag{3.21}$$

where

$$\begin{aligned}
A &= \widehat{\beta}_k^3 \zeta^k \\
B &= -2\widehat{\beta}_k^2 \zeta^k + \widehat{\beta}_k(\beta_k - \widehat{\beta}_k)R + \widehat{\beta}_k \beta_{k+1} S - \widehat{\beta}_k^2 T \\
C &= \widehat{\beta}_k \zeta^k + (\widehat{\alpha}_{k-1}\beta_{k+1} - \beta_k + \widehat{\beta}_k)R - \beta_{k+1} S + 2\widehat{\beta}_k T \\
D &= -T \\
R &= \sum_{j=0}^{k-1} \widehat{\alpha}_j \zeta^j, \quad S = \sum_{j=0}^{k-2} \widehat{\alpha}_j \zeta^{j+1}, \quad T = \sum_{j=0}^{k} \alpha_j \zeta^j.
\end{aligned} \tag{3.22}$$

Inserting $\zeta = e^{i\theta}$, Equation (3.21) gives us three roots $\mu_i(\theta)$ $i = 1,2,3$, which describe the stability domain. These, computed by Cardano's formula, are displayed in Fig. 3.4. The corresponding stability characteristics are given in Table 3.5. The methods are A-stable for $p \le 4$ and are stiffly stable for orders up to 9.

Table 3.5. Stability measures for Cash's modified EBDF methods

k	1	2	3	4	5	6	7	8
p	2	3	4	5	6	7	8	9
α	90°	90°	90°	88.36°	83.07°	74.48°	61.98°	42.87°
D	0.	0.	0.	0.040	0.246	0.684	1.402	2.432

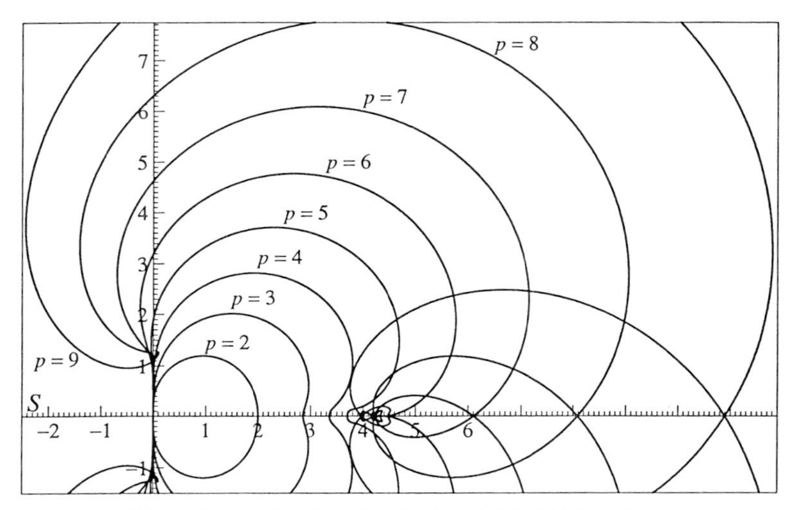

Fig. 3.4. Stability domains for Cash's MEBDF methods

Multistep Collocation Methods

> ... a theorem of great antiquity ... the simple theorem of polynomial interpolation upon which much practical numerical analysis rests ...
>
> (P.J. Davis, Interp. and Approx., Chapter II, 1963)

There are essentially two possibilities to extend the idea of collocation, which is so successful in the Runge-Kutta case (see Sect. II.7, Formulas (II.7.16)), into the multistep scene:

a) In a Nordsieck type manner with given $y_n, hy'_n, h^2y''_n/2, \ldots$ compute y_{n+1}, hy'_{n+1}, $h^2y''_{n+1}/2, \ldots$ The result is a spline function which approximates the solution globally. Butcher's generalized singly-implicit methods (Butcher 1981) are of this type. Extensive studies of these methods are due to Mülthei (1982).

b) In a multistep manner with given y_n, $y_{n-1}, \ldots, y_{n-k+1}$ compute y_{n+1}, then discard, as usual, the last point y_{n-k+1} and continue. This possibility was first proposed and analysed by Guillou & Soulé (1969). It is also the subject of a paper by Lie & Nørsett (1989) and will retain our attention here in more detail. In evident generalization of Definition II.7.6, the method is defined as follows:

Definition 3.2. Let s real numbers $c_1, \ldots, c_s$ (typically between 0 and 1) be given and k solution values $y_n, y_{n-1}, \ldots, y_{n-k+1}$. Then define the corresponding *collocation polynomial* $u(x)$ of degree $s+k-1$ by (see Fig. 3.5)

$$u(x_j) = y_j \qquad j = n-k+1, \ldots, n \tag{3.23a}$$

$$u'(x_n + c_i h) = f(x_n + c_i h, u(x_n + c_i h)) \quad i = 1, \ldots, s. \tag{3.23b}$$

The numerical solution is then

$$y_{n+1} := u(x_{n+1}). \tag{3.23c}$$

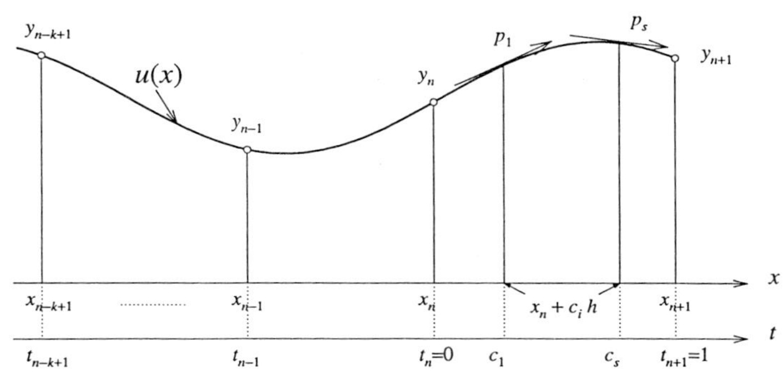

Fig. 3.5. The collocation polynomial

If we suppose the derivatives $u'(x_n + c_i h)$ are known, Eqs. (3.23a) and (3.23b) constitute a Hermite interpolation problem with incomplete data: the function values at $x_n + c_i h$ are missing. We therefore have no nice formulas and reduce the problem to a linear algebraic equation. We introduce the dimensionless coordinate $t = (x - x_n)/h$, $x = x_n + th$, nodes $t_1 = -k+1, \ldots, t_{k-1} = -1$, $t_k = 0$ and define polynomials $\varphi_i(t)$ $(i = 1, \ldots, k)$ of degree $s+k-1$ by

$$\begin{aligned} \varphi_i(t_j) &= \begin{cases} 0 & \text{if } i \neq j \\ 1 & \text{if } i = j \end{cases} \qquad j = 1, \ldots k \\ \varphi_i'(c_j) &= 0 \qquad j = 1, \ldots, s \end{aligned} \tag{3.24}$$

and polynomials $\psi_i(t)$ $(i = 1, \ldots, s)$ by

$$\begin{aligned} \psi_i(t_j) &= 0 \qquad j = 1, \ldots, k \\ \psi_i'(c_j) &= \begin{cases} 1 & \text{if } i = j \\ 0 & \text{if } i \neq j \end{cases} \qquad j = 1, \ldots, s. \end{aligned} \tag{3.25}$$

This makes these polynomials a (generalized) Lagrange basis and the polynomial $u(x)$ is readily written as

$$u(x_n + th) = \sum_{j=1}^{k} \varphi_j(t) y_{n-k+j} + h \sum_{j=1}^{s} \psi_j(t) u'(x_n + c_j h). \tag{3.26}$$

Formulas (3.24) and (3.25) do not always have a solution (Exercise 4 below). A convenient way of computing them is indicated in Exercise 5. Putting $t = c_i$ in (3.26), writing $u(x_n + c_i h) = v_i$ and inserting the collocation condition (3.23b) we obtain

$$v_i = \sum_{j=1}^{k} \varphi_j(c_i) y_{n-k+j} + h \sum_{j=1}^{s} \psi_j(c_i) f(x_n + c_j h, v_j) \qquad i = 1, \ldots, s \tag{3.27a}$$

$$y_{n+1} = \sum_{j=1}^{k} \varphi_j(1) y_{n-k+j} + h \sum_{j=1}^{s} \psi_j(1) f(x_n + c_j h, v_j), \tag{3.27b}$$

a general linear method as defined in (III.8.7).

Theorem 3.3. *The collocation method (3.23) is equivalent to the general linear method*

$$\begin{aligned} v_i &= \sum_{j=1}^{k} a_{ij}\, y_{n-k+j} + h \sum_{j=1}^{s} b_{ij}\, f(x_n + c_j h, v_j) \qquad i = 1, \ldots, s \\ y_{n+1} &= \sum_{j=1}^{k} a_{k+1,j}\, y_{n-k+j} + h \sum_{j=1}^{s} b_{k+1,j}\, f(x_n + c_j h, v_j) \end{aligned} \tag{3.28}$$

where

$$a_{ij} = \varphi_j(c_i), \quad b_{ij} = \psi_j(c_i), \quad a_{k+1,j} = \varphi_j(1), \quad b_{k+1,j} = \psi_j(1) \tag{3.29}$$

and $\varphi_j(t)$, $\psi_j(t)$ are polynomials defined by (3.24) and (3.25). Formula (3.26) provides a continuous output. □

A straightforward extension of the proof of Theorem II.7.9, again using the Gröbner & Alekseev formula (I.14.18), yields

Theorem 3.4 (Guillou & Soulé 1969). *If the quadrature formula (3.27b) is exact for polynomials $g(t)$ of degree $\le s+k+r$, i.e., $\sum_{j=1}^{k} \varphi_j(1) = 1$ and*

$$\sum_{j=1}^{k} \varphi_j(1) \int_{j-k}^{1} g(t)dt = \sum_{i=1}^{s} \psi_i(1) g(c_i),$$

then the multistep collocation method (3.28) also has order $s+k+r$. □

Methods of "Radau" Type

Nous allons maintenant étudier une classe de formules qui généralise les formules ordinaires de Gauss, Radau et Lobatto.

(Guillou & Soulé 1969)

An interesting question is now how to choose the nodes c_i in order to obtain the highest possible order. Using an elegant idea of Krylov (1959) (see the last chapter of his book on integration), Guillou & Soulé (1969) and Lie & Nørsett (1989) constructed such methods of maximal order $p=2s+k-1$. Unfortunately, these methods are not stiffly stable and therefore of no use for stiff problems. Consequently, we fix $c_s = 1$ to achieve stability at infinity and try to determine $c_1, \ldots, c_{s-1}$ so that the order becomes $p = 2s+k-2$. Because of Theorem 3.4, it is sufficient to consider quadrature problems.

And now to Krylov's idea for integrals, adapted to our situation. We fill in the gaps in the data for Hermite interpolation, i.e., we suppose that the *function values* $v_i = u(x_n + c_i h)$ $(i = 1, \ldots, s-1)$ are known and we extend our Lagrange basis accordingly: firstly, we add polynomials $\chi_1(t), \ldots, \chi_{s-1}(t)$ of degree $2s+k-2$ which must satisfy

$$\chi_i(t_j) = 0 \qquad j = 1, \ldots, k \tag{3.30a}$$

$$\chi_i'(c_j) = 0 \qquad j = 1, \ldots, s \tag{3.30b}$$

$$\chi_i(c_j) = \begin{cases} 1 & i = j \\ 0 & i \neq j \end{cases} \qquad j = 1, \ldots, s-1 \tag{3.30c}$$

(Caution: the last condition is *not* for $j = s$, because c_s is not a free node). Secondly, the polynomials $\varphi_i(t)$ and $\psi_i(t)$ are replaced by $\widetilde{\varphi}_i(t)$, $\widetilde{\psi}_i(t)$ of degree $2s+k-2$ which, in addition to (3.24) and (3.25), must satisfy

$$\widetilde{\varphi}_i(c_j) = 0 \quad \text{and} \quad \widetilde{\psi}_i(c_j) = 0 \qquad j = 1, \ldots, s-1. \tag{3.31}$$

Then Eq. (3.26) is replaced by

$$\widetilde{u}(x_n + th) = \sum_{j=1}^{k} \widetilde{\varphi}_j(t) y_{n-k+j} + \sum_{j=1}^{s-1} \chi_j(t) v_j + h \sum_{j=1}^{s} \widetilde{\psi}_j(t) u'(x_n + c_j h), \tag{3.32}$$

and (3.27b) becomes the integration formula

$$y_{n+1} = \sum_{j=1}^{k} \widetilde{\varphi}_j(1) y_{n-k+j} + \sum_{j=1}^{s-1} \chi_j(1) v_j + h \sum_{j=1}^{s} \widetilde{\psi}_j(1) u'(x_n + c_j h) \tag{3.33}$$

which is of order $2s+k-2$. If now, by a miracle, all coefficients

$$\chi_j(1) = 0 \qquad (j = 1, \ldots, s-1) \tag{3.34}$$

were zero, then the quadrature Formula (3.27b) would become equal to (3.33), since by uniqueness the remaining coefficients $\widetilde{\varphi}_j(1)$ and $\widetilde{\psi}_j(1)$ must also be equal to $\varphi_j(1)$ and $\psi_j(1)$.

Theorem 3.5. *If the collocation points* $c_1, \ldots, c_{s-1}$ *(with* $c_s = 1$*) are chosen such that the polynomials* $\varphi_i(t), \psi_i(t)$ *of (3.24), (3.25) exist uniquely and that (3.34) is true, then the collocation method (3.28) is of highest possible order* $2s + k - 2$. □

Computation of the Nodes. Equation (3.34) together with the conditions (3.30) allow us to write the polynomials $\chi_i(t)$ in the simple form

$$\chi_i(t) = C \prod_{j=1}^{k} (t - t_j) \prod_{\substack{j=1 \\ j \neq i}}^{s} (t - c_j)^2. \tag{3.35}$$

where C is determined by $\chi_i(c_i) = 1$. This then satisfies all derivative requirements (3.30b), except at c_i. $\chi_i'(c_i)$ is readily computed from (3.35) by taking logarithms and the conditions $\chi_i'(c_i) = 0$ give

$$\sum_{j=1}^{k} \frac{1}{c_i - t_j} + \sum_{\substack{j=1 \\ j \neq i}}^{s} \frac{2}{c_i - c_j} = 0, \qquad i = 1, \ldots, s-1. \tag{3.36}$$

Example. For the case $s = 3$, Eqs. (3.36) become $(c_3 = 1)$

$$\begin{aligned} \frac{2}{c_2 - c_1} &= \frac{2}{c_1 - 1} + \sum_{j=1}^{k} \frac{1}{c_1 - t_j}, \\ \frac{2}{c_1 - c_2} &= \frac{2}{c_2 - 1} + \sum_{j=1}^{k} \frac{1}{c_2 - t_j}. \end{aligned} \tag{3.37}$$

These two equations can easily be solved for c_2 and c_1 respectively, and lead to the curves displayed for $k = 3$ and $k = 4$ in Fig. 3.6. We see that a huge number of solutions is possible (precisely $\binom{s+k-1}{k-1}$, Krylov imagined charged electrical particles in equilibrium to prove their existence), but most of these lead to totally unstable and therefore useless methods (in the sense of Sect. III.3). Thus the only choice which we retain are the rightmost solutions c_i with $0 < c_1,\ c_2 < 1$, shown in Table 3.6 below. In addition, as Krylov has shown (see Krylov (1959), English translation 1962, p. 329) this choice leads to the smallest error constant (for once, stability and small error are *not* in conflict!)

Stability of the Radau-Type Methods. The stability analysis of the Radau methods is done by inserting $y' = \lambda y$ into (3.28). Since $c_s = 1$ we have $y_{n+1} = v_s$ and thus obtain (for $s = 3$) the characteristic equation

$$\begin{pmatrix} 1-\mu b_{11} & -\mu b_{12} & -\mu b_{13} \\ -\mu b_{21} & 1-\mu b_{22} & -\mu b_{23} \\ -\mu b_{31} & -\mu b_{32} & 1-\mu b_{33} \end{pmatrix} \begin{pmatrix} v_1 \\ v_2 \\ \zeta^3 \end{pmatrix} = \begin{pmatrix} a_{11} & a_{12} & a_{13} \\ a_{21} & a_{22} & a_{23} \\ a_{31} & a_{32} & a_{33} \end{pmatrix} \begin{pmatrix} 1 \\ \zeta \\ \zeta^2 \end{pmatrix},$$

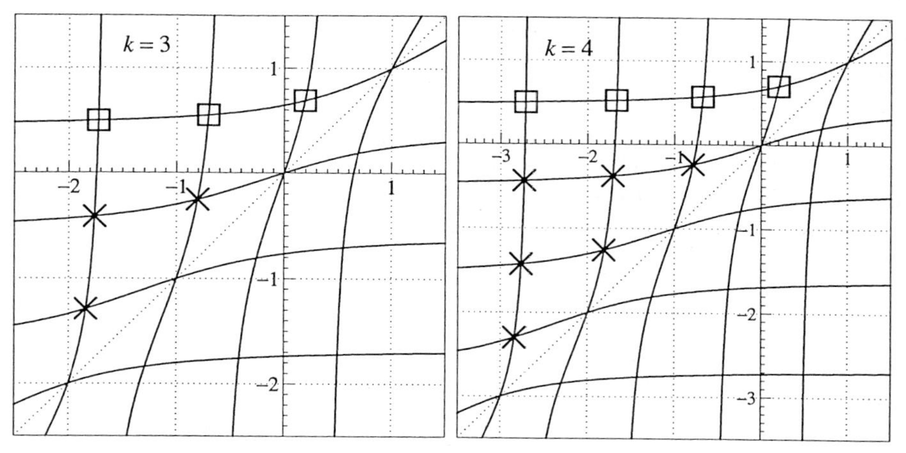

Fig. 3.6. Solutions of (3.37). × unstable, □ stable

or

$$\zeta^3 = (0,0,1)\begin{pmatrix} 1-\mu b_{11} & -\mu b_{12} & -\mu b_{13} \\ -\mu b_{21} & 1-\mu b_{22} & -\mu b_{23} \\ -\mu b_{31} & -\mu b_{32} & 1-\mu b_{33} \end{pmatrix}^{-1} \begin{pmatrix} a_{11} & a_{12} & a_{13} \\ a_{21} & a_{22} & a_{23} \\ a_{31} & a_{32} & a_{33} \end{pmatrix} \begin{pmatrix} 1 \\ \zeta \\ \zeta^2 \end{pmatrix} \tag{3.38}$$

which, when multiplied by $\det(I-\mu B)$, becomes a polynomial of degree 3 in μ. For a general multistep collocation method (3.28) we obtain in this way

$$q_k(\mu)\zeta^k + q_{k-1}(\mu)\zeta^{k-1} + \ldots + q_0(\mu) = 0$$

where $q_k(\mu) = \det(I-\mu B)$ and all $q_i(\mu)$ are polynomials of degree at most s.

The root locus curves of Fig. 3.7 were again obtained by Cardano's formula. Coefficients and stability measures are given in Table 3.6. The methods for $k=1,2$ (orders $p=5$ and 6) are A-stable. The subsequent methods have surprisingly large α-values for very high orders (up to $p \approx 20$), which makes this class very promising.

Exercises

1. Show that the coefficients ν_j in (3.13) for the Enright methods can be computed recursively by

$$\nu_j = -\frac{1}{(j+1)(j+2)} - \sum_{k=0}^{j-1} \nu_k S_{j+1-k} \quad \text{where} \quad S_l = \sum_{k=1}^{l} \frac{1}{k(l+1-k)}.$$

Hint. See the proof of Eq. (III.1.7). The generating function $G(t) = \sum_{j=0}^{\infty} \nu_j t^j$ becomes here $\int_0^1 (s-1)(1-t)^{1-s}\,ds$.

Table 3.6. Coefficients and stability measures for multistep Radau methods ($s = 3$)

k	p	c_1	c_2	c_3	α	D
1	5	0.155051025721682	0.644948974278318	1.	90°	0.000
2	6	0.177891722985607	0.673235257220651	1.	90°	0.000
3	7	0.192169638937766	0.689317969824851	1.	89.73°	0.016
4	8	0.202814874040288	0.700407719104611	1.	89.13°	0.084
5	9	0.211395456069620	0.708798418188500	1.	88.61°	0.178
6	10	0.218626151232186	0.715507419158199	1.	88.14°	0.278
7	11	0.224897548200883	0.721072684914921	1.	87.70°	0.376
8	12	0.230448266933707	0.725812172023161	1.	87.28°	0.467
9	13	0.235435607740434	0.729928926504599	1.	86.89°	0.555
10	14	0.239969169367303	0.733560240031675	1.	86.51°	0.649
11	15	0.244128606044551	0.736803122952198	1.	86.14°	0.763
12	16	0.247973766491964	0.739728565298052	1.	85.79°	0.917
13	17	0.251550844436705	0.742390019356757	1.	85.44°	1.135
14	18	0.254896295040291	0.744828697795402	1.	85.07°	1.462
15	19	0.258039429919700	0.747077018862741	1.	84.68°	1.995
16	20	0.261004194709515	0.749160923778290	1.	84.23°	3.037

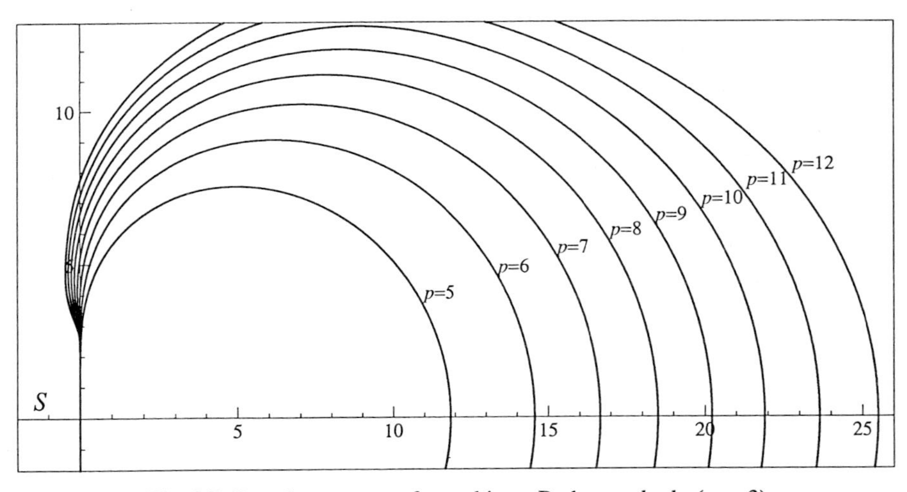

Fig. 3.7. Root locus curves for multistep Radau methods ($s = 3$)

2. The Enright Formulas are stiffly stable for $k \le 7$ and are *not* stiffly stable, as one can easily inspect, e.g. by a computer plot, for $k = 8$, $k = 9, \ldots$ and so on. Hence, everybody agrees that they are not stiffly stable for *any* $k > 7$. However, no rigorous proof has been found for this, as for instance the proof of Theorem III.3.4. Why don't you try to find one?

3. Prove that the second derivative BDF methods (3.14) are unstable (in the sense of Sect. III.3) for $k > 11$.

4. a) Show that for $k=2$, $t_1=-1$, $t_2=0$, $s=1$, $c_1=-1/2$ neither equations (3.24) nor equations (3.25) possess a solution.

   b) Show that (3.24) and (3.25) always admit unique solutions if all c_i are distinct and satisfy $c_i \geq 0$.

   Hint for (b). If φ_i (or ψ_i) are written as $\sum_{l=1}^{s+k} a_l t^{l-1}$, then (3.24) and (3.25) become linear systems with the same matrix and different right-hand sides. The corresponding *homogeneous* system then possesses a non-zero solution iff the interpolation problem

$$\begin{aligned} p(t_j) &= 0 \qquad j=1,\ldots,k \\ p'(c_j) &= 0 \qquad j=1,\ldots,s \end{aligned} \tag{3.39}$$

   has a non-zero solution. Since $p'(t)$ has at most $k+s-2$ real zeros and since (Rolle's theorem) each interval (t_l, t_{l+1}) must contain at least one of these, there can be at most $s-1$ zeros beyond $t_k=0$.

5. A convenient way of computing the polynomials (3.24), (3.25) (written here for the case $s=3$) is to put

$$\varphi_i(t) = (a_1 + a_2 t + a_3 t^2 + a_4 t^3) \prod_{l=1,\, l\neq i}^{k} (t - t_l). \tag{3.40}$$

   Show that Eqs. (3.24) (for $i=j$) and (3.25) then become the following linear system

$$a_1 + t_i a_2 + t_i^2 a_3 + t_i^3 a_4 = 1/r_i, \tag{3.41}$$

$$s_j a_1 + (s_j c_j + 1) a_2 + (s_j c_j^2 + 2c_j) a_3 + (s_j c_j^3 + 3c_j^2) a_4 = 0, \qquad j=1,2,3$$

   where $r_i = \prod_{l=1,\, l\neq i}^{k} (t_i - t_l)$, $s_j = \sum_{l=1,\, l\neq i}^{k} \frac{1}{c_j - t_l}$. Secondly, for

$$\psi_i(t) = (a_1 + a_2 t + a_3 t^2) \prod_{l=1}^{k} (t - t_l) \tag{3.42}$$

   Eq. (3.25) becomes

$$s_j a_1 + (s_j c_j + 1) a_2 + (s_j c_j^2 + 2c_j) a_3 = \begin{cases} 0 & \text{if } j \neq i \\ 1/r_i & \text{if } j = i \end{cases} \qquad j=1,2,3$$

   where $r_i = \prod_{l=1}^{k} (c_i - t_l)$, $s_j = \sum_{l=1}^{k} \frac{1}{c_j - t_l}$.

6. Generalize the proof and the result of Theorem IV.3.10 to multistep collocation methods.

   Hint. Instead of $KM(x)$ in (IV.3.26) we have to insert a linear combination $\sum_{\ell=1}^{k} \alpha_\ell M_\ell(x)$ where $M_\ell(x) = M(x) \cdot x^{\ell-1}$, $M(x) = \frac{1}{s!}\prod_{i=1}^{s}(x - c_i)$ and $\alpha_1, \ldots, \alpha_k$ are arbitrary. Instead of (IV.3.27) we then obtain

$$u(x) = \sum_{\ell=1}^{k} \alpha_\ell \sum_{j=0}^{s} \frac{M_\ell^{(j)}(x)}{\mu^j}. \tag{3.43}$$

   Putting $x = t_1, t_2, \ldots, t_k, t_{k+1}$ and $u(t_i) = y_i$ gives an overdetermined system for $\alpha_1, \ldots, \alpha_k$ which has a solution only if a certain determinant is zero. Setting $y_1 = 1$, $y_2 = \zeta$, $y_3 = \zeta^2, \ldots$ there leads to the characteristic equation

$$\det \begin{pmatrix} \sum_{j=0}^{s} M_1^{(j)}(t_1)\mu^{s-j} & \ldots & \sum_{j=0}^{s} M_k^{(j)}(t_1)\mu^{s-j} & 1 \\ \sum_{j=0}^{s} M_1^{(j)}(t_2)\mu^{s-j} & \ldots & \sum_{j=0}^{s} M_k^{(j)}(t_2)\mu^{s-j} & \zeta \\ \vdots & & \vdots & \vdots \\ \sum_{j=0}^{s} M_1^{(j)}(t_{k+1})\mu^{s-j} & \ldots & \sum_{j=0}^{s} M_k^{(j)}(t_{k+1})\mu^{s-j} & \zeta^k \end{pmatrix}$$

   as a generalization of (IV.3.22,23). Tedious expansions of this determinant into powers of ζ and μ (with many coefficients equal to zero) then leads to an explicit expression (see Theorem 7 of Lie 1990).

7. Prove that the 2-step 2-stage collocation method with $c_2 = 1$ is A-stable iff

$$c_1 \geq (\sqrt{17} - 1)/8.$$

   Hint. a) Show that the characteristic equation is $q_2(\mu)\zeta^2 + q_1(\mu)\zeta + q_0(\mu) = 0$, where

$$\begin{aligned} q_2(\mu) &= -(9c_1 + 5) + \mu(3c_1^2 + 7c_1 + 2) - \mu^2 2c_1(c_1 + 1) \\ q_1(\mu) &= 12c_1 + 4 - \mu 4(c_1^2 - 1) \\ q_0(\mu) &= -3c_1 + 1 + \mu c_1(c_1 - 1). \end{aligned} \tag{3.44}$$

   b) Apply Schur's criterion (1918) to the polynomial (3.46) with $\mu = it$, $t \in \mathbb{R}$.

   Schur's criterion. Let $a(\zeta) = a_k\zeta^k + a_{k-1}\zeta^{k-1} + \ldots + a_0$ $(a_k \neq 0)$ be a polynomial with complex coefficients and set

$$a^*(\zeta) = \overline{a}_0\zeta^k + \overline{a}_1\zeta^{k-1} + \ldots + \overline{a}_k.$$

   Then, all zeros of $a(\zeta)$ lie inside the unit circle, iff

   i) $|a_0| < |a_k|$

   ii) the zeros of $\zeta^{-1}(a^*(0)a(\zeta) - a(0)a^*(\zeta))$, a polynomial of degree $k-1$, are all inside the unit circle.

8. Prove that $c_1 = (\sqrt{17} - 1)/8$ is a super-convergence point for the 2-step 2-stage collocation methods with $c_2 = 1$.

V.4 Order Stars on Riemann Surfaces

Riemann ist der Mann der glänzenden Intuition. Durch seine umfassende Genialität überragt er alle seine Zeitgenossen ... Im Auftreten schüchtern, ja ungeschickt, musste sich der junge Dozent, zu dem wir Nachgeborenen wie zu einem Heiligen aufblicken, mancherlei Neckereien von seinen Kollegen gefallen lassen.

(F. Klein, Entwicklung der Mathematik im 19. Jhd., p. 246, 247)

We have seen in the foregoing sections that the highest possible order of A-stable linear multistep methods is two; furthermore, the second derivative Enright methods as well as the SDBDF methods were seen to be A-stable for $p \le 4$; the three-stage Radau multistep methods were A-stable for $p \le 6$. In this section we shall see that these observations are special cases of a general principle, the so-called "Daniel-Moore conjecture" which says that the order of an A-stable multistep method involving either s derivatives or s implicit stages satisfies $p \le 2s$. Before proceeding to its proof, we should become familiar with Riemann surfaces.

Riemann Surfaces

Für manche Untersuchungen, namentlich für die Untersuchung algebraischer und Abel'scher Functionen ist es vortheilhaft, die Verzweigungsart einer mehrwerthigen Function in folgender Weise geometrisch darzustellen. Man denke sich in der (x, y)-Ebene eine andere mit ihr zusammenfallende Fläche (oder auf der Ebene einen unendlich dünnen Körper) ausgebreitet, welche sich so weit und nur so weit erstreckt, als die Function gegeben ist. Bei Fortsetzung dieser Function wird also diese Fläche ebenfalls weiter ausgedehnt werden. In einem Theile der Ebene, für welchen zwei oder mehrere Fortsetzungen der Function vorhanden sind, wird die Fläche doppelt oder mehrfach sein; sie wird dort aus zwei oder mehreren Blättern bestehen, deren jedes einen Zweig der Function vertritt. Um einen Verzweigungspunkt der Function herum wird sich ein Blatt der Fläche in ein anderes fortsetzen, so dass in der Umgebung eines solchen Punktes die Fläche als eine Schraubenfläche mit einer in diesem Punkte auf der (x, y)-Ebene senkrechten Axe und unendlich kleiner Höhe des Schraubenganges betrachtet werden kann. Wenn die Function nach mehreren Umläufen des z um den Verzweigungswerth ihren vorigen Werth wieder erhält (wie z.B. $(z-a)^{m/n}$, wenn m, n relative Primzahlen sind, nach n Umläufen von z um a), muss man dann freilich annehmen, dass sich das oberste Blatt der Fläche durch die übrigen hindurch in das unterste fortsetzt.

Die mehrwerthige Function hat für jeden Punkt einer solchen ihre Verzweigungsart darstellenden Fläche nur *einen* bestimmten Werth und kann daher als eine völlig bestimmte Function des Orts in dieser Fläche angesehen werden.

(B. Riemann 1857)

We take as example the BDF method (III.1.22") for $k=2$ which has the characteristic equation

$$\Big(\frac{3}{2}-\mu\Big)\zeta^2-2\zeta+\frac{1}{2}=0. \tag{4.1}$$

This quadratic equation expresses ζ as a function of μ, both are complex variables. It is immediately solved to yield

$$\zeta_{1,2}=\frac{2\pm\sqrt{1+2\mu}}{3-2\mu} \tag{4.2}$$

which defines a *two-valued function*, i.e., to each $\mu\in\mathbb{C}$ we have two solutions ζ. These two solutions are displayed in Fig. 4.1 (we have plotted the level curves of $|\zeta_{1,2}(\mu)|$; the region with $|\zeta_1(\mu)|>1$ is in white).

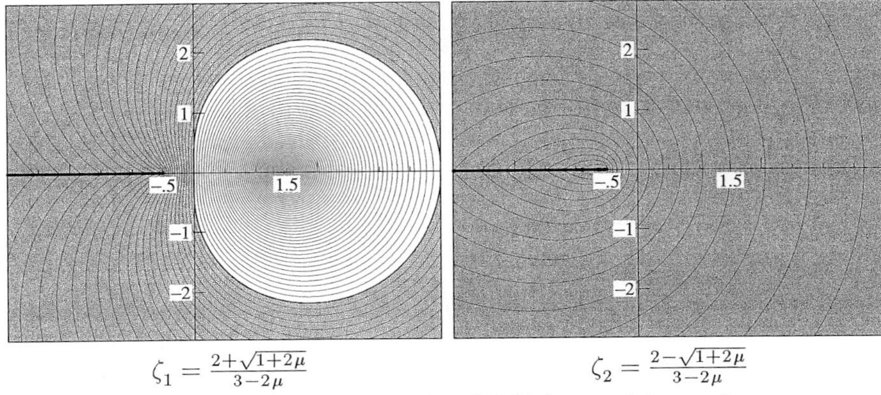

Fig. 4.1. The two solutions of the BDF2 characteristic equation

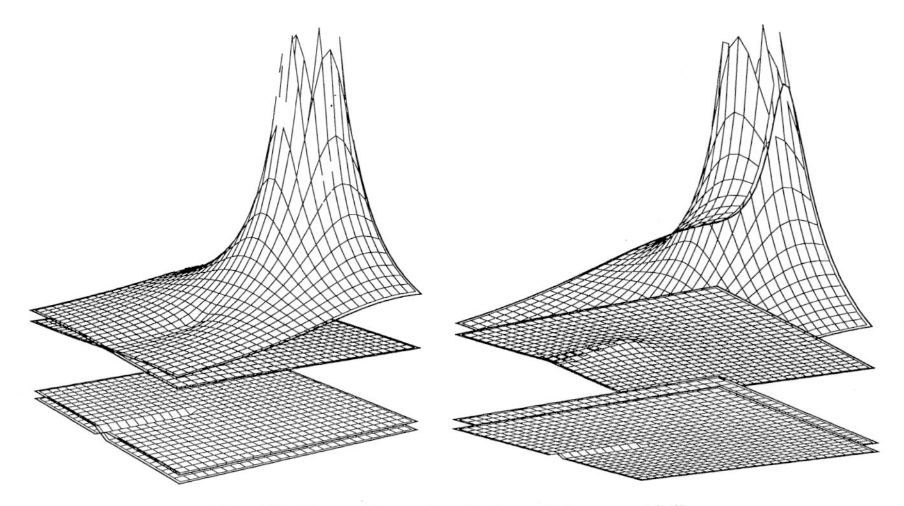

Fig. 4.2. Three dimensional view of the map (4.4)

We observe two essential facts. First, there is a pole of ζ_1, but not of ζ_2, at the point $\mu = 3/2$. This is due to the factor $(3/2-\mu)$ in (4.1) which represents the implicit stage of the method. Second, we observe a curious discontinuity on the negative real axis left of the point $-1/2$, a phenomenon first observed in a famous paper of Puiseux (1850) ("... a encore cet inconvénient, que u devient alors une fonction discontinue ..."). It has its reason in the complex square root $\sqrt{1+2\mu}$ which, while $1+2\mu$ performs a revolution around the origin, only does *half* a revolution and exchanges the two roots. We cannot therefore speak in a natural way of the *two* complex functions $\zeta_1(\mu)$ and $\zeta_2(\mu)$. And here comes the great idea of Riemann (1857): Instead of varying μ in the complex plane $\mathbb{C}$, we imagine it varying in a *double sheet* of (in Riemann's words: infinitely close) complex planes $\mathbb{C} \cup \mathbb{C}$. The $\mu's$ in the upper sheet are mapped to ζ_1, the $\mu's$ in the lower sheet are mapped to ζ_2. The double-valued function then becomes single-valued. At the "cut", left of the point $-1/2$, the two roots ζ_1 and ζ_2 are interchanged, so we must imagine that the upper sheet for ζ_1 continues into the lower sheet for ζ_2 (shaded in Fig. 4.1) and vice-versa. If we denote the manifold obtained in this way by M, then the map

$$\begin{cases} M \longrightarrow \mathbb{C} \\ \mu \longmapsto \zeta \end{cases} \tag{4.3}$$

becomes an everywhere continuous and holomorphic map (with the exception of the pole). M is then called the *Riemann surface* of the algebraic function $\mu \mapsto \zeta$.

A three-dimensional view of the map

$$\begin{cases} M \longrightarrow \mathbb{R} \\ \mu \longmapsto |\zeta| \end{cases} \tag{4.4}$$

is represented in Fig. 4.2.

More General Methods. Most methods of Sect. V.3 are so-called *multistep Runge-Kutta methods* defined by the formulas

$$y_{n+k} = \sum_{j=1}^{k} a_j\, y_{n+j-1} + h \sum_{j=1}^{s} b_j\, f(x_n + c_j h, v_j^{(n)}) \tag{4.5a}$$

$$v_i^{(n)} = \sum_{j=1}^{k} \widetilde{a}_{ij}\, y_{n+j-1} + h \sum_{j=1}^{s} \widetilde{b}_{ij}\, f(x_n + c_j h, v_j^{(n)}). \tag{4.5b}$$

This is *the* subclass of general linear methods (Example III.8.5) for which the external stages represent the solution $y(x)$ on an equidistant grid. The bulk of numerical work for applying the above method are the implicit stages (4.5b).

For the stability analysis we set as now usual $f(x,y) = \lambda y$, $h\lambda = \mu$ and $(y_n, y_{n+1}, \ldots, y_{n+k}) = (1, \zeta, \ldots, \zeta^k)$. Equation (4.5b) then becomes in vector notation (using $\vec{\zeta} = (1, \zeta, \ldots, \zeta^{k-1})^T$)

$$\vec{v} = \left(I - \mu \widetilde{B}\right)^{-1} \widetilde{A}\, \vec{\zeta}, \tag{4.6}$$

which is rational in μ with denominator $\det(I-\mu\widetilde{B})$. Inserting this into (4.5a) and multiplying with this denominator we obtain a characteristic equation of the form

$$Q(\mu,\zeta) \equiv q_k(\mu)\zeta^k + q_{k-1}(\mu)\zeta^{k-1} + \ldots + q_0(\mu) = 0 \tag{4.7}$$

where $q_k(\mu) = \det(I-\mu\widetilde{B})$ and all $q_j(\mu)$ are polynomials in μ of degree $\leq s$.

Multiderivative multistep methods, on the other hand, may be written as (M. Reimer 1967, R. Jeltsch 1976)

$$\sum_{j=0}^{s} h^j \sum_{i=0}^{k} \alpha_{ij} D^j y_{n+i} = 0 \tag{4.8}$$

where the computation of higher derivatives $D^j y$ is done by Eq. (II.12.3). For the equation $y' = \lambda y$ we have $D^j y = \lambda^j y$ and inserting this into (4.8) together with $(y_n, y_{n+1}, \ldots, y_{n+k}) = (1, \zeta, \ldots, \zeta^k)$ we obtain at once a characteristic equation of the form (4.7). Here, the degree s of the polynomials $\varphi_j(\mu)$ is equal to the order of the highest derivative taken. The bulk of numerical work for evaluating (4.8) is the determination of y_{n+k} from an implicit equation containing y_{n+k}, $Dy_{n+k}, \ldots, D^s y_{n+k}$. If the last of these derivatives is present (i.e., if $\alpha_{ks} \neq 0$), then the degree of $q_k(\mu)$ in (4.7) will be s.

The Riemann surface M of (4.7) will consist of k sheets, one for each of the k roots ζ_j. The *branch points* are values of μ for which two or several roots of (4.5) coalesce to an m-fold root. These are the roots of a certain "discriminant" (see any classical book on Algebra, e.g., the famous "Weber", Vol. I, § 50); hence for irreducible $Q(\mu,\zeta)$ there are only a finite number of such points. The movement of the coalescing roots ζ_j, when μ surrounds such a branch point, has been carefully studied by Puiseux: They usually form what Puiseux calls a "système circulaire", i.e., they are cyclically permuted at each revolution like the values of the complex function $\sqrt[m]{z}$ near the origin. The Riemann surface must then follow these "monodromies" and must be cut along certain lines and rejoined appropriately. The location of these cuts is not unique.

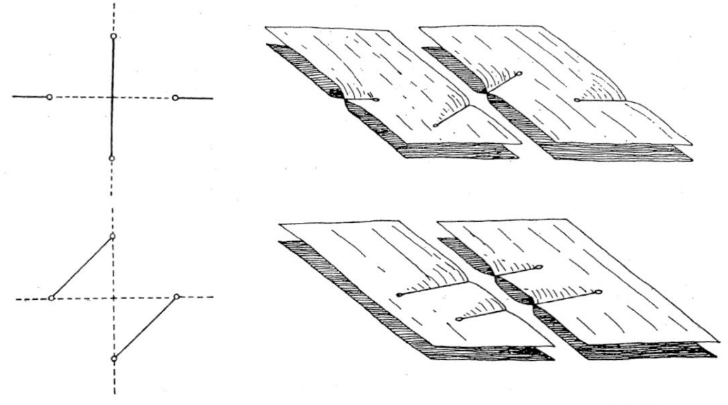

Fig. 4.3. Different cuts for (4.9) (Hurwitz & Courant 1925)

Different possibilities for cutting the Riemann surface of, say, the function

$$\zeta^2 - (1 - \mu^4) = 0 \tag{4.9}$$

with branch points at ± 1 and $\pm i$, are shown in a classical figure reproduced from the book of Hurwitz & Courant, second edition 1925, p. 360 (Fig. 4.3).

Poles Representing Numerical Work

> Only 85 miles (geog.) from the Pole, but it's going to be a stiff pull *both ways* apparently; still we do make progress, which is something. (R.F. Scott, January 10, 1912; first mention of interrelation between poles and stiffness in the literature)

We have just seen that the degree s of $q_k(\mu)$ in (4.7) expresses the numerical work (either the number of implicit stages or the number of derivatives for the implicit solution). Now $q_k(\mu)$ will possess s zeros $\mu_1, \mu_2, \ldots, \mu_s$. What happens if μ approaches one of these zeros? The polynomial (4.7) of degree k (with k roots $\zeta_1(\mu), \ldots, \zeta_k(\mu)$) suddenly becomes a polynomial of degree $k-1$ with only $k-1$ roots. Where does the last one go? Well, by Vieta's Theorem, it must go to infinity. In order to compute its asymptotic behaviour, suppose $q_k(\mu_0) = 0$, $q_k'(\mu_0) \neq 0$, $q_{k-1}(\mu_0) \neq 0$ and that ζ is large. Then all terms $q_{k-2}(\mu)\zeta^{k-2}, \ldots,$ $q_0(\mu)$ are dominated by $q_{k-1}(\mu)\zeta^{k-1}$ and may be neglected. It results that

$$\zeta \sim -\frac{q_{k-1}(\mu_0)}{q_k'(\mu_0)} \frac{1}{\mu - \mu_0} \quad \text{as} \quad \mu \to \mu_0, \tag{4.10}$$

hence the algebraic function $\zeta(\mu)$ possesses a pole on *one* of its sheets. If $q_k(\mu_0) = 0$ is a multiple root, the corresponding pole will be multiple too.

It is also possible that the pole in question coincides with a branch point. This happens when in addition to $q_k(\mu_0) = 0$ also $q_{k-1}(\mu_0) = 0$. In this case *two* roots $\zeta_j(\mu)$ tend to infinity, but *more slowly*, like $\pm C(\mu - \mu_0)^{-1/2}$ (Exercise 1). We therefore count both "half-poles" together as *one* pole again. If c is a boundary curve of a neighbourhood V of μ_0 (which around this branch point surrounds μ_0 twice before closing up), the argument of $\zeta(\mu)$ makes just *one* clockwise revolution on this path. Fig. 4.4 illustrates this fact with an example.

Recapitulating we may state:

Lemma 4.1. *The Riemann surface for the characteristic equation of a multistep Runge-Kutta method with s implicit stages per step (or a multiderivative multistep method with s implicit derivative evaluations) includes at most s poles of the algebraic function $\zeta(\mu)$.* □

We shall see below that Lemma 4.1 remains true for the whole class of general linear methods, but for the moment we are "impatient et joyeux d'aller au combat"

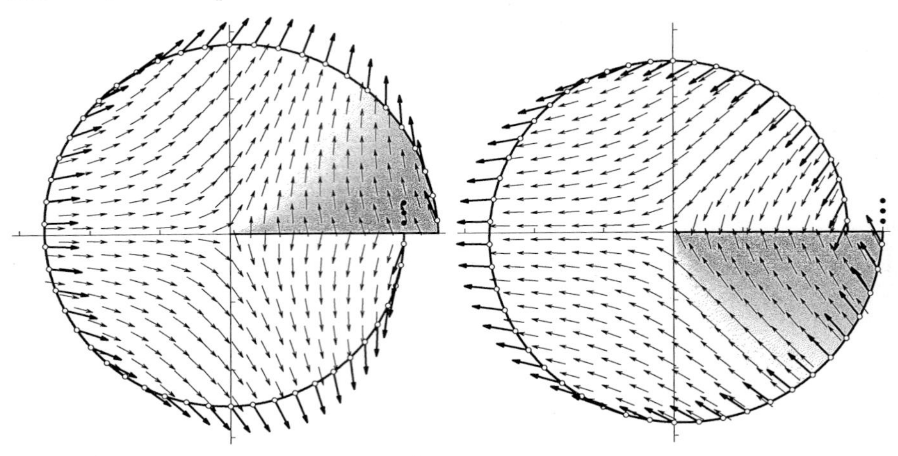

Fig. 4.4. Behaviour of roots of $\mu\zeta^2 + 2\mu\zeta + 2 - \mu = 0$ near the origin $\mu = 0$

(Astérix Légionnaire, pp. 29 and 30). The *argument principle* also remains valid on Riemann surfaces and we state it as follows:

> "On the left, isn't it ?" — "Right."
> "On the right ?" — "Left, leeeft!"
> (John Cleese in "Clockwise")

Lemma 4.2. *Suppose that a domain $F \subset M$ contains no zeros of $\zeta(\mu)$ and that its boundary consists of closed loops $\gamma_1, \ldots, \gamma_\ell$. Then the number of poles of $\zeta(\mu)$ contained in F is equal to the total number of clockwise revolutions of* $\arg(\zeta(\mu))$ *along $\gamma_1, \ldots, \gamma_\ell$, each passed through in* that *direction which leaves F to the* left *of γ_j.*

The *proof* is by cutting F into thousand pieces, each of which is homeomorphic to a disc in $\mathbb{C}$, and by adding up all revolution numbers which cancel along the cuts, because the adjacent edges are traversed in opposite directions. □

Order and Order Stars

> ... denn das Klare und leicht Faßliche zieht uns an, das Verwickelte schreckt uns ab. (D. Hilbert, Paris 1900)

Guided by the ideas of Sect. IV.4, we now compare the absolute values of the characteristic roots $|\zeta_1|$ and $|\zeta_2|$ for the BDF2 scheme (4.2) with the exponential function $|e^{\mu}| = e^{\mathrm{Re}\,\mu}$, hence we define (Wanner, Hairer & Nørsett 1978)

$$A_j = \left\{ \mu \in \mathbb{C}\,;\;\; |\zeta_j(\mu)| > |e^{\mu}| \right\} \qquad j = 1, 2. \tag{4.11}$$

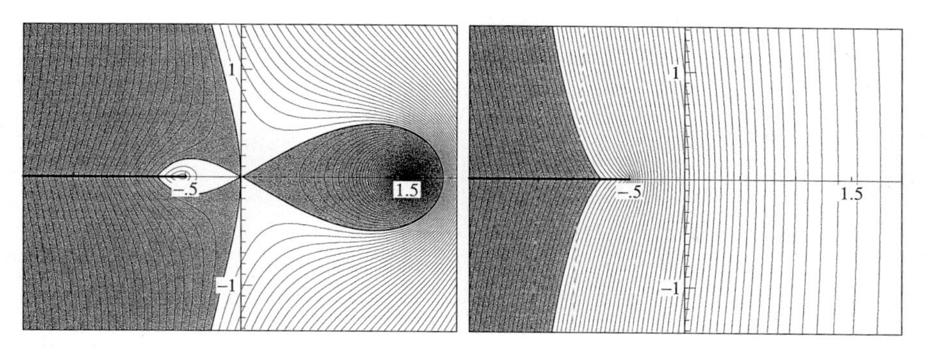

Fig. 4.5. The order star (4.14) for BDF2

These sets, on precisely the same scale as in Fig. 4.1, are represented in Fig. 4.5.

The sets A_j continue across the cuts in the same way as do the roots, it is therefore natural to embed them into the Riemann surface M and define

$$A = \left\{ \mu \in M \; ; \; |\zeta(\mu)| > |e^{\pi(\mu)}| \right\} \tag{4.12}$$

where $\pi : M \to \mathbb{C}$ is the natural projection.

Fig. 4.5 shows clearly an order star with three sectors for $\zeta_1(\mu)$, but none for $\zeta_2(\mu)$, and we guess that this has to do with the order of the method, which is two. Lemma 4.3 below will extend Lemma IV.4.3 to multistep methods.

By putting $h = 0$ in (4.5) (hence $\mu = 0$ in (4.7)), and

$$(y_n, y_{n+1}, \ldots, y_{n+k-1}) = (1, 1, \ldots, 1)$$

(hence $\zeta = 1$ in (4.7)), we must have by consistency that $y_{n+k} = 1$ too, i.e., that $Q(0,1) = 0$. This corresponds to the formula $\varrho(1) = 0$ in the multistep case (see (III.2.6)). But for $h = 0$ the difference equation (4.5a) is stable only if $\zeta = 1$ is a *simple* root of the polynomial equation $Q(0, \zeta) = 0$. Hence we must have

$$Q(0,1) = 0, \qquad \frac{\partial Q}{\partial \zeta}(0,1) \neq 0. \tag{4.13}$$

The analytic continuation $\zeta_1(\mu)$ of this root in the neighbourhood of the origin (as far as it is not embarassed with branch points) will be called the *principal root*, the corresponding surface the *principal sheet* of M.

Lemma 4.3. *For stable multistep Runge-Kutta (or multiderivative) methods of order p the set A possesses a star of $p+1$ sectors on the principal sheet in the neighbourhood of the origin.*

Proof. We fix $\lambda \in \mathbb{C}$, set $y' = \lambda y$ and take for $y_0, \ldots, y_{k-1}$ exact initial values $1, e^{\mu}, \ldots, e^{(k-1)\mu}$. The order of the method then tells us that the *local error* (see Fig. III.2.1), i.e., the difference between $e^{k\mu}$ and the numerical solution y_k computed from (4.5a), must be $\widetilde{C} \cdot h^{p+1}$ for $h \to 0$, hence $\widetilde{C}\lambda^{-p-1}\mu^{p+1}$ for $\mu \to 0$.

Thus, Formula (4.5) with *all* y_j replaced by $e^{j\mu}$ will lead to

$$Q(\mu, e^{\mu}) = \overline{C}\mu^{p+1} + \mathcal{O}(\mu^{p+2}). \tag{4.14}$$

We subtract (4.14) from (4.7), choose for $\zeta(\mu)$ the principal root $\zeta_1(\mu)$ (for which $e^{\mu} - \zeta_1(\mu)$ is small for $|\mu|$ small) and linearize. This gives

$$\frac{\partial Q}{\partial \zeta}(0,1) \cdot (e^{\mu} - \zeta_1(\mu)) = \overline{C}\mu^{p+1} + \ldots$$

and by dividing through by the non-zero constant (4.13)

$$e^{\mu} - \zeta_1(\mu) = C \cdot \mu^{p+1} + \mathcal{O}(\mu^{p+2}) \qquad \text{for} \qquad \mu \to 0. \tag{4.15}$$

The rest of the proof now goes exactly analogously to that of Lemma IV.4.3. There is also not much difference in the case of multiderivative methods. □

The constant C of (4.15) is called the *error constant* of the method. This is consistent with Formula (III.2.6) and (III.2.13) for multistep methods and with (IV.3.5) for Runge-Kutta methods.

The stability domain of multistep Runge-Kutta methods as well as their A-stability is defined in the same way as for multistep methods (see Definition 1.1). One has only to interpret $\zeta_1(\mu), \ldots, \zeta_k(\mu)$ as the roots of (4.7).

The "Daniel and Moore Conjecture"

> It is conjectured here that no A-stable method of the form of Eq. 5-6 can be of order greater than $2J+2$ and that, of those A-stable methods of order $2J+2$, the smallest error constant is exhibited by the *Hermite method* . . .
>
> (Daniel & Moore 1970, p. 80)

At the time when no simple proof for Dahlquist's second barrier was known, a proof of its generalization, the Daniel & Moore conjecture, seemed quite hopeless. Y. Genin (1974) constructed A-stable multistep multiderivative methods with astonishingly high "order" contradicting the conjecture. R. Jeltsch (1976) later cleared up the mystery by showing that Genin's methods had 1 as multiple root of $\varrho(\zeta)$ and hence the "effective" order was lower. The conjecture was finally proved in 1978 with the help of order stars:

Theorem 4.4. *The highest order of an A-stable s-stage Runge-Kutta (or s-derivative) multistep method is $2s$. For the A-stable methods of order $2s$ the error constant satisfies*

$$(-1)^s C \geq \frac{s!\,s!}{(2s)!\,(2s+1)!}. \tag{4.16}$$

Proof. By A-stability, we have for *all* roots $|\zeta_j(iy)| \le 1$ along the imaginary axis; hence the order star A is nowhere allowed to cross the imaginary axis. We consider $A^+ = A \cap \pi^{-1}(\mathbb{C}^+)$, the part of the order star which lies above $\mathbb{C}^+$. As in Lemma IV.4.4, A^+ must be finite on *all* sheets of M. The boundary of A^+ may consist of several closed curves. As in Lemma IV.4.5, the argument of $\zeta(\mu)/e^\mu$ is steadily increasing along ∂A^+. Since at the origin we have a star with $p+1$ sectors (Lemma 4.3), of which at least $[\frac{p+1}{2}]$ lie in $\mathbb{C}^+$, the boundary curves of A^+ must visit the origin at least $[\frac{p+1}{2}]$ times. Hence the total rotation number is at least $[\frac{p+1}{2}]$ and from Lemmas 4.1 and 4.2 we conclude that

$$\left[\frac{p+1}{2}\right] \le s. \tag{4.17}$$

This implies that $p \le 2s$ and the first assertion is proved.

We now need a new idea for the part concerning the error constant. The following reasoning will help: the star A expresses the fact that the surface $|\zeta(\mu)/e^\mu|$ goes up and down around the origin like Montaigne's ruff. There, the *error constant* has to do with the *height* of these waves. So if we want to compare different error constants we must compare $|\zeta(\mu)/e^\mu|$ to $|R(\mu)/e^\mu|$, where $R(\mu)$ is the characteristic function of a second method. By dividing the two expressions, e^μ cancels and we define

$$B = \left\{ \mu \in M\ ;\ \left| \frac{\zeta(\mu)}{R(\pi(\mu))} \right| > 1 \right\}, \tag{4.18}$$

called the *relative order star*. For $R(z)$ we choose the diagonal Padé approximation $R_{ss}(z)$ with s zeros and s poles (see (IV.3.30)). By subtracting (IV.3.31) (with $j = k = s$) from (4.15) (where it is now supposed that $p = 2s$) we obtain

$$R_{ss}(\mu) - \zeta_1(\mu) = \underbrace{\left(C - (-1)^s \frac{s!\, s!}{(2s)!\,(2s+1)!} \right)}_{\widetilde{C}} \mu^{2s+1} + \ldots . \tag{4.19}$$

It is known that $|R_{ss}(iy)| = 1$ for all $y \in \mathbb{R}$ and that all zeros of $R_{ss}(z)$ lie in $\mathbb{C}^-$ (Theorem IV.4.12). Therefore the set B in (4.18) cannot cross the imaginary axis (as before) and the quotient $|\zeta(\mu)/R(\pi(\mu))|$ has no other poles above $\mathbb{C}^+$ than those of $\zeta(\mu)$, of which, we know, there are at most s. Therefore the sectors of the relative order star B must exhibit the same colours as those of the classical order star A for diagonal Padé (see Fig. IV.4.2). Otherwise an extra pole would be needed. We conclude that the error constants must have the same sign (see Lemma IV.4.3), hence (see IV.3.31) $(-1)^s \widetilde{C} > 0$, which leads to (4.16).

Equality $\widetilde{C} = 0$ would produce an order star B of even higher order which is impossible with s poles, unless the two methods are identical. □

Remarks. a) The first half is in fact superfluous, since the inequality (4.16) implies that the $2s$-th order error constant $C \neq 0$, hence necessarily $p \leq 2s$. It has been retained for its beauty and simplicity, and for readers who do not want to study the second half.

b) The proof never uses the full hypothesis of A-stability; the only property used is stability on the imaginary axis (I-stability, see (IV.3.6)). Thus Theorem 4.4 allows the following sharpening, which then extends Theorem IV.4.7 to multistep methods:

Theorem 4.5. *Suppose that an I-stable s-stage Runge-Kutta (or s-derivative) multistep method possesses a characteristic function $\zeta(\mu)$ with s_1 poles in $\mathbb{C}^+$. Then*

$$p \leq 2s_1 \tag{4.20}$$

and the error constant for all such I-stable methods of order $p = 2s_1$ satisfies

$$(-1)^{s_1} C \geq \frac{s_1!\, s_1!}{(2s_1)!\,(2s_1+1)!}. \tag{4.21}$$

□

Another interpretation of this theorem is the following result (compare with Theorem IV.4.8), which in the case $s=1$ is due to R. Jeltsch (1978).

Thorem 4.6. *Suppose that an I-stable method with s poles satisfies $p \geq 2s-1$. Then it is A-stable.*

Proof. If only $s-1$ poles were in $\mathbb{C}^+$, we would have $p \leq 2s-2$, a contradiction. Hence all poles of $\zeta(\mu)$ are in $\mathbb{C}^+$ and A-stability follows from the maximum principle. □

Methods with Property C

It is now tempting to extend the proof of Theorem 4.4 to *any* method other than the diagonal Padé method. But this meets with an essential difficulty in defining (4.18) if $R(\mu)$ is a multistep method defined on *another* Riemann surface, since then the definition of B makes no sense. The following observation will help: The second part of the proof of Theorem 4.4 only took place in $\mathbb{C}^+$, which was the *instability* domain of the "comparison method". This leads to

Definition 4.7 (Jeltsch & Nevanlinna 1982). Let a method be given with characteristic polynomial (4.7) satisfying (4.13) and denote its stability domain by S_R. We say that this method has *Property C* if the principal sheet includes no branch points outside of $\pi^{-1}(S_R)$ (with ∞ included if S_R is bounded), and the principal root $R_1(\mu)$ produces the whole instability of the method, i.e.,

$$\Delta_R := \partial S_R = \{\mu \in \mathbb{C}\,;\ |R_1(\mu)| = 1\}. \tag{4.22}$$

Examples. All one-step methods have Property C, of course. Linear multistep methods whose root locus curve is simply closed have Property C too. In this situation all roots except $R_1(\mu)$ have modulus smaller than one for all $\mu \notin \pi^{-1}(S_R)$. Thus the principal sheet cannot have a branch point there. The explicit 4th order Adams method analyzed in Fig. 1.1 does *not* have Property C. The implicit Adams methods (see Fig. 1.3) have Property C for $k \leq 5$. Also, the 4th order implicit Milne-Simpson method (1.16) has property C.

Definition 4.7 allows us to replace $R_{ss}(\mu)$ in the proof of Theorem 4.4 by $R_1(\mu)$, $\mathbb{C}^+$ by the exterior of S_R, the imaginary axis by Δ_R and to obtain the following theorem (Jeltsch and Nevanlinna the 5th of April, 1979 at 5 a.m. in Champaign; G.W. the 5th of April, 1979 at 4.30 a.m. in Urbana. How was this coincidence possible? E-mail was not yet in general use at that time; was it Psi-mail?)

Theorem 4.8. *Let a method with characteristic function $R(\mu)$, stability domain S_R and order p_R possess Property C. If another method with characteristic function $\zeta(\mu)$, stability domain S_ζ and order p_ζ is* more stable *than R, i.e., if*

$$S_\zeta \supset S_R, \tag{4.23}$$

then

$$p \leq 2s \tag{4.24}$$

where

$$p = \min(p_R\,, p_\zeta) \tag{4.25}$$

and s is the number of poles of $\zeta(\mu)$, each counted with its multiplicity, which are not poles of the principal root $R_1(\mu)$ of $R(\mu)$. □

> ... and tried to optimize the stability boundary. Despite many efforts we were not able to exceed $\sqrt{3}$, the stability boundary of the Milne-Simpson method ... (K. Dekker 1981)

As an illustration of Theorem 4.8 we ask for the largest stability interval on the imaginary axis $I_r = [-ir, ir] \subset \mathbb{C}$ of a 3rd order multistep method (for hyperbolic equations). Since we have $s = 1$ for linear multistep methods, $p = 3$ contradicts (4.24) and we obtain from Theorem 4.8 by using for $R(\mu)$ the Milne-Simpson method (1.16):

Theorem 4.9 (Dekker 1981, Jeltsch & Nevanlinna 1982). *If a linear multistep method of order $p \geq 3$ is stable on I_r, then $r \leq \sqrt{3}$.* □

The second part of Theorem 4.4 also allows an extension, the essential ingredient for its proof has been the sign of the error constant for the diagonal Padé approximation.

Theorem 4.10. *Consider a method with characteristic equation (4.7) satisfying (4.13) and let p denote its order and C its error constant. Suppose*

a) the method possesses Property C,

b) the principal root $R_1(\mu)$ possesses s poles,

c) $\operatorname{sign}(C) = (-1)^s$

d) $p \geq 2s-1$.

Then this method is "optimal" in the sense that every other method with s poles which is stable on Δ_R of (4.22) has either lower order or, for the same order, a larger (in absolute value) error constant. □

Examples. The diagonal and first sub-diagonal Padé approximations satisfy the above hypotheses (see Eq. (IV.3.30)). Also I-stable linear multistep methods with Property C can be applied.

Remark 4.11. Property C allows the extension of Theorem IV.4.17 of Jeltsch & Nevanlinna to explicit multistep methods. Thus explicit methods with comparable numerical work cannot have including stability domains. Exercise 4 below shows that Property C is a necessary condition. Remember that explicit methods have all their poles at infinity.

General Linear Methods

The large class of general linear methods (Example III.8.5) written in obvious matrix notation

$$v_n = \widetilde{A}u_n + h\widetilde{B}f(v_n) \tag{4.26a}$$

$$u_{n+1} = Au_n + hBf(v_n) \tag{4.26b}$$

seems to allow much more freedom to break the Daniel & Moore conjecture. This is not the case as we shall see in the sequel.

The bulk of numerical work for solving (4.26) is represented by the implicit stages (4.26a) and hence depends on the structure of the matrix $\widetilde{B}$. Inserting $y' = \lambda y$ leads to

$$u_{n+1} = S(\mu)u_n \tag{4.27}$$

where

$$S(\mu) = A + \mu B(I - \mu\widetilde{B})^{-1}\widetilde{A}. \tag{4.28}$$

The stability of the numerical method (4.27) is thus governed by the eigenvalues of the matrix $S(\mu)$. The elements of this matrix are seen to be rational functions in μ.

Lemma 4.12. *If the characteristic polynomial of* $S(\mu)$ *is multiplied by* $\det(I-\mu\widetilde{B})$ *then it becomes polynomial in* μ*:*

$$\begin{aligned}\det(\zeta I - S(\mu))\cdot\det(I-\mu\widetilde{B}) &= q_k(\mu)\zeta^k + q_{k-1}(\mu)\zeta^{k-1} + \ldots + q_0(\mu)\\ &=: Q(\mu,\zeta)\end{aligned}\tag{4.29}$$

where $q_0,\ldots,q_k$ *are polynomials of degree* $\le s$ *and* $q_k(\mu)=\det(I-\mu\widetilde{B})$.

Proof. Suppose first that $\widetilde{B}$ is diagonalizable as

$$T^{-1}\widetilde{B}T = \operatorname{diag}(\beta_1,\ldots,\beta_s)\tag{4.30}$$

so that from (4.28)

$$S(\mu) = A + B\,T\operatorname{diag}(w_1,\ldots,w_s)T^{-1}\widetilde{A} = A + \sum_{i=1}^{s} w_i\,\vec{d}_i\,\vec{c}_i^{\,T}\tag{4.31}$$

where

$$\left.\begin{aligned} w_i &= \frac{\mu}{1-\mu\beta_i}\\ \vec{d}_i &= i\text{-th column of } BT\\ \vec{c}_i^{\,T} &= i\text{-th row of } T^{-1}\widetilde{A}.\end{aligned}\right\}\qquad i=1,\ldots s\tag{4.32}$$

We write the matrix $\zeta I - S(\mu)$ in terms of its column vectors

$$\Big(\zeta\vec{e}_1-\vec{a}_1-w_1c_{11}\vec{d}_1-w_2c_{12}\vec{d}_2-\ldots,\ \zeta\vec{e}_2-\vec{a}_2-w_1c_{21}\vec{d}_1-w_2c_{22}\vec{d}_2-\ldots,\ \ldots\Big).$$

Its determinant, the characteristic polynomial of $S(\mu)$, is computed using the multilinearity of det and considering ζ, w_i, c_{ij} as scalars. All terms containing one of the w_j to any power higher than 1 cancel, because the corresponding factor is a determinant with two or more identical columns. Thus, if $\det(\zeta I - S(\mu))$ is multiplied by $\prod_{i=1}^{s}(1-\mu\beta_i)=\det(I-\mu\widetilde{B})$ it becomes a polynomial of the form (4.29).

A non-diagonalizable matrix $\widetilde{B}$ is considered as the limit of diagonalizable matrices. The coefficients of the polynomial $Q(\mu,\zeta)$ depend continuously on $\widetilde{B}$. □

We conclude that Lemma 4.1 again remains valid for general linear methods. The s poles on the Riemann surface for the algebraic function $Q(\mu,\zeta)=0$ are located at the positions $\mu=1/\beta_1,\ldots,\mu=1/\beta_s$ where β_i are the eigenvalues of the matrix $\widetilde{B}$.

We next have to investigate the *order conditions*, i.e., the analogue of Lemma 4.3. Recall that general linear methods must be equipped with a *starting procedure* (see Eq. (III.8.4a)) which for the differential equation $y'=\lambda y$ will be of the form $u_0=\psi(\mu)\cdot y_0$ with $\psi(0)\neq 0$. Here $\mu=h\lambda$ and $\psi(\mu)$ is a k-vector of polynomials or rational functions of μ. Then the diagram of Fig. III.8.1 becomes the one sketched in Fig. 4.6.

The order condition (see Formula (III.8.16) of Lemma III.8.11) then gives:

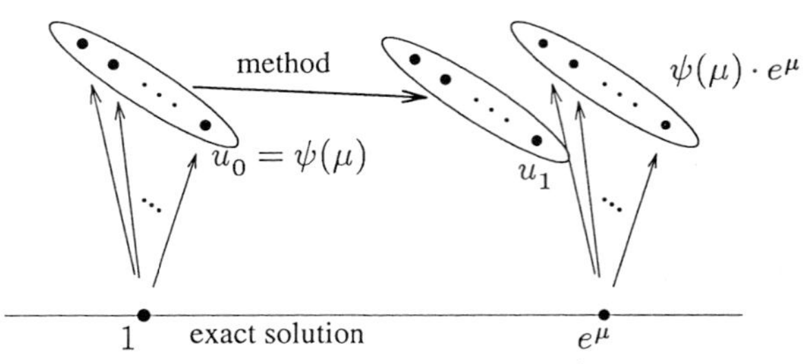

Fig. 4.6. General linear method for $y' = \lambda y$

Lemma 4.13. *If the general linear method (4.26) is of order* p *then*

$$\big(e^{\mu}I - S(\mu)\big)\psi(\mu) = \mathcal{O}(\mu^{p}) \qquad \textit{for } \mu \to 0 \tag{4.33a}$$

$$E\big(e^{\mu}I - S(\mu)\big)\psi(\mu) = \mathcal{O}(\mu^{p+1}) \quad \textit{for } \mu \to 0 \tag{4.33b}$$

where E *is defined in (III.8.12) and* $S(\mu)$ *is given in (4.28).* □

Formula (4.33) tells us, roughly, that $\psi(\mu)$ is an approximate eigenvector of $S(\mu)$ with eigenvalue e^{μ}. We shall now see how this information can be turned into order conditions for the correct eigenvalues of $S(\mu)$.

Definition 4.14. Let ℓ be the *number of principal sheets* of (4.29), i.e., the multiplicity of 1 as eigenvalue of $S(0)$ (which, by stability, must then be a simple root of the minimal polynomial). ℓ is also the dimension of I in (III.8.12) and the rank of E.

Theorem 4.15. *Suppose that there exists* $\psi(\mu)$ *with* $\psi(0) \neq 0$ *such that the general linear method satisfies the conditions (4.33) for order* $p \geq 1$. *Then the* ℓ*-fold eigenvalue* 1 *of* S *continues into* ℓ *eigenvalues* $\zeta_j(\mu)$ *of* $S(\mu)$ *which satisfy*

$$e^{\mu} - \zeta_j(\mu) = \mathcal{O}(\mu^{p_j+1}) \qquad \mu \to 0 \tag{4.34}$$

with

$$p_j \geq 0, \qquad \sum_{j=1}^{\ell} p_j \geq p. \tag{4.35}$$

Examples. a) The matrix

$$S(\mu) = \begin{pmatrix} 1+\mu & \frac{20}{9}\mu^2 \\ 3\mu + \frac{11}{80}\mu^2 & 1 - \frac{37}{3}\mu + \frac{13}{3}\mu^2 \end{pmatrix} \tag{4.36}$$

has $\ell = 2$ so that $E = I$ in (4.33b). There is a vector $\psi(\mu)$ (non-vanishing for $\mu = 0$) such that

$$(e^{\mu} I - S(\mu))\psi(\mu) = \mathcal{O}(\mu^6),$$

i.e., $p = 5$. The eigenvalues

$$\zeta_{1,2}(\mu) = \Big(1 - \frac{17\mu}{3} + \frac{13\mu^2}{6}\Big) \pm \frac{20\mu}{3}\sqrt{1 - \frac{\mu}{2} + \frac{9\mu^2}{80}}$$

satisfy $e^{\mu} - \zeta_1(\mu) = \mathcal{O}(\mu^6)$, $e^{\mu} - \zeta_2(\mu) = \mathcal{O}(\mu)$, which is (4.34) with $p_1 = 5$, $p_2 = 0$.

b) The matrix

$$S(\mu) = \begin{pmatrix} 1 + 2\mu + \frac{\mu^2}{2} & -\mu \\ \mu & 1 + \frac{\mu^2}{2} \end{pmatrix} \tag{4.37}$$

satisfies (4.33) with $\ell = 2$, $p = 4$. Its eigenvalues $\zeta_{1,2}(\mu) = 1 + \mu + \mu^2/2$ fulfil (4.34) with $p_1 = p_2 = 2$.

c) The example

$$S(\mu) = \begin{pmatrix} 1 + 2\mu & -\mu + \mu^2 \\ \mu & 1 \end{pmatrix} \tag{4.38}$$

has $\ell = 2$, $p = 1$ in (4.33). Its eigenvalues $\zeta_{1,2}(\mu) = 1 + \mu \pm \sqrt{\mu^3}$ satisfy (4.34) with $p_1 = p_2 = 1/2$. This example shows that the p_j in (4.34) need not be integers.

Proof of Theorem 4.15. We introduce the matrix

$$\widetilde{S}(\mu) = e^{\mu} I - S(\mu) \tag{4.39}$$

which has the same eigenvectors as $S(\mu)$ and the corresponding eigenvalues

$$\widetilde{\zeta}_j(\mu) = e^{\mu} - \zeta_j(\mu). \tag{4.40}$$

Formulas (4.34) and (4.35) now say simply that

$$\prod_{j=1}^{\ell} \widetilde{\zeta}_j(\mu) = \mathcal{O}(\mu^{p+\ell}) \qquad \mu \to 0. \tag{4.41}$$

Since the product of the eigenvalues is, as we know, the determinant of the matrix, we look for information about $\det \widetilde{S}(\mu)$.

After a suitable change of coordinates (via the transformation matrix T of (III.8.12)) we suppose the matrix $S = S(0)$ in Jordan canonical form. We then separate blocks of dimensions ℓ and $k - \ell$ so that

$$E = \begin{pmatrix} I & 0 \\ 0 & 0 \end{pmatrix}, \quad S(\mu) = \begin{pmatrix} I + \mathcal{O}(\mu) & \mathcal{O}(\mu) \\ \mathcal{O}(\mu) & \mathcal{O}(1) \end{pmatrix}, \quad \psi(\mu) = \begin{pmatrix} \psi_1(\mu) \\ \psi_2(\mu) \end{pmatrix} \tag{4.42}$$

$$\widetilde{S}(\mu) = \begin{pmatrix} \widetilde{S}_{11}(\mu) & \widetilde{S}_{12}(\mu) \\ \widetilde{S}_{21}(\mu) & \widetilde{S}_{22}(\mu) \end{pmatrix} = \begin{pmatrix} \mathcal{O}(\mu) & \mathcal{O}(\mu) \\ \mathcal{O}(\mu) & \mathcal{O}(1) \end{pmatrix} \tag{4.43}$$

where it is important to notice that $\widetilde{S}_{22}(0)$ is invertible; this is because E collects all eigenvalues equal to 1, thus $S_{22}(0)$ has no eigenvalues equal to 1 and $\widetilde{S}_{22}(0)$ has none equal to zero. Conditions (4.33) now read

$$\begin{pmatrix} \widetilde{S}_{11}(\mu) & \widetilde{S}_{12}(\mu) \\ \widetilde{S}_{21}(\mu) & \widetilde{S}_{22}(\mu) \end{pmatrix} \begin{pmatrix} \psi_1(\mu) \\ \psi_2(\mu) \end{pmatrix} = \begin{pmatrix} \mathcal{O}(\mu^{p+1}) \\ \mathcal{O}(\mu^p) \end{pmatrix}. \tag{4.44}$$

Putting $\mu = 0$ in (4.44) we get $\psi_2(0) = 0$. The assumption $\psi(0) \neq 0$ thus implies that at least one component of $\psi_1(0)$, say the j-th component $\psi_{1j}(0)$, does not vanish. Cramer's rule then yields

$$\det \widetilde{S}(\mu) \cdot \psi_{1j}(\mu) = \det T(\mu), \tag{4.45}$$

where $T(\mu)$ is obtained from $\widetilde{S}(\mu)$ by replacing its j-th column by the right-hand side of (4.44). One easily sees that $\det T(\mu) = \mathcal{O}(\mu^{p+\ell})$ (take out a factor μ from each of the first ℓ lines and a factor μ^p from the j-th column). Because of $\psi_{1j}(0) \neq 0$ this implies $\det \widetilde{S}(\mu) = \mathcal{O}(\mu^{p+\ell})$. We have thus proved (4.41) (hence (4.34) and (4.35)), because $\widetilde{\zeta}_{\ell+1}, \ldots, \widetilde{\zeta}_k$ do not converge to zero for $\mu \to 0$. □

The next lemma excludes fractional orders for A-stable methods:

Lemma 4.16. *For I-stable general linear methods the orders p_j in (4.34) must be integers.*

Proof. Divide (4.34) by e^μ, let

$$\frac{\zeta_j(\mu)}{e^\mu} = 1 - C\mu^{m/r} + \ldots \tag{4.46}$$

where $p_j + 1 = m/r$, and suppose that $r > 1$ and m, r are relatively prime. Since $e^\mu - \zeta_j(\mu)$ are the eigenvalues of the matrix (4.39), hence the roots of an analytic equation, the presence of a root $\mu^{m/r}$ involves the occurrence of all branches $\mu^{m/r} \cdot e^{2i\pi j/r}$ $(j = 0, 1, \ldots, r-1)$. For $\mu = \pm iy = e^{\pm i\pi/2} y$ ($y \in \mathbb{R}$ small), inserted into (4.46), we thus obtain $2r$ different values

$$1 - Cy^{m/r} e^{\pm im\pi/2r} e^{2i\pi j/r} + \ldots \qquad j = 0, 1, \ldots r-1$$

which form a regular $2r$-Mercedes star; hence whatever the argument of C is, there are values of $C(\pm iy)^{m/r} e^{2i\pi j/r}$ (for some $0 \leq j \leq r-1$) with negative real part, such that from (4.46) $|\zeta_j(\pm iy)| > 1$. This is a contradiction to I-stability. □

And here is the "Daniel-Moore conjecture" for general linear methods:

Theorem 4.17. *Let the characteristic function $Q(\mu, \zeta)$ of an I-stable general linear method possess s poles in $\mathbb{C}^+$. Then*

$$p \leq 2s. \tag{4.47}$$

Proof. Again we denote by $A^+ = A \cap \pi^{-1}(\mathbb{C}^+)$, the part of the order star lying above $\mathbb{C}^+$. By I-stability A^+ does not intersect the imaginary axis $\pi^{-1}(i\mathbb{R})$ on *any* sheet.

By Theorem 4.15 the boundary curves γ_m of A^+ visit the origin on the different principal sheets at least $[\frac{p_j+1}{2}]$ times $(j = 1, \ldots, \ell)$ (see (4.17)), where the p_j are integers by Lemma 4.16. Thus by Lemma 4.2

$$\sum_{j=1}^{\ell} \Big[\frac{p_j+1}{2}\Big] \le s. \tag{4.48}$$

Multiplying this by 2, using $p_j \le 2[\frac{p_j+1}{2}]$ and (4.35), we get $p \le 2s$. □

Dual Order Stars

Why not interchange the role of the two variables ζ and μ ...?
(J. Butcher,
June 27, 1989, in West Park Hall, Dundee, at midsummernight)

A-stability implies that *for all solutions $\zeta_j(\mu)$ of $Q(\mu,\zeta) = 0$ we have*

$$\operatorname{Re}\mu \le 0 \qquad \Longrightarrow \qquad |\zeta_j(\mu)| \le 1. \tag{4.49}$$

This is logically equivalent to: *For all solutions $\mu_j(\zeta)$ of $Q(\mu,\zeta) = 0$ we have*

$$|\zeta| \ge 1 \qquad \Longrightarrow \qquad \operatorname{Re}\mu_j(\zeta) \ge 0 \tag{4.50}$$

(in fact, pure logic gives us "$>$" on both sides; the "$\ge$" then follow by continuity). Further the order condition (4.15) becomes, by passing to inverse functions for the principal root,

$$\log\zeta - \mu_1(\zeta) = -C(\zeta-1)^{p+1} + \ldots. \tag{4.51}$$

Thus order star theory can be very much dualized by the replacements

$$\begin{array}{llcl}
\text{a)} & \mu & \longleftrightarrow & \zeta \\
\text{b)} & 0 & \longleftrightarrow & 1 \\
\text{c)} & \text{Imag. axis} & \longleftrightarrow & \text{Unit circle} \\
\text{d)} & \operatorname{Re} & \longleftrightarrow & |\cdot| \\
\text{e)} & \operatorname{Im} & \longleftrightarrow & \operatorname{Arg} \\
\text{f)} & \exp & \longleftrightarrow & \log
\end{array} \tag{4.52}$$

The analogue of the star defined in (4.12) becomes

$$A = \Big\{\zeta\,;\ \operatorname{Re}\mu(\zeta) \le \operatorname{Re}(\log\zeta)\Big\} = \Big\{\zeta\,;\ \operatorname{Re}\mu(\zeta) \le \log|\zeta|\Big\} \tag{4.53}$$

and the analogue of the relative order star (4.18) becomes

$$B = \Big\{\zeta\,;\ \operatorname{Re}\mu(\zeta) \le \operatorname{Re}\mu_R(\zeta)\Big\}. \tag{4.54}$$

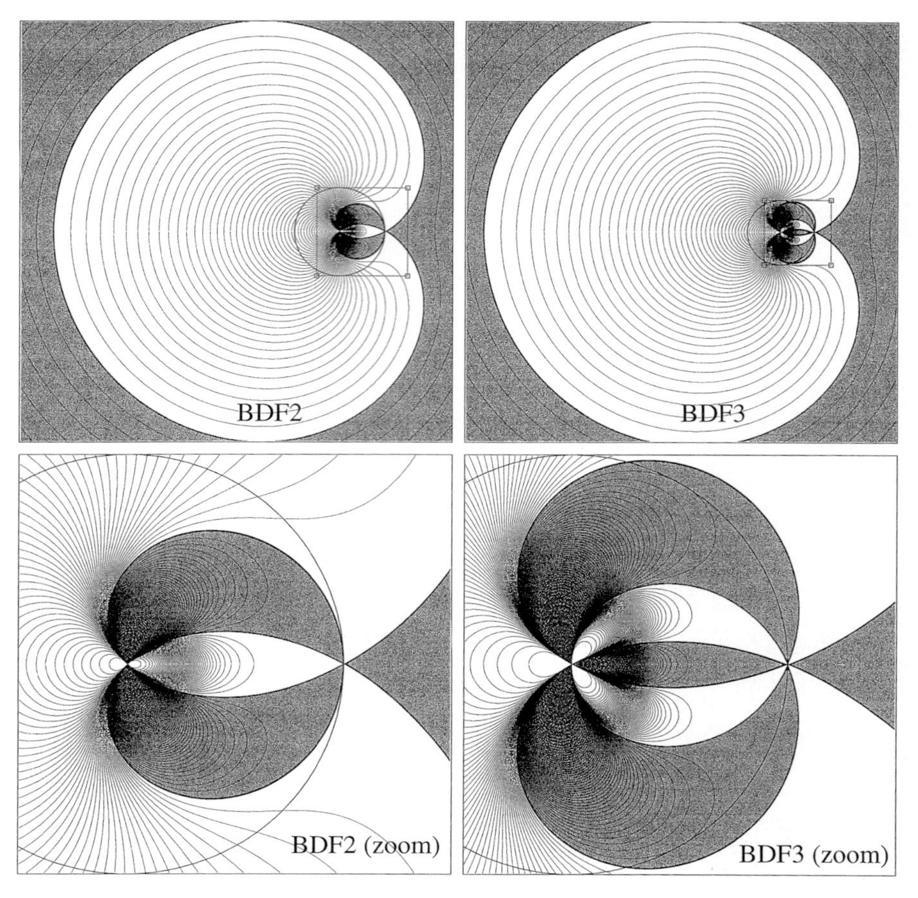

Fig. 4.7. Dual order stars (4.53) for BDF methods

For the special case of the trapezoidal rule this is

$$B = \left\{ \zeta\ ;\ \ \mathrm{Re}\,\mu(\zeta) \leq \mathrm{Re}\left(2\frac{\zeta - 1}{\zeta + 1} \right) \right\}. \tag{4.55}$$

The set A is displayed in Fig. 4.7 for the BDF2 and BDF3 methods. It explains once again why A-stable methods of order $> 2s$ are not possible (see Exercise 5).

Still another possibility is to replace (4.50) by the obviously equivalent condition

$$|\zeta| \geq 1 \qquad \Longrightarrow \qquad \mathrm{Re}\,\frac{1}{\mu_j(\zeta)} \geq 0 \tag{4.56}$$

in which case order condition (4.51) becomes

$$\frac{1}{\log \zeta} - \frac{1}{\mu_1(\zeta)} = C(\zeta - 1)^{p-1} + \ldots \tag{4.57}$$

since $\log\zeta$ as well as $\mu_1(\zeta)$ are $(\zeta-1)+\mathcal{O}((\zeta-1)^2)$. The order stars now become analogously

$$A=\left\{\zeta\,;\ \operatorname{Re}\frac{1}{\mu(\zeta)}\geq \operatorname{Re}\frac{1}{\log\zeta}\right\} \tag{4.58}$$

and

$$B=\left\{\zeta\,;\ \operatorname{Re}\frac{1}{\mu(\zeta)}\geq \operatorname{Re}\frac{1}{\mu_R(\zeta)}\right\}. \tag{4.59}$$

A special advantage of these last definitions is that for linear multistep methods $1/\mu=\sigma(\zeta)/\varrho(\zeta)$, hence the *poles* of the functions involved are the *zeros* of $\varrho(\zeta)$, which play a role in the definition of *ordinary* stability (Sect. III.3). This can be used to obtain a geometric proof of the *first* Dahlquist barrier (Theorem III.3.5), inspired by the paper Iserles & Nørsett (1984) (see Exercise 6).

Also, the proof for Dahlquist's second barrier of Sect. V.1 (Theorem 1.4) can be seen to be nothing else but a study of B of (4.59) where $\mu_R(\zeta)$ represents the trapezoidal rule.

Exercises

1. Analyze the behaviour of the characteristic roots of (4.7) in the neighbourhood of a pole which coincides with a branch point, i.e., solve (4.7) asymptotically for ζ large in the case

$$\varphi_k(\mu_0)=0,\quad \varphi_k'(\mu_0)\neq 0,\quad \varphi_{k-1}(\mu_0)=0,\quad \varphi_{k-2}(\mu_0)\neq 0.$$

   Show that these roots behave like $\pm C(\mu-\mu_0)^{-1/2}$.

2. Compute the approximate eigenvectors $\psi(\mu)$ such that

$$\bigl(e^{\mu}I-S(\mu)\bigr)\psi(\mu)=\mathcal{O}(\mu^{p+1})$$

   for the matrices $S(\mu)$ given in (4.36), (4.37), (4.38). Show that the stated orders are optimal.

3. Explain with the help of order stars, why the 2-step 2-stage collocation method with $c_2=1$ (see Exercise 7 of Sect. V.3) looses A-stability exactly when c_1 crosses the superconvergence point (Exercise 8 of Sect. V.3).

4. Modify the coefficient β in the method

$$y_{n+1}=y_n+h\Bigl(f_n+\frac{1}{2}\nabla f_n+\frac{5}{12}\nabla^2 f_n+\beta\nabla^3 f_n\Bigr),$$

   which for $\beta=3/8$ is the Adams method of order 4, in such a way that the stability domain becomes *strictly* larger. This example shows that the multistep version of Theorem IV.4.17 of Jeltsch & Nevanlinna requires the hypothesis of "Property C".

5. Prove the Daniel & Moore conjecture with the help of the order star A from (4.53).

   Hint. The set A is not allowed to cross the unit circle and along the borderlines of A the imaginary part of $\log\zeta - \mu(\zeta)$ must steadily *decrease* (consult (4.52) and the proof of Lemma IV.4.5). Hence a borderline starting and ending at the origin must either pass through a pole (which is not outside the unit circle) or cross the negative real axis in the upward direction (where $\operatorname{Im}(\log\zeta)$ increases by 2π). Since then the set A must be to the left, this is only possible once on each sheet.

Fig. 4.8. Dual order stars (4.58) and (4.59) for

$$\varrho_R(\zeta) = (\zeta-1)(\zeta+1)^5, \quad \varrho(\zeta) = \zeta^6 - 1,$$

$$\sigma_R(\zeta) = (251\zeta^6 + 2736\zeta^5 + 6957\zeta^4 + 10352\zeta^3 + 6957\zeta^2 + 2736\zeta + 251)/945$$

$$\sigma(\zeta) = (41\zeta^6 + 216\zeta^5 + 27\zeta^4 + 272\zeta^3 + 27\zeta^2 + 216\zeta + 41)/140$$

6. Prove the first Dahlquist barrier by order stars, i.e., prove that stable linear multistep methods satisfy $p \leq k+2$ (k even) and $p \leq k+1$ (k odd). Prove also that for methods with optimal order the smallest error constant is assumed by the method with

$$\varrho_R(\zeta) = (\zeta - 1)(\zeta + 1)^{\widetilde{k}-1}. \tag{4.60}$$

where $\widetilde{k} = k$ (if k is even) and $\widetilde{k} = k - 1$ (if k is odd).

Hint. Study the order stars (4.58) (with $\mu = \mu_R$) and (4.59) where $\mu_R = \sigma_R / \varrho_R$ with ϱ_R from (4.60) (see Fig. 4.8 for the case $k = 6$, $p = 8$, $\varrho(\zeta) = \zeta^6 - 1$). You must show that the two order stars in the vicinity of $\zeta = 1$ have the *same* colours. The following observations will help:

i) The stars in the vicinity of $\zeta = -1$ (produced by the pole $1/(\zeta + 1)^{\widetilde{k}-1}$) have *opposite* colours;

ii) By stability all poles of

$$d_A(\zeta) = \mathrm{Re}\left(\frac{1}{\mu_R(\zeta)} - \frac{1}{\log \zeta}\right), \quad d_B(\zeta) = \mathrm{Re}\left(\frac{1}{\mu_R(\zeta)} - \frac{1}{\mu(\zeta)}\right)$$

lie on or inside the unit circle;

iii) The boundary curves of A and B cannot cross the unit circle arbitrarily often, since $d_A(e^{i\varphi})$ and $d_B(e^{i\varphi})$ are trigonometric polynomials.

iv) Study the behaviour of A and B at infinity.

7. Prove the second Dahlquist barrier for linear multistep methods with the help of the order star (4.55).

8. Compute on a computer for an implicit multistep method of order 3 the order star B of (4.18), where $R(\mu)$ is the maximal root of the Milne-Simpson method (1.17). Understand at once the validity of Theorem 4.9.

V.5 Experiments with Multistep Codes

> ... we know that theory is unable to predict much of what happens in practice at present and software writers need to discover the way ahead by numerical experiment ...
>
> (J.R. Cash, in Aiken 1985)

> A comparison of different codes is a notoriously difficult and inexact area ... but there are some clear conclusions that can ...
>
> (J.R. Cash 1983)

This section presents numerical results of multistep codes on precisely the same problems as in Sect. IV.10. These are, in increasing dimension, VDPOL (the van der Pol equation (IV.10.1)), ROBER (the famous Robertson problem (IV.10.2)), OREGO (the Oregonator (IV.10.3)), HIRES (the physiological problem (IV.10.4)), E5 (the badly scaled chemical reaction (IV.10.5)), PLATE ((IV.10.6), a car moving on a plate, the only linear and non autonomous problem), BEAM (the nonlinear elastic beam equation (IV.1.10') with $n = 40$), CUSP (the cusp catastrophe (IV.10.8)), BRUSS (the brusselator (IV.1.6') with one-dimensional diffusion $\alpha = 1/50$ and $n = 500$), and KS (the one-dimensional Kuramoto-Sivashinsky equation (IV.10.11) with $n = 1022$). We have *not* included here the problems BECKDO and BRUSS-2D, since they require a special treatment of the linear algebra routines.

As in Sect. IV.10, the codes have been applied with tolerances

$$Rtol = 10^{-2-m/4} \qquad m = 0, 1, 2, \ldots$$

and $Atol = Rtol$ (with the exceptions $Atol = 10^{-6} \cdot Rtol$ for OREGO and ROBER, $Atol = 10^{-4} \cdot Rtol$ for HIRES, $Atol = 10^{-3} \cdot Rtol$ for PLATE, and $Atol = 1.7 \cdot 10^{-24}$ for E5). The numerical precisions obtained compared to the CPU times (where all codes are compiled with the same optimization options) are then displayed in Figs. 5.1 and 5.2, again with the symbols representing the required precision $Rtol = 10^{-5}$ displayed in grey tone.

The Codes Used

LSODE — is the "Livermore Solver" of Hindmarsh (1980, 1983). Since we are dealing with stiff equations, we use "stiff" method flags MF $= 21, 22, 24$ or 25, so that the code is based on the Nordsieck representation of the fixed step size BDF methods (see Sections III.6 and III.7). This code emerged from a long development starting with Gear's DIFSUB in 1971. Its exemplary user interface and ease of application has been a model for much subsequent ODE Software (including ours). Most problems were computed with analytical Jacobian and full linear algebra (MF $= 21$), with the exception of BRUSS and KS (analytical banded Jacobian, MF $= 24$), BEAM (numerical full Jacobian, MF $= 22$), and CUSP (numerical banded Jacobian, MF $= 25$).

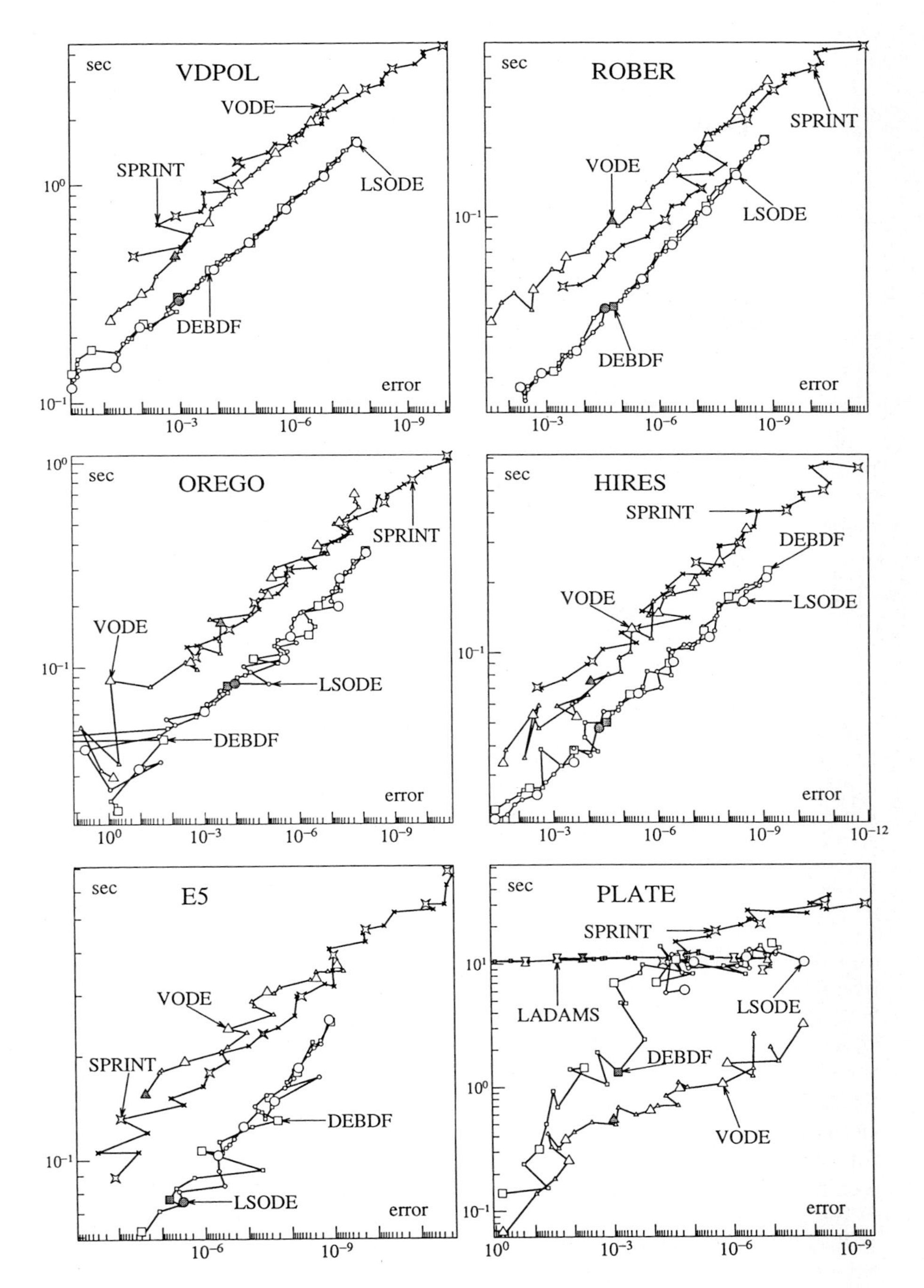

Fig. 5.1. Work-precision diagrams for problems of dimension 2 to 80

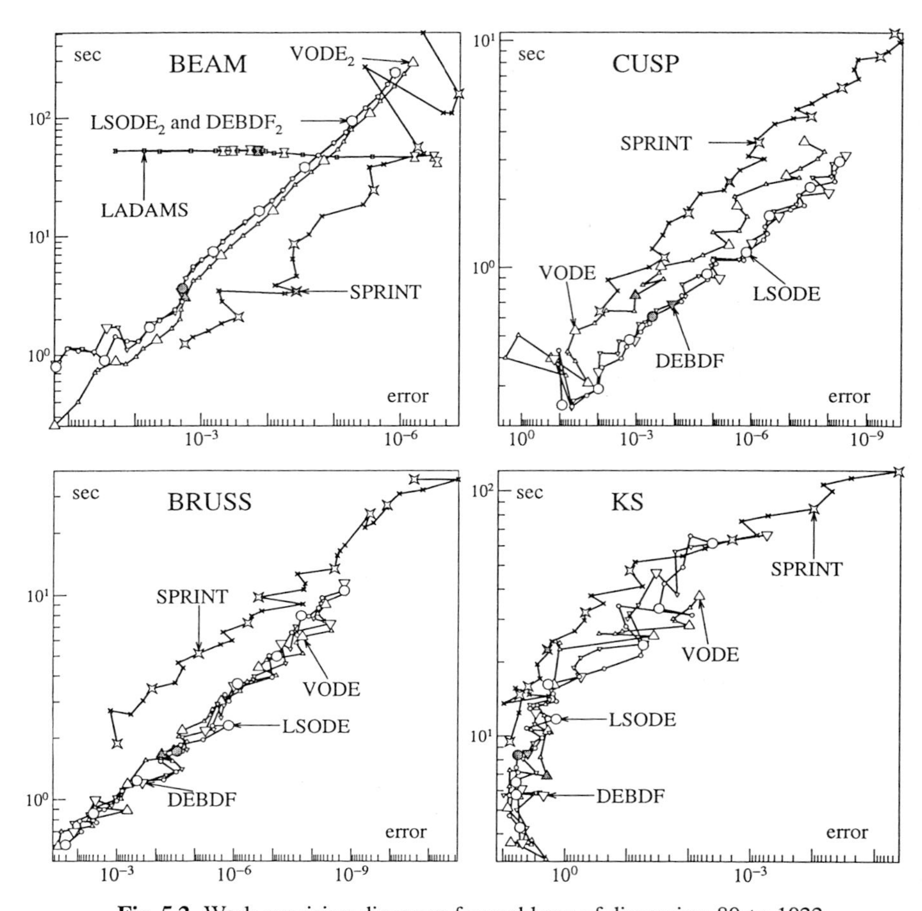

Fig. 5.2. Work-precision diagrams for problems of dimension 80 to 1022

For E5, the code worked correctly only for $Tol \le 10^{-5}$, for PLATE it was necessary to have $Tol \le 10^{-7}$. For the BEAM problem, which has eigenvalues on the imaginary axis, it was necessary to restrict the maximal order to 2 because of the lack of A-stability of the higher order BDF methods. The disastrous effect of the allowance of orders higher than 2 can be seen in Fig. 5.3.

DEBDF — this is Shampine & Watts's driver for a modification of the code LSODE and is included in the "DEPAC" family (Shampine & Watts 1979). As is to be expected, it behaves nearly identically to LSODE (see Figs. 5.1 and 5.2). It also requires a restriction of the order for the BEAM problem (see Fig. 5.3).

VODE — is the "Variable-coefficient Ordinary Differential Equation solver" of Brown, Byrne & Hindmarsh (1989). It is based on the EPISODE and EPISODEB packages (see Sect. III.7) which use BDF methods on a non uniform grid (Byrne

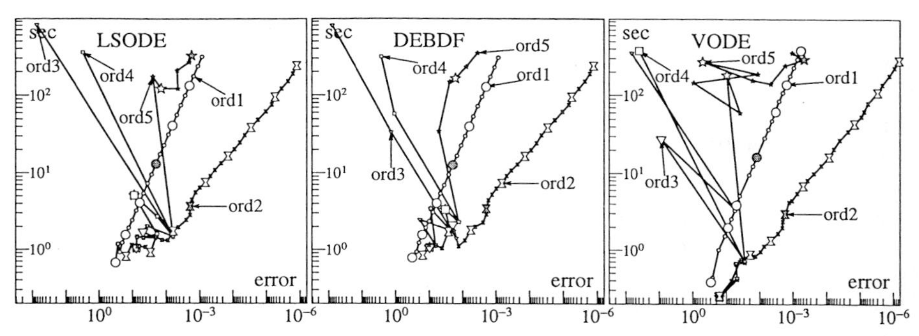

Fig. 5.3. Performance of LSODE, DEBDF and VODE on the BEAM problem with restricted maximal order

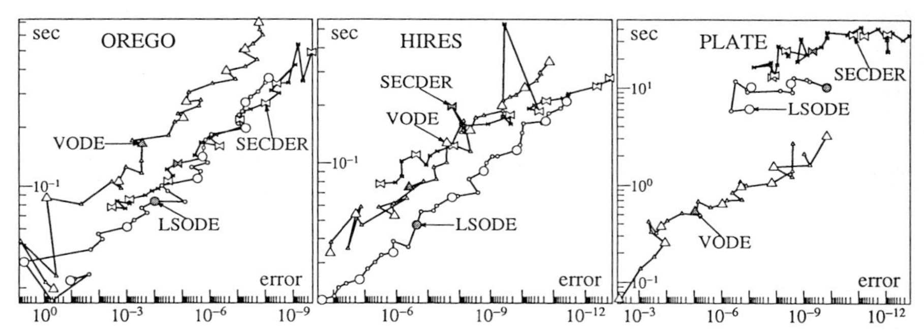

Fig. 5.4. Performance of SECDER, compared to LSODE and VODE

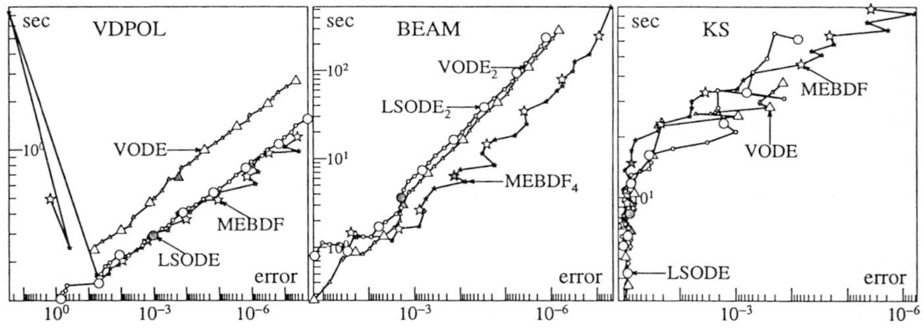

Fig. 5.5. Performance of MEBDF, compared to LSODE and VODE (for the BEAM problem with restricted maximal order)

& Hindmarsh 1975). The user interface is very similar to that of LSODE; the code again allows selection between full or banded linear algebra and between analytical or numerical Jacobian. The numerical results of VODE (see Figs. 5.1 and 5.2) are very similar for the large problems to those of LSODE and DEBDF, the code is, however, considerably slower on the small problems. For problem E5 this code required a tolerance requirement ($Rtol \leq 10^{-5}$). On the PLATE problem, this code was by far the best. On the BEAM problem, one has to restrict the maximal order to two (Fig. 5.3).

SPRINT — this package, written by M. Berzins (see Berzins & Furzeland 1985), which has been incorporated into the NAG library ("subchapter D02N"), contains several modules for the step integrator, one of which is SBLEND. This allows us to study the effect of the blended multistep methods (3.15) of Skeel & Kong (1977). It can be seen from Table 3.4 that these methods are A-stable for orders up to 4. We therefore expect them to be much better on the oscillatory BEAM problem. As can be observed in Fig. 5.2 (as well as in Fig. IV.10.8), this code gives excellent results for this problem. An observation of the grey points for $Tol = 10^{-5}$ (Figs. 5.1 and 5.2) shows that the code gives better values than the other multistep codes for a same given tolerance. From time to time, it is fairly slow (e.g., in the PLATE and KS problems).

SECDER — this code, written in 1979 by C.A. Addison (see Addison 1979), implements the SECond DERivative multistep methods (3.7) of Enright. The high order of the methods accompanied with good stability leads us to expect good performance at high tolerances. This has shown to be true (see Fig. 5.4) for OREGO, HIRES and PLATE; however, for the latter it is very slow. We have not used it on the large problems since it has no built-in banded algebra and requires an analytic Jacobian.

MEBDF — this code by Cash & Considine (1992) implements the modified extended BDF methods (see Eq. (3.17.mod) and Table 3.5). Its good performance is shown on selected examples in Fig. 5.5. For the BEAM problem, the code works well if the maximal order is limited to 4.

LADAMS — this is the "Livermore Adams" code, i.e., LSODE with method flag $MF = 10$, included to demonstrate the performance of an *explicit* multistep method on large and/or mildly stiff problems. One can see that it has its chance on several large problems (PLATE, BEAM). It is, when compared to DOPRI5 in Fig. IV.10.8, a good deal slower when f-evaluations are cheap (CUSP), but not on BEAM.

The codes LSODE, DEBDF, VODE and MEBDF can be obtained by sending an electronic mail (e.g., "send lsode.f from odepack") to "netlib@research.att.com".

Exercises

1. Do your own experiments and draw your own conclusions for the above problems. The authors will be happy to provide you with drivers and function subroutines.

V.6 One-Leg Methods and G-Stability

... the error analysis is simpler to formulate for one-leg methods than for linear multistep methods. (G. Dahlquist 1975)

The first stability results for *nonlinear* differential equations and multistep methods are fairly old (Liniger 1956, Dahlquist 1963), older than similar studies for Runge-Kutta methods. The great break-through occured in 1975 (at the Dundee conference) when Dahlquist proposed considering nonlinear problems

$$y' = f(x, y) \tag{6.1}$$

which satisfy a one-sided Lipschitz condition

$$\langle f(x,y) - f(x,z),\, y - z\rangle \le \nu \|y - z\|^2 \tag{6.2}$$

or, if the functions are complex-valued,

$$\mathrm{Re}\,\langle f(x,y) - f(x,z),\, y - z\rangle \le \nu \|y - z\|^2 \tag{6.2'}$$

(see Sect. IV.12). He also found that the study of nonlinear stability for general multistep methods is simplified, if a related class of methods — the so-called one-leg (multistep) methods — is considered.

One-Leg (Multistep) Methods

... the somewhat crazy name *one-leg methods* ...
(G. Dahlquist 1983)

Je ne suis absolument pas capable de traduire "one-leg" en français ... uni-jambiste? (M. Crouzeix, in 1987)

Signor mio, le gru non hanno se non una coscia ed una gamba ... (Boccacio, Decamerone 1353; quotation suggested by M. Crouzeix)

Suppose that a linear k-step method

$$\sum_{i=0}^{k} \alpha_i y_{m+i} = h \sum_{i=0}^{k} \beta_i f(x_{m+i}, y_{m+i}) \tag{6.3}$$

is given, and that the generating polynomials

$$\varrho(\zeta) = \sum_{i=0}^{k} \alpha_i \zeta^i, \qquad \sigma(\zeta) = \sum_{i=0}^{k} \beta_i \zeta^i \tag{6.4}$$

have real coefficients and no common divisor (see Sect. III.2). We also assume throughout the normalization

$$\sigma(1) = 1. \tag{6.5}$$

Then the associated *one-leg method* is defined by

$$\sum_{i=0}^{k} \alpha_i y_{m+i} = hf\Big(\sum_{i=0}^{k} \beta_i x_{m+i}, \sum_{i=0}^{k} \beta_i y_{m+i}\Big). \tag{6.6}$$

In this new method, the derivative f is evaluated at one point only, which makes it easier to analyze.

It is, of course, interesting to know how the solutions of the one-leg method (6.6) are related to those of its "multistep twin" (6.3). If the differential equation is linear and autonomous, $y' = Ay$, then both formulas — (6.3) and (6.6) — are identical. For the BDF schemes (1.18) there is in any case only one f-value in the multistep-version, hence the equations (6.3) and (6.6) are the same. For general methods and general nonlinear equations, however, the formulas are *not* identical, but the solutions are related by certain transformations (see Exercise 3). We consider, as an example, the trapezoidal rule, which is a two-leg method,

$$y_{m+1} - y_m = \frac{h}{2}\Big(f(x_m, y_m) + f(x_{m+1}, y_{m+1})\Big). \tag{6.7}$$

The corresponding one-leg method is the implicit midpoint rule,

$$y_{m+1} - y_m = hf\Big(\frac{x_m + x_{m+1}}{2}, \frac{y_m + y_{m+1}}{2}\Big). \tag{6.8}$$

If $\{y_m\}$ is a solution of the one-leg formula (6.8), then

$$\widehat{y}_m = \frac{1}{2}(y_m + y_{m+1}), \qquad \widehat{x}_m = \frac{1}{2}(x_m + x_{m+1})$$

satisfies (6.7). On the other hand, if $\{\widehat{y}_m, \widehat{x}_m\}$ satisfy (6.7), then

$$y_m = \widehat{y}_m - \frac{h}{2} f(\widehat{x}_m, \widehat{y}_m), \qquad x_m = \widehat{x}_m - \frac{h}{2}$$

is a solution of (6.8). This relationship has already been extensively exploited in the proof of Theorem IV.15.8.

Existence and Uniqueness

We suppose $\alpha_k \neq 0$ (as always) and $\beta_k \neq 0$ (otherwise the method is explicit). In the case of multistep methods, we write (6.3) in the form

$$y - \eta - h\frac{\beta_k}{\alpha_k} f(x, y) = 0, \tag{6.9}$$

where x is given, η is a vector composed of known quantities, and $y = y_{m+k}$ is the unknown vector. The one-leg Formula (6.6) can also be brought to the form (6.9) by the transformation $y = \beta_k y_{m+k} + \ldots + \beta_0 y_m$, so that all subsequent results on existence and uniqueness will be valid for multistep *and* one-leg methods. To obtain existence results for Eq. (6.9), we replace $h\beta_k/\alpha_k$ by a new "step size" $\widetilde{h}$ and obtain nothing else but implicit Euler. All theorems for implicit Runge-Kutta methods (Theorems 14.2, 14.3, and 14.4 of Sect. IV.14) are immediately applicable and give

Theorem 6.1 (Dahlquist 1975). *Let f be continuously differentiable and satisfy (6.2). If*

$$h\nu < \frac{\alpha_k}{\beta_k} \tag{6.10}$$

then the nonlinear equation (6.9) has a unique solution y. □

Theorem 6.2. *Let y be given by (6.9) and consider a perturbed value $\widehat{y}$ satisfying*

$$\widehat{y} - \eta - h\,\frac{\beta_k}{\alpha_k}\,f(x,\widehat{y}) = \delta. \tag{6.11}$$

Under the assumption (6.10) we then have

$$\|\widehat{y} - y\| \le \frac{1}{1-(\beta_k/\alpha_k)h\nu}\,\|\delta\|. \tag{6.12}$$

□

Remark. Theorems IV.14.2, IV.14.3 and IV.14.4 are for much more general methods than just the implicit Euler needed here. The reader who is not interested in the more general case can rewrite the proofs of Sect. IV.14 nearly word for word. Since there is now only one implicit stage, all tensor products disappear and the formulas, but not the ideas of the proof, simplify considerably.

G-Stability

If the differential equation satisfies the one-sided Lipschitz condition (6.2) (or (6.2')) with $\nu = 0$, then the exact solutions are contractive (Lemma IV.12.1). We shall investigate here, which one-leg (multistep) methods then also have contractive solutions. Since the numerical value y_{m+k} depends on all $y_{m+k-1}, \ldots, y_m$, it makes no sense to require $\|y_{m+k} - \widehat{y}_{m+k}\| \le \|y_{m+k-1} - \widehat{y}_{m+k-1}\|$ as in the one-step case (Definition IV.12.2). We have to consider the method as a mapping $\mathbb{R}^{n\cdot k} \to \mathbb{R}^{n\cdot k}$. For this we introduce the notation

$$Y_m = (y_{m+k-1}, \ldots, y_m)^T \tag{6.13}$$

and consider inner product norms on $\mathbb{R}^{n\cdot k}$

$$\|Y_m\|_G^2 = \sum_{i=1}^{k}\sum_{j=1}^{k} g_{ij}\,\langle y_{m+i-1}, y_{m+j-1}\rangle, \tag{6.14}$$

where $\langle\cdot,\cdot\rangle$ is the inner product on $\mathbb{R}^n$ used in (6.2) and the k-dimensional matrix

$$G=(g_{ij})_{i,j=1,\ldots,k}$$

is assumed to be real, symmetric and positive definite.

Definition 6.3 (Dahlquist 1975). The one-leg method (6.6) is called *G-stable*, if there exists a real, symmetric and positive definite matrix G, such that for two numerical solutions $\{y_m\}$ and $\{\widehat{y}_m\}$ we have

$$\|Y_{m+1}-\widehat{Y}_{m+1}\|_G\le\|Y_m-\widehat{Y}_m\|_G \tag{6.15}$$

for all step sizes $h>0$ and for all differential equations satisfying (6.2) or (6.2') with $\nu=0$.

Since $y'=\lambda y, \mathrm{Re}\,\lambda\le 0$ satisfies (6.2') with $\nu=0$, we immediately get

Theorem 6.4. *G-stability implies A-stability.* □

Example 6.5. Consider the 2-step BDF method

$$\frac{3}{2}y_{m+2}-2y_{m+1}+\frac{1}{2}y_m=hf(x_{m+2},y_{m+2}). \tag{6.16}$$

We take a second numerical solution $\{\widehat{y}_m\}$ and denote its difference to $\{y_m\}$ by $\Delta y_m=y_m-\widehat{y}_m$. If we insert (6.16) into our assumption (6.2')

$$\mathrm{Re}\,\langle f(x_{m+2},y_{m+2})-f(x_{m+2}\,,\,\widehat{y}_{m+2}),\;y_{m+2}-\widehat{y}_{m+2}\rangle\le 0$$

we obtain

$$E=\mathrm{Re}\,\Big\langle\frac{3}{2}\Delta y_{m+2}-2\Delta y_{m+1}+\frac{1}{2}\Delta y_m\,,\,\Delta y_{m+2}\Big\rangle\le 0. \tag{6.17}$$

The main idea is now to subtract from this inequality a well-chosen quadratic term $\|a_2\Delta y_{m+2}+a_1\Delta y_{m+1}+a_0\Delta y_m\|^2$ in order to bring it to the form required by (6.15). With $\Delta Y_m=(\Delta y_{m+1},\Delta y_m)^T$ this means that

$$E=\|\Delta Y_{m+1}\|_G^2-\|\Delta Y_m\|_G^2+\|a_2\Delta y_{m+2}+a_1\Delta y_{m+1}+a_0\Delta y_m\|^2 \tag{6.18}$$

with a positive definite matrix

$$G=\begin{pmatrix} g_{22} & g_{21}\\ g_{21} & g_{11}\end{pmatrix}.$$

Multiplying out and comparing the coefficients of $\mathrm{Re}\,\langle\Delta y_i,\Delta y_j\rangle$ in (6.17) and (6.18) gives the six relations

$$\frac{3}{2}=g_{22}+a_2^2,\qquad 0=g_{11}-g_{22}+a_1^2,\qquad 0=-g_{11}+a_0^2, \tag{6.19a}$$

$$-2=2g_{21}+2a_2a_1,\qquad \frac{1}{2}=2a_2a_0,\qquad 0=-2g_{21}+2a_1a_0. \tag{6.19b}$$

Adding all six equations gives $0=(a_0+a_1+a_2)^2$, so that $a_0+a_1+a_2=0$. This relation together with (6.19b) determines the a_i as $a_0=\pm 1/2$, $a_1=\mp 1$, $a_2=\pm 1/2$. Inserting this into (6.19) yields the positive definite matrix

$$G=\frac{1}{4}\begin{pmatrix} 5 & -2 \\ -2 & 1 \end{pmatrix}. \tag{6.20}$$

Since $E\le 0$ by (6.17), it follows from (6.18) that the 2-step BDF method is G-stable.

An Algebraic Criterion

The algebraic structures of the foregoing computations become much more visible, if we replace formally in (6.17) and (6.18) all

$$\langle \Delta y_{m+i}, \Delta y_{m+j}\rangle \longmapsto \zeta^i\omega^j$$

and use

$$2\mathrm{Re}\,\langle \Delta y_{m+i}, \Delta y_{m+j}\rangle = \langle \Delta y_{m+i}, \Delta y_{m+j}\rangle + \langle \Delta y_{m+j}, \Delta y_{m+i}\rangle.$$

This yields

$$E=\frac{1}{2}\big(\varrho(\zeta)\sigma(\omega)+\varrho(\omega)\sigma(\zeta)\big) \tag{6.17'}$$

$$E=(\zeta\omega-1)\sum_{i,j=1}^{k} g_{ij}\zeta^{i-1}\omega^{j-1}+\Big(\sum_{i=0}^{k} a_i\zeta^i\Big)\Big(\sum_{j=0}^{k} a_j\omega^j\Big). \tag{6.18'}$$

We can now formulate an algebraic criterion which, in a different notation, already appears in Dahlquist (1975).

Theorem 6.6 (Baiocchi & Crouzeix 1989). *Consider a method* (ϱ,σ). *If there exists a real, symmetric and positive definite matrix* G *and real numbers* $a_0,\ldots,a_k$, *such that*

$$\begin{aligned}&\frac{1}{2}\big(\varrho(\zeta)\sigma(\omega)+\varrho(\omega)\sigma(\zeta)\big)\\ &\qquad=(\zeta\omega-1)\sum_{i,j=1}^{k} g_{ij}\zeta^{i-1}\omega^{j-1}+\Big(\sum_{i=0}^{k} a_i\zeta^i\Big)\Big(\sum_{j=0}^{k} a_j\omega^j\Big),\end{aligned} \tag{G}$$

then the corresponding one-leg method is G*-stable.*

Remark. The factor $1/2$ on the left-hand side of (G) is of no significance and can be replaced by any other positive constant, leading to another scaling of the coefficients g_{ij} and a_i.

Proof. We just replace $\zeta^i\omega^j$ by $\langle \Delta y_{m+i}, \Delta y_{m+j}\rangle$ in Eq. (G) and obtain

$$\mathrm{Re}\Big\langle \sum_{i=0}^{k} \alpha_i \Delta y_{m+i}, \sum_{j=0}^{k} \beta_j \Delta y_{m+j}\Big\rangle = \\ \|\Delta Y_{m+1}\|_G^2 - \|\Delta Y_m\|_G^2 + \|\sum_{i=0}^{k} a_i \Delta y_{m+i}\|^2. \tag{6.21}$$

We then insert (6.6) and use (6.2') with $\nu = 0$ and obtain the desired estimate $\|\Delta Y_{m+1}\|_G \le \|\Delta Y_m\|_G$. □

An interesting question is now, for which methods (ϱ, σ) Condition (6.21) is satisfied. By Theorem 6.4 the method is necessarily A-stable. Is this also sufficient?

The Equivalence of A-Stability and G-Stability

Dahlquist struggled for three years to get the answer, which is

Theorem 6.7 (Dahlquist 1978). *If ϱ and σ have no common divisor, then the method (ϱ, σ) is A-stable if and only if the corresponding one-leg method is G-stable.*

Proof. We follow here the presentation of Baiocchi & Crouzeix (1989). Recall first that A-stability of the method (ϱ, σ) implies

$$\mathrm{Re}\, \varrho(\zeta)\overline{\sigma(\zeta)} \ge 0 \qquad \text{for } |\zeta| \ge 1 \tag{A}$$

(see Sect. V.1). Because of Theorems 6.4 and 6.6 it is sufficient to prove that condition (A) implies the existence of a real, symmetric and positive definite matrix G and real numbers $a_0, \ldots, a_k$ such that Property (G) holds. The proof is in three steps:

a) computation of $a_0, \ldots, a_k$;
b) computation of G;
c) show that G is positive definite.

a) The term containing the g_{ij}'s in (G) disappears if we put $\omega = 1/\zeta$. We therefore consider the function

$$E(\zeta) = \frac{1}{2}\big(\varrho(\zeta)\sigma(1/\zeta) + \varrho(1/\zeta)\sigma(\zeta)\big), \tag{6.22}$$

which is of the form

$$
\begin{aligned}
E(\zeta) &= c_r\Big(\zeta^r+\frac{1}{\zeta^r}\Big)+c_{r-1}\Big(\zeta^{r-1}+\frac{1}{\zeta^{r-1}}\Big)+\ldots+c_1\Big(\zeta+\frac{1}{\zeta}\Big)+c_0 \\
&= \frac{c_r}{\zeta^r}\prod_{j=1}^{2r}(\zeta-\zeta_j)
\end{aligned}
\tag{6.23}
$$

with some $r\le k$. Since $E(\zeta)=E(1/\zeta)$, for each root ζ_j of the polynomial $\zeta^r E(\zeta)$ the inverse $1/\zeta_j$ is also a root with the same multiplicity. Therefore there are as many roots *inside* the unit circle as there are *outside*. As to the roots *on* the unit circle, Condition (A) tells us that $E(\zeta)=\mathrm{Re}\,\varrho(\zeta)\sigma(\overline{\zeta})\ge 0$ on the unit circle. Therefore, all roots on the unit circle must have *even multiplicity*, half of them we declare "inside" and half of them we declare "outside". The clever idea is now to collect all roots "outside" the unit circle into a product, so that

$$
\begin{aligned}
E(\zeta) &= \frac{c_r}{\zeta^r}\prod_{\zeta_j\text{ outside}}(\zeta-\zeta_j)\prod_{\zeta_j\text{ inside}}(\zeta-\zeta_j) \\
&= \frac{c_r}{\zeta^r}\prod_{\zeta_j\text{ outside}}(\zeta-\zeta_j)\prod_{\zeta_j\text{ outside}}\Big(\zeta-\frac{1}{\zeta_j}\Big) \\
&= K\prod_{\zeta_j\text{ outside}}(\zeta-\zeta_j)\prod_{\zeta_j\text{ outside}}\Big(\frac{1}{\zeta}-\zeta_j\Big)
\end{aligned}
\tag{6.24}
$$

where K is a constant. But this constant must be non-negative, as can be seen thus: by Condition (A), $E(\zeta)$ is non-negative on the unit circle. The same is true for the function divided by K, since each factor $(e^{i\theta}-\zeta_j)$ from the first product has a complex conjugate brother $(e^{-i\theta}-\overline{\zeta}_j)$ in the second. Therefore $E(\zeta)$ in (6.24) can be factored as

$$
E(\zeta)=a(\zeta)\cdot a(1/\zeta)
\tag{6.25}
$$

where

$$
a(\zeta)=\sqrt{K}\prod_{\zeta_j\text{ outside}}(\zeta-\zeta_j)=:\sum_{i=0}^{k}a_i\zeta^i.
\tag{6.26}
$$

and step (a) is done.

b) It follows from (6.22) and (6.25) that the polynomial

$$
P(\zeta,\omega)=\frac{1}{2}\Big(\varrho(\zeta)\sigma(\omega)+\varrho(\omega)\sigma(\zeta)\Big)-a(\zeta)a(\omega)
\tag{6.27}
$$

vanishes when $\zeta\omega-1=0$. It can therefore be written as

$$
P(\zeta,\omega)=(\zeta\omega-1)\sum_{i,j=1}^{k}g_{ij}\zeta^{i-1}\omega^{j-1}.
\tag{6.28}
$$

The coefficients g_{ij} are real and satisfy $g_{ij}=g_{ji}$, because $P(\zeta,\omega)=P(\omega,\zeta)$.

c) Looking at (6.28), it appears at first sight a difficult task to prove positive definiteness for the matrix $G=(g_{ij})$ defined there. The crucial idea is the following: choose k (at first arbitrary) complex numbers $\zeta_1,\ldots,\zeta_k$ and replace in (6.28) $\zeta\mapsto\overline{\zeta}_q$, $\omega\mapsto\zeta_r$, which gives together with (6.27)

$$\begin{aligned} b_{qr} &= \sum_{i,j=1}^{k}\overline{\zeta}_q^{\,i-1} g_{ij}\zeta_r^{j-1} \\ &= \frac{1}{1-\overline{\zeta}_q\zeta_r}\Big\{-\frac{1}{2}\Big(\varrho(\overline{\zeta}_q)\sigma(\zeta_r)+\varrho(\zeta_r)\sigma(\overline{\zeta}_q)\Big)+a(\overline{\zeta}_q)a(\zeta_r)\Big\}. \end{aligned} \tag{6.29}$$

Here the b_{qr} are the elements of the matrix

$$B=V^*GV$$

where $V=(\zeta_j^{i-1})$ is a Vandermonde matrix. Thus, we now have to prove that B is positive definite, which appears much easier. First, we develop

$$\frac{1}{1-\overline{\zeta}_q\zeta_r}=1+\overline{\zeta}_q\zeta_r+\overline{\zeta}_q^{\,2}\zeta_r^2+\overline{\zeta}_q^{\,3}\zeta_r^3+\ldots \tag{6.30a}$$

which converges if

$$|\zeta_q|<1 \qquad q=1,2,\ldots,k. \tag{6.30b}$$

Next, we require that for all q

$$\varrho(\zeta_q)+\lambda\sigma(\zeta_q)=0 \qquad \text{for some } \lambda>0. \tag{6.31}$$

With the exception of a finite number of $\lambda's$, the k roots of Eq. (6.31) are all different. A-stability (assumption (A)) implies (6.30b), because $-\lambda$ lies in the interior of the stability domain. Inserting (6.31) and (6.30a) into (6.29) gives, for an arbitrary non-zero vector $\vec{v}=(v_1,\ldots,v_k)$,

$$\vec{v}^*B\vec{v}=\sum_{q,r=1}^{k}\overline{v}_q b_{qr}v_r=\sum_{m=0}^{\infty}\Big\{\Big|\sum_{q=1}^{k}v_q\zeta_q^m a(\zeta_q)\Big|^2+\lambda\Big|\sum_{q=1}^{k}v_q\zeta_q^m\sigma(\zeta_q)\Big|^2\Big\},$$

which looks rather positive. This expression cannot be zero for $\vec{v}\neq 0$, because it follows from (6.31) that $\sigma(\zeta_q)\neq 0$ for all q, otherwise ϱ and σ would have a common factor. Therefore $\vec{v}^*B\vec{v}>0$, thus the matrix B, and consequently the matrix G, is positive definite. □

It is worth noting that the above proof provides constructive formulas for the matrix G. As an illustration, we again consider the 2-step BDF method (6.16) with generating polynomials

$$\varrho(\zeta)=\frac{3}{2}\zeta^2-2\zeta+\frac{1}{2}, \qquad \sigma(\zeta)=\zeta^2.$$

The function $E(\zeta)$ (Formula (6.22)) becomes

$$E(\zeta) = \frac{1}{4}(\zeta^2 + \frac{1}{\zeta^2}) - (\zeta + \frac{1}{\zeta}) + \frac{3}{2} = \frac{1}{4}(\zeta - 1)^2 (\frac{1}{\zeta} - 1)^2$$

so that $a(\zeta) = \frac{1}{2}(\zeta - 1)^2$. Inserting this into (6.27) gives

$$P(\zeta, \omega) = (\zeta\omega - 1)(\frac{5}{4}\zeta\omega - \frac{1}{2}\zeta - \frac{1}{2}\omega + \frac{1}{4}),$$

so that $g_{22} = 5/4$, $g_{12} = g_{21} = -1/2$, $g_{11} = 1/4$ is the same as (6.20).

A Criterion for Positive Functions

In the proof of Lemma IV.13.19 we have used the following criterion for positive functions, which is an immediate consequence of the above equivalence result.

Lemma 6.8. *Let $\chi(z) = \alpha(z)/\beta(z)$ be an irreducible rational function with real polynomials $\alpha(z)$ of degree $\leq k-1$ and $\beta(z)$ of degree k. Then $\chi(z)$ is a positive function, i.e.,*

$$\operatorname{Re} \chi(z) > 0 \qquad \textit{for} \qquad \operatorname{Re} z > 0, \tag{6.32}$$

if and only if there exist a real, symmetric and positive definite matrix A and a real, symmetric and non-negative definite matrix B, such that

$$\alpha(z)\beta(w) + \alpha(w)\beta(z) = (z + w) \sum_{i,j=1}^{k} a_{ij} z^{i-1} w^{j-1} + \sum_{i,j=1}^{k} b_{ij} z^{i-1} w^{j-1}. \tag{6.33}$$

Proof. The "if"-part follows immediately by putting $w = \overline{z}$ in (6.33). For the "only if"-part we consider the transformations

$$\zeta = \frac{z+1}{z-1}, \quad z = \frac{\zeta+1}{\zeta-1} \qquad \text{and} \qquad \omega = \frac{w+1}{w-1}, \quad w = \frac{\omega+1}{\omega-1} \tag{6.34}$$

and introduce the polynomials

$$\varrho(\zeta) = \Big(\frac{\zeta-1}{2}\Big)^k \alpha\Big(\frac{\zeta+1}{\zeta-1}\Big), \qquad \sigma(\zeta) = \Big(\frac{\zeta-1}{2}\Big)^k \beta\Big(\frac{\zeta+1}{\zeta-1}\Big).$$

As the transformation (6.34) maps $|\zeta| > 1$ onto the half plane $\operatorname{Re} z > 0$, Condition (6.32) is equivalent to Assumption (A). Therefore, Theorem 6.7 implies the existence of a real, symmetric and positive definite matrix G and of real numbers $a_0, \ldots, a_k$ such that

$$\frac{1}{2}\big(\varrho(\zeta)\sigma(\omega) + \varrho(\omega)\sigma(\zeta)\big) = (\zeta\omega - 1) \sum_{i,j=1}^{k} g_{ij} \zeta^{i-1} \omega^{j-1} + \Big(\sum_{i=0}^{k} a_i \zeta^i\Big)\Big(\sum_{j=0}^{k} a_j \omega^j\Big).$$

Backsubstitution of the old variables yields

$$\frac{1}{2}\big(\alpha(z)\beta(w)+\alpha(w)\beta(z)\big) \tag{6.35}$$
$$= 2(z+w)\sum_{i,j=1}^{k} g_{ij}(z+1)^{i-1}(z-1)^{k-i}(w+1)^{j-1}(w-1)^{k-j}$$
$$+\Big(\sum_{i=0}^{k} a_i(z+1)^i(z-1)^{k-i}\Big)\Big(\sum_{j=0}^{k} a_j(w+1)^j(w-1)^{k-j}\Big).$$

Rearranging into powers of z and w gives Eq. (6.33). Since the polynomials $(z+1)^{i-1}(z-1)^{k-i}$ for $i=1,\dots,k$ are linearly independent, the resulting matrix A is positive definite. The coefficient of $z^k w^k$ in the second term of the right-hand side of (6.35) must vanish, because the degree of $\alpha(z)$ is at most $k-1$. We remark that the matrix B of this construction is only of rank 1. □

Error Bounds for One-Leg Methods

We shall apply the stability results of this section to derive bounds for the global error of one-leg methods. For a differential equation (6.1) with exact (smooth) solution $y(x)$ it is natural to define the discretization error of (6.6) as

$$\delta_{OL}(x)=\sum_{i=0}^{k}\alpha_i y(x+ih)-hf\Big(x+\beta h,\sum_{i=0}^{k}\beta_i y(x+ih)\Big) \tag{6.36}$$

with $\beta=\sigma'(1)=\sum i\beta_i$. For the BDF methods we have $\sum_i \beta_i y(x+ih)=y(x+\beta h)$, so that (6.36) equals

$$\delta_D(x)=\sum_{i=0}^{k}\alpha_i y(x+ih)-hy'(x+\beta h), \tag{6.37}$$

the so-called *differentiation error* of the method. For methods which do not satisfy $\sum_i \beta_i y(x+ih)=y(x+\beta h)$, the right hand side of (6.36) may become very large for stiff problems, even if the derivatives of the solution are bounded by a constant of moderate size. In this case, the expression (6.36) is not a suitable quantity for error estimates. Dahlquist (1983) proposed considering in addition to $\delta_D(x)$ also the *interpolation error*

$$\delta_I(x)=\sum_{i=0}^{k}\beta_i y(x+ih)-y(x+\beta h). \tag{6.38}$$

For nonstiff problems (with bounded derivatives of f) these two error expressions are related to $\delta_{OL}(x)$ by

$$\delta_{OL}(x)=\delta_D(x)-h\frac{\partial f}{\partial y}\big(x,y(x)\big)\,\delta_I(x)+\mathcal{O}(h\|\delta_I(x)\|^2).$$

Taylor expansion of (6.37) and (6.38) shows that

$$\delta_D(x) = \mathcal{O}(h^{p_D+1}), \qquad \delta_I(x) = \mathcal{O}(h^{p_I+1}), \tag{6.39}$$

where the optimal orders p_D and p_I are determined by certain algebraic conditions (see Exercise 1a). From $\beta = \sigma'(1)$ we always have $p_I \geq 1$ and from the consistency conditions it follows that $p_D \geq 1$. However, the orders p_D and p_I may be significantly smaller than the order of the corresponding multistep method (Exercise 1). The constants in the $\mathcal{O}(\ldots)$-terms of (6.39) depend only on bounds for a certain derivative of the solution, but not on the stiffness of the problem.

Using $\delta_D(x)$ and $\delta_I(x)$ it is possible to interpret the exact solution of (6.1) as the solution of the following perturbed one-leg formula

$$\sum_{i=0}^{k} \alpha_i y(x+ih) - \delta_D(x) = hf\Big(x+\beta h, \sum_{i=0}^{k} \beta_i y(x+ih) - \delta_I(x)\Big). \tag{6.40}$$

The next lemma, which extends results of Dahlquist (1975) and of Nevanlinna (1976), investigates the influence of perturbations to the solution of a one-leg method.

Lemma 6.9. *Consider, in addition to the one-leg method (6.6), the perturbed formula*

$$\sum_{i=0}^{k} \alpha_i \widehat{y}_{m+i} - \delta_m = hf\Big(x_m + \beta h, \sum_{i=0}^{k} \beta_i \widehat{y}_{m+i} - \varepsilon_m\Big). \tag{6.41}$$

Suppose that the condition (6.2') holds for the differential equation (6.1) and that the method is G-stable. Then the differences

$$\Delta y_j = \widehat{y}_j - y_j, \qquad \Delta Y_m = (\Delta y_{m+k-1}, \ldots, \Delta y_m)^T$$

satisfy in the norm (6.14)

$$\|\Delta Y_{m+1}\|_G \leq (1+ch\nu)\|\Delta Y_m\|_G + C\big(\|\delta_m\| + \|\varepsilon_m\|\big) \qquad \textit{for } 0 < h\nu \leq \textit{Const.}$$

The constants c, C, and Const depend only on the method, not on the differential equation. If $\nu \leq 0$ we have

$$\|\Delta Y_{m+1}\|_G \leq \|\Delta Y_m\|_G + C\big(\|\delta_m\| + \|\varepsilon_m\|\big) \qquad \textit{for all } h > 0.$$

Proof. We shall make the additional assumption that f is continuously differentiable. A direct proof without this assumption is possible, but leads to a quadratic inequality for $\|\Delta Y_{m+1}\|_G$.

The idea is to subtract (6.6) from (6.41) and to use

$$\begin{aligned} f(x_m+\beta h, \textstyle\sum \beta_i \widehat{y}_{m+i} - \varepsilon_m) - f(x_m+\beta h, \textstyle\sum \beta_i y_{m+i}) \\ = J_m\big(\textstyle\sum \beta_i \Delta y_{m+i} - \varepsilon_m\big) \end{aligned}$$

where

$$J_m = \int_0^1 \frac{\partial f}{\partial y}\Big(x_m + \beta h, t\sum \beta_i y_{m+i} + (1-t)\big(\sum \beta_i \widehat{y}_{m+i} - \varepsilon_m\big)\Big)\, dt.$$

This yields

$$\sum_{i=0}^{k} \alpha_i \Delta y_{m+i} = hJ_m \sum_{i=0}^{k} \beta_i \Delta y_{m+i} + \delta_m - hJ_m \varepsilon_m .$$

Computing Δy_{m+k} from this relation gives

$$\Delta y_{m+k} = \Delta z_{m+k} + (\alpha_k - \beta_k h J_m)^{-1}(\delta_m - hJ_m \varepsilon_m) \tag{6.42}$$

where Δz_{m+k} is defined by

$$\sum_{i=0}^{k} \alpha_i \Delta z_{m+i} = hJ_m \sum_{i=0}^{k} \beta_i \Delta z_{m+i} \tag{6.43}$$

and $\Delta z_j = \Delta y_j$ for $j < m+k$. By our assumption (6.2') the matrix J_m satisfies the one-sided Lipschitz condition $\operatorname{Re}\langle J_m u, u\rangle \le \nu \|u\|^2$ (see Exercise 6 of Sect. I.10). Taking the scalar product of (6.43) with $\sum \beta_i \Delta z_{m+i}$ and using (6.21) we thus obtain in the notation of (6.13)

$$\begin{aligned} \|\Delta Z_{m+1}\|_G^2 - \|\Delta Z_m\|_G^2 &\le c_0 h\nu \big\|\sum \beta_i \Delta z_{m+i}\big\|^2 \\ &\le c_1 h\nu (\|\Delta Z_{m+1}\|_G + \|\Delta Z_m\|_G)^2 \end{aligned}$$

(the second inequality is only valid for $\nu \ge 0$; for negative values of ν we replace ν by 0 in (6.2')). A division by $\|\Delta Z_{m+1}\|_G + \|\Delta Z_m\|_G$ then leads to the estimate

$$\|\Delta Z_{m+1}\|_G \le (1 + ch\nu)\|\Delta Z_m\|_G . \tag{6.44}$$

With the help of von Neumann's theorem (Sect. IV.11) the second term of (6.42) can be bounded by *Const* $(\|\delta_m\| + \|\varepsilon_m\|)$. Inserting this and (6.44) into (6.42) yields the desired estimate. □

The above lemma allows us to derive a convergence result for one-leg methods, which is related to B-convergence for Runge-Kutta methods.

Theorem 6.10. *Consider a G-stable one-leg method with differentiation order $p_D \ge p$ and interpolation order $p_I \ge p-1$. Suppose that the differential equation satisfies the one-sided Lipschitz condition (6.2'). Then there exists $C_0 > 0$ such that for $h\nu \le C_0$*

$$\|y_m - y(x_m)\| \le C \max_{0\le j<k} \|y_j - y(x_j)\| + Mh^p . \tag{6.45}$$

The constant C depends on the method and, for $\nu > 0$, on the length $x_m - x_0$ of the integration interval; the constant M depends in addition on bounds for the p-th and $(p+1)$-th derivative of the exact solution.

Proof. A direct application of Lemma 6.9 to Eq. (6.40) yields the desired error bounds only for $p_I \geq p$. Following Hundsdorfer & Steininger (1991) we therefore introduce $\widehat{y}(x) = y(x) - \delta_I(x)$, so that (6.40) becomes

$$\sum_{i=0}^{k} \alpha_i \widehat{y}(x+ih) - \widehat{\delta}(x) = hf\Big(x+\beta h, \sum_{i=0}^{k} \beta_i \widehat{y}(x+ih) - \widehat{\varepsilon}(x)\Big), \tag{6.46}$$

where

$$\widehat{\delta}(x) = \delta_D(x) - \sum_{i=0}^{k} \alpha_i \delta_I(x+ih), \quad \widehat{\varepsilon}(x) = \delta_I(x) - \sum_{i=0}^{k} \beta_i \delta_I(x+ih). \tag{6.47}$$

Using $\varrho(1) = 0$ and $\sigma(1) = 1$, Taylor expansion of these functions yields

$$\|\widehat{\delta}(x)\| + \|\widehat{\varepsilon}(x)\| \leq C_1 h^p \int_x^{x+kh} \|y^{(p+1)}(\zeta)\| d\zeta.$$

We thus can apply Lemma 6.9 to (6.46) and obtain

$$\|\Delta Y_{m+1}\|_G \leq (1 + ch\nu)\|\Delta Y_m\|_G + M_1 h^{p+1}$$

where $\Delta y_j = \widehat{y}(x_j) - y_j$. Using $(1+ch\nu)^j \leq \exp(c\nu(x_j - x_0))$, a simple induction argument gives

$$\|\Delta Y_{m+1}\|_G \leq C\|\Delta Y_0\|_G + Mh^p.$$

The statement now follows from the equivalence of norms

$$d_0\|\Delta Y_0\|_G \leq \max_{0\leq j<k} \|\Delta y_j\| \leq d_1\|\Delta Y_0\|_G,$$

from the estimate $\|y_m - y(x_m)\| \leq \|y_m - \widehat{y}(x_m)\| + \|\delta_I(x_m)\|$, and from the fact that $\|\delta_I(x_m)\| = \mathcal{O}(h^p)$. □

Convergence of A-Stable Multistep Methods

An interesting equivalence relation between one-leg and linear multistep methods is presented in Dahlquist (1975) (see Exercise 3). This allows us to translate the above convergence result into a corresponding one for multistep methods (Hundsdorfer & Steininger 1991). A different and more direct approach will be presented in Sect. V.8 (Theorem 8.2).

We consider the linear multistep method

$$\sum_{i=0}^{k} \alpha_i \widehat{y}_{m+i} = h \sum_{i=0}^{k} \beta_i f(\widehat{x}_{m+i}, \widehat{y}_{m+i}). \tag{6.48}$$

We let $x_m = \widehat{x}_m - \beta h$, so that $\sum_{i=0}^k \beta_i x_{m+i} = \widehat{x}_m$, and, in view of Eq. (6.54), we define $\{y_0, y_1, \ldots, y_{2k-1}\}$ as the solution of the linear system

$$\sum_{i=0}^{k} \beta_i y_{j+i} = \widehat{y}_j, \qquad \sum_{i=0}^{k} \alpha_i y_{j+i} = hf(\widehat{x}_j, \widehat{y}_j), \qquad j = 0, \ldots, k-1. \tag{6.49}$$

This system is uniquely solvable, because the polynomials $\varrho(\zeta)$ and $\sigma(\zeta)$ are relatively prime. With these starting values we define $\{y_m\}$ as solution of the one-leg relation (for $m \geq k$)

$$\sum_{i=0}^{k} \alpha_i y_{m+i} = hf\Big(\sum_{i=0}^{k} \beta_i x_{m+i}, \sum_{i=0}^{k} \beta_i y_{m+i}\Big). \tag{6.50}$$

By the second relation of (6.49), Eq. (6.50) holds for all $m \geq 0$. Consequently (Exercise 3a) the expression $\sum_{i=0}^k \beta_i y_{m+i}$ is a solution of the multistep method (6.48). Because of (6.49) and the uniqueness of the numerical solution this gives

$$\sum_{i=0}^{k} \beta_i y_{m+i} = \widehat{y}_m \qquad \text{for all} \quad m \geq 0. \tag{6.51}$$

This relation leads to a proof of the following result.

Theorem 6.11. *Consider an A-stable linear multistep method of order p. Suppose the differential equation satisfies (6.2'). Then there exists $C_0 > 0$ such that for $h\nu \leq C_0$,*

$$\|\widehat{y}_m - y(\widehat{x}_m)\| \leq C\big(\max_{0\leq j<k} \|\widehat{y}_j - y(\widehat{x}_j)\| + h \max_{0\leq j<k} \|f(\widehat{x}_j, \widehat{y}_j) - y'(\widehat{x}_j)\|\big) + Mh^p.$$

The constants C and M are as in Theorem 6.10.

Proof. By Theorem 6.7, A-stability implies G-stability of the corresponding one-leg method. Further, Taylor expansion of (6.37) and (6.38) shows that $p_D \geq \min(p, 2)$ and $p_I \geq 1$. Since $p \leq 2$ by Dahlquist's second barrier, all assumptions of Theorem 6.10 are verified. The one-leg solution $\{y_m\}$ thus satisfies (6.45). In order to estimate $\|y_j - y(x_j)\|$ for $j < k$ we subtract the definitions of $\delta_D(x)$ and $\delta_I(x)$ from (6.48) and obtain

$$\sum_{i=0}^{k} \beta_i \big(y_{j+i} - y(x_{j+i})\big) = \widehat{y}_j - y(\widehat{x}_j) - \delta_I(x_j)$$

$$\sum_{i=0}^{k} \alpha_i \big(y_{j+i} - y(x_{j+i})\big) = hf(\widehat{x}_j, \widehat{y}_j) - hy'(\widehat{x}_j) - \delta_D(x_j).$$

Solving these relations for $y_j - y(x_j)$ yields

$$\max_{0\leq j<k} \|y_j - y(x_j)\| \leq C_0\Big(\max_{0\leq j<k} \|\widehat{y}_j - y(\widehat{x}_j)\| + h \max_{0\leq j<k} \|f(\widehat{x}_j, \widehat{y}_j) - y'(\widehat{x}_j)\|\Big) + M_0 h^p.$$

This proves the statement, because by (6.51)

$$\|\widehat{y}_m - y(\widehat{x}_m)\| \le \sum_{i=0}^{k} |\beta_i| \, \|y_{m+i} - y(x_{m+i})\| + \|\delta_I(x_m)\|. \qquad \square$$

Exercises

1. a) Prove that the one-leg method (6.6) satisfies (6.39) iff

$$\sum_{i=0}^{k} \alpha_i i^q = q\beta^{q-1} \qquad \text{for } q = 0, 1, \ldots, p_D \tag{6.52}$$

$$\sum_{i=0}^{k} \beta_i i^q = \beta^q \qquad \text{for } q = 0, \ldots, p_I. \tag{6.53}$$

   Compare this result with Theorem III.2.4.

   b) Compute the orders p_D and p_I for the Adams methods.

2. a) Show that the one-leg method (6.6) can be written in the form of a general linear method (Sect. III.8).

   b) Prove that the order of convergence p of this method is given by

$$p = \min(p_D, p_I + 1)$$

   with p_D, p_I defined in (6.39).

   c) The order of a one-leg method is never larger than the order of the corresponding multistep method.

3. (Dahlquist 1975).

   a) Let $\{y_m\}$ and $\{x_m = x_0 + mh\}$ satisfy the (one-leg) difference relation (6.6); then

$$\widehat{y}_m = \sum_{j=0}^{k} \beta_j y_{m+j}, \qquad \widehat{x}_m = \sum_{j=0}^{k} \beta_j x_{m+j} \tag{6.54}$$

   satisfy the (linear multistep) difference relation (6.3).

   b) Conversely, let

$$P(\zeta) = \sum_{j=0}^{k-1} a_j \zeta^j, \qquad Q(\zeta) = \sum_{j=0}^{k-1} b_j \zeta^j$$

be such that $P(\zeta)\sigma(\zeta) - Q(\zeta)\varrho(\zeta) = \zeta^l$ for some integer l $(0 \le l \le k)$, then

$$y_{m+l} = \sum_{j=0}^{k-1} a_j \widehat{y}_{m+j} - h \sum_{j=0}^{k-1} b_j f(\widehat{x}_{m+j}, \widehat{y}_{m+j})$$

$$x_{m+l} = \sum_{j=0}^{k-1} a_j \widehat{x}_{m+j} - h \sum_{j=0}^{k-1} b_j$$

satisfy (6.6), whenever $\{\widehat{y}_m\}$ and $\{\widehat{x}_m\}$ are a solution of (6.3).

Hint for a). Multiply (6.6) by β_j, replace m by $m+j$, sum from $j=0$ to $j=k$, and interchange the summations.

4. *One-leg collocation* methods (Dahlquist 1983).

   a) For a given β there exists a unique k-step one-leg method with $p_D = k$ and $p_I = k$.

   b) This one-leg method is of order $p = k+1$ iff

$$\sum_{i=0}^{k} \frac{1}{(\beta - i)} = 0.$$

   c) Discuss numerically the zero-stability of these methods.

5. (proposed by M. Crouzeix). a) Let $R(z) = P(z)/Q(z)$ be an irreducible rational function where $\deg P \le k$, $\deg Q \le k$. Show that $R(z)$ is A-stable, if and only if there exist polynomials $\alpha_i(z)$, $\beta(z)$ with real coefficients and with $\deg \alpha_i \le k-1$, $\deg \beta \le k$, such that

$$Q(z)Q(w) - P(z)P(w) = -(z+w) \sum_{i=1}^{k} \alpha_i(z)\alpha_i(w) + \beta(z)\beta(w). \qquad (6.55)$$

   b) Use this characterization to give a new proof of von Neumann's theorem (Corollary IV.11.3).

   Hint. Part (a) can be proved along the lines of the proofs of Theorem 6.7 and Lemma 6.8. Remark that (6.55) reduces to the E-polynomial (IV.3.8) if $z = iy$ and $w = -iy$. For the proof of (b), deduce from (6.55) the identity

$$\|Q(A)u\|^2 - \|P(A)u\|^2 = -\sum_{i=1}^{k} \operatorname{Re}\langle \alpha_i(A)u, A\alpha_i(A)u\rangle + \|\beta(A)u\|^2.$$

V.7 Convergence for Linear Problems

Theorems 6.10 and 6.11 give satisfactory convergence results for G-stable one-leg methods and A-stable multistep methods. But there are only few such methods and their highest order is two (Theorem 1.4). It is therefore interesting to relax the requirement of A-stability and to investigate higher-order multistep and one-leg methods. This section is devoted to linear stiff problems, while Sect. V.8 will treat non-linear problems.

We shall describe two different approaches for convergence results. One is with the help of the discrete variation of constants formula and shall be given at the end of this section (see Lemma 7.9 and Theorem 7.10 below). The other possibility is based on a formulation as a one-step method and on the use of the Kreiss matrix theorem.

Difference Equations for the Global Error

Most of the difficulties can already be seen by studying the one-dimensional problem of Prothero and Robinson

$$y' = \lambda y + g(x), \qquad y(x_0) = y_0. \tag{7.1}$$

We assume $\operatorname{Re}\lambda \le 0$ and the solution $y(x)$ to be smooth in the sense that sufficiently many derivatives are bounded independently of the stiffness parameter λ.

Applying a *linear multistep method* to (7.1) yields

$$\sum_{i=0}^{k} \alpha_i y_{m+i} = h\lambda \sum_{i=0}^{k} \beta_i y_{m+i} + h \sum_{i=0}^{k} \beta_i g(x_{m+i}). \tag{7.2}$$

The global error

$$e_m = y_m - y(x_m) \tag{7.3}$$

is seen to satisfy the difference relation

$$\sum_{i=0}^{k} (\alpha_i - h\lambda\beta_i) e_{m+i} = -\delta_{LM}(x_m) \tag{7.4}$$

with

$$\delta_{LM}(x) = \sum_{i=0}^{k} \alpha_i y(x+ih) - h \sum_{i=0}^{k} \beta_i y'(x+ih) \tag{7.5}$$

(to be compared with Formula III.2.3). We observe that the right-hand side of (7.4) is independent of the stiffness (i.e., of λ). Further, if the classical order of the method is p, then $\delta_{LM}(x) = \mathcal{O}(h^{p+1})$.

If we apply the method in its *one-leg* version, we obtain

$$\sum_{i=0}^{k} \alpha_i y_{m+i} = h\lambda \sum_{i=0}^{k} \beta_i y_{m+i} + hg(x_m + \beta h), \tag{7.6}$$

where $\sum \beta_i = 1$ and $\sum \beta_i i = \beta$. In this case the global error $e_m = y_m - y(x_m)$ satisfies

$$\sum_{i=0}^{k} (\alpha_i - h\lambda\beta_i) e_{m+i} = h\lambda\delta_I(x_m) - \delta_D(x_m) \tag{7.7}$$

with $\delta_D(x)$ and $\delta_I(x)$ defined in (6.37) and (6.38), respectively. Unless $\delta_I(x) = 0$ (which is the case for the BDF methods), Eq. (7.7) is disappointing, because its right-hand side becomes large in the stiff case $(h\lambda \to \infty)$.

In order to overcome this difficulty, Dahlquist (1983) proposes that one consider instead of $e_m = y_m - y(x_m)$ the quantities

$$e_m^* = \sum_{i=0}^{k} \beta_i y_{m+i} - y(x_m + \beta h) \tag{7.8}$$

("... a more adequate measure of the global error than the customary one ...", Dahlquist 1983). Replacing m by $m+j$ in (7.6), multiplying by β_j and summing up gives the error formula

$$\sum_{i=0}^{k} (\alpha_i - h\lambda\beta_i) e_{m+i}^* = -\delta_{LM}(x_m + \beta h) \tag{7.9}$$

with $\delta_{LM}(x)$ of (7.5). This difference relation now has the same strength as (7.4).

It has been pointed out by Hundsdorfer & Steininger (1991) that we usually get better error estimates for one-leg methods by considering $\widehat{e}_m = e_m + \delta_I(x_m)$. We then have

$$\sum_{i=0}^{k} (\alpha_i - h\lambda\beta_i) \widehat{e}_{m+i} = h\lambda\widehat{\varepsilon}(x_m) - \widehat{\delta}(x_m) \tag{7.10}$$

with $\widehat{\varepsilon}(x)$ and $\widehat{\delta}(x)$ given by (6.47). Observe that $\widehat{\varepsilon}(x) = \mathcal{O}(h^{p_I+2})$ and $\widehat{\delta}(x) = \mathcal{O}(h^{\min(p_D+1, p_I+2)})$.

Formulation as a One-Step Method. The error relations (7.4), (7.7), (7.9), and (7.10) are all of the form

$$\sum_{i=0}^{k}(\alpha_i - h\lambda\beta_i)e_{m+i} = \delta_h(x_m). \tag{7.11}$$

In order to estimate e_m it is convenient to introduce, as in Sect. III.4, the vector

$$E_m = (e_{m+k-1}, \dots, e_{m+1}, e_m)^T, \tag{7.12}$$

the companion matrix

$$C(\mu) = \begin{pmatrix} c_{k-1}(\mu) & \dots & c_1(\mu) & c_0(\mu) \\ 1 & & & \\ & \ddots & & \\ & & 1 & 0 \end{pmatrix}, \qquad c_j(\mu) = -\frac{\alpha_j - \mu\beta_j}{\alpha_k - \mu\beta_k} \tag{7.13}$$

and

$$\Delta_m = \big(\delta_h(x_m)/(\alpha_k - \mu\beta_k), 0, \dots, 0\big)^T, \qquad \mu = h\lambda. \tag{7.14}$$

Then, Eq. (7.11) becomes

$$E_{m+1} = C(h\lambda)E_m + \Delta_m, \tag{7.15}$$

which leads to

$$E_{m+1} = C(h\lambda)^{m+1}E_0 + \sum_{j=0}^{m} C(h\lambda)^{m-j}\Delta_j. \tag{7.16}$$

To estimate E_{m+1} we have to bound the powers of $C(h\lambda)$ uniformly in $h\lambda$. This is the subject of the next subsection.

The Kreiss Matrix Theorem

> Als Fakultätsopponent für meine Stockholmer Dissertation brachte Dr. G. Dahlquist die Frage der Stabilitätsdefinition zur Sprache.
> (H.-O. Kreiss 1962)

The following Theorem of Kreiss (1962) is a powerful tool for proving uniform power boundedness of an arbitrary family of matrices.

Theorem 7.1 (Kreiss 1962). *Let $\mathcal{F}$ be a family of $k \times k$ matrices A. Then the "power condition"*

$$\|A^n\| \le M \qquad \textit{for} \quad n = 0, 1, 2, \dots \quad \textit{and} \quad A \in \mathcal{F} \tag{P}$$

is equivalent to the "resolvent condition"

$$\|(A - zI)^{-1}\| \le \frac{C}{|z| - 1} \qquad \textit{for} \quad |z| > 1 \quad \textit{and} \quad A \in \mathcal{F}. \tag{R}$$

Remark. The difficult step is to prove that (R) implies (P). Several mathematicians contributed to a better understanding of this result (Richtmyer & Morton 1967, Tadmor 1981). LeVeque & Trefethen (1984) have given a marvellous version of the proof; the best we can do is to copy it nearly word for word:

Proof. Necessity. If (P) is true, the eigenvalues of A lie within the closed unit disk and therefore $(A-zI)^{-1}$ exists for $|z|>1$. Moreover,

$$\|(A-zI)^{-1}\| = \|\sum_{n=0}^{\infty} A^n z^{-n-1}\| \le M \sum_{n=0}^{\infty} |z|^{-n-1} = \frac{M}{|z|-1}, \tag{7.17}$$

so that (R) holds with $C=M$.

Sufficiency. Assume condition (R), so that all eigenvalues of A lie inside the closed unit disk. The matrix A^n can then be written in terms of the resolvent by means of a Cauchy integral (see Exercise 1)

$$A^n = \frac{1}{2\pi i}\int_\Gamma z^n (zI-A)^{-1}\, dz, \tag{7.18}$$

where the contour of integration is, for example, a circle of radius $\varrho>1$ centred at the origin. Let u and v be arbitrary unit vectors, i.e., $\|u\|=\|v\|=1$. Then,

$$v^*A^n u = \frac{1}{2\pi i}\int_\Gamma z^n q(z)dz \qquad \text{with} \qquad q(z)=v^*(zI-A)^{-1}u.$$

Integration by parts gives

$$v^*A^n u = \frac{-1}{2\pi i(n+1)}\int_\Gamma z^{n+1} q'(z)dz.$$

Now fix as contour of integration the circle of radius $\varrho = 1+1/(n+1)$. On this path one has $|z^{n+1}|\le e$, and therefore

$$|v^*A^n u| \le \frac{e}{2\pi(n+1)}\int_\Gamma |q'(z)|\,|dz|. \tag{7.19}$$

By Cramer's rule, $q(z)$ is a rational function of degree k. Applying Lemma 7.2 below, the integral in (7.19) is bounded by $4\pi k$ times the supremum of $|q(z)|$ on Γ, and by (R) this supremum is at most $(n+1)C$. Hence

$$|v^*A^n u| \le 2ekC.$$

Since $\|A^n\|$ is the supremum of $|v^*A^n u|$ over all unit vectors u and v, this proves the estimate (P) with $M=2ekC$. □

The above proof used the following lemma, which relates the arc length of a rational function on a circle to its maximum value. For the case of a polynomial of degree k the result is a corollary of Bernstein's inequality $\sup_{|z|=1}|q'(z)| \le k\sup_{|z|=1}|q(z)|$ (see e.g., Marden 1966).

Lemma 7.2. *Let $q(z)=p(z)/r(z)$ be a rational function with* $\deg p \le k$, $\deg r \le k$ *and suppose that no poles lie on the circle* $\Gamma : |z| = \varrho$. *Then*

$$\int_\Gamma |q'(z)|\,|dz| \le 4\pi k \sup_{|z|=\varrho} |q(z)|. \tag{7.20}$$

Proof. Replacing $q(z)$ by $q(\varrho z)$ we may assume without loss of generality that $\varrho = 1$. With the parametrization e^{it} of Γ we introduce

$$\gamma(t) = q(e^{it}), \qquad \gamma'(t) = ie^{it}q'(e^{it})$$

so that

$$\gamma'(t) = |q'(e^{it})| \cdot e^{ig(t)} \qquad \text{with} \quad g(t) = \arg(\gamma'(t)).$$

Integration by parts now yields

$$\int_\Gamma |q'(z)|\,|dz| = \int_0^{2\pi} |q'(e^{it})| dt = \int_0^{2\pi} \gamma'(t) e^{-ig(t)}\,dt$$
$$= i\int_0^{2\pi} \gamma(t) g'(t) e^{-ig(t)}\,dt \le \sup|\gamma(t)| \cdot \int_0^{2\pi} |g'(t)| dt.$$

It remains to prove that the total variation of g, i.e., $\text{TV}[g] = \int_0^{2\pi} |g'(t)|dt$, can be bounded by $4\pi k$. To prove this, note that $zq'(z)$ is a rational function of degree $(2k, 2k)$ and can be written as a product

$$zq'(z) = \prod_{j=1}^{2k} \frac{a_j z + b_j}{c_j z + d_j}.$$

This implies for $z = e^{it}$

$$g(t) = \arg(izq'(z)) = \frac{\pi}{2} + \sum_{j=1}^{2k} \arg\Big(\frac{a_j z + b_j}{c_j z + d_j}\Big).$$

Since the Möbius transformation $(az+b)/(cz+d)$ maps the unit circle to some other circle, the total variation of $\arg((az+b)/(cz+d))$ is at most 2π. Consequently,

$$\text{TV}[g] \le \sum_{j=1}^{2k} \text{TV}\Big[\arg\Big(\frac{a_j z + b_j}{c_j z + d_j}\Big)\Big] \le 4\pi k. \qquad \square$$

Remark. It has been conjectured by LeVeque & Trefethen (1984) that the bound (7.20) is valid with a factor 2π instead of 4π. This conjecture has been proved to be true by Spijker (1991).

Some Applications of the Kreiss Matrix Theorem

Following Dahlquist, Mingyou & LeVeque (1983) we now obtain some results on the uniform power boundedness of the matrix $C(\mu)$, defined in (7.13), with the help of the Kreiss matrix theorem. Similar results were found independently by Crouzeix & Raviart (1980) and Gekeler (1979, 1984).

Lemma 7.3. *Let $S \subset \overline{\mathbb{C}}$ denote the stability region of a method (ϱ, σ). If S is closed in $\overline{\mathbb{C}}$, then there exists a constant M such that*

$$\|C(\mu)^n\| \le M \qquad \textit{for} \quad \mu \in S \qquad \textit{and} \quad n = 0, 1, 2, \ldots .$$

Proof. Because of Theorem 7.1 it is sufficient to prove that

$$\|(C(\mu) - zI)^{-1}\| \le \frac{C}{|z| - 1} \qquad \text{for} \quad \mu \in S \quad \text{and} \quad |z| > 1.$$

To show this, we make use of the inequality (Kato (1960), see Exercise 2)

$$\|(C(\mu) - zI)^{-1}\| \le \frac{(\|C(\mu)\| + |z|)^{k-1}}{|\det(C(\mu) - zI)|}.$$

$\|C(\mu)\|$ is uniformly bounded for $\mu \in S$. Therefore it suffices to show that

$$\varphi(\mu) = \inf_{|z|>1} \frac{|\det(C(\mu) - zI)|}{|z|^{k-1}(|z| - 1)} \tag{7.21}$$

is bounded away from zero for all $\mu \in S$. For $|z| \to \infty$ the expression in (7.21) tends to 1 and so poses no problem. Further, observe that

$$|\det(C(\mu) - zI)| = \Big| \prod_{j=1}^{k} (z - \zeta_j(\mu)) \Big|, \tag{7.22}$$

where $\zeta_j(\mu)$ are the eigenvalues of $C(\mu)$, i.e., the roots of

$$\sum_{i=0}^{k} (\alpha_i - \mu\beta_i)\zeta^i = 0. \tag{7.23}$$

By definition of the stability region S, the values $\zeta_j(\mu)$ lie, for $\mu \in S$, inside the closed unit disc and those on the unit circle are well separated. Therefore, for fixed $\mu_0 \in S$, only one of the $\zeta_j(\mu_0)$ can be close to a z with $|z| > 1$. The corresponding factor in (7.22) will be minorized by $|z| - 1$, the other factors are bounded away from zero. By continuity of the $\zeta_j(\mu)$, the same holds for all $\mu \in S$ in a sufficiently small neighbourhood $V(\mu_0)$ of μ_0. Hence $\varphi(\mu) \ge a > 0$ for $\mu \in V(\mu_0) \cap S$. Since S is closed (compact in $\overline{\mathbb{C}}$) it is covered by a finite number of $V(\mu_0)$. Consequently $\varphi(\mu) \ge a > 0$ for all $\mu \in S$, which completes the proof of the theorem. □

Remark. The hypothesis "S is closed in $\overline{\mathbb{C}}$" is usually satisfied. For methods which do *not* satisfy this hypothesis (see e.g., Exercise 2 of Sect. V.1 or Dahlquist, Mingyou & LeVeque (1981)) the above lemma remains valid on closed subsets $D \subset S \subset \overline{\mathbb{C}}$.

The estimate of this lemma can be improved, if we consider closed sets D lying in the interior of S.

Lemma 7.4. *Let S be the stability region of a method (ϱ, σ). If $D \subset \operatorname{Int} S$ is closed in $\overline{\mathbb{C}}$, then there exist constants M and κ $(0 < \kappa < 1)$ such that*

$$\|C(\mu)^n\| \le M\kappa^n \qquad \textit{for} \quad \mu \in D \quad \textit{and} \quad n = 0, 1, 2, \ldots .$$

Proof. If μ lies in the interior of S, all roots of (7.23) satisfy $|\zeta_j(\mu)| < 1$ (maximum principle). Since D is closed, this implies the existence of $\varepsilon > 0$ such that

$$D \subset S_\varepsilon = \{\mu \in \overline{\mathbb{C}}\,;\ |\zeta_j(\mu)| \le 1 - 2\varepsilon,\ j = 1, \ldots, k\}.$$

We now consider $R(\mu) = \kappa^{-1} C(\mu)$ with $\kappa = 1 - \varepsilon$. The eigenvalues of $R(\mu)$ satisfy $|\kappa^{-1}\zeta_j(\mu)| \le (1-2\varepsilon)/(1-\varepsilon) < 1 - \varepsilon$ for $\mu \in S_\varepsilon$. As in the proof of Lemma 7.3 (more easily, because $R(\mu)$ has no eigenvalues of modulus 1) we conclude that $R(\mu)$ is uniformly power bounded for $\mu \in S_\varepsilon$. This implies the statement. □

Since the origin is never in the interior of S, we add the following estimate for its neighbourhood:

Lemma 7.5. *Suppose that the method (ϱ, σ) is consistent and strictly stable (see Sect. III.9, Assumption A1). Then there exists a neighbourhood V of 0 and constants M and a such that*

$$\|C(\mu)^n\| \le M e^{n(\operatorname{Re}\mu + a|\mu|^2)} \qquad \textit{for} \quad \mu \in V \quad \textit{and} \quad n = 0, 1, 2, \ldots .$$

Proof. Since the method is strictly stable there exists a compact neighbourhood V of 0, in which $|\zeta_j(\mu)| < |\zeta_1(\mu)|$ for $j = 2, \ldots, k$ ($\zeta_j(\mu)$ are the roots of (7.23)). The matrix $R(\mu) = \zeta_1(\mu)^{-1} C(\mu)$ then has a simple eigenvalue 1 and all other eigenvalues are strictly smaller than 1. As in the proof of Lemma 7.3 we obtain $\|R(\mu)^n\| \le M$ and consequently $\|C(\mu)^n\| \le M |\zeta_1(\mu)|^n$ for $\mu \in V$. The stated estimate now follows from $\zeta_1(\mu) = e^\mu + \mathcal{O}(\mu^2)$. □

Global Error for Prothero and Robinson Problem

The above lemmas permit us to continue our analysis of Eq. (7.16). Whenever we consider λ and h such that their product λh varies in a closed subset of S, it follows that

$$\|E_{m+1}\| \le M\Big(\|E_0\| + \sum_{j=0}^{m} \|\Delta_j\|\Big) \tag{7.24}$$

(Lemma 7.3). If $h\lambda$ varies in a closed subset of the interior of S, we have the better estimate

$$\|E_{m+1}\| \le M\Big(\kappa^{m+1}\|E_0\| + \sum_{j=0}^{m} \kappa^{m-j}\|\Delta_j\|\Big) \qquad \text{with some } \kappa < 1 \tag{7.25}$$

(Lemma 7.4). The resulting asymptotic estimates for the global errors $e_m = y_m - y(x_m)$ for $mh \le Const$ are presented in Table 7.1 (p denotes the classical order, p_D the differentiation order and p_I the interpolation order of Sect. V.6). We assume that the initial values are exact and that simultaneously $h\lambda \to \infty$ and $h \to 0$. This is the most interesting situation because any reasonable method for stiff problems should integrate the equation with step sizes h such that $h\lambda$ is large. We distinguish two cases:

(A) the half-ray $\{h\lambda\,;\ h>0,\ |h\lambda| \ge c\} \cup \{\infty\}$ lies in S (Lemma 7.3 is applicable, i.e., Eq. (7.24)).

(B) ∞ is an interior point of S (estimate (7.25) is applicable; the global error $\|E_m\|$ is essentially equal to the last term in the sum of (7.25)).

Table 7.1. Error for (7.1) when $h\lambda \to \infty$ and $h \to 0$

Method	error	(A)	(B)
multistep	e_m	$\mathcal{O}(\lvert\lambda\rvert^{-1}h^{p-1})$	$\mathcal{O}(\lvert\lambda\rvert^{-1}h^{p})$
one-leg	e_m	$\mathcal{O}(h^{p_I+1} + \lvert\lambda\rvert^{-1}h^{p_D-1})$	$\mathcal{O}(h^{p_I+1} + \lvert\lambda\rvert^{-1}h^{p_D})$

We remark that the global error of the multistep method contains a factor $|\lambda|^{-1}$, so that the error decreases if $|\lambda|$ increases ("the stiffer the better"). The estimate in case (A) for one-leg methods is obtained by the use of Recursion (7.10).

Convergence for Linear Systems with Constant Coefficients

The extension of the above results to linear systems

$$y' = Ay + g(x), \qquad y(x_0) = y_0 \tag{7.26}$$

is straightforward, if we assume that the matrix A is diagonalizable. The following results have been derived by Crouzeix & Raviart (1980).

Theorem 7.6. *Suppose that the multistep method* (ϱ, σ) *is of order* p, $A(\alpha)$*-stable and stable at infinity. If the matrix* A *of (7.26) is diagonalizable (i.e.,* $T^{-1}AT = \operatorname{diag}(\lambda_1, \ldots, \lambda_n)$*) with eigenvalues satisfying*

$$|\arg(-\lambda_i)| \le \alpha \qquad \textit{for} \quad i = 1, \ldots, n,$$

then there exists a constant M *(depending only on the method) such that for all* $h > 0$ *the global error satisfies*

$$\|y(x_m) - y_m\| \le M \cdot \|T\| \cdot \|T^{-1}\| \Big(\max_{0 \le j < k} \|y(x_j) - y_j\| + h^p \int_{x_0}^{x_m} \|y^{(p+1)}(\xi)\| d\xi \Big).$$

Proof. The transformation $y = Tz$ decouples the system (7.26) into n scalar equations

$$z_i' = \lambda_i z_i + (T^{-1}g)_i(x). \tag{7.27}$$

Since this transformation leaves the numerical solution invariant, it suffices to consider Eq. (7.27). Lemma 7.3 yields the power boundedness

$$\|C(h\lambda_i)^m\| \le M_0 \qquad \text{for} \quad h > 0, \quad i = 1, \ldots, n \quad \text{and} \quad m \ge 0. \tag{7.28}$$

The discretization error $\delta_{LM}(x)$ (Eq. (7.5)) can be written as

$$\delta_{LM}(x) = h^{p+1} \int_0^k K_p(s) z_i^{(p+1)}(x + sh) ds, \tag{7.29}$$

where $K_p(s)$ is the Peano-kernel of the multistep method (Theorem III.2.8). By $A(\alpha)$-stability we have $\alpha_k \cdot \beta_k > 0$, so that $|\alpha_k - h\lambda_i \beta_k|^{-1} \le |\alpha_k|^{-1}$. This together with (7.29) implies that

$$\|\Delta_j\| \le Ch^p \int_{x_j}^{x_{j+k}} |z_i^{(p+1)}(\xi)| d\xi, \tag{7.30}$$

where C depends only on the method. The estimates (7.28) and (7.30) inserted into (7.16) yield a bound for the global error of (7.27), which, by backsubstitution into the original variables, proves the statement. □

Because of its exponentially decaying term, the following estimate is especially useful in the case when large time intervals are considered (or when the starting values do not lie on the exact solution).

Theorem 7.7. *Let the multistep method* (ϱ, σ) *be of order* $p \geq 1$, $A(\alpha)$*-stable and strictly stable at zero and at infinity (i.e.,* $\sigma(\zeta) = 0$ *implies* $|\zeta| < 1$*). If the matrix* A *of (7.26) is diagonalizable (*$T^{-1}AT = \operatorname{diag}(\lambda_1, \ldots, \lambda_n)$*) with eigenvalues* λ_i *satisfying*

$$|\arg(-\lambda_i)| \leq \gamma < \alpha, \qquad \operatorname{Re}\lambda_i \leq -\widehat{\lambda} < 0$$

then, for given $h_0 > 0$*, there exist constants* M *and* $\nu > 0$ *such that for* $0 < h \leq h_0$

$$\|y(x_m) - y_m\| \leq M \cdot \|T\| \cdot \|T^{-1}\| \cdot \Big(e^{-\nu(x_m - x_0)} \cdot \max_{0 \leq j < k} \|y(x_j) - y_j\| + h^p \int_{x_0}^{x_m} e^{-\nu(x_m - \xi)} \|y^{(p+1)}(\xi)\| d\xi \Big).$$

Remark. The constants M and ν may depend on $\gamma, \widehat{\lambda}, h_0$ and on the method, but they are independent of the eigenvalues λ_i and of the length $x_m - x_0$ of the integration interval.

Proof. By Lemma 7.5 there exists an $r > 0$ such that

$$\|C(h\lambda_i)^m\| \leq M_0 e^{-mh\widehat{\lambda}/2} \qquad \text{for} \quad |h\lambda_i| \leq r \tag{7.31}$$

(observe that $|\mu| \leq \mathit{Const} \cdot |\operatorname{Re}\mu|$, if $|\arg(-\mu)| \leq \gamma < \pi/2$). Since

$$D = \{\mu\,;\ |\arg(-\mu)| \leq \gamma,\ |\mu| \geq r\} \cup \{\infty\}$$

lies in the interior of the stability region S, it follows from Lemma 7.4 that

$$\|C(h\lambda_i)^m\| \leq M_1 \varrho^m \qquad \text{for} \quad |h\lambda_i| \geq r \tag{7.32}$$

with some $\varrho < 1$. Combining the estimates (7.31) and (7.32) we get

$$\|C(h\lambda_i)^m\| \leq M e^{-mh\nu} \qquad \text{for} \quad 0 < h \leq h_0, \tag{7.33}$$

where $M = \max(M_0, M_1)$ and $\nu = \min(\widehat{\lambda}/2, -\ln\varrho/h_0)$. Using (7.33) instead of (7.28) and $mh = x_m - x_0$, the statement now follows as in the proof of Theorem 7.6. □

Matrix Valued Theorem of von Neumann

An interesting contractivity result is obtained by the following matrix valued version of a theorem of von Neumann (Theorem IV.11.2).

We consider the Euclidean scalar product $\langle \cdot, \cdot \rangle$ on $\mathbb{R}^n$, the norm $\|\cdot\|_G$ on $\mathbb{R}^k$ which is defined by a symmetric, positiv definite matrix G, and

$$|||u|||_G = \sqrt{\sum_{i,j=1}^{k} g_{ij} \langle u_i, u_j \rangle} \qquad \text{for } u = (u_1, \ldots, u_k)^T \in \mathbb{R}^{nk}. \tag{7.34}$$

The corresponding operator norms are denoted by the same symbols.

Theorem 7.8 (O. Nevanlinna 1985). *Let* $C(\mu)=(c_{ij}(\mu))_{i,j=1}^{k}$ *be a matrix whose elements are rational functions of* μ. *If*

$$\|C(\mu)\|_G \le 1 \qquad \text{for} \quad \operatorname{Re}\mu \le 0, \tag{7.35}$$

then

$$|||C(A)|||_G \le 1 \tag{7.36}$$

for all matrices A *such that*

$$\operatorname{Re}\langle y, Ay\rangle \le 0 \qquad \text{for} \quad y\in\mathbb{C}^n. \tag{7.37}$$

Remark. If $C(\mu)$ is the companion matrix of a G-stable method (ϱ,σ), the result follows from Theorem 6.7 and Exercise 3 below ("It would be interesting to have a more operator-theoretical proof of this." Dahlquist & Söderlind 1982).

Proof. This is a straight-forward extension of Crouzeix's proof of Theorem IV.11.2. We first suppose that A is normal, so that $A=QDQ^*$ with a unitary matrix Q and a diagonal matrix $D=\operatorname{diag}(\lambda_1,\ldots,\lambda_n)$. In this case we have

$$|||C(A)|||_G = |||(I\otimes Q)C(D)(I\otimes Q^*)|||_G = |||C(D)|||_G. \tag{7.38}$$

With the permutation matrix $P=(I\otimes e_1,\ldots,I\otimes e_n)$ (where I is the k-dimensional identity and e_j is the n-dimensional j-th unit vector) the matrix $C(D)$ is transformed to block-diagonal form according to

$$P^*C(D)P = \text{blockdiag}\,(C(\lambda_1),\ldots,C(\lambda_n)).$$

We further have $P^*(G\otimes I)P = I\otimes G$. This implies that

$$P^*C(D)^*(G\otimes I)C(D)P = \text{blockdiag}\,(C(\lambda_1)^*GC(\lambda_1),\ldots)$$

and hence also

$$|||C(D)|||_G = \max_{i=1,\ldots,n}\|C(\lambda_i)\|_G. \tag{7.39}$$

The statement now follows from (7.38) and (7.39), because $\operatorname{Re}\lambda_i\le 0$ by (7.37).

For a general A we consider $A(\omega)=\frac{\omega}{2}(A+A^*)+\frac{1}{2}(A-A^*)$ and define the rational function

$$\varphi(\omega)=\langle u, C(A(\omega))v\rangle_G = u^*(G\otimes I)C(A(\omega))v.$$

The statement of the theorem can then be deduced exactly as in the proof of Theorem IV.11.2. □

This theorem can be used to derive convergence results for differential equations (7.26) with A satisfying (7.37). Indeed, if the method (ϱ,σ) is A-stable, the companion matrix (7.13) satisfies $\|C(\mu)\|_G\le 1$ for $\operatorname{Re}\mu\le 0$ in some suitable norm (Exercise 3). The above theorem then implies $|||C(hA)|||_G \le 1$ and Formula (7.16) with λ replaced by A yields the estimate

$$|||E_{m+1}|||_G \le |||E_0|||_G + \sum_{j=0}^{m} |||\Delta_j|||_G. \tag{7.40}$$

This proves convergence, because Δ_j can be bounded as in (7.30).

Discrete Variation of Constants Formula

A second approach to convergence results of linear multistep methods is by the use of a discrete variation of constants formula. This is an extension of the classical proofs for nonstiff problems (Dahlquist 1956, Henrici 1962) to the case $\mu \neq 0$. It has been developed by Crouzeix & Raviart (1976), and more recently by Lubich (1988, 1991).

We consider the error equation (cf. (7.13))

$$\sum_{i=0}^{k} (\alpha_i - \mu\beta_i) e_{m+i} = d_{m+k} \qquad \text{for} \quad m \ge 0, \tag{7.41}$$

and extend this relation to negative m by putting $e_j = 0$ (for $j < 0$) and by defining $d_0, \ldots, d_{k-1}$ according to (7.41). The main idea is now to introduce the generating power series

$$e(\zeta) = \sum_{j\ge 0} e_j \zeta^j, \qquad d(\zeta) = \sum_{j\ge 0} d_j \zeta^j$$

so that (7.41) becomes the m-th coefficient of the identity

$$\big(\varrho(\zeta^{-1}) - \mu\sigma(\zeta^{-1})\big) e(\zeta) = \zeta^{-k} d(\zeta). \tag{7.42}$$

This gives

$$e(\zeta) = \big(\varrho(\zeta^{-1}) - \mu\sigma(\zeta^{-1})\big)^{-1} \zeta^{-k} d(\zeta) = r(\zeta,\mu) d(\zeta) \tag{7.43}$$

and allows to compute easily e_m in terms of d_j as

$$e_m = \sum_{j=0}^{m} r_{m-j}(\mu) d_j. \tag{7.43'}$$

Here $r_j(\mu)$ are the coefficients of the *discrete resolvent*

$$r(\zeta,\mu) = \big(\delta(\zeta) - \mu\big)^{-1} \frac{\zeta^{-k}}{\sigma(\zeta^{-1})} = \sum_{j\ge 0} r_j(\mu)\zeta^j, \tag{7.44}$$

where

$$\delta(\zeta) = \frac{\varrho(\zeta^{-1})}{\sigma(\zeta^{-1})} = \frac{\alpha_0\zeta^k + \ldots + \alpha_{k-1}\zeta + \alpha_k}{\beta_0\zeta^k + \ldots + \beta_{k-1}\zeta + \beta_k}. \tag{7.45}$$

Since $(\varrho(\zeta^{-1})-\mu\sigma(\zeta^{-1}))r(\zeta,\mu)=\zeta^{-k}$, the coefficients $r_j(\mu)$ can be interpreted as the numerical solution y_j of the multistep method applied to the homogeneous equation $y'=\mu y$ with step size $h=1$, and with starting values $y_{-k+1}=\ldots=y_{-1}=0$, $y_0=(\alpha_k-\mu\beta_k)^{-1}$.

Formula (7.43') can be used to estimate e_m, if appropriate bounds for the coefficients $r_j(\mu)$ of the discrete resolvent are known. Such bounds are given in the following lemma.

Lemma 7.9. *Let $S\subset\overline{\mathbb{C}}$ denote the stability region of the multistep method.*

a) If S is closed in $\overline{\mathbb{C}}$ then

$$|r_j(\mu)|\le\frac{M}{1+|\mu|}\qquad \textit{for}\quad \mu\in S\quad \textit{and}\quad j=0,1,2,\ldots$$

b) If $D\subset \operatorname{Int} S$ is closed in $\overline{\mathbb{C}}$ then there exists a constant κ $(0<\kappa<1)$ such that

$$|r_j(\mu)|\le\frac{M\kappa^j}{1+|\mu|}\qquad \textit{for}\quad \mu\in D\quad \textit{and}\quad j=0,1,2,\ldots$$

c) If the method is strictly stable then there exists a neighbourhood V of 0 such that

$$|r_j(\mu)|\le Me^{j(\operatorname{Re}\mu+a|\mu|^2)}\qquad \textit{for}\quad \mu\in V\quad \textit{and}\quad j=0,1,2,\ldots$$

The constants M, κ, and a are independent of j and μ.

Proof. The estimates for $|r_j(\mu)|$ in (a), (b), and (c) can easily be deduced from Lemmas 7.3, 7.4, and 7.5 because $r_j(\mu)$ is the numerical solution for the problem $y'=\mu y$ with step size $h=1$ and starting values $y_{-k+1}=\ldots=y_{-1}=0$, $y_0=(\alpha_k-\mu\beta_k)^{-1}$.

As noted by Crouzeix & Raviart (1976) and Lubich (1988) the estimates of Lemma 7.9 can be proved *directly*, without any use of the Kreiss matrix theorem. We illustrate these ideas by proving statement (b) (for a proof of statement (a) see Exercise 4).

By definition of the stability region the function $\zeta^k(\varrho(\zeta^{-1})-\mu\sigma(\zeta^{-1}))$ does not vanish for $|\zeta|\le 1$ if $\mu\in\operatorname{Int}S$. Therefore there exists a κ $(0<\kappa<1)$ such that $\zeta^k(\varrho(\zeta^{-1})-\mu\sigma(\zeta^{-1}))$ has no zeros in the disk $|\zeta|\le 1/\kappa$. Hence, for $\mu\in D$

$$\sup_{|\zeta|\le 1/\kappa}\left|\big(\varrho(\zeta^{-1})-\mu\sigma(\zeta^{-1})\big)^{-1}\zeta^{-k}\right|\le\frac{M}{1+|\mu|},$$

and Cauchy's integral formula

$$r_j(\mu)=\frac{1}{2\pi i}\int_{|\zeta|=1/\kappa}\big(\varrho(\zeta^{-1})-\mu\sigma(\zeta^{-1})\big)^{-1}\zeta^{-k}\zeta^{-j-1}d\zeta \tag{7.46}$$

yields the desired estimate. □

The use of the discrete resolvent allows elegant convergence proofs for linear multistep methods. We shall demonstrate this for the linear problem (7.26) where the matrix A satisfies

$$\|(sI-A)^{-1}\| \leq \frac{M}{1+|s|} \qquad \text{for} \quad |\arg(s-c)| \leq \pi - \alpha' \tag{7.47}$$

with some $c \in \mathbb{R}$. This is a common assumption in the theory of holomorphic semigroups for parabolic problems (see e.g., Kato (1966) or Pazy (1983)). If all eigenvalues λ_i of A satisfy $|\arg(\lambda_i - c) - \pi| < \alpha'$ then Condition (7.47) is satisfied with a constant M depending on the matrix A (Exercise 2). The following theorem, which was communicated to us by Ch. Lubich, is an improvement of results of Crouzeix & Raviart (1976).

Theorem 7.10. *Let the multistep method be of order $p \geq 1$, $A(\alpha)$-stable and strictly stable at zero and at infinity. If the matrix A of (7.26) satisfies (7.47) with $\alpha' < \alpha$, then there exist constants C, h_0, and γ (γ of the same sign as c in (7.47)), which depend only on M, c, α' and the method, such that for $h \leq h_0$ the global error satisfies*

$$\begin{aligned}&\|y(x_m)-y_m\| \\ &\qquad \leq C\Big(e^{\gamma x_m} \max_{0\leq j<k} \|y(x_j)-y_j\| + h^p \int_{x_0}^{x_m} e^{\gamma(x_m-\xi)} \|y^{(p+1)}(\xi)\| d\xi\Big).\end{aligned}$$

Moreover, if $c \leq 0$, then h_0 can be chosen arbitrarily.

Proof. The global error $e_m = y(x_m) - y_m$ satisfies

$$\sum_{i=0}^{k} (\alpha_i - hA\beta_i) e_{m+i} = d_{m+k}$$

where

$$\|d_{m+k}\| \leq Ch^p \int_{x_m}^{x_{m+k}} \|y^{(p+1)}(\xi)\| d\xi, \qquad m \geq 0 \tag{7.48}$$

and $d_0, \ldots, d_{k-1}$ are linear combinations of the e_j and hAe_j with $j<k$. We split these expressions into

$$d_\ell = d'_\ell + hAd''_\ell \qquad \text{for} \quad \ell < k,$$

so that d'_ℓ and d''_ℓ are linear combinations of the e_j $(j<k)$ only. We also put $d'_\ell = d_\ell$ and $d''_\ell = 0$ for $\ell \geq k$. The analysis at the beginning of this subsection (Eq. (7.43)) then shows that

$$e(\zeta) = r(\zeta, hA) d'(\zeta) + r(\zeta, hA) hAd''(\zeta), \tag{7.49}$$

where as in the scalar case

$$r(\zeta, hA) = (\delta(\zeta)I - hA)^{-1} \frac{\zeta^{-k}}{\sigma(\zeta^{-1})} = \sum_{j\geq 0} r_j(hA)\zeta^j. \tag{7.50}$$

We now apply Lemma 7.11 below with $\Phi(s) = (sI - A)^{-1}$. By assumption the estimate (7.57) holds with $\beta = 1$ so that

$$\|r_j(hA)\| \le C e^{\gamma j h}. \tag{7.51}$$

The second term in (7.49) can be written as

$$r(\zeta, hA) hA (\delta(\zeta))^{-1} \delta(\zeta) d''(\zeta) = r'(\zeta, hA) \widehat{d}(\zeta) \tag{7.52}$$

where

$$\begin{aligned} r'(\zeta, hA) &= (\delta(\zeta) I - hA)^{-1} hA (\delta(\zeta))^{-1} \frac{\zeta^{-k}}{\sigma(\zeta^{-1})} = \sum_{j\ge 0} r'_j(hA)\zeta^j \\ \widehat{d}(\zeta) &= \delta(\zeta) d''(\zeta) = \sum_{j\ge 0} \widehat{d}_j \zeta^j. \end{aligned} \tag{7.53}$$

We apply Lemma 7.11 again, this time to

$$\Phi(s) = (sI - A)^{-1} A s^{-1} = (sI - A)^{-1} - s^{-1} I.$$

Condition (7.57) is satisfied with $\beta = 1$ so that

$$\|r'_j(hA)\| \le C' e^{\gamma j h}. \tag{7.54}$$

The coefficients δ_j of $\delta(\zeta)$ are exponentially decaying because all zeros of $\sigma(\zeta)$ lie in $|\zeta| < 1$. Consequently, we have

$$\|\widehat{d}_j\| \le \kappa^j \widehat{C} \max_{0 \le \ell < k} \|e_\ell\| \tag{7.55}$$

with some $\kappa < 1$. The coefficient of ζ^m in (7.49) gives

$$e_m = \sum_{j=0}^{m} r_{m-j}(hA) d'_j + \sum_{j=0}^{m} r'_{m-j}(hA) \widehat{d}_j.$$

Inserting the estimates (7.48), (7.51), (7.54), and (7.55) proves the statement. □

We still have to prove the estimates for $r_j(hA)$ and $r'_j(hA)$. For this we let $\Phi(s)$ be some analytic (scalar-, vector-, or matrix-valued) function and consider the coefficients of

$$\Phi(\delta(\zeta)/h) \cdot \frac{\zeta^{-k}}{\sigma(\zeta^{-1})} = h \sum_{j \ge 0} \varphi_j(h) \zeta^j. \tag{7.56}$$

We then have the following result.

Lemma 7.11 (Lubich 1991). *Assume that the multistep method is $A(\alpha)$-stable and strictly stable at zero and at infinity. Further suppose that $\Phi(s)$ is analytic in a sector $|\arg(s - c)| < \pi - \alpha'$ with $\alpha' < \alpha$, $c \in \mathbb{R}$ and there satisfies*

$$\|\Phi(s)\| \le M \cdot |s|^{-\beta} \qquad \textit{for some} \quad \beta > 0. \tag{7.57}$$

Then the coefficients $\varphi_j(h)$ *of (7.56) are bounded for* $h \le h_0$ *(sufficiently small) by*

$$\|\varphi_j(h)\| \le C \cdot (jh)^{\beta-1} e^{\gamma jh} \qquad \text{for} \quad j \ge 1, \tag{7.58}$$

and for $j=0$ *the same bound holds as for* $j=1$. *The constants* C, γ, *and* h_0 *depend only on* α', c, M, β, *and the multistep method. Moreover, if* $c<0$, *then also* $\gamma<0$, *and the result holds for arbitrary* h_0.

Proof. By $A(\alpha)$-stability we have $\beta_k/\alpha_k > 0$, so that $\delta(0)/h$ lies in the region of analyticity of Φ for $h \le h_0$. Cauchy's integral formula thus gives

$$\Phi(\delta(\zeta)/h) = \frac{1}{2\pi i}\int_\Gamma (\delta(\zeta)/h - \lambda)^{-1}\Phi(\lambda)d\lambda \tag{7.59}$$

where Γ is a suitable contour from "$\infty \cdot e^{-i(\pi-\alpha')}$" to "$\infty \cdot e^{i(\pi-\alpha')}$" within the sector of analyticity of Φ and does not meet the origin (see Fig. 7.1; observe that $\Phi(s)$ decays sufficiently rapidly at infinity). Multiplying (7.59) by $\zeta^{-k}/\sigma(\zeta^{-1})$ and comparing coefficients of equal powers of ζ yields the representation

$$\varphi_j(h) = \frac{1}{2\pi i}\int_\Gamma r_j(h\lambda)\Phi(\lambda)d\lambda, \qquad j \ge 0, \tag{7.60}$$

which is a discrete analogue of the Laplace inversion formula. We next substitute $\omega = jh\lambda$ (for $j=0$ we put $\omega = h\lambda$) so that with $\Gamma_j = jh\cdot\Gamma$ Eq. (7.60) becomes

$$\varphi_j(h) = \frac{1}{2\pi i}\int_{\Gamma_j} r_j\Big(\frac{\omega}{j}\Big)\Phi\Big(\frac{\omega}{jh}\Big)\frac{d\omega}{jh}, \qquad j\ge 1, \tag{7.61}$$

and the use of (7.57) yields

$$\|\varphi_j(h)\| \le \frac{M}{2\pi}(jh)^{\beta-1}\int_{\Gamma_j}\Big|r_j\Big(\frac{\omega}{j}\Big)\Big|\cdot|\omega|^{-\beta}\cdot|d\omega|. \tag{7.62}$$

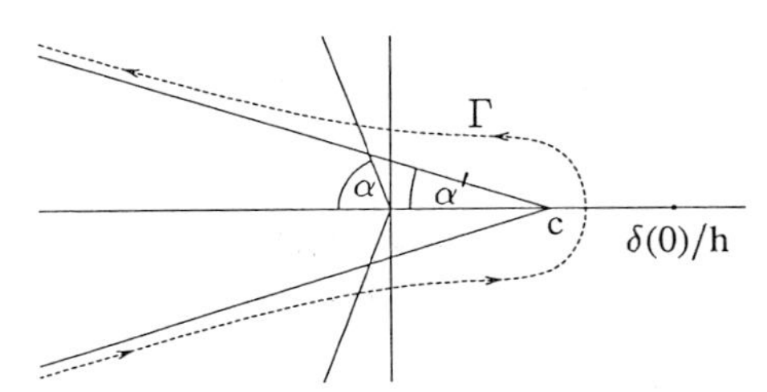

Fig. 7.1. Contour Γ in Formula (7.59)

We still have to show that the integral in (7.62) is bounded by $C\cdot e^{\gamma jh}$. For this we split it into two parts: the first one corresponds to those ω such that ω/j lies in a closed subset of the interior of the stability domain of the method. There we can use Lemma 7.9b so that the corresponding part of the integral in (7.62) is bounded by

$$j\cdot\kappa^j\int|\omega|^{-\beta-1}|d\omega| \le Ce^{\gamma jh} \qquad \text{for} \quad h\le h_0.$$

For the remaining part, the argument $\omega/j = h\lambda$ of r_j in (7.62) lies, for sufficiently small h_0, in a neighbourhood V of the origin, where the estimate of Lemma 7.9c holds. For $jh \geq 1$ we thus obtain the bound

$$\int e^{\mathrm{Re}\,\omega + a|\omega|^2/j} |\omega|^{-\beta} |d\omega| \leq C e^{\gamma jh},$$

because $\mathrm{Re}\,\omega = jh\mathrm{Re}\,\lambda$, $|\omega|^2/j \leq jh \cdot \mathit{Const}$ and $|\omega| \geq |\lambda|$ is bounded away from the origin. For small jh the contour Γ_j comes arbitrarily close to the origin so that a more refined estimate is required. The idea is to replace the corresponding part of Γ_j (in (7.61) and hence also in (7.62)) by an equivalent contour which is independent of $jh \in [h, 1]$, has a positive distance to the origin and remains in the neighbourhood V. The corresponding integral is thus bounded by some constant.

□

Remark 7.12. In Lemma 7.11 it is sufficient to require the analyticity of $\Phi(s)$ and the estimate (7.57) in a sector $|\arg(s-c)| < \pi - \alpha'$, where some compact neighbourhood of the origin is removed. We just have to take the contour Γ in (7.59) so that it lies outside this compact neighbourhood of 0. In this situation, the constant γ may be positive also if $c < 0$.

Exercises

1. Prove the Cauchy integral formula (7.18) in the case where all eigenvalues λ of A satisfy $|\lambda| \leq 1$ and the contour of integration is the circle $|z| = \varrho$ with $\varrho > 1$.
   Hint. Integrate the identity

$$z^n (zI - A)^{-1} = \sum_{j=0}^{\infty} A^j z^{n-j-1}.$$

2. (Kato 1960). For a non-singular $k \times k$-matrix B show that in the Euclidean norm

$$\|B^{-1}\| \leq \frac{\|B\|^{k-1}}{|\det B|}.$$

   Hint. Use the singular value decomposition of B, i.e., $B = U^T \Lambda V$, where U and V are orthogonal and $\Lambda = \mathrm{diag}\,(\sigma_1, \ldots, \sigma_k)$ with $\sigma_1 \geq \sigma_2 \geq \ldots > \sigma_k > 0$.

3. A method (ϱ, σ) is called *A-contractive* in the norm $\|\cdot\|_G$ (Nevanlinna & Liniger 1978-79, Dahlquist & Söderlind 1982), if

$$\|C(\mu)\|_G \leq 1 \qquad \text{for} \quad \mathrm{Re}\,\mu \leq 0$$

   where $C(\mu)$ is the companion matrix (7.13).

a) Prove that a method (ϱ, σ) is A-contractive for some positive definite matrix G, if and only if it is A-stable.

b) Compute the contractivity region

$$\{\mu \in \mathbb{C}\,;\ \|C(\mu)\|_G \leq 1\}$$

for the 2-step BDF method with G given in (6.20). Observe that it is strictly smaller than the stability domain.

Result. The contractivity region is $\{\mu \in \mathbb{C}\,;\ \mathrm{Re}\,\mu \leq 0\}$.

4. Give a direct proof for the statement of Lemma 7.9a.

Hint. Observe that

$$r(\zeta,\mu) = \frac{1}{\alpha_k - \mu\beta_k} \prod_{i=1}^{k} \frac{1}{(1-\zeta\cdot\zeta_i(\mu))}, \tag{7.63}$$

where $\zeta_1(\mu),\ldots,\zeta_k(\mu)$ are the k zeros of $\varrho(\zeta)-\mu\sigma(\zeta)$. If $\mu_0 \in \mathrm{Int}\, S$ then there exists a neighbourhood $\mathcal{U}$ of μ_0 such that $|\zeta_i(\mu)| \leq a < 1$ for all i and $\mu \in \mathcal{U}$. Hence the coefficients $r_j(\mu)$ are bounded. For $\mu_0 \in \partial S$ we have $|\zeta_i(\mu_0)| = 1$ for, say, $i = 1,\ldots,\ell$ with $1 \leq \ell \leq k$. These ℓ zeros are simple for all μ in a sufficiently small neighbourhood $\mathcal{U}$ of μ_0 and the other zeros satisfy $|\zeta_i(\mu)| \leq a < 1$ for $\mu \in \mathcal{U} \cap S$. A partial fraction decomposition

$$r(\zeta,\mu) = \frac{1}{\alpha_k - \mu\beta_k}\Big(\sum_{i=1}^{\ell} \frac{c_i(\mu)}{1-\zeta\cdot\zeta_i(\mu))} + s(\zeta,\mu)\Big)$$

shows that

$$r_j(\mu) = \frac{1}{\alpha_k - \mu\beta_k}\Big(\sum_{i=1}^{\ell} c_i(\mu)(\zeta_i(\mu))^j + s_j(\mu)\Big), \tag{7.64}$$

where $s_j(\mu)$ are the coefficients of $s(\zeta,\mu)$. Since the function $s(\zeta,\mu)$ is uniformly bounded for $|\zeta| \leq 1$ and $\mu \in \mathcal{U} \cap S$, it follows from Cauchy's integral formula with integration along $|\zeta| = 1$ that $s_j(\mu)$ is bounded. The statement thus follows from (7.64) and the fact that a finite set of the family $\{\mathcal{U}\}_{\mu_0 \in S}$ covers S (Heine-Borel).

V.8 Convergence for Nonlinear Problems

In Sect. V.6 we have seen a convergence result for one-leg methods (Theorem 6.10) applied to nonlinear problems satisfying a one-sided Lipschitz condition. An extension to linear multistep methods has been given in Theorem 6.11. A different and direct proof of this result will be the first goal of this section. Unfortunately, such a result is valid only for A-stable methods (whose order cannot exceed two). The subsequent parts of this section are then devoted to convergence results for nonlinear problems, where the assumptions on the method are relaxed (e.g., $A(\alpha)$-stability), but the class of problems considered is restricted. We shall present two different theories: the multiplier technique of Nevanlinna & Odeh (1981) and Lubich's perturbation approach via the discrete variation of constants formula (Lubich 1991).

Problems Satisfying a One-Sided Lipschitz Condition

Suppose that the differential equation $y' = f(x,y)$ satisfies

$$\mathrm{Re}\,\langle f(x,y) - f(x,z)\,,\ y - z\rangle \le \nu \|y - z\|^2 \tag{8.1}$$

for some inner product. We consider the linear multistep method

$$\sum_{i=0}^{k} \alpha_i\, y_{m+i} = h \sum_{i=0}^{k} \beta_i\, f(x_{m+i}, y_{m+i}) \tag{8.2}$$

together with its perturbed formula

$$\sum_{i=0}^{k} \alpha_i\, \widehat{y}_{m+i} = h \sum_{i=0}^{k} \beta_i\, f(x_{m+i}, \widehat{y}_{m+i}) + d_{m+k}. \tag{8.3}$$

The perturbations d_{m+k} can be interpreted as the influence of round-off, as the error due to the iterative solution of the nonlinear equation, or as the local discretization error (compare Eq. (7.5)). Taking the difference of (8.3) and (8.2) we obtain (for $m \ge 0$)

$$\sum_{i=0}^{k} \alpha_i\, \Delta y_{m+i} = h \sum_{i=0}^{k} \beta_i\, \Delta f_{m+i} + d_{m+k}, \tag{8.4}$$

where we have introduced the notation

$$\Delta y_j = \widehat{y}_j - y_j, \qquad \Delta f_j = f(x_j, \widehat{y}_j) - f(x_j, y_j). \tag{8.5}$$

The one-sided Lipschitz condition cannot be used directly, because several Δf_j appear in (8.4) (in contrast to one-leg methods). In order to express *one* Δf_m in terms of Δy_j only we introduce the formal power series

$$\Delta y(\zeta) = \sum_{j\geq 0} \Delta y_j \, \zeta^j, \quad \Delta f(\zeta) = \sum_{j\geq 0} \Delta f_j \, \zeta^j, \quad d(\zeta) = \sum_{j\geq 0} d_j \, \zeta^j.$$

It is convenient to assume that $\Delta y_j = 0$, $\Delta f_j = 0$, $d_j = 0$ for negative indices and that $d_0, \ldots, d_{k-1}$ are defined by Eq. (8.4) with $m \in \{-k, \ldots, -1\}$. Then Eq. (8.4) just compares the coefficient of ζ^m in the identity

$$\varrho(\zeta^{-1})\, \Delta y(\zeta) = h\sigma(\zeta^{-1})\, \Delta f(\zeta) + \zeta^{-k} d(\zeta). \tag{8.4'}$$

Dividing (8.4') by $\sigma(\zeta^{-1})$ and comparing the coefficients of ζ^m yields

$$\sum_{j=0}^{m} \delta_{m-j}\, \Delta y_j = h \Delta f_m + \widetilde{d}_m, \tag{8.6}$$

where

$$\frac{\varrho(\zeta^{-1})}{\sigma(\zeta^{-1})} = \delta(\zeta) = \sum_{j\geq 0} \delta_j \zeta^j \tag{8.7}$$

as in (7.45) and

$$\frac{\zeta^{-k}}{\sigma(\zeta^{-1})}\, d(\zeta) = \widetilde{d}(\zeta) = \sum_{j\geq 0} \widetilde{d}_j \zeta^j. \tag{8.8}$$

In (8.6) Δf_m is now isolated as desired and we can take the scalar product of (8.6) with Δy_m. We then exploit the assumption (8.1) and obtain

$$\sum_{j=0}^{m} \delta_{m-j} \mathrm{Re}\, \langle \Delta y_j\, , \, \Delta y_m \rangle \leq h\nu \|\Delta y_m\|^2 + \mathrm{Re}\, \langle \widetilde{d}_m\, , \, \Delta y_m \rangle. \tag{8.9}$$

This allows us to prove the following estimate.

Lemma 8.1. *Let* $\{\Delta y_j\}$ *and* $\{\Delta f_j\}$ *satisfy (8.6) with* δ_j *given by (8.7). If*

$$\mathrm{Re}\, \langle \Delta f_m\, , \, \Delta y_m \rangle \leq \nu \|\Delta y_m\|^2, \qquad m \geq 0,$$

and the method is A*-stable, then there exist constants* C *and* $C_0 > 0$ *such that for* $mh \leq x_{\text{end}} - x_0$ *and* $h\nu \leq C_0$,

$$\|\Delta y_m\| \leq C \sum_{j=0}^{m} \|\widetilde{d}_j\|.$$

Proof. We first reformulate the left-hand side of (8.9). For this we introduce $\{\Delta z_j\}$ by the relation

$$\sum_{i=0}^{k} \beta_i \Delta z_{m+i} = \Delta y_m, \qquad m \geq 0 \tag{8.10}$$

and assume that $\Delta z_j = 0$ for $j < k$. With $\Delta z(\zeta) = \sum_j \Delta z_j \zeta^j$ this means that $\sigma(\zeta^{-1})\Delta z(\zeta) = \Delta y(\zeta)$. Consequently we also have

$$\delta(\zeta)\,\Delta y(\zeta) = \varrho(\zeta^{-1})\,\Delta z(\zeta),$$

which is equivalent to

$$\sum_{j=0}^{m} \delta_{m-j}\,\Delta y_j = \sum_{i=0}^{k} \alpha_i\,\Delta z_{m+i}. \tag{8.11}$$

Inserting (8.11) and (8.10) into (8.9) yields

$$\mathrm{Re}\Big\langle \sum_{i=0}^{k} \alpha_i\,\Delta z_{m+i}, \sum_{i=0}^{k} \beta_i\,\Delta z_{m+i}\Big\rangle$$
$$\leq h\nu \Big\| \sum_{i=0}^{k} \beta_i\,\Delta z_{m+i}\Big\|^2 + \mathrm{Re}\Big\langle \widetilde{d}_m, \sum_{i=0}^{k} \beta_i\,\Delta z_{m+i}\Big\rangle.$$

By Theorem 6.7 the method (ϱ,σ) is G-stable, so that Eq. (6.21) can be applied. As in the proof of Lemma 6.9 this yields for $\Delta Z_m = (\Delta z_{m+k-1},\ldots,\Delta z_m)^T$ and $\nu \geq 0$

$$\|\Delta Z_{m+1}\|_G \leq (1 + C_1 h\nu)\|\Delta Z_m\|_G + C_2\|\widetilde{d}_m\|,$$

(if $\nu < 0$ replace ν by $\nu = 0$). But this implies

$$\|\Delta Z_{m+1}\|_G \leq C_3\Big(\|\Delta Z_0\|_G + \sum_{j=0}^{m} \|\widetilde{d}_j\|\Big).$$

By definition of Δz_j we have $\Delta Z_0 = 0$. The statement now follows from the fact that $\|\Delta y_m\| \leq C_4\,(\|\Delta Z_{m+1}\|_G + \|\Delta Z_m\|_G)$. □

This lemma allows a direct proof for the convergence of A-stable multistep methods which are strictly stable at infinity (compare Theorem 6.11).

Theorem 8.2. *Consider an A-stable multistep method of order p which is strictly stable at infinity. Suppose that the differential equation satisfies (8.1). Then there exists $C_0 > 0$ such that for $h\nu \leq C_0$*

$$\|y_m - y(x_m)\| \leq C\big(\max_{0\leq j<k}\|y_j - y(x_j)\| + h\max_{0\leq j<k}\|f(x_j,y_j) - y'(x_j)\|\big) + Mh^p.$$

The constant C depends on the method and, for $\nu > 0$, on the length $x_m - x_0$ of the integration interval; the constant M depends in addition on bounds for the $(p+1)$-th derivative of the exact solution.

Proof. We put $\widehat{y}_m = y(x_m)$ in (8.3). The perturbations thus become the local truncation errors $d_{m+k} = \delta_{LM}(x_m)$, where

$$\delta_{LM}(x) = \sum_{i=0}^{k} \alpha_i\, y(x+ih) - h \sum_{i=0}^{k} \beta_i\, y'(x+ih). \tag{8.12}$$

If the zeros of $\sigma(\zeta)$ all lie in the open unit disk, the coefficients of $\zeta^{-k}/\sigma(\zeta^{-1})$ are absolutely summable and by (8.8) we have

$$\sum_{j=0}^{m} \|\tilde{\hat{d}}_j\| \le C_1 \sum_{j=0}^{m} \|d_j\|.$$

The statement then follows from Lemma 8.1, from $\|\delta_{LM}(x)\| \le Mh^{p+1}$, and from the fact that $d_0, \ldots, d_{k-1}$ are linear combinations of the $y_j - y(x_j)$ and $h\big(f(x_j, y_j) - y'(x_j)\big)$ for $j < k$. □

Multiplier Technique

> ... the best of all multipliers would be $\{1, -\eta\}$ with a very small $\eta > 0$; ...
> (Nevanlinna & Odeh 1981)

The above convergence proof is based on Eq. (8.6) and on the A-stability of the multistep method. How can we modify this proof in order to get convergence results also for methods which are not A-stable? This can be done by the so-called "multiplier technique", introduced by Nevanlinna & Odeh (1981) and based on previous ideas of Nevanlinna (1977) and Odeh & Liniger (1977).

The main idea is the following: instead of multiplying scalarly the identity (8.6) by Δy_m, we multiply it by

$$\sum_{j=0}^{m} \mu_{m-j}\, \Delta y_j$$

where $\{\mu_j\}$ are the coefficients of a rational function (the multiplier)

$$\mu(\zeta) = \sum_{j\ge 0} \mu_j \zeta^j = \frac{\eta(\zeta^{-1})}{\tau(\zeta^{-1})} \tag{8.13}$$

(η and τ are polynomials). We obtain

$$\begin{aligned} \mathrm{Re}\Big\langle \sum_{j=0}^{m} \delta_{m-j}\, \Delta y_j, \sum_{j=0}^{m} \mu_{m-j}\, \Delta y_j \Big\rangle &= h\mathrm{Re}\Big\langle \Delta f_m, \sum_{j=0}^{m} \mu_{m-j}\, \Delta y_j \Big\rangle \\ &\quad + \Big\langle \tilde{d}_m, \sum_{j=0}^{m} \mu_{m-j}\, \Delta y_j \Big\rangle. \end{aligned} \tag{8.14}$$

Our next aim is to introduce new variables Δz_j such that the left-hand side of (8.14) becomes

$$\Big\langle \sum_{j=0}^{m} \delta_{m-j}\,\Delta y_j\,,\ \sum_{j=0}^{m} \mu_{m-j}\,\Delta y_j \Big\rangle = \Big\langle \sum_{i=0}^{\ell} \widetilde{\alpha}_i\,\Delta z_{m+i}\,,\ \sum_{i=0}^{\ell} \widetilde{\beta}_i\,\Delta z_{m+i} \Big\rangle. \tag{8.15}$$

Denoting

$$\widetilde{\varrho}(\zeta) = \sum_{i=0}^{\ell} \widetilde{\alpha}_i \zeta^i, \qquad \widetilde{\sigma}(\zeta) = \sum_{i=0}^{\ell} \widetilde{\beta}_i \zeta^i, \tag{8.16}$$

the identity (8.15) certainly holds, if

$$\begin{aligned} \varrho(\zeta^{-1})\,\Delta y(\zeta) &= \sigma(\zeta^{-1})\,\widetilde{\varrho}(\zeta^{-1})\,\Delta z(\zeta) \\ \eta(\zeta^{-1})\,\Delta y(\zeta) &= \tau(\zeta^{-1})\,\widetilde{\sigma}(\zeta^{-1})\,\Delta z(\zeta). \end{aligned} \tag{8.17}$$

Dividing these two relations motivates the following definition of the new generating polynomials

$$\widetilde{\varrho}(\zeta) = \varrho(\zeta)\tau(\zeta)/\chi(\zeta), \qquad \widetilde{\sigma}(\zeta) = \sigma(\zeta)\eta(\zeta)/\chi(\zeta). \tag{8.18}$$

Here $\chi(\zeta)$ denotes the greatest common divisor of $\varrho(\zeta)\tau(\zeta)$ and $\sigma(\zeta)\eta(\zeta)$. If we define $\Delta z_j = 0$ for $j < 0$ and the remaining Δz_j by

$$\chi(\zeta^{-1})\,\Delta y(\zeta) = \sigma(\zeta^{-1})\tau(\zeta^{-1})\,\Delta z(\zeta) \tag{8.19}$$

the identity (8.15) holds for all m. Suppose now that the multistep method $(\widetilde{\varrho}, \widetilde{\sigma})$ is A-stable, then the left hand side of (8.14) can be minorized by the G-stability estimate (6.21) and we shall be able to derive convergence results. This motivates the following

Definition 8.3. The rational function $\mu(\zeta)$ of (8.13) is called a *multiplier* for (ϱ, σ) if $\mu(\zeta) \not\equiv \varrho(\zeta^{-1})/\sigma(\zeta^{-1})$ and if the method $(\widetilde{\varrho}, \widetilde{\sigma})$, given by (8.18) is A-stable, i.e., if

$$\mathrm{Re}\left(\frac{1}{\mu(\zeta^{-1})} \cdot \frac{\varrho(\zeta)}{\sigma(\zeta)} \right) > 0 \qquad \text{for} \quad |\zeta| > 1. \tag{8.20}$$

A continuation of the above analysis yields the following convergence result.

Lemma 8.4. *Let $\{\Delta y_j\}$ and $\{\Delta f_j\}$ satisfy (8.6) with δ_j given by (8.7). If*

$$\sum_{m=0}^{N} \sum_{j=0}^{m} \mu_{m-j} \mathrm{Re}\,\langle \Delta f_m\,,\ \Delta y_j \rangle \le 0 \qquad \textit{for all} \quad N \ge 0$$

and if $\mu(\zeta)$ is a multiplier for the method, then there exists a constant C such that for $mh \le x_{\mathrm{end}} - x_0$

$$\|\Delta y_m\| \le C \sum_{j=0}^{m} \|\widetilde{d}_j\|.$$

Proof. Inserting (8.15) into (8.14) and using the estimate (6.21) for the A-stable method $(\widetilde{\varrho},\widetilde{\sigma})$ yields for $\Delta Z_m = (\Delta z_{m+\ell-1},\ldots,\Delta z_m)^T$

$$\begin{aligned}\|\Delta Z_{m+1}\|_G^2 - \|\Delta Z_m\|_G^2 \le h\mathrm{Re}\Big\langle \Delta f_m\,,\, \sum_{j=0}^{m} \mu_{m-j}\,\Delta y_j\Big\rangle \\ + \|\widetilde{d}_m\|\cdot\Big\|\sum_{i=0}^{\ell}\widetilde{\beta}_i\,\Delta z_{m+i}\Big\|.\end{aligned} \tag{8.21}$$

Summing up this inequality from $m=0$ to $m=N$ gives

$$\|\Delta Z_{N+1}\|_G^2 \le C_1 \sum_{m=0}^{N} \|\widetilde{d}_m\|\cdot\big(\|\Delta Z_{m+1}\|_G + \|\Delta Z_m\|_G\big),$$

because $\Delta Z_0 = 0$ by (8.19). This also implies

$$\max_{N\le M} \|\Delta Z_{N+1}\|_G^2 \le 2C_1 \sum_{m=0}^{M} \|\widetilde{d}_m\|\cdot \max_{m\le M}\|\Delta Z_{m+1}\|_G.$$

A division by $\max_{N\le M}\|\Delta Z_{N+1}\|_G$ yields the desired estimate, because Δy_M is a linear combination of the elements of ΔZ_{M+1}. □

The proof of Theorem 8.2 applied to the A-stable method $(\widetilde{\varrho},\widetilde{\sigma})$ now yields:

Theorem 8.5 (Nevanlinna & Odeh 1981). *Consider a linear multistep method (8.2) of order p, which is strictly stable at infinity and has a multiplier $\mu(\zeta)$. Suppose that the differential equation satisfies*

$$\sum_{m=0}^{N}\sum_{j=0}^{m} \mu_{m-j}\mathrm{Re}\,\big\langle f(x_m,u_m) - f(x_m,v_m)\,,\, u_j - v_j\big\rangle \le 0 \tag{8.22}$$

for all $N\ge 0$ and for all sequences $\{u_j\}$ and $\{v_j\}$. Then we have

$$\|y_m - y(x_m)\| \le C\Big(\max_{0\le j<k}\|y_j - y(x_j)\| + h\max_{0\le j<k}\|f(x_j,y_j) - y'(x_j)\|\Big) + Mh^p,$$

where the constants C and M are as in Theorem 8.2. □

In the next two subsections we shall study the existence and construction of multipliers, and try to better understand the condition (8.22).

Construction of Multipliers. Obviously $\mu(\zeta)=1$ is a multiplier iff the method itself is A-stable. Moreover, the limit $|\zeta|\to\infty$ in (8.20) shows that $\mu(0)$ must have the same sign as α_k/β_k (which we always assume to be positive). Therefore, the simplest (and most important) nontrivial multiplier has the form

$$\mu(\zeta) = 1 - \eta\zeta. \tag{8.23}$$

Suppose now that the method (ϱ, σ) is stable at infinity. The maximum principle for harmonic functions then implies that (8.23) is a multiplier for (ϱ, σ) iff $|\eta| \le 1$ and

$$\mathrm{Re}\left((1-\eta e^{it})\frac{\varrho(e^{it})}{\sigma(e^{it})}\right) \ge 0 \qquad \text{for all} \quad t \in \mathbb{R}.$$

This condition motivates the study of

$$\gamma(t) = \left(\mathrm{Re}\left(\frac{\varrho(e^{it})}{\sigma(e^{it})}\right), -\mathrm{Re}\left(e^{it}\frac{\varrho(e^{it})}{\sigma(e^{it})}\right)\right), \tag{8.24}$$

which is called the *modified root-locus* curve by Nevanlinna & Odeh (1981). We then have:

Criterion 8.6. Consider a method which is stable at infinity. The function (8.23) is a *multiplier for* (ϱ, σ) iff $|\eta| \le 1$ and the modified root-locus curve lies to the right of the straight line through the origin with slope $-1/\eta$.

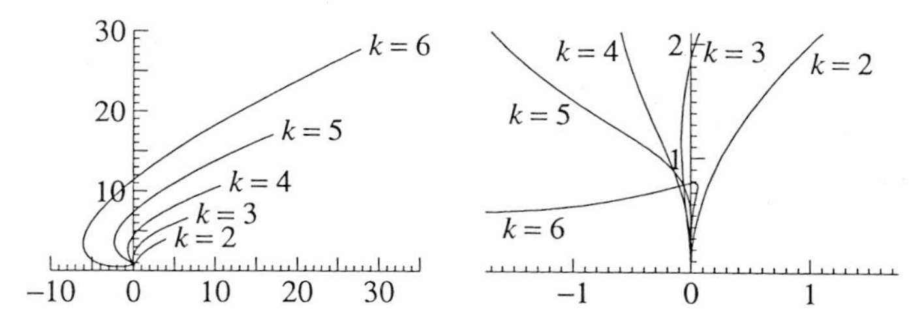

Fig. 8.1. Modified root-locus curve for BDF schemes

Fig. 8.1 shows the modified root-locus curves for the BDF schemes for $2 \le k \le 6$. The optimal values for η are given in Table 8.1.

Table 8.1. Multiplier for BDF schemes

k	η	$\arccos\eta$	$A(\alpha)$-stable
2	0	$\pi/2$	$\pi/2$
3	0.0836	85.20°	86.03°
4	0.2878	73.27°	73.35°
5	0.8160	35.32°	51.84°
6	5.0130	—	17.84°

Proposition 8.7. *If* $\mu(\zeta)$ *is a multiplier for* (ϱ, σ) *and we have*

$$|\arg\mu(\zeta)| \le \frac{\pi}{2} - \alpha \qquad \textit{for} \quad |\zeta| \le 1 \tag{8.25}$$

then the method is $A(\alpha)$*-stable.*

Proof. Condition (8.20) together with (8.25) implies that

$$\left|\arg\Big(\frac{\varrho(\zeta)}{\sigma(\zeta)}\Big)-\pi\right|\geq\alpha \qquad \text{for} \quad |\zeta|\geq 1.$$

But this condition implies $A(\alpha)$-stability. □

A simple calculation shows that the multiplier (8.23) satisfies (8.25) with $\alpha = \arccos\eta$. For the BDF schemes we have included these values in Table 8.1 together with the α-values for linear stability.

Multipliers and Nonlinearities

We still have to investigate the problem under what conditions on the multiplier $\mu(\zeta)$ and on the function $f(x,y)$ one has (8.22) for all sequences $\{u_j\}$ and $\{v_j\}$. To get an idea of the nature of (8.22) we first look, following Nevanlinna & Odeh (1981), at the linear problem $y' = Ay$.

Proposition 8.8. *If the multiplier $\mu(\zeta)$ satisfies (8.25) and if the range of the matrix A lies in the sector $|\arg\langle Au,u\rangle-\pi|\leq\alpha$ for all $u\in\mathbb{C}^n$, then we have*

$$\sum_{m=0}^{N}\sum_{j=0}^{m}\mu_{m-j}\,\mathrm{Re}\,\langle Au_m,u_j\rangle\leq 0 \tag{8.26}$$

for all $N\geq 0$ and all sequences $\{u_j\}$.

Proof. A direct computation shows that the expression in (8.26) equals

$$\mathrm{Re}\left(\frac{1}{2\pi}\int_0^{2\pi}\mu(e^{it})\big\langle A\widehat{u}_N(t),\widehat{u}_N(t)\big\rangle\,dt\right) \tag{8.27}$$

where

$$\widehat{u}_N(t)=\sum_{j=0}^{N}e^{-ijt}u_j$$

denotes the Fourier transform of $(u_0,u_1,\ldots,u_N)$. The assumptions on $\mu(\zeta)$ and on A imply that the integrand in (8.27) has non-positive real part. This proves (8.26). □

Problems which satisfy (8.22) for some multiplier $\mu(\zeta)$ must also satisfy the one-sided Lipschitz condition (8.1) with $\nu=0$ (this is seen by putting $N=0$ in (8.22)). A class of nonlinear problems, for which (8.22) holds, is given by the following perturbation result.

Proposition 8.9. *Let $f(x,y) = -Ay + Ag(x,y)$ where A is a symmetric and positive semi-definite matrix. With $\|u\|_A^2 = u^T Au$ suppose that*

$$\|g(x,y) - g(x,z)\|_A \le L\|y - z\|_A. \tag{8.28}$$

Then Condition (8.22) holds if

$$L \cdot \max_{|\zeta|=1} |\mu(\zeta)| \le \min_{|\zeta|=1} \operatorname{Re} \mu(\zeta). \tag{8.29}$$

Remark. For the multiplier (8.23) Condition (8.29) is equivalent to $L \cdot (1+\eta) \le (1-\eta)$.

Proof. As in the proof of Proposition 8.8 we get for $w_j = u_j - v_j$

$$\begin{aligned} -\sum_{m=0}^{N} \sum_{j=0}^{m} \mu_{m-j} \operatorname{Re} \langle Aw_m, w_j \rangle &= -\operatorname{Re} \Big(\frac{1}{2\pi} \int_0^{2\pi} \mu(e^{it}) \big\langle A\widehat{w}_N(t), \widehat{w}_N(t) \big\rangle dt \Big) \\ &\le -m_0 \frac{1}{2\pi} \int_0^{2\pi} \langle A\widehat{w}_N(t), \widehat{w}_N(t) \rangle dt = -m_0 \sum_{j=0}^{N} \langle Aw_j, w_j \rangle \end{aligned} \tag{8.30}$$

where $m_0 = \min \operatorname{Re} \mu(e^{it})$. On the other hand, the inequality of Cauchy-Schwarz gives

$$\begin{aligned} &\sum_{m=0}^{N} \operatorname{Re} \Big\langle A\big(g(x_m, u_m) - g(x_m, v_m)\big),\ \sum_{j=0}^{m} \mu_{m-j}(u_j - v_j) \Big\rangle \\ &\le \Big(\sum_{m=0}^{N} \|g(x_m, u_m) - g(x_m, v_m)\|_A^2 \Big)^{1/2} \cdot \Big(\sum_{m=0}^{N} \Big\| \sum_{j=0}^{m} \mu_{m-j}(u_j - v_j) \Big\|_A^2 \Big)^{1/2}. \end{aligned} \tag{8.31}$$

The last term in (8.31) can be estimated as (for the moment put $w_j = 0$ for $j > N$)

$$\begin{aligned} \sum_{m=0}^{N} \Big\| \sum_{j=0}^{m} \mu_{m-j} w_j \Big\|_A^2 &\le \frac{1}{2\pi} \int_0^{2\pi} \Big\| \sum_{m\ge 0} e^{-imt} \sum_{j=0}^{m} \mu_{m-j} w_j \Big\|_A^2 dt \\ &= \frac{1}{2\pi} \int_0^{2\pi} |\mu(e^{-it})|^2 \cdot \Big\| \sum_{j\ge 0} e^{-ijt} w_j \Big\|_A^2 dt \le M^2 \sum_{j=0}^{N} \|w_j\|_A^2 \end{aligned}$$

where $M = \max |\mu(e^{-it})|$. These estimates together with (8.28) show that the expression in (8.22) is majorized by

$$(L \cdot M - m_0) \sum_{j=0}^{N} \|u_j - v_j\|_A^2.$$

This is non-positive if (8.29) holds. □

Discrete Variation of Constants and Perturbations

We now turn our attention to the perturbation approach of Lubich (1991), which extends the ideas of Sect. V.7 (discrete variation of constants) to nonlinear problems. For this we consider nonlinear differential equations written in the form

$$y' = Ay + g(t, y). \tag{8.32}$$

Inserting this equation into Formulas (8.2), (8.3), and (8.4) we get

$$\sum_{i=0}^{k} (\alpha_i I - hA\beta_i)\Delta y_{m+i} = h\Delta g_{m+k} + d_{m+k}, \tag{8.33}$$

where

$$\Delta g_{m+k} = \sum_{i=0}^{k} \beta_i \Big(g(x_{m+i}, \widehat{y}_{m+i}) - g(x_{m+i}, y_{m+i}) \Big) \tag{8.34}$$

for $m \geq 0$. We further put $\Delta g_j = 0$ for $j < k$. Recall that d_j (for $j \geq k$) are usually the local truncation errors and $d_0, \ldots, d_{k-1}$ are defined by (8.33) with $m \in \{-1, \ldots, -k\}$. The differences Δy_j are then the global errors of the method. If we introduce the formal power series

$$\Delta y(\zeta) = \sum_{j\geq 0} \Delta y_j \zeta^j, \quad \Delta g(\zeta) = \sum_{j\geq 0} \Delta g_j \zeta^j, \quad d(\zeta) = \sum_{j\geq 0} d_j \zeta^j$$

then the recursion (8.33) can be written as

$$\Delta y(\zeta) = r(\zeta, hA) \big(h\Delta g(\zeta) + d(\zeta) \big). \tag{8.35}$$

The resolvent $r(\zeta, hA)$ was introduced in (7.44) and (7.50). The coefficient of ζ^m in (8.35) then yields

$$\Delta y_m = h \sum_{j=0}^{m} r_{m-j}(hA)\Delta g_j + \sum_{j=0}^{m} r_{m-j}(hA) d_j. \tag{8.36}$$

The second sum on the right-hand side of (8.36) can be estimated as in Sect. V.7. In order to estimate the first term we have to combine estimates for $r_j(hA)$ with a Lipschitz condition for $g(x, y)$. This will lead to a Gronwall-type inequality, whose resolution gives the desired estimates for Δy_m. Let us illustrate this procedure in a simple situation.

Theorem 8.10. *Let the multistep method and the matrix A satisfy the assumptions of Theorem 7.10. If the nonlinearity $g(x, y)$ satisfies*

$$\|g(x, y) - g(x, z)\| \leq L\|y - z\| \tag{8.37}$$

then there exist constants C, h_0 and γ as in Theorem 7.10, and Λ (h_0 and Λ

depend on L) such that

$$\|y(x_m)-y_m\| \\ \le Ce^{\Lambda x_m}\Big(\max_{0\le j<k}\|y(x_j)-y_j\|+h^p\int_{x_0}^{x_m}e^{\gamma(x_m-\xi)}\|y^{(p+1)}(\xi)\|d\xi\Big).$$

Proof. It follows from the proof of Theorem 7.10 and from (8.36) that

$$\|\Delta y_m\|\le hLC_1\sum_{j=0}^{m}e^{\gamma(m-j)h}\|\Delta y_j\|+C_2\sum_{j=0}^{m}e^{\gamma(m-j)h}\varepsilon_j, \tag{8.38}$$

where (with $0\le\kappa<1$)

$$\varepsilon_m=C_0\Big(\kappa^m\max_{0\le j<k}\|\Delta y_j\|+h^p\int_{x_{m-k}}^{x_m}\|y^{(p+1)}(\xi)\|d\xi\Big).$$

Application of Exercise 1 to the sequence $\{e^{-\gamma mh}\|\Delta y_m\|\}$ yields the statement of the theorem. □

Lubich (1991) has shown how the above estimates can be improved to obtain convergence results for singularly perturbed problems (see Sect. VI.2) and for discretized nonlinear parabolic equations, as we shall see in the sequel.

Convergence for Nonlinear Parabolic Problems

We consider the initial value problem

$$y'+Ay=g(t,y),\qquad y(0)=y_0 \tag{8.39}$$

obtained by space discretization of a parabolic differential equation. The matrix A is assumed to satisfy for some $\alpha'\in(0,\pi/2)$

$$\|(sI+A)^{-1}\|\le\frac{M}{1+|s|}\qquad\text{for}\qquad|\arg s|\le\pi-\alpha' \tag{8.40}$$

(compare (7.47)). In order to motivate our assumptions on $g(t,y)$ we begin with two examples.

Burgers' Equation. For this problem (Burgers 1948)

$$u_t+uu_x=\mu u_{xx}\qquad\text{or}\qquad u_t+\Big(\frac{u^2}{2}\Big)_x=\mu u_{xx},\qquad\mu>0,$$

we consider the discretization

$$\dot u_i=-\frac{u_{i+1}^2-u_{i-1}^2}{4\Delta x}+\mu\frac{u_{i+1}-2u_i+u_{i-1}}{(\Delta x)^2}.$$

It is of the form (8.39) with

$$A=\frac{\mu}{\Delta x^2}\begin{pmatrix} 2 & -1 & & & \\ -1 & 2 & -1 & & \\ & -1 & 2 & \ddots & \\ & & \ddots & \ddots & -1 \\ & & & -1 & 2 \end{pmatrix}, \qquad g(t,y)=\frac{1}{4\Delta x}\begin{pmatrix} y_2^2-y_0^2 \\ y_3^2-y_1^2 \\ \vdots \\ y_{n+1}^2-y_{n-1}^2 \end{pmatrix}, \tag{8.41}$$

where $\mu>0$ is a given constant, $\Delta x=1/(n+1)$ and, due to homogeneous boundary conditions, $y_0=y_{n+1}=0$. In this situation we work with the scaled norm (on $\mathbb{R}^n$)

$$\|u\|=\sqrt{\Delta x\sum_{i=1}^n |u_i|^2}, \tag{8.42}$$

which tends to that of $L^2(0,1)$ for $n\to\infty$. As the eigenvalues of the symmetric matrix A are real and positive, condition (8.40) is verified for every $\alpha'>0$, uniformly in Δx.

The presence of the denominator Δx in $g(t,y)$ of (8.41) does not allow a Lipschitz condition (8.37) uniformly in $\Delta x>0$ (not even in a neighbourhood of the exact solution). However, using the energy norm $\|A^{1/2}u\|$, which already contains the factor $1/\Delta x$, we show that

$$\|g(t,y)-g(t,z)\|\le \mu^{-1}r\cdot\|A^{1/2}(y-z)\| \qquad \text{for } \|A^{1/2}y\|+\|A^{1/2}z\|\le r. \tag{8.43}$$

For the proof of this relation we consider the bilinear map $b:\mathbb{R}^n\times\mathbb{R}^n\to\mathbb{R}^n$, whose ith component is defined by

$$b_i(u,v)=(4\Delta x)^{-1}(u_{i+1}+u_{i-1})(v_{i+1}-v_{i-1})$$

(again we put $u_0=v_0=u_{n+1}=v_{n+1}=0$). Then

$$g(t,y)-g(t,z)=b(y,y)-b(z,z)=b(y,y-z)+b(y-z,z), \tag{8.44}$$

and we need an estimate for $\|b(u,v)\|$. Using

$$|(u_{i+1}+u_{i-1})(v_{i+1}-v_{i-1})|\le 2\cdot|v_{i+1}-v_{i-1}|\cdot\max_j|u_j|,$$

and the estimates of Exercise 3 we obtain

$$\|b(u,v)\|\le\|u\|_\infty\cdot\|Dv\|\le\mu^{-1}\cdot\|A^{1/2}u\|\cdot\|A^{1/2}v\|. \tag{8.45}$$

where

$$D=\frac{1}{2\Delta x}\begin{pmatrix} 0 & 1 & & & \\ -1 & 0 & 1 & & \\ & -1 & 0 & \ddots & \\ & & \ddots & \ddots & 1 \\ & & & -1 & 0 \end{pmatrix} \tag{8.46}$$

represents the first central difference operator. The estimate (8.45) applied to (8.44) proves (8.43).

Incompressible Navier-Stokes Equation. The motion of a viscous incompressible fluid in a domain $\Omega \subset \mathbb{R}^d$ is governed by the equations of Navier (1823) and Stokes (1845)

$$\frac{\partial \mathbf{u}}{\partial t} + \sum_{i=1}^{d} u_i \frac{\partial \mathbf{u}}{\partial x_i} = \Delta \mathbf{u} - \mathbf{grad}\, p, \qquad \operatorname{div} \mathbf{u} = 0 \tag{8.47}$$

where $\mathbf{u} = (u_1, \ldots, u_d)^T$. We denote by P the orthogonal projection from $L^2(\Omega)^d$ onto X, where X is the subspace of functions with $\operatorname{div} \mathbf{u} = 0$ (more precisely: the closure of the set of smooth functions with vanishing divergence and support contained in Ω). If we apply P to Eq. (8.47), $\mathbf{grad}\, p$ is eliminated and we obtain

$$\frac{\partial \mathbf{u}}{\partial t} - P\Delta \mathbf{u} = -P\Big(\sum_{i=1}^{d} u_i \frac{\partial \mathbf{u}}{\partial x_i}\Big). \tag{8.48}$$

These equations are now precisely of the form (8.39), where $A = -P\Delta$ (or some discretization of it) and $g(t,y)$ is the right-hand side of (8.48). Lipschitz estimations for this nonlinear term have been obtained by Sobolevskiĭ (1959) and Fujita & Kato (1964). They are of the form

$$\|g(t,u) - g(t,v)\|_{\beta-\gamma} \le \ell(r) \cdot \|u - v\|_\beta \qquad \text{for} \quad \|u\|_\beta + \|v\|_\beta \le r \tag{8.49}$$

where $\|\cdot\|_\beta$ denotes the norm

$$\|u\|_\beta = \|A^\beta u\|. \tag{8.50}$$

In particular, for $d = 3$, condition (8.49) is true for $\beta = 1/2$, $\gamma \ge 3/4$ as well as for $\beta = \gamma > 3/4$ (Fujita & Kato 1964, pp. 272-273).

Motivated by these examples we consider the initial value problem (8.39) on $\mathbb{R}^n$, where A is supposed to satisfy (8.40) for some $\alpha' \in (0, \pi/2)$ and the nonlinearity $g(t,y)$ is assumed to satisfy the Lipschitz condition (8.49).

Application of a linear multistep method to (8.39) yields

$$\sum_{i=0}^{k} \alpha_i y_{m+i} + hA \sum_{i=0}^{k} \beta_i y_{m+i} = h \sum_{i=0}^{k} \beta_i g(t_{m+i}, y_{m+i}). \tag{8.51}$$

Instead of comparing the numerical solution $\{y_m\}$ with the analytic solution $y(t)$ of (8.39), it is more interesting to compare it with the exact solution of the original partial differential equation. We therefore denote by $\eta(t)$ a projection of the solution of the PDE into the finite-dimensional space under consideration. In this way we obtain

$$\eta' + A\eta = g(t,\eta) + s(t)$$

where $s(t)$ is the spatial discretization error.

Theorem 8.11 (Lubich 1991). *Consider the problem (8.39) with A and $g(t,y)$ satisfying (8.40) and (8.49) with $\gamma < 1$, respectively. Assume that the multistep method is of order p, $A(\alpha)$-stable for some $\alpha > \alpha'$, and strictly stable at infinity. Then, the full discretization error is bounded by*

$$\|y_m - \eta(t_m)\|_\beta \le C \cdot \Big(\max_{0 \le j < k} \|y_j - \eta(t_j)\|_\beta + h^p \int_0^{t_m} \|\eta^{(p+1)}(t)\|_\beta \, dt$$
$$+ \|A^{-1}s(0)\|_\beta + \int_0^{t_m} \|A^{-1}s'(t)\|_\beta \, dt \Big). \tag{8.52}$$

The estimate holds for $t_m = mh \le T$ provided that $h \le h_0$ and the expression in brackets on the right-hand side is bounded by ε, where h_0 and ε are sufficiently small. The constants C, h_0 and ε depend on $\max_{0\le t \le T} \|\eta(t)\|_\beta$ and M of (8.40), but are otherwise independent of A and the dimension of the system, and independent of m and h.

Proof. a) The projected solution $\eta(t)$ of the PDE, inserted into (8.51), gives

$$\sum_{i=0}^k \alpha_i \eta(t_{m+i}) = h \sum_{i=0}^k \beta_i \Big(-A\eta(t_{m+i}) + g(t_{m+i}, \eta(t_{m+i})) + s(t_{m+i}) \Big) + d_{m+k}$$

where

$$\|d_{m+k}\|_\beta \le C_0 h^p \int_{t_m}^{t_{m+k}} \|\eta^{(p+1)}(t)\|_\beta \, dt , \qquad m \ge 0. \tag{8.53}$$

The same analysis which was necessary for (8.36) now gives for the error $\Delta y_m = \eta(t_m) - y_m$ the relation

$$\Delta y_m = h \sum_{j=0}^m r_{m-j}(-hA)\Delta g_j + h \sum_{j=0}^m r_{m-j}(-hA)\Delta s_j + h \sum_{j=0}^m r_{m-j}(-hA) d_j . \tag{8.54}$$

As in (8.34) the quantities Δg_j and Δs_j are defined by

$$\Delta g_{m+k} = \sum_{i=0}^k \beta_i \Big(g(t_{m+i}, \eta(t_{m+i})) - g(t_{m+i}, y_{m+i}) \Big)$$
$$\Delta s_{m+k} = \sum_{i=0}^k \beta_i s(t_{m+i})$$

for $m \ge 0$, and $\Delta g_j = 0$, $\Delta s_j = 0$ for $j < k$. The values $d_0, \ldots, d_{k-1}$ are defined as usual (see their definition before (8.4')). The following three parts of the proof treat the three terms in the right-hand side of (8.54) separately.

b) The Lipschitz condition (8.49) can be written as

$$\big\| A^{-\gamma}\big(g(t,y) - g(t,z)\big) \big\|_\beta \le \ell(r) \cdot \|y - z\|_\beta \qquad \text{for} \qquad \|y\|_\beta + \|z\|_\beta \le r.$$

We put $\varrho = \max_{0 \le t \le T} \|\eta(t)\|_\beta$ and assume that for $hm \le T$ the numerical solution y_m exists and is bounded by $\|y_m\| \le \varrho + 1$ (this will be verified recursively in part

(f) of the proof) so that

$$\|A^{-\gamma}\Delta g_{m+k}\|_\beta \le \ell(2\varrho+1)\cdot\sum_{i=0}^{k}|\beta_i|\cdot\|\Delta y_{m+i}\|_\beta. \tag{8.55}$$

Consequently we have to find an estimate for $\|r_{m-j}(-hA)A^\gamma\|_\beta$ (for the matrix norm corresponding to the vector norm $\|\cdot\|_\beta$; see Sect. I.9). We note that $\|r_{m-j}(-hA)A^\gamma\|_\beta = \|A^\gamma r_{m-j}(-hA)\|$ and recall that $A^\gamma r_j(-hA)$ is the coefficient of ζ^j in the series for

$$A^\gamma r(\zeta,-hA) = A^\gamma\left(\delta(\zeta)I+hA\right)^{-1}\frac{\zeta^{-k}}{\sigma(\zeta^{-1})}.$$

In order to apply Lemma 7.11 we have to estimate $\Phi(s)=A^\gamma(sI+A)^{-1}$. If A can be transformed to diagonal form with an orthogonal matrix (as it is the case for (8.41)), we have for $|\arg s|\le\pi-\alpha'$ $(0<\alpha'<\alpha)$

$$\|A^\gamma(sI+A)^{-1}\|\le\sup_{a\ge0}\frac{a^\gamma}{|s+a|}\le M_1\cdot|s|^{\gamma-1}.$$

For the general case we refer the reader to Henry (1981, pp. 26-28). Application of Lemma 7.11 (see also Remark 7.12) yields

$$\|r_j(-hA)A^\gamma\|_\beta\le C_1\left((j+1)h\right)^{-\gamma}\qquad\text{for } j\ge0.$$

Together with the Lipschitz condition (8.55) this gives with $L=C_1\cdot\ell(2\varrho+1)$

$$h\Big\|\sum_{j=0}^{m}r_{m-j}(-hA)\,\Delta g_j\Big\|_\beta\le h^{1-\gamma}L\sum_{j=0}^{m}(m-j+1)^{-\gamma}\,\|\Delta y_j\|_\beta. \tag{8.56}$$

c) The second term in (8.54) is the coefficient of ζ^m in

$$hr(\zeta,-hA)\Delta s(\zeta)=\widetilde r(\zeta)\widetilde{\Delta s}(\zeta)$$

where we have introduced

$$\widetilde r(\zeta)=\left(\delta(\zeta)I+hA\right)^{-1}hA\,\delta(\zeta)^{-1}\frac{\zeta^{-k}}{\sigma(\zeta^{-1})}=\sum_{j\ge0}\widetilde r_j\,\zeta^j$$

$$\widetilde{\Delta s}(\zeta)=\delta(\zeta)\,A^{-1}\,\Delta s(\zeta)=\sum_{j\ge0}\widetilde{\Delta s}_j\,\zeta^j.$$

In order to estimate $\|\widetilde r_j\|_\beta$ (matrix norm) we note that $\|\widetilde r_j\|_\beta=\|\widetilde r_j\|$. In view of an application of Lemma 7.11 we have to consider $\Phi(s)=(sI+A)^{-1}As^{-1}=s^{-1}I-(sI+A)^{-1}$ which, because of (8.40), is bounded by $(M+1)/|s|$. Lemma 7.11 thus yields $\|\widetilde r_j\|_\beta\le C_2$. Further we have

$$\widetilde{\Delta s}(\zeta)=\frac{\delta(\zeta)}{1-\zeta}\cdot\Big(A^{-1}\Delta s_k\,\zeta^k+\sum_{j\ge k+1}A^{-1}(\Delta s_j-\Delta s_{j-1})\zeta^j\Big)$$

where the coefficients of $\delta(\zeta)/(1-\zeta)$ are absolutely summable, because the zeros of $\sigma(\zeta)$ lie all inside $|\zeta|<1$. Combining all these estimates we get

$$\begin{aligned} h\Big\| \sum_{j=0}^{m} r_{m-j}(-hA)\,\Delta s_j \Big\|_\beta &= \Big\| \sum_{j=0}^{m} \widetilde{r}_{m-j}\,\widetilde{\Delta s}_j \Big\|_\beta \\ &\le C_3\Big(\|A^{-1}\,\Delta s_k\|_\beta + \sum_{j=k+1}^{m} \|A^{-1}\,(\Delta s_j - \Delta s_{j-1})\|_\beta \Big) \\ &\le C_4\Big(\|A^{-1} s(0)\|_\beta + \int_0^{t_m} \|A^{-1}\, s'(t)\|_\beta\, dt \Big). \end{aligned} \tag{8.57}$$

d) The last term in (8.54) can be estimated in the same way as the corresponding term in the proof of Theorem 7.10. We just have to take the norm (8.50) and get

$$h\Big\| \sum_{j=0}^{m} r_{m-j}(-hA) d_j \Big\|_\beta \le C_5\Big(\max_{0\le j<k} \|y_j - \eta(t_j)\|_\beta + h^p \int_0^{t_m} \|\eta^{(p+1)}(t)\|_\beta\, dt \Big). \tag{8.58}$$

e) Inserting (8.56), (8.57), and (8.58) into (8.54) gives

$$\|\Delta y_m\|_\beta \le h^{1-\gamma} L \sum_{j=0}^{m} (m-j+1)^{-\gamma}\, \|\Delta y_j\|_\beta + C_6 \varepsilon_m \tag{8.59}$$

where $C_6 = \max(C_4, C_5)$ and ε_m denotes the expression in brackets on the right-hand side of (8.52). For $h \le h_0$ and $h_0^{1-\gamma} L < 1$ this Gronwall-type inequality can be solved (Exercise 2) and gives $\|\Delta y_m\| \le C_7 \varepsilon_m$, the desired result.

f) We now justify recurively our assumption $\|y_m\|_\beta \le \varrho + 1$ used in (b). Suppose that $\|y_j\|_\beta \le \varrho + 1$ for $j = 0, 1, \ldots, m-1$, then it follows from $h^{1-\gamma} L < 1$ and the contraction mapping theorem that a unique solution y_m of (8.54) exists. This solution verifies $\|y_m\|_\beta \le \|\eta(t_m)\|_\beta + \|\Delta y_m\|_\beta \le \varrho + 1$ if ε is small enough, more precisely, if $C_7 \varepsilon < 1$. □

Remark. A different approach to convergence results of multistep methods for nonlinear parabolic equations is given by Le Roux (1980). A corresponding theorem for Runge-Kutta methods is proved in Lubich & Ostermann (1993).

Exercises

1. Let $L \ge 0$ and consider two sequences $\{u_j\}$ and $\{\varepsilon_j\}$ of nonnegative numbers which satisfy

$$u_m \le hL \sum_{j=0}^{m} u_j + \sum_{j=0}^{m} \varepsilon_j \qquad \text{for} \quad m \ge 0.$$

Prove that for $hL \le 1 - C^{-1}$

$$u_m \le C e^{LCmh} \sum_{j=0}^{m} \varepsilon_j .$$

Hint. Show by induction that $v_m \le h\Lambda \sum_{j=0}^{m-1} v_j + M$ implies $v_m \le M(1+h\Lambda)^m \le Me^{\Lambda mh}$.

2. Consider the inequality (8.59) with $\gamma < 1$, $L \ge 0$, $\varepsilon_m \ge 0$ and $h > 0$. Under the assumptions $h \le h_0$ and $h_0^{1-\gamma} L < 1$ prove that there exists a constant C such that $\|\Delta y_m\|_\beta \le C\varepsilon_m$ for $mh \le T$.

   Hint. Move the term $h^{1-\gamma} L \|\Delta y_m\|_\beta$ to the left and divide the inequality by $(1 - h^{1-\gamma}L)$. This yields

$$\|\Delta y_m\|_\beta \le h^{1-\gamma} \widehat{L} \sum_{j=0}^{m-1} (m-j)^{-\gamma} \|\Delta y_j\|_\beta + \widehat{\varepsilon} \quad \text{for} \quad m \ge 0.$$

   Show that $\|\Delta y_m\|_\beta \le \widehat{\varepsilon} u(mh)$, where $u(x)$ is the solution of

$$u(x) = 1 + \widehat{L} \int_0^x (x-t)^{-\gamma} u(t)\, dt. \tag{8.60}$$

   Estimate the solution of (8.60) (see Henry 1981, pp. 188-190).

3. Let A and D be the matrices of (8.41) (suppose $\mu = 1$) and (8.46). Prove that for all $u \in \mathbb{R}^n$

$$\text{a)} \quad \|u\|_\infty \le \|A^{1/2} u\|, \qquad \text{b)} \quad \|Du\| \le \|A^{1/2} u\|,$$

   where $\|u\|_\infty = \max_i |u_i|$ and $\|\cdot\|$ is the norm of (8.42).
   Hint. a) Let $u_0 = 0$ and apply the inequality of Cauchy-Bunyakovski-Schwarz to $u_i = \sum_{j=1}^{i} (u_j - u_{j-1})$. This gives

$$\|u\|_\infty^2 \le \sum_{j=1}^{n} \Big(\frac{u_j - u_{j-1}}{\Delta x} \Big)^2 = u^T A u.$$

   b) The inequality $u^T A u \ge \|Du\|^2$ is a consequence of the algebraic identity $(u_0 = u_{n+1} = 0)$

$$4 \sum_{i=1}^{n} (2u_i^2 - u_i u_{i+1} - u_i u_{i-1}) - \sum_{i=1}^{n} (u_{i+1} - u_{i-1})^2 = \sum_{i=1}^{n} (u_{i+1} - 2u_i + u_{i-1})^2 + 2u_1^2 + 2u_n^2.$$

V.9 Algebraic Stability of General Linear Methods

General linear methods were originally introduced as a means of unifying and generalizing existing theories for traditional methods. (J.C. Butcher 1987)

In Sections IV.12 and V.6 we have studied the nonlinear stability of Runge-Kutta methods (B-stability) and of one-leg methods (G-stability). It is natural to ask whether these theories can be combined within the class of general linear methods. This work was initiated by Burrage & Butcher (1980).

We consider the differential equation $y' = f(x, y)$ where y and f are complex-valued vectors and we assume the one-sided Lipschitz condition

$$\operatorname{Re}\langle f(x,y) - f(x,z),\ y - z\rangle \le \nu \|y - z\|^2. \tag{9.1}$$

General linear methods are defined by (see Example 8.5 of Sect. III.8)

$$u_i^{(n+1)} = \sum_{j=1}^{k} a_{ij} u_j^{(n)} + h \sum_{j=1}^{s} b_{ij} f(x_n + c_j h, v_j^{(n)}), \quad i = 1, \ldots, k \tag{9.2a}$$

$$v_i^{(n)} = \sum_{j=1}^{k} \widetilde{a}_{ij} u_j^{(n)} + h \sum_{j=1}^{s} \widetilde{b}_{ij} f(x_n + c_j h, v_j^{(n)}), \quad i = 1, \ldots, s. \tag{9.2b}$$

Here, $u_n = (u_1^{(n)}, \ldots, u_k^{(n)})^T$ contains the necessary information from the previous step. The internal stages $(v_1^{(n)}, \ldots, v_s^{(n)})$, defined by (9.2b), serve for the computation of u_{n+1} in (9.2a).

G-Stability

As in Sect. V.6, we consider inner product norms

$$\|u_n\|_G^2 = \sum_{i=1}^{k} \sum_{j=1}^{k} g_{ij} \langle u_i^{(n)}, u_j^{(n)} \rangle, \tag{9.3}$$

where $G = (g_{ij})$ is a real, symmetric and positive definite matrix.

Definition 9.1. The general linear method (9.2) is called *G-stable*, if there exists a real, symmetric and positive definite matrix G, such that for two numerical solutions $\{u_n\}$ and $\{\widehat{u}_n\}$,

$$\|u_{n+1} - \widehat{u}_{n+1}\|_G \le \|u_n - \widehat{u}_n\|_G \tag{9.4}$$

for all step sizes $h > 0$ and for all differential equations satisying (9.1) with $\nu = 0$.

For Runge-Kutta methods (where $k=1$ and apart from a scaling factor $G=(1)$) this definition reduces to B-stability as introduced in Definition IV.12.2. For one-leg methods (where $s=1$ and $u_n=(y_{n+k-1},\dots,y_n)^T$) it is equivalent to Definition 6.3.

Many methods can be written in different ways as general linear methods and the above definition of G-stability may depend on the particular formulation. For example, the trapezoidal rule

$$y_{n+1}=y_n+\frac{h}{2}\big(f(x_n,y_n)+f(x_{n+1},y_{n+1})\big)$$

can be considered as a Runge-Kutta method (with $u_n=y_n$). In this case it is not G-stable (because it is not B-stable, see Theorem IV.12.12). However, if we let $u_n=(y_n,hy_n')$ where $y_n'=f(x_n,y_n)$, then the trapezoidal rule satisfies (9.4) with

$$G=\begin{pmatrix}1 & 1/2\\ 1/2 & 1/4\end{pmatrix}. \tag{9.5}$$

This follows from the fact that whenever $\{y_n\}$ is the solution obtained by the trapezoidal rule, then $z_n=y_n+\frac{h}{2}y_n'$ is a solution of the implicit midpoint rule, which is known to be B-stable (see Example IV.12.3 or Theorem IV.12.9). Therefore

$$\|y_{n+1}+\frac{h}{2}y_{n+1}'\|\le\|y_n+\frac{h}{2}y_n'\|$$

which proves the statement. The matrix G in (9.5) is singular and thus not strictly positive definite. Burrage & Butcher (1980), however, admit non-zero non-negative definite matrices G in their definition of G-stability (which they call *monotonicity*). Therefore the trapezoidal rule is G-stable in their definition.

Algebraic Stability

In addition to (9.2) we consider a second numerical solution (marked with hats) produced by the same method using different starting values. We denote the differences by

$$\begin{aligned}\Delta u_i^{(n)}&=u_i^{(n)}-\widehat{u}_i^{(n)}, &\qquad \Delta u_n&=u_n-\widehat{u}_n\\ \Delta v_i^{(n)}&=v_i^{(n)}-\widehat{v}_i^{(n)}, &\qquad \Delta f_i^{(n)}&=h\big(f(x_n+c_ih,v_i^{(n)})-f(x_n+c_ih,\widehat{v}_i^{(n)})\big).\end{aligned}$$

The following lemma states an identity which will be essential in the study of G-stability.

Lemma 9.2 (Burrage & Butcher 1980). *Let G be a real, symmetric matrix and $D=\operatorname{diag}(d_1,\dots,d_s)$ be a real diagonal matrix. The difference of two solutions of (9.2) then satisfies*

$$\|\Delta u_{n+1}\|_G^2-\|\Delta u_n\|_G^2=2\sum_{i=1}^{s}d_i\operatorname{Re}\langle\Delta f_i^{(n)},\Delta v_i^{(n)}\rangle-\sum_{i,j=1}^{s+k}m_{ij}\langle w_i,w_j\rangle$$

where $(w_1,\dots,w_{s+k}) = (\Delta u_1^{(n)},\dots,\Delta u_k^{(n)},\ \Delta f_1^{(n)},\dots,\Delta f_s^{(n)})$ *and the matrix* $M=(m_{ij})$ *is given by*

$$M = \begin{pmatrix} G - A^T G A & \widetilde{A}^T D - A^T G B \\ D\widetilde{A} - B^T G A & D\widetilde{B} + \widetilde{B}^T D - B^T G B \end{pmatrix}. \tag{9.6}$$

Proof. We consider the identity

$$\begin{aligned} \|\Delta u_{n+1}\|_G^2 - \|\Delta u_n\|_G^2 - 2\sum_{i=1}^{s} d_i \mathrm{Re}\,\langle \Delta f_i^{(n)}, \Delta v_i^{(n)}\rangle \\ = \sum_{i,j=1}^{k} g_{ij}\langle \Delta u_i^{(n+1)}, \Delta u_j^{(n+1)}\rangle - \sum_{i,j=1}^{k} g_{ij}\langle \Delta u_i^{(n)}, \Delta u_j^{(n)}\rangle \\ - \sum_{i=1}^{s} d_i\langle \Delta f_i^{(n)}, \Delta v_i^{(n)}\rangle - \sum_{i=1}^{s} d_i\langle \Delta v_i^{(n)}, \Delta f_i^{(n)}\rangle \end{aligned}$$

and insert the formulas (9.2). This gives

$$\begin{aligned} \ldots = \sum_{i,j=1}^{k} g_{ij}\Big\langle \sum_{\ell=1}^{k} a_{i\ell}\Delta u_\ell^{(n)} + h\sum_{\ell=1}^{s} b_{i\ell}\Delta f_\ell^{(n)},\ \sum_{\ell=1}^{k} a_{j\ell}\Delta u_\ell^{(n)} + h\sum_{\ell=1}^{s} b_{j\ell}\Delta f_\ell^{(n)}\Big\rangle \\ - \sum_{i,j=1}^{k} g_{ij}\langle \Delta u_i^{(n)}, \Delta u_j^{(n)}\rangle - \sum_{i=1}^{s} d_i\Big\langle \Delta f_i^{(n)}, \sum_{\ell=1}^{k} \widetilde{a}_{i\ell}\Delta u_\ell^{(n)} + h\sum_{\ell=1}^{s} \widetilde{b}_{i\ell}\Delta f_\ell^{(n)}\Big\rangle \\ - \sum_{i=1}^{s} d_i\Big\langle \sum_{\ell=1}^{k} \widetilde{a}_{i\ell}\Delta u_\ell^{(n)} + h\sum_{\ell=1}^{s} \widetilde{b}_{i\ell}\Delta f_\ell^{(n)},\ \Delta f_i^{(n)}\Big\rangle. \end{aligned}$$

Multiplying out and collecting suitable terms proves the statement. □

Definition 9.3. The general linear method (9.2) is called *algebraically stable*, if there exist a real, symmetric and positive definite matrix G and a real non-negative definite diagonal matrix D, such that the matrix M of (9.6) is non-negative definite.

An immediate consequence of our assumption (9.1) with $\nu = 0$ and of Lemma 9.2 is the following result.

Theorem 9.4. *Algebraic stability implies* G*-stability.* □

For a given method it may be difficult to find matrices D and G such that M of (9.6) is non-negative definite. The following lemma shows some useful relations, which hold if the method is assumed to be *preconsistent*, i.e., if there exists a vector $\xi_0 \in \mathbb{R}^k$ such that

$$A\xi_0 = \xi_0, \qquad \widetilde{A}\xi_0 = 1\!\!1 \tag{9.7}$$

(cf. Eq. (8.25) of Sect. III.8).

Lemma 9.5. *If a general linear method is preconsistent and algebraically stable, then the matrices D and G satisfy*

i) $(d_1,\ldots,d_s)^T = D1\!\!1 = B^T G\xi_0$,

ii) $(I-A^T)G\xi_0 = 0$, *i.e., $G\xi_0$ is a left-eigenvector of A corresponding to the eigenvalue 1.*

Proof. i) Let $\eta \in \mathbb{R}^s$ and $\varepsilon \in \mathbb{R}$ be arbitrary. The non-negativity of M, given by (9.6), implies

$$(\xi_0^T, \varepsilon\eta^T)M\begin{pmatrix}\xi_0\\ \varepsilon\eta\end{pmatrix} \geq 0$$

so that

$$\xi_0^T(G-A^TGA)\xi_0 + 2\varepsilon\eta^T(D\widetilde{A}-B^TGA)\xi_0 + \varepsilon^2\eta^T(D\widetilde{B}+\widetilde{B}^TD-B^TGB)\eta \geq 0.$$

Since the ε-independent term vanishes (due to $A\xi_0 = \xi_0$), the coefficient of ε must be zero and since this holds for all η, the result follows.

ii) A similar argument applied to

$$(\xi_0+\varepsilon\xi_1)^T(G-A^TGA)(\xi_0+\varepsilon\xi_1) \geq 0 \quad \text{for all} \quad \xi_1 \in \mathbb{R}^k,\ \varepsilon \in \mathbb{R}$$

implies the second statement. □

AN-Stability and Equivalence Results

It is interesting to study in which situation algebraic stability is also necessary for G-stability. For this we consider the differential equation

$$y' = \lambda(x)y \qquad \text{with} \qquad \operatorname{Re}\lambda(x) \leq 0.$$

If we apply the general linear method (9.2) to this problem, we obtain

$$u_{n+1} = S(Z)u_n \tag{9.8}$$

where $Z = \operatorname{diag}(z_1,\ldots,z_s)$, $z_j = h\lambda(x_n + c_j h)$ and

$$S(Z) = A + BZ(I-\widetilde{B}Z)^{-1}\widetilde{A}. \tag{9.9}$$

In the sequel we assume that the abscissae c_j are related to the other coefficients of the method by (see also Remark III.8.17)

$$(c_1,\ldots,c_s)^T = c = \widetilde{A}\xi_1 + \widetilde{B}1\!\!1, \tag{9.10}$$

where $\xi_1 \in \mathbb{R}^k$ is the second coefficient vector of the exact value function

$$z(x,h) = y(x)\xi_0 + hy'(x)\xi_1 + \mathcal{O}(h^2).$$

This means that the internal stages approximate the exact solution as $v_j^{(n)} = y(x_n + c_j h) + \mathcal{O}(h^2)$.

Definition 9.6. A general linear method is called *AN-stable*, if there exists a real, symmetric and positive definite matrix G such that

$$\|S(Z)u\|_G \le \|u\|_G \qquad \begin{array}{l} \text{for all } Z = \operatorname{diag}(z_1,\ldots,z_s) \text{ satisfying } \operatorname{Re} z_j \le 0 \\ (j=1,\ldots,s) \text{ and } z_j = z_k \text{ whenever } c_j = c_k. \end{array}$$

Other possible definitions of AN-stability are given in Butcher (1987). For example, if the condition $\|S(Z)u\|_G \le \|u\|_G$ is replaced by the powerboundedness of the matrix $S(Z)$, the method is called *weakly AN-stable*. This definition, however, does not allow the values $z_j = h\lambda(x_n + c_j h)$ to change at each step. Another modification is to consider arbitrary norms (instead of inner product norms only) in the definition of AN-stability. Butcher (1987) has shown that this does not lead to a larger class of AN-stable methods, but makes the analysis much more difficult.

We are now interested in the relations between the various stability definitions: the implications

$$\text{algebraically stable} \implies G\text{-stable} \implies AN\text{-stable} \implies A\text{-stable}$$

are either trivial or follow from Theorem 9.4. We also know that A-stability does not, in general, imply AN-stability (see e.g., Theorem IV.12.12). The following result shows that the other two implications are (nearly always) reversible.

Theorem 9.7 (Butcher 1987). *For preconsistent and non-confluent general linear methods (i.e., methods with distinct c_j) we have*

$$\textit{algebraically stable} \iff G\textit{-stable} \iff AN\textit{-stable}.$$

Proof. It is sufficient to prove that AN-stability implies algebraic stability. For this we take the matrix G, whose existence is known by the definition of AN-stability, and show that the matrices D and M, given by Lemma 9.5i and (9.6), are non-negative definite.

In order to prove $d_j \ge 0$ we put $z_j = -\varepsilon$ ($\varepsilon > 0$) and $z_k = 0$ for $k \ne j$. We further let $\Delta u_n = \xi_0$ (the preconsistency vector of (9.7)) and $\Delta f_\ell^{(n)} = z_\ell \Delta v_\ell^{(n)}$, so that $\Delta u_{n+1} = S(Z)\xi_0$ and $\Delta v_\ell^{(n)} = 1 + \mathcal{O}(\varepsilon)$. Using

$$M\begin{pmatrix} \xi_0 \\ 0 \end{pmatrix} = 0, \tag{9.11}$$

which is a consequence of Lemma 9.5, the identity of Lemma 9.2 yields

$$\|S(Z)\xi_0\|_G^2 - \|\xi_0\|_G^2 = -2\varepsilon d_j + \mathcal{O}(\varepsilon^2).$$

Since the left-hand side of this equation is non-positive by AN-stability, we obtain $d_j \geq 0$.

We next put $z_\ell = i\varepsilon\eta_\ell$ where $\eta = (\eta_1, \ldots, \eta_s)^T \in \mathbb{R}^s$ is arbitrary and ε is a small real parameter. We further put $\Delta u_n = \xi_0 + i\varepsilon\mu$ with $\mu \in \mathbb{R}^k$ and $\Delta f_\ell^{(n)} = z_\ell \Delta v_\ell^{(n)}$. This again implies $\Delta v_\ell^{(n)} = 1 + \mathcal{O}(\varepsilon)$. The identity of Lemma 9.2 together with (9.11) gives

$$\begin{aligned}\|S(Z)\xi_0\|_G^2 - \|\xi_0\|_G^2 &= -(\xi_0 - i\varepsilon\mu, i\varepsilon\eta + \mathcal{O}(\varepsilon^2))M\begin{pmatrix}\xi_0 + i\varepsilon\mu \\ i\varepsilon\eta + \mathcal{O}(\varepsilon^2)\end{pmatrix} = \\ &= -\varepsilon^2(\mu,\eta)^T M\begin{pmatrix}\mu \\ \eta\end{pmatrix} + \mathcal{O}(\varepsilon^3).\end{aligned}$$

Since this relation holds for all μ and η, the matrix M has to be non-negative definite. □

Example 9.8. Let us investigate the G-stability of *multistep collocation methods* as introduced in Sect. V.3. We consider here the case $k = 2$ and $s = 2$, and fix one collocation point at $c_2 = 1$. The method is then given by

$$\begin{aligned}\begin{pmatrix} y_{n+1} \\ y_n \end{pmatrix} &= \overbrace{\begin{pmatrix} 1-\varphi(1) & \varphi(1) \\ 1 & 0 \end{pmatrix}}^{A} \begin{pmatrix} y_n \\ y_{n-1} \end{pmatrix} \\ &\quad + h \overbrace{\begin{pmatrix} \psi_1(1) & \psi_2(1) \\ 0 & 0 \end{pmatrix}}^{B} \begin{pmatrix} f(x_n + c_1 h, v_1) \\ f(x_n + h, v_2) \end{pmatrix} \\ \begin{pmatrix} v_1 \\ v_2 \end{pmatrix} &= \underbrace{\begin{pmatrix} 1-\varphi(c_1) & \varphi(c_1) \\ 1-\varphi(1) & \varphi(1) \end{pmatrix}}_{\widetilde{A}} \begin{pmatrix} y_n \\ y_{n-1} \end{pmatrix} \\ &\quad + h \underbrace{\begin{pmatrix} \psi_1(c_1) & \psi_2(c_1) \\ \psi_1(1) & \psi_2(1) \end{pmatrix}}_{\widetilde{B}} \begin{pmatrix} f(x_n + c_1 h, v_1) \\ f(x_n + h, v_2) \end{pmatrix}\end{aligned} \tag{9.12}$$

where

$$\begin{aligned}\varphi(x) &= -\frac{6}{5+9c_1}\Big(\frac{x^3}{3} - \frac{x^2}{2}(1+c_1) + xc_1\Big) \\ \psi_1(x) &= \frac{x(x+1)}{(1-c_1)(5+9c_1)}(5-3x) \\ \psi_2(x) &= \frac{x(x+1)}{(1-c_1)(5+9c_1)}\big((2c_1+1)x - c_1(3c_1+2)\big).\end{aligned}$$

We know from Exercise V.3.7 that the method is A-stable if and only if $c_1 \geq (\sqrt{17}-1)/8$. For the study of its G-stability we assume that after an appropriate scaling of G, $g_{11} = 1$. By Lemma 9.5ii the matrix G must then be of the form (recall that $\xi_0 = (1,1)^T$)

$$G = \begin{pmatrix} 1 & \gamma - 1 \\ \gamma - 1 & (\varphi(1) - 1)\gamma + 1 \end{pmatrix}. \tag{9.13}$$

A necessary condition for G to be positive definite is that $\det G > 0$. For $c_1 \geq 0$ this is equivalent to

$$0 < \gamma < \frac{6(1+c_1)}{5+9c_1}. \tag{9.14}$$

Next we use Lemma 9.5i which implies that

$$d_1 = \gamma\psi_1(1), \qquad d_2 = \gamma\psi_2(1). \tag{9.15}$$

Inserting (9.13) and (9.15) into the matrix M of (9.6) yields for its lower right block

$$\begin{pmatrix} \psi_1(1) & 0 \\ 0 & \psi_2(1) \end{pmatrix} \begin{pmatrix} 2\gamma\chi_1 - 1 & (\chi_2+1)\gamma - 1 \\ (\chi_2+1)\gamma - 1 & 2\gamma - 1 \end{pmatrix} \begin{pmatrix} \psi_1(1) & 0 \\ 0 & \psi_2(1) \end{pmatrix} \tag{9.16}$$

where

$$\chi_1 = \frac{\psi_1(c_1)}{\psi_1(1)} = \frac{1}{4}c_1(c_1+1)(5-3c_1), \qquad \chi_2 = \frac{\psi_2(c_1)}{\psi_2(1)} = \frac{c_1^2(c_1+1)^2}{2(3c_1^2-1)}.$$

A direct computation (see Exercise 2) shows that this 2×2 matrix can not be non-negative definite for $c_1 \geq (\sqrt{17}-1)/8$ and γ satisfying (9.14). Consequently the considered methods are never G-stable.

In the next subsections we shall show how high-order algebraically stable general linear methods can be constructed.

Multistep Runge-Kutta Methods

An interesting extension of multistep collocation methods are the so-called multistep Runge-Kutta methods. They are defined by the formulas

$$\begin{aligned} y_{n+1} &= \sum_{j=1}^{k} \alpha_j y_{n+1-j} + h\sum_{j=1}^{s} b_j f(x_n + c_j h, v_j^{(n)}) \\ v_i^{(n)} &= \sum_{j=1}^{k} \widetilde{a}_{ij} y_{n+1-j} + h\sum_{j=1}^{s} \widetilde{b}_{ij} f(x_n + c_j h, v_j^{(n)}). \end{aligned} \tag{9.17}$$

They obviously form a subclass of the general linear methods (9.2). This is seen by putting $u_n = (y_n, y_{n-1}, \ldots, y_{n-k+1})^T$ so that the exact value function is

$$z(x,h) = \big(y(x), y(x-h), \ldots, y(x-(k-1)h)\big)^T.$$

Further, the matrices A and B have the special form

$$A=\begin{pmatrix} \alpha_1 & \dots & \dots & \alpha_k \\ 1 & & & 0 \\ & \ddots & & \vdots \\ & & 1 & 0 \end{pmatrix}, \qquad B=\begin{pmatrix} b_1 & \dots & b_s \\ 0 & \dots & 0 \\ \vdots & & \vdots \\ 0 & \dots & 0 \end{pmatrix}. \tag{9.18}$$

The order conditions for such methods were derived in Theorem III.8.14. It follows from this theorem that the method (9.17) is of order p, iff

$$1=\sum_{j=1}^{k}\alpha_j(1-j)^{\varrho(t)}+\sum_{j=1}^{s}b_j\mathbf{v}_j'(t) \qquad \text{for} \quad t\in T,\ \ \varrho(t)\le p. \tag{9.19}$$

The values $\mathbf{v}_j'(t)$ are given recursively by

$$\mathbf{v}_i(t)=\sum_{j=1}^{k}\widetilde{a}_{ij}(1-j)^{\varrho(t)}+\sum_{j=1}^{s}\widetilde{b}_{ij}\mathbf{v}_j'(t). \tag{9.20}$$

Recall from Corollary II.12.7 that

$$\begin{aligned} &\mathbf{v}_j'(\emptyset)=0, \qquad \mathbf{v}_j'(\tau)=1 \\ &\mathbf{v}_j'(t)=\varrho(t)\mathbf{v}_j(t_1)\cdot\ldots\cdot\mathbf{v}_j(t_m) \quad \text{if} \quad t=[t_1,\ldots,t_m]. \end{aligned} \tag{9.21}$$

The order conditions (9.19) constitute a system of nonlinear equations in the coefficients of the method. Without any preparation, solving them may be difficult. We therefore introduce additional assumptions which simplify the construction of multistep Runge-Kutta methods.

Simplifying Assumptions

The conditions $B(p)$, $C(\eta)$ and $D(\xi)$ of Sect. IV.5 were useful for the construction of high-order implicit Runge-Kutta methods. Burrage (1988) showed how these simplifying assumptions can be extended to general linear methods. In the sequel we specialize his approach to multistep Runge-Kutta methods. We consider the assumptions

$$B(p): \qquad q\sum_{j=1}^{s}b_jc_j^{q-1}+\sum_{j=1}^{k}\alpha_j(1-j)^q=1 \qquad q=1,\ldots,p;$$

$$C(\eta): \qquad q\sum_{j=1}^{s}\widetilde{b}_{ij}c_j^{q-1}+\sum_{j=1}^{k}\widetilde{a}_{ij}(1-j)^q=c_i^q \qquad q=1,\ldots,\eta,\ \text{all } i;$$

$$D_A(\xi): \qquad q\sum_{i=1}^{s}b_ic_i^{q-1}\widetilde{a}_{ij}=\alpha_j\big(1-(1-j)^q\big) \qquad q=1,\ldots,\xi,\ \text{all } j;$$

$$D_B(\xi): \qquad q\sum_{i=1}^{s}b_ic_i^{q-1}\widetilde{b}_{ij}=b_j(1-c_j^q) \qquad q=1,\ldots,\xi,\ \text{all } j.$$

Condition $B(p)$ is equivalent to the order conditions (9.19) for bushy trees. Condition $C(\eta)$ means that $\mathbf{v}_j(t)$, defined by (9.20), satisfies

$$\mathbf{v}_j(t) = c_j^{\varrho(t)} \quad \text{for} \quad \varrho(t) \le \eta. \tag{9.22}$$

We remark that the preconsistency condition (9.7) with $\xi_0 = (1,\dots,1)^T$,

$$\sum_{j=1}^{k} \alpha_j = 1, \qquad \sum_{j=1}^{k} \widetilde{a}_{ij} = 1 \quad \text{for} \quad i = 1,\dots,s, \tag{9.23}$$

is obtained by putting $q = 0$ in $B(p)$ and $C(\eta)$. The condition $D(\xi)$ for Runge-Kutta methods splits into $D_A(\xi)$ and $D_B(\xi)$. However, under certain assumptions one of these conditions is automatically satisfied.

Lemma 9.9. *Suppose that the coefficients $c_1,\dots,c_s$ of a multistep Runge-Kutta method are distinct and $b_i \neq 0$. Then,*

i) $B(\xi + k - 1),\ C(k-1),\ D_B(\xi) \Longrightarrow D_A(\xi)$,

ii) $B(\xi + s),\ C(s),\ D_A(\xi) \Longrightarrow D_B(\xi)$,

iii) $B(\eta + s),\ D_A(s),\ D_B(s) \Longrightarrow C(\eta)$.

Proof. The first two implications are a consequence of the identity

$$\sum_{j=1}^{k}\Big(q\sum_{i=1}^{s} b_i c_i^{q-1}\widetilde{a}_{ij} - \alpha_j\big(1-(1-j)^q\big)\Big)(1-j)^\ell$$
$$= -\ell\sum_{j=1}^{s}\Big(q\sum_{i=1}^{s} b_i c_i^{q-1}\widetilde{b}_{ij} - b_j(1-c_j^q)\Big)c_j^{\ell-1}$$

which holds under the assumptions $C(\ell)$ and $B(q+\ell)$. The last implication can be proved similarly. □

The fundamental theorem, which generalizes Theorem IV.5.1, is

Theorem 9.10 (Burrage 1988). *If the coefficients of a multistep Runge-Kutta method (9.17) satisfy the simplifying assumptions $B(p)$, $C(\eta)$, $D_A(\xi)$, $D_B(\xi)$ with $p \le \eta + \xi + 1$ and $p \le 2\eta + 2$, then the method is of order p.*

Proof. The conditions $C(\eta)$ and $D_A(\xi)$, $D_B(\xi)$ allow the reduction of order conditions of trees as sketched in Fig. 7.1 and Fig. 7.2 of Sect. II.7, respectively. Under the restrictions $p \le \eta + \xi + 1$ and $p \le 2\eta + 2$ all order conditions reduce to those for bushy trees which are satisfied by $B(p)$. □

Remember that we are searching for high-order algebraically stable methods. Due to the Daniel-Moore conjecture (Theorem V.4.4) the order is restricted by $p \le 2s$. It is therefore natural to look for methods satisfying $B(2s)$, $C(s)$ and

$D_A(s)$, $D_B(s)$. They will be of order $2s$ by Theorem 9.10 and are an extension of the Runge-Kutta methods based on Gauss quadrature. Let us begin by studying the condition $B(2s)$.

Quadrature Formulas

Because of (9.23) condition $B(p)$ of the preceding subsection is equivalent to

$$\sum_{j=1}^{s} b_j f(c_j) = \sum_{j=1}^{k} \alpha_j \int_{1-j}^{1} f(x)dx, \qquad \deg f \le p-1, \tag{9.24}$$

where f stands for a polynomial of degree at most $p-1$. For the construction of such quadrature formulas it is useful to consider the bilinear form

$$\langle f, g \rangle = \sum_{j=1}^{k} \alpha_j \int_{1-j}^{1} f(x)g(x)\,dx = \int_{1-k}^{1} \omega(x)f(x)g(x)\,dx, \tag{9.25}$$

where $\omega(x)$ is the step-function sketched in Fig. 9.1. Under the assumption

$$\alpha_k \ge 0, \quad \alpha_k + \alpha_{k-1} \ge 0, \ldots, \quad \alpha_k + \ldots + \alpha_2 \ge 0, \quad \alpha_k + \ldots + \alpha_1 = 1, \tag{9.26}$$

$\omega(x)$ is non-negative and (9.25) becomes an inner product on the space of real polynomials. We call the quadrature formula (9.24) *interpolatory* if $B(s)$ holds. This implies that

$$b_i = \int_{1-k}^{1} \omega(x)\ell_i(x)\,dx, \qquad \ell_i(x) = \prod_{\substack{l=1 \\ l \ne i}}^{s} \frac{(x-c_l)}{(c_i-c_l)}. \tag{9.27}$$

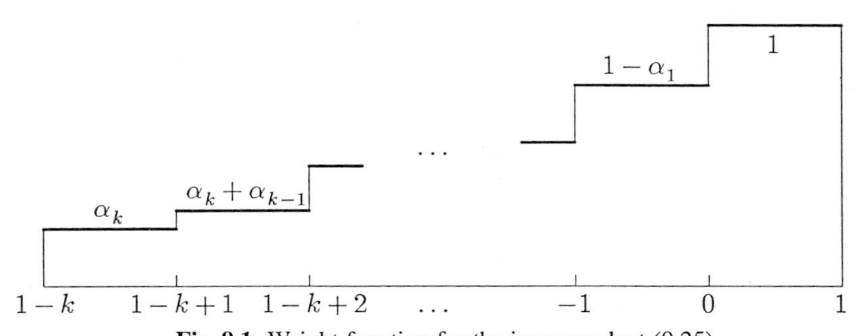

Fig. 9.1. Weight function for the inner product (9.25)

The following results on Gaussian quadrature and orthogonal polynomials are classical.

Lemma 9.11. *Let $M(x)=(x-c_1)\cdot\ldots\cdot(x-c_s)$. An interpolatory quadrature formula satisfies $B(s+m)$ if and only if*

$$\sum_{j=1}^{k}\alpha_j\int_{1-j}^{1}M(x)\,x^{q-1}\,dx=0\quad \textit{for}\quad q=1,\ldots,m. \qquad \square$$

Let $p_s(x)$ be the polynomial of degree s which is orthogonal with respect to (9.25) to all polynomials of degree $s-1$. Lemma 9.11 then states that a quadrature formula (9.24) is of order $2s$ iff $M(x)$ is a scalar multiple of $p_s(x)$. The polynomials $p_s(x)$ which depend on $\alpha_1,\ldots,\alpha_k$ via the bilinear form (9.25) can be computed from a standard three term recursion

$$\begin{aligned} &p_0(x)=1, \qquad p_1(x)=x-\beta_0\\ &p_{s+1}(x)=(x-\beta_s)p_s(x)-\gamma_s p_{s-1}(x)\end{aligned} \tag{9.28}$$

where

$$\beta_s=\frac{\langle xp_s,p_s\rangle}{\langle p_s,p_s\rangle},\qquad \gamma_s=\frac{\langle p_s,p_s\rangle}{\langle p_{s-1},p_{s-1}\rangle}. \tag{9.29}$$

Obviously this is only possible if $\langle p_j,p_j\rangle\neq 0$ for $j=1,\ldots,s$. This is certainly the case under the assumption (9.26).

Lemma 9.12. *If $\alpha_1,\ldots,\alpha_k$ satisfy (9.26) then all zeros of $p_s(x)$ are real, simple and lie in the open interval $(1-k,1)$.* $\square$

For the construction of algebraically stable methods, quadrature formulas with positive weights will be of particular interest. Sufficient conditions for this property are given in the following theorem.

Theorem 9.13. *If the quadrature formula (9.24) is of order $p\geq 2s-1$ and if $\alpha_1,\ldots,\alpha_k$ satisfy (9.26), then*

$$b_i>0\qquad \textit{for}\qquad i=1,\ldots,s. \qquad \square$$

Algebraically Stable Methods of Order $2s$

> ... the analysis of the algebraic stability properties of multivalue methods ... is not as difficult as was generally thought ...
> (Burrage 1987)

Following Burrage (1987) we consider the following class of multistep Runge-Kutta methods.

Definition 9.14. Let $\alpha_1,\ldots,\alpha_k$ with $\sum\alpha_j=1$ and $\alpha_k\neq 0$ be given such that the zeros $c_1,\ldots,c_s$ of $p_s(x)$ (Formula (9.28)) are real and simple. We then denote by $E(\alpha_1,\ldots,\alpha_k)$ the multistep Runge-Kutta method (9.17) whose coefficients are given by

$$b_i = \sum_{j=1}^{k} \alpha_j \int_{1-j}^{1} \ell_i(x)dx, \qquad i=1,\ldots,s,$$

$$\widetilde{a}_{ij} = \frac{\alpha_j}{b_j} \int_{1-j}^{1} \ell_i(x)dx, \qquad i=1,\ldots,s;\ j=1,\ldots k$$

$$\widetilde{b}_{ij} = \frac{b_j}{b_i} \int_{c_j}^{1} \ell_i(x)dx, \qquad i=1,\ldots,s;\ j=1,\ldots s$$

where $\ell_i(x)$ is the function of (9.27).

The definitions of c_i and b_i imply $B(2s)$ by Lemma 9.11. The formulas for $\widetilde{a}_{ij}$ and $\widetilde{b}_{ij}$ are equivalent to $D_A(s)$ and $D_B(s)$, respectively. Lemma 9.9iii thus implies $C(s)$ and Theorem 9.10 finally proves that the considered methods are of order $2s$. The following theorem gives sufficient conditions for the algebraic stability of these methods.

Theorem 9.15 (Burrage 1987). *If $\alpha_j \ge 0$ for $j=1,\ldots,k$, then the method $E(\alpha_1,\ldots,\alpha_k)$ is G-stable with*

$$G = \operatorname{diag}(1, \alpha_2+\ldots+\alpha_k, \ldots, \alpha_{k-1}+\alpha_k, \alpha_k). \tag{9.30}$$

Proof. For multistep Runge-Kutta methods the preconsistency vector is given by $\xi_0 = (1,1,\ldots,1)^T$. With the matrix G of (9.30) it therefore follows from Lemma 9.5 that

$$d_i = b_i \qquad \text{for} \qquad i=1,\ldots,s. \tag{9.31}$$

By Theorem 9.13 this implies $d_i > 0$ so that the first condition for algebraic stability is satisfied. In order to verify that the matrix M of (9.6) is non-negative definite, we transform it by a suitable matrix. We put

$$V = \left(c_i^{j-1}\right)_{i,j=1,\ldots,s} \qquad \text{and} \qquad \alpha = (\alpha_1,\ldots,\alpha_k)^T. \tag{9.32}$$

A straightforward calculation using the simplifying assumptions $D_A(s)$, $D_B(s)$ and $B(2s)$ shows that

$$\begin{pmatrix} I & 0 \\ 0 & V^T \end{pmatrix} M \begin{pmatrix} I & 0 \\ 0 & V \end{pmatrix} = \begin{pmatrix} I & 0 \\ 0 & W^T \end{pmatrix} \widehat{M} \begin{pmatrix} I & 0 \\ 0 & W \end{pmatrix} \tag{9.33}$$

where

$$W = \left(\frac{1}{j}(1-i)^j\right)_{\substack{i=1,\ldots,k \\ j=1,\ldots,s}}$$

and the $2k \times 2k$ matrix $\widehat{M}$ is given by

$$\widehat{M} = \begin{pmatrix} Z & Z \\ Z & Z \end{pmatrix}, \qquad Z = \operatorname{diag}(\alpha_1,\ldots,\alpha_k) - \alpha\alpha^T. \tag{9.34}$$

Since $\alpha_j \geq 0$ and $\sum \alpha_j = 1$ it follows from the Cauchy-Schwarz inequality that

$$x^T Z x = \sum_{j=1}^{k} \alpha_j x_j^2 - \Big(\sum_{j=1}^{k} \alpha_j x_j \Big)^2 \geq 0$$

Therefore the matrix Z, and hence also $\widehat{M}$, are non-negative definite matrices. This completes the proof of the theorem. □

One can ask what are the advantages of the methods $E(\alpha_1, \ldots, \alpha_k)$ with $k > 1$ over the s-stage Gauss Runge-Kutta methods of order $2s$. All these methods have the same order and are algebraically stable for $\alpha_j \geq 0$.

- The Gauss methods have a stability function whose value at infinity satisfies $|R(\infty)| = 1$. In contrast, the new methods allow the spectral radius $\varrho(S(\infty))$ to be smaller than 1, which improves stability at infinity. For example, numerical investigations of the case $s = 2$, $k = 2$ show that $\varrho(S(\infty))$ has the minimal value $\sqrt{2} - 1 \approx 0.41421$ for $\alpha_1 = 12\sqrt{2} - 16$ and $\alpha_2 = 1 - \alpha_1$ (see Exercise 7). There are some indications that L-stable methods do not exist: if we could find methods with an internal stage, say $v_s^{(n)}$, equal to y_{n+1}, then the method would be L-stable. Unfortunately, this would imply $c_s = 1$, which is in contradiction to Lemma 9.12 and to $\alpha_j \geq 0$.
- The eigenvalues of the Runge-Kutta matrix of the Gauss methods are complex (with the exception of one real eigenvalue, if s is odd). Can we hope that, for a suitable choice of $\alpha_j \geq 0$, all eigenvalues of $\widetilde{B}$ become real? Numerical computations for $s = 2$ and $k = 2$ indicate that this is not possible.

B-Convergence

Many results of Sections IV.14 and IV.15 have a straightforward extension to general linear methods. The following theorem corresponds to Theorems IV.14.2, IV.14.3, and IV.14.4 and is proved in the same way:

Theorem 9.16. *Let f be continuously differentiable and satisfy (9.1). If the matrix $\widetilde{B}$ of method (9.2) is invertible and if*

$$h\nu < \alpha_0(\widetilde{B}^{-1}),$$

then the nonlinear system (9.2b) has a unique solution. □

The next results give estimates of the local and global errors. We formulate these results only for multistep Runge-Kutta methods, because in this case the definitions of $C(\eta)$ and $B(p)$ are already available. In analogy to Runge-Kutta methods we say that method (9.17) has *stage order* q, if $C(q)$ and $B(q)$ are satisfied. Recall that for the definition of the local error

$$\delta_h(x) = y_1 - y(x+h)$$

one assumes that $y_i = y(x+ih)$ for $i = 1-k, \dots, 0$ lie on the exact solution.

Theorem 9.17. *Suppose that the differential equation satisfies (9.1). If the matrix $\widetilde{B}$ is invertible, if $\alpha_0(\widetilde{B}^{-1}) > 0$ and if the stage order is q, then the local error of method (9.17) satisfies*

$$\|\delta_h(x)\| \le Ch^{q+1} \max_{\xi\in[x-(k-1)h,x+h]} \|y^{(q+1)}(\xi)\| \quad \textit{for} \quad h\nu \le \alpha < \alpha_0(\widetilde{B}^{-1})$$

where C depends only on the coefficients of the method and on α. □

This result, which corresponds to Proposition IV.15.1, is of particular interest for multistep collocation methods, for which the stage order $q = s+k-1$ is maximal. The global error allows the following estimate, which extends Theorem IV.15.3.

Theorem 9.18. *Suppose, in addition to the assumptions of Theorem 9.17, that the method (9.17) is algebraically stable.*

a) If $\nu > 0$ then the global error satisfies for $\quad h\nu \le \alpha < \alpha_0(\widetilde{B}^{-1})$

$$\|y_n - y(x_n)\| \le h^q \, \frac{e^{C_1\nu(x_n-x_0)} - 1}{C_1\nu} \, C_2 \max_{x\in[x_0,x_n]} \|y^{(q+1)}(x)\|.$$

b) If $\nu \le 0$ then (for all $h > 0$)

$$\|y_n - y(x_n)\| \le h^q (x_n - x_0)\, C_2 \max_{x\in[x_0,x_n]} \|y^{(q+1)}(x)\|.$$

The constants C_1 and C_2 depend only on the coefficients of the method and (for case a) on α. □

In contrast to the results of Sect. IV.15 the above theorem holds only for a constant step size implementation.

Exercises

1. Show that for Runge-Kutta methods, where $A=(1)$, $\widetilde{A}=1\!\!1$, both definitions of algebraic stability (IV.12.5 and V.9.3) are the same.

2. Prove in detail the statement of Example 9.8, that the 2-step 2-stage collocation methods with $c_2=1$ (and $c_1\neq 1$) are not G-stable.

   Hint. The non-negativity of the matrix (9.16) implies $\gamma\geq 1/2$ and by considering its determinant,
$$\gamma\big(4\chi_1-(1+\chi_2)^2\big)\geq 2(\chi_1-\chi_2).$$
   This inequality contradicts (9.14).

3. If a multistep Runge-Kutta method with distinct c_i and $c_i\geq 0$ satisfies the assumptions $B(s+k+\xi)$ and $C(s+k-1)$, then it also satisfies $D_B(\xi)$.

   Hint. Show that
$$\sum_{j=1}^{k}\Big(q\sum_{i=1}^{s}b_ic_i^{q-1}\widetilde{a}_{ij}-\alpha_j(1-(1-j)^q)\Big)\big(r(1)-r(1-j)\big)=0$$
   for all polynomials $r(x)$ of degree $\leq s+k-1$ which satisfy $r(c_1)=\ldots=r(c_s)=0$. For given j, construct such a polynomial which also satisfies
$$r(1-j)=1,\quad r(1-i)=0\quad\text{for}\quad i=1,\ldots,k\quad\text{and}\quad i\neq j\,.$$

4. Disprove the conjecture of Burrage (1988) that for every k and s there exist zero-stable multistep Runge-Kutta methods of order $2s+k-1$.

   Hint. Consider the case $s=1$ so that these methods are equivalent to one-leg methods and consult a result of Dahlquist (1983).

5. (Burrage 1988). Show that there exists a zero-stable multistep Runge-Kutta method with $s=2$ and $k=2$ which is of order 5.

   Result. $c_{1,2}=(\sqrt{7}\pm\sqrt{2})/5$

6. (Stability at infinity). If a multistep Runge-Kutta method satisfies $D_A(s)$ and $D_B(s)$ then we have, e.g., for $s=2$ and $k=2$,
$$S(\infty)=\begin{pmatrix}\alpha_1&\alpha_2\\1&0\end{pmatrix}-\begin{pmatrix}1&1\\0&0\end{pmatrix}\begin{pmatrix}1-c_1&1-c_2\\1-c_1^2&1-c_2^2\end{pmatrix}^{-1}\begin{pmatrix}\alpha_1&2\alpha_2\\\alpha_1&0\end{pmatrix}.$$
   Formulate this result also for general s and k.

7. Verify that for the method $E(\alpha_1,\alpha_2)$ with $0\leq\alpha_1\leq 1$, $\alpha_2=1-\alpha_1$, the spectral radius $\varrho(S(\infty))$ is minimal for $\alpha_1=12\sqrt{2}-16$.

Chapter VI. Singular Perturbation Problems and Index 1 Problems

(Drawing by G. Di Marzo)

Singular perturbation problems (SPP) form a special class of problems containing a parameter ε. When this parameter is small, the corresponding differential equation is stiff; when ε tends to zero, the differential equation becomes differential algebraic. This chapter investigates the numerical solution of such singular perturbation problems. This allows us to understand many phenomena observed for very stiff problems. Much insight is obtained by studying the limit case $\varepsilon = 0$ ("the reduced system" or "problem of index 1") which is usually much easier to analyze.

We start by considering the limit case $\varepsilon = 0$. Two numerical approaches – the ε-embedding method and the state space form method – are investigated in Sect. VI.1. We then analyze multistep methods in Sect. VI.2, Runge-Kutta methods in Sect. VI.3, Rosenbrock methods in Sect. VI.4 and extrapolation methods in Sect. VI.5. Convergence is studied for singular perturbation problems and for semi-explicit differential-algebraic systems of "index 1".

VI.1 Solving Index 1 Problems

Singular perturbation problems (SPP) have several origins in applied mathematics. One comes from fluid dynamics and results in linear boundary value problems containing a small parameter ε (the coefficient of viscosity) such that for $\varepsilon \to 0$ the differential equation loses the highest derivative (see Exercise 1 below). Others originate in the study of nonlinear oscillations with *large* parameters (van der Pol 1926, Dorodnicyn 1947) or in the study of chemical kinetics with slow and fast reactions (see e.g., Example (IV.1.4)).

Asymptotic Solution of van der Pol's Equation

The classical paper of Dorodnicyn (1947) studied the van der Pol Equation (IV.1.5') with large μ, i.e., with small ε. The investigation becomes a little easier if we use Liénard's coordinates (see Exercise I.16.8). In Eq. (IV.1.5'), written here as

$$\varepsilon z'' + (z^2 - 1)z' + z = 0, \tag{1.1}$$

we insert the identity

$$\varepsilon z'' + (z^2 - 1)z' = \frac{d}{dx}\underbrace{\Big(\varepsilon z' + \big(\frac{z^3}{3} - z\big)\Big)}_{:= y}$$

so that (1.1) becomes

$$\begin{aligned} y' &= -z & &=: f(y,z) \\ \varepsilon z' &= y - \Big(\frac{z^3}{3} - z\Big) & &=: g(y,z). \end{aligned} \tag{1.2}$$

Fig. 1.1 shows solutions of Eq. (1.2) with $\varepsilon = 0.03$ in the (y,z)-plane. One observes rapid movements towards the manifold M defined by $y = z^3/3 - z$, close to which the solution becomes smooth. In order to approximate the solution for very small ε, we set $\varepsilon = 0$ in (1.2) and obtain the so-called *reduced* system

$$\begin{aligned} y' &= -z & &= f(y,z) \\ 0 &= y - \Big(\frac{z^3}{3} - z\Big) & &= g(y,z). \end{aligned} \tag{1.2'}$$

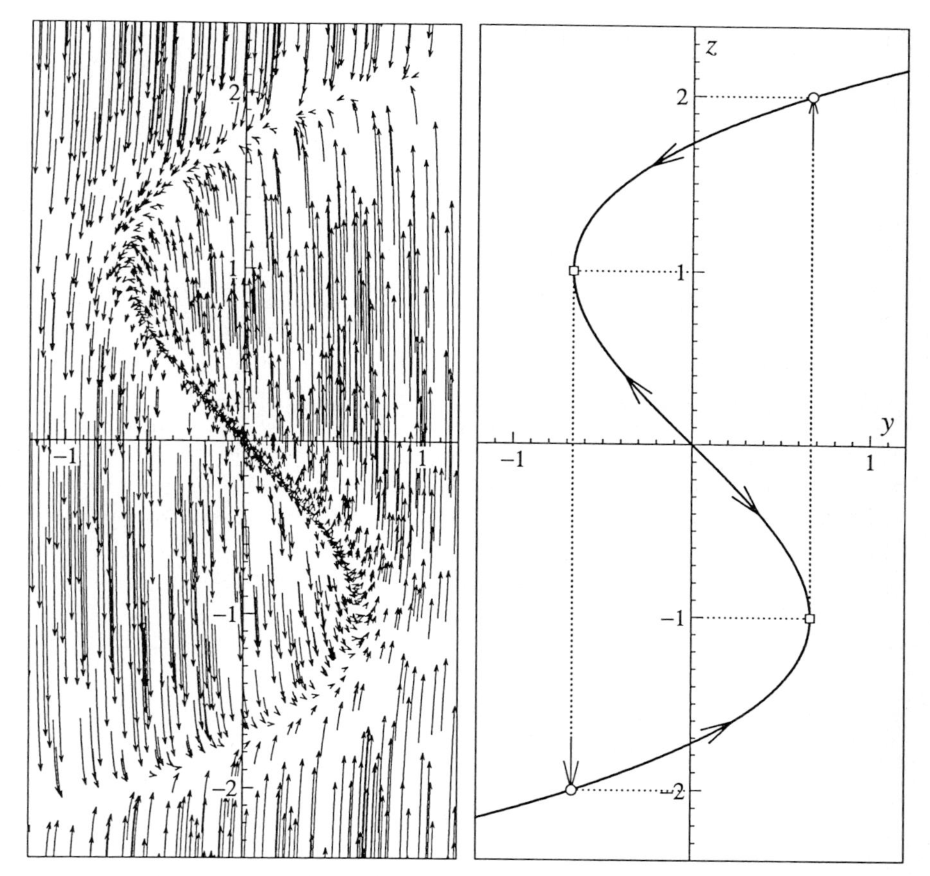

Fig. 1.1. Solutions of SPP (1.2)

Fig. 1.2. Reduced problem (1.2')

While (1.2) has no analytic solution, (1.2') can easily be solved to give

$$y' = -z = (z^2 - 1)z' \quad \text{or} \quad \ln|z| - \frac{z^2}{2} = x + C. \tag{1.3}$$

Equation (1.2') is called a *differential algebraic equation* (DAE), since it combines a differential equation (first line) with an algebraic equation (second line). Such a problem only makes sense if the initial values are *consistent*, i.e., lie on the manifold M. The points of M with coordinates $y = \pm 2/3$, $z = \mp 1$ are of special interest (Fig. 1.2): at these points the partial derivative $g_z = \partial g/\partial z$ vanishes and the defining manifold is no longer "transversal" to the direction of the fast movement. Here the solutions of (1.2') cease to exist, while the solutions of the full problem (1.2) for $\varepsilon \to 0$ jump with "infinite" speed to the opposite manifold. For $-1 < z < 1$ the manifold M is *unstable* for the solution of (1.2) (here $g_z > 0$), otherwise M is *stable* ($g_z < 0$).

We demonstrate the power of the reduced equation by answering the question:

what is the period T of the limit cycle solution of van der Pol's equation for $\varepsilon \to 0$? Fig. 1.2 shows that the asymptotic value of T is just twice the time which $z(x)$ of (1.3) needs to advance from $z=-2$ to $z=-1$, i.e.,

$$T = 3 - 2\ln 2. \tag{1.4}$$

This is the first term of Dorodnicyn's asymptotic formula. We also see that $z(x)$ reaches its largest values (i.e., crosses the Poincaré cut $z'=0$, see Fig. I.16.2) at $z=\pm 2$. We thus have the curious result that the limit cycle of van der Pol's equation (1.1) has the same asymptotic initial value $z=2$ and $z'=0$ for $\varepsilon \to 0$ and for $\varepsilon \to \infty$ (see Eq. (I.16.10)).

The ε-Embedding Method for Problems of Index 1

We now want to study the behaviour of the *numerical solution* for $\varepsilon \to 0$. This will give us insight into many phenomena encountered for very stiff equations and also suggest advantageous numerical procedures for stiff and differential-algebraic equations. Let an arbitrary singular perturbation problem be given,

$$y' = f(y,z) \tag{1.5a}$$
$$\varepsilon z' = g(y,z), \tag{1.5b}$$

where y and z are vectors; suppose that f and g are sufficiently often differentiable vector functions of the same dimensions as y and z, respectively. The corresponding *reduced* equation is the DAE

$$y' = f(y,z) \tag{1.6a}$$
$$0 = g(y,z), \tag{1.6b}$$

whose initial values are *consistent* if $0 = g(y_0, z_0)$. A general assumption of the present chapter will be that the Jacobian

$$g_z(y,z) \quad \text{is invertible} \tag{1.7}$$

in a neighbourhood of the solution of (1.6). Equation (1.6b) then possesses a locally unique solution $z = G(y)$ ("Implicit Function Theorem") which inserted into (1.6a) gives

$$y' = f\big(y, G(y)\big), \tag{1.8}$$

the so-called "state space form", an ordinary differential system. Under the assumption (1.7), Eq. (1.6) is said to be a differential-algebraic equation of *index 1*.

An interesting approach for solving (1.6) is to apply some numerical method to the SPP (1.5) and to put $\varepsilon = 0$ in the resulting formulas. Let us illustrate this approach for Runge-Kutta methods. Applied to the system (1.5) we obtain

$$Y_{ni} = y_n + h \sum_{j=1}^{s} a_{ij}\, f\,(Y_{nj}, Z_{nj}) \tag{1.9a}$$

$$\varepsilon Z_{ni} = \varepsilon z_n + h \sum_{j=1}^{s} a_{ij}\, g\,(Y_{nj}, Z_{nj}) \tag{1.9b}$$

$$y_{n+1} = y_n + h \sum_{i=1}^{s} b_i\, f\,(Y_{ni}, Z_{ni}) \tag{1.9c}$$

$$\varepsilon z_{n+1} = \varepsilon z_n + h \sum_{i=1}^{s} b_i\, g\,(Y_{ni}, Z_{ni}). \tag{1.9d}$$

We now suppose that the RK matrix (a_{ij}) is invertible and obtain from (1.9b)

$$hg(Y_{ni}, Z_{ni}) = \varepsilon \sum_{j=1}^{s} \omega_{ij}(Z_{nj} - z_n), \tag{1.10}$$

where the ω_{ij} are the elements of the inverse of (a_{ij}). Inserting this into (1.9d) makes the definition of z_{n+1} independent of ε. We thus put without more ado $\varepsilon = 0$ and obtain

$$Y_{ni} = y_n + h \sum_{j=1}^{s} a_{ij}\, f\,(Y_{nj}, Z_{nj}) \tag{1.11a}$$

$$0 = g(Y_{ni}, Z_{ni}) \tag{1.11b}$$

$$y_{n+1} = y_n + h \sum_{i=1}^{s} b_i\, f\,(Y_{ni}, Z_{ni}) \tag{1.11c}$$

$$z_{n+1} = \Big(1 - \sum_{i,j=1}^{s} b_i\, \omega_{ij}\Big) z_n + \sum_{i,j=1}^{s} b_i\, \omega_{ij}\, Z_{nj}. \tag{1.11d}$$

Here

$$1 - \sum_{i,j=1}^{s} b_i\, \omega_{ij} = R(\infty) \tag{1.11e}$$

(see Eq. (IV.3.15)), where $R(z)$ is the stability function of the method.

State Space Form Method

The numerical solution (y_{n+1}, z_{n+1}) of the above approach will usually *not* lie on the manifold $g(y,z) = 0$. However, this can easily be repaired by replacing (1.11d) by the condition

$$0 = g(y_{n+1}, z_{n+1}). \tag{1.12}$$

Then, we do not only have $Z_{nj} = G(Y_{nj})$ (see (1.11b)), but also $z_{n+1} = G(y_{n+1})$. In this case the method (1.11a–c), (1.12) is *identical* to the solution of the state space form (1.8) with the same Runge-Kutta method. This will be called the *state space form method*. The whole situation is summarized in the following diagram:

$$\begin{array}{ccccc}
\text{SPP (1.5)} & \xleftarrow{\varepsilon \leftarrow 0} & \text{DAE (1.6)} & \xrightarrow{z=G(y)} & \text{ODE (1.8)} \\
\text{RK} \downarrow & & \downarrow & & \text{RK} \downarrow \\
\text{Sol. (1.9)} & \xrightarrow{\varepsilon \to 0} & \begin{array}{c}\varepsilon\text{-embedding method} \\ \text{state space form method}\end{array} & \longleftarrow & \text{Sol. (1.12)}
\end{array}$$

Of special importance here are *stiffly accurate* methods, i.e., methods which satisfy

$$a_{si} = b_i \qquad \text{for} \quad i = 1, \ldots, s. \tag{1.13}$$

This means that $y_{n+1} = Y_{ns}$, $z_{n+1} = Z_{ns}$ and (1.12) is satisfied anyway. Hence for stiffly accurate methods the ε-embedding method and the state space form method are identical. For this reason, Griepentrog & März (1986) denote such methods IRK(DAE).

Both approaches have their own merits. Theoretical results for the ε-embedding method yield insight into the method when applied to singular perturbation problems. Moreover, this approach can easily be extended to more general situations, where the algebraic relation is not explicitly separated from the differential equation (see below). The state space form method, on the other hand, has the advantage that it is not restricted to implicit methods. Applying an explicit Runge-Kutta method or a multistep method to Eq. (1.8) is certainly a method of choice for semi-explicit index 1 equations. No new theory is necessary in this case.

A Transistor Amplifier

> ... auf eine merkwürdige Tatsache aufmerksam machen, das ist die außerordentlich grosse Zahl berühmter Mathematiker, die aus Königsberg stammen ...: Kant 1724, Richelot 1808, Hesse 1811, Kirchhoff 1824, Carl Neumann 1832, Clebsch 1833, Hilbert 1862.
> (F. Klein, Entw. der Math., p. 159)

Very often, differential-algebraic problems arising in practice are not at once in the semi-explicit form (1.6), but rather in the form $Mu' = \varphi(u)$ where M is a constant *singular* matrix.

As an example we compute the amplifier of Fig. 1.3, where $U_e(t)$ is the entry voltage, $U_b = 6$ the operating voltage, $U_i(t)$ $(i = 1, 2, 3, 4, 5)$ the voltages at the nodes $1, 2, 3, 4, 5$, and $U_5(t)$ the output voltage. The current through a resistor satisfies $I = U/R$ (Ohm 1827), the current through a capacitor $I = C \cdot dU/dt$, where R and C are constants and U the voltage. The transistor acts as amplifier in that the current from node 4 to node 3 is 99 times larger than that from node 2 to node 3 and depends on the voltage difference $U_3 - U_2$ in a nonlinear way. Kirchhoff's law (a Königsberg discovery) says that the sum of currents entering a node vanishes. This law applied to the 5 nodes of Fig. 1.3 leads to the following equations:

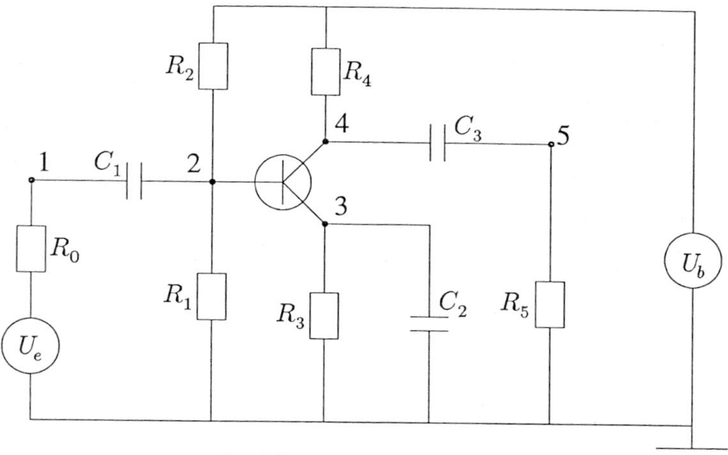

Fig. 1.3. A transistor amplifier

$$
\begin{aligned}
&\text{node 1:} && \frac{U_e(t)}{R_0} - \frac{U_1}{R_0} + C_1(U_2' - U_1') = 0 \\
&\text{node 2:} && \frac{U_b}{R_2} - U_2\Big(\frac{1}{R_1} + \frac{1}{R_2}\Big) + C_1(U_1' - U_2') - 0.01\, f(U_2 - U_3) = 0 \\
&\text{node 3:} && f(U_2 - U_3) - \frac{U_3}{R_3} - C_2 U_3' = 0 \\
&\text{node 4:} && \frac{U_b}{R_4} - \frac{U_4}{R_4} + C_3(U_5' - U_4') - 0.99 f(U_2 - U_3) = 0 \\
&\text{node 5:} && -\frac{U_5}{R_5} + C_3(U_4' - U_5') = 0.
\end{aligned}
\tag{1.14}
$$

As constants we adopt the values reported (for a similar problem) by Rentrop, Roche & Steinebach (1989)

$$
\begin{aligned}
f(U) &= 10^{-6}\Big(\exp\Big(\frac{U}{0.026}\Big) - 1\Big) \\
R_0 &= 1000, \quad R_1 = \ldots = R_5 = 9000 \\
C_k &= k \cdot 10^{-6}, \quad k = 1, 2, 3,
\end{aligned}
$$

and the initial signal is chosen as

$$
U_e(t) = 0.4 \cdot \sin(200\pi t). \tag{1.15}
$$

Equations (1.14) are of the form $Mu' = \varphi(u)$ where

$$
M = \begin{pmatrix}
-C_1 & C_1 & & & \\
C_1 & -C_1 & & & \\
& & -C_2 & & \\
& & & -C_3 & C_3 \\
& & & C_3 & -C_3
\end{pmatrix}
$$

is obviously a singular matrix of rank 3. The sum of the first two and of the last two equations leads directly to two algebraic equations. Introducing e.g.,

$$U_1 - U_2 = y_1, \quad U_3 = y_2, \quad U_4 - U_5 = y_3, \quad U_1 = z_1, \quad U_4 = z_2,$$

transforms equations (1.14) to the form (1.6). *Consistent initial values* must thus satisfy $\varphi_1(u) + \varphi_2(u) = 0$ and $\varphi_4(u) + \varphi_5(u) = 0$. If we put $U_2(0) = U_3(0)$, we have $f(U_2(0) - U_3(0)) = 0$. Since $U_e(0) = 0$, we then easily find consistent initial values, e.g., as

$$U_1(0) = 0, \quad U_2(0) = U_3(0) = \frac{U_b R_1}{R_1 + R_2}, \quad U_4(0) = U_b, \quad U_5(0) = 0. \tag{1.16}$$

Problems of the Form $Mu' = \varphi(u)$

Numerical methods for problems of the form

$$Mu' = \varphi(u), \tag{1.17}$$

where M is a constant matrix, can be derived as follows: we assume that M is regular, apply an ODE method to $u' = M^{-1}\varphi(u)$ and multiply the resulting formulas by M. For Runge-Kutta methods we obtain in this way

$$M(U_{ni} - u_n) = h \sum_{j=1}^{s} a_{ij} \varphi(U_{nj}) \tag{1.18a}$$

$$u_{n+1} = \Big(1 - \sum_{i,j=1}^{s} b_i \omega_{ij}\Big) u_n + \sum_{i,j=1}^{s} b_i \omega_{ij} U_{nj}, \tag{1.18b}$$

where again (ω_{ij}) is the inverse of (a_{ij}). The second formula was obtained from

$$M(u_{n+1} - u_n) = h \sum_{i=1}^{s} b_i \varphi(U_{ni}) \tag{1.18c}$$

in exactly the same way as above (see (1.10)).

Formulas (1.18) also make sense formally when M is a *singular* matrix. In this case, problem (1.17) is mathematically equivalent to a semi-explicit system (1.6) and method (1.18) corresponds to method (1.11). This can be seen as follows: we decompose the matrix M (e.g., by Gaussian elimination with total pivoting) as

$$M = S \begin{pmatrix} I & 0 \\ 0 & 0 \end{pmatrix} T, \tag{1.19}$$

where S and T are invertible matrices and the dimension of I represents the rank of M. Inserting this into (1.17), multiplying by S^{-1}, and using the transformed variables

$$Tu = \begin{pmatrix} y \\ z \end{pmatrix} \tag{1.20}$$

gives

$$\begin{pmatrix} y' \\ 0 \end{pmatrix} = S^{-1}\varphi\Big(T^{-1}\begin{pmatrix} y \\ z \end{pmatrix}\Big) =: \begin{pmatrix} f(y,z) \\ g(y,z) \end{pmatrix}, \tag{1.21}$$

a problem of type (1.6). An initial value u_0 is *consistent* if $\varphi(u_0)$ lies in the range of the matrix M.

Similarly, if (1.19) is inserted into (1.18), and the variables

$$TU_{nj} = \begin{pmatrix} Y_{nj} \\ Z_{nj} \end{pmatrix}, \qquad Tu_n = \begin{pmatrix} y_n \\ z_n \end{pmatrix} \tag{1.22}$$

are introduced, Eq. (1.18b) (for Z_{n+1}) and Eq. (1.18c) (for Y_{n+1}) lead precisely to equations (1.11). This means that the diagram

$$\begin{array}{ccc} \text{Problem (1.17)} & \xrightarrow{\textit{Transf. (1.20)}} & \text{Problem (1.6)} \\ \textit{Meth.}\ \Big\downarrow\ \textit{(1.18)} & & \textit{Meth.}\ \Big\downarrow\ \textit{(1.11)} \\ \{u_n\} & \xrightarrow{\textit{Transf. (1.22)}} & \{y_n\},\ \{z_n\} \end{array} \tag{1.23}$$

commutes. An important consequence of this commutativity is that all results for semi-explicit systems (1.6) and the ε-embedding method (1.11) (existence of a numerical solution, convergence, asymptotic expansions, ...) also apply to implicit problems (1.17) with singular M and method (1.18).

All codes, such as RADAU5, which have an option for implicit differential equations (1.17) can thus be applied directly. This has been done for problem (1.14) with initial values (1.16), integration interval $0 \le x \le 0.2$, and $Tol = 10^{-4}$. The code computed the solution $U_5(t)$ displayed in Fig. 1.4 in 556 (accepted) steps. The comparison with the entry voltage $U_e(t)$ shows that our amplifier is working. See also Hairer, Lubich & Roche (1989), p. 108-111 for a more elaborate example.

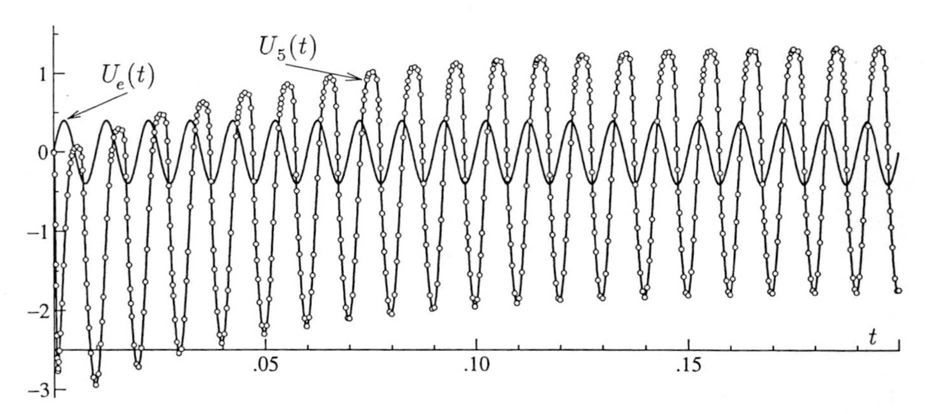

Fig. 1.4. Computed solution of amplifier problem (1.14)

Convergence of Runge-Kutta Methods

If the method is stiffly accurate, the numerical solutions (1.11) are equivalent to those of the *ordinary* equation (1.8). Therefore the convergence of the solutions is described by Theorems II.3.4 and II.3.6 as

$$y_n - y(x_n) = \mathcal{O}(h^p), \qquad z_n - z(x_n) = \mathcal{O}(h^p), \tag{1.24}$$

where p is the *classical* order of the method (the second formula follows from a Lipschitz condition for G). For *general* methods, the estimate (1.24) remains valid for y_n, because (1.11a,b,c) are independent of z_n and do not change if (1.11d) is replaced by (1.12). Thus we only have to prove a convergence result for z_n. An essential ingredient of the following theorem is the *stage order* q of the method, i.e., condition $C(q)$ of Sect. II.7 or IV.5.

Theorem 1.1. *Suppose that the system (1.6) satisfies (1.7) in a neighbourhood of the exact solution $(y(x), z(x))$ and assume the initial values are consistent. Consider a Runge-Kutta method of order p, stage order q and with invertible matrix A. Then the numerical solution of (1.11a–d) has global error*

$$z_n - z(x_n) = \mathcal{O}(h^r) \quad \textit{for} \quad x_n - x_0 = nh \le \textit{Const}, \tag{1.25}$$

where
a) $r = p$ *for stiffly accurate methods,*
b) $r = \min(p, q+1)$ *if the stability function satisfies* $-1 \le R(\infty) < 1$,
c) $r = \min(p-1, q)$ *if* $R(\infty) = +1$.
d) If $|R(\infty)| > 1$, *the numerical solution diverges.*

Proof. Part (a) has already been discussed. For the remaining cases we proceed as follows: we first observe that Condition $C(q)$ and order p imply

$$z(x_n + c_i h) = z(x_n) + h \sum_{j=1}^{s} a_{ij} z'(x_n + c_j h) + \mathcal{O}(h^{q+1}) \tag{1.26a}$$

$$z(x_{n+1}) = z(x_n) + h \sum_{i=1}^{s} b_i z'(x_n + c_i h) + \mathcal{O}(h^{p+1})\,. \tag{1.26b}$$

Since A is invertible we can compute $z'(x_n + c_j h)$ from (1.26a) and insert it into (1.26b). This gives

$$z(x_{n+1}) = \varrho z(x_n) + b^T A^{-1} \widehat{Z}_n + \mathcal{O}(h^{p+1}) + \mathcal{O}(h^{q+1}) \tag{1.27}$$

where $\varrho = 1 - b^T A^{-1} \mathbb{1} = R(\infty)$ and $\widehat{Z}_n = (z(x_n + c_1 h), \ldots, z(x_n + c_s h))^T$. We then denote the global error by $\Delta z_n = z_n - z(x_n)$, and $\Delta Z_n = Z_n - \widehat{Z}_n$. Subtracting (1.27) from (1.11d) yields

$$\Delta z_{n+1} = \varrho \Delta z_n + b^T A^{-1} \Delta Z_n + \mathcal{O}(h^{p+1}) + \mathcal{O}(h^{q+1}). \tag{1.28}$$

Our next aim is to estimate ΔZ_n. For this we have to consider the y-component of the system. Due to (1.11a–c) the values y_n, Y_{ni} are those of the Runge-Kutta method applied to (1.8). It thus follows from Theorem II.8.1 that $y_n - y(x_n) = e_p(x_n)h^p + \mathcal{O}(h^{p+1})$. Since Eq. (1.26a) also holds with $z(x)$ replaced by $y(x)$, we can subtract this formula from (1.11a) and so obtain

$$Y_{ni} - y(x_n + c_i h) = y_n - y(x_n)$$
$$+ h\sum_{j=1}^{s} a_{ij}\Big(f(Y_{nj}, G(Y_{nj})) - f\big(y(x_n + c_j h), G(y(x_n + c_j h))\big)\Big) + \mathcal{O}(h^{q+1}).$$

This implies that

$$Y_{ni} - y(x_n + c_i h) = \mathcal{O}(h^\nu) \qquad \text{with} \qquad \nu = \min(p, q+1).$$

Because of (1.11b) we get

$$Z_{ni} - z(x_n + c_i h) = G(Y_{ni}) - G\big(y(x_n + c_i h)\big) = \mathcal{O}(h^\nu)$$

and Eq. (1.28) becomes

$$\Delta z_{n+1} = \varrho \Delta z_n + \delta_{n+1}, \qquad \text{where} \qquad \delta_{n+1} = \mathcal{O}(h^\nu). \tag{1.29}$$

Repeated insertion of this formula gives

$$\Delta z_n = \sum_{i=1}^{n} \varrho^{n-i}\delta_i, \tag{1.30}$$

because $\Delta z_0 = 0$. This proves the statement for $\varrho \neq -1$. For the case $\varrho = -1$ the error Δz_n is a sum of differences $\delta_{j+1} - \delta_j$. Since δ_{n+1} is actually of the form $\delta_{n+1} = d(x_n)h^\nu + \mathcal{O}(h^{\nu+1})$ we have $\delta_{j+1} - \delta_j = \mathcal{O}(h^{\nu+1})$ and the statement also follows in this situation. □

The order reduction in the z-component (for non stiffly accurate methods) was first studied by Petzold (1986) in a more general context.

Exercises

1. Compute the solutions of the boundary value problems

$$\varepsilon y'' + y' + y = 1 \quad \text{respectively} \quad \varepsilon y'' - y' + y = 1 \tag{1.31}$$
$$y(0) = y(1) = 0, \qquad \text{for} \quad \varepsilon > 0.$$

Observe that the solutions possess, for $\varepsilon \to 0$, a "boundary layer" on one of the two sides of $[0,1]$ and that the limit solutions for $\varepsilon = 0$ satisfy

$$y' + y = 1 \qquad \text{respectively} \qquad -y' + y = 1$$

with one of the two boundary conditions being lost.

VI.2 Multistep Methods

The aim of this section is to study convergence of multistep methods when applied to singular perturbation problems (Runge-Kutta methods will be treated in Sect. VI.3). We are interested in estimates that hold uniformly for $\varepsilon \to 0$. The results of the previous chapters cannot be applied. Since the Lipschitz constant of the singular perturbation problem (1.5) is of size $\mathcal{O}(\varepsilon^{-1})$, the estimates of Sect. III.4 are useless. Also the one-sided Lipschitz constant is in general $\mathcal{O}(\varepsilon^{-1})$, so that the convergence results of Sect. V.8 can neither be applied. Let us start by considering the reduced problem.

Methods for Index 1 Problems

A multistep method applied to the system $y' = f(y,z)$, $\varepsilon z' = g(y,z)$ gives

$$\sum_{i=0}^{k} \alpha_i \, y_{n+i} = h \sum_{i=0}^{k} \beta_i \, f(y_{n+i}, z_{n+i}) \tag{2.1a}$$

$$\varepsilon \sum_{i=0}^{k} \alpha_i \, z_{n+i} = h \sum_{i=0}^{k} \beta_i \, g(y_{n+i}, z_{n+i}). \tag{2.1b}$$

By putting $\varepsilon = 0$ we obtain (ε*-embedding method*)

$$\sum_{i=0}^{k} \alpha_i \, y_{n+i} = h \sum_{i=0}^{k} \beta_i \, f(y_{n+i}, z_{n+i}) \tag{2.2a}$$

$$0 = \sum_{i=0}^{k} \beta_i \, g(y_{n+i}, z_{n+i}) \tag{2.2b}$$

which allows us to apply a multistep method to the differential-algebraic system (1.6). This approach was first proposed (for the BDF methods) by Gear (1971).

Theorem 2.1. *Suppose that the system (1.6) satisfies (1.7). Consider a multistep method of order* p *which is stable at the origin and at infinity* (0 *and* ∞ *are in the stability region) and suppose that the error of the starting values* y_j, z_j *for* $j = 0, \ldots, k-1$ *is* $\mathcal{O}(h^p)$. *Then the global error of (2.2) satisfies*

$$y_n - y(x_n) = \mathcal{O}(h^p), \qquad z_n - z(x_n) = \mathcal{O}(h^p)$$

for $x_n - x_0 = nh \le Const.$

Proof. Formula (2.2b) is a stable recursion for $\delta_n = g(y_n, z_n)$, because ∞ lies in the stability region of the method. This together with the assumption on the starting values implies that $\delta_n = \mathcal{O}(h^p)$ for all $n \geq 0$. By the Implicit Function Theorem $g(y_n, z_n) = \delta_n$ can be solved for z_n and yields

$$z_n = G(y_n) + \mathcal{O}(h^p) \tag{2.3}$$

with $G(y)$ as in (1.8). Inserting (2.3) into (2.2a) gives the multistep formula for the differential equation (1.8) with an $\mathcal{O}(h^{p+1})$ perturbation. The statement then follows from the convergence proof of Sect. III.4. □

For the implicit index 1 problem (1.17) the multistep method becomes

$$M\sum_{i=0}^{k} \alpha_i u_{n+i} = h\sum_{i=0}^{k} \beta_i \varphi(u_{n+i}) \tag{2.4}$$

and convergence without any order reduction for methods satisfying the hypotheses of Theorem 2.1 follows from the transformation (1.20) and the diagram (1.23).

The *state space from approach* is also possible for multistep methods. We just have to replace (2.2b) by

$$g(y_{n+k}, z_{n+k}) = 0. \tag{2.2c}$$

Method (2.2a,c) is equivalent to the solution of (1.8) by the above multistep method. Hence, we have convergence as for nonstiff ordinary differential equations. The assumption "$\infty \in S$" is no longer necessary and even explicit methods can be applied.

Convergence for Singular Perturbation Problems

The error propagation has been studied by Söderlind & Dahlquist (1981) using G-stability estimates. Convergence results were first obtained by Lötstedt (1985) for BDF methods. The following convergence result by Lubich (1991), based on the smoothness of the exact solution and thus uniform in ε as long as we stay away from transient phases, gives optimal error bounds for arbitrary multistep methods.

The Jacobian of the system (1.5) is of the form

$$\begin{pmatrix} f_y & f_z \\ \varepsilon^{-1} g_y & \varepsilon^{-1} g_z \end{pmatrix}$$

and its dominant eigenvalues are seen to be close to $\varepsilon^{-1}\lambda$ where λ represents the eigenvalues of g_z. For reasons of stability we assume throughout this subsection that the eigenvalues of g_z have negative real part. More precisely, we assume that

$$\text{the eigenvalues } \lambda \text{ of } g_z(y,z) \text{ lie in } |\arg\lambda - \pi| < \alpha \tag{2.5}$$

for (y, z) in a neighbourhood of the considered solution. We then have the following result for method (2.1a,b):

Theorem 2.2 (Lubich 1991). *Suppose that the multistep method is of order p, $A(\alpha)$-stable and strictly stable at infinity. If the problem (1.5) satisfies (2.5), then the error is bounded for $h \geq \varepsilon$ and $nh \leq \overline{x} - x_0$ by*

$$\begin{aligned}\|y_n - y(x_n)\| + \|z_n - z(x_n)\| \\ \leq C\Big(\max_{0\leq j<k} \|y_j - y(x_j)\| + h^p \int_{x_0}^{x_n} \|y^{(p+1)}(x)\|\, dx \\ + (h+\varrho^n) \max_{0\leq j<k} \|z_j - z(x_j)\| + \varepsilon h^p \max_{x_0\leq x\leq x_n} \|z^{(p+1)}(x)\|\Big)\end{aligned}$$

with $0<\varrho<1$. This estimate holds for $h \leq h_0$ (h_0 sufficiently small, but independent of ε), and provided that the starting values are in a sufficiently small, h- and ε-independent neighbourhood of the exact solution. The constants C and ϱ are independent of ε and h.

Proof. The proof is divided into several parts: in part (a) we shall derive recursive estimates for the global error, these will be solved in part (b); part (c) proves an inequality which is needed in (a).

a) First we insert the exact solution of (1.5) into the method (2.1) and so obtain

$$\sum_{i=0}^{k} \alpha_i\, y(x_{n+i}) = h \sum_{i=0}^{k} \beta_i\, f\big(y(x_{n+i}), z(x_{n+i})\big) + d_{n+k} \tag{2.6a}$$

$$\sum_{i=0}^{k} \alpha_i\, z(x_{n+i}) = \frac{h}{\varepsilon} \sum_{i=0}^{k} \beta_i\, g\big(y(x_{n+i}), z(x_{n+i})\big) + e_{n+k}, \tag{2.6b}$$

where the perturbations d_{n+k}, e_{n+k} can be estimated (for $n \geq 0$) as

$$\|d_{n+k}\| \leq C_1 h^p \int_{x_n}^{x_{n+k}} \|y^{(p+1)}(x)\| dx \tag{2.7a}$$

$$\|e_{n+k}\| \leq C_2 h^{p+1} \max_{x_n\leq x\leq x_{n+k}} \|z^{(p+1)}(x)\|. \tag{2.7b}$$

We then denote the global errors by $\Delta y_n = y_n - y(x_n)$, $\Delta z_n = z_n - z(x_n)$ and introduce the differences

$$\Delta f_{n+k} = \sum_{i=0}^{k} \beta_i \Big(f(y_{n+i}, z_{n+i}) - f\big(y(x_{n+i}), z(x_{n+i})\big)\Big), \qquad n \geq 0,$$

$\Delta f_j = 0$ for $j<k$. Subtraction of (2.6a) from (2.1a) yields for $n \geq 0$

$$\sum_{i=0}^{k} \alpha_i \Delta y_{n+i} = h\Delta f_{n+k} - d_{n+k}. \tag{2.8}$$

Guided by previous experience (see (V.7.41)), we define $d_0, \ldots, d_{k-1}$ so that (2.8) also holds for negative n. Solving for Δy_n gives

$$\Delta y_n = h\sum_{j=0}^{n} r_{n-j}(0)\,\Delta f_j - \sum_{j=0}^{n} r_{n-j}(0)\,d_j$$

where $r_j(0)$ is defined in (V.7.44). These numbers are the coefficients of $r(\zeta,0) = \zeta^{-k}/\varrho(\zeta^{-1})$. By zero-stability of the method, the sequence $\{r_j(0)\}$ is bounded, so that a Lipschitz condition for $f(y,z)$ implies the estimate

$$\|\Delta y_n\| \le h\sum_{j=0}^{n}\big(M\|\Delta y_j\| + N\|\Delta z_j\|\big) + C_3\sum_{j=0}^{n}\|d_j\|. \tag{2.9}$$

A more refined estimate is necessary for the z-component. We take the difference of (2.1b) and (2.6b) and then subtract from both sides the quantity

$$\frac{h}{\varepsilon}\sum_{i=0}^{k}\beta_i J\,\Delta z_{n+i} \qquad \text{where} \qquad J = g_z(y_0, z_0). \tag{2.10}$$

This yields

$$\sum_{i=0}^{k}\Big(\alpha_i I - \beta_i\frac{h}{\varepsilon}J\Big)\Delta z_{n+i} = \frac{h}{\varepsilon}\,\Delta g_{n+k} - e_{n+k} \tag{2.11}$$

where

$$\Delta g_{n+k} = \sum_{i=0}^{k}\beta_i\Big(g(y_{n+i}, z_{n+i}) - g\big(y(x_{n+i}), z(x_{n+i})\big) - J\,\Delta z_{n+i}\Big), \tag{2.12}$$

and $\Delta g_j = 0$ for $j<k$. We again define $e_0,\ldots,e_{k-1}$ such that (2.11) holds for negative n, and we then solve (2.11) for Δz_n. This gives

$$\Delta z_n = \frac{h}{\varepsilon}\sum_{j=0}^{n} r_{n-j}\Big(\frac{h}{\varepsilon}J\Big)\Delta g_j - \sum_{j=0}^{n} r_{n-j}\Big(\frac{h}{\varepsilon}J\Big)e_j \tag{2.13}$$

where the matrices $r_j(\frac{h}{\varepsilon}J)$ are defined by (see Formula (V.7.50))

$$\frac{h}{\varepsilon}\sum_{j\ge 0} r_j\Big(\frac{h}{\varepsilon}J\Big)\zeta^j = \Big(\frac{\varepsilon}{h}\,\delta(\zeta)I - J\Big)^{-1}\frac{\zeta^{-k}}{\sigma(\zeta^{-1})} \tag{2.14}$$

with $\delta(\zeta)$ given in (V.7.45). In part (c) below we shall prove that

$$\frac{h}{\varepsilon}\Big\|r_j\Big(\frac{h}{\varepsilon}J\Big)\Big\| \le C\kappa^j \qquad \text{with} \qquad 0<\kappa<1. \tag{2.15}$$

Inserted into (2.13) we thus get

$$\|\Delta z_n\| \le \sum_{j=0}^{n}\kappa^{n-j}\big(L\|\Delta y_j\| + \ell\|\Delta z_j\|\big) + C_4\frac{\varepsilon}{h}\sum_{j=0}^{n}\kappa^{n-j}\|e_j\|. \tag{2.16}$$

It is important to remark that the Lipschitz constant ℓ can be made arbitrarily small by shrinking the considered interval.

b) In order to solve the inequalities (2.9) and (2.16) we define sequences $\{u_n\}$ and $\{v_n\}$ by

$$\begin{aligned} u_n &= h\sum_{j=0}^{n}(Mu_j + Nv_j) + C_3\sum_{j=0}^{n}\|d_j\|, \\ v_n &= \sum_{j=0}^{n}\kappa^{n-j}(Lu_j + \ell v_j) + C_4\frac{\varepsilon}{h}\sum_{j=0}^{n}\kappa^{n-j}\|e_j\|. \end{aligned} \tag{2.17}$$

An induction argument shows that

$$\|\Delta y_n\| \le u_n, \qquad \|\Delta z_n\| \le v_n$$

provided $\ell < 1$ and $h \le h_0$. We then rewrite (2.17) as

$$\begin{aligned} u_n &= u_{n-1} + hMu_n + hNv_n + C_3\|d_n\|, & u_{-1} &= 0, \\ v_n &= \kappa v_{n-1} + Lu_n + \ell v_n + C_4\frac{\varepsilon}{h}\|e_n\|, & v_{-1} &= 0. \end{aligned}$$

Solving for u_n, v_n we get (with $\varrho = \kappa/(1-\ell)$)

$$\begin{pmatrix} u_n \\ v_n \end{pmatrix} = A(h)\begin{pmatrix} u_{n-1} \\ v_{n-1} \end{pmatrix} + \begin{pmatrix} \widehat{d}_n \\ \widehat{e}_n \end{pmatrix}, \qquad A(h) = \begin{pmatrix} 1+\mathcal{O}(h) & \mathcal{O}(h) \\ \mathcal{O}(1) & \varrho + \mathcal{O}(h) \end{pmatrix} \tag{2.18}$$

where

$$|\widehat{d}_n| \le C_5\big(\|d_n\| + \varepsilon\|e_n\|\big), \qquad |\widehat{e}_n| \le C_6\big(\|d_n\| + \frac{\varepsilon}{h}\|e_n\|\big). \tag{2.19}$$

Inserting (2.18) repeatedly we obtain

$$\begin{pmatrix} u_n \\ v_n \end{pmatrix} = \sum_{j=0}^{n} A(h)^{n-j}\begin{pmatrix} \widehat{d}_j \\ \widehat{e}_j \end{pmatrix}. \tag{2.20}$$

If ℓ is small enough so that $\varrho = \kappa/(1-\ell) < 1$ and if $h \le h_0$, then the eigenvalues of $A(h)$ are distinct and $A(h)$ can be diagonalized as

$$A(h) = T^{-1}(h)\begin{pmatrix} 1+\mathcal{O}(h) & 0 \\ 0 & \varrho + \mathcal{O}(h) \end{pmatrix} T(h), \quad T(h) = \begin{pmatrix} 1 & \mathcal{O}(h) \\ \mathcal{O}(1) & 1 \end{pmatrix}.$$

Inserted into (2.20) this yields

$$u_n + v_n \le Const.\Big(\sum_{j=1}^{n}\widehat{d}_j + \sum_{j=1}^{n}(h + \varrho^{n-j})\,\widehat{e}_j\Big).$$

Since $d_0,\ldots,d_{k-1}$ are linear combinations of the values $\Delta y_0,\ldots,\Delta y_{k-1}$, and $e_0,\ldots,e_{k-1}$ are linear combinations of the Δz_j and $\frac{h}{\varepsilon}\Delta z_j$, the statement of the theorem follows from (2.19) and (2.7). Because of our assumption on ℓ (that $\varrho = \kappa/(1-\ell) < 1$) we have proved the theorem for sufficiently small (but ε-independent) intervals. Compact intervals $[x_0, \overline{x}]$ can be covered by repeated application of the above estimates.

c) It still remains to prove (2.15). More generally, we shall show that

$$\frac{h}{\varepsilon}\Big\| r_j\Big(\frac{h}{\varepsilon}\, g_z(y,z)\Big)\Big\| \le C\kappa^j \qquad \text{with} \quad 0<\kappa<1 \tag{2.21}$$

holds uniformly in a compact neighbourhood of the solution. This is necessary, if the above estimates are applied to several subintervals. In order to prove (2.21) we remember that $r_j(\frac{h}{\varepsilon}J)$ is defined by (2.14). If we are able to show that

$$\Big\|\Big(\frac{\varepsilon}{h}\,\delta(\zeta)I - g_z(y,z)\Big)^{-1}\frac{\zeta^{-k}}{\sigma(\zeta^{-k})}\Big\| \le C \qquad \text{for} \quad |\zeta|\le 1/\kappa \tag{2.22}$$

then the estimate (2.21) follows immediately from Cauchy's integral formula

$$\frac{h}{\varepsilon}\, r_j\Big(\frac{h}{\varepsilon}J\Big) = \frac{1}{2\pi i}\int_{|\zeta|=1/\kappa}\Big(\frac{\varepsilon}{h}\,\delta(\zeta)I - J\Big)^{-1}\frac{\zeta^{-k}}{\sigma(\zeta^{-1})}\cdot \zeta^{-j-1}\,d\zeta.$$

By definition of the stability region S of a multistep method, the value $\delta(\zeta)$ lies outside of S whenever $|\zeta|<1$. Recall that the method is $A(\alpha)$-stable and strictly stable at infinity, and the differential equation satisfies (2.5). Therefore the set of eigenvalues of $g_z(y,z)$ (with (y,z) varying in a compact neighbourhood of the solution) is well separated from $\{\gamma\delta(\zeta)\ ;\ \gamma\le 1,\ |\zeta|\le 1\}$. It is even separated from $\{\gamma\delta(\zeta)\ ;\ \gamma\le 1,\ |\zeta|\le 1/\kappa\}$ with some $\kappa<1$. Together with Exercise 2 of Sect. V.7 this proves (2.22). □

Exercises

1. (Lubich 1991). Prove that for the BDF-schemes the estimate of Theorem 2.1 (for $n\ge k$) is valid with $(h+\varrho^n)$ replaced by $\varepsilon(1+\varrho^n/h)$ in the factor multiplying the z-component of the errors in the starting values.

   Hint. Give a direct proof for $n\in\{k,\dots,2k-1\}$; then apply Theorem 2.1 to shifted starting values.

VI.3 Epsilon Expansions for Exact and RK Solutions

In the preceding section we have proved convergence of multistep methods for singular perturbation problems. The same techniques do not yield optimal estimates for Runge-Kutta methods. We therefore investigate more thoroughly the structure of the solutions of singular perturbation problems. A first systematic study of the qualitative aspects of such problems is due to Tikhonov (1952). Asymptotic expansions were then analyzed by Vasil'eva (1963). Classical books on this subject are Wasow (1965), O'Malley (1974), and Tikhonov, Vasil'eva & Sveshnikov (1985).

Expansion of the Smooth Solution

> Tihonov's theorem is only the first step ... The actual approximate solution of such problems in series form is still a difficult question. It has been analyzed in a series of papers by Vasil'eva ...
> (W. Wasow 1965)

We consider the singular perturbation problem

$$\begin{aligned} y' &= f(y,z) \\ \varepsilon z' &= g(y,z), \qquad 0 < \varepsilon \ll 1 \end{aligned} \tag{3.1}$$

where f and g are sufficiently differentiable. The functions f, g and the initial values $y(0)$, $z(0)$ may depend smoothly on ε. For simplicity of notation we suppress this dependence. The corresponding equation for $\varepsilon = 0$,

$$\begin{aligned} y' &= f(y,z) \\ 0 &= g(y,z), \end{aligned} \tag{3.2}$$

is the *reduced problem*. In order to guarantee the solvability of (3.2), we assume that $g_z(y,z)$ is invertible (in a neighbourhood of the solution of (3.2)).

We are mainly interested in smooth solutions of (3.1), which are of the form

$$\begin{aligned} y(x) &= y_0(x) + \varepsilon y_1(x) + \varepsilon^2 y_2(x) + \ldots \\ z(x) &= z_0(x) + \varepsilon z_1(x) + \varepsilon^2 z_2(x) + \ldots . \end{aligned} \tag{3.3}$$

Inserting (3.3) into (3.1) and collecting equal powers of ε yields

$$\varepsilon^0 : \quad \left.\begin{aligned} y_0' &= f(y_0, z_0) \\ 0 &= g(y_0, z_0) \end{aligned}\right\} \tag{3.4a}$$

$$\varepsilon^1: \quad \left.\begin{aligned} y_1' &= f_y(y_0,z_0)y_1 + f_z(y_0,z_0)z_1 \\ z_0' &= g_y(y_0,z_0)y_1 + g_z(y_0,z_0)z_1 \end{aligned}\right\} \tag{3.4b}$$

$\dots$

$$\varepsilon^\nu: \quad \left.\begin{aligned} y_\nu' &= f_y(y_0,z_0)y_\nu + f_z(y_0,z_0)z_\nu + \varphi_\nu(y_0,z_0,\ldots,y_{\nu-1},z_{\nu-1}) \\ z_{\nu-1}' &= g_y(y_0,z_0)y_\nu + g_z(y_0,z_0)z_\nu + \psi_\nu(y_0,z_0,\ldots,y_{\nu-1},z_{\nu-1}) \end{aligned}\right\} \tag{3.4c}$$

As expected, we see from (3.4a) that $y_0(x)$, $z_0(x)$ is a solution of the reduced system. Since g_z is invertible, the second equation of (3.4b) can be solved for z_1. By inserting z_1 into the upper relation of (3.4b) we obtain a linear differential equation for $y_1(x)$. Hence, $y_1(x)$ and $z_1(x)$ are determined. Similarly, we get $y_2(x)$, $z_2(x)$ from (3.4c), etc.

This construction of the coefficients of (3.3) shows that we can choose the initial values $y_j(0)$ arbitrarily, but that there is no freedom in the choice of $z_j(0)$. Consequently, not every solution of (3.1) can be written in the form (3.3).

Expansions with Boundary Layer Terms

> To construct a uniform asymptotic expansion we must combine the Maclaurin expansion with another expansion of special form. The terms in this expansion are exponential functions that are appreciable inside the boundary layer, but negligibly small outside it. (A.B. Vasil'eva 1963)

Example 3.1. We consider the problem (IV.1.1), written in the form

$$\varepsilon z' = -z + \cos x. \tag{3.5}$$

Its analytic solution

$$\begin{aligned} z(x) &= (1+\varepsilon^2)^{-1}(\cos x + \varepsilon \sin x) + Ce^{-x/\varepsilon} \\ &= \cos x + \varepsilon \sin x - \varepsilon^2 \cos x - \varepsilon^3 \sin x + \ldots + Ce^{-x/\varepsilon} \end{aligned}$$

is a superposition of a smooth solution of the form (3.3) and of a rapidly decaying function. This additional term (transient phase, boundary layer) compensates the missing freedom in the choice of the initial values $z_j(0)$.

Motivated by this example, we seek solutions of the general problem (3.1) which are of the form

$$\begin{aligned} y(x) &= \sum_{j\ge 0} \varepsilon^j y_j(x) + \varepsilon \sum_{j\ge 0} \varepsilon^j \eta_j(x/\varepsilon) \\ z(x) &= \sum_{j\ge 0} \varepsilon^j z_j(x) + \sum_{j\ge 0} \varepsilon^j \zeta_j(x/\varepsilon), \end{aligned} \tag{3.6}$$

where $y_j(x)$, $z_j(x)$ are determined by (3.4) and the ε-independent functions η_j, ζ_j are assumed to satisfy

$$\|\eta_j(\xi)\| \le e^{-\kappa\xi}, \qquad \|\zeta_j(\xi)\| \le e^{-\kappa\xi} \tag{3.7}$$

with some $\kappa > 0$. Inserting (3.6) into (3.1) and using (3.4) we obtain formally

$$\sum_{j\ge 0}\varepsilon^j\eta_j'\Big(\frac{x}{\varepsilon}\Big) = f\Big(\sum_{j\ge 0}\varepsilon^j y_j(x)+\varepsilon\sum_{j\ge 0}\varepsilon^j\eta_j\Big(\frac{x}{\varepsilon}\Big),\ \sum_{j\ge 0}\varepsilon^j z_j(x)+\sum_{j\ge 0}\varepsilon^j\zeta_j\Big(\frac{x}{\varepsilon}\Big)\Big) - f\Big(\sum_{j\ge 0}\varepsilon^j y_j(x),\ \sum_{j\ge 0}\varepsilon^j z_j(x)\Big) \tag{3.8a}$$

$$\sum_{j\ge 0}\varepsilon^j\zeta_j'\Big(\frac{x}{\varepsilon}\Big) = g\Big(\sum_{j\ge 0}\varepsilon^j y_j(x)+\varepsilon\sum_{j\ge 0}\varepsilon^j\eta_j\Big(\frac{x}{\varepsilon}\Big),\ \sum_{j\ge 0}\varepsilon^j z_j(x)+\sum_{j\ge 0}\varepsilon^j\zeta_j\Big(\frac{x}{\varepsilon}\Big)\Big) - g\Big(\sum_{j\ge 0}\varepsilon^j y_j(x),\ \sum_{j\ge 0}\varepsilon^j z_j(x)\Big). \tag{3.8b}$$

We then replace x by the stretched variable

$$\xi = x/\varepsilon \tag{3.9}$$

and compare like powers of ε in (3.8). This gives for ε^0

$$\eta_0'(\xi) = f\big(y_0(0), z_0(0)+\zeta_0(\xi)\big) - f\big(y_0(0), z_0(0)\big) \tag{3.10a}$$

$$\zeta_0'(\xi) = g\big(y_0(0), z_0(0)+\zeta_0(\xi)\big) - g\big(y_0(0), z_0(0)\big). \tag{3.10b}$$

At this point it is necessary to introduce some stability assumption for (3.1) in order to obtain (3.7). We shall require that the logarithmic norm of g_z satisfy

$$\mu\big(g_z(y,z)\big) \le -1 \tag{3.11}$$

in an ε-independent neighbourhood of the solution of (3.2) (any negative bound other than -1 can be normalized by re-scaling ε). By Theorem I.10.6 Eqs. (3.10b) and (3.11) imply

$$\|\zeta_0(\xi)\| \le \|\zeta_0(0)\| e^{-\xi}.$$

Since $f(y,z)$ satisfies locally a Lipschitz condition, the right-hand side of (3.10a), denoted by $\varphi(\xi)$, is bounded by $\|\varphi(\xi)\| \le L\|\zeta_0(0)\|e^{-\xi}$. Consequently, there is only one solution of (3.10a) which satisfies (3.7), namely

$$\eta_0(\xi) = \int_0^\xi \varphi(s)\,ds - \int_0^\infty \varphi(s)\,ds. \tag{3.12}$$

A comparison of the powers of ε^1 in (3.8) yields

$$\begin{aligned}
\eta_1'(\xi) &= f_y\big(y_0(0), z_0(0)+\zeta_0(\xi)\big)\big(y_1(0)+\xi y_0'(0)+\eta_0(\xi)\big)\\
&\quad + f_z\big(y_0(0), z_0(0)+\zeta_0(\xi)\big)\big(z_1(0)+\xi z_0'(0)+\zeta_1(\xi)\big)\\
&\quad - f_y\big(y_0(0), z_0(0)\big)\big(y_1(0)+\xi y_0'(0)\big)\\
&\quad - f_z\big(y_0(0), z_0(0)\big)\big(z_1(0)+\xi z_0'(0)\big)
\end{aligned} \tag{3.13a}$$

$$\begin{aligned}
\zeta_1'(\xi) &= g_y\big(y_0(0), z_0(0)+\zeta_0(\xi)\big)\big(y_1(0)+\xi y_0'(0)+\eta_0(\xi)\big)\\
&\quad + g_z\big(y_0(0), z_0(0)+\zeta_0(\xi)\big)\big(z_1(0)+\xi z_0'(0)+\zeta_1(\xi)\big)
\end{aligned}$$

$$
\begin{aligned}
&- g_y\big(y_0(0), z_0(0)\big)\big(y_1(0) + \xi y_0'(0)\big) \\
&- g_z\big(y_0(0), z_0(0)\big)\big(z_1(0) + \xi z_0'(0)\big). \qquad (3.13b)
\end{aligned}
$$

Eq. (3.13b) is a linear differential equation for $\zeta_1(\xi)$. Its defect, for ζ_1 replaced by 0, is bounded by $Ce^{-\xi}$. Therefore, an application of Theorem I.10.6 yields

$$
\|\zeta_1(\xi)\| \le e^{-\xi}\big(\|\zeta_1(0)\| + C\xi\big),
$$

which implies (3.7) for any $\kappa < 1$. The right-hand side of (3.13a) is then bounded by $C_1 e^{-\kappa\xi}$. As in (3.12) we obtain a unique solution to (3.13a), which satisfies (3.7). This procedure can be continued to construct all further $\eta_j(\xi)$, $\zeta_j(\xi)$. At each step, the value of κ in (3.7) may become smaller. This is no serious difficulty, because we are only interested in a finite part of the series (3.6).

We point out that for the construction of $\eta_j(\xi)$, $\zeta_j(\xi)$ we can choose $\zeta_j(0)$ arbitrarily, but that there is no freedom in the choice of $\eta_j(0)$.

As a consequence, for an arbitrary initial value for (3.1) with expansion

$$
\begin{aligned}
y(0) &= y_0^0 + \varepsilon y_1^0 + \varepsilon^2 y_2^0 + \ldots \\
z(0) &= z_0^0 + \varepsilon z_1^0 + \varepsilon^2 z_2^0 + \ldots,
\end{aligned}
\qquad (3.14)
$$

the coefficients of the series (3.6) can be constructed as follows: put $x = 0$ in (3.6) to obtain the necessary relations

$$
y_0(0) = y_0^0, \qquad y_j(0) + \eta_{j-1}(0) = y_j^0, \qquad z_j(0) + \zeta_j(0) = z_j^0. \qquad (3.15)
$$

This initial value $y_0(0) = y_0^0$ determines $z_0(0)$ by (3.4a), $\zeta_0(0)$ is then given by (3.15), $\eta_0(0)$ by (3.12), $y_1(0)$ by (3.15), $z_1(0)$ by (3.4b), $\zeta_1(0)$ by (3.15), $\eta_1(0)$ by (3.13a) and (3.7), $y_2(0)$ by (3.15), etc.

Estimation of the Remainder

The following result gives a rigorous estimate of the remainder in (3.6), when only a truncated series is considered.

Theorem 3.2. *Consider the initial value problem (3.1), (3.14), and suppose that (3.11) holds in an ε-independent neighbourhood of the solution $y_0(x)$, $z_0(x)$ $(0 \le x \le \overline{x})$ of the reduced problem $(y_0(0) = y_0^0)$. If (y_0^0, z_0^0) lies in this neighbourhood, then the problem (3.1), (3.14) has a unique solution for ε sufficiently small and for $0 \le x \le \overline{x}$, which is of the form*

$$
\begin{aligned}
y(x) &= \sum_{j=0}^{N} \varepsilon^j y_j(x) + \varepsilon \sum_{j=0}^{N-1} \varepsilon^j \eta_j(x/\varepsilon) + \mathcal{O}(\varepsilon^{N+1}) \\
z(x) &= \sum_{j=0}^{N} \varepsilon^j z_j(x) + \sum_{j=0}^{N} \varepsilon^j \zeta_j(x/\varepsilon) + \mathcal{O}(\varepsilon^{N+1}).
\end{aligned}
\qquad (3.16)
$$

The coefficients $y_j(x)$, $z_j(x)$, $\eta_j(\xi)$, $\zeta_j(\xi)$ are given by (3.4), (3.10), (3.13), and satisfy (3.7).

Proof. We denote the truncated series by

$$\begin{aligned}\widehat{y}(x) &= \sum_{j=0}^{N} \varepsilon^j y_j(x) + \varepsilon \sum_{j=0}^{N} \varepsilon^j \eta_j(x/\varepsilon) \\ \widehat{z}(x) &= \sum_{j=0}^{N} \varepsilon^j z_j(x) + \sum_{j=0}^{N} \varepsilon^j \zeta_j(x/\varepsilon).\end{aligned} \tag{3.17}$$

By our construction of $y_j(x)$, $z_j(x)$, $\eta_j(\xi)$, $\zeta_j(\xi)$ we have

$$\begin{aligned}\widehat{y}'(x) &= f\big(\widehat{y}(x), \widehat{z}(x)\big) + \mathcal{O}(\varepsilon^{N+1}) \\ \varepsilon \widehat{z}'(x) &= g\big(\widehat{y}(x), \widehat{z}(x)\big) + \mathcal{O}(\varepsilon^{N+1}).\end{aligned} \tag{3.18}$$

Subtracting (3.1) from (3.18) and exploiting Lipschitz conditions for f and g we obtain

$$\begin{aligned}D_+\|\widehat{y}(x) - y(x)\| &\le L_1\|\widehat{y}(x) - y(x)\| + L_2\|\widehat{z}(x) - z(x)\| + C_1\varepsilon^{N+1} \\ \varepsilon D_+\|\widehat{z}(x) - z(x)\| &\le L_3\|\widehat{y}(x) - y(x)\| - \|\widehat{z}(x) - z(x)\| + C_2\varepsilon^{N+1}.\end{aligned} \tag{3.19}$$

Here, D_+ denotes the Dini derivate introduced in Section I.10. We have used $D_+\|w(x)\| \le \|w'(x)\|$ (see Eq. (I.10.4)) and, for the second inequality of (3.19), Formula (I.10.17) together with (3.11).

In order to solve inequality (3.19) we replace $\le$ by $=$ and so obtain

$$\begin{aligned}u' &= L_1 u + L_2 v + C_1\varepsilon^{N+1}, \qquad & u_0 &= \|\widehat{y}(0) - y(0)\| = \mathcal{O}(\varepsilon^{N+1}) \\ \varepsilon v' &= L_3 u - v + C_2\varepsilon^{N+1}, \qquad & v_0 &= \|\widehat{z}(0) - z(0)\| = \mathcal{O}(\varepsilon^{N+1}).\end{aligned} \tag{3.20}$$

This system is quasimonotone, it thus follows from Exercise 7 (Sect. I.10) that

$$\|\widehat{y}(x) - y(x)\| \le u(x), \qquad \|\widehat{z}(x) - z(x)\| \le v(x). \tag{3.21}$$

Transforming (3.20) to diagonal form one easily finds its analytic solution and verifies that $u(x) = \mathcal{O}(\varepsilon^{N+1})$, $v(x) = \mathcal{O}(\varepsilon^{N+1})$ on compact intervals. Inserted into (3.21) this proves the statement. □

Expansion of the Runge-Kutta Solution

After having understood the structure of the analytic solution of (3.1), we turn our attention to its numerical counterpart. We consider the Runge-Kutta method

$$\begin{pmatrix} y_{n+1} \\ z_{n+1} \end{pmatrix} = \begin{pmatrix} y_n \\ z_n \end{pmatrix} + h \sum_{i=1}^{s} b_i \begin{pmatrix} k_{ni} \\ \ell_{ni} \end{pmatrix} \tag{3.22}$$

where

$$\begin{pmatrix} k_{ni} \\ \varepsilon \ell_{ni} \end{pmatrix} = \begin{pmatrix} f(Y_{ni}, Z_{ni}) \\ g(Y_{ni}, Z_{ni}) \end{pmatrix} \tag{3.23}$$

and the internal stages are given by

$$\begin{pmatrix} Y_{ni} \\ Z_{ni} \end{pmatrix} = \begin{pmatrix} y_n \\ z_n \end{pmatrix} + h \sum_{j=1}^{s} a_{ij} \begin{pmatrix} k_{nj} \\ \ell_{nj} \end{pmatrix}. \tag{3.24}$$

For arbitrary initial values, the solution possesses a transient phase (as described by Theorem 3.2), and the numerical method has anyway to take small step sizes of magnitude $\mathcal{O}(\varepsilon)$. We shall therefore focus on the situation where the transient phase is over and the method has reached the smooth solution within the given tolerance. We thus suppose that the initial values lie on the smooth solution (i.e., that an expansion of the form (3.3) holds) and that the step size h is large compared to ε. Our first goal is an ε-expansion of the numerical solution. To this end, we formally expand all occuring quantities into powers of ε with ε-independent coefficients (see Hairer, Lubich & Roche 1988)

$$y_n = y_n^0 + \varepsilon y_n^1 + \varepsilon^2 y_n^2 + \ldots \tag{3.25a}$$
$$Y_{ni} = Y_{ni}^0 + \varepsilon Y_{ni}^1 + \varepsilon^2 Y_{ni}^2 + \ldots \tag{3.25b}$$
$$k_{ni} = k_{ni}^0 + \varepsilon k_{ni}^1 + \varepsilon^2 k_{ni}^2 + \ldots \tag{3.25c}$$

and similarly for z_n, Z_{ni}, ℓ_{ni}. Because of the linearity of the relations (3.22) and (3.24) we have

$$\begin{pmatrix} y_{n+1}^\nu \\ z_{n+1}^\nu \end{pmatrix} = \begin{pmatrix} y_n^\nu \\ z_n^\nu \end{pmatrix} + h \sum_{i=1}^{s} b_i \begin{pmatrix} k_{ni}^\nu \\ \ell_{ni}^\nu \end{pmatrix} \tag{3.26}$$

and

$$\begin{pmatrix} Y_{ni}^\nu \\ Z_{ni}^\nu \end{pmatrix} = \begin{pmatrix} y_n^\nu \\ z_n^\nu \end{pmatrix} + h \sum_{j=1}^{s} a_{ij} \begin{pmatrix} k_{nj}^\nu \\ \ell_{nj}^\nu \end{pmatrix}. \tag{3.27}$$

Inserting (3.25b, c) into (3.23) and comparing equal powers of ε we obtain

$$\varepsilon^0: \quad \left.\begin{aligned} k_{ni}^0 &= f(Y_{ni}^0, Z_{ni}^0) \\ 0 &= g(Y_{ni}^0, Z_{ni}^0) \end{aligned}\right\} \tag{3.28a}$$

$$\varepsilon^1: \quad \left.\begin{aligned} k_{ni}^1 &= f_y(Y_{ni}^0, Z_{ni}^0)Y_{ni}^1 + f_z(Y_{ni}^0, Z_{ni}^0)Z_{ni}^1 \\ \ell_{ni}^0 &= g_y(Y_{ni}^0, Z_{ni}^0)Y_{ni}^1 + g_z(Y_{ni}^0, Z_{ni}^0)Z_{ni}^1 \end{aligned}\right\} \tag{3.28b}$$

$$\ldots$$

$$\varepsilon^\nu: \quad \left.\begin{aligned} k_{ni}^\nu &= f_y(Y_{ni}^0, Z_{ni}^0)Y_{ni}^\nu + f_z(Y_{ni}^0, Z_{ni}^0)Z_{ni}^\nu + \varphi_\nu(Y_{ni}^0, Z_{ni}^0, \ldots, Y_{ni}^{\nu-1}, Z_{ni}^{\nu-1}) \\ \ell_{ni}^{\nu-1} &= g_y(Y_{ni}^0, Z_{ni}^0)Y_{ni}^\nu + g_z(Y_{ni}^0, Z_{ni}^0)Z_{ni}^\nu + \psi_\nu(Y_{ni}^0, Z_{ni}^0, \ldots, Y_{ni}^{\nu-1}, Z_{ni}^{\nu-1}) \end{aligned}\right\} \tag{3.28c}$$

Since (3.23) has the same form as the differential equation (3.1), it is obvious that the formulas of (3.28) are exactly the same as those of (3.4). An interesting interpretation of this fact is the following: the coefficients y_n^0, z_n^0, y_n^1, $z_n^1, \ldots$ represent the numerical solution of the Runge-Kutta method applied to the differential-algebraic system (3.4) (ε-embedding method of Sect. VI.1). This can be expressed by the commutativity of the following diagram:

$$\begin{array}{ccc} \text{Problem (3.1)} & \xrightarrow{(3.3)} & \text{DAE (3.4)} \\ RK \downarrow \text{method} & & RK \downarrow \text{method} \\ \{y_n, z_n\} & \xrightarrow{(3.25)} & \{y_n^0, z_n^0, y_n^1, z_n^1, \ldots\} \end{array}$$

Subtracting (3.25a) from (3.3) we get formally

$$\begin{aligned} y_n - y(x_n) &= \sum_{\nu \geq 0} \varepsilon^\nu \left(y_n^\nu - y_\nu(x_n)\right) \\ z_n - z(x_n) &= \sum_{\nu \geq 0} \varepsilon^\nu \left(z_n^\nu - z_\nu(x_n)\right). \end{aligned} \tag{3.29}$$

In order to study this error we first investigate the differences $y_n^\nu - y_\nu(x_n)$, $z_n^\nu - z_\nu(x_n)$ (next subsection). A rigorous estimate of the remainder in (3.29) will then follow. The presentation follows that of Hairer, Lubich & Roche (1988).

Convergence of RK-Methods for Differential-Algebraic Systems

The first differences $y_n^0 - y_0(x_n)$, $z_n^0 - z_0(x_n)$ in the expansions of (3.29) are just the global errors of the Runge-Kutta method applied to the reduced system (3.4a). By assumption (3.11) this system is of index 1. Therefore, the following result is an immediate consequence of Theorem 1.1.

Theorem 3.3. *Consider a Runge-Kutta method of (classical) order p, with invertible coefficient matrix (a_{ij}). Suppose that Problem (3.4a) satisfies (3.11) and that the initial values are consistent.*
a) If the method is stiffly accurate (i.e., $a_{si} = b_i$ for $i = 1, \ldots, s$) then the global error satisfies

$$y_n^0 - y_0(x_n) = \mathcal{O}(h^p), \qquad z_n^0 - z_0(x_n) = \mathcal{O}(h^p). \tag{3.30}$$

b) If the stability function satisfies $|R(\infty)| < 1$, and the stage order is q $(q < p)$, then

$$y_n^0 - y_0(x_n) = \mathcal{O}(h^p), \qquad z_n^0 - z_0(x_n) = \mathcal{O}(h^{q+1}). \tag{3.31}$$

In both cases the estimates hold uniformly for $nh \leq Const$. □

Estimating the second differences $y_n^1 - y_1(x_n)$, $z_n^1 - z_1(x_n)$ is not as simple, because the enlarged system (3.4a,b) with differential variables y_0, z_0, y_1 and algebraic variable z_1, is no longer of index 1. It is actually of index 2, as will become clear in Sect. VII.1 below (Exercise 5). In principle it is possible to consult the results of Sect. VII.4 (Theorems VII.4.5 and VII.4.6). For the special system (3.4a,b), however, a simpler proof is possible. It also extends more easily to the higher-index problems (3.4a-c).

Theorem 3.4 (Hairer, Lubich & Roche 1988). *Consider a Runge-Kutta method of order p, stage order q $(q<p)$, such that (a_{ij}) is invertible and the stability function satisfies $|R(\infty)|<1$. If (3.11) holds and if the initial values of the differential-algebraic system (3.4a-c) are consistent, then the global error of method (3.26)–(3.28) satisfies for $1 \le \nu \le q+1$*

$$y_n^\nu - y_\nu(x_n) = \mathcal{O}(h^{q+2-\nu}), \qquad z_n^\nu - z_\nu(x_n) = \mathcal{O}(h^{q+1-\nu}).$$

Proof. We denote the differences to the exact solution values by

$$\begin{aligned} \Delta y_n^\nu &= y_n^\nu - y_\nu(x_n), & \Delta z_n^\nu &= z_n^\nu - z_\nu(x_n), \\ \Delta Y_{ni}^\nu &= Y_{ni}^\nu - y_\nu(x_n+c_ih), & \Delta Z_{ni}^\nu &= Z_{ni}^\nu - z_\nu(x_n+c_ih), \\ \Delta k_{ni}^\nu &= k_{ni}^\nu - y_\nu'(x_n+c_ih), & \Delta \ell_{ni}^\nu &= \ell_{ni}^\nu - z_\nu'(x_n+c_ih). \end{aligned} \tag{3.32}$$

Since the quadrature formula with nodes c_i and weights b_i is of order p, we have from (3.26)

$$\begin{pmatrix} \Delta y_{n+1}^\nu \\ \Delta z_{n+1}^\nu \end{pmatrix} = \begin{pmatrix} \Delta y_n^\nu \\ \Delta z_n^\nu \end{pmatrix} + h\sum_{i=1}^s b_i \begin{pmatrix} \Delta k_{ni}^\nu \\ \Delta \ell_{ni}^\nu \end{pmatrix} + \mathcal{O}(h^{p+1}). \tag{3.33}$$

Similarly, the definition of the stage order implies

$$\begin{pmatrix} \Delta Y_{ni}^\nu \\ \Delta Z_{ni}^\nu \end{pmatrix} = \begin{pmatrix} \Delta y_n^\nu \\ \Delta z_n^\nu \end{pmatrix} + h\sum_{j=1}^s a_{ij} \begin{pmatrix} \Delta k_{nj}^\nu \\ \Delta \ell_{nj}^\nu \end{pmatrix} + \mathcal{O}(h^{q+1}). \tag{3.34}$$

It follows from Theorem 3.3 (see also the proof of Theorem 1.1) that

$$\begin{aligned} \Delta y_n^0 &= \mathcal{O}(h^p), & \Delta Y_{ni}^0 &= \mathcal{O}(h^{q+1}), & \Delta k_{ni}^0 &= \mathcal{O}(h^{q+1}) \\ \Delta z_n^0 &= \mathcal{O}(h^{q+1}), & \Delta Z_{ni}^0 &= \mathcal{O}(h^{q+1}), & \Delta \ell_{ni}^0 &= \mathcal{O}(h^q). \end{aligned} \tag{3.35}$$

a) We first consider the case $\nu=1$. Replacing in (3.28b) Y_{ni}^0, Z_{ni}^0 by $y_0(x_n + c_ih)+\Delta Y_{ni}^0$, $z_0(x_n+c_ih)+\Delta Z_{ni}^0$ and subtracting Equation (3.4b) at the position $x=x_n+c_ih$, we obtain with the help of (3.35)

$$\begin{aligned} \Delta k_{ni}^1 &= f_y(x_n+c_ih)\Delta Y_{ni}^1 + f_z(x_n+c_ih)\Delta Z_{ni}^1 \\ &\qquad + \mathcal{O}(h^{q+1}+h^{q+1}\|\Delta Y_{ni}^1\| + h^{q+1}\|\Delta Z_{ni}^1\|) \\ \Delta \ell_{ni}^0 &= g_y(x_n+c_ih)\Delta Y_{ni}^1 + g_z(x_n+c_ih)\Delta Z_{ni}^1 \\ &\qquad + \mathcal{O}(h^{q+1}+h^{q+1}\|\Delta Y_{ni}^1\| + h^{q+1}\|\Delta Z_{ni}^1\|). \end{aligned} \tag{3.36}$$

Here we have used the abbreviations $f_y(x) = f_y(y_0(x), z_0(x))$, etc. Computing ΔZ_{ni}^1 from the second relation of (3.36) and inserting it into the first one yields

$$\begin{aligned} &\Delta k_{ni}^1 - (f_z g_z^{-1})(x_n+c_ih)\Delta \ell_{ni}^0 \\ &\qquad = (f_y - f_z g_z^{-1} g_y)(x_n+c_ih)\Delta Y_{ni}^1 + \mathcal{O}(h^{q+1}+h^{q+1}\|\Delta Y_{ni}^1\|). \end{aligned}$$

Using (3.34) we can eliminate ΔY_{ni}^1 and obtain (with (3.35))

$$\Delta k_{ni}^1 - (f_z g_z^{-1})(x_n + c_i h)\Delta \ell_{ni}^0 = \mathcal{O}(\|\Delta y_n^1\|) + \mathcal{O}(h^{q+1}). \tag{3.37}$$

Since $\Delta \ell_{ni}^0$ is of size $\mathcal{O}(h^q)$, we only have $\Delta k_{ni}^1 = \mathcal{O}(\|\Delta y_n^1\|) + \mathcal{O}(h^q)$, and a direct estimation of Δy_n^1 in (3.33) would lead to $\Delta y_n^1 = \mathcal{O}(h^q)$, which is not optimal. We therefore introduce the new variable

$$\Delta u_n^1 = \Delta y_n^1 - (f_z g_z^{-1})(x_n)\Delta z_n^0. \tag{3.38}$$

From (3.33) we get

$$\begin{aligned}\Delta u_{n+1}^1 = \Delta u_n^1 &+ h\sum_{i=1}^{s} b_i\big(\Delta k_{ni}^1 - (f_z g_z^{-1})(x_n)\Delta \ell_{ni}^0\big) \\ &- \big((f_z g_z^{-1})(x_n+h) - (f_z g_z^{-1})(x_n)\big)\Delta z_{n+1}^0 + \mathcal{O}(h^{p+1}).\end{aligned} \tag{3.39}$$

The estimates (3.35), (3.37) and the fact that $\Delta y_n^1 = \Delta u_n^1 + \mathcal{O}(h^{q+1})$ imply that

$$\|\Delta u_{n+1}^1\| \le (1 + Ch)\|\Delta u_n^1\| + \mathcal{O}(h^{q+2}). \tag{3.40}$$

Standard techniques now show that $\Delta u_n^1 = \mathcal{O}(h^{q+1})$ for $nh \le$ *Const* (observe that the initial values are assumed to be consistent, i.e., $\Delta u_0^1 = 0$), so that by (3.38) and (3.35) also $\Delta y_n^1 = \mathcal{O}(h^{q+1})$. This implies $\Delta k_{ni}^1 = \mathcal{O}(h^q)$ by (3.37) and $\Delta Y_{ni}^1 = \mathcal{O}(h^{q+1})$ by (3.34). The second relation of (3.36) then proves that $\Delta Z_{ni}^1 = \mathcal{O}(h^q)$. In order to estimate Δz_n^1, we compute $\Delta \ell_{ni}^1$ from (3.34) and insert it into (3.33). Using $\Delta Z_{ni}^1 = \mathcal{O}(h^q)$ this gives

$$\Delta z_{n+1}^1 = (1 - b^T A^{-1}\mathbb{1})\Delta z_n^1 + \mathcal{O}(h^q), \tag{3.41}$$

and it follows from $|1 - b^T A^{-1}\mathbb{1}| = |R(\infty)| < 1$ that $\Delta z_n^1 = \mathcal{O}(h^q)$. We have thus proved the case $\nu = 1$.

b) The proof for general ν is by induction. We shall show that

$$\begin{aligned}\Delta y_n^\nu &= \mathcal{O}(h^{q+2-\nu}), \qquad & \Delta Y_{ni}^\nu &= \mathcal{O}(h^{q+2-\nu}) \\ \Delta z_n^\nu &= \mathcal{O}(h^{q+1-\nu}), \qquad & \Delta Z_{ni}^\nu &= \mathcal{O}(h^{q+1-\nu})\end{aligned} \tag{3.42}$$

holds for $\nu = 1,\ldots,q+1$. The main difference to the case $\nu = 1$ consists in the additional inhomogeneities φ_ν and ψ_ν in (3.4c). Using their Lipschitz continuity one obtains an additional term of size $\mathcal{O}(h^{q+2-\nu})$ in (3.36). Otherwise the proof is identical to that for $\nu = 1$. □

We next study the existence and local uniqueness of the solution of the Runge-Kutta method (3.22)–(3.24). Further, we investigate the influence of perturbations in (3.24) to the numerical solution. This will be important for the estimation of the remainder in the expansion (3.29).

Existence and Uniqueness of the Runge-Kutta Solution

For h small compared to ε, the existence of a unique numerical solution of (3.23), (3.24) follows from standard fixed point iteration (e.g., Theorem II.7.2). For the (more interesting) case where the step size h is large compared to ε, we suppose that (y_n, z_n) are known, denote it by (η, ζ), and prove the existence of (y_{n+1}, z_{n+1}) as follows:

Theorem 3.5 (Hairer, Lubich & Roche 1988). *Assume that $g(\eta,\zeta) = \mathcal{O}(h)$, $\mu(g_z(\eta,\zeta)) \le -1$ and that the eigenvalues of the Runge-Kutta matrix (a_{ij}) have positive real part. Then, the nonlinear system*

$$\begin{pmatrix} Y_i - \eta \\ \varepsilon(Z_i - \zeta) \end{pmatrix} = h \sum_{j=1}^{s} a_{ij} \begin{pmatrix} f(Y_j, Z_j) \\ g(Y_j, Z_j) \end{pmatrix} \tag{3.43}$$

possesses a locally unique solution for $h \le h_0$, where h_0 is sufficiently small but independent of ε. This solution satisfies

$$Y_i - \eta = \mathcal{O}(h), \qquad Z_i - \zeta = \mathcal{O}(h). \tag{3.44}$$

Proof. We apply Newton's method to the nonlinear system (3.43), whose second equation is divided by h. The existence and uniqueness statement can then be deduced from the theorem of Newton-Kantorovich (Kantorovich & Akilov 1959, Ortega & Rheinboldt 1970) as follows: for the starting values $Y_i^{(0)} = \eta$, $Z_i^{(0)} = \zeta$ the Jacobian of the system is of the form

$$\begin{pmatrix} I + \mathcal{O}(h) & \mathcal{O}(h) \\ \mathcal{O}(1) & (\varepsilon/h)I - A \otimes g_z(\eta,\zeta) \end{pmatrix}. \tag{3.45}$$

Since $\mu(g_z(\eta,\zeta)) \le -1$ it follows from the matrix-valued theorem of von Neumann (Theorem V.7.8) that

$$\left\| \left(\kappa I - A \otimes g_z(\eta,\zeta) \right)^{-1} \right\| \le \max_{\operatorname{Re}\mu \le -1} \|(\kappa I - \mu A)^{-1}\|. \tag{3.46}$$

The right-hand side of (3.46) is bounded by a constant independent of $\kappa \ge 0$, because the eigenvalues of A are assumed to have positive real part. Consequently, also the inverse of (3.45) is uniformly bounded for $\varepsilon > 0$ and $h \le h_0$. This together with $g(\eta,\zeta) = \mathcal{O}(h)$ implies that the first increment (of Newton's method) is of size $\mathcal{O}(h)$. Hence, for sufficiently small h, the Newton-Kantorovich assumptions are fulfilled. □

Influence of Perturbations

For the perturbed Runge-Kutta method

$$\begin{pmatrix} \widehat{Y}_i - \widehat{\eta} \\ \varepsilon(\widehat{Z}_i - \widehat{\zeta}) \end{pmatrix} = h \sum_{j=1}^{s} a_{ij} \begin{pmatrix} f(\widehat{Y}_j, \widehat{Z}_j) \\ g(\widehat{Y}_j, \widehat{Z}_j) \end{pmatrix} + h \begin{pmatrix} \delta_i \\ \theta_i \end{pmatrix} \tag{3.47}$$

we have the following result.

Theorem 3.6 (Hairer, Lubich & Roche 1988). *Let Y_i, Z_i be given by (3.43) and consider perturbed values $\widehat{Y}_i$, $\widehat{Z}_i$ satisfying (3.47). In addition to the assumptions of Theorem 3.5 suppose that $\widehat{\eta} - \eta = \mathcal{O}(h)$, $\widehat{\zeta} - \zeta = \mathcal{O}(h)$, $\delta_i = \mathcal{O}(1)$, and $\theta_i = \mathcal{O}(h)$. Then we have for $h \le h_0$*

$$\begin{aligned} \|\widehat{Y}_i - Y_i\| &\le C(\|\widehat{\eta} - \eta\| + \varepsilon\|\widehat{\zeta} - \zeta\|) + hC(\|\delta\| + \|\theta\|) \\ \|\widehat{Z}_i - Z_i\| &\le C(\|\widehat{\eta} - \eta\| + \frac{\varepsilon}{h}\|\widehat{\zeta} - \zeta\|) + C(h\|\delta\| + \|\theta\|). \end{aligned} \tag{3.48}$$

Here $\delta = (\delta_1, \ldots, \delta_s)^T$ and $\theta = (\theta_1, \ldots, \theta_s)^T$.

Proof. The essential idea is to consider the homotopy

$$\begin{pmatrix} Y_i - \eta \\ \varepsilon(Z_i - \zeta) \end{pmatrix} - h \sum_{j=1}^{s} a_{ij} \begin{pmatrix} f(Y_j, Z_j) \\ g(Y_j, Z_j) \end{pmatrix} = \tau \begin{pmatrix} \widehat{\eta} - \eta + h\delta_i \\ \varepsilon(\widehat{\zeta} - \zeta) + h\theta_i \end{pmatrix} \tag{3.49}$$

which relates the system (3.43) for $\tau = 0$ to the perturbed system (3.47) for $\tau = 1$. The solutions Y_i and Z_i of (3.49) are functions of τ. If we differentiate (3.49) with respect to τ and divide its second formula by h, we obtain the differential equation

$$\begin{pmatrix} I + \mathcal{O}(h) & \mathcal{O}(h) \\ \mathcal{O}(1) & M(\varepsilon/h, Y, Z) \end{pmatrix} \begin{pmatrix} \dot{Y} \\ \dot{Z} \end{pmatrix} = \begin{pmatrix} 1\!\!1 \cdot (\widehat{\eta} - \eta) + h\delta \\ (\varepsilon/h) 1\!\!1 \cdot (\widehat{\zeta} - \zeta) + \theta \end{pmatrix} \tag{3.50}$$

where $1\!\!1 = (1, \ldots, 1)^T$, $Y = (Y_1, \ldots, Y_s)^T$, $Z = (Z_1, \ldots, Z_s)^T$ and

$$M(\kappa, Y, Z) = \kappa I - A \otimes I \cdot \begin{pmatrix} g_z(Y_1, Z_1) & & 0 \\ & \ddots & \\ 0 & & g_z(Y_s, Z_s) \end{pmatrix}. \tag{3.51}$$

Whenever $\|Y_i - \eta\| \le d$ and $\|Z_i - \zeta\| \le d$ for all i, we have

$$M(\kappa, Y, Z) = \kappa I - A \otimes g_z(\eta, \zeta) + \mathcal{O}(d) \tag{3.52}$$

and it follows from (3.46) that $M^{-1}(\kappa, Y, Z)$ is uniformly bounded for $\kappa \ge 0$, if d is sufficiently small. Hence, the inverse of the matrix in (3.50) satisfies

$$\begin{pmatrix} I + \mathcal{O}(h) & \mathcal{O}(h) \\ \mathcal{O}(1) & M(\varepsilon/h, Y, Z) \end{pmatrix}^{-1} = \begin{pmatrix} I + \mathcal{O}(h) & \mathcal{O}(h) \\ \mathcal{O}(1) & \mathcal{O}(1) \end{pmatrix}$$

and the statement (3.48) follows from the fact that

$$\widehat{Y} - Y = \int_0^1 \dot{Y}(\tau)d\tau, \qquad \widehat{Z} - Z = \int_0^1 \dot{Z}(\tau)d\tau. \qquad \square$$

Remark 3.7. If the Runge-Kutta matrix A is only assumed to be invertible, the results of Theorems 3.5 and 3.6 still hold for $\varepsilon \le Kh$, where K is any constant smaller than the modulus of the smallest eigenvalue of A (i.e., $K < |\lambda_{\min}|$). In this situation, the right-hand side of (3.48) is also bounded, and the same conclusions hold.

Estimation of the Remainder in the Numerical Solution

We are now in the position to estimate the remainder in (3.29). The result is the following.

Theorem 3.8 (Hairer, Lubich & Roche 1988). *Consider the stiff problem (3.1), (3.11) with initial values $y(0)$, $z(0)$ admitting a smooth solution. Apply the Runge-Kutta method (3.22)–(3.24) of classical order p and stage order q $(1 \le q < p)$. Assume that the method is A-stable, that the stability function satisfies $|R(\infty)| < 1$, and that the eigenvalues of the coefficient matrix A have positive real part. Then for any fixed constant $c > 0$ the global error satisfies for $\varepsilon \le ch$ and $\nu \le q+1$*

$$\begin{aligned} y_n - y(x_n) &= \Delta y_n^0 + \varepsilon \Delta y_n^1 + \ldots + \varepsilon^\nu \Delta y_n^\nu + \mathcal{O}(\varepsilon^{\nu+1}) \\ z_n - z(x_n) &= \Delta z_n^0 + \varepsilon \Delta z_n^1 + \ldots + \varepsilon^\nu \Delta z_n^\nu + \mathcal{O}(\varepsilon^{\nu+1}/h). \end{aligned} \tag{3.53}$$

Here $\Delta y_n^0 = y_n^0 - y_0(x_n)$, $\Delta z_n^0 = z_n^0 - z_0(x_n), \ldots$ (see Formula (3.32)) are the global errors of the method applied to the system (3.4). The estimates (3.53) hold uniformly for $h \le h_0$ and $nh \le Const$.

Proof. By Theorem 3.4 it suffices to prove the result for $\nu = q+1$. We denote the truncated series of (3.25) by

$$\begin{aligned} \widehat{y}_n &= y_n^0 + \varepsilon y_n^1 + \ldots + \varepsilon^\nu y_n^\nu \\ \widehat{Y}_{ni} &= Y_{ni}^0 + \varepsilon Y_{ni}^1 + \ldots + \varepsilon^\nu Y_{ni}^\nu \\ \widehat{k}_{ni} &= k_{ni}^0 + \varepsilon k_{ni}^1 + \ldots + \varepsilon^\nu k_{ni}^\nu \end{aligned} \tag{3.54}$$

and similarly $\widehat{z}_n$, $\widehat{Z}_{ni}$, $\widehat{\ell}_{ni}$. Further we denote

$$\Delta y_n = \widehat{y}_n - y_n, \qquad \Delta Y_{ni} = \widehat{Y}_{ni} - Y_{ni}, \qquad \Delta k_{ni} = \widehat{k}_{ni} - k_{ni}, \ldots \tag{3.55}$$

Using (3.3) and Theorem 3.4 the statement (3.53) is then equivalent to

$$\Delta y_n = \mathcal{O}(\varepsilon^{\nu+1}), \qquad \Delta z_n = \mathcal{O}(\varepsilon^{\nu+1}/h). \tag{3.56}$$

a) We first estimate the differences ΔY_{ni}, ΔZ_{ni} of the internal stages. For this we investigate the defect when (3.54) is inserted into (3.23). By our construction (3.28) it follows from (3.42) and $\nu = q+1$ that

$$\begin{aligned} \widehat{k}_{ni} &= f(\widehat{Y}_{ni}, \widehat{Z}_{ni}) + \mathcal{O}(\varepsilon^{\nu+1}) \\ \varepsilon\widehat{\ell}_{ni} &= g(\widehat{Y}_{ni}, \widehat{Z}_{ni}) + \varepsilon^{\nu+1}\ell^{\nu}_{ni} + \mathcal{O}(\varepsilon^{\nu+1}). \end{aligned} \tag{3.57}$$

From (3.42) and (3.27) we know that $\ell^{\nu}_{ni} = \mathcal{O}(h^{-1})$. Together with (3.27) this implies

$$\begin{pmatrix} \widehat{Y}_{ni} - \widehat{y}_n \\ \varepsilon(\widehat{Z}_{ni} - \widehat{z}_n) \end{pmatrix} = h\sum_{j=1}^{s} a_{ij} \begin{pmatrix} f(\widehat{Y}_{nj}, \widehat{Z}_{nj}) \\ g(\widehat{Y}_{nj}, \widehat{Z}_{nj}) \end{pmatrix} + \begin{pmatrix} \mathcal{O}(h\varepsilon^{\nu+1}) \\ \mathcal{O}(\varepsilon^{\nu+1}) \end{pmatrix}, \tag{3.58}$$

which is of the form (3.47). Application of Theorem 3.6 yields

$$\begin{aligned} \|\Delta Y_{ni}\| &\le C(\|\Delta y_n\| + \varepsilon\|\Delta z_n\|) + \mathcal{O}(\varepsilon^{\nu+1}) \\ \|\Delta Z_{ni}\| &\le C(\|\Delta y_n\| + \frac{\varepsilon}{h}\|\Delta z_n\|) + \mathcal{O}(\varepsilon^{\nu+1}/h) \end{aligned} \tag{3.59}$$

provided that Δy_n and Δz_n are of size $\mathcal{O}(h)$. This will be justified in part (c).

b) Our next aim is to prove the recursion

$$\begin{pmatrix} \|\Delta y_{n+1}\| \\ \|\Delta z_{n+1}\| \end{pmatrix} \le \begin{pmatrix} 1+\mathcal{O}(h) & \mathcal{O}(\varepsilon) \\ \mathcal{O}(1) & \alpha + \mathcal{O}(\varepsilon) \end{pmatrix} \begin{pmatrix} \|\Delta y_n\| \\ \|\Delta z_n\| \end{pmatrix} + \begin{pmatrix} \mathcal{O}(\varepsilon^{\nu+1}) \\ \mathcal{O}(\varepsilon^{\nu+1}/h) \end{pmatrix} \tag{3.60}$$

where we assume again that Δy_n and Δz_n are of size $\mathcal{O}(h)$. The value $\alpha < 1$ will be given in Formula (3.63) below. The upper relation of (3.60) follows from

$$\Delta y_{n+1} = \Delta y_n + h\sum_{i=1}^{s} b_i\Big(f(\widehat{Y}_{ni}, \widehat{Z}_{ni}) - f(Y_{ni}, Z_{ni})\Big) + \mathcal{O}(h\varepsilon^{\nu+1})$$

by the use of (3.59) and a Lipschitz condition for f.

For the verification of the second relation in (3.60) we subtract (3.57) from (3.23), and use (3.59) and (3.42) to obtain

$$\varepsilon\Delta\ell_{ni} = g_z(x_n)\Delta Z_{ni} + \mathcal{O}(\|\Delta Y_{ni}\| + h\|\Delta Z_{ni}\|) + \mathcal{O}(\varepsilon^{\nu+1}/h). \tag{3.61}$$

Here we use the notation $g_z(x) = g_z(y_0(x), z_0(x))$. Inserting $\Delta Z_{ni} = \Delta z_n + h\sum a_{ij}\Delta\ell_{nj}$ into this relation and using (3.59) again we obtain

$$\varepsilon\Delta\ell_{ni} - h\sum_{j=1}^{s} a_{ij} g_z(x_n)\Delta\ell_{nj} = g_z(x_n)\Delta z_n + \mathcal{O}(\|\Delta y_n\| + \varepsilon\|\Delta z_n\|) + \mathcal{O}(\varepsilon^{\nu+1}/h).$$

We now solve for $h\Delta\ell_{ni}$ and insert it into $\Delta z_{n+1} = \Delta z_n + h\sum b_i\Delta\ell_{ni}$. Since the matrix $(\varepsilon/h)I - A\otimes g_z(x_n)$ has a bounded inverse by (3.46), this gives

$$\Delta z_{n+1} = R\Big(\frac{h}{\varepsilon}\, g_z(x_n)\Big)\Delta z_n + \mathcal{O}(\|\Delta y_n\| + \varepsilon\|\Delta z_n\|) + \mathcal{O}(\varepsilon^{\nu+1}/h), \tag{3.62}$$

where $R(\mu)$ is the stability function of the method. Because of (3.11) we can apply von Neumann's theorem (Corollary IV.11.4) to estimate

$$\Big\| R\Big(\frac{h}{\varepsilon}\, g_z(x_n)\Big)\Big\| \le \sup\{|R(\mu)|\;;\;\mathrm{Re}\,\mu \le -h/\varepsilon\} \le \alpha < 1. \tag{3.63}$$

The bound α is strictly smaller than 1, because $|R(\infty)|<1$ and $-h/\varepsilon \le -1/c<0$. The triangle inequality applied to (3.62) completes the proof of Formula (3.60).

c) Applying Lemma 3.9 below to the difference inequality (3.60) gives

$$\Delta y_n = \mathcal{O}(\varepsilon^{\nu+1}/h), \qquad \Delta z_n = \mathcal{O}(\varepsilon^{\nu+1}/h) \tag{3.64}$$

for $nh \le Const$. We are now in a position to justify the assumption $\Delta y_n = \mathcal{O}(h)$ and $\Delta z_n = \mathcal{O}(h)$ of the beginning of the proof. Indeed, this follows by induction on n ($\Delta y_0 = \mathcal{O}(\varepsilon^{\nu+1})$, $\Delta z_0 = \mathcal{O}(\varepsilon^{\nu+1})$) and from (3.64), because $\nu = q+1 \ge 2$.

d) Formula (3.64) proves the desired result (3.56) for the z-component. However, the estimate (3.64) is not yet optimal for the y-component. The proof for the correct estimate is similar to that of Theorem 3.4. We have to treat more carefully the expression which gives rise to the $\mathcal{O}(\varepsilon^{\nu+1}/h)$ term in (3.61). Using (3.59) and (3.64) the same calculations which gave (3.61), now yield

$$\Delta k_{ni} = f_y(x_n)\Delta Y_{ni} + f_z(x_n)\Delta Z_{ni} + \mathcal{O}(\varepsilon^{\nu+1}) \tag{3.65a}$$

$$\varepsilon \Delta \ell_{ni} = g_y(x_n)\Delta Y_{ni} + g_z(x_n)\Delta Z_{ni} + \varepsilon^{\nu+1}\ell^{\nu}_{ni} + \mathcal{O}(\varepsilon^{\nu+1}). \tag{3.65b}$$

We compute ΔZ_{ni} from (3.65b) and insert it into (3.65a). This gives

$$\begin{aligned}\Delta k_{ni} - (f_z g_z^{-1})(x_n)\big(\varepsilon\Delta \ell_{ni} - \varepsilon^{\nu+1}\ell^{\nu}_{ni}\big)& \\ = (f_y - f_z g_z^{-1} g_y)(x_n)\Delta Y_{ni} + \mathcal{O}(\varepsilon^{\nu+1}).&\end{aligned} \tag{3.66}$$

Guided by this formula we put

$$\Delta u_n = \Delta y_n - (f_z g_z^{-1})(x_n)\big(\varepsilon \Delta z_n - \varepsilon^{\nu+1} z^{\nu}_n\big). \tag{3.67}$$

Since

$$\begin{aligned}\Delta u_{n+1} = \Delta u_n + h\sum_{i=1}^{s} b_i\Big(\Delta k_{ni} - (f_z g_z^{-1})(x_n)\big(\varepsilon\Delta\ell_{ni} - \varepsilon^{\nu+1}\ell^{\nu}_{ni}\big)\Big)& \\ - \Big((f_z g_z^{-1})(x_n+h) - (f_z g_z^{-1})(x_n)\Big)\big(\varepsilon\Delta z_{n+1} - \varepsilon^{\nu+1} z^{\nu}_{n+1}\big)&\end{aligned}$$

it follows from (3.66), (3.64), and (3.42) that

$$\|\Delta u_{n+1}\| \le (1+ch)\|\Delta u_n\| + \mathcal{O}(h\varepsilon^{\nu+1}). \tag{3.68}$$

As in the proof of Theorem 3.4 we deduce $\Delta u_n = \mathcal{O}(\varepsilon^{\nu+1})$ and $\Delta y_n = \mathcal{O}(\varepsilon^{\nu+1})$. □

In the above proof we used the following result.

Lemma 3.9. *Let $\{u_n\}$, $\{v_n\}$ be two sequences of non-negative numbers satisfying (componentwise)*

$$\begin{pmatrix} u_{n+1} \\ v_{n+1} \end{pmatrix} \le \begin{pmatrix} 1+\mathcal{O}(h) & \mathcal{O}(\varepsilon) \\ \mathcal{O}(1) & \alpha+\mathcal{O}(\varepsilon) \end{pmatrix} \begin{pmatrix} u_n \\ v_n \end{pmatrix} + M \begin{pmatrix} h \\ 1 \end{pmatrix} \tag{3.69}$$

with $0 \le \alpha < 1$ and $M \ge 0$. Then the following estimates hold for $\varepsilon \le ch$, $h \le h_0$ and $nh \le Const$

$$\begin{aligned} u_n &\le C(u_0 + \varepsilon v_0 + M) \\ v_n &\le C(u_0 + (\varepsilon + \alpha^n) v_0 + M). \end{aligned} \tag{3.70}$$

Proof. We transform the matrix in (3.69) to diagonal form and so obtain

$$\begin{pmatrix} u_n \\ v_n \end{pmatrix} \le T^{-1} \begin{pmatrix} \lambda_1^n & 0 \\ 0 & \lambda_2^n \end{pmatrix} T \begin{pmatrix} u_0 \\ v_0 \end{pmatrix} + M \sum_{j=1}^{n} T^{-1} \begin{pmatrix} \lambda_1^{n-j} & 0 \\ 0 & \lambda_2^{n-j} \end{pmatrix} T \begin{pmatrix} h \\ 1 \end{pmatrix}$$

where $\lambda_1 = 1 + \mathcal{O}(h)$, $\lambda_2 = \alpha + \mathcal{O}(\varepsilon)$ are the eigenvalues and the transformation matrix T (composed of eigenvectors) satisfies

$$T = \begin{pmatrix} 1 & \mathcal{O}(\varepsilon) \\ \mathcal{O}(1) & 1 \end{pmatrix}.$$

The statement now follows from the fact that $(\alpha + \mathcal{O}(\varepsilon))^n = \mathcal{O}(\alpha^n) + \mathcal{O}(\varepsilon)$ for $\varepsilon \le ch$ and $nh \le Const$. □

By combining Theorems 3.3, 3.4 and 3.8 we get the following result.

Corollary 3.10 (Hairer, Lubich & Roche 1988). *Under the assumptions of Theorem 3.8 the global error of a Runge-Kutta method satisfies*

$$y_n - y(x_n) = \mathcal{O}(h^p) + \mathcal{O}(\varepsilon h^{q+1}), \qquad z_n - z(x_n) = \mathcal{O}(h^{q+1}). \tag{3.71}$$

If in addition $a_{si} = b_i$ for all i, we have

$$z_n - z(x_n) = \mathcal{O}(h^p) + \mathcal{O}(\varepsilon h^q). \tag{3.72}$$

Remarks. a) If the A-stability assumption is dropped and the coefficient matrix A is only assumed to be invertible, then the estimates of Corollary 3.10 still hold for $\varepsilon \le Kh$ where K is a method-dependent constant (see Remark 3.7).

b) A-stability and the invertibility of the matrix A imply in general that the eigenvalues of A have positive real part. Otherwise the stability function would have to be reducible.

c) For several Runge-Kutta methods satisfying $a_{si} = b_i$ the estimate (3.71) for the y-component can be improved. E.g., for Radau IIA and for Lobatto IIIC one has $y_n - y(x_n) = \mathcal{O}(h^p) + \mathcal{O}(\varepsilon^2 h^q)$. This follows from Table VII.4.1 below.

d) A completely different proof of the estimates (3.71) is given by Nipp & Stoffer (1995). They show that the Runge-Kutta method, considered as a discrete dynamical system, admits an attractive invariant manifold $M_{h,\varepsilon}$, which is close to the invariant manifold M_ε of the problem (3.1). Studying the closeness of the two manifolds, they obtain the error estimates (3.71) without considering ε-expansions.

e) The analogues of Theorem 3.8 and Corollary 3.10 for Rosenbrock methods are given in Hairer, Lubich & Roche (1989).

f) Estimates for $p = q$ are given in Exercise 3 below.

Numerical Confirmation

The estimates of Corollary 3.10 can be observed numerically. As an example of (3.1) we choose the van der Pol equation

$$\begin{aligned} y' &= z \\ \varepsilon z' &= (1-y^2)z - y \end{aligned} \tag{3.73}$$

with $\varepsilon = 10^{-5}$ and initial values

$$y(0) = 2, \qquad z(0) = -0.6666654321121172 \tag{3.74}$$

on the smooth solution (Exercise 2).

Table 3.1 shows the methods of our experiment together with the theoretical error bounds. In Fig. 3.1 we have plotted the relative global error at $x_{end} = 0.5$ as a function of the step size h, which was taken constant over the considered interval. The use of logarithmic scales in both directions makes the curves appear as straight lines of slope r, whenever the leading term of the global error behaves like $Const \cdot h^r$. The figures show complete agreement with our theoretical results.

Table 3.1. Global errors predicted by Corollary 3.10

Method	$a_{si} = b_i$	y-comp.	z-comp.
Radau IA	no	$h^{2s-1} + \varepsilon h^s$	h^s
Radau IIA	yes	$h^{2s-1} + \varepsilon^2 h^s$	$h^{2s-1} + \varepsilon h^s$
Lobatto IIIC	yes	$h^{2s-2} + \varepsilon^2 h^{s-1}$	$h^{2s-2} + \varepsilon h^{s-1}$
SDIRK (IV.6.16)	yes	$h^4 + \varepsilon h^2$	$h^4 + \varepsilon h$
SDIRK (IV.6.18)	no	$h^4 + \varepsilon h^2$	h^2

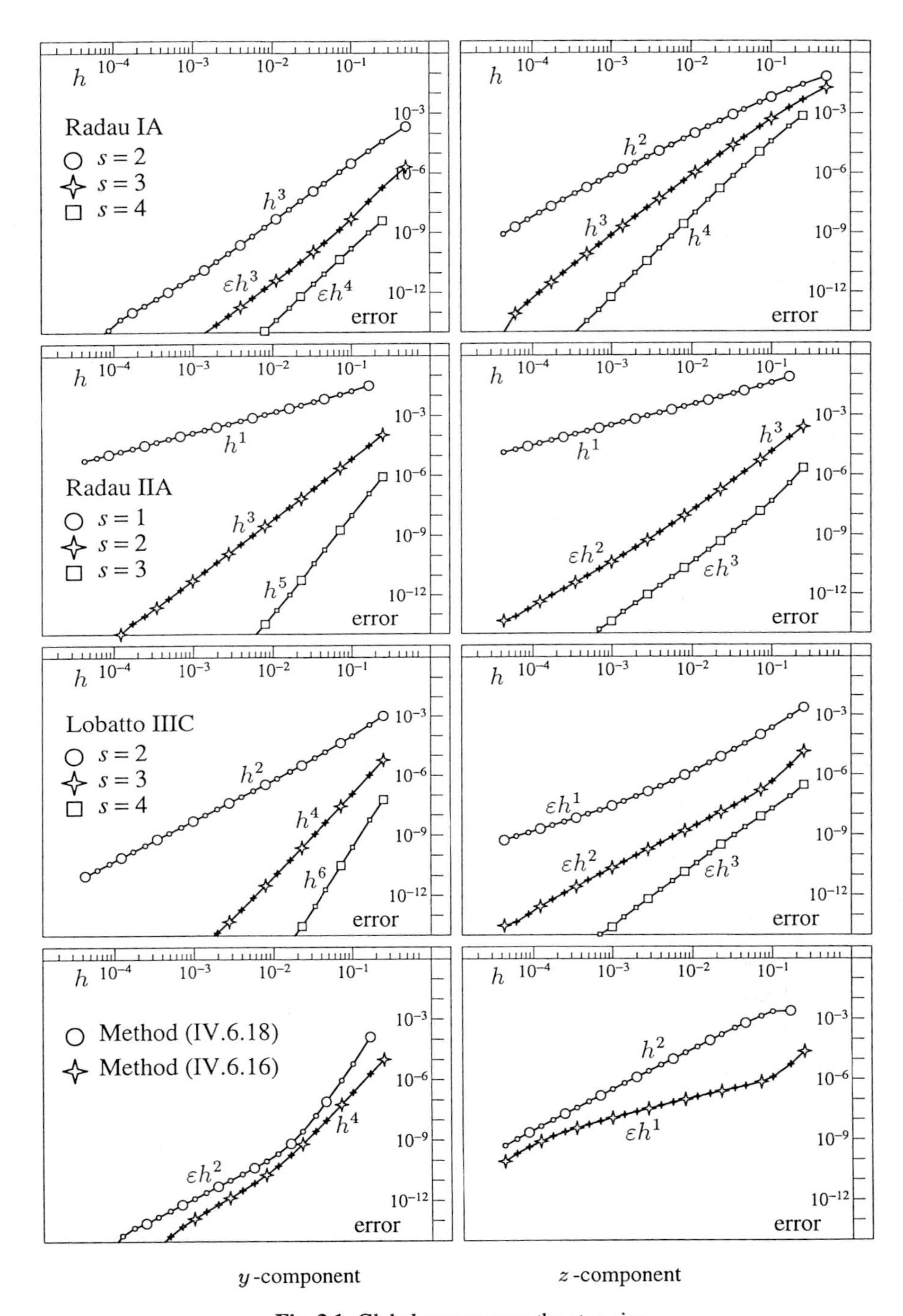

Fig. 3.1. Global error versus the step size

Perturbed Initial Values

When integrating a singular perturbation problem, the numerical solution approximates the smooth solution only within the given tolerance *Tol*. It is therefore interesting to investigate the influence of perturbations in the initial values on the global and local errors of the method. Let us begin with a numerical experiment. We perturb the $z(0)$ value of (3.74) by an amount of 10^{-6} and apply the Radau IIA methods to the problem (3.73). For the global error at $x_{end} = 0.5$ we obtain exactly the same results as in Fig. 3.1. This shows that the perturbation is completely damped out during integration. The results for the local error show a different behaviour and are displayed in Fig. 3.3. We observe the presence of a "hump", exactly as in Fig. IV.7.4 and in Fig. IV.8.2.

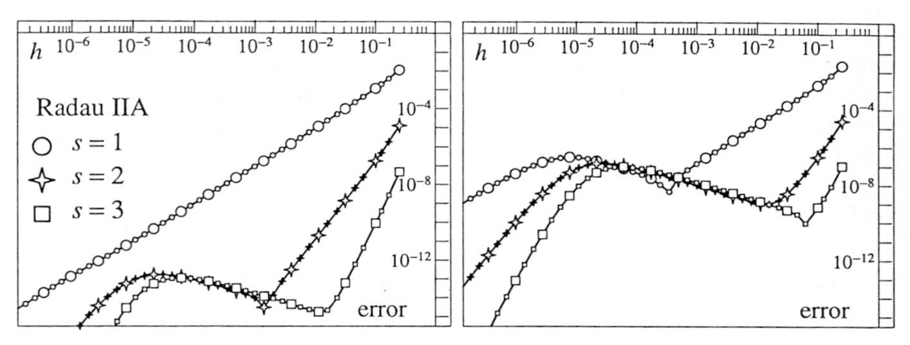

Fig. 3.2. Local error of Radau IIA (perturbed initial value)

In order to explain this phenomenon we denote by (y_0, z_0) the considered initial value, and by (y_1, z_1) the numerical solution after one step with step size h. The exact solution $y(x)$, $z(x)$ passing through (y_0, z_0) will have a boundary layer, and (under suitable assumptions, see Theorem 3.2) can be written as

$$y(x) = \widetilde{y}(x) + \mathcal{O}(\varepsilon e^{-x/\varepsilon}), \qquad z(x) = \widetilde{z}(x) + \mathcal{O}(e^{-x/\varepsilon}). \tag{3.75}$$

Here $\widetilde{y}(x)$, $\widetilde{z}(x)$ represents a smooth solution of (3.1). We denote by $\widetilde{y}_0 = \widetilde{y}(0)$, $\widetilde{z}_0 = \widetilde{z}(0)$ the initial values on this smooth solution, and by $(\widetilde{y}_1, \widetilde{z}_1)$ the numerical approximation obtained by the same method with step size h and initial values $(\widetilde{y}_0, \widetilde{z}_0)$. The local error can now be written as

$$z_1 - z(h) = (z_1 - \widetilde{z}_1) + (\widetilde{z}_1 - \widetilde{z}(h)) + (\widetilde{z}(h) - z(h)) \tag{3.76}$$

and similarly for the y-component. The last term in (3.76), which is of size $\mathcal{O}(Tol \cdot e^{-h/\varepsilon})$, can be neglected if the step size h is significantly larger than ε. The term $\widetilde{z}_1 - \widetilde{z}(h)$ represents the local error in the "smooth" situation and is bounded by at least $\mathcal{O}(h^{q+1})$ (apply Corollary 3.10 with $n = 1$). It can be observed in Fig. 3.2 whenever h or the error is large. The difference $z_1 - \widetilde{z}_1$ is the term which causes the irregularity in Fig. 3.2. Using Theorem 3.6 (with $\delta = 0$,

$\theta = 0$, $\widehat{\eta} - \eta = \mathcal{O}(\varepsilon \cdot Tol)$, $\widehat{\zeta} - \zeta = \mathcal{O}(Tol)$) and the ideas of the proof of Theorem 3.8 (in particular Eq. (3.62)) we obtain

$$\begin{aligned} z_1 - \widetilde{z}_1 &= R\Big(\frac{h}{\varepsilon}\, g_z(0)\Big)(z_0 - \widetilde{z}_0) + \mathcal{O}(\varepsilon \cdot Tol) \\ y_1 - \widetilde{y}_1 &= \mathcal{O}(\varepsilon \cdot Tol). \end{aligned} \tag{3.77}$$

For $\varepsilon < h$ we develop

$$R\Big(\frac{h}{\varepsilon}\, g_z(0)\Big) = R(\infty) + C\frac{\varepsilon}{h}\, g_z^{-1}(0) + \mathcal{O}\Big(\Big(\frac{\varepsilon}{h}\Big)^2\Big). \tag{3.78}$$

This shows that an h-independent expression $R(\infty)(z_0 - \widetilde{z}_0) = \mathcal{O}(Tol)$ will be observed in the local error, if $R(\infty) \neq 0$. For methods with $R(\infty) = 0$ (such as Radau IIA) the dominant part in $z_1 - \widetilde{z}_1$ is $C(\varepsilon/h)g_z^{-1}(0)(z_0 - \widetilde{z}_0) = \mathcal{O}(Tol \cdot \varepsilon/h)$. This term can be observed in Fig. 3.2 as a straight line of slope -1. Thus in this region the local error increases like h^{-1} when h decreases. A similar perturbation, multiplied however by ε, is observed for the y-component.

This is not a serious drawback for a numerical implementation, because the phenomenon appears only for step sizes where the local error is smaller than Tol.

Exercises

1. Prove that the statement of Theorem 3.2 remains valid, if the assumption (3.11) is replaced by

   the eigenvalues λ of $g_z(y,z)$ satisfy $\operatorname{Re}\lambda \le -1$

   for all y, z in a neighbourhood of the solution $y_0(x)$, $z_0(x)$ of the reduced system.

   Hint. Split the interval into a finite number of small subintervals and construct for each of them an inner product norm such that, after a rescaling of ε, (3.11) holds (see Nevanlinna 1976).

2. Let $y(0) = 2$; find the corresponding $z(0)$ for the van der Pol equation (3.73), such that its solution is smooth.

   Result.
   $$z(0) = -\frac{2}{3} + \frac{10}{81}\varepsilon - \frac{292}{2187}\varepsilon^2 - \frac{1814}{19683}\varepsilon^3 + \mathcal{O}(\varepsilon^4).$$

3. If the assumption $q < p$ (p classical order, q stage order) is dropped in Corollary 3.10, we still have

   $$y_n - y(x_n) = \mathcal{O}(h^p), \qquad z_n - z(x_n) = \mathcal{O}(h^p).$$

   Prove this statement. The implicit Euler method and the SIRK methods of Lemma IV.8.1 are typical examples with $p = q$.

   Hint. Apply Corollary 3.10 with q reduced by 1.

VI.4 Rosenbrock Methods

This section is devoted to the extension of Rosenbrock methods (see Sect. IV.7) to differential-algebraic equations in semi-explicit form

$$y' = f(y,z), \qquad y(x_0) = y_0 \tag{4.1a}$$

$$0 = g(y,z), \qquad z(x_0) = z_0. \tag{4.1b}$$

We suppose that g_z is invertible (see (1.7)), so that the problem is of index 1. We shall obtain new methods for the numerical solution of such problems, and at the same time get more insight into the behaviour of Rosenbrock methods for stiff differential equations. In particular, the phenomenon of Fig. IV.7.4 will be explained.

Definition of the Method

The main advantage of Rosenbrock methods over implicit Runge-Kutta methods is that nonlinear systems are completely avoided. The state space form method (transforming (4.1) to $y' = f(y, G(y))$) would destroy this advantage. This is one more reason for considering the ε-embedding method. For the problem (1.5) a Rosenbrock method reads

$$\begin{pmatrix} k_i \\ \ell_i \end{pmatrix} = h \begin{pmatrix} f(v_i, w_i) \\ \varepsilon^{-1} g(v_i, w_i) \end{pmatrix} + h \begin{pmatrix} f_y & f_z \\ \varepsilon^{-1} g_y & \varepsilon^{-1} g_z \end{pmatrix} (y_0, z_0) \sum_{j=1}^{i} \gamma_{ij} \begin{pmatrix} k_j \\ \ell_j \end{pmatrix} \tag{4.2}$$

$$\begin{pmatrix} v_i \\ w_i \end{pmatrix} = \begin{pmatrix} y_0 \\ z_0 \end{pmatrix} + \sum_{j=1}^{i-1} \alpha_{ij} \begin{pmatrix} k_j \\ \ell_j \end{pmatrix}, \qquad \begin{pmatrix} y_1 \\ z_1 \end{pmatrix} = \begin{pmatrix} y_0 \\ z_0 \end{pmatrix} + \sum_{i=1}^{s} b_i \begin{pmatrix} k_i \\ \ell_i \end{pmatrix}. \tag{4.3a}$$

If we multiply the second line of (4.2) by ε and then put $\varepsilon = 0$ we obtain

$$\begin{pmatrix} k_i \\ 0 \end{pmatrix} = h \begin{pmatrix} f(v_i, w_i) \\ g(v_i, w_i) \end{pmatrix} + h \begin{pmatrix} f_y & f_z \\ g_y & g_z \end{pmatrix} (y_0, z_0) \sum_{j=1}^{i} \gamma_{ij} \begin{pmatrix} k_j \\ \ell_j \end{pmatrix}. \tag{4.3b}$$

Formulas (4.3a) and (4.3b) together constitute the extension of a Rosenbrock method to the problem (4.1). This type of method was first considered by Michelsen (1976) (quoted by Feng, Holland & Gallun (1984)). Further studies are due to Roche

(1988). We remark that the computation of (k_i, ℓ_i) from (4.3b) requires the solution of a linear system with matrix

$$\begin{pmatrix} I-\gamma h f_y & -\gamma h f_z \\ -\gamma h g_y & -\gamma h g_z \end{pmatrix} \tag{4.4}$$

where all derivatives are evaluated at (y_0, z_0). For nonsingular g_z, nonzero γ, and small enough $h > 0$, this matrix is invertible. This can be seen by dividing the lower blocks by γh and then putting $h = 0$.

Non-autonomous equations. If the functions f and g in (4.1) also depend on x, we replace (4.3b) by

$$\begin{pmatrix} k_i \\ 0 \end{pmatrix} = h \begin{pmatrix} f(x_0+\alpha_i h, v_i, w_i) \\ g(x_0+\alpha_i h, v_i, w_i) \end{pmatrix} + h \begin{pmatrix} f_y & f_z \\ g_y & g_z \end{pmatrix} \sum_{j=1}^{i} \gamma_{ij} \begin{pmatrix} k_j \\ \ell_j \end{pmatrix} + h^2 \gamma_i \begin{pmatrix} f_x \\ g_x \end{pmatrix} \tag{4.5}$$

(compare with (IV.7.4a) and recall the definition of α_i and γ_i in (IV.7.5)). All derivatives are evaluated at the initial value (x_0, y_0, z_0).

Problems of the form $Mu' = \varphi(u)$. Rosenbrock formulas for these problems have been developed in Sect. IV.7 (Formula (IV.7.4b)) in the case of regular M. This formula is also applicable for singular M, and can be justified as follows: It is theoretically possible to apply the transformation (1.20) so that M becomes the block-diagonal matrix with entries I and 0. The method (IV.7.4b) is then identical to method (4.3). Therefore, the theory to be developed in this section will also be valid for Rosenbrock method (IV.7.4b) applied to index 1 problems of the form $Mu' = \varphi(u)$.

Having introduced a new class of methods, we must study their order conditions. As usual, this is done by Taylor expansion of both the exact and the numerical solution (similar to Section II.2). A nice correspondence between the order conditions and certain rooted trees with two different kinds of vertices will be obtained (Roche 1988).

Derivatives of the Exact Solution

In contrast to Sect. II.2, where we used "hordes of indices" (see Dieudonné's preface to his "Foundations of Modern Analysis") to show us the way through the "woud met bomen" (Hundsdorfer), we here write higher derivatives as multilinear mappings. For example, the expression

$$\sum_{j,k} \frac{\partial^2 g_i}{\partial y_j \partial z_k} \cdot u_j v_k \quad \text{is written as} \quad g_{yz}(u, v),$$

which simplifies the subsequent formulas.

We differentiate (4.1b) to obtain $0 = g_y \cdot y' + g_z \cdot z'$ and, equivalently,

$$z' = (-g_z^{-1}) g_y f. \tag{4.6}$$

We now differentiate successively (4.1a) and (4.6) with respect to x. We use the formula

$$(-g_z^{-1})' u = (-g_z^{-1}) \Big(g_{zy}\big((-g_z^{-1})u, f\big) + g_{zz}\big((-g_z^{-1})u, (-g_z^{-1}) g_y f\big)\Big) \tag{4.7}$$

which is a consequence of $(A^{-1}(x))' = -A^{-1}(x) A'(x) A^{-1}(x)$ and the chain rule. This gives

$$y'' = f_y \cdot y' + f_z \cdot z' = f_y f + f_z (-g_z^{-1}) g_y f \tag{4.8}$$

$$\begin{aligned} z'' = (-g_z^{-1}) \Big(& g_{zy}\big((-g_z^{-1}) g_y f, f\big) + g_{zz}\big((-g_z^{-1}) g_y f, (-g_z^{-1}) g_y f\big)\Big) \\ & + (-g_z^{-1}) \Big(g_{yy}(f, f) + g_{yz}\big(f, (-g_z^{-1}) g_y f\big)\Big) \\ & + (-g_z^{-1}) g_y \Big(f_y f + f_z (-g_z^{-1}) g_y f \Big). \end{aligned} \tag{4.9}$$

Clearly, these expressions soon become very complicated and a graphical representation of the terms in (4.8) and (4.9) is desirable.

Trees and Elementary Differentials

We shall identify each occuring f with a meagre vertex, and each of its derivatives with an upward leaving branch. The expression $(-g_z^{-1})g$ is identified with a fat vertex. The derivatives of g therein are again indicated by upwards leaving branches. For example, the second expression of (4.8) and the first one of (4.9) correspond to the trees in Fig. 4.1.

The above formulas for y', z', y'', z'' thus become

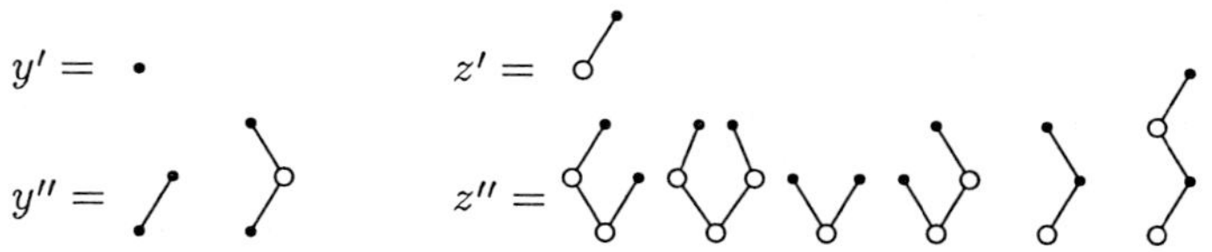

$y' =$ $\qquad$ $z' =$

$y'' =$ $\qquad$ $z'' =$ $\qquad$ (4.10)

The first and fourth expressions in (4.9) are identical, because $g_{zy}(u, v) = g_{yz}(v, u)$. This is in nice accordance with the fact that the corresponding trees are topologically equivalent. The lowest vertex of a tree will be called its *root*.

We see that derivatives of y are characterized by trees with a *meagre root*. These trees will be denoted by t or t_i, the tree consisting only of the root (for y') being τ_y. Derivatives of z have trees with a *fat root*. These will be written as u or u_i, the tree for z' being τ_z.

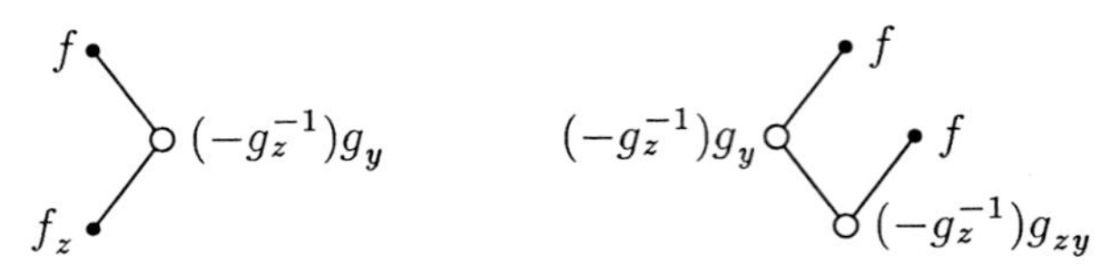

Fig. 4.1. Graphical representation of elementary differentials

Definition 4.1. Let $DAT = DAT_y \cup DAT_z$ denote the set of (differential algebraic rooted) *trees* defined recursively by

a) $\tau_y \in DAT_y$, $\tau_z \in DAT_z$;

b) $[t_1,\ldots,t_m,u_1,\ldots u_n]_y \in DAT_y$
if $t_1,\ldots,t_m \in DAT_y$ and $u_1,\ldots u_n \in DAT_z$;

c) $[t_1,\ldots,t_m,u_1,\ldots,u_n]_z \in DAT_z$
if $t_1,\ldots,t_m \in DAT_y$, $u_1,\ldots,u_n \in DAT_z$, and $(m,n) \neq (0,1)$.

Here $[t_1,\ldots,t_m,u_1,\ldots,u_n]_y$ and $[t_1,\ldots,t_m,u_1,\ldots,u_n]_z$ represent unordered $(m+n)$-tuples.

The graphical representation of these trees is as follows: if we connect the roots of $t_1,\ldots,t_m$, $u_1,\ldots,u_n$ by $m+n$ branches to a new meagre vertex (the new root) we obtain $[t_1,\ldots,t_m,\ u_1,\ldots,u_n]_y$; if we connect them to a new fat vertex we obtain $[t_1,\ldots,t_m,\ u_1,\ldots,u_n]_z$. For example, the two trees of Fig. 4.1 can be written as $[\tau_z]_y$ and $[\tau_z,\tau_y]_z$.

Definition 4.2. The *order* of a tree $t \in DAT_y$ or $u \in DAT_z$, denoted by $\varrho(t)$ or $\varrho(u)$, is the number of its meagre vertices.

We see in (4.10) that this definition of order coincides with the derivative order of $y^{(i)}$ or $z^{(i)}$ as far as they are computed there.

We next give a recursive definition of the one-to-one correspondence between the trees in (4.10) and the expressions in (4.8) and (4.9).

Definition 4.3. The *elementary differentials* $F(t)$ (or $F(u)$) corresponding to trees in DAT are defined as follows:

a) $F(\tau_y) = f,\quad F(\tau_z) = (-g_z^{-1})g_y f,$

b) $$F(t) = \frac{\partial^{m+n} f}{\partial y^m \partial z^n}\Big(F(t_1),\ldots,F(t_m),\ F(u_1),\ldots,F(u_n)\Big)$$
if $t = [t_1,\ldots,t_m,\ u_1,\ldots,u_n]_y \in DAT_y$,

c) $$F(u) = (-g_z)^{-1}\frac{\partial^{m+n} g}{\partial y^m \partial z^n}\Big(F(t_1),\ldots,F(t_m),\ F(u_1),\ldots,F(u_n)\Big)$$
if $u = [t_1,\ldots,t_m,\ u_1,\ldots,u_n]_z \in DAT_z$.

Because of the symmetry of partial derivatives, this definition is unaffected by a permutation of $t_1,\ldots,t_m$, $u_1,\ldots,u_n$ and therefore the functions $F(t)$ and $F(u)$ are well defined.

Taylor Expansion of the Exact Solution

In order to get more insight into the process of (4.8) and (4.9) we study the differentiation of an elementary differential with respect to x. By Leibniz' rule the differentiation of $F(t)$ (or $F(u)$) gives a sum of new elementary differentials which are obtained by the following four rules:

i) attach to each vertex a branch with τ_y (derivative of f or g with respect to y and addition of the factor $y' = f$);

ii) attach to each vertex a branch with τ_z (derivative of f or g with respect to z and addition of the factor $z' = (-g_z^{-1})g_y f$);

iii) split each fat vertex into two new fat vertices (linked by a new branch) and attach to the lower of these fat vertices a branch with τ_y;

iv) as in (iii) split each fat vertex into two new fat vertices, but attach this time to the lower of the new fat vertices a branch with τ_z.

The rules (iii) and (iv) correspond to the differentiation of $(-g_z^{-1})$ and follow at once from (4.7). We observe that the differentiation of a tree of order q (or, more precisely, of its corresponding elementary differential) generates trees of order $q+1$.

As was the case in Sect. II.2, some of these trees appear *several times* in the derivative (as the first and fourth tree for z'' in (4.10)). In order to distinguish all these trees, we indicate the *order of generation* of the meagre vertices by labels. This is demonstrated, for thé first derivatives of y, in Fig. 4.2. Since in the above differentiation process the new meagre vertex is always an end-vertex of the tree, the labelling thus obtained is necessarily increasing from the root upwards along each branch.

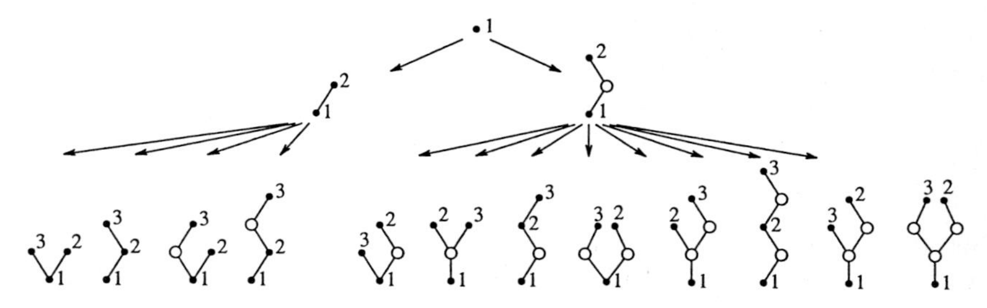

Fig. 4.2. Monotonically labelled trees ($LDAT_y$)

Definition 4.4. A tree $t \in DAT_y$ (or $u \in DAT_z$) together with a monotonic labelling of its meagre vertices is called a *monotonically labelled tree*. The sets of all such monotonically labelled trees are denoted by $LDAT_y$, $LDAT_z$ and $LDAT$.

Definition 4.2 (order of a tree) and Definition 4.3 (elementary differential) are extended in a natural way to monotonically labelled trees. We can therefore write the derivatives of the exact solution as follows:

Theorem 4.5 (Roche 1988). *For the exact solution of (4.1) we have:*

$$y^{(q)}(x_0) = \sum_{t \in LDAT_y, \varrho(t)=q} F(t)(y_0, z_0) = \sum_{t \in DAT_y, \varrho(t)=q} \alpha(t) F(t)(y_0, z_0)$$

$$z^{(q)}(x_0) = \sum_{u \in LDAT_z, \varrho(u)=q} F(u)(y_0, z_0) = \sum_{u \in DAT_z, \varrho(u)=q} \alpha(u) F(u)(y_0, z_0).$$

The integer coefficients $\alpha(t)$ and $\alpha(u)$ indicate the number of possible monotonic labellings of a tree.

Proof. For $q=1$ and $q=2$ this is just (4.1a), (4.6), (4.8) and (4.9). For general q the above differentiation process of trees generates all elements of *LDAT*, each element exactly once. If the sum is taken over DAT_y and DAT_z , the factors $\alpha(t)$ and $\alpha(u)$ must be added. □

Taylor Expansion of the Numerical Solution

Our next aim is to prove an analogue of Theorem 4.5 for the numerical solution of a Rosenbrock method. We consider y_1, z_1 as functions of the step size h and compute their derivatives. From (4.3a) it follows that

$$y_1^{(q)}(0) = \sum_{i=1}^{s} b_i k_i^{(q)}(0), \qquad z_1^{(q)}(0) = \sum_{i=1}^{s} b_i \ell_i^{(q)}(0). \tag{4.11}$$

Consequently we have to compute the derivatives of k_i and ℓ_i. This is done as for Runge-Kutta methods (Sect. II.2) or for Rosenbrock methods applied to ordinary differential equations (Sect. IV.7).

We differentiate the first line of (4.3b) with respect to h. Using Leibniz' rule (II.2.4) this yields for $h=0$

$$k_i^{(q)} = q\big(f(v_i, w_i)\big)^{(q-1)} + (f_y)_0\, q \sum_{j=1}^{i} \gamma_{ij} k_j^{(q-1)} + (f_z)_0\, q \sum_{j=1}^{i} \gamma_{ij} \ell_j^{(q-1)}. \tag{4.12}$$

The index 0 in $(f_y)_0$ and $(f_z)_0$ indicates that the derivatives are evaluated at (y_0, z_0). The second line of (4.3b) is divided by h before differentiation. This gives (again for $h=0$)

$$0 = \big(g(v_i, w_i)\big)^{(q)} + (g_y)_0 \sum_{j=1}^{i} \gamma_{ij} k_j^{(q)} + (g_z)_0 \sum_{j=1}^{i} \gamma_{ij} \ell_j^{(q)}. \tag{4.13}$$

The derivatives of f and g can be computed by Faà di Bruno's formula (Lemma II.2.8). This yields

$$\big(f(v_i, w_i)\big)^{(q-1)} = \sum \frac{\partial^{m+n} f(v_i, w_i)}{\partial y^m \partial z^n} \big(v_i^{(\mu_1)}, \ldots, v_i^{(\mu_m)}, w_i^{(\nu_1)}, \ldots, w_i^{(\nu_n)}\big) \tag{4.14}$$

where the sum is over all "special $LDAT_y$'s" of order q. These are monotonically labelled trees $[t_1,\ldots,t_m,u_1,\ldots,u_n]_y$ where t_j and u_j do not have any ramification and all their vertices are meagre with the exception of the roots of $u_1,\ldots,u_n$. The integers μ_j and ν_j are the orders of t_j and u_j, respectively. They satisfy $\mu_1+\ldots+\mu_m+\nu_1+\ldots+\nu_n=q-1$. Similarly we apply Faà di Bruno's formula to g and obtain

$$\big(g(v_i,w_i)\big)^{(q)}=\sum\frac{\partial^{m+n}g(v_i,w_i)}{\partial y^m\partial z^n}\big(v_i^{(\mu_1)},\ldots,v_i^{(\mu_m)},w_i^{(\nu_1)},\ldots,w_i^{(\nu_n)}\big)+g_z(v_i,w_i)w_i^{(q)}. \tag{4.15}$$

Here the sum is over all "special $LDAT_z$'s" of order q. They are defined as above but have a fat vertex. The integers μ_j,ν_j satisfy $\mu_1+\ldots+\mu_m+\nu_1+\ldots+\nu_n=q$. The term with g_z is written separately, because (by the definition of $LDAT_z$) $[u_1]_z$ is not an admissible tree.

We are now in a position to compute the derivatives of k_i and ℓ_i. For this it is convenient to introduce the notation

$$\beta_{ij}=\alpha_{ij}+\gamma_{ij} \tag{4.16}$$

(with $\alpha_{ii}=0$) as in (IV.7.12). We also need the inverse of the matrix (β_{ij}), whose coefficients we denote by ω_{ij}:

$$(\omega_{ij})=(\beta_{ij})^{-1}. \tag{4.17}$$

Theorem 4.6. *The derivatives of k_i and ℓ_i satisfy*

$$\begin{aligned} k_i^{(q)}(0)&=\sum_{t\in LDAT_y,\varrho(t)=q}\gamma(t)\Phi_i(t)F(t)(y_0,z_0)\\ \ell_i^{(q)}(0)&=\sum_{u\in LDAT_z,\varrho(u)=q}\gamma(u)\Phi_i(u)F(u)(y_0,z_0), \end{aligned} \tag{4.18}$$

where the coefficients $\Phi_i(t)$ and $\Phi_i(u)$ are given by $\Phi_i(\tau_y)=1$, $\Phi_i(\tau_z)=1$ and

$$\Phi_i(t)=\begin{cases} \displaystyle\sum_{\mu_1,..,\mu_m,\nu_1,..,\nu_n}\alpha_{i\mu_1}\cdots\alpha_{i\mu_m}\alpha_{i\nu_1}\cdots\alpha_{i\nu_n} \\ \qquad\qquad\cdot\Phi_{\mu_1}(t_1)\cdots\Phi_{\mu_m}(t_m)\Phi_{\nu_1}(u_1)\cdots\Phi_{\nu_n}(u_n) \\ \qquad\qquad \text{if } t=[t_1,\ldots,t_m,u_1,\ldots,u_n]_y \text{ and } m+n\ge 2 \\ \displaystyle\sum_j\beta_{ij}\Phi_j(t_1) \qquad \text{if } t=[t_1]_y \\ \displaystyle\sum_j\beta_{ij}\Phi_j(u_1) \qquad \text{if } t=[u_1]_y, \end{cases}$$

$$\Phi_i(u) = \begin{cases} \displaystyle\sum_{j,\mu_1,..,\mu_m,\nu_1,..,\nu_n} \omega_{ij}\alpha_{j\mu_1}\cdots\alpha_{j\mu_m}\alpha_{j\nu_1}\cdots\alpha_{j\nu_n} \\ \qquad\qquad \cdot\Phi_{\mu_1}(t_1)\cdots\Phi_{\mu_m}(t_m)\Phi_{\nu_1}(u_1)\cdots\Phi_{\nu_n}(u_n) \\ \qquad\qquad \textit{if } u=[t_1,\ldots,t_m,u_1,\ldots,u_n]_z \textit{ and } m+n\geq 2 \\ \Phi_i(t_1) \qquad \textit{if } u=[t_1]_z \end{cases}$$

and the integer coefficients $\gamma(t)$ and $\gamma(u)$ are defined by $\gamma(\tau_y)=1,\ \gamma(\tau_z)=1$ and

$$\gamma(t)=\varrho(t)\gamma(t_1)\ldots\gamma(t_m)\gamma(u_1)\ldots\gamma(u_n) \quad \textit{if } t=[t_1,\ldots,t_m,u_1,\ldots,u_n]_y$$
$$\gamma(u)=\gamma(t_1)\ldots\gamma(t_m)\,\gamma(u_1)\ldots\gamma(u_n) \quad \textit{if } u=[t_1,\ldots,t_m,u_1,\ldots,u_n]_z.$$

Proof. By (4.3a) we have

$$v_i^{(\mu)}=\sum_{j=1}^{i-1}\alpha_{ij}k_j^{(\mu)}, \qquad w_i^{(\nu)}=\sum_{j=1}^{i-1}\alpha_{ij}\ell_j^{(\nu)}. \tag{4.19}$$

We now insert (4.19) into (4.14) and the resulting formula for $(f(v_i,w_i))^{(q-1)}$ into (4.12). This yields (all expressions have to be evaluated at $h=0$)

$$\begin{aligned} k_i^{(q)} = q\sum_{m+n\geq 2}\frac{\partial^{m+n}f(y_0,z_0)}{\partial y^m\partial z^n}\Big(\sum_{j=1}^{i-1}\alpha_{ij}k_j^{(\mu_1)},\ldots,\sum_{j=1}^{i-1}\alpha_{ij}\ell_j^{(\nu_1)},\ldots\Big) \\ +q(f_y)_0\sum_{j=1}^{i}\beta_{ij}k_j^{(q-1)}+q(f_z)_0\sum_{j=1}^{i}\beta_{ij}\ell_j^{(q-1)}. \end{aligned} \tag{4.20}$$

The same analysis for the second component leads to

$$\begin{aligned} 0 = \sum_{m+n\geq 2}\frac{\partial^{m+n}g(y_0,z_0)}{\partial y^m\partial z^n}\Big(\sum_{j=1}^{i-1}\alpha_{ij}k_j^{(\mu_1)},\ldots,\sum_{j=1}^{i-1}\alpha_{ij}\ell_j^{(\nu_1)},\ldots\Big) \\ +(g_y)_0\sum_{j=1}^{i}\beta_{ij}k_j^{(q)}+(g_z)_0\sum_{j=1}^{i}\beta_{ij}\ell_j^{(q)}. \end{aligned} \tag{4.21}$$

The sums in (4.20) and (4.21) are over elements of *LDAT* exactly as in (4.14) and (4.15). Equation (4.21) allows us to extract $\ell_i^{(q)}$ if we use the inverse of (β_{ij}). This gives

$$\begin{aligned} \ell_i^{(q)} = (-g_z)_0^{-1}\sum_{j=1}^{i}\omega_{ij}\sum_{m+n\geq 2}\frac{\partial^{m+n}g(y_0,z_0)}{\partial y^m\partial z^n}\Big(\sum_{\kappa=1}^{j-1}\alpha_{j\kappa}k_\kappa^{(\mu_1)},\ldots,\sum_{\kappa=1}^{j-1}\alpha_{j\kappa}\ell_\kappa^{(\nu_1)},\ldots\Big) \\ +\big((-g_z^{-1})g_y\big)_0k_i^{(q)}. \end{aligned} \tag{4.22}$$

The proof of Formula (4.18) is now by induction on q. The case $q=1$ follows immediately from (4.12) and (4.13). For general q, we insert the induction hypothesis into (4.20) and (4.22), exploit the multilinearity of the derivatives, and arrange the summations as in the proof of Theorem II.2.11. □

Finally, Eq. (4.11) yields the derivatives of the numerical solution.

Theorem 4.7 (Roche 1988). *The numerical solution of (4.3) satisfies:*

$$y_1^{(q)}|_{h=0} = \sum_{t\in LDAT_y,\varrho(t)=q} \gamma(t) \sum_{i=1}^{s} b_i\Phi_i(t)F(t)(y_0,z_0)$$

$$z_1^{(q)}|_{h=0} = \sum_{u\in LDAT_z,\varrho(u)=q} \gamma(u) \sum_{i=1}^{s} b_i\Phi_i(u)F(u)(y_0,z_0)$$

where the coefficients γ and Φ_i are given in Theorem 4.6. □

Order Conditions

Comparing Theorem 4.5 and 4.7 we obtain

Theorem 4.8. *For the Rosenbrock method (4.3) we have:*

$$y(x_0+h)-y_1 = \mathcal{O}(h^{p+1}) \quad \textit{iff}$$

$$\sum_{i=1}^{s} b_i\Phi_i(t) = \frac{1}{\gamma(t)} \qquad \textit{for} \quad t\in DAT_y,\ \varrho(t)\le p;$$

$$z(x_0+h)-z_1 = \mathcal{O}(h^{q+1}) \quad \textit{iff}$$

$$\sum_{i=1}^{s} b_i\Phi_i(u) = \frac{1}{\gamma(u)} \qquad \textit{for} \quad u\in DAT_z,\ \varrho(u)\le q,$$

where the coefficients Φ_i and γ are those of Theorem 4.6. □

Repeated application of the recursive definition of Φ_i in Theorem 4.6 yields the following algorithm:

Forming the Order Condition for a Given Tree: attach to each meagre vertex one summation index, and to each fat vertex two indices (one above the other). Then the left hand side of the order condition is a sum over all indices of a product with factors

- b_i if "i" is the index of the root (the lower index if the root is fat);
- α_{ij} if "j" lies directly above "i" and "i" is multiply branched;
- β_{ij} if "j" lies directly above "i" and "i" is singly branched;
- ω_{ij} if "i,j" are the two indices of a fat vertex ("i" below "j").

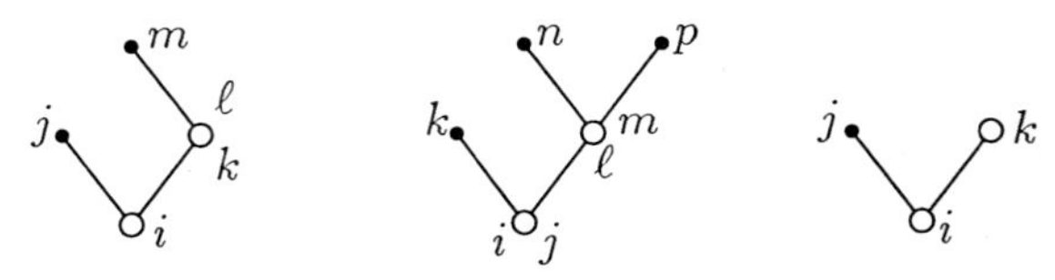

Fig. 4.3. Trees with labelling

As an example, we present the order conditions for the first two trees of Fig. 4.3.

$$\sum_{i,j,k,\ell,m} b_i \alpha_{ij} \alpha_{ik} \omega_{k\ell} \beta_{\ell m} = \frac{1}{3} \tag{4.23}$$

$$\sum_{i,j,k,\ell,m,n,p} b_i \omega_{ij} \alpha_{jk} \alpha_{j\ell} \omega_{\ell m} \alpha_{mn} \alpha_{mp} = 1. \tag{4.24}$$

The condition (4.23) can be further simplified if we use the fact that (ω_{ij}) is the inverse of the matrix (β_{ij}). Indeed, (4.23) is equivalent to

$$\sum_{i,j,k} b_i \alpha_{ij} \alpha_{ik} = \frac{1}{3}$$

which is the order condition for the third tree in Fig. 4.3. Exploiting this reduction systematically we arrive at the following result.

Lemma 4.9. *For a Rosenbrock method (4.3) the order conditions corresponding to one of the following situations are redundant:*

a) a fat vertex is singly branched.

b) a singly branched vertex is followed by a fat vertex. □

The subset of DAT_y which consists of trees with only meagre vertices, is simply T (the set of trees of Sect. II.2). The corresponding order conditions are those given in Sect. IV.7. Consequently, a p-th order Rosenbrock method has to satisfy all "classical" order conditions and, in addition, several "algebraic" conditions. The first of these new order conditions are given in Table 4.1. We have included the polynomial $p_t(\gamma)$ in its last column, which is the right-hand side of the order condition, when written in the form (IV.7.11').

Convergence

Before we proceed to the actual construction of a new Rosenbrock method, we still have to study its convergence property. The following result will also involve

$$R(\infty) = 1 - b^T B^{-1} 1\!\!1 = 1 - \sum_{i,j} b_i \omega_{ij} \tag{4.25}$$

where $R(z)$ is the stability function (IV.7.14).

Table 4.1. Trees and elementary differentials

$\varrho(t)$	t	graph	$\gamma(t)$	$\Phi_j(t)$	$p_t(\gamma)$
4	t_{45}	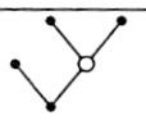	4	$\sum \alpha_{jk}\alpha_{j\ell}\omega_{\ell m}\alpha_{mn}\alpha_{mp}$	1/4
2	u_{21}		1	$\sum \omega_{jk}\alpha_{k\ell}\alpha_{km}$	1
3	u_{31}		1	$\sum \omega_{jk}\alpha_{k\ell}\alpha_{km}\alpha_{kn}$	1
3	u_{32}		2	$\sum \omega_{jk}\alpha_{k\ell}\alpha_{km}\beta_{mn}$	$1/2-\gamma$
3	u_{33}	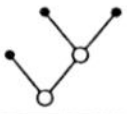	1	$\sum \omega_{jk}\alpha_{k\ell}\alpha_{km}\omega_{mn}\alpha_{np}\alpha_{nq}$	1

We denote the local error of the Rosenbrock method (4.3) by

$$\delta y_h(x) = y_1 - y(x+h), \qquad \delta z_h(x) = z_1 - z(x+h). \tag{4.26}$$

Here y_1, z_1 is the numerical solution obtained with the exact initial values $y_0 = y(x)$, $z_0 = z(x)$.

Theorem 4.10. *Suppose that g_z is regular in a neighbourhood of the solution $(y(x), z(x))$ of (4.1) and that the initial values (y_0, z_0) are consistent. If the stability function is such that $|R(\infty)| < 1$, and the local error satisfies*

$$\delta y_h(x) = \mathcal{O}(h^{p+1}), \qquad \delta z_h(x) = \mathcal{O}(h^p), \tag{4.27}$$

then the Rosenbrock method (4.3) is convergent of order p; i.e.,

$$y_n - y(x_n) = \mathcal{O}(h^p), \quad z_n - z(x_n) = \mathcal{O}(h^p) \quad \textit{for} \quad x_n - x_0 = nh \le \textit{Const.}$$

Proof. Since g_z is regular we have

$$\|g_z^{-1}(y,z)g(y,z)\| \le \delta \tag{4.28}$$

for (y,z) in a compact neighbourhood $\mathcal{U}$ of the solution. The h-independent value of δ can be made arbitrarily small by shrinking $\mathcal{U}$. We also suppose for the moment that the numerical solution and all its internal stages remain in this neighbourhood. The propagation of local errors will be studied in part (a), and their accumulation over the whole interval in part (b).

a) We consider two pairs of initial values, (y_0, z_0) and $(\widehat{y}_0, \widehat{z}_0)$, and apply the method to each (these values may be inconsistent, but they are assumed to lie in $\mathcal{U}$). We shall prove that

$$\begin{aligned} \|y_1 - \widehat{y}_1\| &\le (1+hL)\|y_0 - \widehat{y}_0\| + hM\|z_0 - \widehat{z}_0\| \\ \|z_1 - \widehat{z}_1\| &\le N\|y_0 - \widehat{y}_0\| + \kappa\|z_0 - \widehat{z}_0\| \end{aligned} \tag{4.29}$$

where $\kappa < 1$. For this we fix a sufficiently small step size h, and consider y_1, z_1,

k_i, ℓ_i as functions of (y_0, z_0). We shall show that

$$\begin{aligned} \frac{\partial y_1}{\partial y_0} &= I + \mathcal{O}(h), & \frac{\partial y_1}{\partial z_0} &= \mathcal{O}(h), \\ \frac{\partial z_1}{\partial y_0} &= \mathcal{O}(1), & \frac{\partial z_1}{\partial z_0} &= R(\infty)I + \mathcal{O}(h+\delta). \end{aligned} \tag{4.30}$$

The mean value theorem then implies (4.29).

We first estimate k_i and ℓ_i, defined in (4.3b). Using (4.28) we compute ℓ_i from the second line and insert it into the first one. This yields successively $k_i = \mathcal{O}(h)$ and $\ell_i = \mathcal{O}(h+\delta)$ for all internal stages. We then differentiate (4.3b) once with respect to y_0 and once with respect to z_0. An analysis similar to that for k_i and ℓ_i yields

$$\begin{aligned} \frac{\partial k_i}{\partial y_0} &= \mathcal{O}(h), & \frac{\partial k_i}{\partial z_0} &= \mathcal{O}(h) \\ \frac{\partial \ell_i}{\partial y_0} &= \mathcal{O}(1), & \frac{\partial \ell_i}{\partial z_0} &= -\sum_j \omega_{ij} I + \mathcal{O}(h+\delta) \end{aligned} \tag{4.31}$$

and the estimates (4.30) follow from (4.3a) and (4.25).

b) As a consequence of Lemma 3.9 (see Exercise 8), the propagation of the local errors $\delta y_h(x_{j-1})$, $\delta z_h(x_{j-1})$ to the solution at x_n can be bounded by

$$C\big(\|\delta y_h(x_{j-1})\| + (h+\kappa^{n-j})\|\delta z_h(x_{j-1})\|\big). \tag{4.32}$$

Summing up these terms from $j=1$ to $j=n$ and using (4.27) gives the stated bounds for the global error, because $\sum_{j=1}^{n}(h+\kappa^{n-j}) \le Const$.

Our assumption that the numerical solution and the internal stages lie in $\mathcal{U}$ can now easily be justified by induction on the step number. The numerical solution remains $\mathcal{O}(h^p)$-close to the exact solution and thus remains in $\mathcal{U}$ for sufficiently small h. This implies $g(y_j, z_j) = \mathcal{O}(h^p)$ for all j and hence also $\ell_i = \mathcal{O}(h)$. Consequently (v_i, w_i) are also as close to the exact solution as we want. □

Stiffly Accurate Rosenbrock Methods

We have already had several occasions to admire the beneficial effect of stiffly accurate Runge-Kutta methods (methods with $a_{si} = b_i$ for all i; see Theorem 1.1 and Corollary 3.10). What is the corresponding condition for Rosenbrock methods?

Definition 4.11. A Rosenbrock method is called *stiffly accurate*, if

$$\alpha_{si} + \gamma_{si} = b_i \quad (i=1,\dots,s) \quad \text{and} \quad \alpha_s = 1. \tag{4.33}$$

Recall that $\alpha_i = \sum_j \alpha_{ij}$. It has already been remarked at the end of Sect. IV.15 that methods satisfying (4.33) yield asymptotically exact results for the problem

$y' = \lambda(y - \varphi(x)) + \varphi'(x)$. A further interesting interpretation of this condition has been given by C. Schneider (1991). He argues that DAE's are combinations of differential equations and algebraic equations; hence methods should be equally valuable for both extreme cases, either a purely differential equation, or a purely algebraic equation

$$x' = 1, \qquad 0 = g(x, z), \qquad g_z \text{ invertible} . \tag{4.34}$$

Proposition 4.12. *A stiffly accurate Rosenbrock method, applied to (4.34), yields*

$$z_1 = w_s - g_z^{-1}(x_0, z_0) \cdot g(x_0 + h, w_s).$$

The numerical solution z_1 is thus the result of one simplified Newton iteration for $0 = g(x_0 + h, z)$ (with starting value w_s).

Proof. Condition (4.33) together with $\sum_i b_i = 1$ implies that $\gamma_s = \sum_j \gamma_{sj} = 0$. Therefore, the second line of (4.5) gives (observe that $k_i = h$ for the problem (4.34))

$$0 = g(x_0 + h, w_s) + g_z(x_0, z_0) \sum_{j=1}^{i} \gamma_{ij} \ell_j .$$

Inserting the expression thus obtained for $\sum_j \gamma_{ij} \ell_j$ into

$$z_1 = z_0 + \sum_{j=1}^{s} b_j \ell_j = w_s + \sum_{j=1}^{s} \gamma_{sj} \ell_j$$

proves the statement. □

The values (v_s, w_s) of the last stage are often used as an embedded solution for step size control. If this is the case for a stiffly accurate method, then many of the algebraic order condition are automatically satisfied. This is a consequence of the following result.

Proposition 4.13. *Consider a stiffly accurate Rosenbrock method. For sufficiently regular problems (4.1) we have*

$$z_1 - z(x_0 + h) = \mathcal{O}(h^{q+1}) \tag{4.35}$$

if and only if

$$v_s - y(x_0 + h) = \mathcal{O}(h^q) \qquad \text{and} \qquad w_s - z(x_0 + h) = \mathcal{O}(h^q). \tag{4.36}$$

Proof. We use the characterization of Theorem 4.8 and the fact that (with ω_{ij} defined in (4.17))

$$\sum_i b_i \omega_{ij} = \begin{cases} 1 & \text{if } j = s \\ 0 & \text{else} . \end{cases} \tag{4.37}$$

Suppose first that (4.35) holds. For a tree $u=[\tau_y,t_2]_z$ with arbitrary $t_2\in DAT_y$ we have, by definition of $\Phi_j(u)$ and $\gamma(u)$,

$$\sum_i b_i\Phi_i(u)=\sum_{i,j,k} b_i\omega_{ij}\alpha_j\alpha_{jk}\Phi_k(t_2)=\sum_k \alpha_{sk}\Phi_k(t_2) \tag{4.38}$$

and $\gamma(u)=\gamma(t_2)$. Consequently, the order condition is satisfied for u iff it is satisfied for t_2. Since $\varrho(t_2)=\varrho(u)-1$, we see that $v_s-y(x_0+h)=\mathcal{O}(h^q)$ is a consequence of (4.35). By considering $u=[\tau_y,u_1]_z$ with $u_1\in DAT_z$ we deduce the second relation of (4.36). The "if" part is proved in a similar way. □

Finally we remark that because of (4.25) and (4.37) the stability function of a stiffly accurate Rosenbrock method always satisfies $R(\infty)=0$. This is a desirable property when solving stiff or differential algebraic equations.

Construction of RODAS, a Stiffly Accurate Embedded Method

We want to construct an embedded Rosenbrock method (where $\widehat{y}_1=v_s$, $\widehat{z}_1=w_s$), such that both methods are stiffly accurate. This gives the following conditions

$$\begin{aligned} &b_i=\beta_{si} \qquad (i=1,\ldots,s), && \alpha_s=1\\ &\widehat{b}_i=\alpha_{si}=\beta_{s-1,i} \qquad (i=1,\ldots,s-1), && \alpha_{s-1}=1 \end{aligned} \tag{4.39}$$

(as usual $\beta_{ij}=\alpha_{ij}+\gamma_{ij}$). It follows from Proposition 4.12 that the last *two* stages represent simplified Newton iterations. Further, both methods have a stability function which vanishes at infinity. The construction of such a method of order 4(3) seems to be impossible with $s=5$. We therefore put $s=6$.

Here is the list of order conditions which have to be solved. We use the abbreviations α_i,β_i' defined in (IV.7.16), and the coefficients ω_{ij} from (4.17). We shall require that

$$y_1-y(x_0+h)=\mathcal{O}(h^5), \qquad \widehat{y}_1-y(x_0+h)=\mathcal{O}(h^4). \tag{4.40}$$

Since we have sufficiently many parameters we also require

$$v_{s-1}-y(x_0+h)=\mathcal{O}(h^3), \qquad w_{s-1}-z(x_0+h)=\mathcal{O}(h^3). \tag{4.41}$$

By Proposition 4.13 this implies

$$\widehat{z}_1-z(x_0+h)=\mathcal{O}(h^4), \qquad z_1-z(x_0+h)=\mathcal{O}(h^5), \tag{4.42}$$

which is more than sufficient to ensure convergence of order 4 (see Theorem 4.10). The conditions for (4.40) and (4.41) are (see Table IV.7.1 and Table 4.1)

$$b_1+b_2+b_3+b_4+(b_5+b_6)=1 \tag{4.43a}$$

$$b_2\beta_2'+b_3\beta_3'+b_4\beta_4'+(b_5+b_6)(1-\gamma)=\tfrac{1}{2}-\gamma \tag{4.43b}$$

$$b_2\alpha_2^2+b_3\alpha_3^2+b_4\alpha_4^2+(b_5+b_6)=\tfrac{1}{3} \tag{4.43c}$$

$$b_3\beta_{32}\beta_2' + b_4 \sum{}' \beta_{4i}\beta_i' + (b_5+b_6)\big(\tfrac{1}{2} - 2\gamma + \gamma^2\big) = \tfrac{1}{6} - \gamma + \gamma^2 \tag{4.43d}$$

$$b_2\alpha_2^3 + b_3\alpha_3^3 + b_4\alpha_4^3 + (b_5+b_6) = \tfrac{1}{4} \tag{4.43e}$$

$$b_3\alpha_3\alpha_{32}\beta_2' + b_4\alpha_4 \sum{}' \alpha_{4i}\beta_i' + (b_5+b_6)\big(\tfrac{1}{2} - \gamma\big) = \tfrac{1}{8} - \tfrac{\gamma}{3} \tag{4.43f}$$

$$b_3\beta_{32}\alpha_2^2 + b_4 \sum{}' \beta_{4i}\alpha_i^2 + (b_5+b_6)\big(\tfrac{1}{3} - \gamma\big) = \tfrac{1}{12} - \tfrac{\gamma}{3} \tag{4.43g}$$

$$b_4\beta_{43}\beta_{32}\beta_2' + (b_5+b_6)\big(\tfrac{1}{6} - \tfrac{3}{2}\gamma + 3\gamma^2 - \gamma^3\big) = \tfrac{1}{24} - \tfrac{\gamma}{2} + \tfrac{3}{2}\gamma^2 - \gamma^3 \tag{4.43h}$$

$$b_3\alpha_3\alpha_{32}\omega_{22}\alpha_2^2 + b_4\alpha_4 \sum_{i,j} \alpha_{4i}\omega_{ij}\alpha_j^2 + (b_5+b_6) = \tfrac{1}{4} \tag{4.43i}$$

$$\alpha_{62}\beta_2' + \alpha_{63}\beta_3' + \alpha_{64}\beta_4' = \tfrac{1}{2} - 2\gamma + \gamma^2 \tag{4.43j}$$

$$\alpha_{62}\alpha_2^2 + \alpha_{63}\alpha_3^2 + \alpha_{64}\alpha_4^2 = \tfrac{1}{3} - \gamma \tag{4.43k}$$

$$\alpha_{63}\beta_{32}\beta_2' + \alpha_{64} \sum{}' \beta_{4i}\beta_i' = \tfrac{1}{6} - \tfrac{3}{2}\gamma + 3\gamma^2 - \gamma^3 \tag{4.43l}$$

$$\alpha_{52}\beta_2' + \alpha_{53}\beta_3' + \alpha_{54}\beta_4' = \tfrac{1}{2} - \gamma \tag{4.43m}$$

$$\sum_{i=1}^4 \alpha_{5i} \sum_{j=1}^i \omega_{ij}\alpha_j^2 = 1 \tag{4.43n}$$

In order to solve the system (4.39), (4.43a–n) we can take γ, α_2, α_3, α_4, $\beta_2' = \beta_{21}$, β_3', β_4' as free parameters. The remaining coefficients can then be computed as follows:

Step 1. We have $b_6 = \gamma$ by (4.39). The remaining b_i can be chosen such that (4.43a,b,c,e) are satisfied. We have one degree of freedom which can be exploited to fulfill the additional order condition $\sum_i b_i\alpha_i^4 = 1/5$. This step also yields $\beta_{6i} = b_i$ for $i = 1,\ldots,6$.

Step 2. Compute the two expressions $b_3\beta_{32} + b_4\beta_{42}$ and $b_4\beta_{43}$ from (4.43d,g), and then β_{32} from (4.43h). Because of $\beta_i' = \sum_{j=1}^{i-1}\beta_{ij}$ this determines all β_{ij} with $i \le 4$. Observe that $\beta_{ii} = \gamma$ for all i.

Step 3. Solve the linear system (4.43j,k,l) for α_{62}, α_{63}, α_{64}. We have $\alpha_{65} = \gamma$ by (4.39) and compute α_{61} from $\alpha_6 = \sum_i \alpha_{6i} = 1$. This also yields $\beta_{5i} = \alpha_{6i}$ by (4.39). Hence all β_{ij} and ω_{ij}, and also $\widehat{b}_i = \beta_{5i}$ $(i = 1,\ldots,5)$ are determined at this stage.

Step 4. The conditions (4.43m,n) and $\alpha_5 = 1$ constitute 3 linear equations in the four unknown parameters α_{51}, α_{52}, α_{53}, α_{54}. We have one degree of freedom in this step.

Step 5. The remaining two conditions (4.43f,i) are linear equations in α_{32}, α_{42}, α_{43}. We have one more degree of freedom which can be exploited to fulfill the order condition for the tree $[\tau_y, \tau_y, [\tau_y]_y]_y$. The values of α_{i1} are then determined by $\alpha_i = \sum_{j=1}^{i-1}\alpha_{ij}$, and those of γ_{ij} are given by $\gamma_{ij} = \beta_{ij} - \alpha_{ij}$.

The coefficients for the code RODAS of the appendix were computed with the above procedure. In step 4 we have added the condition

$$\sum_{i,j} \alpha_{5i}\omega_{ij} = 1 \tag{4.44}$$

which will be explained in Exercise 3 below. The free parameters were chosen in order to get an A-stable method with small error constants. The result is

$$\begin{aligned} &\gamma = 0.25 \\ &\alpha_2 = 0.386 \qquad \alpha_3 = 0.21 \qquad \alpha_4 = 0.63 \\ &\beta_2' = 0.0317 \qquad \beta_3' = 0.0635 \qquad \beta_4' = 0.3438 \end{aligned} \tag{4.45}$$

We do not claim that these values are optimal. Nevertheless, the numerical results of Sect. IV.10 (Fig. IV.10.8, IV.10.9 and IV.10.12) are encouraging. Although the new method needs 6 function evaluations per step, it is in general superior to the classical methods of Table IV.7.2 which need only 3 evaluations per step.

A different set of coefficients, based on the same construction, has been proposed by Steinebach (1995). The free parameters are chosen in order to satisfy the Scholz conditions $C_2(z) \equiv 0$ and $C_3(z) \equiv 0$ (see Eq. (15.41) of Sect. IV.15).

Dense Output. A natural way to define a continuous numerical solution for $y(x_0 + \theta h)$, $z(x_0 + \theta h)$ is

$$y_1(\theta) = y_0 + \sum_{i=1}^{s} b_i(\theta) k_i, \qquad z_1(\theta) = z_0 + \sum_{i=1}^{s} b_i(\theta) \ell_i, \tag{4.46}$$

where the $b_i(\theta)$ are polynomials which satisfy $b_i(0) = 0$, $b_i(1) = b_i$. In complete analogy to Theorem 4.8 we have

$$\begin{aligned} y(x_0 + \theta h) - y_1(\theta) = \mathcal{O}(h^{p+1}) \quad &\text{iff} \quad \sum_{i=1}^{s} b_i(\theta)\Phi_i(t) = \frac{\theta^{\varrho(t)}}{\gamma(t)} \\ &\text{for } t \in DAT_y,\ \varrho(t) \le p, \\ z(x_0 + \theta h) - z_1(\theta) = \mathcal{O}(h^{q+1}) \quad &\text{iff} \quad \sum_{i=1}^{s} b_i(\theta)\Phi_i(u) = \frac{\theta^{\varrho(u)}}{\gamma(u)} \\ &\text{for } u \in DAT_z,\ \varrho(u) \le q. \end{aligned} \tag{4.47}$$

In our situation ($s = 6$) it is easy to fulfill these conditions with $p = 3$ and $q = 2$. The additional condition $b_s(\theta) = \gamma\theta$ makes the solution unique.

Methods of Order 5. C. Schneider (1991b) first constructed stiffly accurate Rosenbrock methods of order 5 with $s = 8$ stages. Di Marzo (1992) then determined carefully the free parameters to obtain A-stability and small error constants. The resulting method, implemented in the code RODAS5, gives excellent results (see Sect. IV.10).

Inconsistent Initial Values

Even if we start the computation with consistent initial values, the numerical solution (y_n, z_n) of a Rosenbrock method does not, in general, satisfy $g(y_n, z_n) = 0$. It is therefore of interest to investigate the local error also for inconsistent initial values (y_0, z_0). But what is the local error? To which solution of (4.1) should we compare the numerical values? If

$$\|(g_z^{-1}g)(y_0, z_0)\| \le \delta \tag{4.48}$$

with sufficiently small δ, we can find (because of (1.7)) a locally unique $\widehat{z}_0$ which satisfies $g(y_0, \widehat{z}_0) = 0$. It is natural to compare the numerical solution (y_1, z_1) to that solution of (4.1) which passes through $(y_0, \widehat{z}_0)$.

Our first aim is to write this solution in terms of elementary differentials evaluated at (y_0, z_0). Using

$$\widehat{z}_0 - z_0 = (-g_z^{-1}g)(y_0, z_0) + \mathcal{O}(\delta^2),$$

which is a consequence of $0 = g(y_0, z_0) + g_z(y_0, z_0)(\widehat{z}_0 - z_0) + \ldots$, we get

$$\begin{aligned} y(x_0 + h) &= y_0 + hf(y_0, \widehat{z}_0) + \mathcal{O}(h^2) \qquad (4.49)\\ &= y_0 + hf(y_0, z_0) + h(f_z(-g_z^{-1})g)(y_0, z_0) + \mathcal{O}(h^2 + h\delta^2) \end{aligned}$$

$$\begin{aligned} z(x_0 + h) &= \widehat{z}_0 + h(-g_z^{-1}g_y f)(y_0, \widehat{z}_0) + \mathcal{O}(h^2) \qquad (4.50)\\ &= z_0 + (-g_z^{-1}g)(y_0, z_0) + h(-g_z^{-1}g_y f)(y_0, z_0)\\ &\quad + h(-g_z^{-1}g_{zz}(-g_z^{-1}g, -g_z^{-1}g_y f))(y_0, z_0)\\ &\quad + h(-g_z^{-1}g_{yz}(f, -g_z^{-1}g))(y_0, z_0)\\ &\quad + h(-g_z^{-1}g_y f_z(-g_z^{-1})g)(y_0, z_0) + \mathcal{O}(h^2 + \delta^2) \end{aligned}$$

The expressions so obtained allow a nice interpretation using trees. We only have to add in the recursive Definition 4.1 a tree of order 0, which consists of a fat root. We denote this tree by $\emptyset_z$, and extend Definition 4.3 by setting $F(\emptyset_z)(y, z) = (-g_z^{-1}g)(y, z)$. Then, the expressions of (4.49) and (4.50) correspond to the trees of Fig. 4.4.

Fig. 4.4. Trees, to be considered for inconsistent initial values

The numerical solution also possesses an expansion of the form (4.49), (4.50) with additional method-dependent coefficients. The first few terms are as follows:

$$\begin{aligned} y_1 &= y_0 + \Big(\sum_i b_i\Big) hf(y_0, z_0) + \Big(\sum_{i,j,k} b_i\beta_{ij}\omega_{jk}\Big) h(f_z(-g_z^{-1})g)(y_0, z_0)\\ &\qquad + \mathcal{O}(h^2 + h\delta^2)\\ z_1 &= z_0 + \Big(\sum_{i,j} b_i\omega_{ij}\Big)(-g_z^{-1}g)(y_0, z_0) + \mathcal{O}(h + \delta^2). \end{aligned}$$

In order to understand the form of these new coefficients we have to extend the proof of Theorem 4.6. It turns out that the elementary differentials are multiplied by $\gamma(t)\sum_i b_i\Phi_i(t)$ or $\gamma(u)\sum_i b_i\Phi_i(u)$, where γ and Φ_i are defined by $\gamma(\emptyset_z) =$

1, $\Phi_i(\emptyset_z) = \sum_j \omega_{ij}$ and the recursion of Theorem 4.6. Equating the coefficients of the exact and numerical solutions yields new order conditions for the case of inconsistent initial values. The first of these (to be added to those of Table IV.7.1 and Table 4.1) are presented in Table 4.2.

Table 4.2. Order conditions for inconsistent initial values

tree	order condition	size of error term
 [tree]	$\sum b_i \alpha_i \alpha_{ij} \omega_{jk} = 1/2$	$\mathcal{O}(h^2\delta)$
○	$\sum b_i \omega_{ij} = 1$	$\mathcal{O}(\delta)$
 [tree]	$\sum b_i \omega_{ij} \alpha_j \alpha_{jk} \omega_{k\ell} = 1$	$\mathcal{O}(h\delta)$

Remarks. a) The first condition of Table 4.2 is exactly the same as that found by van Veldhuizen (1984) in a different context. It implies that the local error of the y-component is of size $\mathcal{O}(h^{p+1} + h^3\delta + h\delta^2)$.

b) Condition $\sum_{i,j} b_i \omega_{ij} = 1$ means that the stability function satisfies $R(\infty) = 0$. Unless this condition is satisfied, the local error of the z-component contains an h-independent term of size δ (which usually is near to *Tol*). This was observed numerically in Fig. IV.7.4 and explains the phenomenon of Fig. IV.7.3.

c) For Rosenbrock methods which satisfy (4.39), the second and third conditions of Table 4.2 are automatically fulfilled. For such methods the local error of the z-component is of size $\mathcal{O}(h^{q+1} + h^2\delta + \delta^2)$.

Exercises

1. (Roche 1989). Consider the implicit Runge-Kutta method (1.11) applied to (1.6).

   a) Prove that $z_1 - z(x_0 + h) = \mathcal{O}(h^{q+1})$ iff
$$\sum_{i=1}^{s} b_i \Phi_i(u) = \frac{1}{\gamma(u)} \qquad \text{for} \quad u \in DAT_z,\ \varrho(u) \le q,$$
   where $\gamma(u)$ and $\Phi_i(u)$ are defined as in Theorem 4.6, but all coefficients α_{ij} and β_{ij} are replaced by the Runge-Kutta coefficients a_{ij}.

   b) Show that those trees in DAT_z which have more than one fat vertex, are redundant.

2. The simplifying assumptions (4.39) imply that many of the (algebraic) order conditions are automatically satisfied. Characterize the corresponding trees.

3. State the order condition for the tree $[\tau_y, [\tau_y, \emptyset_z]_z]_z$.

   a) Show that the corresponding error term is of size $\mathcal{O}(h^2\delta)$ with δ given in (4.48).

   b) For methods satisfying (4.39), this condition is equivalent to (4.44).

4. (Ostermann 1990). Suppose that the Rosenbrock method (4.3) satisfies (4.27). Define polynomials $b_i(\theta)$ of degree $q=[(p+1)/2]$ by $b_i(0)=0$, $b_i(1)=b_i$, and

$$\int_0^1 b_i(\theta)\theta^{\ell-1}\,d\theta = \begin{cases} \sum_j b_j(\alpha_{ji}+\gamma_{ji}) & \text{if } \ell=1 \\ \sum_j b_j\alpha_j^{\ell-1}\alpha_{ji} & \text{if } \ell=2,\dots,q-1. \end{cases}$$

   Prove that the error of the dense output formulas (4.46) is $\mathcal{O}(h^{q+1})$.

   Hint. Extend the ideas of Exercise II.17.5 to Rosenbrock methods.

5. Suppose that a Rosenbrock method is implemented in the form (IV.7.25). If it satisfies (4.39), then its last two stages allow a very simple implementation

   Hint. Prove that

$$m_i = \begin{cases} a_{si} & i=1,\dots,s-1 \\ 1 & i=s\,, \end{cases} \qquad a_{si} = \begin{cases} a_{s-1,i} & i=1,\dots,s-2 \\ 1 & i=s-1\,. \end{cases}$$

6. *Partitioned Rosenbrock methods* (Rentrop, Roche & Steinebach 1989). Consider the method (4.3) with f_y and f_z replaced by 0. Derive necessary and sufficient conditions that it be of order p.

   Remark. Case (a) of Lemma 4.9 remains valid in this new situation. However, the trees of Lemma 4.9b give rise to new conditions.

7. What is the "algebraic order" of the classical 4th order Rosenbrock methods of Section IV.7?

8. Let $\{u_n\}$, $\{v_n\}$ be two sequences of non-negative numbers satisfying (componentwise)

$$\begin{pmatrix} u_{n+1} \\ v_{n+1} \end{pmatrix} \le \begin{pmatrix} 1+hL & hM \\ N & \kappa \end{pmatrix} \begin{pmatrix} u_n \\ v_n \end{pmatrix}$$

   with $0 \le \kappa < 1$ and positive constants L, M, N. Prove that for $h \le h_0$ and $nh \le Const$

$$u_n \le C(u_0 + hv_0), \qquad v_n \le C(u_0 + (h+\kappa^n)v_0).$$

   Hint. Apply Lemma 3.9 with $\varepsilon = h$ and $M = 0$.

VI.5 Extrapolation Methods

The numerical computations of Sect. IV.10 have revealed the extrapolation code SEULEX as one of the best method for very stringent tolerances. The aim of the present section is to justify theoretically the underlying numerical method, the extrapolated linearly implicit Euler method, for singular perturbation problems as a representative of stiff equations.

Linearly Implicit Euler Discretization

The linearly implicit Euler method (IV.9.25) applied to the singular perturbation problem (1.5) reads

$$\begin{pmatrix} I-hf_y(0) & -hf_z(0) \\ -hg_y(0) & \varepsilon I-hg_z(0) \end{pmatrix}\begin{pmatrix} y_{i+1}-y_i \\ z_{i+1}-z_i \end{pmatrix} = h\begin{pmatrix} f(y_i,z_i) \\ g(y_i,z_i) \end{pmatrix}. \tag{5.1}$$

Here we have used abbreviations such as $f_y(0)=f_y(y_0,z_0)$ for the partial derivatives. We recall that the numerical approximations at x_0+H $(H=nh)$ are extrapolated according to (IV.9.26).

For the differential algebraic problem (1.6) we just put $\varepsilon=0$ in (5.1). This yields

$$\begin{pmatrix} I-hf_y(0) & -hf_z(0) \\ -hg_y(0) & -hg_z(0) \end{pmatrix}\begin{pmatrix} y_{i+1}-y_i \\ z_{i+1}-z_i \end{pmatrix} = h\begin{pmatrix} f(y_i,z_i) \\ g(y_i,z_i) \end{pmatrix}. \tag{5.2}$$

Possible extensions to non-autonomous problems have been presented in Sect. IV.9. For problems $Mu'=\varphi(u)$ we use the formulation (IV.9.34) also for singular M. Due to the invariance of the method with respect to the transformation (1.23), all results of this section are equally valid for $Mu'=\varphi(u)$ of index 1.

The performance of extrapolation methods relies heavily on the existence of an asymptotic expansion of the global error. Such expansions are well understood, if the differential equation is nonstiff (see Sections II.8 and IV.9). But what happens if the problem is stiff or differential-algebraic?

> Continued study of special problems is still a commendable way towards greater insight ... (E. Hopf 1950)

Example 5.1. Consider the test problem

$$y' = 1, \qquad \varepsilon z' = -z + g(y). \tag{5.3}$$

Method (5.1) yields the exact result $y_i = x_i = x_0 + ih$ for the y-component, and the recursion

$$(\varepsilon + h)z_{i+1} = \varepsilon z_i + hg(x_i) + h^2 g'(x_0) \tag{5.4}$$

for the z-component. In order to compute the coefficients of the asymptotic expansion (Theorem II.8.1), we insert

$$z_i = z(x_i) + hb_1(x_i) + h^2 b_2(x_i) + h^3 b_3(x_i) + \ldots \tag{5.5}$$

into (5.4), expand into a Taylor series and compare the coefficients of h^j. This yields the differential equation

$$\varepsilon b_1'(x) + b_1(x) = -\frac{\varepsilon}{2} z''(x) - z'(x) + g'(x_0)$$

for $b_1(x)$, and similar ones for $b_2(x)$, $b_3(x)$, etc. Putting $i = 0$ in (5.5) we get the initial values $b_i(x_0) = 0$ (all i). In general, the computation of the functions $b_1(x)$, $b_2(x), \ldots$ is rather tedious. We therefore continue this example for the special case $x_0 = 0$, $g(x) = x^2 + 2\varepsilon x$, and $z_0 = 0$, so that the exact solution of (5.3) is $z(x) = x^2$. In this situation we get

$$\begin{aligned} b_1(x) &= -3\varepsilon e^{-x/\varepsilon} + 3\varepsilon - 2x \\ b_2(x) &= -\Big(1 + \frac{3x}{2\varepsilon}\Big) e^{-x/\varepsilon} + 1 \\ b_3(x) &= \Big(\frac{x}{2\varepsilon^2} - \frac{3x^2}{8\varepsilon^3}\Big) e^{-x/\varepsilon} \end{aligned} \tag{5.6}$$

etc. We observe that for $\varepsilon \to 0$, the function $b_2(x)$ becomes discontinuous at $x = 0$, and $b_3(x)$ is even not uniformly bounded. Hence, the expansion (5.5) is not useful for the study of extrapolation, if ε is small compared to the step size H.

The idea is now to omit in (5.6) the terms containing the factor $e^{-x/\varepsilon}$ by requiring that the functions $b_i(x)$ be *smooth* uniformly in ε and, instead, to add a discrete perturbation β_i to (5.5). For our example, this then becomes

$$z_i = x_i^2 + h(3\varepsilon - 2x_i) + h^2 + \beta_i. \tag{5.7a}$$

Inserting (5.7a) into (5.4) gives the relation $(\varepsilon + h)\beta_{i+1} = \varepsilon \beta_i$. The value of β_0 is obtained from (5.7a) with $i = 0$. We thus get

$$\beta_i = -\Big(1 + \frac{h}{\varepsilon}\Big)^{-i} (3\varepsilon h + h^2). \tag{5.7b}$$

If the numerical solution is extrapolated, the smooth terms in (5.7) are eliminated one after the other. It remains to study the effect of extrapolation on the perturbation terms β_i. If the differential equation is very stiff ($\varepsilon \ll h$), these terms are very small and may be neglected over a wide range of h (observe that $i \geq n_1$).

Example 5.2. For the differential-algebraic problem

$$y' = 1, \qquad 0 = -z + g(y) \tag{5.8}$$

with initial values $y_0 = x_0$, $z_0 = g(x_0)$ the numerical solution, given by (5.2), is

$$z_i = \begin{cases} g(x_0) & \text{for } i = 0 \\ g(x_{i-1}) + hg'(x_0) & \text{for } i \geq 1\,. \end{cases}$$

Developing its second formula (for $i \geq 1$) yields

$$z_i = g(x_i) + h\big(g'(x_0) - g'(x_i)\big) + \frac{h^2}{2} g''(x_i) - \frac{h^3}{6} g'''(x_i) + \mathcal{O}(h^4).$$

If we add the perturbation

$$\beta_i = h\beta_i^1 + h^2\beta_i^2 + h^3\beta_i^3 + \ldots \tag{5.9}$$

(which is different from zero only for $i = 0$) we get for *all* i

$$z_i - g(x_i) = \sum_{j=1}^{3} h^j (b_j(x_i) + \beta_i^j) + \mathcal{O}(h^4) \tag{5.10}$$

where

$$b_1(x) = g'(x_0) - g'(x), \quad b_2(x) = \frac{1}{2} g''(x), \quad b_3(x) = -\frac{1}{6} g'''(x)$$

are smooth functions and the perturbations are given by

$$\beta_0^1 = 0, \quad \beta_0^2 = -\frac{1}{2} g''(x_0), \quad \beta_0^3 = \frac{1}{6} g'''(x_0).$$

If we add a further algebraic equation to (5.8), e.g., $0 = u - k(z)$, and again apply Method (5.2), we get three different formulas for u_i, one for $i = 0$, one for $i = 1$, and a different one for $i \geq 2$. In an expansion of the type (5.10) for $u_i - k(g(x_i))$, perturbation terms will be present for $i = 0$ and for $i = 1$.

Perturbed Asymptotic Expansion

For general differential algebraic problems we have the following result.

Theorem 5.3 (Deuflhard, Hairer & Zugck 1987). *Consider the problem (1.6) with consistent initial values (y_0, z_0), and suppose that (1.7) is satisfied. The global error of the linearly implicit Euler method (5.2) then has an asymptotic h-expansion of the form*

$$\begin{aligned} y_i - y(x_i) &= \sum_{j=1}^{M} h^j \big(a_j(x_i) + \alpha_i^j\big) + \mathcal{O}(h^{M+1}) \\ z_i - z(x_i) &= \sum_{j=1}^{M} h^j \big(b_j(x_i) + \beta_i^j\big) + \mathcal{O}(h^{M+1}) \end{aligned} \tag{5.11}$$

where $a_j(x)$, $b_j(x)$ are smooth functions and the perturbations satisfy (see Table 5.1 and 5.2)

$$\alpha_i^1 = 0, \quad \alpha_i^2 = 0, \quad \alpha_i^3 = 0, \quad \beta_i^1 = 0 \qquad \text{for } i \geq 0 \tag{5.12a}$$

$$\beta_i^2 = 0 \qquad \text{for } i \geq 1 \tag{5.12b}$$

$$\alpha_i^j = 0 \qquad \text{for } i \geq j-4 \text{ and } j \geq 4 \tag{5.12c}$$

$$\beta_i^j = 0 \qquad \text{for } i \geq j-2 \text{ and } j \geq 3\,. \tag{5.12d}$$

The error terms in (5.11) are uniformly bounded for $x_i = ih \leq H$, if H is sufficiently small.

Table 5.1. Non-zero α's

	h	h^2	h^3	h^4	h^5	h^6	h^7
y_0	0	0	0	0	*	*	*
y_1	0	0	0	0	0	*	*
y_2	0	0	0	0	0	0	*
y_3	0	0	0	0	0	0	0
y_4	0	0	0	0	0	0	0
y_5	0	0	0	0	0	0	0

Table 5.2. Non-zero β's

	h	h^2	h^3	h^4	h^5	h^6	h^7
z_0	0	*	*	*	*	*	*
z_1	0	0	0	*	*	*	*
z_2	0	0	0	0	*	*	*
z_3	0	0	0	0	0	*	*
z_4	0	0	0	0	0	0	*
z_5	0	0	0	0	0	0	0

Proof. In part (a) we shall recursively construct truncated expansions

$$\begin{aligned} \widehat{y}_i &= y(x_i) + \sum_{j=1}^{M} h^j \big(a_j(x_i) + \alpha_i^j\big) + h^{M+1}\alpha_i^{M+1} \\ \widehat{z}_i &= z(x_i) + \sum_{j=1}^{M} h^j \big(b_j(x_i) + \beta_i^j\big) \end{aligned} \tag{5.13}$$

such that the defect of $\widehat{y}_i$, $\widehat{z}_i$ inserted into the method is small; more precisely, we require that

$$\begin{pmatrix} I - hf_y(0) & -hf_z(0) \\ -hg_y(0) & -hg_z(0) \end{pmatrix} \begin{pmatrix} \widehat{y}_{i+1} - \widehat{y}_i \\ \widehat{z}_{i+1} - \widehat{z}_i \end{pmatrix} = h \begin{pmatrix} f(\widehat{y}_i, \widehat{z}_i) \\ g(\widehat{y}_i, \widehat{z}_i) \end{pmatrix} + \mathcal{O}(h^{M+2}). \tag{5.14}$$

For the initial values we require $\widehat{y}_0 = y_0$, $\widehat{z}_0 = z_0$, or equivalently

$$a_j(0) + \alpha_0^j = 0, \qquad b_j(0) + \beta_0^j = 0, \tag{5.15}$$

and the perturbation terms are assumed to satisfy

$$\alpha_i^j \to 0, \qquad \beta_i^j \to 0 \qquad \text{for} \quad i \to \infty, \tag{5.16}$$

otherwise, these limits could be added to the smooth parts. The result will then follow from a stability estimate derived in part (b).

a) For the construction of $a_j(x)$, $b_j(x)$, α_i^j, β_i^j we insert (5.13) into (5.14), and develop

$$\begin{aligned} f(\widehat{y}_i, \widehat{z}_i) &= f(y(x_i), z(x_i)) + f_y(x_i)\big(ha_1(x_i) + h\alpha_i^1 + \ldots\big) \\ &\quad + f_z(x_i)\big(hb_1(x_i) + h\beta_i^1 + \ldots\big) \\ &\quad + f_{yy}(x_i)\big(ha_1(x_i) + h\alpha_i^1 + \ldots\big)^2 + \ldots, \end{aligned}$$

$$\begin{aligned} \widehat{y}_{i+1} - \widehat{y}_i &= y(x_{i+1}) - y(x_i) + h\big(a_1(x_{i+1}) - a_1(x_i) + \alpha_{i+1}^1 - \alpha_i^1\big) + \ldots \\ &= hy'(x_i) + \frac{h^2}{2} y''(x_i) + \ldots + h^2 a_1'(x_i) + h(\alpha_{i+1}^1 - \alpha_i^1) + \ldots, \end{aligned}$$

where $f_y(x) = f_y(y(x), z(x))$, etc. Similarly, we develop $g(\widehat{y}_i, \widehat{z}_i)$ and $\widehat{z}_{i+1} - \widehat{z}_i$, and compare coefficients of h^{j+1} (for $j = 0, \ldots, M$). Each power of h will lead to *two* conditions — one containing the smooth functions and the other containing the perturbation terms.

First step. Equating the coefficients of h^1 yields the equations (1.6) for the smooth part (due to consistency of the method), and $\alpha_{i+1}^1 - \alpha_i^1 = 0$ for $i \geq 0$. Because of (5.16) we get $\alpha_i^1 = 0$ for all $i \geq 0$ (compare (5.12a)).

Second step. The coefficents of h^2 give

$$a_1'(x) + \frac{1}{2} y''(x) - f_y(0)y'(x) - f_z(0)z'(x) = f_y(x)a_1(x) + f_z(x)b_1(x) \tag{5.17a}$$

$$-g_y(0)y'(x) - g_z(0)z'(x) = g_y(x)a_1(x) + g_z(x)b_1(x) \tag{5.17b}$$

$$\alpha_{i+1}^2 - \alpha_i^2 - f_z(0)(\beta_{i+1}^1 - \beta_i^1) = f_z(0)\beta_i^1 \tag{5.17c}$$

$$-g_z(0)(\beta_{i+1}^1 - \beta_i^1) = g_z(0)\beta_i^1. \tag{5.17d}$$

Observe that the coefficients α_i^ℓ, β_i^ℓ have to be independent of h, so that $f_z(0)$, $g_z(0)$ cannot be replaced by $f_z(x_i)$, $g_z(x_i)$ in the right-hand sides of (5.17c, d). The system (5.17) can be solved as follows. Compute $b_1(x)$ from (5.17b) and insert it into (5.17a). This gives a linear differential equation for $a_1(x)$. Because of (5.15) and $\alpha_0^1 = 0$ the initial value is $a_1(0) = 0$. Therefore $a_1(x)$ and $b_1(x)$ are uniquely determined by (5.17a, b). Differentiating $g(y(x), z(x)) = 0$ and putting $x = 0$ implies that the left-hand side of (5.17b) vanishes at $x = 0$. Consequently, we have $b_1(0) = 0$ and by (5.15), also $\beta_0^1 = 0$. Condition (5.17d) then implies $\beta_i^1 = 0$ (all i), and (5.17c) together with (5.16) give $\alpha_i^2 = 0$ (all i).

Third step. As in the second step we get (for $j = 2$)

$$a_j'(x) = f_y(x)a_j(x) + f_z(x)b_j(x) + r(x) \tag{5.18a}$$

$$0 = g_y(x)a_j(x) + g_z(x)b_j(x) + s(x), \tag{5.18b}$$

where $r(x)$, $s(x)$ are known functions depending on derivatives of $y(x)$, $z(x)$, and on $a_\ell(x)$, $b_\ell(x)$ with $\ell \leq j-1$. We further get

$$\alpha_{i+1}^3 - \alpha_i^3 = f_z(0)\beta_{i+1}^2 \tag{5.18c}$$

$$0 = g_z(0)\beta_{i+1}^2. \tag{5.18d}$$

We compute $a_2(x)$, $b_2(x)$ as in step 2. However, $b_2(0) \neq 0$ in general, and for the first time, we are forced to introduce a perturbation term $\beta_0^2 \neq 0$. From (5.18c, d) we then get $\beta_i^2 = 0$ (for $i \geq 1$) and $\alpha_i^3 = 0$ (for all i).

Fourth step. Comparing the coefficients of h^4 we just get (5.18a,b) with $j = 3$ and (5.18c,d) with the upper index raised by 1. As above we conclude $\beta_i^3 = 0$ (for $i \geq 1$) and $\alpha_i^4 = 0$ (for all i).

General step. The conditions for the smooth functions are (5.18a,b). For the perturbation terms we get

$$\alpha_{i+1}^{j+1} - \alpha_i^{j+1} = f_z(0)\beta_{i+1}^j + \varrho_i^j \tag{5.19c}$$

$$0 = g_z(0)\beta_{i+1}^j + \sigma_i^j, \tag{5.19d}$$

where ϱ_i^j, σ_i^j are linear combinations of expressions which contain as factors α_{i+1}^ℓ, $\alpha_i^{\ell-1}$, $\beta_i^{\ell-1}$ with $\ell \leq j$. For example, we have $\varrho_i^4 = f_{zz}(0)(\beta_i^2)^2$ and $\sigma_i^4 = g_{zz}(0)(\beta_i^2)^2$. The proof of (5.12) is now by induction on j. By the induction hypothesis we have $\varrho_i^j = 0$, $\sigma_i^j = 0$ for $i \geq j-3$. Formula (5.19d) hence implies $\beta_{i+1}^j = 0$ (for $i \geq j-3$) and (5.19c) together with (5.16) gives $\alpha_i^{j+1} = 0$ (for $i \geq j-3$). But this is simply the statement (5.12c,d).

b) We still have to estimate the remainder term, i.e., differences $\Delta y_i = y_i - \widehat{y}_i$, $\Delta z_i = z_i - \widehat{z}_i$. Subtracting (5.14) from (5.2) and eliminating Δy_{i+1}, Δz_{i+1} yields

$$\begin{pmatrix} \Delta y_{i+1} \\ \Delta z_{i+1} \end{pmatrix} = \begin{pmatrix} \Delta y_i \\ \Delta z_i \end{pmatrix} + \begin{pmatrix} I + \mathcal{O}(h) & \mathcal{O}(h) \\ \mathcal{O}(1) & -g_z^{-1}(0) \end{pmatrix} \begin{pmatrix} h\big(f(y_i, z_i) - f(\widehat{y}_i, \widehat{z}_i)\big) \\ g(y_i, z_i) - g(\widehat{y}_i, \widehat{z}_i) \end{pmatrix} + \begin{pmatrix} \mathcal{O}(h^{M+2}) \\ \mathcal{O}(h^{M+1}) \end{pmatrix}.$$

The application of a Lipschitz condition for $f(y,z)$ and $g(y,z)$ then gives

$$\begin{pmatrix} \|\Delta y_{i+1}\| \\ \|\Delta z_{i+1}\| \end{pmatrix} \leq \begin{pmatrix} 1 + \mathcal{O}(h) & \mathcal{O}(h) \\ \mathcal{O}(1) & \varrho \end{pmatrix} \begin{pmatrix} \|\Delta y_i\| \\ \|\Delta z_i\| \end{pmatrix} + \begin{pmatrix} \mathcal{O}(h^{M+2}) \\ \mathcal{O}(h^{M+1}) \end{pmatrix}, \tag{5.20}$$

where $|\varrho| < 1$ if H is sufficiently small. Applying Lemma 3.9 we deduce $\|\Delta y_i\| + \|\Delta z_i\| = \mathcal{O}(h^{M+1})$. □

Order Tableau

We consider (5.2) as our basic method for extrapolation, i.e., we take some step number sequence $n_1 < n_2 < \ldots$, put $h_j = H/n_j$, and define

$$Y_{j1} = y_{h_j}(x_0 + H), \qquad Z_{j1} = z_{h_j}(x_0 + H), \tag{5.21}$$

the numerical solution of (1.6) after n_j steps with step size h_j. We then extrapolate these values according to (IV.9.26) and obtain Y_{jk}, Z_{jk}. What is the order of the approximations thus obtained?

Theorem 5.4 (Deuflhard, Hairer & Zugck 1987). *If we consider the harmonic sequence* $\{1, 2, 3, 4, \ldots\}$, *then the extrapolated values* Y_{jk}, Z_{jk} *satisfy*

$$Y_{jk} - y(x_0 + h) = \mathcal{O}(H^{r_{jk}+1}), \qquad Z_{jk} - z(x_0 + H) = \mathcal{O}(H^{s_{jk}}) \tag{5.22}$$

where the differential-algebraic orders r_{jk}, s_{jk} *are given in Tables 5.3 and 5.5.*

Table 5.3. orders r_{jk}.

1									
1	2								
1	2	3							
1	2	3	4						
1	2	3	4	4					
1	2	3	4	4	5				
1	2	3	4	4	5	5			
1	2	3	4	4	5	6	5		
1	2	3	4	4	5	6	6	5	
1	2	3	4	4	5	6	7	6	5

Table 5.4. orders s_{jk}.

2									
2	2								
2	2	3							
2	2	3	4						
2	2	3	4	4					
2	2	3	4	5	4				
2	2	3	4	5	5	4			
2	2	3	4	5	6	5	4		
2	2	3	4	5	6	6	5	4	
2	2	3	4	5	6	7	6	5	4

Proof. We use the expansion (5.11). It follows from $\alpha_i^1 = \beta_i^1 = 0$ (for all $i \geq 0$) and from (5.15) that $a_1(x_0) = b_1(x_0) = 0$. Since $a_j(x)$ and $b_j(x)$ are smooth functions we obtain $a_1(x_0 + H) = \mathcal{O}(H)$, $b_1(x_0 + H) = \mathcal{O}(H)$ and the errors of Y_{j1}, Z_{j1} are seen to be of size $\mathcal{O}(H^2)$. This verifies the entries of the first columns of Tables 5.3 and 5.4. In the same way we deduce that $a_2(x_0 + H) = \mathcal{O}(H)$. However, since $\beta_0^2 \neq 0$ in general, we have $b_2(x_0) \neq 0$ by (5.15) and the term $b_2(x_0 + H)$ is only of size $\mathcal{O}(1)$. One extrapolation of the numerical solution eliminates the terms with $j = 1$ in (5.11). The error is thus of size $\mathcal{O}(H^3)$ for Y_{j2} but only $\mathcal{O}(H^2)$ for Z_{j2}, verifying the second columns of Tables 5.3 and 5.4. If we continue the extrapolation process, the smooth parts of the error expansion (5.11) are eliminated one after the other. The perturbation terms, however, are *not* eliminated.

For the y-component the first non-vanishing perturbation for $i \geq n_1 = 1$ is α_1^6. Therefore, the diagonal elements of the extrapolation tableau for the y-component (Table 5.3) contain an error term of size $\mathcal{O}(H^6)$ (observe that α_1^6 is multiplied by h^6 in (5.11)). The elements $Y_{j,j-1}$ of the first subdiagonal depend only on $Y_{\ell 1} = y_{n_\ell}$ for $\ell \geq 2$. Since $n_2 \geq 2$, only the perturbations α_i^j with $i \geq 2$ can have an influence. We see from (5.12) that the first non-vanishing perturbation for $i \geq 2$ is α_2^7. This explains the $\mathcal{O}(H^7)$ error term in the first subdiagonal of Table 5.3.

For the z-component, β_1^4 is the first perturbation term for $i \geq 1$. Hence the diagonal entries of the extrapolation tableau for the z-component contain an error of size $\mathcal{O}(H^4)$. All other entries of Tables 5.3 and 5.4 can be verified analogously. □

If we consider a step number sequence $\{n_j\}$ which is different from the harmonic sequence, we obtain the corresponding order tableaux as follows: the jth diagonal of the new tableau is the n_jth diagonal of Table 5.3 and 5.4, respectively.

Theorem 5.4 then remains valid with r_{jk}, s_{jk} given by these new tableaux. This implies that a larger n_1, say $n_1 = 2$ increases, the order of the extrapolated values. Numerical computations have shown that the sequence

$$\{2, 3, 4, 5, 6, \ldots\} \tag{5.23}$$

is superior to the harmonic sequence. It is therefore recommended for SEULEX.

It is interesting to study the influence of the perturbation terms on the extrapolated values. Suppose that $\alpha^j_{n_1}$ (or $\beta^j_{n_1}$) is the leading perturbation term in Y_{11} (or Z_{11}). Because of the recursion (IV.9.26) all Y_{kk} then contain an error term of the form $C_k H^j \alpha^j_{n_1}$, whereas the Y_{jk} (for $j > k$) do not depend on $\alpha^j_{n_1}$. The error constants C_k are given recursively by

$$C_1 = \frac{1}{n_1^j}, \qquad C_k = -\frac{n_1}{n_k - n_1} C_{k-1} \tag{5.24}$$

and tend to zero exponentially, if k increases.

Error Expansion for Singular Perturbation Problems

Our aim is to extend the analysis of Example 5.1 to general singular perturbation problems

$$\begin{aligned} y' &= f(y, z), \qquad & y(0) &= y_0 \\ \varepsilon z' &= g(y, z), \qquad & z(0) &= z_0, \quad 0 < \varepsilon \ll 1, \end{aligned} \tag{5.25}$$

where the solution $y(x)$, $z(x)$ is assumed to be sufficiently smooth (i.e., its derivatives up to a certain order are bounded independently of ε). An important observation in Example 5.1 was the existence of smooth solutions of the (linear) differential equations for the coefficients $b_i(x)$. In the general situation we shall be concerned with equations of the form

$$\begin{aligned} a' &= f_y(x)a + f_z(x)b + c(x, \varepsilon) \\ \varepsilon b' &= g_y(x)a + g_z(x)b + d(x, \varepsilon) \end{aligned} \tag{5.26}$$

(the coefficients $f_y(x) = f_y(y(x), z(x))$, etc. depend smoothly on ε because the solution of (5.25) itself depends on ε, even if f and g are ε-independent).

Lemma 5.5. *Suppose that the logarithmic norm of $g_z(x)$ satisfies*

$$\mu(g_z(x)) \le -1 \qquad \text{for} \quad 0 \le x \le \bar{x}. \tag{5.27}$$

For a given value

$$a(0) = a_0^0 + \varepsilon a_0^1 + \ldots + \varepsilon^N a_0^N + \mathcal{O}(\varepsilon^{N+1})$$

there exists a unique (up to $\mathcal{O}(\varepsilon^{N+1})$)

$$b(0) = b_0^0 + \varepsilon b_0^1 + \ldots + \varepsilon^N b_0^N + \mathcal{O}(\varepsilon^{N+1})$$

such that the solutions $a(x)$, $b(x)$ of (5.26) and their first N derivatives are bounded independently of ε.

Proof. We insert the finite expansions

$$\widehat{a}(x) = \sum_{i=0}^{N} \varepsilon^i a_i(x), \quad \widehat{b}(x) = \sum_{i=0}^{N} \varepsilon^i b_i(x)$$

with ε-independent coefficients $a_i(x)$ $b_i(x)$ into (5.26) and compare powers of ε (see Section VI.2). This leads to the differential-algebraic system (2.4). Consequently, a_0^0 determines b_0^0; these two together with a_0^1 determine b_0^1, etc. The remainders $a(x)-\widehat{a}(x)$, $b(x)-\widehat{b}(x)$ are then estimated as in the proof of Theorem 2.1. □

The next result exhibits the dominant perturbation terms in an asymptotic expansion of the error of the linearly implicit Euler method, when it is applied to a singular perturbation problem.

Theorem 5.6 (Hairer & Lubich 1988). *Assume that the solution of (5.25) is smooth. Under the condition*

$$\|(I-\gamma g_z(0))^{-1}\| \le \frac{1}{1+\gamma} \qquad \text{for all} \quad \gamma \ge 1 \tag{5.28}$$

(which is a consequence of (5.27) and Theorem IV.11.2), the numerical solution of (5.1) possesses for $\varepsilon \le h$ a perturbed asymptotic expansion of the form

$$\begin{aligned} y_i &= y(x_i) + ha_1(x_i) + h^2 a_2(x_i) + \mathcal{O}(h^3) \\ &\quad - \varepsilon f_z(0) g_z^{-1}(0) \Big(I - \frac{h}{\varepsilon} g_z(0)\Big)^{-i} \big(hb_1(0) + h^2 b_2(0)\big) \end{aligned} \tag{5.29}$$

$$\begin{aligned} z_i &= z(x_i) + hb_1(x_i) + h^2 b_2(x_i) + \mathcal{O}(h^3) \\ &\quad - \Big(I - \frac{h}{\varepsilon} g_z(0)\Big)^{-i} \big(hb_1(0) + h^2 b_2(0)\big) \end{aligned} \tag{5.30}$$

where $x_i = ih \le H$ with H sufficiently small (but independent of ε). The smooth functions $a_j(x)$, $b_j(x)$ satisfy

$$a_1(0) = \mathcal{O}(\varepsilon^2), \quad a_2(0) = \mathcal{O}(\varepsilon), \quad b_1(0) = \mathcal{O}(\varepsilon), \quad b_2(0) = \mathcal{O}(1).$$

Proof. This proof is organized like that of Theorem 5.3. In part (a) we recursively construct truncated expansions (for $M \le 2$)

$$\begin{aligned} \widehat{y}_i &= y(x_i) + \sum_{j=1}^{M} h^j \big(a_j(x_i) + \alpha_i^j\big) \\ \widehat{z}_i &= z(x_i) + \sum_{j=1}^{M} h^j \big(b_j(x_i) + \beta_i^j\big) \end{aligned} \tag{5.31}$$

such that

$$\begin{pmatrix} I - hf_y(0) & -hf_z(0) \\ -hg_y(0) & \varepsilon I - hg_z(0) \end{pmatrix} \begin{pmatrix} \widehat{y}_{i+1} - \widehat{y}_i \\ \widehat{z}_{i+1} - \widehat{z}_i \end{pmatrix} = h \begin{pmatrix} f(\widehat{y}_i, \widehat{z}_i) \\ g(\widehat{y}_i, \widehat{z}_i) \end{pmatrix} + \mathcal{O}(h^{M+2}). \tag{5.32}$$

The smooth functions $a_j(x)$, $b_j(x)$ clearly depend on ε, but are independent of h. The perturbation terms α_i^j, β_i^j (for $i \ge 1$), however, will depend smoothly on ε and on ε/h. As in the case $\varepsilon = 0$, we shall require that (5.15) and (5.16) hold. The differences $y_i - \widehat{y}_i$ and $z_i - \widehat{z}_i$ will then be estimated in part b).

a) The case $M = 0$ is obvious. Indeed, the values $\widehat{y}_i = y(x_i)$, $\widehat{z}_i = z(x_i)$ satisfy (5.32) with $M = 0$. The construction of the coefficients in (5.31) is done in two steps.

First step ($M = 1$). We insert (5.31) into (5.32) and compare the smooth coefficients of h^2. This gives

$$a_1'(x) + \frac{1}{2}y''(x) - f_y(0)y'(x) - f_z(0)z'(x) = f_y(x)a_1(x) + f_z(x)b_1(x) \tag{5.33a}$$

$$\varepsilon b_1'(x) + \frac{\varepsilon}{2}z''(x) - g_y(0)y'(x) - g_z(0)z'(x) = g_y(x)a_1(x) + g_z(x)b_1(x) \tag{5.33b}$$

By Lemma 5.5 the initial value $b_1(0)$ is uniquely determined by $a_1(0)$. Differentiation of $\varepsilon z' = g(y,z)$ with respect to x gives $\varepsilon z''(x) = g_y(x)y'(x) + g_z(x)z'(x)$. Inserted into (5.33b) this yields the relation

$$g_y(0)a_1(0) + g_z(0)b_1(0) = \mathcal{O}(\varepsilon) \tag{5.34}$$

with known right-hand side.

As to the perturbation terms, we obtain by collecting everything up to $\mathcal{O}(h^2)$

$$\begin{aligned} \alpha_{i+1}^1 - \alpha_i^1 - hf_y(0)(\alpha_{i+1}^1 - \alpha_i^1) - hf_z(0)(\beta_{i+1}^1 - \beta_i^1) \\ = hf_y(x_i)\alpha_i^1 + hf_z(x_i)\beta_i^1 \\ \varepsilon(\beta_{i+1}^1 - \beta_i^1) - hg_y(0)(\alpha_{i+1}^1 - \alpha_i^1) - hg_z(0)(\beta_{i+1}^1 - \beta_i^1) \\ = hg_y(x_i)\alpha_i^1 + hg_z(x_i)\beta_i^1 \end{aligned} \tag{5.35}$$

and try to determine the most important parts of this. We firstly replace $hf_y(x_i)\alpha_i^1$ by $hf_y(0)\alpha_i^1$ and similarly for three other terms. This is motivated by the fact that we search for exponentially decaying α_i. Therefore with $x_i = ih$,

$$\big(f_y(x_i) - f_y(0)\big)\alpha_i^1 = \mathcal{O}(h).$$

Then many terms cancel in (5.35). We next observe that $\beta_{i+1}^1 - \beta_i^1$ is multiplied by ε, but not $\alpha_{i+1}^1 - \alpha_i^1$. This suggests that the β_{i+1}^1 are an order of magnitude larger than α_{i+1}^1. Neglecting therefore α_{i+1}^1 where it competes with β_{i+1}^1, we are led to define

$$\alpha_{i+1}^1 - \alpha_i^1 = hf_z(0)\beta_{i+1}^1 \tag{5.33c}$$

$$\varepsilon(\beta_{i+1}^1 - \beta_i^1) = hg_z(0)\beta_{i+1}^1. \tag{5.33d}$$

It remains to verify a posteriori, that there exist solutions of (5.33a,b,c,d) which produce an error term $\mathcal{O}(h^3)$ in (5.32): from (5.33d) we obtain

$$\beta_i^1 = \Big(I - \frac{h}{\varepsilon}\, g_z(0)\Big)^{-i} \beta_0^1. \tag{5.36a}$$

Since we require $\alpha_i^1 \to 0$ for $i \to \infty$, the solution of (5.33c) is given by

$$\alpha_i^1 = \varepsilon f_z(0) g_z^{-1}(0)\Big(I - \frac{h}{\varepsilon}\, g_z(0)\Big)^{-i} \beta_0^1. \tag{5.36b}$$

For $i = 0$ this implies the relation

$$\alpha_0^1 = \varepsilon f_z(0) g_z^{-1}(0) \beta_0^1. \tag{5.37}$$

The assumption (5.15) together with (5.34) and (5.37) uniquely determine the coefficients $a_1(0)$, $b_1(0)$, α_0^1, β_0^1. We remark that $b_1(0) = \mathcal{O}(\varepsilon)$ and $a_1(0) = \mathcal{O}(\varepsilon^2)$. Using the fact that $\alpha_i^1 = \mathcal{O}(\varepsilon^2)$ and $\varepsilon \le h$, one easily verifies that the quantities (5.31) with $M = 1$ satisfy (5.32).

Second step $(M=2)$. Comparing the smooth coefficients of h^3 in (5.32) gives two differential equations for $a_2(x)$, $b_2(x)$ which are of the form (5.26). It follows from Lemma 5.5 that the initial values have to satisfy a relation

$$g_y(0) a_2(0) + g_z(0) b_2(0) = \mathcal{O}(1) \tag{5.38}$$

with known right-hand side. As in the first step we require for the perturbations

$$\begin{aligned} \alpha_{i+1}^2 - \alpha_i^2 &= h f_z(0) \beta_{i+1}^2 \\ \varepsilon(\beta_{i+1}^2 - \beta_i^2) &= h g_z(0) \beta_{i+1}^2. \end{aligned} \tag{5.39}$$

and obtain the formulas (5.36) and (5.37) with α_i^1, β_i^1 replaced by α_i^2, β_i^2. Again the values $a_2(0)$, $b_2(0)$, α_0^2, β_0^2 are uniquely determined by (5.15), (5.38), and (5.37). Due to the $\mathcal{O}(1)$ term in (5.38) we only have $b_2(0) = \mathcal{O}(1)$ and $a_2(0) = \mathcal{O}(\varepsilon)$.

We still have to verify (5.32) with $M = 2$. In the left-hand side we have neglected terms of the form $h f_y(0)(h\alpha_i^1 + h^2\alpha_i^2)$. This is justified, because $\alpha_i^1 = \mathcal{O}(\varepsilon^2)$, $\alpha_i^2 = \mathcal{O}(\varepsilon)$ and $\varepsilon \le h$. The most dangerous term, neglected in the right-hand side of (5.32) is

$$h\big(f_z(x_i) - f_z(0)\big)(h\beta_i^1 + h^2\beta_i^2). \tag{5.40}$$

However, $f_z(x_i) - f_z(0) = \mathcal{O}(ih)$, and $\beta_i^1 = \mathcal{O}(\varepsilon 2^{-i})$, $\beta_i^2 = \mathcal{O}(2^{-i})$ by (5.28) and $\varepsilon \le h$. This shows that the term (5.40) is also of size $\mathcal{O}(h^4)$, so that (5.32) holds with $M = 2$.

b) In order to estimate the remainder term, i.e., the differences $\Delta y_i = y_i - \widehat{y}_i$, $\Delta z_i = z_i - \widehat{z}_i$ we subtract (5.32) from (5.1) and eliminate Δy_{i+1} and Δz_{i+1}. This gives

$$\begin{pmatrix} \Delta y_{i+1} \\ \Delta z_{i+1} \end{pmatrix} = \begin{pmatrix} \Delta y_i \\ \Delta z_i \end{pmatrix}$$
$$+ \begin{pmatrix} I + \mathcal{O}(h) & \mathcal{O}(h) \\ \mathcal{O}(1) & \left(\frac{\varepsilon}{h} I - g_z(0)\right)^{-1} \end{pmatrix} \begin{pmatrix} h\big(f(y_i, z_i) - f(\widehat{y}_i, \widehat{z}_i)\big) \\ g(y_i, z_i) - g(\widehat{y}_i, \widehat{z}_i) \end{pmatrix} + \begin{pmatrix} \mathcal{O}(h^{M+2}) \\ \mathcal{O}(h^{M+1}) \end{pmatrix}.$$

Due to (5.28) and $\varepsilon \le h$ we have

$$\Big\| I + \Big(\frac{\varepsilon}{h} I - g_z(0)\Big)^{-1} g_z(0) \Big\| = \Big\| \Big(I - \frac{h}{\varepsilon} g_z(0)\Big)^{-1} \Big\| \le \frac{\varepsilon}{\varepsilon + h} \le \frac{1}{2}. \qquad (5.41)$$

We therefore again obtain (5.20) with some $|\varrho| < 1$, if H is sufficiently small. We then deduce the result as in the proof of Theorem 5.3. □

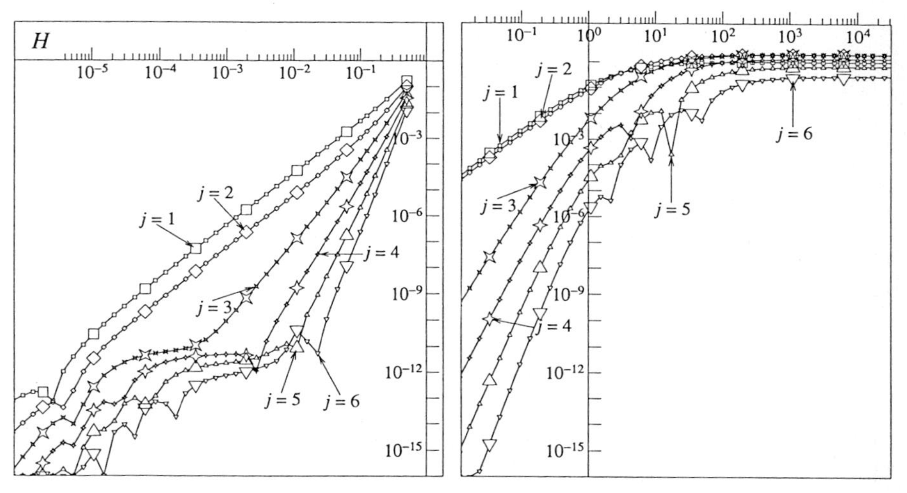

Fig. 5.1. Step size/error diagram **Fig. 5.2.** T_{jj} as functions of H/ε

Of course, it is possible to add a third step to the above proof. However, the recursions for α_i^3, β_i^3 are no longer as simple as in (5.33) or (5.39). Moreover, the perturbations of (5.29) and (5.30) already describe very well the situation encountered in practice. We shall illustrate this with the following numerical example (see also Hairer & Lubich 1988).

Consider van der Pol's equation (2.73) with $\varepsilon = 10^{-5}$ and with initial values (2.74) on the smooth solution. We take the step number sequence (5.23) and apply Method (5.1) n_j times with step size $h = H/n_j$. The numerical result Y_{j1}, Z_{j1} is then extrapolated according to (IV.9.26). In Fig. 5.1 we show in logarithmic scale the errors $|Z_{jj} - z(H)|$ for $j = 1, 2, \ldots, 6$ as functions of H. We observe that whenever the error is larger than $\varepsilon^2 = 10^{-10}$, the curves appear as straight lines with slopes $2, 2, 3, 4, 5$, and 6, respectively. If its slope is q, we have $\log(\mathit{error}) \approx$

$q \log H + Const$, or equivalently $error \approx CH^q$. This corresponds (with exception of the last one) to the orders predicted by the subdiagonal entries of Table 5.4 for the case $\varepsilon = 0$.

In order to understand the irregular behaviour of the curves when the error becomes smaller than $\varepsilon^2 = 10^{-10}$, we study the influence of extrapolation on the perturbation terms in (5.30). Since $b_1(0)$ contains a factor ε, the dominant part of the perturbation in Z_{j1} is $(I - (h/\varepsilon)g_z(0))^{-n_j} h^2 b_2(0)$, where $b_2(0)$ is some constant and $h = H/n_j$. We assume the matrix $g_z(0)$ to be diagonalized and put $g_z(0) = -1$. The dominant perturbation in Z_{j1} is therefore $\varepsilon^2 T_{j1} b_2(0)$, where

$$T_{j1} = \Big(\frac{H}{\varepsilon n_j}\Big)^2 \Big(1 + \frac{H}{\varepsilon n_j}\Big)^{-n_j}. \tag{5.42}$$

Due to the linearity of extrapolation, the dominant perturbation in Z_{jj} will be $\varepsilon^2 T_{jj} b_2(0)$, where T_{jj} is obtained from (5.42) and (IV.9.26). For the step number sequence (5.23) the values of T_{jj} are plotted as functions of H/ε in Fig. 5.2. For large values of H/ε the curves appear as horizontal lines. This is a consequence of our choice $n_1 = 2$ and of the fact that

$$T_{jj} = C_j \cdot \Big(\frac{H}{\varepsilon}\Big)^{2-n_1} + \mathcal{O}\Big(\Big(\frac{H}{\varepsilon}\Big)\Big)^{1-n_1} \qquad \text{for} \quad \frac{H}{\varepsilon} \to \infty,$$

where $C_1 = 1$ and the other C_j are given by the recursion (5.24).

The errors of Fig. 5.1 are now seen to be a superposition of the errors, predicted from the case $\varepsilon = 0$ (Theorem 5.4), and of the perturbations of Fig. 5.2 scaled by a factor $\mathcal{O}(\varepsilon^2)$.

Remark. As mentioned in Sect. VI.1, the *implicit Euler* discretization possesses a classical asymptotic expansion for differential-algebraic problems (1.6) of index 1 (case $\varepsilon = 0$). However, for singular perturbation problems, perturbations of the same type as in (5.29) and (5.30) are present. The only difference is that all $b_i(0)$ contain a factor ε for the implicit Euler method. For details and numerical experiments we refer to Hairer & Lubich (1988). A related analysis for a slightly different class of singular perturbation problems is presented in Auzinger, Frank & Macsek (1990).

Dense Output

Extrapolation methods typically take very large (basic) step sizes during integration. This makes it important that the method possess a continuous numerical solution. The first attempt to get a dense output for extrapolation methods is due to Lindberg (1972). His approach, however, imposes severe restrictions on the step number sequence. We present here the dense output of Hairer & Ostermann (1990), which exists for any step number sequence.

The main idea (due to Ch. Lubich) is the following: when computing the j-th entry of the extrapolation tableau, we consider not only $Y_{j1} = y_{n_j}$, but also compute the difference $(y_{n_j} - y_{n_j-1})/h_j$. Since these expressions possess an h-expansion, their extrapolation gives an accurate approximation to $y'(x_0 + H)$. By considering higher differences, we get also approximations to higher derivatives of $y(x)$ at $x_0 + H$. They are then used for Hermite interpolation. The reason for computing the derivatives only at the right end of the basic interval, is the presence of perturbation terms as described in Theorems 5.3 and 5.6. These perturbations are large at the beginning (near the initial value), but decrease exponentially for increasing i. For the same reason, one must not use differences of a too high an order. We thus choose an integer λ (usually 0 or 1) and avoid the values $y_0, \dots, y_{n_1+\lambda-2}$ for the computation of the finite differences. We remark that a similar idea was used by Deuflhard & Nowak (1987) to construct consistent initial values for differential-algebraic problems.

An algorithmic description of the dense output for the linearly implicit Euler method is as follows (we suppose that the value $Y_{\kappa\kappa}$ has been accepted as a numerical approximation to $y(x_0 + H)$).

Step 1. For each $j \in \{1, \dots, \kappa\}$ we compute

$$r_j^{(k)} = \frac{\nabla^k y_{n_j}^{(j)}}{h_j^k} \qquad \text{for} \quad k = 1, \dots, j - \lambda. \tag{5.43}$$

Here $y_i^{(j)}$ is the approximation of $y(x_i)$, obtained during the computation of Y_{j1}, and $\nabla y_i = y_i - y_{i-1}$ is the backward difference operator.

Step 2. We extrapolate $r_j^{(k)}$, $(\kappa - k - \lambda)$ times. This yields the improved approximation $r^{(k)}$ to $y^{(k)}(x_0 + H)$.

Step 3. We define the polynomial $P(\theta)$ of degree κ by

$$\begin{aligned} P(0) &= y_0, & P(1) &= Y_{\kappa\kappa} \\ P^{(k)}(1) &= H^k r^{(k)} & &\text{for } k = 1, \dots, \kappa - 1. \end{aligned} \tag{5.44}$$

The following theorem shows to which order these polynomials approximate the exact solution.

Theorem 5.7 (Hairer & Ostermann 1990). *Consider a nonstiff differential equation and let* $\lambda \in \{0, 1\}$. *Then, the error of the interpolation polynomial* $P(\theta)$ *satisfies*

$$P(\theta) - y(x_0 + \theta H) = \mathcal{O}(H^{\kappa+1-\lambda}) \qquad \textit{for} \quad H \to 0.$$

Proof. Since $P(\theta)$ is a polynomial of degree κ, the error due to interpolation is of size $\mathcal{O}(H^{\kappa+1})$. We know that $Y_{\kappa\kappa} - y(x_0 + H) = \mathcal{O}(H^{\kappa+1})$. Therefore it suffices to prove that

$$r^{(k)} = y^{(k)}(x_0 + H) + \mathcal{O}(H^{\kappa-k-\lambda+1}) \qquad \text{for} \quad k = 1, \dots, \kappa - 1. \tag{5.45}$$

Due to the asymptotic expansion of the global error $y_i - y(x_i)$, the approximations $r_j^{(k)}$ also have an expansion of the form

$$r_j^{(k)} = y^{(k)}(x_0 + H) + h_j a_1^{(k)} + h_j^2 a_2^{(k)} + \dots. \tag{5.46}$$

The statement (5.45) now follows from the fact that each extrapolation eliminates one power of h in (5.46). □

It is now natural to investigate the error of the dense output $P(\theta)$ also for stiff differential equations, such as singular perturbation problems. We shall treat here the limit case $\varepsilon = 0$ which is easier to analyse and, nevertheless, gives much insight into the structure of the error for very stiff problems.

For the differential-algebraic system (1.6) one defines the dense output in exactly the same way as for ordinary differential equations. As the system (1.6) is partitioned into y- and z-components, it is convenient to denote the corresponding interpolation polynomials by $P(\theta)$ and $Q(\theta)$, respectively.

Theorem 5.8 (Hairer & Ostermann 1990). *Let $y(x)$, $z(x)$ be the solution of (1.6). Suppose that the step number sequence satisfies $n_1 + \lambda \geq 2$ with $\lambda \in \{0,1\}$. We then have*

$$\begin{aligned} P(\theta) - y(x_0 + \theta H) &= \mathcal{O}(H^{\kappa+1-\lambda}) + \mathcal{O}(H^{r+1}), \\ Q(\theta) - z(x_0 + \theta H) &= \mathcal{O}(H^{\kappa+1-\lambda}) + \mathcal{O}(H^{s}), \end{aligned} \tag{5.47}$$

where r and s are the $(\kappa + n_1 + \lambda - 2, \kappa)$-entries of Table 5.3 and Table 5.4, respectively.

Proof. We use the perturbed asymptotic error expansions of Theorem 5.3. Their smooth terms are treated exactly as in the proof of Theorem 5.7 and yield the $\mathcal{O}(H^{\kappa+1-\lambda})$ error term in (5.47). The second error terms in (5.47) are due to the perturbations in (5.11). We observe that the computation of $r_j^{(k)}$ involves only y_i (or z_i) with $i \geq n_j - j + \lambda$. Since $n_j - j \geq n_1 - 1$, the values $y_0, \ldots, y_{n_1+\lambda-2}$ do not enter into the formulas for $r_j^{(k)}$, so that the dominant perturbation comes from $y_{n_1+\lambda-1}$ (or $z_{n_1+\lambda-1}$). □

It is interesting to note that for $\lambda = 1$, the second errror term in (5.47) is of the same size as that in the numerical solution $Y_{\kappa\kappa}$, $Z_{\kappa\kappa}$ (see Theorem 5.4). However, one power of H is lost in the first term of (5.47). On the other hand, one H may be lost in the second error term, if $\lambda = 0$. Both choices lead to a cheap (no additional function evaluations) and accurate dense output. Its order for $\theta \in (0,1)$ is at most one lower than the order obtained for $\theta = 1$.

Exercises

1. The linearly implicit mid-point rule, applied to the differential-algebraic system (1.6), reads

$$\begin{pmatrix} I-hf_y(0) & -hf_z(0) \\ -hg_y(0) & -hg_z(0) \end{pmatrix} \begin{pmatrix} y_{i+1}-y_i \\ z_{i+1}-z_i \end{pmatrix} \tag{5.48}$$
$$= -\begin{pmatrix} I+hf_y(0) & hf_z(0) \\ hg_z(0) & hg_z(0) \end{pmatrix} \begin{pmatrix} y_i-y_{i-1} \\ z_i-z_{i-1} \end{pmatrix} + 2h \begin{pmatrix} f(y_i,z_i) \\ g(y_i,z_i) \end{pmatrix}.$$

If we compute y_1, z_1 from (5.2), and if we define the numerical solution at x_0+H $(H=2mh)$ by

$$y_h(x_0+H)=\tfrac{1}{2}(y_{2m+1}+y_{2m-1}), \qquad z_h(x_0+H)=\tfrac{1}{2}(z_{2m+1}+z_{2m-1}),$$

this algorithm constitutes an extension of (IV.9.16) to differential-algebraic problems.

a) Show that this method integrates the problem (5.8) exactly.

b) Apply the algorithm to

$$y'=1, \quad 0=u-y^2, \quad 0=v-yu, \quad 0=w-yv, \quad 0=z-yw$$

with zero initial values and verify the formula

$$\tfrac{1}{2}\left(z_{2m+1}+z_{2m-1}\right)-z(x_{2m}) = -10x_{2m}^3h^2+9x_{2m}h^4 - (-1)^m\left(\tfrac{1}{8}\,x_{2m}^5+x_{2m}^3h^2+9x_{2m}h^4\right).$$

Remark. The error of the z-component thus contains an h-independent term of size $\mathcal{O}(H^5)$, which is not affected by extrapolation.

2. Consider the method of Exercise 1 as the basis of an h^2-extrapolation method. Prove that for the step number sequence (IV.9.22) the extrapolated values satisfy

$$Y_{jk}-y(x_0+H)=\mathcal{O}(H^{r_{jk}+1}), \qquad Z_{jk}-z(x_0+H)=\mathcal{O}(H^{s_{jk}})$$

with r_{jk}, s_{jk} given in Tables 5.5 and 5.6.

Hint. Interpret Y_{j1}, Z_{j1} as numerical solution of a Rosenbrock method (Exercise 3 of Sect. IV.9) and verify the order condition derived in Sect. VI.3 (see also Hairer & Lubich (1988b) and C. Schneider (1993)).

Table 5.5. orders r_{jk}.

1						
1	3					
1	3	5				
1	3	5	7			
1	3	5	7	7		
1	3	5	7	7	7	
1	3	5	7	7	7	7

Table 5.6. orders s_{jk}.

2						
2	4					
2	4	5				
2	4	5	5			
2	4	5	5	5		
2	4	5	5	5	5	
2	4	5	5	5	5	5

VI.6 Quasilinear Problems

Quasilinear differential equations are usually understood to be equations in which the highest derivative appears linearly. In the case of first order ODE systems, they are of the form

$$C(y) \cdot y' = f(y), \tag{6.1}$$

where $C(y)$ is a $n \times n$-matrix. In the regions where $C(y)$ is invertible, Eq. (6.1) can be written as

$$y' = C(y)^{-1} \cdot f(y) \tag{6.1'}$$

and every ODE-code can be applied by solving at every function call a linear system. But this would destroy, for example, a banded structure of the Jacobian and it is therefore often preferable to treat Eq. (6.1) directly. If the matrix C is everywhere of rank m $(m < n)$, Eq. (6.1) represents a quasilinear differential-algebraic system.

Example: Moving Finite Elements

As an example, we present the classical idea of "moving finite elements", described in K. Miller & R.N. Miller (1981): the solution $u(x,t)$ of a nonlinear partial differential equation

$$\frac{\partial u}{\partial t} = L\big(u(x,t)\big), \quad u(0,t) = u(1,t) = 0, \tag{6.2}$$

where $L(u)$ is an unbounded nonlinear differential operator, is approximated by finite element polygons $v(x, a_1, s_1, \ldots, a_n, s_n)$ which satisfy $v(s_j, \ldots) = a_j$ (see Fig. 6.1). These polygons form a $2n$-dimensional manifold in the Hilbert space $L^2(0,1)$ parametrized by $a_1, s_1, \ldots, a_n, s_n$. The idea is now to *move simultaneously* $a(t)$ and $s(t)$ in oder to adapt at any time the finite element solution as best as possible to Eq. (6.2).

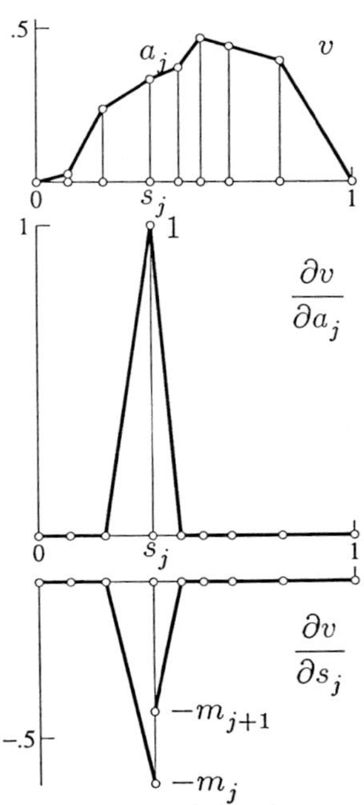

Fig. 6.1. Moving finite elements

We thus require that the defect $\dot{v} - L(v)$ remains always orthogonal to the tangent space. The conditions

$$\int_0^1 \big(\dot{v} - L(v)\big) \cdot \frac{\partial v}{\partial a_j}\, dx = 0 \qquad \int_0^1 \big(\dot{v} - L(v)\big) \cdot \frac{\partial v}{\partial s_j}\, dx = 0 \tag{6.3}$$

lead to a system of type (6.1) with

$$\begin{aligned} c_{2j-1,2k-1} &= \textstyle\int_0^1 \frac{\partial v}{\partial a_j} \cdot \frac{\partial v}{\partial a_k}\, dx, & c_{2j-1,2k} &= \textstyle\int_0^1 \frac{\partial v}{\partial a_j} \cdot \frac{\partial v}{\partial s_k}\, dx, \\ c_{2j,2k-1} &= \textstyle\int_0^1 \frac{\partial v}{\partial s_j} \cdot \frac{\partial v}{\partial a_k}\, dx, & c_{2j,2k} &= \textstyle\int_0^1 \frac{\partial v}{\partial s_j} \cdot \frac{\partial v}{\partial s_k}\, dx, \\ f_{2j-1} &= \textstyle\int_0^1 L(v) \cdot \frac{\partial v}{\partial a_j}\, dx, & f_{2j} &= \textstyle\int_0^1 L(v) \cdot \frac{\partial v}{\partial s_j}\, dx. \end{aligned} \tag{6.4}$$

For the partial derivatives of v, sketched in Fig. 6.1, the non-zero of these scalar products become

$$\begin{aligned} c_{2j-1,2j-1} &= \tfrac{1}{3}(\Delta_j + \Delta_{j+1}) & c_{2j-1,2j} &= -\tfrac{1}{3}(m_j\Delta_j + m_{j+1}\Delta_{j+1}) \\ c_{2j,2j-1} &= -\tfrac{1}{3}(m_j\Delta_j + m_{j+1}\Delta_{j+1}) & c_{2j,2j} &= \tfrac{1}{3}(m_j^2\Delta_j + m_{j+1}^2\Delta_{j+1}) + 2\varepsilon^2 \end{aligned} \tag{6.5a}$$

$$\begin{aligned} c_{2j-1,2j+1} &= c_{2j+1,2j-1} = \tfrac{1}{6}\Delta_{j+1} & c_{2j-1,2j+2} &= c_{2j+2,2j-1} = -\tfrac{1}{6}m_{j+1}\Delta_{j+1} \\ c_{2j,2j+1} &= c_{2j+1,2j} = -\tfrac{1}{6}m_{j+1}\Delta_{j+1} & c_{2j,2j+2} &= c_{2j+2,2j} = \tfrac{1}{6}m_{j+1}^2\Delta_{j+1} - \varepsilon^2 \end{aligned} \tag{6.5b}$$

where

$$\Delta_j = s_j - s_{j-1}, \quad m_j = (a_j - a_{j-1})/\Delta_j, \qquad j = 1, \ldots, n+1.$$

The matrix $C(y)$ is banded with bandwidth $3+1+3$. The ε^2-terms in (6.5) come from an "internodal viscosity" penalty term, explained in Miller & Miller (1981), which aims to regularize the relative movement of the nodes s_j whenever their position is ill-conditioned, which happens to appear in the vicinity of inflection points (see Fig. 6.2).

It is then hoped that the nodes move automatically into the critical regions of the solutions, move with shocks which may appear, and that $a(t)$ and $s(t)$ become smooth functions.

Application to Burgers' Equation. Burgers' Equation is given by

$$u_t = -uu_x + \mu u_{xx} \qquad \text{or} \qquad u_t = -\Big(\frac{u^2}{2}\Big)_x + \mu u_{xx} \tag{6.6}$$

where $\mu = 1/R$ and R is called te Reynolds number. This is one of the equations originally designed by Burgers (1948) as "a mathematical model illustrating the theory of turbulence". However, soon afterwards, E. Hopf (1950) presented an analytical solution (see Exercise 1 below) and concluded that "we doubt that Burgers' equation fully illustrates the statistics of free turbulence. (...) Equation (1) is too simple a model to display chance fluctuations ...". Nowadays it remains interesting as a nonlinear equation resembling the Navier-Stokes' equations in fluid dynamics which possesses, for R large, shock waves and, for $R \to \infty$, discontinuous solutions. Here, the integrals in (6.4) become

$$f_{2j-1} = A_j + \mu B_j, \quad f_{2j} = C_j + \mu D_j, \qquad j = 1, \dots n. \tag{6.5c}$$

where

$$\begin{aligned}
A_j &= (a_{j-1} - a_j)(\tfrac{1}{3}a_j + \tfrac{1}{6}a_{j-1}) + (a_j - a_{j+1})(\tfrac{1}{3}a_j + \tfrac{1}{6}a_{j+1}),\\
B_j &= (m_{j+1} - m_j),\\
C_j &= -m_j(a_{j-1} - a_j)(\tfrac{1}{3}a_j + \tfrac{1}{6}a_{j-1}) - m_{j+1}(a_j - a_{j+1})(\tfrac{1}{3}a_j + \tfrac{1}{6}a_{j+1}),\\
D_j &= (m_{j+1} - m_j)(\tfrac{1}{2}m_{j+1} + \tfrac{1}{2}m_j),
\end{aligned} \tag{6.5d}$$

(in the case of D_j appears the product of a Dirac δ function with a discontinuous function; these must be suitably "mollified"). We choose as initial function

$$u(x,0) = (\sin(3\pi x))^2 \cdot (1-x)^{3/2}, \qquad \mu = 0.0003 \tag{6.7}$$

and as initial positions

$$s_j = j/(n+1), \quad a_j = u(s_j, 0), \qquad j = 1, \dots, n, \quad n = 100,$$

and solve the problem with smoothing parameter $\varepsilon = 10^{-2}$ for $0 \le t \le 1.9$. Two shock waves arise which later fuse into one (see Fig. 6.2).

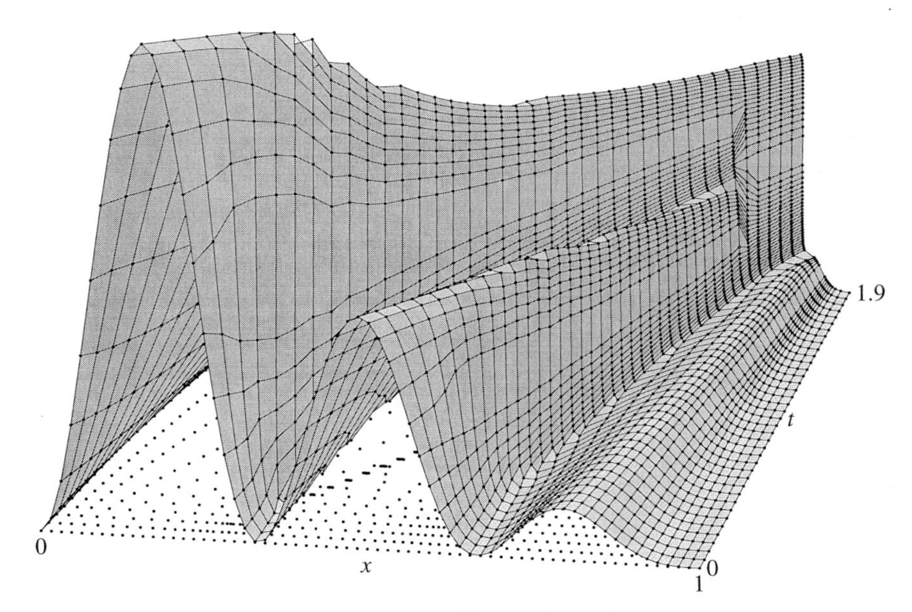

Fig. 6.2. Moving finite element solution of Burgers' equation

Problems of Index One

For invertible $C(y)$, Eq. (6.1) is an ordinary differential equation and standard theory (for existence and uniqueness results) can be applied. If the matrix is ill-conditioned or even singular, new investigations are necessary. In order to exclude equations with singularities, such as $xy' = (q+bx)y$ (see Sect. I.5), we assume that

$$C(y) \text{ has constant rank } m \ (m < n) \tag{6.8}$$

in a neighbourhood of the solution. Then the columns of $C(y)$ span an m-dimensional subspace $\mathcal{I}m\, C(y)$ which moves with y. Clearly, in order that (6.1) can make sense, we need consistent initial values, i.e., we need

$$f(y_0) \in \mathcal{I}m\, C(y_0). \tag{6.9}$$

We shall now show, how, under a certain condition, this property can be satisfied for all x and determines uniquely the solution: choose a nonsingular matrix

$$T(y) = \begin{pmatrix} T_1(y) \\ T_2(y) \end{pmatrix} \qquad \text{such that} \qquad T(y)C(y) = \begin{pmatrix} B_1(y) \\ 0 \end{pmatrix}; \tag{6.10}$$

this means that the rows of $T_2(y)$ must span the $(n-m)$-dimensional orthogonal complement of $\mathcal{I}m\, C(y)$. Then we multiply Eq. (6.1) by $T(y)$ and obtain

$$\begin{pmatrix} B_1(y) \\ 0 \end{pmatrix} y' = \begin{pmatrix} T_1(y)f(y) \\ T_2(y)f(y) \end{pmatrix}, \tag{6.11}$$

so that the condition corresponding to (6.9) becomes visible in the form $T_2(y)f(y) = 0$. Differentiating this relation and inserting the derivative into the second part of (6.11), we obtain

$$\begin{pmatrix} B_1(y) \\ (T_2 f)'(y) \end{pmatrix} y' = \begin{pmatrix} (T_1 f)(y) \\ 0 \end{pmatrix}, \tag{6.12}$$

which is a *regular* quasilinear equation if the matrix

$$\begin{pmatrix} B_1(y) \\ (T_2 f)'(y) \end{pmatrix} \qquad \text{is invertible.} \tag{6.13}$$

Lemma 6.1. *Let the matrix $C(y)$ satisfy (6.8) and (6.13), and let the initial values y_0 fulfill (6.9). Then, the quasilinear problem $C(y)y' = f(y)$, $y(x_0) = y_0$ possesses a locally unique solution.*

Proof. Condition (6.9) means that $T_2(y_0)f(y_0) = 0$ and the second part of (6.12) assures that $(T_2(y(x))f(y(x)))' = 0$. Therefore we have $(T_2 f)(y(x)) = 0$ for all x, and the solution of (6.12) solves also (6.11) and (6.1). □

The following result gives a consequence of condition (6.13) which shall be essential in the later discussions of feasibility of numerical procedures.

Lemma 6.2. *Assume that $C(y)$ satisfies (6.8) and (6.13). If $f(y_0) = C(y_0)y_0'$, then the matrix*

$$C(y) + \lambda\big(f'(y_0) - \Gamma(y_0, y_0')\big)$$

is invertible for sufficiently small $\lambda \neq 0$ and for y sufficiently close to y_0. Here,

$$\Gamma(y, y') = \frac{\partial}{\partial y}\big(C(y)y'\big).$$

Proof. Condition (6.13) implies that

$$T(y)C(y) + \lambda(Tf)'(y_0) \qquad \text{is invertible} \tag{6.14}$$

for small $\lambda \neq 0$ and y close to y_0. Using $T'C + TC' = B'$ we have

$$(T'f)(y_0) = T'(y_0)C(y_0)y_0' = -T(y_0)\Gamma(y_0, y_0') + B'(y_0)y_0'. \tag{6.15}$$

Since $B'(y_0)y_0'$ does not contribute to the lower block of the matrix (6.14), it can be neglected after insertion of $(Tf)' = Tf' + T'f$ and (6.15) into (6.14). This implies that

$$T(y)C(y) + \lambda T(y_0)\big(f'(y_0) - \Gamma(y_0, y_0')\big) \qquad \text{is invertible.}$$

The statement of the Lemma now follows from a continuity argument. □

Numerical Treatment of $C(y)y' = f(y)$

As has been said above, in the case of invertible matrices $C(y)$, one can eventually apply an explicit numerical method to (6.1'). However, if Eq. (6.1') is stiff, implicit methods have to be applied. In this case it may be advantageous to have methods that avoid the computation of the Jacobian of $C(y)^{-1}f(y)$.

Transformation to Semi-Explicit Form. In the case where (6.1') is stiff or where $C(y)$ is singular and satisfies (6.8) and (6.13) we introduce $z = y'$ as new variable, such that system (6.1) becomes of the semi-explicit form

$$\begin{aligned} y' &= z \\ 0 &= C(y)z - f(y) \end{aligned} \tag{6.16}$$

Here, all methods of the preceding sections can be applied (at least formally). The study of convergence, however, needs further investigation, because Condition (1.7) is no longer satisfied here.

Implicit Runge-Kutta and Multistep Methods. With the ε-embedding approach (see (1.11) for Runge-Kutta methods and (2.2) for multistep methods) we are led to nonlinear equations, which, when solved by Newton iterations, require the solution of linear systems of the form

$$\begin{pmatrix} -\alpha I & I \\ \Gamma(y_0,z_0)-f'(y_0) & C(y_0) \end{pmatrix}. \tag{6.17}$$

Here $\alpha=(\gamma h)^{-1}$, and γ is an eigenvalue of the Runge-Kutta matrix. By Lemma 6.2 this matrix is invertible for small enough $h>0$. Convergence follows from the results of Sections VII.3 and VII.4 (see Exercise 2).

Rosenbrock Methods. Method (4.3) applied to system (6.16) leads to

$$\begin{pmatrix} v_i \\ w_i \end{pmatrix} = \begin{pmatrix} y_0 \\ z_0 \end{pmatrix} + \sum_{j=1}^{i-1} \alpha_{ij} \begin{pmatrix} k_j \\ \ell_j \end{pmatrix}, \qquad \begin{pmatrix} y_1 \\ z_1 \end{pmatrix} = \begin{pmatrix} y_0 \\ z_0 \end{pmatrix} + \sum_{i=1}^{s} b_i \begin{pmatrix} k_i \\ \ell_i \end{pmatrix}. \tag{6.18a}$$

$$\begin{pmatrix} k_i \\ 0 \end{pmatrix} = h \begin{pmatrix} w_i \\ C(v_i)w_i - f(v_i) \end{pmatrix} + h \begin{pmatrix} 0 & I \\ \Gamma_0 - f_0' & C(y_0) \end{pmatrix} \sum_{j=1}^{i} \gamma_{ij} \begin{pmatrix} k_j \\ \ell_j \end{pmatrix}. \tag{6.18b}$$

Again, it can be seen that (6.18b) represents a linear system whose regularity is assured by Lemma 6.2. However, since Condition (1.7) is not satisfied, a new theory for the order conditions of the local error as well as for convergence of the global error is necessary. This theory reveals, for example, that new order conditions for the coefficients are necessary and explains why, say, the code RODAS, directly applied to (6.16), does not give precise results. For full details we refer the reader to the original publication Lubich & Roche (1990).

Extrapolation Methods

The first problem is to find suitable linearly implicit Euler discretizations for (6.1), to serve as basic method for the extrapolation algorithm (see Sect. IV.9).

Method of Deuflhard & Nowak. Applying the linearly implicit Euler method (IV.9.15) to the differential equation (6.1') we obtain

$$(I-hA)(y_{i+1}-y_i)=hC(y_i)^{-1}f(y_i) \tag{6.19}$$

where

$$A\approx(C^{-1}f)'(y_0)=C(y_0)^{-1}\big(f'(y_0)-\Gamma(y_0,y_0')\big)$$

with $\Gamma(y,y')$ as in Lemma 6.2. Multiplication of (6.19) with $C(y_i)$ yields

$$\big(C(y_i)-hC(y_i)C(y_0)^{-1}J\big)(y_{i+1}-y_i)=hf(y_i)$$

with $J=f'(y_0)-\Gamma(y_0,y_0')$. Deuflhard & Nowak (1987) suggest to replace $C(y_i)C(y_0)^{-1}$ by the identity matrix, which "may be interpreted as just introducing an approximation error into the Jacobian matrix". This leads to the discretization

$$\big(C(y_i)-hJ\big)(y_{i+1}-y_i)=hf(y_i) \tag{6.20}$$

which represents the basic step for the code LIMEX described in Deuflhard & Nowak (1987). The regularity of the matrix of this linear system is again assured by Lemma 6.2.

The computation of J requires an approximation to $z_0 = y_0'$. Such consistent initial values must be computed explicitely for the first basic steps, and are obtained by extrapolation of

$$z_n = (y_n - y_{n-1})/h \tag{6.21}$$

in the subsequent steps.

Linearly-Implicit Euler to Semi-Explicit Model. Another possibility is to apply the linearly-implicit Euler discretization (5.2) to the differential-algebraic system (6.16). This gives

$$\begin{pmatrix} I & -hI \\ -hJ & hC(y_0) \end{pmatrix} \begin{pmatrix} y_{i+1} - y_i \\ z_{i+1} - z_i \end{pmatrix} = h \begin{pmatrix} z_i \\ f(y_i) - C(y_i) z_i \end{pmatrix} \tag{6.22}$$

with $z_0 = y_0' = y'(x_0)$. The first line yields $z_{i+1} = (y_{i+1} - y_i)/h$ and the second line becomes

$$\bigl(C(y_0) - hJ\bigr)(y_{i+1} - y_i) = hf(y_i) - \bigl(C(y_i) - C(y_0)\bigr)(y_i - y_{i-1}). \tag{6.23}$$

The right-most term vanishes for $i = 0$, so that y_{-1} does not enter the algorithm.

Asymptotic Expansions. The theoretical justification of the use of either (6.20) or (6.23) as basic step for an extrapolation process requires the investigation of the asymptotic expansion of their global errors.

In the situation where $C(y)$ is invertible, the discretization (6.20) is a consistent one-step discretization of (6.1') and possesses therefore, by standard theory (Theorem II.8.1), an asymptotic expansion, the terms of which, however, depend on the stiffness. Since the system (6.16) is of the form (1.6) with assumption (1.7) satisfied, we can conclude from Theorem 5.3 the existence of a *perturbed* asymptotic expansion for the second discretization (6.23).

In the situation where $C(y)$ is singular, Lubich (1989) reveiled the existence of a perturbed asymptotic expansion for both discretizations (6.20) and (6.23). We refer to this original publication for further details, in particular to the study of the influence of these perturbations to the extrapolated numerical approximation.

Exercises

1. Reconstruct E. Hopf's analytic solution of Burgers' equation (6.6).

   Hint. Introduce the new dependent variable

$$\varphi(x,t) = \exp\Bigl\{-\frac{1}{2\mu}\int_0^x u(\xi,t)\,d\xi - \int_0^t c(\tau)\,d\tau\Bigr\}.$$

Show that for a suitably chosen $c(t)$ the function $\varphi(x,t)$ satisfies the one dimensional heat equation. The solution $u(x,t)$ of (6.6) can then be recovered from $\varphi(x,t)$ by

$$u = -2\mu(\log\varphi)_x = -2\mu(\varphi_x/\varphi) .$$

2. Assume that (6.8) and (6.13) hold. By eliminating from $0 = C(y)z - f(y)$ as many components of z as possible, transform the system (6.16) into an equivalent one of the form

$$y' = F(y,u), \qquad 0 = G(y),$$

where u collects the remaining components of z.

a) Prove that Runge-Kutta methods and multistep methods are invariant with respect to this transformation.

b) Show that $G_y(y)F_u(y,u)$ is invertible, so that the convergence results of Sections VII.3 and VII.4 can be applied.

3. (Quasilinear problems with gradient-type mass-matrix, see HLR89, page 111). Consider the electrical circuit (1.14), but suppose this time that the capacities depend on the voltages, e.g., as

$$C_k = C_{k0}/(1-(U_i-U_j)/U_b)^{1/2}$$

so that the expressions $C_k(U_i'-U_j')$ in (1.14) must be replaced by $(C_k(U_i-U_j))'$. Show that then the corresponding equations are of the form (6.1) with

$$C(y) = Aq'(y)$$

where A is a constant matrix and $q(y)$ a known function of y. Show that such problems can be efficiently solved by introducing $q(y) = z$ as a new variable such that the problem becomes semi-explicit as

$$Az' = f(y), \qquad 0 = z - q(y).$$

Chapter VII. Differential-Algebraic Equations of Higher Index

(Drawing by K. Wanner)

In the preceding chapter we considered the simplest special case of differential-algebraic equations – the so-called index 1 problem. Many problems of practical interest are, however, of higher index, which makes them more and more difficult for their numerical treatment.

We start by classifying differential-algebraic equations (DAE) by the index (index of nilpotency for linear problems with constant coefficients; differentiation and perturbation index for general nonlinear problems) and present some examples arising in applications (Sect. VII.1). Several different approaches for solving numerically higher index problems are discussed in Sect. VII.2: index reduction by differentiation combined with suitable projections, state space form methods, and treatment as overdetermined or unstructured systems. Sections VII.3 and VII.4 study the convergence properties of multistep methods and Runge-Kutta methods when they are applied directly to index 2 systems. It may happen that the order of convergence is lower than for ordinary differential equations ("order reduction"). The study of conditions which guarantee a certain order is the subject of Sect. VII.5. Half-explicit methods for index 2 problems are especially suited for constrained mechanical systems (Sect. VII.6). A multibody mechanism and its numerical treatment are detailed in Sect. VII.7. Finally, we discuss symplectic methods for constrained Hamiltonian systems (Sect. VII.8), and explain their long-term behaviour by a backward error analysis for differential equations on manifolds.

VII.1 The Index and Various Examples

The most general form of a differential-algebraic system is that of an implicit differential equation

$$F(u',u)=0 \tag{1.1}$$

where F and u have the same dimension. We always assume F to be sufficiently differentiable. A non-autonomous system is brought to the form (1.1) by appending x to the vector u, and by adding the equation $x'=1$.

If $\partial F/\partial u'$ is invertible we can formally solve (1.1) for u' to obtain an ordinary differential equation. In this chapter we are interested in problems (1.1) where $\partial F/\partial u'$ is singular.

Linear Equations with Constant Coefficients

> Uebrigens kann ich die Meinung des Hrn. *Jordan* nicht theilen, dass es ziemlich schwer sei, der *Weierstrass*-schen Analyse zu folgen; sie scheint mir im Gegentheil vollkommen durchsichtig zu sein, ... (L. Kronecker 1874)

The simplest and best understood problems of the form (1.1) are linear differential equations with constant coefficients

$$Bu'+Au=d(x). \tag{1.2}$$

In looking for solutions of the form $e^{\lambda x}u_0$ (if $d(x)\equiv 0$) we are led to consider the "matrix pencil" $A+\lambda B$. When $A+\lambda B$ is singular for all values of λ, then (1.2) has either no solution or infinitely many solutions for a given initial value (Exercise 1). We shall therefore deal only with *regular matrix pencils*, i.e., with problems where the polynomial $\det(A+\lambda B)$ does not vanish identically. The key to the solution of (1.2) is the following simultaneous transformation of A and B to canonical form.

Theorem 1.1 (Weierstrass 1868, Kronecker 1890). *Let $A+\lambda B$ be a regular matrix pencil. Then there exist nonsingular matrices P and Q such that*

$$PAQ=\begin{pmatrix} C & 0\\ 0 & I\end{pmatrix},\qquad PBQ=\begin{pmatrix} I & 0\\ 0 & N\end{pmatrix} \tag{1.3}$$

where $N = \operatorname{blockdiag}(N_1, \ldots, N_k)$, *each* N_i *is of the form*

$$N_i = \begin{pmatrix} 0 & 1 & & 0 \\ & \ddots & \ddots & \\ & & 0 & 1 \\ 0 & & & 0 \end{pmatrix}, \quad \textit{of dimension } m_i \,, \tag{1.4}$$

and C *can be assumed to be in Jordan canonical form.*

Proof (Gantmacher 1954 (Chapter XII), see also Exercises 2 and 3). We fix some c such that $A + cB$ is invertible. If we multiply

$$A + \lambda B = A + cB + (\lambda - c)B$$

by the inverse of $A + cB$ and then transform $(A + cB)^{-1}B$ to Jordan canonical form (Theorem I.12.2) we obtain

$$\begin{pmatrix} I & 0 \\ 0 & I \end{pmatrix} + (\lambda - c)\begin{pmatrix} J_1 & 0 \\ 0 & J_2 \end{pmatrix}. \tag{1.5}$$

Here, J_1 contains the Jordan blocks with non-zero eigenvalues, J_2 those with zero eigenvalues (the dimension of J_1 is just the degree of the polynomial $\det(A + \lambda B)$). Consequently, J_1 and $I - cJ_2$ are both invertible and multiplying (1.5) from the left by $\operatorname{blockdiag}(J_1^{-1}, (I - cJ_2)^{-1})$ gives

$$\begin{pmatrix} J_1^{-1}(I - cJ_1) & 0 \\ 0 & I \end{pmatrix} + \lambda \begin{pmatrix} I & 0 \\ 0 & (I - cJ_2)^{-1}J_2 \end{pmatrix}.$$

The matrices $J_1^{-1}(I - cJ_1)$ and $(I - cJ_2)^{-1}J_2$ can then be brought to Jordan canonical form. Since all eigenvalues of $(I - cJ_2)^{-1}J_2$ are zero, we obtain the desired decomposition (1.3). □

Theorem 1.1 allows us to solve (1.2) as follows: we premultiply (1.2) by P and use the transformation

$$u = Q\begin{pmatrix} y \\ z \end{pmatrix}, \qquad Pd(x) = \begin{pmatrix} \eta(x) \\ \delta(x) \end{pmatrix}.$$

This decouples the differential-algebraic system (1.2) into

$$y' + Cy = \eta(x), \qquad Nz' + z = \delta(x). \tag{1.6}$$

The equation for y is just an ordinary differential equation. The relation for z decouples again into k subsystems, each of the form (with $m = m_i$)

$$\begin{aligned} z_2' + z_1 &= \delta_1(x) \\ &\vdots \\ z_m' + z_{m-1} &= \delta_{m-1}(x) \\ z_m &= \delta_m(x). \end{aligned} \tag{1.7}$$

Here z_m is determined by the last equation, and the other components are obtained recursively by repeated differentiation. Thus z_1 depends on the $(m-1)$-th derivative of $\delta_m(x)$. Since numerical differentiation is an unstable procedure, the largest m_i appearing in (1.4) is a measure of numerical difficulty for solving problem (1.2). This integer $(\max m_i)$ is called the *index of nilpotency* of the matrix pencil $A + \lambda B$. It does not depend on the particular transformation used to get (1.3) (see Exercise 4).

Linear Equations with Variable Coefficients. In the case, where the matrices A and B in (1.2) depend on x, the study of the solutions is much more complicated. Multiplying the equation by $P(x)$ and substituting $u = Q(x)v$, yields the system

$$PBQv' + (PAQ + PBQ')v = 0, \tag{1.8}$$

which shows that the transformation (1.3) is no longer relevant. With the use of transformations of the form (1.8), Kunkel & Mehrmann (1995) derive a canonical form for linear systems with variable coefficients.

Differentiation Index

A lot of English cars have steering wheels.
(*Fawlty Towers*, Cleese and Booth 1979)

Let us start with the following example:

$$\begin{aligned} y_1' &= 0.7 \cdot y_2 + \sin(2.5 \cdot z) = f_1(y, z) \\ y_2' &= 1.4 \cdot y_1 + \cos(2.5 \cdot z) = f_2(y, z) \end{aligned} \tag{1.9a}$$

$$0 = y_1^2 + y_2^2 - 1 = g(y). \tag{1.9b}$$

The "control variable" z in (1.9a) can be interpreted as the position of a "steering wheel" keeping the vector field (y_1', y_2') tangent to the circle $y_1^2 + y_2^2 = 1$, so that condition (1.9b) remains continually satisfied (see Fig. 1.1a). By differentiating (1.9b) and substituting (1.9a) we therefore must have

$$g_y(y)f(y, z) = 0. \tag{1.9c}$$

This defines a "hidden" submanifold of the cylinder, on which all solutions of (1.9a,b) must lie (see Fig. 1.1b). We still do not know how, with increasing x, the variable z changes. This is obtained by differentiating (1.9c) with respect to x: $g_{yy}(f, f) + g_y f_y f + g_y f_z z' = 0$. From this relation we can extract

$$z' = -(g_y f_z)^{-1}\Big(g_{yy}(f, f) + g_y f_y f\Big) \tag{1.9d}$$

if

$$g_y(y)f_z(y, z) \qquad \text{is invertible.} \tag{1.10}$$

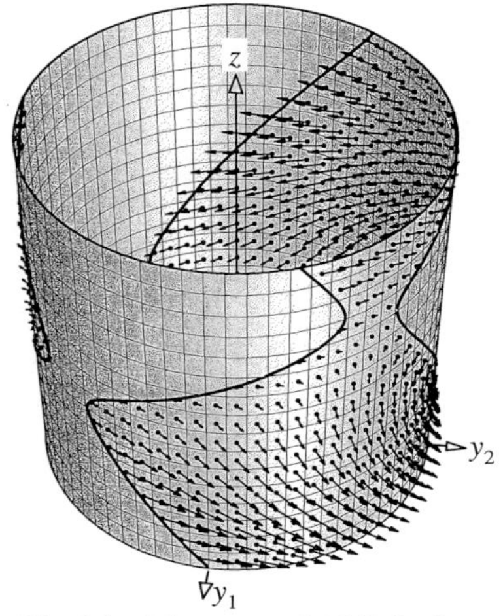

Fig. 1.1a. The vector field (1.9a,d)

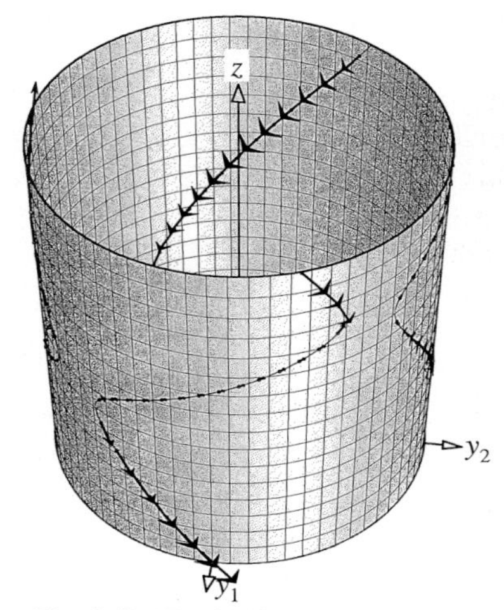

Fig. 1.1b. The hidden submanifold

We have been able to transform the above differential-algebraic equation (1.9a,b) into an ordinary differential system (1.9a,d) by *two analytic differentiations* of the constraint (1.9c). This fact is used for the following definition, which has been developed in several papers (Gear & Petzold 1983, 1984; Gear, Gupta & Leimkuhler 1985, Gear 1990, Campbell & Gear 1995).

Definition 1.2. Equation (1.1) has *differentiation index* $di = m$ if m is the minimal number of analytical differentiations

$$F(u',u)=0, \quad \frac{dF(u',u)}{dx}=0\,, \quad \ldots\,, \quad \frac{d^m F(u',u)}{dx^m}=0 \tag{1.11}$$

such that equations (1.11) allow us to extract by algebraic manipulations an explicit ordinary differential system $u' = \varphi(u)$ (which is called the "*underlying ODE*").

Examples. *Linear Equations with Constant Coefficients.* The following problem

$$\begin{aligned} z_2' + z_1 &= \delta_1 \\ z_3' + z_2 &= \delta_2 \\ z_3 &= \delta_3 \end{aligned} \quad \Rightarrow \quad \begin{aligned} z_2'' + z_1' &= \delta_1' \\ z_3''' + z_2'' &= \delta_2'' \\ z_3''' &= \delta_3''' \end{aligned} \quad \Rightarrow \quad z_1' = \delta_1' - \delta_2'' + \delta_3''' \tag{1.12}$$

can be seen to have differentiation index 3. For linear equations with constant coefficients the differentiation index and the index of nilpotency are therefore the same.

Systems of Index 1. The differential-algebraic systems already seen in Chapter VI

$$y' = f(y,z) \tag{1.13a}$$
$$0 = g(y,z) \tag{1.13b}$$

have no z'. We therefore differentiate (1.13b) to obtain

$$z' = -g_z^{-1}(y,z)g_y(y,z)f(y,z) \tag{1.13c}$$

which is possible if g_z is invertible in a neighbourhood of the solution. The problem (1.13a,b), for invertible g_z, is thus of differentiation index 1.

Systems of Index 2. In the system (see example (1.9))

$$y' = f(y,z) \tag{1.14a}$$
$$0 = g(y), \tag{1.14b}$$

where the variable z is absent in the algebraic constraint, we obtain by differentiation of (1.14b) the "hidden constraint"

$$0 = g_y(y)f(y,z). \tag{1.14c}$$

If (1.10) is satisfied in a neighbourhood of the solution, then (1.14a) and (1.14c) constitute an index 1 problem. Differentiation of (1.14c) yields the missing differential equation for z, so that the problem (1.14a,b) is of differentiation index 2. If the initial values satisfy $0 = g(y_0)$ and $0 = g_y(y_0)f(y_0, z_0)$, we call them *consistent*. In this case, and only in this case, the system (1.14a,b) possesses a (locally) unique solution.

System (1.14a,b) is a representative of the larger class of problems of type (1.13a,b) with *singular* g_z. If we assume that g_z has constant rank in a neighbourhood of the solution, we can eliminate certain algebraic variables from $0 = g(y,z)$ until the system is of the form (1.14). This can be done as follows: from the constant rank assumption it follows that either there exists a component of g such that $\partial g_i/\partial z_1 \neq 0$ locally, or $\partial g/\partial z_1$ vanishes identically so that g is already independent of z_1. In the first case we can express z_1 as a function of y and the remaining components of z, and then we can eliminate z_1 from the system. Repeating this procedure with z_2, z_3, etc., will lead to a system of the form (1.14). This transformation does not change the index. Moreover, most numerical methods are invariant under this transformation. Therefore, theoretical work done for systems of the form (1.14) will also be valid for more general problems.

Systems of Index 3. Problems of the form

$$y' = f(y,z) \tag{1.15a}$$
$$z' = k(y,z,u) \tag{1.15b}$$
$$0 = g(y) \tag{1.15c}$$

are of differentiation index 3, if

$$g_y f_z k_u \qquad \text{is invertible} \tag{1.16}$$

in a neighbourhood of the solution. Differentiating (1.15c) twice gives

$$0 = g_y f \tag{1.15d}$$
$$0 = g_{yy}(f,f) + g_y f_y f + g_y f_z k. \tag{1.15e}$$

Equations (1.15a,b), (1.15e) together with Condition (1.16) are of the index 1 form (1.13a,b). Consistent inital values must satisfy the three conditions (1.15c,d,e).

An extensive study of the solution space of general differential-algebraic systems is done by Griepentrog & März (1986), März (1989, 1990). These authors try to avoid assumptions on the smoothness on the problem as far as possible and replace the above differentiations by a careful study of suitable projections depending only on the first derivatives of F.

Differential Equations on Manifolds

In the language of differentiable manifolds, whose use in DAE theory was urged by Rheinboldt (1984), a constraint (such as $g(y)=0$) represents a manifold, which we denote by

$$\mathcal{M}=\{y\in\mathbb{R}^n\mid g(y)=0\}. \tag{1.17}$$

We assume that $g:\mathbb{R}^n\to\mathbb{R}^m$ (with $m<n$) is a sufficiently differentiable function whose Jacobian $g_y(y)$ has full rank for $y\in\mathcal{M}$. For a fixed $y\in\mathcal{M}$ we denote by

$$T_y\mathcal{M}=\{v\in\mathbb{R}^n\mid g_y(y)v=0\}, \tag{1.18}$$

the tangent space of $\mathcal{M}$ at y. This is a linear space and has the same dimension $n-m$ as the manifold $\mathcal{M}$.

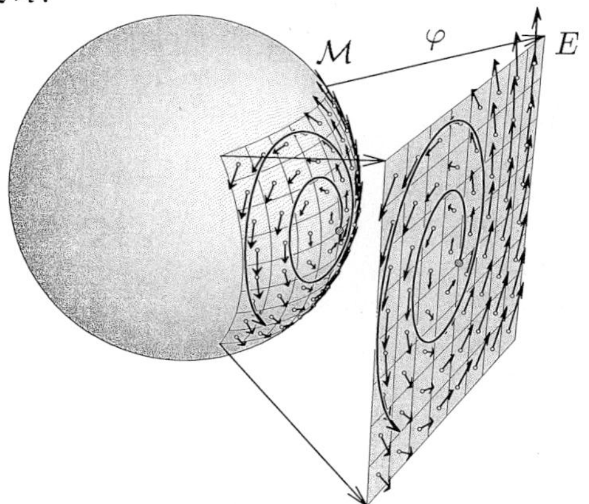

Fig. 1.2. A manifold with a tangent vector field, a chart, and a solution curve

A *vector field on* $\mathcal{M}$ is a mapping $v:\mathcal{M}\to\mathbb{R}^n$, which satisfies $v(y)\in T_y\mathcal{M}$ for all $y\in\mathcal{M}$. For such a vector field we call

$$y'=v(y),\qquad y\in\mathcal{M} \tag{1.19}$$

a *differential equation on the manifold* $\mathcal{M}$. Differentiation on an $(n-m)$-dimensional manifold is described by so-called *charts* $\varphi_i:U_i\to E_i$, where the U_i cover the manifold $\mathcal{M}$ and the E_i are open subsets of $\mathbb{R}^{n-m}$ (Fig. 1.2; see also Lang (1962), Chap. II and Abraham, Marsden & Ratiu (1983), Chap. III). The local theory of ordinary differential equations can be extended to vector fields on manifolds in a straightforward manner:

> Project the vectors $v(y)$ onto E_i via a chart φ_i by multiplying $v(y)$ with the Jacobian of φ_i at y. Then apply standard results to the projected vector field in $\mathbb{R}^{n-m}$, and pull the solution back to $\mathcal{M}$.

(see Fig. 1.2). The local existence of solutions of (1.19) can be shown in this way. The obtained solution is independent of the chosen chart. Where the solution leaves the domain of a chart, the integration must be continued via another one.

Index 2 Problems. Consider the system (1.14a,b) and suppose that (1.10) is satisfied. This condition implies that $g_y(y)$ is of full rank, so that (1.17) is a smooth manifold. Moreover, the Implicit Function Theorem implies that the differentiated constraint (1.14c) can be solved for z (in a neighbourhood of the solution), i.e., there exists a smooth function $h(y)$ such that

$$g_y(y)f(y,z)=0 \qquad \Longleftrightarrow \qquad z=h(y). \tag{1.20}$$

Inserting this relation into (1.10a) yields

$$y'=f\big(y,h(y)\big), \qquad y\in\mathcal{M} \tag{1.21}$$

which is a differential equation on the manifold (1.17), because $f(y,h(y))\in T_y\mathcal{M}$ by (1.20). The differential equation (1.21) is equivalent to (1.14a,b).

Example. The manifold $\mathcal{M}$ for problem (1.9) is one-dimensional (circle). In points, where $y_1\neq\pm 1$, we can solve (1.9b) to obtain locally $y_2=\pm\sqrt{1-y_1^2}$. The map $(y_1,y_2)\mapsto y_1$ consitutes a chart φ, which is bijective in a neighbourhood of the considered point. Inserting z from (1.9c) and the above y_2 into (1.9a), yields an equation $y_1'=G(y_1)$, which is the projected vector field in $\mathbb{R}^1$.

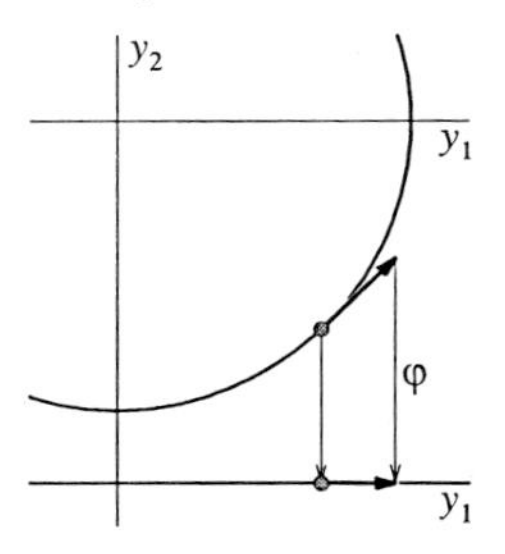

Index 3 Problems. For the system (1.15a,b,c) the solutions lie on the manifold

$$\mathcal{M}=\{(y,z)\mid g(y)=0,\;\; g_y(y)f(y,z)=0\}. \tag{1.22}$$

The assumption (1.16) implies that $g_y(y)$ and $g_y(y)f_z(y,z)$ have full rank, so that $\mathcal{M}$ is a manifold. Its tangent space at (y,z) is

$$\begin{aligned}T_{(y,z)}\mathcal{M}=\{(v,w)\mid g_y(y)v=0,\;\; & g_{yy}(y)\big(f(y,z),v\big)\\ & +g_y(y)\big(f_y(y,z)v+f_z(y,z)w\big)=0\}.\end{aligned} \tag{1.23}$$

Solving Eq. (1.15e) for u and inserting the result into (1.15b) yields a differential equation on the manifold $\mathcal{M}$. Because of (1.15d,e), the obtained vector field lies in the tangent space $T_{(y,z)}\mathcal{M}$ for all $(y,z)\in\mathcal{M}$.

The Perturbation Index

Now fills thy sleep with perturbations.
(The *Ghost of Anne* in Shakespeare's *Richard III*, act V, sc. III)

A second concept of index, due to HLR89 [1], interprets the index as a measure of sensitivity of the solutions with respect to perturbations of the given problem.

Definition 1.3. Equation (1.1) has *perturbation index pi* $=m$ along a solution $u(x)$ on $[0,\overline{x}]$, if m is the smallest integer such that, for all functions $\widehat{u}(x)$ having a defect

$$F(\widehat{u}',\widehat{u})=\delta(x), \tag{1.24}$$

there exists on $[0,\overline{x}]$ an estimate

$$\|\widehat{u}(x)-u(x)\|\leq C\Big(\|\widehat{u}(0)-u(0)\|+\max_{0\leq\xi\leq x}\|\delta(\xi)\|+\ldots+\max_{0\leq\xi\leq x}\|\delta^{(m-1)}(\xi)\|\Big) \tag{1.25}$$

whenever the expression on the right-hand side is sufficiently small.

Remark. We deliberately do not write "Let $\widehat{u}(x)$ be the solution of $F(\widehat{u}',\widehat{u})=\delta(x)$ ..." in this definition, because the existence of such a solution $\widehat{u}(x)$ for an arbitrarily given $\delta(x)$ is not assured. We start with $\widehat{u}$ and then compute δ as defect of (1.1).

Systems of Index 1. For the computation of the perturbation index of (1.13a,b) we consider the perturbed system

$$\widehat{y}'=f(\widehat{y},\widehat{z})+\delta_1(x) \tag{1.26a}$$
$$0=g(\widehat{y},\widehat{z})+\delta_2(x). \tag{1.26b}$$

The essential observation is that the difference $\widehat{z}-z$ can be estimated with the help of the Implicit Function Theorem, without any differentiation of the equation. Since g_z is invertible by hypothesis, this theorem gives from (1.26b) compared to (1.13b)

$$\|\widehat{z}(x)-z(x)\|\leq C_1\big(\|\widehat{y}(x)-y(x)\|+\|\delta_2(x)\|\big) \tag{1.27}$$

as long as the right-hand side of (1.27) is sufficiently small. We now subtract (1.26a) from (1.13a), integrate from 0 to x, use a Lipschitz condition for f and the above estimate for $\widehat{z}(x)-z(x)$. This gives for $e(x)=\|\widehat{y}(x)-y(x)\|$:

$$e(x)\leq e(0)+C_2\int_0^x e(t)dt+C_3\int_0^x\|\delta_2(t)\|dt+\Big\|\int_0^x\delta_1(t)dt\Big\|.$$

In this estimate the norm is *inside* the integral for δ_2, but *outside* the integral for δ_1. This is due to the fact that perturbations of the algebraic equation (1.13b) are more

[1] The "Lecture Notes" of Hairer, Lubich & Roche (1989) will be cited frequently in the subsequent sections. Reference to this publication will henceforth be denoted by HLR89.

serious than perturbations of the differential equation (1.13a). We finally apply Gronwall's Lemma (Exercise I.10.2) to obtain on a bounded interval $[0, \overline{x}]$

$$\begin{aligned} \|\widehat{y}(x) - y(x)\| &\le C_4 \Big(\|\widehat{y}(0) - y(0)\| + \int_0^x \|\delta_2(t)\| dt + \max_{0\le\xi\le x} \Big\| \int_0^\xi \delta_1(t) dt \Big\| \Big) \\ &\le C_5 \Big(\|\widehat{y}(0) - y(0)\| + \max_{0\le\xi\le x} \|\delta_2(\xi)\| + \max_{0\le\xi\le x} \|\delta_1(\xi)\| \Big). \end{aligned}$$

This inequality, together with (1.27), shows that the perturbation index of the problem is 1.

Systems of Index 2. We consider the following perturbation of system (1.14a,b)

$$\widehat{y}' = f(\widehat{y}, \widehat{z}) + \delta(x) \qquad (1.28a)$$

$$0 = g(\widehat{y}) + \theta(x). \qquad (1.28b)$$

Differentiation of (1.28b) gives

$$0 = g_y(\widehat{y}) f(\widehat{y}, \widehat{z}) + g_y(\widehat{y}) \delta(x) + \theta'(x). \qquad (1.29)$$

Under the assumption (1.10) we can use the estimates of the index 1 case (with $\delta_2(x)$ replaced by $g_y(\widehat{y}(x))\delta(x) + \theta'(x)$) to obtain

$$\begin{aligned} \|\widehat{y}(x) - y(x)\| &\le C \Big(\|\widehat{y}(0) - y(0)\| + \int_0^x \big(\|\delta(\xi)\| + \|\theta'(\xi)\| \big) d\xi \Big) \\ \|\widehat{z}(x) - z(x)\| &\le C \Big(\|\widehat{y}(0) - y(0)\| + \max_{0\le\xi\le x} \|\delta(\xi)\| + \max_{0\le\xi\le x} \|\theta'(\xi)\| \Big). \end{aligned} \qquad (1.30)$$

Since these estimates depend on the first derivative of θ, the perturbation index of this problem is 2. A sharper estimate for the y-component is given in Exercise 6.

Example. Fig. 1.3 presents an illustration for the index 2 problem (1.9a,b). Small perturbations of $g(y)$, once discontinuous in the first derivative (left), the other of oscillatory type (right), results in discontinuities or violent oscillations of z, respectively.

The above examples might give the impression that the differentiation index and the perturbation index are always equal. The following counter-examples show that this is not true.

Counterexamples. The first counterexample of type $M(y)y' = f(y)$ is given by Lubich (1989):

$$\begin{aligned} y_1' - y_3 y_2' + y_2 y_3' &= 0 \qquad & \widehat{y}_1' - \widehat{y}_3 \widehat{y}_2' + \widehat{y}_2 \widehat{y}_3' &= 0 \\ y_2 &= 0 & \widehat{y}_2 &= \varepsilon \sin \omega x \\ y_3 &= 0 & \widehat{y}_3 &= \varepsilon \cos \omega x \end{aligned} \qquad (1.31)$$

with $y_i(0) = 0$ $(i = 1, 2, 3)$. Inserting $\widehat{y}_2 = \varepsilon \sin \omega x$ and $\widehat{y}_3 = \varepsilon \cos \omega x$ into the first equation gives $\widehat{y}_1' = \varepsilon^2 \omega$ which makes, for ε fixed and $\omega \to \infty$, an estimate

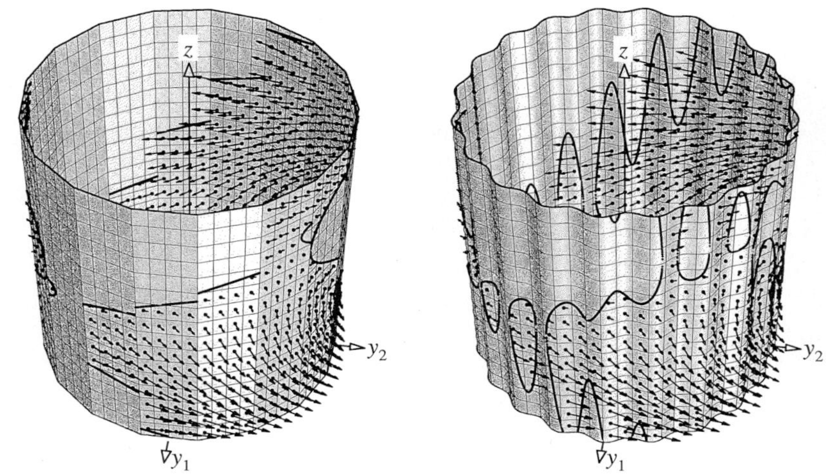

Fig. 1.3. Perturbations of an index 2 problem

(1.25) with $m=1$ impossible. However, for $m=2$ the estimate (1.25) is clearly satisfied. This problem, which is obviously of differentiation index 1, is thus of perturbation index 2.

It was believed for some time (see the first edition, p. 479), that the differentiation and perturbation indices can differ at most by 1. The following example, due to Campbell & Gear (1995), was therefore a big surprise:

$$y_m N y' + y = 0, \tag{1.32}$$

where N is a $m \times m$ upper triangular nilpotent Jordan block. Since the last row of N is zero, we have $y_m = 0$, and the differentiation index is 1. On the other hand, adding a perturbation makes y_m different from zero. This is the reason why the perturbation index of (1.32) is m.

Control Problems

Many problems of control theory lead to ordinary differential equations of the form $y' = f(y,u)$, where u represents a set of controls. Similar as in example (1.9) above, these controls must be applied so that the solution satisfies some constraints $0 = g(y,u)$. For numerical examples of such control problems we refer to Brenan (1983) (space shuttle simulation) and Brenan, Campbell & Petzold (1989).

Optimal Control Problems are differential equations $y' = f(y,u)$ formulated in such a way that the control $u(x)$ has to minimize some cost functional. The Euler–Lagrange equation then often becomes a differential-algebraic system (Pontryagin, Boltyanskij, Gamkrelidze & Mishchenko 1961, Athans & Falb 1966, Campbell 1982). We demonstrate this on the problem

$$y' = f(y,u), \qquad y(0) = y_0 \tag{1.33a}$$

with cost functional

$$J(u)=\int_0^1 \varphi\big(y(x),u(x)\big)\,dx. \tag{1.33b}$$

For a given function $u(x)$ the solution $y(x)$ is determined by (1.33a). In order to find conditions for $u(x)$ which minimize $J(u)$ of (1.33b), we consider the perturbed control $u(x)+\varepsilon\delta u(x)$ where $\delta u(x)$ is an arbitrary function and ε a small number. To this control there corresponds a solution $y(x)+\varepsilon\delta y(x)+\mathcal{O}(\varepsilon^2)$ of (1.33a); hence (by comparing powers of ε)

$$\delta y'(x)=f_y(x)\delta y(x)+f_u(x)\delta u(x),\qquad \delta y(0)=0, \tag{1.34}$$

where, as usual, $f_y(x)=f_y(y(x),u(x))$, etc. Linearization of (1.33b) shows that

$$J(u+\varepsilon\delta u)-J(u)=\varepsilon\int_0^1\Big(\varphi_y(x)\delta y(x)+\varphi_u(x)\delta u(x)\Big)\,dx+\mathcal{O}(\varepsilon^2)$$

so that

$$\int_0^1\Big(\varphi_y(x)\delta y(x)+\varphi_u(x)\delta u(x)\Big)\,dx=0 \tag{1.35}$$

is a necessary condition for $u(x)$ to be an optimal solution of our problem. In order to express δy in terms of δu in (1.35), we introduce the adjoint differential equation

$$v'=-f_y(x)^T v-\varphi_y(x)^T,\qquad v(1)=0 \tag{1.36}$$

with inhomogeneity $\varphi_y(x)^T$. Hence we have (see Exercise 7)

$$\int_0^1 \varphi_y(x)\delta y(x)dx=\int_0^1 v^T(x)f_u(x)\delta u(x)dx. \tag{1.37}$$

Inserted into (1.35) this gives the necessary condition

$$\int_0^1\Big(v^T(x)f_u(x)+\varphi_u(x)\Big)\delta u(x)dx=0. \tag{1.38}$$

Since this relation has to be satisfied for all δu we obtain the necessary relation $v^T(x)f_u(x)+\varphi_u(x)=0$ by the so-called "fundamental lemma of variational calculus".

In summary, we have proved that a solution of the above optimal control problem has to satisfy the system

$$\begin{aligned} y'&=f(y,u), & y(0)&=y_0\\ v'&=-f_y(y,u)^T v-\varphi_y(y,u)^T, & v(1)&=0\\ 0&=v^T f_u(y,u)+\varphi_u(y,u). \end{aligned} \tag{1.39}$$

This is a boundary value differential-algebraic problem. It can also be obtained directly from the Pontryagin minimum principle (see Pontryagin et al. 1961, Athans & Falb 1966).

Differentiation of the algebraic relation in (1.39) shows that the system (1.39) has index 1 if the matrix

$$\sum_{i=1}^{n} v_i \frac{\partial^2 f_i}{\partial u^2}(y,u) + \frac{\partial^2 \varphi}{\partial u^2}(y,u) \tag{1.40}$$

is invertible along the solution. A situation where the system (1.39) has index 3 is presented in Exercise 8. An index 5 problem of this type is given in "Example 3.1" of Clark (1988). Other control problems with a large index are discussed in Campbell (1995).

Mechanical Systems

> ... berechnen wir T, V, L. Mehr brauchen wir von der Geometrie und Mechanik unseres Systems nicht zu wissen. Alles übrige besorgt ohne unser Zutun der Formalismus von LAGRANGE.
>
> (Sommerfeld 1942, §35)

An interesting class of differential-algebraic systems appears in mechanical modeling of constrained systems. A choice method for deriving the equations of motion of mechanical systems is the Lagrange-Hamilton principle, whose long history goes back to merely theological ideas of Leibniz and Maupertuis. Let $q_1, \ldots, q_n$ be position coordinates of a system and $u_i = \dot{q}_i$ the velocities. Suppose a function $L(q, \dot{q})$ is given; then the Euler equations of the variational problem

$$\int_{t_1}^{t_2} L(q,\dot{q})dt = \min! \tag{1.41}$$

are given by

$$\frac{d}{dt}\left(\frac{\partial L}{\partial \dot{q}_k}\right) - \frac{\partial L}{\partial q_k} = 0, \qquad k = 1, \ldots, n \tag{1.42}$$

or

$$\sum_{\ell=1}^{n} L_{\dot{q}_k \dot{q}_\ell} \ddot{q}_\ell = L_{q_k} - \sum_{\ell=1}^{n} L_{\dot{q}_k q_\ell} \dot{q}_\ell . \tag{1.43}$$

The great discovery of Lagrange (1788) is that for $L = T - U$, where T is the *kinetic energy* and U the *potential energy*, the differential equations (1.43) describe the movement of the corresponding "conservative system". For a proof and various generalizations, consult any book on mechanics e.g., Sommerfeld (1942), vol. I, §§ 33–37, or Arnol'd (1979), part II.

Example 1. For the mathematical pendulum of length ℓ we choose as position coordinate the angle $\theta = q_1$ such that $T = m\ell^2\dot{\theta}^2/2$ and $U = -\ell mg\cos\theta$. Then (1.43) becomes $\ell\ddot{\theta} = -g\sin\theta$, the well-known pendulum equation.

Movement with Constraints. Suppose now that we have some constraints $g_1(q) = 0, \dots, g_m(q) = 0$ on our movement. Another great idea of Lagrange is to vary the "Lagrange function" as follows in this case

$$L = T - U - \lambda_1 g_1(q) - \dots - \lambda_m g_m(q) \tag{1.44}$$

where the "Lagrange multipliers" λ_i are appended to the coordinates. The important fact is that, since L is independent of $\dot{\lambda}_i$, the equation (1.43), for the derivatives with respect to λ_k, just becomes $0 = g_k(q)$, the desired side conditions.

Example 2. We now describe the pendulum in Cartesian coordinates x, y with constraint $x^2 + y^2 - \ell^2 = 0$. This gives for (1.44)

$$L = \frac{m}{2}(\dot{x}^2 + \dot{y}^2) - mgy - \lambda(x^2 + y^2 - \ell^2)$$

and (1.43) becomes

$$\begin{aligned} m\ddot{x} &= -2x\lambda \\ m\ddot{y} &= -mg - 2y\lambda \\ 0 &= x^2 + y^2 - \ell^2. \end{aligned} \tag{1.45}$$

In this example the physical meaning of λ is the tension in the rod which maintains the mass point on the desired orbit.

The general form of a constrained mechanical system (1.43) is in vector notation (after replacing dots by primes)

$$q' = u \tag{1.46a}$$

$$M(q)u' = f(q, u) - G^T(q)\lambda \tag{1.46b}$$

$$0 = g(q) \tag{1.46c}$$

where $M(q) = T_{\dot{q}\dot{q}} = T_{uu}$ is a positive definite matrix, $G(q) = \partial g / \partial q$ and $q = (q_1, \dots, q_n)^T$, $u = (\dot{q}_1, \dots, \dot{q}_n)^T$, $\lambda = (\lambda_1, \dots, \lambda_m)^T$. Various formulations are possible for such a problem, each of which leads to a different numerical approach.

Index 3 Formulation (position level, descriptor form). If we formally multiply (1.46b) by M^{-1}, the system (1.46) becomes of the form (1.15) with (q, u, λ) in the roles of (y, z, u). The condition (1.16), written out for (1.46), is

$$GM^{-1}G^T \qquad \text{is invertible}\,. \tag{1.47}$$

This is satisfied, if the constraints (1.46c) are independent, i.e., if the rows of the matrix G are linearly independent. Under this assumption, the system (1.46a,b,c) is thus an index 3 problem.

Index 2 Formulation (velocity level). Differentiation of (1.46c) gives

$$0 = G(q)u. \tag{1.46d}$$

If we replace (1.46c) by (1.46d) we obtain a system of the form (1.14a,b) with (q,u) in the role of y and λ that of z. One verifies that Condition (1.10) is equivalent to (1.47), so that (1.46a,b,d) represents a problem of index 2.

Index 1 Formulation (acceleration level). If we differentiate twice the constraint (1.46c), the resulting equation together with (1.46b) yield

$$\begin{pmatrix} M(q) & G^T(q) \\ G(q) & 0 \end{pmatrix} \begin{pmatrix} u' \\ \lambda \end{pmatrix} = \begin{pmatrix} f(q,u) \\ -g_{qq}(q)(u,u) \end{pmatrix}. \tag{1.46e}$$

This allows us to express u' and λ as functions of q,u, provided that the matrix in (1.46e) is invertible. Hence, (1.46a,e) consitute an index 1 problem. The assumption on the matrix in Eq. (1.46e) is weaker than (1.47), because $M(q)$ need not be regular.

All these formulations are mathematically equivalent, if the initial values are consistent, i.e., if (1.46c,d,e) are satisfied. However, if for example the index 2 system (1.46a,b,d) is integrated numerically, the constraints of the original problem will no longer be exactly satisfied. For this reason Gear, Gupta & Leimkuhler (1985) introduced another index 2 formulation ("... an interesting way of reducing the problem to index two and adding variables so that the constraint continues to be satisfied".).

GGL Formulation. The idea is to add the constraint (1.46d) to the original system and to introduce an additional Lagrange multiplier μ in (1.46a). For the sake of symmetry we also multiply (1.46a) by $M(q)$, so that the whole system becomes

$$\begin{aligned} M(q)q' &= M(q)u - G^T(q)\mu \\ M(q)u' &= f(q,u) - G^T(q)\lambda \\ 0 &= g(q) \\ 0 &= G(q)u. \end{aligned} \tag{1.48}$$

Here the differential variables are (q,u) and the algebraic variables are (μ,λ). System (1.48) is of the form (1.14a,b) and the index 2 assumption is satisfied if (1.47) holds.

A concrete mechanical system is described in detail, together with numerical results for all the above formulations, in Sect. VII.7.

Exercises

1. Prove that the initial value problem

$$Bu' + Au = 0, \qquad u(0) = u_0 \tag{1.49}$$

has a unique solution if and only if the matrix pencil $A + \lambda B$ is regular.

Hint for the "only if" part. If n is the dimension of u, choose arbitrarily $n+1$ distinct λ_i and vectors $v_i \neq 0$ satisfying $(A+\lambda_i B)v_i = 0$. Then take a linear combination, such that $\sum \alpha_i v_i = 0$, but $\sum \alpha_i e^{\lambda_i x} v_i \neq 0$.

2. (Stewart 1972). Let $A+\lambda B$ be a regular matrix pencil. Show that there exist unitary matrices Q and Z such that

$$QAZ = \begin{pmatrix} A_{11} & A_{12} \\ 0 & A_{22} \end{pmatrix}, \qquad QBZ = \begin{pmatrix} B_{11} & B_{12} \\ 0 & B_{22} \end{pmatrix} \tag{1.50}$$

are both triangular. Further, the diagonal elements of A_{22} and B_{11} are all 1, those of B_{22} are all 0.

Hint (Compare with the Schur decomposition of Theorem I.12.1). Let λ_1 be a zero of $\det(A+\lambda B)$ and $v_1 \neq 0$ be such that $(A+\lambda_1 B)v_1 = 0$. Verify that $Bv_1 \neq 0$ and that

$$AZ_1 = Q_1 \begin{pmatrix} -\lambda_1 & * \\ 0 & \widetilde{A} \end{pmatrix}, \qquad BZ_1 = Q_1 \begin{pmatrix} 1 & * \\ 0 & \widetilde{B} \end{pmatrix}$$

where Q_1, Z_1 are unitary matrices whose first columns are Bv_1 and v_1, respectively. The matrix pencil $\widetilde{A}+\lambda\widetilde{B}$ is again regular and this procedure can be continued until $\det(\widetilde{A}+\lambda\widetilde{B}) = Const$ which implies that $\det\widetilde{B} = 0$. In this case we take a vector $v_2 \neq 0$ such that $\widetilde{B}v_2 = 0$ and transform $\widetilde{A}+\lambda\widetilde{B}$ with unitary matrices Q_2, Z_2, whose first columns are $\widetilde{A}v_2$ and v_2, respectively. For a practical computation of the decomposition (1.50) see Golub & Van Loan (1989), Sect. 7.7.

3. Under the assumptions of Exercise 2 show that there exist matrices S and T such that

$$\begin{pmatrix} I & S \\ 0 & I \end{pmatrix}\begin{pmatrix} A_{11} & A_{12} \\ 0 & A_{22} \end{pmatrix}\begin{pmatrix} I & T \\ 0 & I \end{pmatrix} = \begin{pmatrix} A_{11} & 0 \\ 0 & A_{22} \end{pmatrix},$$

$$\begin{pmatrix} I & S \\ 0 & I \end{pmatrix}\begin{pmatrix} B_{11} & B_{12} \\ 0 & B_{22} \end{pmatrix}\begin{pmatrix} I & T \\ 0 & I \end{pmatrix} = \begin{pmatrix} B_{11} & 0 \\ 0 & B_{22} \end{pmatrix}.$$

Hint. These matrices have to satisfy

$$A_{11}T + A_{12} + SA_{22} = 0 \tag{1.51a}$$

$$B_{11}T + B_{12} + SB_{22} = 0 \tag{1.51b}$$

and can be computed as follows: the first column of T is obtained from (1.51b) because B_{11} is invertible and the first column of SB_{22} vanishes; then the first column of S is given by (1.51a) because A_{22} is invertible; the second column of SB_{22} is then known and we can compute the second column of T from (1.51b), etc.

4. Prove that the index of nilpotency of a regular matrix pencil $A+\lambda B$ does not depend on the choice of P and Q in (1.3).

Hint. Consider two different decompositions of the form (1.3) and denote the matrices which appear by C_1, N_1 and C_2, N_2, respectively. Show the existence of a regular matrix T such that $N_2 = T^{-1}N_1T$.

5. Prove that the system (VI.3.4a,b) has index 2 (it is of the form (1.14a,b) and satisfies (1.10)). The full system (VI.3.4) has perturbation index k.

6. (Arnold 1993). Consider the index 2 problem (1.14) and its perturbation (1.28). Prove that the difference $\Delta y(x) = \widehat{y}(x) - y(x)$ satisfies

$$\|\Delta y(x)\| \le C\Big(\|\Delta y(0)\| + \max_{0\le\xi\le x}\Big(\Big\|\int_0^\xi P(t)\delta(t)\,dt\Big\| + \|\theta(\xi)\| + \big(\|\delta(\xi)\| + \|\theta'(x)\|\big)^2\Big)\Big)$$

with the projector $P(t) = I - \big(f_z(g_yf_z)^{-1}g_y\big)\big(y(t), z(t)\big)$, provided that the right hand side is sufficiently small.

Hint. Linearize Eq. (1.29) around (y, z), extract $\widehat{z} - z$, and insert it into the difference of (1.28a) and (1.14a). The term $\big(f_z(g_yf_z)^{-1}\big)\big(y(x), z(x)\big)\,\theta'(x)$ can be replaced by $\frac{d}{dx}\big(f_z(g_yf_z)^{-1}\big(y(x), z(x)\big)\,\theta(x)\big) + \mathcal{O}(\|\theta(x)\|)$ before integration.

7. For the linear initial value problem

$$y' = A(x)y + f(x), \qquad y(0) = 0$$

consider the *adjoint* problem

$$v' = -A(x)^Tv - g(x), \qquad v(1) = 0.$$

Prove that $\displaystyle\int_0^1 g(x)^Ty(x)dx = \int_0^1 v(x)^Tf(x)dx.$

8. Consider a linear optimal control problem with quadratic cost functional

$$y' = Ay + Bu + f(x), \qquad y(0) = y_0$$

$$J(u) = \frac{1}{2}\int_0^1 \Big(y(x)^TCy(x) + u(x)^TDu(x)\Big)\,dx,$$

where C and D are assumed to be positive semi-definite.

a) Prove that $J(u)$ is minimal if and only if

$$\begin{aligned} y' &= Ay + Bu + f(x), \qquad & y(0) &= y_0 \\ v' &= -A^Tv - Cy, & v(1) &= 0 \\ 0 &= B^Tv + Du. \end{aligned} \tag{1.52}$$

b) If D is positive definite, then (1.52) has index 1.

c) If $D = 0$ and B^TCB is positive definite, then (1.52) has index 3.

VII.2 Index Reduction Methods

We have seen in Sect. VI.1 that the numerical treatment of problems of index 1, which are either in the half-explicit form (1.13) or in the form $Mu' = \varphi(u)$, is not much more difficult than that of ordinary differential equations. For higher index problems the situation changes completely. This section is devoted to the study of several approaches that are all based on the idea of modifying the problem in such a way that the index is reduced.

Index Reduction by Differentiation

The most apparent way of reducing the index is to differentiate repeatedly the algebraic constraints (see Definition 1.2). In general, it is recommended to differentiate until having obtained an index 1 problem. For example, the index 2 problem (1.14a,b) is replaced by (1.14a,c), or the constrained mechanical system (1.46a,b,c) by (1.46a,b,e). The resulting problem is then solved by the methods of Chapter VI.

We illustrate this approach at the "pendulum example"

$$x' = u, \qquad u' = -x\lambda \tag{2.1a}$$

$$y' = v, \qquad v' = -1 - y\lambda \tag{2.1b}$$

$$0 = x^2 + y^2 - 1. \tag{2.1c}$$

In this form it has index 3. Differentiating the algebraic constraint twice yields

$$0 = xu + yv, \tag{2.2}$$

$$0 = -\lambda(x^2 + y^2) - y + u^2 + v^2. \tag{2.3}$$

Equations (2.1a,b) together with (2.3) represent an index 1 problem. We can extract λ from (2.3) and insert it into (2.1a,b) to get a differential equation for x, y, u, v, which can be solved by standard methods.

Drift-off Phenomenon. As an example we apply the code DOPRI5 to the index 1 problem (2.1a,b), (2.3) with initial values $x_0 = 1$, $y_0 = 0$, $u_0 = 0$, $v_0 = 0$. We are interested, how well the constraints (2.1c) and (2.2) are preserved by the numerical solution. The result presented in Fig. 2.1 shows that the error in the constraint (2.2) grows linearly, that in (2.1c) grows even quadratically. This phenomenon is explained as follows:

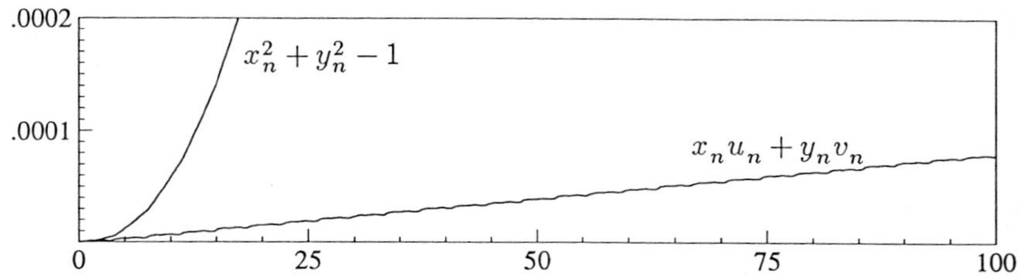

Fig. 2.1. Error in the constraints for DOPRI5 ($Atol = Rtol = 10^{-6}$)

Consider a constrained mechanical system (see (1.46))

$$q' = u \tag{2.4a}$$
$$M(q)u' = f(q,u) - G^T(q)\lambda \tag{2.4b}$$
$$0 = g(q). \tag{2.4c}$$

Differentiating (2.4c) twice we get

$$\begin{pmatrix} M(q) & G^T(q) \\ G(q) & 0 \end{pmatrix} \begin{pmatrix} u' \\ \lambda \end{pmatrix} = \begin{pmatrix} f(q,u) \\ -g_{qq}(q)(u,u) \end{pmatrix} \tag{2.5}$$

which, together with (2.4a), is the corresponding index 1 problem. The important observation is now that the index 1 problem possesses a solution for arbitrary initial values q_0 and u_0. Due to the fact that the second derivative of $g(q(t))$ vanishes (this is a consequence of the lower relation of (2.5)), the solution of the index 1 problem satisfies

$$g\big(q(t)\big) = g(q_0) + (t - t_0)G(q_0)u_0, \tag{2.6a}$$
$$G\big(q(t)\big)u(t) = G(q_0)u_0. \tag{2.6b}$$

Theorem 2.1. *If we apply a pth order numerical method to the index 1 problem (2.4a), (2.5) with consistent initial values at $t_0 = 0$, then the numerical solution (q_n, u_n) at time t_n satisfies (for $t_n - t_0 \le Const$)*

$$\|g(q_n)\| \le h^p(At_n + Bt_n^2), \qquad \|G(q_n)u_n\| \le h^p C t_n.$$

The value h represents the maximal step size used.

Proof. Denote by $q(t, t_0, q_0, u_0)$ the solution of the index 1 problem with initial value (q_0, u_0) at $t = t_0$. Since the local error $q_{j+1} - q(t_{j+1}, t_j, q_j, u_j)$ is of size $\mathcal{O}(h_j^{p+1})$ (and similarly for the u-component), it follows from (2.6a) that

$$\big\|g\big(q(t_n, t_{j+1}, q_{j+1}, u_{j+1})\big) - g\big(q(t_n, t_j, q_j, u_j)\big)\big\| \le h_j^{p+1}\big(A + 2B(t_n - t_{j+1})\big).$$

Adding up these inequalities from $j = 0$ to $j = n-1$ gives the desired bound for $g(q_n)$, because the initial values are consistent, i.e., $g(q(t_n, t_0, q_0, u_0)) = 0$. The second estimate of Theorem 2.1 is proved in the same way. □

Baumgarte Stabilization. The historically first remedy for this drift-off is due to Baumgarte (1972). Instead of replacing the constraint (2.4c) by its second time derivative, he proposes to replace (2.4c) by the linear combination

$$0 = \ddot{g} + 2\alpha\dot{g} + \beta^2 g, \tag{2.7}$$

where $\dot{g}$, $\ddot{g}$ are time derivatives of (2.4c), i.e.,

$$g = g(q), \qquad \dot{g} = G(q)u, \qquad \ddot{g} = g_{qq}(q)(u,u) + G(q)\big(f(q,u) - G^T(q)\lambda\big).$$

Eq. (2.7) together with (2.4b) determines u' and λ as functions of (q,u), and the resulting differential equation can be solved numerically. The idea is now to choose the free parameters α and β in such a way that (2.7) is an asymptotically stable differential equation, e.g., $\beta = \alpha$ and $\alpha > 0$. Consequently, the functions $g(q(t))$ and $G(q(t))u(t)$ are exponentially decreasing, in contrast to (2.6). The difficulty of this approach lies in a good choice of α. For small values of α the damping will not be sufficiently strong, whereas for large α the resulting differential equation becomes stiff and explicit methods are no longer efficient. A careful investigation on the choice of α can be found in Ascher, Chin & Reich (1994).

Stabilization by Projection

We shall now analyze another possibility for avoiding the instability of the preceding example, namely the repeated projection of the numerical solution onto the solution manifold.

Index 2 Problems. Consider the system (1.14a,b). Suppose that (y_{n-1}, z_{n-1}) is an approximation to the solution at time t_{n-1} which satisfies $g(y_{n-1}) = 0$ and $g_y(y_{n-1})f(y_{n-1}, z_{n-1}) = 0$. Applying a numerical one-step method (state space form method of Sect. VI.1) with these values to the index 1 system (1.14a,c) yields an approximation $\widetilde{y}_n, \widetilde{z}_n$ that, in general, does not satisfy the constraint (1.14b). A natural way of projecting the approximation $\widetilde{y}_n$ to the solution manifold $\mathcal{M}$ of Eq. (1.17) is along the image of f_z (see also the projected Runge-Kutta methods of Sect. VII.4). We therefore define y_n as the solution of

$$y - \widetilde{y}_n = f_z(\widetilde{y}_n, \widetilde{z}_n)\mu, \qquad g(y) = 0, \tag{2.8}$$

and then we adjust z_n by solving the equation $g_y(y_n)f(y_n, z_n) = 0$. Applying simplified Newton iterations to the nonlinear system (2.8) requires the decomposition of the matrix

$$\begin{pmatrix} I & f_z(\widetilde{y}_n, \widetilde{z}_n) \\ g_y(\widetilde{y}_n) & 0 \end{pmatrix}. \tag{2.9}$$

Block elimination shows that the invertibility of (2.9) is a consequence of (1.10), and that only the matrix $g_y f_z$ has to be decomposed. Such a decomposition is usually already available from the application of the numerical method, so that the projection (2.8) is very cheap.

It is now natural to ask, whether this projection procedure can distroy the convergence properties of the underlying method. For a pth order one-step method the local error is of size $\mathcal{O}(h^{p+1})$. Since the solution of (1.14a,c) passing through (y_{n-1}, z_{n-1}) satisfies $g(y(t)) = 0$, it holds $g(\widetilde{y}_n) = \mathcal{O}(h^{p+1})$. Hence, the solution of (2.8) satisfies $\mu = \mathcal{O}(h^{p+1})$, $y_n - \widetilde{y}_n = \mathcal{O}(h^{p+1})$, and $z_n - \widetilde{z}_n = \mathcal{O}(h^{p+1})$. By the Implicit Function Theorem this solution depends smoothly on $(\widetilde{y}_n, \widetilde{z}_n)$, so that the mapping $(y_{n-1}, z_{n-1}) \mapsto (y_n, z_n)$ represents a pth order one-step method for (1.14a,c). Convergence of order p thus follows from the standard theory (see Sects. VI.1 and II.3). This proof also applies to multistep methods.

Constrained Mechanical Systems. For the index 3 system (2.4a,b,c) the situation is slightly more complex. We assume consistent values $(q_{n-1}, u_{n-1}, \lambda_{n-1})$ at time t_{n-1} and apply a one-step method to the index 1 system (2.4a), (2.5) to obtain $(\widetilde{q}_n, \widetilde{u}_n)$. Since the position constraint (2.4c) only depends on q, the projections for q and u can be done sequentially.

Projection on Position Constraint. We define q_n as solution of the nonlinear system

$$\begin{aligned} M(\widetilde{q}_n)(q_n - \widetilde{q}_n) + G^T(\widetilde{q}_n)\mu &= 0 \\ g(q_n) &= 0. \end{aligned} \tag{2.10}$$

Projection on Velocity Constraint. With the value q_n obtained from the above projection we let u_n be the solution of

$$\begin{aligned} M(q_n)(u_n - \widetilde{u}_n) + G^T(q_n)\mu &= 0 \\ G(q_n)u_n \qquad\qquad &= 0. \end{aligned} \tag{2.11}$$

Lubich (1991) introduced this kind of projection, because "it is invariant under affine transformations of coordinates". We remark that the system (2.11) is linear, whereas (2.10) is nonlinear and has to be solved by (simplified) Newton iterations. The index 3 assumption that the matrix in Eq. (2.5) is invertible, implies the existence of the projected values q_n and u_n (at least for sufficiently small step size). It is possible to alter slightly the arguments of M and G^T in the upper lines of (2.10) and (2.11) or to solve the system (2.11) iteratively, if this is computationally advantageous. Convergence of this method is proved in the same way as in the index 2 case.

Velocity Stabilization. It can be seen from (2.6) that errors in the velocity constraint $G(q)u = 0$ are more critical for the numerical solution than errors in the position constraint $g(q) = 0$. It is therefore interesting to study the method, where the numerical solution is projected only to the velocity constraint. Alishenas & Ólafsson (1994) come to the conclusion that "*velocity projection* is the most efficient projection with regard to improvement of the numerical integration".

We have applied the code DOPRI5 in four different variants to the index 1 formulation of the pendulum equation (2.1): (i) standard application without any projection, (ii) only projection on the position constraint, (iii) only projection on the velocity constraint, (iv) sequential position and velocity projections. The the global

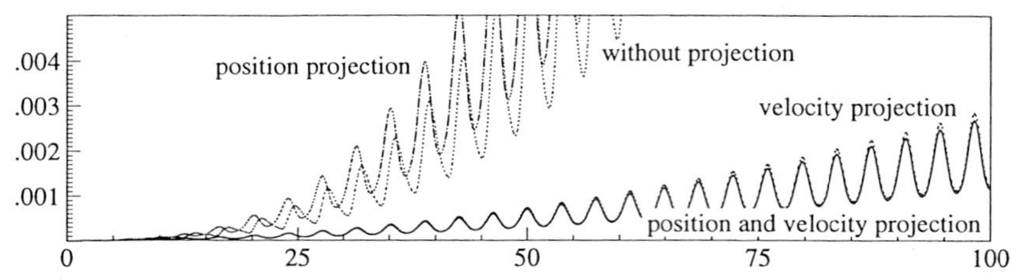

Fig. 2.2. Global error of DOPRI5 with various projections ($Atol = Rtol = 10^{-6}$)

error (in position and velocity) during integration is shown in Fig. 2.2. We conclude that a projection on the position constraint without projection on the velocity constraint does not improve the global error (it makes it even worse in our example). On the other hand, velocity stabilization is as efficient as the complete projection (position and velocity). Nearly no difference can be observed in Fig. 2.2.

Differential Equations with Invariants

Closely related to the above techniques is the numerical treatment of differential equations with invariants. Consider the initial value problem

$$y' = f(y), \qquad y(0) = y_0, \tag{2.12}$$

and suppose that the solution is known to have the invariant

$$\varphi(y) = 0. \tag{2.13}$$

For example, the differential equation (1.46a,e) for (q, u) has the invariants (1.46c) and (1.46d). Conservation laws (total energy,...) may also be written in the form (2.13). The invariant (2.13) is called a *first integral*, if $\varphi_y(y)f(y) \equiv 0$ in a neighbourhood of the solution.

Linear first integrals of the form $\varphi(y) = c + d^T y$ are preserved exactly by most integration methods (e.g., Runge-Kutta and multistep methods). Quadratic first integrals are preserved exactly by symplectic Runge-Kutta methods (see Theorem II.16.7). More complicated invariants are in general not preserved.

The above projection techniques can be adapted to the treatment of the problem (2.12-13) (see Shampine (1986), Eich (1993), Ascher, Chin & Reich (1994)). We apply a numerical method to (2.12) and project (orthogonally or somehow else) the numerical solution onto the manifold defined by (2.13). As discussed above, this precedure retains the order of convergence of the basic method.

Hamiltonian Systems. Differential equations of the form

$$p_i' = -\frac{\partial H}{\partial q_i}(p,q), \qquad q_i' = \frac{\partial H}{\partial p_i}(p,q), \qquad i = 1,\ldots,n, \tag{2.14}$$

where $H : \mathbb{R}^{2n} \to \mathbb{R}$ is a smooth function, always have $H(p,q) = Const$ as first integral. It is tempting to exploit this information and project the numerical solution

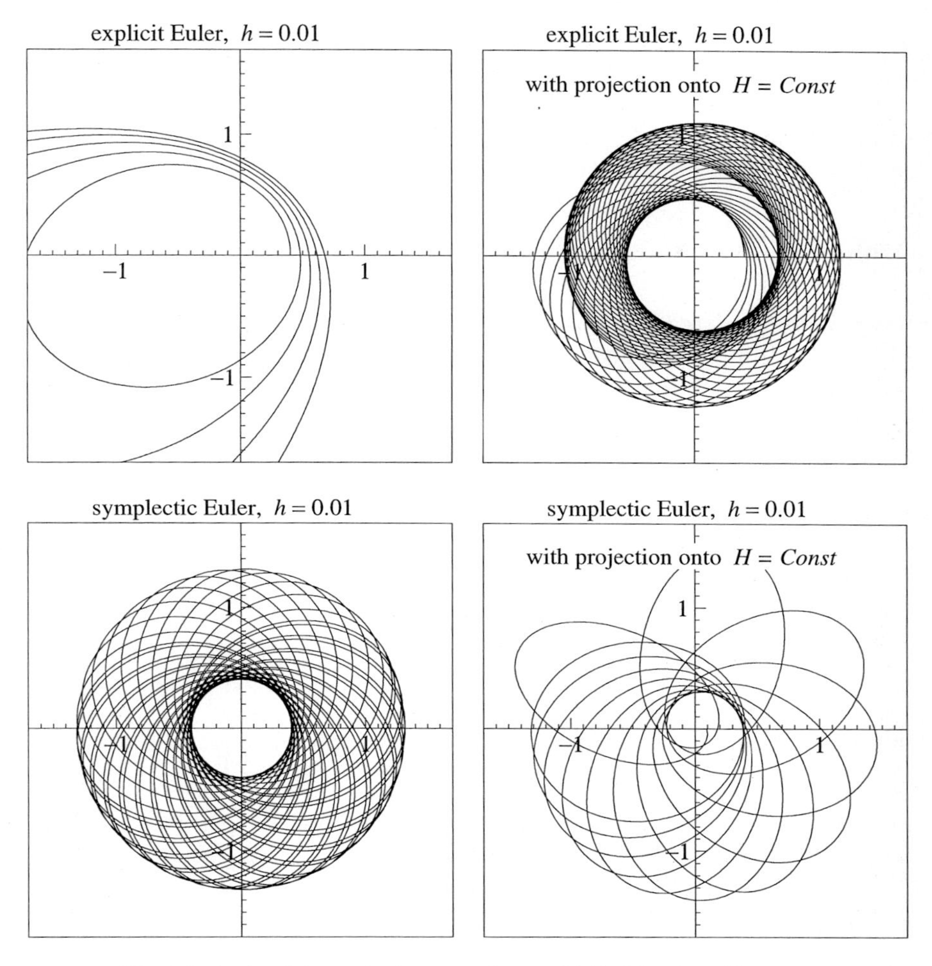

Fig. 2.3. Study of the projection onto the manifold $H(p,q) = H(p_0,q_0)$

onto the manifold $H(p,q)=H(p_0,q_0)$. Consider for example the perturbed Kepler problem with Hamiltonian

$$H(p,q) = \frac{p_1^2+p_2^2}{2} - \frac{1}{\sqrt{q_1^2+q_2^2}} - \frac{0.005}{\sqrt{(q_1^2+q_2^2)^3}} \tag{2.15}$$

and initial values $q_1(0) = 1-e$, $q_2(0) = 0$, $p_1(0) = 0$, $p_2(0) = \sqrt{(1+e)/(1-e)}$ (eccentricity $e = 0.6$). The upper pictures of Fig. 2.3 show the numerical solution obtained by the explicit Euler method with step size $h = 0.01$; to the left without any projection, and to the right with projection onto $H = Const$. An improvement can be observed, but the numerical solution still does not reflect the geometric structure of the exact solution (invariant torus). We also have applied the symplectic Euler method (see Eq. (16.54) of Sect. II.16). Here we see that the numerical solution (without projection) shows the correct qualitative behaviour (this can be explained by a backward error analysis, see Sect. II.16), whereas the projection onto $H = Const$ destroys this property. A remedy could be the following: apply a symplectic method to the problem, project the numerical solution to $H = Const$, but continue the integration with the unprojected values.

Methods Based on Local State Space Forms

This method is also called *differential-geometric approach* by Potra & Rheinboldt (1990). The idea is to regard the differential-algebraic system as a differential equation on a manifold (see Sect. VII.1) and to solve the equation in this manifold by introducing suitable local coordinates.

Let us illustrate this approach at the pendulum example. The equations, formulated in cartesian coordinates, are given in the beginning of this section. The solution manifold is (compare with Eq. (1.22))

$$\mathcal{M} = \big\{(x,y,u,v) \mid x^2+y^2=1,\ \ xu+yv=0\big\}.$$

This is a 2-dimensional manifold in $\mathbb{R}^4$ and can be parametrized by (φ,η) as follows:

$$\begin{aligned} x &= \cos\varphi, \qquad & u &= -\eta\sin\varphi, \\ y &= \sin\varphi, \qquad & v &= \eta\cos\varphi. \end{aligned} \tag{2.16}$$

A short calculation shows that the system (2.1a,b), (2.3), written in the new coordinates, leads to the well-known equation

$$\varphi' = \eta, \qquad \eta' = -\cos\varphi. \tag{2.17}$$

This differential equation can be solved numerically without any difficulties. The numerical approximation in the original coordinates is then obtained via (2.16). Obviously, the position and velocity constraints are satisfied exactly.

Although this example nicely illustrates the main ideas, it may be misleading. First of all, in typical applications it is not possible to use one and the same parametrization throughout the whole integration. Secondly, the choice of coordinates is usually not obvious and the transformed differential equation can be much more complicated than the original one (see for example Alishenas (1992)).

Local State Space Form. Suppose that the differential-algebraic system, which we want to solve, can be written as a differential equation

$$y' = v(y), \qquad y \in \mathcal{M} \tag{2.18}$$

on a smooth d-dimensional manifold $\mathcal{M} \subset \mathbb{R}^n$. Consider a coordinate function $\omega : U \to V$ (sufficiently differentiable, bijective, and $\omega'(\eta)$ of full rank) between the open set $U \subset \mathbb{R}^d$ and $V \subset \mathcal{M}$, and denote the coordinates in U by $\eta \in \mathbb{R}^d$. Under the transformation $y = \omega(\eta)$ the equation (2.18) becomes

$$\omega'(\eta)\eta' = v\big(\omega(\eta)\big). \tag{2.19}$$

Since $v(y) \in T_y\mathcal{M}$ for all $y \in \mathcal{M}$ (see Eq. (1.19)), there exists η' such that (2.19) holds. Moreover η' is unique, because $\omega'(\eta)$ is of full rank. Using the notation $\omega'(\eta)^+ = \big(\omega'(\eta)^T\omega'(\eta)\big)^{-1}\omega'(\eta)^T$ for the pseudo-inverse of $\omega'(\eta)$ we therefore obtain

$$\eta' = \omega'(\eta)^+ v\big(\omega(\eta)\big), \tag{2.20}$$

which is an ordinary differential equation in $\mathbb{R}^d$ and is called *local state space form* of (2.18). Observe that different coordinate functions lead to different state space forms.

The *numerical procedure* for solving (2.18) is the following: suppose that an approximation $y_k \in \mathcal{M}$ of $y(t_k)$ is given. We then choose a coordinate function and apply a standard method (e.g., Runge-Kutta) with initial value $\eta_k = \omega^{-1}(y_k)$ to the state space form (2.20). This yields an approximation η_{k+1} at time t_{k+1}. Finally, we put $y_{k+1} = \omega(\eta_{k+1})$. By definition of this procedure, the numerical approximation y_{k+1} again lies in $\mathcal{M}$.

If one uses one and the same local state space form for the whole integration (as it is the case for the pendulum example, Eq. (2.17)), the convergence properties for (2.20) carry immediately over to (2.18) via the coordinate function $y = \omega(\eta)$. In more complex situations it may be necessary to change the coordinates several times, and from a computational point of view it may even be more advantageous to change them in every integration step.

Theorem 2.2. *Consider the above procedure for the numerical solution of (2.18), and denote by $y = \omega_k(\eta)$ the coordinate transformation of the kth step. If, in a neighbourhood of $\omega_k^{-1}(y_k)$, the matrices $\omega_k'(\eta)$ and $\omega_k'(\eta)^+$ are uniformly bounded in k, then the convergence properties for standard ordinary differential equations carry over to the problem (2.18) on a manifold $\mathcal{M}$.*

Proof. In the case of one-step methods we have

$$y_{k+1} = \omega_k\Big(\omega_k^{-1}(y_k) + h\Phi_k\big(\omega_k^{-1}(y_k), h\big)\Big),$$

where $\Phi_k(\eta, h)$ is the increment function of the method when applied to (2.20) with ω replaced by ω_k. Due to the regularity assumptions on $\omega_k(\eta)$, this formula can be written as

$$y_{k+1} = y_k + h\Psi_k(y_k, h)$$

and takes the form of a standard one-step method. The assumptions guarantee that the functions Ψ_k have a uniform Lipschitz constant with respect to the first argument. Therefore the convergence proofs of Sect. II.3 apply. For multistep methods the situation is analogous. □

Choice of Local Coordinates. Let us explain two choices for the constrained mechanical system (2.4), whose solution manifold is given by

$$\mathcal{M} = \{(q, u) \mid g(q) = 0, \;\; G(q)u = 0\}. \tag{2.21}$$

Here $q, u \in \mathbb{R}^n$ are generalized coordinates, $g(q) \in \mathbb{R}^m$ and $G(q) = g_q(q)$. The adaptation to other differential-algebraic systems with known solution manifold is more or less straightforward.

Generalized Coordinate Partitioning (Wehage & Haug 1982). Assuming that the Jacobian $G(q)$ has full row rank, there exists a partitioning $q = (\eta, \widehat{\eta})$ such that $g_{\widehat{\eta}}(\eta, \widehat{\eta})$ is invertible ($\eta \in \mathbb{R}^{n-m}$, $\widehat{\eta} \in \mathbb{R}^m$). By the Implicit Function Theorem the constraint $g(q) = 0$ can be solved for $\widehat{\eta}$ in a neighbourhood of a consistent value $q_0 = (\eta_0, \widehat{\eta}_0)$. Hence, there exists a function $\widehat{\eta} = h(\eta)$ (defined for η close to η_0) such that $g(\eta, h(\eta)) = 0$. With a corresponding partitioning $u = (\nu, \widehat{\nu})$ the velocity constraint becomes $g_\eta(\eta, \widehat{\eta})\nu + g_{\widehat{\eta}}(\eta, \widehat{\eta})\widehat{\nu} = 0$ and allows us to express $\widehat{\nu}$ in terms of η, ν as $\widehat{\nu} = k(\eta, \nu)$. A coordinate function is thus given by $\omega(\eta, \nu) = ((\eta, h(\eta)), (\nu, k(\eta, \nu)))$, and the differential equation in these local coordinates is

$$\eta' = \nu, \qquad \nu' = \nu'(\omega(\eta, \nu)), \tag{2.22}$$

where $\nu'(q, u)$ collects the ν-components of the solution $u'(q, u)$ of the linear system (1.38e). We emphasize that for a numerical implementation the differential equation (2.22) need not be known analytically. However, a nonlinear system has to be solved each time when the right-hand side of (2.22) has to be evaluated.

Tangent Space Parametrization (Potra & Rheinboldt 1991, Yen 1993). Instead of partitioning the components of q and u we split the vectors $q - q_0$ and $u - u_0$ according to

$$q - q_0 = Q_0\eta + Q_1\widehat{\eta}, \qquad u - u_0 = Q_0\nu + Q_1\widehat{\nu}, \tag{2.23}$$

where the columns of Q_0 form a basis of the tangent space $\{v \mid G(q_0)v = 0\}$ to the manifold $g(q) = 0$, which is completed by the columns of Q_1 to a basis of the whole space. The condition $g(q) = 0$ together with the first relation of (2.23) define (locally) q and $\widehat{\eta}$ as functions of η. Similarly, $G(q)u = 0$ and the second relation of (2.23) define u and $\widehat{\nu}$ as functions of ν and q. Denoting these relationships by $\widehat{\eta} = h(\eta)$, $\widehat{\nu} = k(\eta, \nu)$, we get formally the same coordinate function as in the previous example, and the state space form is given by

$$\eta' = \nu, \qquad \nu' = Q_0^+ u'(\omega(\eta, \nu)), \tag{2.24}$$

where $Q_0^+ = (Q_0^T Q_0)^{-1} Q_0^T$ is the pseudo-inverse of Q_0, and $u'(q, u)$ denotes the solution of the linear system (2.5).

The evaluation of $h(\eta)$ requires the solution of a nonlinear system, whose Jacobian is

$$\begin{pmatrix} I & -Q_1 \\ G(q_0) & 0 \end{pmatrix}.$$

This suggests to take $-Q_1 = G^T(q_0)$ or better $-Q_1 = M^{-1}(q_0)G^T(q_0)$, so that simplified Newton iterations lead to linear systems with a matrix that already appears in (2.5). The linear system for the computation of $k(\eta, \nu)$ has the same structure.

Due to the fact that the evaluation of the right-hand side of (2.24) requires the solution of a nonlinear system, the authors of this approach prefer the use of multistep methods which, in general, use less function evaluations than one-step methods. In connection with Runge-Kutta methods, Potra (1995) suggests the use of certain predicted values instead of the exact solutions of these nonlinear systems, and requires that only the approximation at the end of every step lies on the manifold $\mathcal{M}$. The resulting algorithm is then equivalent to solving the index 1 problem combined with projections onto $\mathcal{M}$ at the end of each step.

Overdetermined Differential-Algebraic Equations

In contrast to the approach at the beginning of this section, where the constraint is replaced by one of its derivatives, we consider the original system and one or more derivatives of the constraints as a unity. For example, the equations of motion of a constrained mechanical system become

$$\begin{aligned} q' &= u && (2.25a)\\ M(q)u' &= f(q,u) - G^T(q)\lambda && (2.25b)\\ 0 &= g(q) && (2.25c)\\ 0 &= G(q)u && (2.25d)\\ 0 &= g_{qq}(q)(u,u) + G(q)M(q)^{-1}\big(f(q,u) - G^T(q)\lambda\big). && (2.25e) \end{aligned}$$

This system is overdetermined, because we are concerned with more equations than unknowns. Nevertheless, it possesses a unique solution, if (1.47) is satisfied and consistent initial values are prescribed.

We illustrate the numerical solution of (2.25) with the BDF method. A formal application (see Sect. VI.2) gives

$$\begin{aligned} & q_k - \widehat{q} - h\gamma u_k = 0 && (2.26a)\\ & M(q_k)(u_k - \widehat{u}) - h\gamma\big(f(q_k,u_k) - G^T(q_k)\lambda_k\big) = 0 && (2.26b)\\ & g(q_k) = 0 && (2.26c)\\ & G(q_k)u_k = 0 && (2.26d)\\ & g_{qq}(q_k)(u_k,u_k) + G(q_k)M(q_k)^{-1}\big(f(q_k,u_k) - G^T(q_k)\lambda_k\big) = 0, && (2.26e) \end{aligned}$$

where $\gamma = \beta_k/\alpha_k$, $\widehat{q} = \big(\sum_{i=0}^{k-1}\alpha_i q_i\big)/\alpha_k$, and $\widehat{u} = \big(\sum_{i=0}^{k-1}\alpha_i u_i\big)/\alpha_k$ are known quantities. The system (2.26) is overdetermined and does not have a solution, in general. A natural idea (Führer 1988) is to search for a least square solution of (2.26). There are several ways to do this. One can consider different norms, or one can require some of the equations to be exactly satisfied and the remaining ones in a least square sense. Führer & Leimkuhler (1991) impose all constraints (2.26c,d,e), and treat the remaining equations by the use of a special pseudoinverse. This can be achieved by introducing Lagrange multipliers μ_k, η_k in the first two equations of (2.26) as follows:

$$M(q_k)(q_k - \widehat{q} - h\gamma u_k) + h\gamma\big(G^T(q_k)\mu_k + (G_q(q_k)u_k)^T\eta_k\big) = 0 \qquad (2.27a)$$

$$M(q_k)(u_k - \widehat{u}) - h\gamma\big(f(q_k, u_k) - G^T(q_k)\lambda_k\big) + h\gamma G^T(q_k)\eta_k = 0. \qquad (2.27b)$$

For sufficiently small h, the system (2.27a,b), (2.26c,d,e) has a locally unique solution, if (1.47) is satisfied.

Connection with GGL-Formulation. If we omit the acceleration constraint (2.26e), there is no need for two Lagrange multipliers, and we can put $\eta_k = 0$. The resulting system (2.27a,b), (2.26c,d) is then nothing else than the standard BDF discretization of the system (1.48).

Unstructured Higher Index Problems

We consider a general differential-algebraic system

$$F(u', u) = 0. \qquad (2.28)$$

For its numerical solution we shall construct an 'underlying ODE' (see Definition 1.2) and solve it by any integration method. This approach has been developed in several papers by Campbell (1989, 1993). We shall explain the main ideas following the presentation of Campbell & Moore (1995).

Inspired by the definition of the differentiation index we consider the *derivative array equations*

$$F(u', u) = 0, \quad \frac{dF(u', u)}{dx} = 0\,, \quad \ldots\,, \quad \frac{d^m F(u', u)}{dx^m} = 0$$

which we write in compact form as

$$G(u', w, u) = 0, \qquad (2.29)$$

where $w = (u'', u''', \ldots, u^{(m+1)})$ collects the higher derivatives of u. In Eq. (2.29) we consider w, u, and also u' as independent variables. Besides the usual differentiability assumptions we assume that

(A1) the matrix $(G_{u'}, G_w)$ is 1-full with respect to u'; this means that the relation $G_{u'}\Delta u' + G_w \Delta w = 0$ implies $\Delta u' = 0$;

(A2) the matrix $(G_{u'}, G_w)$ has constant rank;

(A3) the matrix $(G_{u'}, G_w, G_u)$ has full row rank.

These assumptions are required to hold in a neighbourhood of a particular solution of (2.28). The construction of the underlying ODE is based on the following lemma and on its proof.

Lemma 2.3 (Campbell & Moore 1995). *Consider a sufficiently smooth problem (2.28) and assume that (A1), (A2), and (A3) hold. Then there exist coordinate partitions* $w = (w_a, w_b)$, $u = (u_a, u_b)$ *(and also* $u' = (u'_a, u'_b)$ *with the same partition* *as for* u*), such that the derivative array equations (2.29) are equivalent to*

$$\begin{aligned} u'_a &= f_a(u_b), \qquad & w_a &= \varphi_2(w_b, u_b) \\ u'_b &= f_b(u_b), \qquad & u_a &= \varphi_3(u_b) \end{aligned} \tag{2.30}$$

in a neighbourhood of the consistent initial value (u'_0, w_0, u_0).

Proof. We consider the matrix $(G_{u'}, G_w, G_u)$ evaluated at (u'_0, w_0, u_0) and perform a QR factorization, where column permutations are restricted to components within the vectors u', w, and u. This yields

$$Q^T(G_{u'}, G_w, G_u)P = \left(\begin{array}{c|cc|cc} B_1 & C_1 & C_2 & D_1 & D_2 \\ 0 & C_3 & C_4 & D_3 & D_4 \\ 0 & 0 & 0 & D_5 & D_6 \end{array}\right), \tag{2.31}$$

where B_1, C_3, D_5 are nonsingular by Assumption (A3), Q is an orthogonal matrix, and $P = \operatorname{diag}(P_1, P_2, P_3)$ with suitable permutation matrices P_1, P_2, P_3. Fixing the permutation P, we apply the above factorization also to $(G_{u'}, G_w, G_u)$ evaluated at an arbitrary point (u', w, u) close to (u'_0, w_0, u_0). Because of Assumption (A2) this gives (2.31) with smooth matrices Q, B_i, C_i, and D_i. The decomposition (2.31) defines the partitions $w = (w_a, w_b)$ and $u = (u_a, u_b)$. The first, second and fourth block-columns in (2.31) form an invertible matrix. The Implicit Function Theorem thus implies that (2.29) can be solved for u', w_a, u_a, and we obtain the equivalent system

$$u' = \varphi_1(w_b, u_b), \qquad w_a = \varphi_2(w_b, u_b), \qquad u_a = \varphi_3(w_b, u_b).$$

We still have to show that the functions φ_1 and φ_3 are independent of w_b. By definition of the φ_i we have

$$G\Big(\varphi_1(w_b, u_b), (\varphi_2(w_b, u_b), w_b), (\varphi_3(w_b, u_b), u_b)\Big) = 0.$$

Differentiating with respect to w_b yields

$$G_{u'} \cdot \frac{\partial \varphi_1}{\partial w_b} + G_{w_a} \cdot \frac{\partial \varphi_2}{\partial w_b} + G_{w_b} + G_{u_a} \cdot \frac{\partial \varphi_3}{\partial w_b} = 0. \tag{2.32}$$

Multiplying this relation by Q^T, we see from Eq. (2.31) that $D_5(\partial\varphi_3/\partial w_b) = 0$. Since D_5 is nonsingular, this implies $(\partial\varphi_3/\partial w_b) = 0$, so that φ_3 is independent of w_b. Assumption (A1) now implies from (2.32) that also $(\partial\varphi_1/\partial w_b)$ vanishes. This completes the proof of the lemma. □

Suppose that we know how to compute $f_a(u_b)$, $f_b(u_b)$ and $\varphi_3(u_b)$ for a given value u_b. From (2.30) we then have an ordinary differential equation for u_b, which can be solved by any integration method (Runge-Kutta or multistep, explicit or implicit, ...), and the remaining components are given by $u_a = \varphi_3(u_b)$. The numerical solution of this method thus preserves all constraints (also the hidden ones).

Computation of the Values $f_a(u_b)$, $f_b(u_b)$ *and* $\varphi_3(u_b)$. It follows from Assumption (A3) that $(G_{u'}, G_w, G_u)^T G = 0$ is equivalent to $G = 0$. Thus, for given u_b, any method of finding the minimum (u', w, u_a) of the function $G^T G$ may be used. Campbell & Moore (1995) propose the use of Gauss-Newton iterations.

Remark. A closely related algorithm has been proposed by Kunkel & Mehrmann (1996). Instead of extracting from the derivative array equations an ordinary differential equation for all variables, they extract an equivalent index 1 problem and solve it by standard integration methods. This modification usually requires one differentiation less of the original system (2.28).

Exercises

1. Repeat the experiment of Fig. 2.1 with other numerical methods (explicit Euler method, multistep methods, constant and variable step sizes, ...). You will observe that in some situations the error in $g(q_n)$ grows only linearly, and the error in $G(q_n)u_n$ remains bounded. Try to explain this observation.

2. a) Prove that the matrix in (2.5) is 1-full with respect to u' if and only if the restriction of M to the kernel of G is injective (this is exactly the condition that is needed in order to be able to apply the methods of this section).

   b) Show by examples that neither M needs to be nonsingular nor G has to be of full rank in order that the condition of part (a) is satisfied.

VII.3 Multistep Methods for Index 2 DAE

BDF is so beautiful that it is hard to imagine something else could be better. (L. Petzold 1988, heard by P. Deuflhard)

Convergence results of multistep methods for problems of index at least 2 are harder to obtain than for semi-explicit index 1 problems (see Section VI.2). A first convergence result for BDF schemes, valid for linear constant coefficient DAE's of arbitrary index, was given by Sincovec, Erisman, Yip & Epton (1981). Convergence of BDF for nonlinear DAE systems was then studied by Gear, Gupta & Leimkuhler (1985), Lötstedt & Petzold (1986) and Brenan & Engquist (1988). An independent convergence analysis was given by Griepentrog & März (1986), März (1990). They considered general linear multistep methods and problems, where the differential and algebraic equations (and/or variables) are not explicitly separated.

There are several implementations of the BDF schemes for differential-algebraic systems. The most widely used code is DASSL of Petzold (1982). It is described in detail in the book of Brenan, Campbell & Petzold (1989). Further implementations are LSODI of Hindmarsh (1980) and SPRINT of Berzins & Furzeland (1985).

In this section we consider semi-explicit problems

$$\begin{aligned} y' &= f(y,z) \\ 0 &= g(y). \end{aligned} \tag{3.1}$$

We assume that f and g are sufficiently differentiable and that

$$g_y(y) f_z(y,z) \quad \text{is invertible} \tag{3.2}$$

in a neighbourhood of the solution, so that the problem has index 2. A linear multistep method for (3.1) reads

$$\sum_{i=0}^{k} \alpha_i y_{n+i} = h \sum_{i=0}^{k} \beta_i f(y_{n+i}, z_{n+i}) \tag{3.3a}$$

$$0 = g(y_{n+k}). \tag{3.3b}$$

This is not the only meaningful definition of a multistep method for (3.1). One could as well replace (3.3b) by

$$0 = \sum_{i=0}^{k} \beta_i g(y_{n+i}), \tag{3.4}$$

which is obtained by putting $\varepsilon = 0$ in (VI.2.1). The following results can be extended without any difficulty to the second approach. For BDF schemes (where $\beta_0 = \ldots = \beta_{k-1} = 0$) both definitions are equivalent.

The convergence results of this section are also valid for index 2 systems of the form $y' = f(y,z)$, $0 = g(y,z)$, if they can be transformed to (3.1) without any differentiation (see the discussion after Eq. (1.14)). This is because the multistep method (3.3) is invariant with respect to these transformations. The same is true for problems of the form $M(u)u' = \varphi(u)$, if the multistep method is defined by

$$\sum_{i=0}^{k} \alpha_i u_{n+i} = h \sum_{i=0}^{k} \beta_i v_{n+i}, \qquad M(u_{n+k})v_{n+k} = \varphi(u_{n+k}). \tag{3.5}$$

Existence and Uniqueness of Numerical Solution

Equations (3.3) constitute a nonlinear system for y_{n+k}, z_{n+k}. We have the following result about the existence of its solution.

Theorem 3.1. *Suppose that for a solution $y(x), z(x)$ of (3.1) the starting values satisfy for $j = 0, \ldots, k-1$ and $x_j = x_0 + jh$*

$$y_j - y(x_j) = \mathcal{O}(h), \qquad z_j - z(x_j) = \mathcal{O}(h), \qquad g(y_j) = \mathcal{O}(h^2). \tag{3.6}$$

If (3.2) holds in a neighbourhood of this solution and if $\beta_k \neq 0$, then the nonlinear system

$$\sum_{i=0}^{k} \alpha_i y_i = h \sum_{i=0}^{k} \beta_i f(y_i, z_i) \tag{3.7a}$$

$$0 = g(y_k) \tag{3.7b}$$

has a solution for $h \leq h_0$. This solution is locally unique and satisfies

$$y_k - y(x_k) = \mathcal{O}(h), \qquad z_k - z(x_k) = \mathcal{O}(h). \tag{3.8}$$

Proof. We put

$$\eta = -\sum_{i=0}^{k-1} \frac{\alpha_i}{\alpha_k} y_i + h \sum_{i=0}^{k-1} \frac{\beta_i}{\alpha_k} f(y_i, z_i) \tag{3.9}$$

and define ζ close to $z(x_k)$ such that $g_y(\eta) f(\eta, \zeta) = 0$. We further replace $h(\beta_k/\alpha_k)$ by a new step size which we again denote by h. Then the system (3.7) is equivalent to

$$y_k = \eta + hf(y_k, z_k) \tag{3.10a}$$

$$0 = g(y_k) \tag{3.10b}$$

which is simply the implicit Euler method.

We next show that

$$\eta - y(x_k) = \mathcal{O}(h), \qquad \zeta - z(x_k) = \mathcal{O}(h), \qquad g(\eta) = \mathcal{O}(h^2). \tag{3.11}$$

The first relation follows from $y_j - y(x_j) = \mathcal{O}(h)$ and from $\sum_{i=0}^k \alpha_i = 0$; the second is a consequence of the definition of ζ and of (3.2). The last relation of (3.11) can be seen as follows: we replace all $f(y_i, z_i)$ in (3.9) by $f(y(x_k), z(x_k))$, introducing an error of size $\mathcal{O}(h^2)$ in η. Hence

$$\eta - y(x_k) = -\sum_{i=0}^{k-1} \frac{\alpha_i}{\alpha_k}\big(y_i - y(x_k)\big) + h\Big(\sum_{i=0}^{k-1} \frac{\beta_i}{\alpha_k}\Big) f\big(y(x_k), z(x_k)\big) + \mathcal{O}(h^2).$$

Because of (1.14b,c) this implies

$$g(\eta) = -\sum_{i=0}^{k-1} \frac{\alpha_i}{\alpha_k} g_y\big(y(x_k)\big)\big(y_i - y(x_k)\big) + \mathcal{O}(h^2). \tag{3.12}$$

The last statement of (3.11) now follows from the fact that $g_y(y(x_k))(y_i - y(x_k)) = g(y_i) + \mathcal{O}(h^2)$ and from (3.6).

To show the existence of a locally unique solution of (3.10), it is possible to adapt the proof of "Theorem 4.1" of HLR89 to the implicit Euler method. We shall, however, reformulate (3.10) in such a way that the implicit function theorem is applicable. We write (3.10b) as

$$\begin{aligned} 0 = g(y_k) &= g(y_k) - g\big(\eta(h)\big) + g\big(\eta(h)\big) \\ &= \int_0^1 g_y\Big(\eta(h) + \tau\big(y_k - \eta(h)\big)\Big)d\tau \cdot \big(y_k - \eta(h)\big) + g\big(\eta(h)\big) \end{aligned} \tag{3.13}$$

where we have explicitly indicated the dependence of η on h. Replacing the factor $y_k - \eta(h)$ by $hf(y_k, z_k)$ from (3.10a) and dividing by h we get the system

$$y_k - \eta(h) - hf(y_k, z_k) = 0 \tag{3.14a}$$

$$\int_0^1 g_y\Big(\eta(h) + \tau\big(y_k - \eta(h)\big)\Big)d\tau \cdot f(y_k, z_k) + \frac{1}{h} g\big(\eta(h)\big) = 0 \tag{3.14b}$$

which is the discrete analogue of system (1.14a,c). For $h = 0$ the values $y_k = \eta(0)$ and $z_k = \zeta(0)$ satisfy (3.14) because $g(\eta(h)) = \mathcal{O}(h^2)$ and $g_y(\eta)f(\eta, \zeta) = 0$. Further, the derivative of (3.14) with repect to (y_k, z_k) is of the form

$$\begin{pmatrix} I + \mathcal{O}(h) & \mathcal{O}(h) \\ \mathcal{O}(1) & (g_y f_z)(\eta, \zeta) + \mathcal{O}(h) \end{pmatrix}, \tag{3.15}$$

which has a bounded inverse for $h \le h_0$. Therefore the implicit function theorem (Ortega & Rheinboldt 1970, p. 128) yields the existence of a locally unique solution of (3.14) and hence also of (3.10) and (3.7). □

Influence of Perturbations

The influence of perturbations in the multistep formula (3.3) on the numerical solution will be studied in the next theorem.

Theorem 3.2. *Let y_k, z_k be given by (3.7) and consider perturbed values $\widehat{y}_k, \widehat{z}_k$ satisfying*

$$\sum_{i=0}^{k} \alpha_i \widehat{y}_i = h \sum_{i=0}^{k} \beta_i f(\widehat{y}_i, \widehat{z}_i) + h\delta \tag{3.16a}$$

$$0 = g(\widehat{y}_k) + \theta. \tag{3.16b}$$

In addition to the assumptions of Theorem 3.1 suppose that for $j = 0, \ldots, k-1$

$$\widehat{y}_j - y_j = \mathcal{O}(h^2), \qquad \widehat{z}_j - z_j = \mathcal{O}(h), \qquad \delta = \mathcal{O}(h), \qquad \theta = \mathcal{O}(h^2). \tag{3.17}$$

Then, for $h \le h_0$ we have the estimates

$$\begin{aligned} \|\widehat{y}_k - y_k\| &\le C\Big(\|\widehat{Y}_0 - Y_0\| + h\|\widehat{Z}_0 - Z_0\| + h\|\delta\| + \|\theta\|\Big) \\ \|\widehat{z}_k - z_k\| &\le \frac{C}{h}\Big(\sum_{j=0}^{k-1} \|g_y(\widehat{y}_k)(\widehat{y}_j - y_j)\| + h\|\widehat{Y}_0 - Y_0\| \\ &\qquad + h\|\widehat{Z}_0 - Z_0\| + h\|\delta\| + \|\theta\|\Big) \end{aligned} \tag{3.18}$$

where $\widehat{Y}_0 - Y_0 = (\widehat{y}_{k-1} - y_{k-1}, \ldots, \widehat{y}_0 - y_0)^T$, $\|\widehat{Y}_0 - Y_0\| = \max_{0 \le j \le k-1} \|\widehat{y}_j - y_j\|$, and likewise for the z-component.

Proof. In analogy to the proof of Theorem 3.1 we put

$$\widehat{\eta} = -\sum_{i=0}^{k-1} \frac{\alpha_i}{\alpha_k} \widehat{y}_i + h \sum_{i=0}^{k-1} \frac{\beta_i}{\alpha_k} f(\widehat{y}_i, \widehat{z}_i)$$

and rescale h and δ, so that (3.16) becomes

$$\widehat{y}_k = \widehat{\eta} + hf(\widehat{y}_k, \widehat{z}_k) + h\delta \tag{3.19a}$$

$$0 = g(\widehat{y}_k) + \theta. \tag{3.19b}$$

As in the proof of Theorem 3.1 we conclude from (3.17) that $\widehat{y}_k - \widehat{\eta} = \mathcal{O}(h)$ and $\widehat{z}_k - \widehat{\zeta} = \mathcal{O}(h)$, where $\widehat{\zeta}$ is such that $g_y(\widehat{\eta}) f(\widehat{\eta}, \widehat{\zeta}) = 0$. Inspired by Eq. (3.14) we rewrite (3.19b) as

$$0 = \int_0^1 g_y\Big(\widehat{\eta} + \tau(\widehat{y}_k - \widehat{\eta})\Big) d\tau \cdot \big(f(\widehat{y}_k, \widehat{z}_k) + \delta\big) + \frac{1}{h} g(\widehat{\eta}) + \frac{1}{h}\theta, \tag{3.20}$$

which is now a discrete analogue of Eq. (1.29). Subtracting (3.20) from (3.14b) and exploiting the fact that the matrix $g_y f_z$ is invertible, we deduce the estimate

$$\|\widehat{z}_k - z_k\| \le C\Big(\|\widehat{y}_k - y_k\| + \|\widehat{\eta} - \eta\| + \|\delta\| + \frac{1}{h}\|g(\widehat{\eta}) - g(\eta)\| + \frac{1}{h}\|\theta\|\Big). \tag{3.21}$$

A Lipschitz condition for f applied to the difference of (3.19a) and (3.14a) yields

$$\|\widehat{y}_k - y_k\| \le \|\widehat{\eta} - \eta\| + hL\big(\|\widehat{y}_k - y_k\| + \|\widehat{z}_k - z_k\|\big) + h\|\delta\|.$$

Combining the last two estimates we get

$$\begin{aligned} \|\widehat{y}_k - y_k\| &\le C\big(\|\widehat{\eta} - \eta\| + h\|\delta\| + \|\theta\|\big) \\ \|\widehat{z}_k - z_k\| &\le \frac{C}{h}\big(\|g_y(\widehat{\eta})(\widehat{\eta} - \eta)\| + h\|\widehat{\eta} - \eta\| + h\|\delta\| + \|\theta\|\big). \end{aligned} \tag{3.22}$$

The conclusion now follows from the definitions of η and ζ and from $\widehat{y}_k - \widehat{\eta} = \mathcal{O}(h)$. □

Remark 3.3. a) The above proof shows that the constant C in (3.18) depends on bounds for certain derivatives of f and g, but not on the constants implied by the $\mathcal{O}(\ldots)$ terms in (3.17) (if h is sufficiently small). This observation will be used in the convergence proof below.

b) For one-step methods (e.g., implicit Euler method, trapezoidal rule) the term $\|\sum_{j=0}^{k-1} g_y(\widehat{y}_k)(\widehat{y}_j - y_j)\|$ can be omitted in (3.18), if we require $g(y_0) = g(\widehat{y}_0) = 0$. Indeed, it follows from $\widehat{y}_1 = \widehat{y}_0 + \mathcal{O}(h)$ that $g_y(\widehat{y}_1)(\widehat{y}_0 - y_0) = g_y(\widehat{y}_0)(\widehat{y}_0 - y_0) + \mathcal{O}(h\|\widehat{y}_0 - y_0\|)$. Further we have

$$g_y(\widehat{y}_0)(\widehat{y}_0 - y_0) = g(\widehat{y}_0) - g(y_0) + \mathcal{O}(\|\widehat{y}_0 - y_0\|^2),$$

so that the term in question is estimated by $\mathcal{O}(h\|\widehat{y}_0 - y_0\|)$ if h is sufficiently small.

The Local Error

Consider initial values $y_j = y(x_j)$, $z_j = z(x_j)$ $(j = 0, \ldots, k-1)$ on the exact solution of (3.1) and apply the multistep formula (3.7) once. The differences $y_k - y(x_k)$ and $z_k - z(x_k)$ are then called the *local errors* of the method.

Lemma 3.4. *Suppose that the DAE (3.1) satisfies (3.2) and that the multistep method (3.7) has order p (in the sense of Sect. III.2). Then its local error satisfies*

$$y_k - y(x_k) = \mathcal{O}(h^{p+1}), \qquad z_k - z(x_k) = \mathcal{O}(h^p). \tag{3.23}$$

Proof. We put $\widehat{y}_j = y(x_j)$, $\widehat{z}_j = z(x_j)$ for $j = 0, \ldots, k$. These values satisfy (3.16) with $\delta = \mathcal{O}(h^p)$ and $\theta = 0$. Since $\widehat{y}_j = y_j$ and $\widehat{z}_j = z_j$ for $j < k$, the statement follows immediately from Theorem 3.2. □

Convergence for BDF

The study of convergence is simpler for BDF schemes than for general multistep methods, because y_{n+k} depends only on $y_n, \ldots, y_{n+k-1}$, but not on $z_n, \ldots, z_{n+k-1}$ (due to $\beta_0 = \ldots = \beta_{k-1} = 0$). Therefore the y- and z-components can be treated separately. The following convergence result was obtained by Gear, Gupta & Leimkuhler (1985), Lötstedt & Petzold (1986) and Brenan & Engquist (1988).

Theorem 3.5. *Consider an index 2 problem (3.1) which satisfies (3.2). Then the k-step BDF scheme (III.1.22') is convergent of order $p = k$, if $k \le 6$; i.e.,*

$$y_n - y(x_n) = \mathcal{O}(h^p), \qquad z_n - z(x_n) = \mathcal{O}(h^p) \quad \textit{for} \quad x_n = nh \le \textit{Const}, \tag{3.24}$$

whenever the initial values satisfy

$$y_j - y(x_j) = \mathcal{O}(h^{p+1}) \qquad \textit{for} \quad j = 0, \ldots, k-1. \tag{3.25}$$

Remark. The assumption (3.25) can be relaxed to $y_j - y(x_j) = \mathcal{O}(h^p)$ for $k \ge 3$, but not for $k = 1$ (see Exercise 1).

Proof. We combine the convergence proof for Runge-Kutta methods (HLR89, Theorem 4.4) with the techniques of Sect. III.4. Inspired by Lady Windermere's Fan (Fig. III.4.1) we first study the propagation of the local errors and their accumulation over the whole interval for the y-component (part a). The z-component is treated in part (b) and technical details are given in part (c).

a) In addition to the numerical solution $\{y_n, z_n\}$, which we now also denote by $\{y_n^0, z_n^0\}$, we consider for $\ell = 1, 2, \ldots$ the multistep solutions $\{y_n^\ell, z_n^\ell\}$ with starting values $y_j^\ell = y(x_j)$, $z_j^\ell = z(x_j)$ for $j = \ell-1, \ldots, \ell+k-2$ on the exact solution. Our first aim is to estimate $y_n^\ell - y_n^{\ell+1}$ in terms of the local errors $y_{\ell+k-1}^\ell - y_{\ell+k-1}^{\ell+1}$ (or starting errors if $\ell = 0$). For simplicity we omit the upper index and consider two neighbouring multistep solutions $\{\widehat{y}_n, \widehat{z}_n\}$ and $\{\widetilde{y}_n, \widetilde{z}_n\}$. In order to be able to apply Theorem 3.2 we fix three sufficiently large constants C_0, C_1, C_2 and suppose that for $nh \le \textit{Const}$

$$\|\widehat{y}_n - y(x_n)\| \le C_0 h, \qquad \|\widetilde{y}_n - \widehat{y}_n\| \le C_1 h^2, \qquad \|\widehat{z}_n - z(x_n)\| \le C_2 h. \tag{3.26}$$

This will be justified in part (c) below. We introduce the notation $\Delta y_n = \widetilde{y}_n - \widehat{y}_n$, $\Delta z_n = \widetilde{z}_n - \widehat{z}_n$ and $\Delta Y_n = (\Delta y_{n+k-1}, \ldots, \Delta y_n)^T$. Observing that y_{n+k}, z_{n+k} do not depend on $z_n, \ldots, z_{n+k-1}$ for the BDF schemes, it follows from Theorem 3.2 with $\delta = 0$ and $\theta = 0$ that

$$\|\Delta y_{n+k}\| \le C \|\Delta Y_n\| \tag{3.27a}$$

$$\|\Delta z_{n+k}\| \le \frac{C}{h} \Big(\sum_{j=0}^{k-1} \|g_y(\widehat{y}_{n+k}) \Delta y_{n+j}\| + h \|\Delta Y_n\| \Big). \tag{3.27b}$$

Here C does not depend on the choice of C_0, C_1, C_2, if h is sufficiently small (see Remark 3.3a). Our assumption (3.26) together with (3.27) implies $\Delta y_{n+k} = \mathcal{O}(h^2)$

and $\Delta z_{n+k} = \mathcal{O}(h)$. We therefore obtain by linearization of the multistep formula

$$\sum_{i=0}^{k} \alpha_i \Delta y_{n+i} = h\beta_k f_z(\widehat{y}_{n+k}, \widehat{z}_{n+k})\Delta z_{n+k} + \mathcal{O}(h\|\Delta Y_n\|) \tag{3.28a}$$

$$0 = g_y(\widehat{y}_{n+k})\Delta y_{n+k} + \mathcal{O}(h\|\Delta Y_n\|). \tag{3.28b}$$

We next use the projections (see also Definition 4.3 below)

$$Q_n = \big(f_z(g_y f_z)^{-1} g_y\big)(\widehat{y}_{n+k}, \widehat{z}_{n+k}), \qquad P_n = I - Q_n \tag{3.29}$$

for which

$$P_n^2 = P_n, \quad Q_n^2 = Q_n, \quad P_n Q_n = Q_n P_n = 0, \quad Q_{n+1} = Q_n + \mathcal{O}(h). \tag{3.30}$$

The last relation of (3.30) follows from (3.26) and the smoothness of the solution $y(x), z(x)$. We then multiply (3.28a) by P_{n+k} (which eliminates Δz_{n+k}) and (3.28b) by $f_z(g_y f_z)^{-1}$. This yields with (3.30)

$$\sum_{i=0}^{k} \alpha_i P_{n+i} \Delta y_{n+i} = \mathcal{O}(h\|\Delta Y_n\|), \qquad Q_{n+k}\Delta y_{n+k} = \mathcal{O}(h\|\Delta Y_n\|). \tag{3.31}$$

Introducing the vectors

$$U_n = (P_{n+k-1}\Delta y_{n+k-1}, \ldots, P_n \Delta y_n)^T,$$
$$V_n = (Q_{n+k-1}\Delta y_{n+k-1}, \ldots, Q_n \Delta y_n)^T,$$

we have $\Delta Y_n = U_n + V_n$ and the relations (3.31) become

$$U_{n+1} = (A \otimes I)U_n + \mathcal{O}(h\|U_n\| + h\|V_n\|) \tag{3.32a}$$
$$V_{n+1} = (N \otimes I)V_n + \mathcal{O}(h\|U_n\| + h\|V_n\|) \tag{3.32b}$$

where (with $\alpha_j' = \alpha_j/\alpha_k$)

$$A = \begin{pmatrix} -\alpha_{k-1}' & \ldots & -\alpha_1' & -\alpha_0' \\ 1 & & 0 & 0 \\ & \ddots & \vdots & \vdots \\ & & 1 & 0 \end{pmatrix}, \qquad N = \begin{pmatrix} 0 & \ldots & 0 & 0 \\ 1 & & 0 & 0 \\ & \ddots & \vdots & \vdots \\ & & 1 & 0 \end{pmatrix}. \tag{3.33}$$

According to Lemma III.4.4 we now choose a norm $\|U\|$ such that $\|A \otimes I\| \le 1$. We then choose a (possibly different) norm $\|V\|$, for which $\|N \otimes I\| \le \varrho < 1$. Consequently it follows from (3.32) that

$$\begin{pmatrix} \|U_{n+1}\| \\ \|V_{n+1}\| \end{pmatrix} \le \begin{pmatrix} 1 + \mathcal{O}(h) & \mathcal{O}(h) \\ \mathcal{O}(h) & \varrho + \mathcal{O}(h) \end{pmatrix} \begin{pmatrix} \|U_n\| \\ \|V_n\| \end{pmatrix}. \tag{3.34}$$

As in the proof of Lemma VI.3.9 we diagonalize the matrix in (3.34) and so obtain

$$\|\Delta Y_n\| \le Const_1\big(\|U_n\| + \|V_n\|\big)$$
$$\le Const_2\big(\|U_0\| + (\varrho^n + h)\|V_0\|\big), \tag{3.35a}$$
$$\|V_n\| \le Const_3\big(h\|U_0\| + (\varrho^n + h)\|V_0\|\big). \tag{3.35b}$$

The vectors U_0 and V_0 are composed of local errors (of the y-component) or of errors in the starting values, which are of size $\mathcal{O}(h^{p+1})$ by (3.23) and (3.25). Hence, it follows from (3.35) that the propagated errors satisfy

$$\begin{aligned} \|\Delta y_n\| &\le C_3 h^{p+1}, \\ \|g_y(\widehat{y}_{n+k})\Delta y_{n+j}\| &\le C_4(\varrho^n + h)h^{p+1} \quad \text{for } j = 0, \ldots, k-1. \end{aligned} \tag{3.36}$$

Summing up we obtain

$$\|y_n - y(x_n)\| \le \sum_{\ell=0}^{n-k+1} \|y_n^\ell - y_n^{\ell+1}\| \le C_5 h^p, \tag{3.37}$$

the desired estimate for the y-component.

b) Since z_n depends only on $y_{n-k}, \ldots, y_{n-1}$ but not on the previous z-values, we can apply Theorem 3.2 with $\widehat{y}_i = y(x_i)$, $\widehat{z}_i = z(x_i)$, $\delta = \mathcal{O}(h^p)$ and $\theta = 0$. This yields

$$\|z_n - z(x_n)\| \le \frac{C}{h} \sum_{j=1}^{k} \|g_y\big(y(x_n)\big)\big(y_{n-j} - y(x_{n-j})\big)\| + \mathcal{O}(h^p). \tag{3.38}$$

Using (3.36) and $y_n^\ell = y(x_n) + \mathcal{O}(h^p)$, which follows as in (3.37), we obtain

$$\begin{aligned} \|g_y\big(y(x_n)\big)\big(y_{n-j} - y(x_{n-j})\big)\| &= \Big\| \sum_{\ell=0}^{n-k+1} g_y\big(y(x_n)\big)\big(y_{n-j}^\ell - y_{n-j}^{\ell+1}\big) \Big\| \\ &\le \sum_{\ell=0}^{n-k+1} \Big(\|g_y(y_n^\ell)(y_{n-j}^\ell - y_{n-j}^{\ell+1})\| + \mathcal{O}(h^{2p+1}) \Big) = \mathcal{O}(h^{p+1}) \end{aligned}$$

and hence also

$$\|z_n - z(x_n)\| \le C_6 h^p. \tag{3.39}$$

c) In general, the constants C_3, C_5 and C_6 will depend on C_0, C_1, C_2 of our assumption (3.26). For $p \ge 2$ we can restrict the step size h so that

$$C_5 h^{p-1} \le C_0, \qquad C_3 h^{p-1} \le C_1, \qquad C_6 h^{p-1} \le C_2$$

and the numerical solutions will never violate the conditions (3.26) on the considered interval.

For $p = 1$ (the implicit Euler method) we know from Remark 3.3b that the estimate (3.27b) can be replaced by

$$\|\Delta z_{n+k}\| \le C \|\Delta Y_n\|. \tag{3.40}$$

Instead of (3.28a) we thus immediately get

$$\Delta y_{n+1} - \Delta y_n = \mathcal{O}(h\|\Delta y_n\|) \tag{3.41}$$

where the constant implied by the $\mathcal{O}(\ldots)$ term is independent of C_0, C_1, C_2, if h is sufficiently small. Standard techniques (without considering the projections (3.29)) then yield the convergence result. □

With the ideas of Sect. III.5 the above proof can be extended to cover variable step sizes as well. Originally, such a convergence result was given by Gear, Gupta & Leimkuhler (1985).

General Multistep Methods

For a general multistep method (3.3) with generating polynomials

$$\varrho(\zeta)=\sum_{i=0}^{k}\alpha_i\zeta^i,\qquad \sigma(\zeta)=\sum_{i=0}^{k}\beta_i\zeta^i$$

we have the following convergence result.

Theorem 3.6. *Consider an index 2 problem (3.1) which satisfies (3.2). Assume that the multistep method is stable (Definition III.3.2) and strictly stable at infinity (the zeros of $\sigma(\zeta)$ lie inside the unit disc $|\zeta|<1$). If its order is $p\geq 2$, then the global error satisfies*

$$y_n-y(x_n)=\mathcal{O}(h^p),\qquad z_n-z(x_n)=\mathcal{O}(h^p)\quad \text{for}\quad x_n=nh\leq \text{Const}$$

whenever the initial values satisfy (for $j=0,\dots,k-1$)

$$y_j-y(x_j)=\mathcal{O}(h^{p+1}),\qquad z_j-z(x_j)=\mathcal{O}(h^p).\tag{3.42}$$

Proof. The proof is essentially the same as for the BDF schemes. Due to the dependence of y_{n+k},z_{n+k} on $y_n,\dots,y_{n+k-1}$ *and* on $z_n,\dots,z_{n+k-1}$ the following modifications are necessary.

In addition to (3.26) we assume $\|\widetilde{z}_n-\widehat{z}_n\|\leq C_3h$. Instead of (3.27) we have (from Theorem 3.2)

$$\begin{aligned}\|\Delta y_{n+k}\|&\leq C(\|\Delta Y_n\|+h\|\Delta Z_n\|)\\ \|\Delta z_{n+k}\|&\leq \frac{C}{h}\Big(\sum_{j=0}^{k-1}\|g_y(\widehat{y}_{n+k})\Delta y_{n+j}\|+h\|\Delta Y_n\|+h\|\Delta Z_n\|\Big)\end{aligned}$$

and (3.28) becomes

$$\begin{aligned}\sum_{i=0}^{k}\alpha_i\Delta y_{n+i}&=h\sum_{i=0}^{k}\beta_i f_z(\widehat{y}_{n+k},\widehat{z}_{n+k})\Delta z_{n+i}+\mathcal{O}(h\|\Delta Y_n\|+h^2\|\Delta Z_n\|)\\ 0&=g_y(\widehat{y}_{n+k})\Delta y_{n+k}+\mathcal{O}(h\|\Delta Y_n\|+h^2\|\Delta Z_n\|).\end{aligned}\tag{3.43}$$

A recursion for Δz_n is obtained as follows: we multiply the upper line of (3.43) by $\big((g_yf_z)^{-1}g_y\big)(\widehat{y}_{n+k},\widehat{z}_{n+k})$ and so get

$$\begin{aligned}h\sum_{i=0}^{k}\beta_i\,\Delta z_{n+i}=\sum_{i=0}^{k}\alpha_i\,&\big((g_yf_z)^{-1}g_y\big)(\widehat{y}_{n+k},\widehat{z}_{n+k})\,\Delta y_{n+i}\\ &+\mathcal{O}\big(h\|\Delta Y_n\|+h^2\|\Delta Z_n\|\big).\end{aligned}\tag{3.44}$$

With the projections P_n, Q_n of (3.29) and the vectors U_n, V_n we thus obtain (3.32) with an additional $\mathcal{O}(h^2\|\Delta Z_n\|)$ term. From (3.44) we get

$$h\,\Delta Z_{n+1} = (B\otimes I)\,h\,\Delta Z_n + \mathcal{O}\Big(h\|U_n\| + \|V_n\| + h^2\|\Delta Z_n\|\Big),$$

where

$$B = \begin{pmatrix} -\beta'_{k-1} & \cdots & -\beta'_1 & -\beta'_0 \\ 1 & & 0 & 0 \\ & \ddots & \vdots & \vdots \\ & & 1 & 0 \end{pmatrix}$$

with $\beta'_j = \beta_j/\beta_k$. For this equation we use a norm for which $\|B\otimes I\| \le \kappa < 1$. This is possible, because the method is strictly stable at infinity. Summarizing, we get the inequality

$$\begin{pmatrix} \|U_{n+1}\| \\ \|V_{n+1}\| \\ h\,\|\Delta Z_{n+1}\| \end{pmatrix} \le \begin{pmatrix} 1+\mathcal{O}(h) & \mathcal{O}(h) & \mathcal{O}(h) \\ \mathcal{O}(h) & \varrho+\mathcal{O}(h) & \mathcal{O}(h) \\ \mathcal{O}(h) & \mathcal{O}(1) & \kappa+\mathcal{O}(h) \end{pmatrix} \begin{pmatrix} \|U_n\| \\ \|V_n\| \\ h\,\|\Delta Z_n\| \end{pmatrix} \tag{3.45}$$

which can be solved as before and yields

$$\begin{aligned} &\|\Delta y_n\| \le C_3 h^{p+1}, \qquad \|\Delta z_n\| \le C_7(\varrho^n + \kappa^n + h)h^p, \\ &\|g_y(\widehat{y}_{n+k})\Delta y_{n+j}\| \le C_4(\varrho^n + \kappa^n + h)h^{p+1} \qquad \text{for } j = 0,\ldots,k-1. \end{aligned} \tag{3.46}$$

Summing up the propagated errors as in (3.37) we obtain the desired estimates for the y-and z-component. □

Solution of the Nonlinear System by Simplified Newton

The nonlinear system (3.3) is usually solved by a simplified Newton iteration and it is interesting to study its convergence. As in the proof of Theorem 3.1 we introduce η by (3.9) and rescale h so that the nonlinear system becomes (omitting the indices)

$$\begin{aligned} y - \eta - hf(y,z) &= 0 \\ g(y) &= 0. \end{aligned} \tag{3.47}$$

This is just the implicit Euler method and we can apply the discussion of HLR89, Chapter 7. The Jacobian of the nonlinear system (3.47) is

$$J = \begin{pmatrix} I - hf_y & -hf_z \\ g_y & 0 \end{pmatrix} \tag{3.48}$$

and its inverse has the form

$$J^{-1} = \begin{pmatrix} P + \mathcal{O}(h) & f_z(g_y f_z)^{-1} + \mathcal{O}(h) \\ -h^{-1}(g_y f_z)^{-1} g_y + \mathcal{O}(1) & h^{-1}(g_y f_z)^{-1} + \mathcal{O}(1) \end{pmatrix} \tag{3.49}$$

where $P = I - f_z(g_y f_z)^{-1} g_y$ is the projection of (3.29). We now consider the simplified Newton method as a fixed point iteration with the function

$$\Phi(y,z)=\begin{pmatrix} y\\ z\end{pmatrix}-J_0^{-1}\begin{pmatrix} y-\eta-hf(y,z)\\ g(y)\end{pmatrix}. \tag{3.50}$$

The subscript 0 in J_0 indicates that the arguments of the derivatives in (3.48) are evaluated at some *fixed* approximation $(\widehat{\eta},\widehat{\zeta})$ to the solution of (3.47). We shall use the notation $\{f_y\}_0$ for $f_y(\widehat{\eta},\widehat{\zeta})$, etc. Direct calculation of $\Phi'(y,z)$ gives

$$\begin{pmatrix} \{f_z(g_yf_z)^{-1}\}_0(\{g_y\}_0-g_y)+\mathcal{O}(h) & h\{P\}_0f_z+\mathcal{O}(h^2)\\ h^{-1}\{(g_yf_z)^{-1}\}_0(\{g_y\}_0-g_y)+\mathcal{O}(1) & \{(g_yf_z)^{-1}g_y\}_0(\{f_z\}_0-f_z)+\mathcal{O}(h)\end{pmatrix}.$$

If we assume that $(\widehat{\eta},\widehat{\zeta})$ approximates the fixed point of (3.50) with an error of $\mathcal{O}(h)$, then we have at this fixed point

$$\Phi'(y,z)=\begin{pmatrix} \mathcal{O}(h) & \mathcal{O}(h^2)\\ \mathcal{O}(1) & \mathcal{O}(h)\end{pmatrix}. \tag{3.51}$$

With the scaling matrix $D=\operatorname{diag}(I,hI)$ (this corresponds to a multiplication of the z-variables by h) we have $\|D\Phi'(y,z)D^{-1}\|=\mathcal{O}(h)$. In the norm $\|y\|+h\|z\|$ we therefore gain a factor h in each simplified Newton iteration.

Remark. The above analysis remains valid if f_y or parts of it are replaced by zero in J_0. For mechanical problems such an algorithm was proposed by Gear, Gupta & Leimkuhler (1985).

Exercises

1. Show that the assumption $g(y_j)=\mathcal{O}(h^2)$ for $j=0,\ldots,k-1$ cannot be omitted in Theorem 3.1.

   Counterexample. Consider the system $x'=1, y'=k(z), 0=y-x$, where $k(z)=(e^{z-1}+1)/2$. Apply the implicit Euler method with initial values $x_0=0$, $y_0=h$, $z_0=1$.

2. (Gear, Hsu & Petzold 1981, Gear & Petzold 1984). Consider the problem

$$\begin{pmatrix} 0 & 0\\ 1 & \eta x\end{pmatrix}\begin{pmatrix} y'\\ z'\end{pmatrix}+\begin{pmatrix} 1 & \eta x\\ 0 & 1+\eta\end{pmatrix}\begin{pmatrix} y\\ z\end{pmatrix}=\begin{pmatrix} f(x)\\ g(x)\end{pmatrix}. \tag{3.52}$$

   a) Prove that the system (3.52) has index 2 for all values of η.

   b) The z-component of the exact solution is $z(x)=g(x)-f'(x)$.

   c) The implicit Euler method, applied to (3.52) in an obvious manner, yields the recursion

$$z_{n+1}=\frac{\eta}{1+\eta}z_n+\frac{1}{1+\eta}\Big(g(x_{n+1})-\frac{f(x_{n+1})-f(x_n)}{h}\Big).$$

   Hence, the method is convergent for $\eta>-1/2$, but unstable for $\eta<-1/2$. For $\eta=-1$ the numerical solution does not exist.

VII.4 Runge-Kutta Methods for Index 2 DAE

RK methods prove popular at IMA conference on numerical ODEs.
(Byrne & Hindmarsh, SIAM News, March 1990)

This section is devoted to the convergence of implicit Runge-Kutta methods for semi-explicit index 2 systems (3.1) which satisfy (3.2). The ε-embedding method of Sect. VI.1 defines the numerical solution by

$$y_{n+1} = y_n + h\sum_{i=1}^{s} b_i k_{ni}, \qquad z_{n+1} = z_n + h\sum_{i=1}^{s} b_i \ell_{ni} \tag{4.1a}$$

where

$$k_{ni} = f(Y_{ni}, Z_{ni}), \qquad 0 = g(Y_{ni}) \tag{4.1b}$$

and the internal stages are given by

$$Y_{ni} = y_n + h\sum_{j=1}^{s} a_{ij} k_{nj}, \qquad Z_{ni} = z_n + h\sum_{j=1}^{s} a_{ij} \ell_{nj} \tag{4.1c}$$

(the state space form method (VI.1.12) does not make sense here, because the algebraic conditions do not depend on z).

The first convergence results for this situation are due to Petzold (1986). They are formulated for general problems $F(y', y) = 0$ under the assumption of "uniform index one". Since the system (3.1) becomes "uniform index one" if we replace z by u' (Gear 1988, see also Exercise 1), the results of Petzold can be applied to (3.1). A further study for the semi-explicit system (3.1) is given by Brenan & Petzold (1989). Their main result is that for (4.1) the global error of the y-component is $\mathcal{O}(h^{q+1})$, and that of the z-component is $\mathcal{O}(h^q)$ (where q denotes the stage order of the method). This result was improved by HLR89, using a different approach (local and global error are studied separately).

The Nonlinear System

We first investigate existence, uniqueness and the influence of perturbations to the solution of the nonlinear system (4.1). In order to simplify the notation we write (η, ζ) for (y_n, z_n), which we assume h-dependent, and we suppress the index n in Y_{ni}, etc. The nonlinear system then reads

$$\left.\begin{aligned} Y_i &= \eta + h\sum_{j=1}^{s} a_{ij} f(Y_j, Z_j) \\ 0 &= g(Y_i) \end{aligned}\right\} \qquad i = 1, \ldots, s \tag{4.2}$$

Once a solution to (4.2) is known, we can compute ℓ_{ni} from (4.1c) (whenever (a_{ij}) is an invertible matrix) and then y_{n+1}, z_{n+1} from (4.1a).

Theorem 4.1 (HLR89, p. 31). *Suppose that* (η, ζ) *satisfy*

$$g(\eta) = \mathcal{O}(h^2), \qquad g_y(\eta) f(\eta, \zeta) = \mathcal{O}(h) \tag{4.3}$$

and that (3.2) holds in a neighbourhood of (η, ζ). *If the Runge-Kutta matrix* (a_{ij}) *is invertible, then the nonlinear system (4.2) possesses for* $h \le h_0$ *a locally unique solution which satisfies*

$$Y_i - \eta = \mathcal{O}(h), \qquad Z_i - \zeta = \mathcal{O}(h). \tag{4.4}$$

Remark. Condition (4.3) expresses the fact that (η, ζ) is close to consistent initial values. We also see from (4.2) that the solution (Y_i, Z_i) does not depend on ζ. The value of ζ in (4.3) only specifies the solution branch of $g_y(y) f(y, z) = 0$ to which the numerical solution is close.

The *proof* of Theorem 4.1 for the implicit Euler method was given in Sect. VII.3 (proof of Theorem 3.1). If we replace (3.14) by

$$Y_i - \eta(h) - h \sum_{j=1}^{s} a_{ij} f(Y_j, Z_j) = 0 \tag{4.5a}$$

$$\int_0^1 g_y\Big(\eta(h) + \tau\big(Y_i - \eta(h)\big)\Big)\, d\tau \cdot \sum_{j=1}^{s} a_{ij} f(Y_j, Z_j) + \frac{1}{h} g\big(\eta(h)\big) = 0 \tag{4.5b}$$

it extends in a straightforward manner to general Runge-Kutta methods. □

Influence of Perturbations. Besides (4.2) we also consider the perturbed system

$$\left.\begin{aligned} \widehat{Y}_i &= \widehat{\eta} + h\sum_{j=1}^{s} a_{ij} f(\widehat{Y}_j, \widehat{Z}_j) + h\delta_i \\ 0 &= g(\widehat{Y}_i) + \theta_i \end{aligned}\right\} \qquad i = 1, \ldots, s \tag{4.6}$$

and we investigate the influence of the perturbations δ_i and θ_i on the numerical solution.

Theorem 4.2 (HLR89, p. 33). *Let* Y_i, Z_i *be a solution of (4.2) and consider perturbed values* $\widehat{Y}_i, \widehat{Z}_i$ *satisfying (4.6). In addition to the assumptions of Theorem 4.1 suppose that*

$$\widehat{\eta} - \eta = \mathcal{O}(h^2), \qquad \widehat{Z}_i - \zeta = \mathcal{O}(h), \qquad \delta_i = \mathcal{O}(h), \qquad \theta_i = \mathcal{O}(h^2). \tag{4.7}$$

Then we have for $h \le h_0$ *the estimates*

$$\|\widehat{Y}_i - Y_i\| \le C\Big(\|\widehat{\eta} - \eta\| + h\|\delta\| + \|\theta\|\Big) \tag{4.8a}$$

$$\|\widehat{Z}_i - Z_i\| \le \frac{C}{h}\Big(\|g_y(\eta)(\widehat{\eta} - \eta)\| + h\|\widehat{\eta} - \eta\| + h\|\delta\| + \|\theta\|\Big) \tag{4.8b}$$

where $\|\delta\| = \max_i \|\delta_i\|$ *and* $\|\theta\| = \max_i \|\theta_i\|$. *If the initial values satisfy* $g(\eta) = 0$ *and* $g(\widehat{\eta}) = 0$, *then we have the stronger estimate*

$$\|\widehat{Z}_i - Z_i\| \le \frac{C}{h}\Big(h\|\widehat{\eta} - \eta\| + h\|\delta\| + \|\theta\|\Big). \tag{4.9}$$

The constant C *in (4.8) and (4.9) depends only on bounds for certain derivatives of* f *and* g, *but not on the constants implied by the* $\mathcal{O}(\ldots)$ *terms in (4.3) and (4.7).*

Proof. The estimates (4.8) are obtained by extending the proof of Theorem 3.2. When both initial values, η and $\widehat{\eta}$, lie on the manifold $g(y) = 0$, we have by Taylor expansion $0 = g(\widehat{\eta}) - g(\eta) = g_y(\eta)(\widehat{\eta} - \eta) + \mathcal{O}(\|\widehat{\eta} - \eta\|^2)$. In this situation the term $g_y(\eta)(\widehat{\eta} - \eta)$ in (4.8b) is of size $\mathcal{O}(h^2\|\widehat{\eta} - \eta\|)$ and may be neglected. □

Estimation of the Local Error

We begin by defining two projections which will be important for the study of local errors for index 2 problems (3.1).

Definition 4.3. For given y_0, z_0 for which $(g_y f_z)(y_0, z_0)$ is invertible we define the *projections*

$$Q = \big(f_z(g_y f_z)^{-1} g_y\big)(y_0, z_0), \qquad P = I - Q. \tag{4.10}$$

Geometric interpretation. Let $\mathcal{U}$ be the manifold defined by $\mathcal{U} = \{y; g(y) = 0\}$ and let $T_{y_0}\mathcal{U} = \ker(g_y(y_0))$ be the tangent space at a point $y_0 \in \mathcal{U}$. Further let $\mathcal{V} = \{f(y_0, z)\ ;\ z \text{ arbitrary}\}$ and let $T_{f_0}\mathcal{V} = \mathrm{Im}\,(f_z(y_0, z_0))$ be its tangent space at $f_0 = f(y_0, z_0)$. Here, z_0 is the value for which $f(y_0, z_0)$ lies in $T_{y_0}\mathcal{U}$ (i.e., for which the condition $g_y(y_0)f(y_0, z_0) = 0$ is satisfied (see 1.14c)). By considering the arrows $f(y_0, z)$ with varying z (see Fig. 4.1), the space $T_{f_0}\mathcal{V}$ can be interpreted as the directions in which the control variables z bring the solution to the manifold $\mathcal{U}$. By (3.2) these two spaces are transversal and their direct sum generates the y-space. It follows from (4.10) that P projects onto $T_{y_0}\mathcal{U}$ parallel to $T_{f_0}\mathcal{V}$ and Q projects onto $T_{f_0}\mathcal{V}$ parallel to $T_{y_0}\mathcal{U}$.

Consider now initial values $y_0 = y(x)$, $z_0 = z(x)$ on the exact solution and denote by y_1, z_1 the numerical solution of the Runge-Kutta method (4.1). The *local error*

$$\delta y_h(x) = y_1 - y(x+h), \qquad \delta z_h(x) = z_1 - z(x+h) \tag{4.11}$$

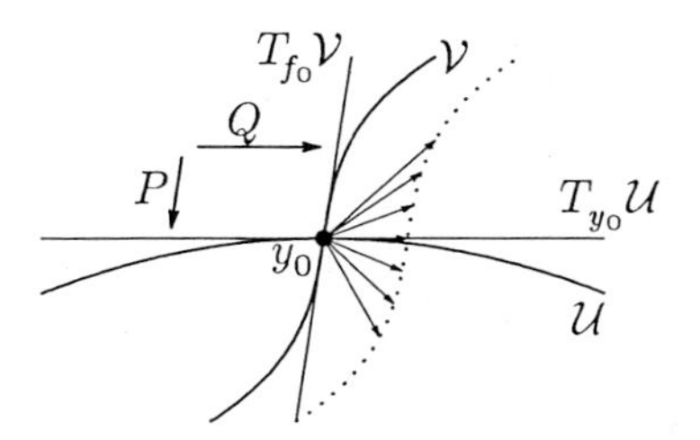

Fig. 4.1. Projections P and Q

can be estimated as follows:

Lemma 4.4 (HLR89, p. 34). *Suppose that a Runge-Kutta method with invertible coefficient matrix (a_{ij}) satisfies the assumptions $B(p)$ and $C(q)$ of Sect. IV.5 with $p \geq q$. Then we have*

$$\begin{aligned} \delta y_h(x) &= \mathcal{O}(h^{q+1}), \qquad P(x)\delta y_h(x) = \mathcal{O}(h^{\min(p+1,q+2)}) \\ \delta z_h(x) &= \mathcal{O}(h^q), \end{aligned} \tag{4.12}$$

where $P(x)$ is the projection (4.10) evaluated at $(y(x), z(x))$. If, in addition, the Runge-Kutta method is stiffly accurate (i.e., satisfies $a_{si} = b_i$ for all i), then

$$\delta y_h(x) = \mathcal{O}(h^{\min(p+1,q+2)}). \tag{4.13}$$

Proof. The exact solution values $\widehat{\eta} = y(x)$, $\widehat{Y}_i = y(x+c_ih)$, $\widehat{Z}_i = z(x+c_ih)$ satisfy (4.6) with $\theta_i = 0$ and

$$\delta_i = \frac{h^q}{q!}\, y^{(q+1)}(x) \Big(\frac{c_i^{q+1}}{q+1} - \sum_{j=1}^{s} a_{ij} c_j^q \Big) + \mathcal{O}(h^{q+1}).$$

The difference to the numerical solution ((4.2) with $\eta = y(x)$) can thus be estimated with Theorem 4.2, yielding

$$Y_i - y(x+c_ih) = \mathcal{O}(h^{q+1}), \qquad Z_i - z(x+c_ih) = \mathcal{O}(h^q). \tag{4.14}$$

Since the quadrature formula $\{b_i, c_i\}$ is of order p, we have

$$y(x+h) - y(x) - h\sum_{i=1}^{s} b_i f\big(y(x+c_ih),\, z(x+c_ih)\big) = \mathcal{O}(h^{p+1}).$$

Subtracting this formula from (4.1a) we get

$$y_1 - y(x+h) = hf_z(y(x), z(x)) \sum_{i=1}^{s} b_i\big(Z_i - z(x+c_ih)\big) + \mathcal{O}(h^{p+1}) + \mathcal{O}(h^{q+2}).$$

Because of $P(x) f_z(y(x), z(x)) \equiv 0$, this proves (4.12) for the y-component. The estimate for the z-component follows from (see (1.28))

$$z_1 - z(x+h) = \sum_{i,j=1}^{s} b_i \omega_{ij} \big(Z_j - z(x+c_j h)\big) + \mathcal{O}(h^{q+1})$$

and (4.14).

Under the assumption $a_{si} = b_i$ (for all i) we have $g(y_1) = 0$ so that by Taylor expansion

$$0 = g(y_1) - g\big(y(x+h)\big) = g_y(y(x))\delta y_h(x) + \mathcal{O}(h\|\delta y_h(x)\|). \tag{4.15}$$

This implies that $Q(x)\delta y_h(x) = \mathcal{O}(h\|\delta y_h(x)\|)$, and (4.13) is a consequence of (4.12) and (4.10). □

For some important Runge-Kutta methods (such as Radau IIA and Lobatto IIIC) the estimates of Lemma 4.4 are not optimal. Sharp estimates will be given in Theorem 4.9 for collocation methods and in Sect. VII.5 for general Runge-Kutta methods.

Convergence for the y-Component

The numerical solution $\{y_n\}$, defined by (4.1), does not depend on $\{z_n\}$. Consequently, the convergence for the y-component can be treated independently of estimates for the z-component.

Theorem 4.5 (HLR89, p. 36). *Suppose that (3.2) holds in a neighbourhood of the solution $(y(x), z(x))$ of (3.1) and that the initial values are consistent. Suppose further that the Runge-Kutta matrix (a_{ij}) is invertible, that $|R(\infty)| < 1$ (see (VI.1.11e)) and that the local error satisfies*

$$\delta y_h(x) = \mathcal{O}(h^r), \qquad P(x)\delta y_h(x) = \mathcal{O}(h^{r+1}) \tag{4.16}$$

with $P(x)$ as in Lemma 4.4. Then the method (4.1) is convergent of order r, i.e.,

$$y_n - y(x_n) = \mathcal{O}(h^r) \qquad \textit{for} \quad x_n - x_0 = nh \le Const.$$

If in addition $\delta y_h(x) = \mathcal{O}(h^{r+1})$, then $g(y_n) = \mathcal{O}(h^{r+1})$.

Proof. A complete proof of this result is given in (HLR89, pp. 36-39). We restrict our presentation to stiffly accurate Runge-Kutta methods (i.e., $a_{si} = b_i$ for all i). This considerably simplifies several parts of the proof, and nevertheless covers many important Runge-Kutta methods (such as Radau IIA, Lobatto IIIC and the SDIRK method (IV.6.16)). The assumption $a_{si} = b_i$ (for all i) implies that $g(y_n) = 0$ for all n and, as a consequence of (4.15) and (4.16), that

$$\delta y_h(x) = \mathcal{O}(h^{r+1}). \tag{4.17}$$

The following proof is similar to that of Theorem 3.5 and uses, once again, Lady Windermere's Fan of Fig. II.3.2.

In addition to the numerical solution $\{y_n, z_n\}$, also denoted by $\{y_n^0, z_n^0\}$, we consider the Runge-Kutta solutions $\{y_n^\ell, z_n^\ell\}$ with initial values $y_\ell^\ell = y(x_\ell)$, $z_\ell^\ell = z(x_\ell)$ on the exact solution. We first estimate $y_n^\ell - y_n^{\ell+1}$ for $n \geq \ell+1$ in terms of the local error $\delta y_h(x_\ell) = y_{\ell+1}^\ell - y_{\ell+1}^{\ell+1}$. In order to simplify the notation we denote two neighbouring Runge-Kutta solutions by $\{\widetilde{y}_n\}$, $\{\widehat{y}_n\}$ and their difference by $\Delta y_n = \widetilde{y}_n - \widehat{y}_n$. We suppose for the moment that

$$\|\widehat{y}_n - y(x_n)\| \leq C_0 h, \qquad \|\Delta y_n\| \leq C_1 h^2 \tag{4.18}$$

(this will be justified below). Theorem 4.2 with $\delta_i = 0$ and $\theta_i = 0$ then yields

$$\|\widetilde{Y}_{ni} - \widehat{Y}_{ni}\| \leq C\|\Delta y_n\|, \qquad \|\widetilde{Z}_{ni} - \widehat{Z}_{ni}\| \leq C\|\Delta y_n\| \tag{4.19}$$

where C is some constant independent of C_0 and C_1. A Lipschitz condition for $f(y,z)$ implies that

$$\|\Delta y_{n+1}\| \leq \|\Delta y_n\| + h\sum_{i=1}^{s} |b_i|\Big(L_1\|\widetilde{Y}_{ni} - \widehat{Y}_{ni}\| + L_2\|\widetilde{Z}_{ni} - \widehat{Z}_{ni}\|\Big).$$

Inserting (4.19) we get $\|\Delta y_{n+1}\| \leq (1+hL)\|\Delta y_n\|$ and hence also

$$\|\Delta y_n\| \leq C_2\|\Delta y_0\| \qquad \text{for} \quad nh \leq Const. \tag{4.20}$$

For our situation in Lady Windermere's Fan the use of (4.17) yields

$$\|y_n^\ell - y_n^{\ell+1}\| \leq C_2\|\delta y_h(x_\ell)\| \leq C_3 h^{r+1} \qquad \text{for } n \geq \ell+1 \text{ and } nh \leq Const.$$

Summing up we obtain the desired estimate

$$\|y_n - y(x_n)\| \leq \sum_{\ell=0}^{n-1} \|y_n^\ell - y_n^{\ell+1}\| \leq C_4 h^r \qquad \text{for } nh \leq Const.$$

Since C_3 and C_4 do not depend on C_0 or C_1 (if h is sufficiently small), the assumption (4.18) is justified by induction on n provided the constants C_0, C_1 are chosen sufficiently large. □

Convergence for the z-Component

Theorem 4.6 (HLR89, p. 40). *Consider the index 2 problem (3.1)–(3.2) with consistent initial values and assume that the Runge-Kutta matrix (a_{ij}) is invertible and $|R(\infty)| < 1$. If the global error of the y-component is $\mathcal{O}(h^r)$, $g(y_n) = \mathcal{O}(h^{r+1})$ and the local error of the z-component is $\mathcal{O}(h^r)$, then we have for the global error*

$$z_n - z(x_n) = \mathcal{O}(h^r) \qquad \textit{for} \quad x_n - x_0 = nh \leq Const.$$

Remark. If, in addition to the invertibility of (a_{ij}) and $|R(\infty)| < 1$, the conditions $B(q)$ and $C(q)$ are satisfied then we have $z_n - z(x_n) = \mathcal{O}(h^q)$ (see Lemma 4.4).

Proof. We write the global error as

$$z_{n+1} - z(x_{n+1}) = z_{n+1} - \widehat{z}_{n+1} + \delta z_h(x_n) \tag{4.21}$$

where $(\widehat{y}_{n+1}, \widehat{z}_{n+1})$ denotes the numerical solution obtained from the starting values $(y(x_n), z(x_n))$ and $\delta z_h(x_n)$ is the local error. From (VI.1.11d) we have

$$z_{n+1} - \widehat{z}_{n+1} = R(\infty)\,\big(z_n - z(x_n)\big) + \sum_{i,j=1}^{s} b_i \omega_{ij} \big(Z_{nj} - \widehat{Z}_{nj}\big). \tag{4.22}$$

The assumption $g(y_n) = \mathcal{O}(h^{r+1})$ implies that $g_y(y_n)(y_n - y(x_n)) = \mathcal{O}(h^{r+1})$ and, together with $y_n - y(x_n) = \mathcal{O}(h^r)$, it follows from Theorem 4.2 that $Z_{nj} - \widehat{Z}_{nj} = \mathcal{O}(h^r)$. Inserting (4.22) into (4.21) we obtain

$$z_{n+1} - z(x_{n+1}) = R(\infty)\,\big(z_n - z(x_n)\big) + \mathcal{O}(h^r),$$

which proves the statement. □

Collocation Methods

An important subclass of implicit Runge-Kutta methods are the collocation methods as introduced in Sect. II.7. For the index 2 problem (3.1) they can be defined as follows.

Definition 4.7. Let $c_1, \ldots, c_s$ be s distinct real numbers and denote by $u(x), v(x)$ the polynomials of degree s (*collocation polynomials*) which satisfy

$$u(x_0) = y_0\,, \qquad v(x_0) = z_0 \tag{4.23a}$$

$$\left.\begin{aligned} u'(x_0 + c_i h) &= f\big(u(x_0 + c_i h), v(x_0 + c_i h)\big) \\ 0 &= g\big(u(x_0 + c_i h)\big) \end{aligned}\right\} \qquad i = 1, \ldots, s. \tag{4.23b}$$

Then, the numerical solution is given by

$$y_1 = u(x_0 + h), \qquad z_1 = v(x_0 + h). \tag{4.23c}$$

A straightforward extension of Theorems II.7.7 and II.7.8 to index 2 problems shows that (4.23) is equivalent to the s-stage Runge-Kutta method (4.1) whose coefficients are defined by $B(s)$ and $C(s)$ (see Sect. IV.5 for their definition). This equivalence allows us to deduce from Theorem 4.1 the existence and local uniqueness of the collocation polynomials provided that the corresponding Runge-Kutta matrix is invertible. Hence we assume in the sequel that $c_i \neq 0$ for all i. The case of a singular Runge-Kutta matrix is considered in Exercises 2 and 3.

The quality of $u(x), v(x)$ as approximations to $y(x), z(x)$ is described by the next theorem, which extends Theorem II.7.10.

Theorem 4.8. *Consider a collocation method (4.23) with all* $c_i \neq 0$. *Then we have for* $k = 0, 1, \ldots, s$ *and* $x \in [x_0, x_0 + h]$

$$\|u^{(k)}(x) - y^{(k)}(x)\| \leq C\,h^{s+1-k},$$
$$\|v^{(k)}(x) - z^{(k)}(x)\| \leq C\,h^{s-k}.$$

Proof. We exploit the fact that $u(x_0 + c_i h) = Y_i$, $v(x_0 + c_i h) = Z_i$ are the internal stages of the Runge-Kutta method (4.1). Consequently the collocation polynomials can be written as

$$u(x_0 + th) = y_0\,\ell_0(t) + \sum_{i=1}^{s} Y_i\,\ell_i(t) \tag{4.24a}$$

$$v(x_0 + th) = z_0\,\ell_0(t) + \sum_{i=1}^{s} Z_i\,\ell_i(t) \tag{4.24b}$$

where the $\ell_i(t)$ are the Lagrange polynomials defined by

$$\ell_0(t) = \prod_{j=1}^{s} \frac{(t - c_j)}{(-c_j)}, \qquad \ell_i(t) = \frac{t}{c_i} \prod_{\substack{j=1\\ j\neq i}}^{s} \frac{(t - c_j)}{(c_i - c_j)}.$$

Familiar estimates of the interpolation error imply that the exact solution $y(x)$ satisfies

$$y(x_0 + th) = y_0 \ell_0(t) + \sum_{i=1}^{s} y(x_0 + c_i h)\ell_i(t) + \mathcal{O}(h^{s+1}). \tag{4.25}$$

The factor h^{s+1} in the interpolation error comes from the $(s+1)$-th derivative of $y(x_0 + th)$ with respect to t. Obviously, the interpolation error is differentiable as often as the function $y(x)$. If we differentiate (4.25) k times, then by Rolle's theorem, the difference

$$h^k y^{(k)}(x_0 + th) - \Big(y_0 \ell_0^{(k)}(t) + \sum_{i=1}^{s} y(x_0 + c_i h)\ell_i^{(k)}(t)\Big) \tag{4.25'}$$

vanishes at least at $s + 1 - k$ points. Hence, the polynomial enclosed in brackets in (4.25') can be interpreted as an interpolation polynomial of degree $s - k$ for the function $h^k y^{(k)}(x_0 + th)$. Its error is thus again of size $\mathcal{O}(h^{s+1})$. Subtracting (4.25) from (4.24a) and differentiating k times thus yields

$$h^k \big(u^{(k)}(x_0 + th) - y^{(k)}(x_0 + th)\big) = \sum_{i=1}^{s} \big(Y_i - y(x_0 + c_i h)\big)\,\ell_i^{(k)}(t) + \mathcal{O}(h^{s+1})$$

and a similar formula for the z-component. The conclusion now follows from (4.14) with $q = s$. □

Superconvergence of Collocation Methods

It is now natural to ask whether superconvergence takes place at x_0+h (as for ordinary differential equations; see Theorem II.7.9). The answer is affirmative, if the method is stiffly accurate, i.e., if $c_s=1$.

Theorem 4.9. *If $c_i \neq 0$ for all i and $c_s=1$, then the y-component of the local error of the collocation method (4.23) satisfies*

$$y_1 - y(x_0+h) = \mathcal{O}(h^{p+1}),$$

where p is the order of the underlying quadrature formula.

Proof. We insert the collocation polynomials into the differential-algebraic problem and define the defect by

$$u'(x) = f\big(u(x), v(x)\big) + \delta(x) \tag{4.26a}$$

$$0 = g\big(u(x)\big) + \theta(x). \tag{4.26b}$$

By Definition 4.7 we have

$$\delta(x_0+c_ih)=0, \qquad \theta(x_0)=0, \qquad \theta(x_0+c_ih)=0. \tag{4.27}$$

We next differentiate (4.26b) with respect to x and use (4.26a):

$$0 = g_y(u(x))\big(f(u(x),v(x))+\delta(x)\big)+\theta'(x). \tag{4.28}$$

This motivates the use of the equation

$$0 = g_y(u)\big(f(u,v)+\delta(x)\big)+\theta'(x) \tag{4.29}$$

for arbitrary (u,v) in a neighbourhood of the solution of (3.1). Because of (3.2) we can extract v from (4.29) so that (4.29) can be written as

$$v = G\big(u, \delta(x), \theta'(x)\big). \tag{4.30}$$

Inserting into (4.26a) and into (3.1) this yields

$$u'(x) = f\Big(u(x), G\big(u(x), \delta(x), \theta'(x)\big)\Big) + \delta(x) \tag{4.31a}$$

$$y'(x) = f\Big(y(x), G\big(y(x), 0, 0\big)\Big). \tag{4.31b}$$

In order to compute $u(x)-y(x)$ we now apply the nonlinear variation-of-constants formula (Theorem I.14.5). This requires the computation of the defect of $u(x)$ inserted into (4.31b)

$$\begin{aligned} &u'(x) - f\Big(u(x), G\big(u(x),0,0\big)\Big) \\ &= f\Big(u(x), G\big(u(x), \delta(x), \theta'(x)\big)\Big) + \delta(x) - f\Big(u(x), G\big(u(x),0,0\big)\Big) \\ &= \Phi(x,1) - \Phi(x,0) + \delta(x) \end{aligned} \tag{4.32}$$

where

$$\Phi(x,\tau) = f\Big(u(x), G\big(u(x), \tau\cdot\delta(x), \tau\cdot\theta'(x)\big)\Big).$$

Then the formula $\Phi(x,1)-\Phi(x,0) = \int_0^1 \partial\Phi/\partial\tau\,(x,\tau)\,d\tau$ shows that the defect (4.32) can be written as

$$Q_1(x)\delta(x) + Q_2(x)\theta'(x). \tag{4.32'}$$

We now insert this into Eq. (I.14.18) and obtain

$$\begin{aligned} u(x)-y(x) &= \int_{x_0}^{x} \text{resolvent}(x,t)\cdot\text{defect}(t)\,dt \\ &= \int_{x_0}^{x} \Big(S_1(x,t)\delta(t) + S_2(x,t)\theta'(t)\Big)\,dt. \end{aligned}$$

Integrating the second term by parts we get (since $\theta(x_0)=0$)

$$\begin{aligned} y_1 - y(x_0+h) = \int_{x_0}^{x_0+h} &\Big(S_1(x_0+h,t)\,\delta(t) - \frac{\partial S_2}{\partial t}(x_0+h,t)\,\theta(t)\Big)\,dt \\ &+ S_2(x_0+h,x_0+h)\,\theta(x_0+h). \end{aligned} \tag{4.33}$$

The assumption $c_s = 1$ implies that $\theta(x_0+h)=0$ so that the last expression in (4.33) vanishes. The main idea is now to integrate the expression in (4.33) with the quadrature formula $\{b_i, c_i\}$ (see also the proof of Theorem II.7.9). With the abbreviation

$$\sigma(t) = S_1(x_0+h,t)\,\delta(t) - \frac{\partial S_2}{\partial t}(x_0+h,t)\,\theta(t) \tag{4.34}$$

this gives

$$y_1 - y(x_0+h) = \int_{x_0}^{x_0+h} \sigma(t)dt = h\sum_{i=1}^{s} b_i\sigma(x_0+c_ih) + err(\sigma). \tag{4.35}$$

Because of (4.27) we have $\sigma(x_0+c_ih)=0$ for all i and the quadrature error is estimated by

$$\|err(\sigma)\| \le Ch^{p+1} \max_{t\in[x_0,x_0+h]} \|\sigma^{(p)}(t)\|. \tag{4.36}$$

The p-th derivative of $\sigma(t)$ contains derivatives of f, g and of $\delta(x)$, $\theta(x)$. By Theorem 4.8 they are uniformly bounded for $h \le h_0$. Hence $y_1 - y(x_0+h) = err(\sigma) = \mathcal{O}(h^{p+1})$, proving the theorem. □

Projected Runge-Kutta Methods

For collocation methods which are not stiffly accurate it is possible to prove superconvergence (as in Theorem 4.9) if the method is combined with a certain projection. We start with a more careful study of the local error of the y-component in (4.33).

Lemma 4.10. *If $c_i \neq 0$ for all i, then the y-component of the local error of the collocation method (4.23) satisfies*

$$y_1 - y(x_0+h) = -\Big(f_z(g_y f_z)^{-1}\Big)\big(y(x_0+h), z(x_0+h)\big)\theta(x_0+h) + \mathcal{O}(h^{p+1}) \tag{4.37}$$

where θ is the defect given by (4.26b) and p is the order of the underlying quadrature formula.

Proof. The above proof of Theorem 4.9 (see Eq. (4.33)) shows that the local error satisfies

$$y_1 - y(x_0+h) = S_2(x_0+h, x_0+h)\,\theta(x_0+h) + \mathcal{O}(h^{p+1}).$$

Hence, we only have to compute $S_2(x,x)$. Since any resolvent equals the identity matrix if both of its arguments are equal, it follows from the definition of $S_2(x,t)$ and from (4.32') that

$$S_2(x,x) = \int_0^1 f_z\Big(u(x), G\big(u(x), \tau\delta(x), \tau\theta'(x)\big)\Big)\frac{\partial G}{\partial \theta'}\Big(u(x), \tau\delta(x), \tau\theta'(x)\Big)\,d\tau.$$

Differentiating (4.29) with respect to θ' gives

$$\frac{\partial G}{\partial \theta'} = \frac{\partial v}{\partial \theta'} = -(g_y f_z)^{-1}(u,v).$$

Furthermore, it follows from (4.27) that $\delta(x) = \mathcal{O}(h^s)$ and $\theta'(x) = \mathcal{O}(h^s)$ for $x = x_0 + h$. Using $u(x) - y(x) = \mathcal{O}(h^{s+1})$ (from Theorem 4.8) we thus obtain for $x = x_0 + h$

$$S_2(x,x) = \Big(f_z(g_y f_z)^{-1}\Big)\big(y(x), z(x)\big) + \mathcal{O}(h^s).$$

The statement now follows from $p \le 2s$ and from $\theta(x_0+h) = \mathcal{O}(h^{s+1})$. □

The geometric interpretation of Lemma 4.10 is as follows: if we split the local error $\delta y_h(x_0)$ according to the projections of Fig. 4.1 then the component $Q(x_0+h)\delta y_h(x_0)$ is of size $\mathcal{O}(h^{s+1})$, whereas the component $P(x_0+h)\delta y_h(x_0)$ is $\mathcal{O}(h^{p+1})$. This suggests to project after every step the numerical solution of a Runge-Kutta method onto the manifold $g(y) = 0$ with the help of the projection operator $P(x_0+h)$ as follows:

Definition 4.11 (Ascher & Petzold 1991). Let y_1, z_1 be the numerical solution of an implicit Runge-Kutta method (4.1) and define $\widehat{y}_1, \lambda$ as the solution of the system

$$\begin{aligned} \widehat{y}_1 &= y_1 + f_z(\widehat{y}_1, z_1)\lambda \\ 0 &= g(\widehat{y}_1). \end{aligned} \tag{4.38}$$

If the value $\widehat{y}_1$ (and z_1) is used for the step by step integration of (3.1), then we call this procedure *projected Runge-Kutta method.*

Remarks. 1) If $g(y_1)$ is sufficiently small, then the nonlinear system (4.38) possesses a locally unique solution. A Newton-type iteration with starting values $\widehat{y}_1^{(0)} = y_1$, $\lambda^{(0)} = 0$ will converge to this solution. This follows at once from the theorem of Newton-Kantorovich (Ortega & Rheinboldt 1970) because the Jacobian of (4.38) evaluated at the starting values

$$\begin{pmatrix} I & -f_z(y_1, z_1) \\ g_y(y_1) & 0 \end{pmatrix}$$

has a bounded inverse by (3.2).

2) For stiffly accurate Runge-Kutta methods (i.e., if $a_{si} = b_i$ for all i) the projected and unprojected Runge-Kutta methods coincide.

3) The proof of the next theorem shows that the argument in $f_z(\widehat{y}_1, z_1)$ may be replaced by some other approximation to $y(x_0+h)$, $z(x_0+h)$ whose error is at most $\mathcal{O}(h^s)$.

The following theorem proves superconvergence for projected collocation methods (also if the corresponding Runge-Kutta method is not stiffly accurate). Superconvergence results for general Runge-Kutta methods are given in Sect. VI.8.

Theorem 4.12 (Ascher & Petzold 1991). *If $c_i \neq 0$ for all i, then the y-component of the local error of the projected collocation method (4.23), (4.38) satisfies*

$$\widehat{y}_1 - y(x_0+h) = \mathcal{O}(h^{p+1})$$

where p is the order of the underlying quadrature formula.

Proof. We write $\widehat{e}_1 = \widehat{y}_1 - y(x_0+h)$, $e_1 = y_1 - y(x_0+h)$ for the local errors and denote the projections of Definition 4.3 by

$$Q = \big(f_z(g_y f_z)^{-1} g_y\big)(\widehat{y}_1, z_1), \qquad P = I - Q.$$

The idea is to split $\widehat{e}_1$ according to

$$\widehat{e}_1 = P\,\widehat{e}_1 + Q\,\widehat{e}_1 \tag{4.39}$$

and to estimate both components separately. The first formula of (4.38) together with (4.37) and $\theta(x_0+h) = \mathcal{O}(h^{s+1})$ imply that

$$P\,\widehat{e}_1 = Pe_1 = \mathcal{O}(h^{p+1}) + \mathcal{O}(h^{s+1}\|\widehat{e}_1\|). \tag{4.40}$$

Further we have $0 = g(\widehat{y}_1) - g(y(x_0 + h)) = g_y(\widehat{y}_1)\widehat{e}_1 + \mathcal{O}(\|\widehat{e}_1\|^2)$, implying

$$Q\,\widehat{e}_1 = \mathcal{O}(\|\widehat{e}_1\|^2). \tag{4.41}$$

Formulas (4.40) and (4.41) inserted into (4.39) give

$$\widehat{e}_1 = \mathcal{O}(h^{p+1}) + \mathcal{O}(h^{s+1}\|\widehat{e}_1\|) + \mathcal{O}(\|\widehat{e}_1\|^2)$$

and the statement of the theorem is an immediate consequence. □

Global convergence of order $\mathcal{O}(h^p)$ of the projected collocation methods is obtained exactly as in the proof of Theorem 4.5. We observe that the numerical solution always remains on the manifold $g(y)=0$ so that the estimate (4.9) applies.

Summary of Convergence Results

Table 4.1 collects the optimal error estimates for some important Runge-Kutta methods when applied to the index 2 problem (3.1)–(3.2). The local error estimates can be verified as follows: Gauss, Radau IA and SDIRK by Lemma 4.4, Radau IIA by Theorem 4.9, Lobatto IIIC by Theorem 5.10 below and Lobatto IIIA with the help of Exercise 4. For the projected methods the estimates follow from Theorem 4.12 and the considerations of Sect. VII.5. Because there are several ways of defining the z-component of the numerical solution, we do not present their convergence behaviour. The global convergence result follows from Theorems 4.5 and 4.6 for the Radau IA, Radau IIA, Lobatto IIIC and SDIRK methods. The remaining methods (Gauss and Lobatto IIIA) require some more effort because their stability function only satisfies $|R(\infty)| = 1$. For a detailed discussion of these methods we refer to HLR89 and Jay (1993).

Table 4.1. Error estimates for the index 2 problem (3.1)-(3.2)

Method	stages	local error		global error	
		y	z	y	z
Gauss	s odd s even	h^{s+1}	h^s	h^{s+1} h^s	h^{s-1} h^{s-2}
projected Gauss	s	h^{2s+1}		h^{2s}	
Radau IA	s	h^s	h^{s-1}	h^s	h^{s-1}
projected Radau IA	s	h^{2s-1}		h^{2s-2}	
Radau IIA	s	h^{2s}	h^s	h^{2s-1}	h^s
Lobatto IIIA	s odd s even	h^{2s-1}	h^s	h^{2s-2}	h^{s-1} h^s
Lobatto IIIC	s	h^{2s-1}	h^{s-1}	h^{2s-2}	h^{s-1}
SDIRK (IV.6.16)	5	h^3	h^1	h^2	h^1
SDIRK (IV.6.18)	3	h^2	h^1	h^2	h^1

Exercises

1. Consider the index 2 problem $y' = f(y,z)$, $0 = g(y)$. Put $z = u'$, $v = (y,u)^T$ so that the problem becomes

$$F(v',v) = \begin{pmatrix} y' - f(y,u') \\ g(y) \end{pmatrix} = 0.$$

Prove that the matrix pencil $F_v + \lambda F_{v'}$ is of index 1 whenever $(g_y f_z)^{-1}$ exists.
Hint. Consider the transformation

$$\begin{pmatrix} I & a \\ 0 & I \end{pmatrix} (F_v + \lambda F_{v'}) \begin{pmatrix} I & b \\ 0 & I \end{pmatrix} \tag{4.42}$$

where $a = f_y f_z (g_y f_z)^{-1}$ and $b = f_z$ are chosen such that the upper right block in (4.42) vanishes.

2. Consider Runge-Kutta methods whose coefficients satisfy:

$$a_{1i} = 0 \text{ for all } i \text{ and } (a_{ij})_{i,j\geq 2} \text{ is invertible.}$$

(Examples are collocation methods with $c_1 = 0$, such as Lobatto IIIA).
If $g(\eta) = 0$ then the nonlinear system (4.2) has a locally unique solution which satisfies $Y_1 = \eta$, $Z_1 = \zeta$.

3. Let $c_1 = 0, c_2, \ldots, c_s$ be s distinct real numbers. Show that there exist unique polynomials $u(x)$ and $v(x)$ ($\deg u = s$, $\deg v = s-1$) such that (4.23a,b) holds.

Hint. Apply the ideas of the proof of Theorem II.7.7 and Exercise 2.

4. Investigate the validity of the conclusions of Theorems 4.8 and 4.9 for the situation where $c_1 = 0$.

5. (Computation of the algebraic variable z by *piecewise discontinuous interpolation*, see Ascher (1989)). Modify the definition of z_{n+1} in the Runge-Kutta method (4.1) as follows: let $v(x)$ be the polynomial of degree $s-1$ satisfying $v(x_n + c_i h) = Z_{ni}$ for all i, then define $z_{n+1} = v(x_n + h)$. In the case of *collocation* methods (4.23) this definition removes the condition $v(x_0) = z_0$ while lowering the degree of $v(x)$ by 1.

a) Verify: z_{n+1} does not depend on z_n, also if the stability function of the method does not vanish at infinity.

b) Prove that for projected collocation methods with $c_i \neq 0$ for all i we have $z_n - z(x_n) = \mathcal{O}(h^s)$.

c) For the projected Gauss methods compare this result with that of the standard approach.

6. The statement of Theorem 4.8 still holds, if one omits the condition $v(x_0) = z_0$ in Definition 4.7 and if one lets $v(x)$ be a polynomial of degree $s-1$.

VII.5 Order Conditions for Index 2 DAE

For an application of the convergence result of the preceding section (Theorem 4.5) it is desirable to know the optimal values of r in (4.16). Comparing the Taylor expansions of the exact and numerical solutions we derive conditions for c_i, a_{ij}, b_j which are equivalent to (4.16). For collocation methods we recover the result of Theorem 4.9. For other methods (such as Lobatto IIIC) the estimates of Lemma 4.4 are substantially improved.

The theory of this section is given in HLR89 (Sect. 5). Our presentation is slightly different and is in complete analogy to the derivation of the index 1 order conditions of Sect. VI.4. The results of this section are here applied to Runge-Kutta methods only; analogous formulas for Rosenbrock methods can be found in Roche (1988). An independent investigation, conducted for the index 2 problem $f(y,z')=0,\ z=g(y)$ by A. Kværnø (1990), leads to the same order conditions for Runge-Kutta methods.

Derivatives of the Exact Solution

We consider the index 2 problem

$$y' = f(y,z) \tag{5.1a}$$

$$0 = g(y) \tag{5.1b}$$

and assume consistent initial values y_0, z_0. The first derivative of the solution $y(x)$ is given by (5.1a). Differentiating this equation we get

$$y'' = f_y(y,z)y' + f_z(y,z)z'. \tag{5.2}$$

In order to compute z' we differentiate (5.1b) twice

$$0 = g_y(y)y' \tag{5.3a}$$

$$0 = g_{yy}(y)(y',y') + g_y(y)y'' \tag{5.3b}$$

and insert (5.2) and (5.1a). This yields (omitting the obvious function arguments)

$$0 = g_{yy}(f,f) + g_y f_y f + g_y f_z z' \tag{5.4}$$

or equivalently

$$z' = (-g_y f_z)^{-1} g_{yy}(f,f) + (-g_y f_z)^{-1} g_y f_y f. \tag{5.5}$$

Here we have used the index 2 assumption (3.2), that $g_y f_z$ is invertible in a neighbourhood of the solution. We now differentiate (5.1a) and (5.5) with respect to x, and replace the appearing y' and z' by (5.1a) and (5.5). We use (for a constant vector u)

$$\begin{aligned}
&\frac{d}{dx}(-g_y f_z)^{-1}u \\
&\quad = (-g_y f_z)^{-1}\Big(g_{yy}\big(f_z(-g_y f_z)^{-1}u, f\big) + g_y f_{zy}\big((-g_y f_z)^{-1}u, f\big) \\
&\qquad + g_y f_{zz}\Big((-g_y f_z)^{-1}u, (-g_y f_z)^{-1}g_{yy}(f,f) + (-g_y f_z)^{-1}g_y f_y f\Big)\Big)
\end{aligned} \tag{5.6}$$

(cf. Formula (VI.4.7)) and thus obtain

$$y'' = f_y f + f_z(-g_y f_z)^{-1}g_{yy}(f,f) + f_z(-g_y f_z)^{-1}g_y f_y f \tag{5.7}$$

$$\begin{aligned}
z'' = {} & (-g_y f_z)^{-1}g_{yyy}(f,f,f) + 3(-g_y f_z)^{-1}g_{yy}(f, f_y f) \\
& + 3(-g_y f_z)^{-1}g_{yy}(f, f_z(-g_y f_z)^{-1}g_{yy}(f,f)) \\
& + 3(-g_y f_z)^{-1}g_{yy}(f, f_z(-g_y f_z)^{-1}g_y f_y f) + (-g_y f_z)^{-1}g_y f_{yy}(f,f) \\
& + 2(-g_y f_z)^{-1}g_y f_{yz}(f, (-g_y f_z)^{-1}g_{yy}(f,f)) \\
& + 2(-g_y f_z)^{-1}g_y f_{yz}(f, (-g_y f_z)^{-1}g_y f_y f) + (-g_y f_z)^{-1}g_y f_y f_y f \\
& + (-g_y f_z)^{-1}g_y f_y f_z(-g_y f_z)^{-1}g_{yy}(f,f) \\
& + (-g_y f_z)^{-1}g_y f_y f_z(-g_y f_z)^{-1}g_y f_y f \\
& + (-g_y f_z)^{-1}g_y f_{zz}((-g_y f_z)^{-1}g_{yy}(f,f), (-g_y f_z)^{-1}g_{yy}(f,f)) \\
& + 2(-g_y f_z)^{-1}g_y f_{zz}((-g_y f_z)^{-1}g_{yy}(f,f), (-g_y f_z)^{-1}g_y f_y f) \\
& + (-g_y f_z)^{-1}g_y f_{zz}((-g_y f_z)^{-1}g_y f_y f, (-g_y f_z)^{-1}g_y f_y f)\,.
\end{aligned} \tag{5.8}$$

Obviously, a graphical representation of these expressions will be of great help.

Trees and Elementary Differentials

As in Sect. VI.4 we identify each occuring f with a meagre vertex, each of its derivatives with an upwards leaving branch, the expression $(-g_y f_z)^{-1}g$ with a fat vertex and the derivatives of g therein again with upwards leaving branches. The corresponding graphs for y', z', y'', z'' (see Formulas (5.1a), (5.5), (5.7), (5.8)) are given in Fig. 5.1.

The derivatives of y are characterized by trees with a *meagre root* (the lowest vertex). These trees will be denoted by t or t_i, the tree consisting of the root only (for y') being τ. Derivatives of z have trees with a *fat root*. They will be denoted by u or u_i.

Definition 5.1. Let $DAT2 = DAT2_y \cup DAT2_z$ denote the set of (*differential algebraic index* 2) *trees* defined recursively by

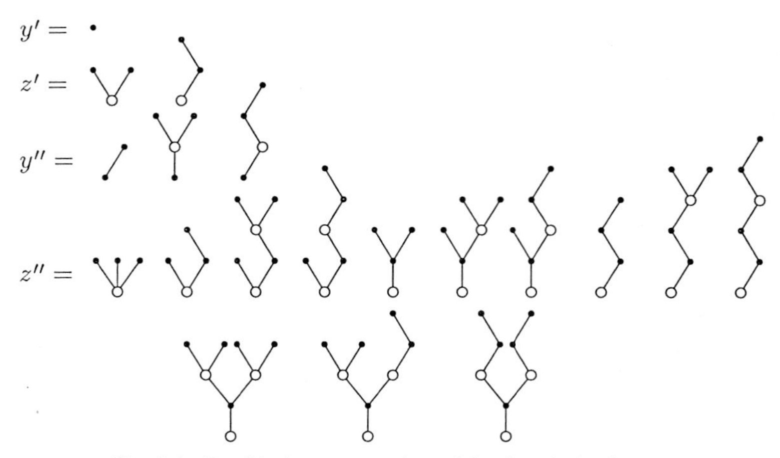

Fig. 5.1. Graphical representation of the first derivatives

a) $\tau \in DAT2_y$,

b) $[t_1, \ldots, t_m, u_1, \ldots u_n]_y \in DAT2_y$
if $t_1, \ldots, t_m \in DAT2_y$ and $u_1, \ldots u_n \in DAT2_z$;

c) $[t_1, \ldots, t_m]_z \in DAT2_z$ if $t_1, \ldots, t_m \in DAT2_y$ and either $m > 1$ or $m = 1$ and $t_1 \neq [u]_y$ with $u \in DAT2_z$.

Definition 5.2. The *order* of a tree $t \in DAT2_y$ or $u \in DAT2_z$, denoted by $\varrho(t)$ or $\varrho(u)$, is the number of meagre vertices minus the number of fat vertices.

Definition 5.3. The *elementary differentials* $F(t)$ (or $F(u)$) corresponding to trees in $DAT2$ are defined as follows:

a) $F(\tau) = f$,

b) $F(t) = \dfrac{\partial^{m+n} f}{\partial y^m \partial z^n} \Big(F(t_1), \ldots, F(t_m), F(u_1), \ldots, F(u_n) \Big)$
if $t = [t_1, \ldots, t_m, u_1, \ldots, u_n]_y \in DAT2_y$,

c) $F(u) = (-g_y f_z)^{-1} \dfrac{\partial^m g}{\partial y^m} \Big(F(t_1), \ldots, F(t_m) \Big)$
if $u = [t_1, \ldots, t_m]_z \in DAT2_z$.

Taylor Expansion of the Exact Solution

In order to continue the process which led to (5.7) and (5.8) we need the differentiation of elementary differentials $F(t)$ and $F(u)$. This is described by the following rules:

i) attach to each vertex a branch with τ (derivative of f or g with respect to y and addition of the factor $y' = f$);

ii) attach to each meagre vertex a branch with $[\tau,\tau]_z$; attach to each meagre vertex a branch with $[[\tau]_y]_z$ (this yields two trees and corresponds to the derivative of f with respect to z and to the addition of the factors $(-g_y f_z)^{-1} g_{yy}(f,f)$ and $(-g_y f_z)^{-1} g_y f_y f$ of (5.5));

iii) split each fat vertex into two new fat vertices (one above the other) and link them via a new meagre vertex. Then four new trees are obtained as follows: attach a branch with τ to the lower of these fat vertices; attach a branch with $\tau, [\tau,\tau]_z$ or $[[\tau]_y]_z$ to the new meagre vertex (this corresponds to the derivation of $(-g_y f_z)^{-1}$ and follows at once from Eq. (5.6)).

Some of the elementary differentials in (5.8) appear more than once. In order to understand how often such an expression (or the corresponding tree) appears in the derivatives of y or z, we indicate the order of generation of the vertices as follows (see Fig. 5.2): for the trees of order 1, namely $\tau, [\tau,\tau]_z$ and $[[\tau]_y]_z$, we add the label 1 to a meagre vertex such that

each fat vertex is followed by at least one unlabelled meagre vertex. (5.9)

Each time a tree is "differentiated" according to the above rules we provide the newly attached tree (of order 1) with a new label such that (5.9) still holds. The labelling so obtained is obviously increasing along each branch.

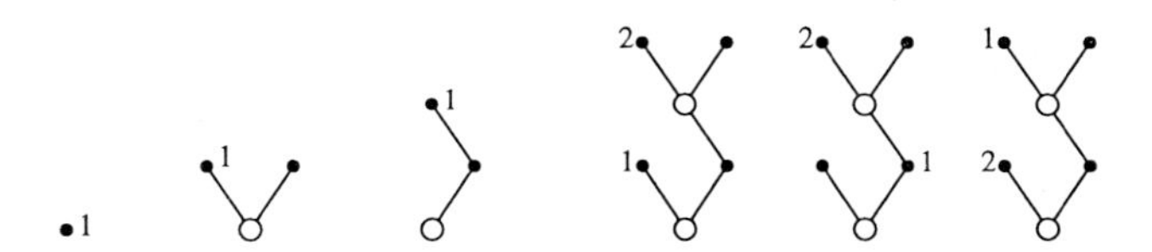

Fig. 5.2. Examples of monotonically labelled trees

Definition 5.4. A tree $t \in DAT2_y$ (or $u \in DAT2_z$), together with a monotonic labelling of $\varrho(t)$ (or $\varrho(u)$) among its meagre vertices such that (5.9) holds, is called a *monotonically labelled tree*. The sets of such monotonically labelled trees are denoted by $LDAT2_y$, $LDAT2_z$, and $LDAT2$.

Since the differentiation process of trees described above generates all elements of $LDAT2$, and each of them exactly once, and since each differentiation increases the order of the trees by one, we have the following result.

Theorem 5.5 (HLR89, p. 58). *For the exact solution of (5.1) we have:*

$$y^{(q)}(x_0) = \sum_{t\in LDAT2_y,\varrho(t)=q} F(t)(y_0,z_0) = \sum_{t\in DAT2_y,\varrho(t)=q} \alpha(t)F(t)(y_0,z_0)$$

$$z^{(q)}(x_0) = \sum_{u\in LDAT2_z,\varrho(u)=q} F(u)(y_0,z_0) = \sum_{u\in DAT2_z,\varrho(u)=q} \alpha(u)F(u)(y_0,z_0).$$

The integer coefficients $\alpha(t)$ and $\alpha(u)$ indicate the number of possible monotonic labellings of a tree. □

Derivatives of the Numerical Solution

For the problem (5.1) with consistent inital values (y_0, z_0) we write one step of a Runge-Kutta method in the form

$$y_1 = y_0 + \sum_{i=1}^{s} b_i k_i, \qquad z_1 = z_0 + \sum_{i=1}^{s} b_i \ell_i \tag{5.10a}$$

where

$$k_i = hf(Y_i, Z_i), \qquad 0 = g(Y_i) \tag{5.10b}$$

and

$$Y_i = y_0 + \sum_{j=1}^{s} a_{ij} k_j, \qquad Z_i = z_0 + \sum_{j=1}^{s} a_{ij} \ell_j. \tag{5.10c}$$

We have replaced $hk_{ni}, h\ell_{ni}$ of Formula (4.1) by k_i, ℓ_i. This is not essential, but adjusts the derivation of the order conditions to the presentation of Sect. VI.4. Since the following derivation is very similar to the one given in Sect. VI.4, we restrict ourselves to the main ideas.

We consider $y_1, z_1, k_i, \ell_i, Y_i, Z_i$ as functions of h and compute their derivatives at $h = 0$. From (5.10a) we get

$$y_1^{(q)}(0) = \sum_{i=1}^{s} b_i k_i^{(q)}(0), \tag{5.11}$$

and (5.10b) yields

$$k_i^{(q)}(0) = q\Big(f(Y_i, Z_i)\Big)^{(q-1)}\Big|_{h=0}, \qquad 0 = \Big(g(Y_i)\Big)^{(q)}\Big|_{h=0}. \tag{5.12}$$

The total derivatives of $f(Y_i, Z_i)$ and $g(Y_i)$ can be computed by Faà di Bruno's formula (see (VI.4.14) and (VI.4.15)). This gives

$$\Big(f(Y_i, Z_i)\Big)^{(q-1)} = \sum \frac{\partial^{m+n} f(Y_i, Z_i)}{\partial y^m \partial z^n} \Big(Y_i^{(\mu_1)}, \ldots, Y_i^{(\mu_m)}, Z_i^{(\nu_1)}, \ldots, Z_i^{(\nu_n)}\Big) \tag{5.13}$$

with $\mu_1 + \ldots + \mu_m + \nu_1 + \ldots + \nu_n = q - 1$, and

$$\Big(g(Y_i)\Big)^{(q)} = \sum \frac{\partial^m g(Y_i)}{\partial y^m} \Big(Y_i^{(\mu_1)}, \ldots, Y_i^{(\mu_m)}\Big) \tag{5.14}$$

with $\mu_1 + \ldots + \mu_m = q$. The summations in (5.13) and (5.14) are over sets of suitable "special labelled trees". We next insert

$$Y_i^{(\mu)} = \sum_{j=1}^{s} a_{ij} k_j^{(\mu)} \tag{5.15}$$

into (5.13) and (5.14) and so obtain from (5.12)

$$k_i^{(q)}(0) = q \sum \frac{\partial^{m+n} f(y_0, z_0)}{\partial y^m \partial z^n} \Big(\sum_{j=1}^{s} a_{ij} k_j^{(\mu_1)}(0), \ldots, Z_i^{(\nu_1)}(0), \ldots\Big) \tag{5.16}$$

and

$$0 = g_y(y_0)\sum_{j=1}^{s} a_{ij}k_j^{(q)}(0) + \sum_{m\geq 2} \frac{\partial^m g(y_0)}{\partial y^m}\Big(\sum_{j=1}^{s} a_{ij}k_j^{(\mu_1)}(0),\ldots\Big). \tag{5.17}$$

Inserting (5.16) into the first term of (5.17) and extracting $Z_j^{(q-1)}(0)$ we get

$$(-g_y f_z)(y_0,z_0)\sum_{j=1}^{s} a_{ij}Z_j^{(q-1)}(0) \tag{5.18}$$

$$= \sum_{j=1}^{s} a_{ij} \sum_{(m,n)\neq(0,1)} g_y(y_0)\frac{\partial^{m+n} f(y_0,z_0)}{\partial y^m \partial z^n}\Big(\sum_{l=1}^{s} a_{jl}k_l^{(\mu_1)}(0),\ldots,Z_j^{(\nu_1)}(0),\ldots\Big)$$
$$+\frac{1}{q}\sum_{m\geq 2}\frac{\partial^m g(y_0)}{\partial y^m}\Big(\sum_{j=1}^{s} a_{ij}k_j^{(\mu_1)}(0),\ldots\Big).$$

This formula allows us to compute $Z_i^{(q-1)}$, whenever $(g_y f_z)$ and (a_{ij}) are invertible. We denote the coefficients of the inverse of (a_{ij}) by ω_{ij}, i.e.,

$$(\omega_{ij}) = (a_{ij})^{-1}. \tag{5.19}$$

The following result then follows by induction on q from (5.16) and (5.18).

Theorem 5.6 (HLR89). *The derivatives of k_i and Z_i satisfy*

$$k_i^{(q)}(0) = \sum_{t\in LDAT2_y,\varrho(t)=q} \gamma(t)\Phi_i(t)F(t)(y_0,z_0)$$
$$Z_i^{(q)}(0) = \sum_{u\in LDAT2_z,\varrho(u)=q} \gamma(u)\Phi_i(u)F(u)(y_0,z_0),$$

where the coefficients $\Phi_i(t)$ and $\Phi_i(u)$ are given by $\Phi_i(\tau)=1$ and

$$\Phi_i(t) = \sum_{\mu_1,..,\mu_m} a_{i\mu_1}\cdots a_{i\mu_m}\cdot\Phi_{\mu_1}(t_1)\cdots\Phi_{\mu_m}(t_m)\Phi_i(u_1)\cdots\Phi_i(u_n)$$
$$\textit{if } t=[t_1,\ldots,t_m,u_1,\ldots,u_n]_y$$

$$\Phi_i(u) = \sum_{j,\mu_1,..,\mu_m} \omega_{ij}a_{j\mu_1}\cdots a_{j\mu_m}\cdot\Phi_{\mu_1}(t_1)\cdots\Phi_{\mu_m}(t_m)$$
$$\textit{if } u=[t_1,\ldots,t_m]_z$$

and the rational coefficients $\gamma(t)$ and $\gamma(u)$ are defined by $\gamma(\tau)=1$ and

$$\gamma(t) = \varrho(t)\gamma(t_1)\ldots\gamma(t_m)\gamma(u_1)\ldots\gamma(u_n) \quad \textit{if } t=[t_1,\ldots,t_m,u_1,\ldots,u_n]_y$$
$$\gamma(u) = \frac{1}{\varrho(u)+1}\gamma(t_1)\ldots\gamma(t_m) \quad \textit{if } u=[t_1,\ldots,t_m]_z.$$

□

The derivatives of the numerical solution y_1 are now obtained from (5.11). In order to get those of z_1, we compute ℓ_i from (5.10c) and insert it into (5.10a). This yields

$$z_1 = z_0 + \sum_{i,j=1}^{s} b_i \omega_{ij} (Z_j - z_0) \tag{5.20}$$

and its derivatives are given by

$$z_1^{(q)}(0) = \sum_{i,j=1}^{s} b_i \omega_{ij} Z_j^{(q)}(0). \tag{5.21}$$

We thus obtain the following result.

Theorem 5.7. *The numerical solution of (5.10) satisfies*

$$y_1^{(q)}|_{h=0} = \sum_{t \in LDAT2_y, \varrho(t)=q} \gamma(t) \sum_{i=1}^{s} b_i \Phi_i(t) F(t)(y_0, z_0),$$

$$z_1^{(q)}|_{h=0} = \sum_{u \in LDAT2_z, \varrho(u)=q} \gamma(u) \sum_{i,j=1}^{s} b_i \omega_{ij} \Phi_j(u) F(u)(y_0, z_0),$$

where the coefficients γ and Φ_i are given in Theorem 5.6. □

Order Conditions

A comparison of Theorem 5.7 with Theorem 5.5 gives

Theorem 5.8 (HLR89). *For the Runge-Kutta method (5.10) we have*

$$y(x_0+h) - y_1 = \mathcal{O}(h^{p+1}) \qquad \textit{iff}$$

$$\sum_{i=1}^{s} b_i \Phi_i(t) = \frac{1}{\gamma(t)} \quad \textit{for} \;\; t \in DAT2_y, \;\; \varrho(t) \le p,$$

$$z(x_0+h) - z_1 = \mathcal{O}(h^{q+1}) \qquad \textit{iff}$$

$$\sum_{i,j=1}^{s} b_i \omega_{ij} \Phi_j(u) = \frac{1}{\gamma(u)} \quad \textit{for} \;\; u \in DAT2_z, \;\; \varrho(u) \le q,$$

where the coefficients γ and Φ_i are those of Theorem 5.6 and ω_{ij} is given by (5.19). □

Remark 5.9. Let $P(x_0) = I - (f_z(g_y f_z)^{-1} g_y)(y_0, z_0)$ be the projection introduced in Definition 4.3. Since $P(x_0) f_z(y_0, z_0) = 0$ we have

$$P(x_0) F(t)(y_0, z_0) = 0 \tag{5.22}$$

for all trees $t \in DAT2_y$ of the form $t = [u]_y$ with $u \in DAT2_z$. Consequently, such trees of order p need not be considered for the construction of Runge-Kutta methods of order p (see Theorem 4.5).

Applying repeatedly the definition of Φ_i in Theorem 5.6 we get the following algorithm:

Forming the Order Condition for a Given Tree. Attach to each vertex one summation index; if the root is fat, attach three indices to this root. Then the left hand side of the order condition is a sum over all indices of a product with factors

b_i	if "i" is the index of a meagre root;
$b_i \omega_{ij} \omega_{jk}$	if "i, j, k" are the three indices of a fat root;
a_{ij}	if "j" lies directly above "i" and "j" is meagre;
ω_{ij}	if "j" lies directly above "i" and "j" is fat.

In Table 5.1 we collect the order conditions for some trees of $DAT2$. We have not included the trees which have only meagre vertices, because their order condition is exactly the same as that of Sect. II.2 (Table 2.2). Several trees of $DAT2$ lead to the same order condition (Exercise 2). We also observe that some of the order conditions for the trees $[u]_y$ with $u \in DAT2_z$ are identical to those for index 1 problems (see Exercise 1 of Sect. VI.4).

Table 5.1. Trees and order conditions

$\varrho(t)$	graph	order condition
2		$\sum b_i \omega_{ij} c_j^2 = 1$
3	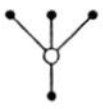	$\sum b_i \omega_{ij} c_j^3 = 1$
3	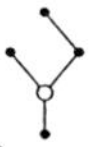	$\sum b_i \omega_{ij} c_j a_{jk} c_k = \frac{1}{2}$
3		$\sum b_i c_i \omega_{ij} c_j^2 = \frac{2}{3}$
3	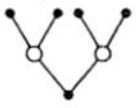	$\sum b_i \omega_{ij} c_j^2 \omega_{ik} c_k^2 = \frac{4}{3}$
$\varrho(u)$	graph	order condition
1		$\sum b_i \omega_{ij} \omega_{jk} c_k^2 = 2$
2		$\sum b_i \omega_{ij} \omega_{jk} c_k^3 = 3$
2		$\sum b_i \omega_{ij} \omega_{jk} c_k a_{k\ell} c_\ell = \frac{3}{2}$
2	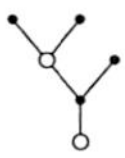	$\sum b_i \omega_{ij} c_j \omega_{jk} c_k^2 = 2$

Simplifying Assumptions

For the construction of implicit Runge-Kutta methods the simplifying conditions $B(p)$, $C(\eta)$, $D(\xi)$ of Sect. IV.5 play an important role. The following result extends Theorem IV.5.1 to index 2 problems.

Theorem 5.10 (HLR89, p. 67). *Suppose that the Runge-Kutta matrix (a_{ij}) is invertible and that $b_i = a_{si}$ for $i = 1, \ldots, s$. Then the conditions $B(p)$, $C(\eta)$, $D(\xi)$ with $p \le 2\eta$ and $p \le \eta + \xi + 1$ imply that the y-component of the local error of (5.1) satisfies*

$$y_1 - y(x_0 + h) = \mathcal{O}(h^{p+1}).$$

Proof. We just outline the main ideas; details are given in (HLR89, pp. 64-67). As in Sect. II.7 (Fig. II.7.1) we first simplify the order conditions with the help of $C(\eta)$. This implies that trees with a branch ending with $[\tau, \ldots, \tau]_y$ (the number of τ's is $k-1$) where $k \le \eta$ need no longer be considered. If we write $C(\eta)$ in the form

$$\sum_{j=1}^{s} \omega_{ij} c_j^k = k c_i^{k-1} \qquad \text{for} \quad k = 1, \ldots, \eta, \tag{5.23}$$

we observe that trees ending with $[\tau, \ldots, \tau]_z$ can also be reduced if the number of τ's is between 1 and η.

The simplifying condition $D(\xi)$ allows us to remove trees $[\tau, \ldots, \tau, t]_y$ with $t \in DAT_y$, where the number of τ's is $\le \xi$. Writing $D(\xi)$ as

$$\sum_{i=1}^{s} b_i c_i^k \omega_{ij} = \sum_{i=1}^{s} b_i \omega_{ij} - k b_j c_j^{k-1} \qquad \text{for} \quad k = 1, \ldots, \xi \tag{5.24}$$

it follows that the trees $[\tau, \ldots, \tau, u]_y$ with $u \in DAT_z$ (number of τ's is k) can also be eliminated for $1 \le k \le \xi$. Since $p \le 2\eta$ and $p \le \eta + \xi + 1$ all that remains after these reductions are the bushy trees $[\tau, \ldots, \tau]_y$ whose order conditions are satisfied by $B(p)$, and trees of the form $[u]_y$ with $u \in DAT_z$. Because of the assumption $b_i = a_{si}$ we have

$$\sum_{i=1}^{s} b_i \omega_{ij} = \begin{cases} 0 & \text{if} \quad j = 1, \ldots, s-1 \\ 1 & \text{if} \quad j = s, \end{cases} \tag{5.25}$$

and these trees can also be reduced to the bushy trees. □

Remark. If the function f of (5.1a) is linear in z, i.e.,

$$f(y, z) = f_0(y) + f_z(y) z, \tag{5.26}$$

then the elementary differentials for trees $[t_1, \ldots, t_m, u_1, \ldots, u_n]_y$ with $n \ge 2$ vanish identically and the corresponding order conditions need not be considered.

In this situation the assumption $p \le 2\eta$ can be relaxed to $p \le 2\eta + 1$. An important class of problems satisfying (5.26) are constrained mechanical systems in the index 2 formulation (1.46a,b,d).

As an illustration of Theorem 5.10 we consider the Lobatto IIIC methods. They satisfy $B(p), C(\eta), D(\xi)$ with $p = 2s-2$, $\eta = s-1$ and $\xi = s-1$ (see Table IV.5.13) and also $a_{si} = b_i$. It therefore follows from Theorem 5.10 that the local error satisfies $\delta y_h(x) = \mathcal{O}(h^{2s-1})$.

The following result shows that for methods which do not satisfy $a_{si} = b_i$ it is unlikely that the estimates of Lemma 4.4 can be improved.

Lemma 5.11. *Let p be the largest integer such that the y-component of the local error satisfies*

$$\delta y_h(x) = \mathcal{O}(h^{p+1}).$$

If the Runge-Kutta matrix is invertible and $c_i \neq 1$ for all i, then

$$p \le s^*$$

where s^* is the number of distinct non-zero values among $c_1, \dots, c_s$.

Proof. The order conditions for the trees $[[\tau, \dots, \tau]_z]_y$ imply that

$$\sum_{i,j=1}^{s} b_i \omega_{ij} \int_0^{c_j} q(t)dt = \int_0^1 q(t)dt \tag{5.27}$$

for all polynomials $q(t)$ of degree $\le p-1$. Put $q(t) = d'(t)$, where $d(t)$ is a polynomial of minimal degree such that $d(c_i) = 0$ for all i, $d(0) = 0$ and $d(1) \neq 0$. Condition (5.27) is violated by this polynomial. The inequality $p \le s^*$ now follows because the degree of this polynomial $q(t)$ is s^*. □

Projected Runge-Kutta Methods

It is, of course, interesting to study the convergence order of projected Runge-Kutta methods (Definition 4.11) which are not yet covered by Theorem 4.12. The main tool for the subsequent study is the following interpretation of projected Runge-Kutta methods.

Table 5.2. Original and extended Runge-Kutta methods

c	A
	b^T

c	A	0
$1+\varepsilon$	b^T	ε
	b^T	ε

Lemma 5.12 (Lubich 1991). *Consider an s-stage Runge-Kutta method with invertible coefficient matrix A and the extended ($s+1$)-stage method defined in Table 5.2. For an initial value y_0 satisfying $g(y_0)=0$ denote their numerical solutions after one step by y_1 and y_1^ε, respectively. If the function f in (5.1a) is linear in z (i.e., (5.26) is satisfied), then the numerical solution $\widehat{y}_1$ of the projected Runge-Kutta method (4.1), (4.38) satisfies*

$$\widehat{y}_1 - y_1^\varepsilon = \mathcal{O}(h\varepsilon) \tag{5.28}$$

for h sufficiently small and $\varepsilon \to 0$.

Proof. The last stage of the extended ($s+1$)-stage Runge-Kutta method reads

$$\begin{aligned} Y_{s+1} &= y_1 + h\varepsilon f(Y_{s+1}, Z_{s+1}) \\ 0 &= g(Y_{s+1}) \end{aligned} \tag{5.29}$$

and we have $y_1^\varepsilon = Y_{s+1}$ (note that this is the result of an implicit Euler step with step size $h\varepsilon$ starting from y_1). Using the linearity of f with respect to z and putting $\lambda = h\varepsilon Z_{s+1}$ we obtain

$$\begin{aligned} y_1^\varepsilon &= y_1 + h\varepsilon f_0(y_1^\varepsilon) + f_z(y_1^\varepsilon)\lambda \\ 0 &= g(y_1^\varepsilon). \end{aligned} \tag{5.30}$$

Comparing (5.30) with (4.38) the implicit function theorem implies that (5.28) is satisfied for sufficiently small h and ε. □

The implicit function theorem, applied to (5.30), also shows that y_1^ε is as often differentiable with respect to h and ε as the right-hand side of the problem (5.1) is. Hence, the Taylor series expansion of y_1^ε with respect to h has coefficients which converge to a finite limit as $\varepsilon \to 0$.

The order conditions for a projected Runge-Kutta method (applied to (5.1), (5.26)) can thus be obtained by considering the limit $\varepsilon \to 0$ in the order conditions for the extended Runge-Kutta method (Exercise 5). Let us illustrate this by extending the statement of Theorem 5.10 to projected Runge-Kutta methods.

Theorem 5.13 (Lubich 1991). *Suppose that the Runge-Kutta matrix A is invertible and that the index 2 problem satisfies (5.26). Then the conditions $B(p)$, $C(\eta)$, $D(\xi)$ with $p \le 2\eta+1$ and $p \le \eta+\xi+1$ imply that the local error of the projected Runge-Kutta method satisfies*

$$\widehat{y}_1 - y(x_0+h) = \mathcal{O}(h^{p+1}). \tag{5.31}$$

If in addition $p \le 2\eta$ then (5.31) holds also when f is nonlinear in z.

Proof. One verifies that the conditions $B(p)$, $C(\eta)$, $D(\xi)$, (5.23), (5.24) and (5.25) are, in the limit $\varepsilon \to 0$, also satisfied for the extended method of Table 5.2. Let us demonstrate this for the Condition (5.23). The inverse of the extended

Runge-Kutta matrix is given by

$$\begin{pmatrix} A & 0 \\ b^T & \varepsilon \end{pmatrix}^{-1} = \begin{pmatrix} A^{-1} & 0 \\ -\varepsilon^{-1} b^T A^{-1} & \varepsilon^{-1} \end{pmatrix}. \tag{5.32}$$

Therefore (5.23) is seen to be satisfied for $i = 1, \ldots, s$. For $i = s+1$ one gets

$$\sum_{j=1}^{s+1} \omega_{s+1,j} c_j^k = -\varepsilon^{-1} \sum_{i,j=1}^{s} b_i \omega_{ij} c_j^k + \varepsilon^{-1}(1+\varepsilon)^k. \tag{5.33}$$

Using (5.23) for $i \leq s$ and $B(p)$ the right-hand expression of (5.33) becomes $-\varepsilon^{-1} + \varepsilon^{-1}(1+\varepsilon)^k$ and tends to k for $\varepsilon \to 0$. Hence, Condition (5.23) is, in the limit $\varepsilon \to 0$, also satisfied for $i = s+1$. As in the proof of Theorem 5.10 (see also the remark after that proof) we deduce the statement for the case where $f(y,z)$ is linear in z.

The generalization to nonlinear problems can be proved by a perturbation argument. We let $z(x)$ be the exact solution of (5.1) and consider the problem (Lubich 1991)

$$\begin{aligned} u' &= f(u, z(x)) + f_z(u, z(x))\lambda \\ 0 &= g(u) \end{aligned} \tag{5.34}$$

in the variables u and λ. This new problem is of index 2 again and has obviously the solution $u(x) = y(x)$ and $\lambda(x) = 0$. Since (5.34) is linear in the algebraic variable λ, the theorem can be applied and we get for the projected Runge-Kutta solution

$$\widehat{u}_1 - y(x_0 + h) = \mathcal{O}(h^{p+1}). \tag{5.35}$$

We still have to estimate $\widehat{y}_1 - \widehat{u}_1$. This is possible with the help of Theorem 4.2. In addition to the nonlinear system (4.2) (with $\eta = y_0$) we consider the method applied to (5.34):

$$\begin{aligned} U_i &= y_0 + h \sum_{j=1}^{s} a_{ij} \Big(f(U_j, z(x_0 + c_j h)) + f_z(U_j, z(x_0 + c_j h)) \Lambda_j \Big) \\ 0 &= g(U_i). \end{aligned} \tag{5.36}$$

Its first line can be written as

$$U_i = y_0 + h \sum_{j=1}^{s} a_{ij} f(U_j, z(x_0 + c_j h) + \Lambda_j) + \mathcal{O}(h \|\Lambda\|^2)$$

where $\|\Lambda\| = \max_j \|\Lambda_j\|$. Theorem 4.2 thus yields

$$\|U_i - Y_i\| \leq Ch\|\Lambda\|^2 \tag{5.37a}$$

$$\|\Lambda_i + z(x_0 + c_i h) - Z_i\| \leq C\|\Lambda\|^2. \tag{5.37b}$$

Since $C(\eta)$ holds, the estimate (4.14) together with (5.37b) proves $\Lambda_i = \mathcal{O}(h^\eta)$. We therefore obtain $y_1 - u_1 = \mathcal{O}(h^{2\eta+1})$ with the help of (5.37), and $\widehat{y}_1 - \widehat{u}_1 = \mathcal{O}(h^{2\eta+1})$ as a consequence of $z_1 - z(x_0 + h) = \mathcal{O}(h^\eta)$. □

Examples. 1) Collocation methods satisfy $B(p)$, $C(s)$ and $D(p-s)$ where s is the number of stages and p the order of the underlying quadrature formula (consult Lemma IV.5.4). Hence, the above presentation provides an alternative proof of Theorem 4.12.

2) The projected s-stage Radau IA method (see Table IV.5.13) has order $2s-1$ for problems which are linear in z, and order $2s-2$ for general nonlinear index 2 problems.

Exercises

1. Denote by r the largest number such that the local error of the z-component satisfies $\delta z_h(x) = \mathcal{O}(h^r)$. For implicit Runge-Kutta methods with invertible coefficient matrix, $R(\infty) = 0$ and $c_j \leq 1$ (all j) prove that

$$r \leq s^*$$

where s^* is the number of distinct non-zero values among $c_1, \ldots, c_s$.
Hint. The order conditions for the bushy trees $[\tau, \ldots, \tau]_z$ imply that

$$\sum_{i,j,k} b_i \omega_{ij} \omega_{jk} \int_0^{c_k} q(t)dt = q(1)$$

for all polynomials $q(t)$ of degree $\leq r-1$.

2. If a tree of *DAT*2 satisfies one of the following two conditions

a) a fat vertex (different from the root) is singly branched

b) a singly branched meagre vertex ($\neq$ root) is followed by a fat vertex

then the corresponding order condition is equivalent to that of a tree of the same order but with fewer fat vertices. Consequently, trees satisfying either (a) or (b) need not be considered for the construction of Runge-Kutta methods.

3. Suppose that the function $f(y,z)$ in (5.1) is linear in z. Characterize the trees of *DAT*2 for which the elementary differentials vanish identically.

4. With the help of Theorem 5.10 and Lemma IV.5.4 give a new (algebraic) proof of Theorem 4.9.

5. (Lubich 1991). Consider a projected Runge-Kutta method for index 2 problems which are linear in z. Prove that $\widehat{y}_1 - y(x_0+h) = \mathcal{O}(h^4)$ iff the condition

$$\sum_{i,j=1}^{s} b_i(1-c_i)\omega_{ij}c_j^2 = \frac{1}{3}$$

is satisfied in addition to the four order conditions already needed for ordinary differential equations.

VII.6 Half-Explicit Methods for Index 2 Systems

The methods of Sects. VII.3 and VII.4 do not use the semi-explicit structure of the differential-algebraic equation

$$y' = f(y,z), \qquad 0 = g(y) \tag{6.1}$$

($y \in \mathbb{R}^n, z \in \mathbb{R}^m$) and can as well be applied to more general situations. Here we shall show how this structure can be exploited for the derivation of new, efficient integration methods. The main idea is to discretize the differential variables y in an explicit manner, and the algebraic variables z in an implicit manner.

The most simple method of this type is the half-explicit Euler method

$$y_1 = y_0 + hf(y_0, z_0) \tag{6.2a}$$

$$0 = g(y_1). \tag{6.2b}$$

Inserting (6.2a) into (6.2b) yields the nonlinear system $0 = g\big(y_0 + hf(y_0, z_0)\big)$ for z_0. It possesses a locally unique solution, if

$$g_y(y) f_z(y,z) \qquad \text{is invertible} \tag{6.3}$$

at (y_0, z_0). Once z_0 is computed, the value y_1 is determined explicitly by (6.2a).

This example shows some interesting features of half-explicit methods. Compared to the implicit Euler discretization, it can be implemented more efficiently, because the nonlinear system is of reduced dimension (m instead of $n+m$). Compared to the explicit Euler method in the mode "index reduction and projection" (see Sect. VII.2), it avoids an accurate computation of the derivative $g_y(y)$. The numerical approximation y_1 only depends on an initial value of the y-component, as does the exact solution of (6.1).

In this section we shall develop half-explicit Runge-Kutta methods, extrapolation methods, and multistep methods. They are in particular very efficient for constrained mechanical systems in their index 2 formulation, because nonlinear systems are completely avoided in this situation (see below).

Half-Explicit Runge-Kutta Methods

In HLR89, the following extension of (6.2) to explicit Runge-Kutta methods is proposed:

$$Y_i = y_0 + h\sum_{j=1}^{i-1} a_{ij} f(Y_j, Z_j), \qquad i = 1, \ldots, s \tag{6.4a}$$

$$0 = g(Y_i) \tag{6.4b}$$

$$y_1 = y_0 + h\sum_{i=1}^{s} b_i f(Y_i, Z_i), \tag{6.4c}$$

$$0 = g(y_1). \tag{6.4d}$$

We have $Y_1 = y_0$, and Eq. (6.4b) is automatically satisfied for $i = 1$, because the initial value is assumed to be consistent. We next insert Y_2 from (6.4a) into (6.4b) and obtain a nonlinear equation for Z_1, which has a (locally) unique solution, if $a_{21} \neq 0$ and the usual index 2 assumption (6.3) is satisfied. We thus obtain Z_1 and Y_2. The next step allows us to compute Z_2 and Y_3, etc.

The local error and convergence properties of (6.4) are studied in HLR89 and Brasey & Hairer (1993). It turns out that the coefficients a_{ij}, b_i have to satisfy additional order conditions. As a consequence, 8 stages are needed for a 5th order method (Brasey 1992), compared to only 6 stages for classical Runge-Kutta methods (see Sect. II.5). Arnold (1995) and Murua (1995) have independently proposed a modification, which simplifies the order conditions and makes the approach more efficient. Their main idea is to introduce an explicit stage $Y_1 = y_0$, $Z_1 = z_0$, $Y_2 = y_0 + ha_{21} f(y_0, z_0)$, and to suppress the condition $g(Y_2) = 0$ in the second stage. We follow here the approach of Murua (1995), because it is slightly more general. For consistent initial values (y_0, z_0) we define

$$Y_1 = y_0, \qquad Z_1 = z_0 \tag{6.5a}$$

$$Y_i = y_0 + h\sum_{j=1}^{i-1} a_{ij} f(Y_j, Z_j), \qquad i = 2, \ldots, s \tag{6.5b}$$

$$\widehat{Y}_i = y_0 + h\sum_{j=1}^{i} \widehat{a}_{ij} f(Y_j, Z_j), \qquad 0 = g(\widehat{Y}_i), \qquad i = 2, \ldots, s \tag{6.5c}$$

$$y_1 = \widehat{Y}_s. \tag{6.5d}$$

The value z_1 can either be computed from the hidden constraint $g_y(y_1) f(y_1, z_1) = 0$, or from the additional stage

$$\widehat{Y}_{s+1} = y_0 + h\sum_{j=1}^{s+1} \widehat{a}_{s+1,j} f(Y_j, Z_j), \qquad 0 = g(\widehat{Y}_{s+1}) \tag{6.5e}$$

as $z_1 = Z_{s+1}$. Here we have put $Y_{s+1} = y_1$, so that the value $f(Y_{s+1}, Z_{s+1})$

can be reused as $f(y_0, z_0)$ for the next step. A significant difference compared to the original approach (6.4) is that the numerical solution (y_1, z_1) depends on both initial values (y_0 and z_0).

Existence of the Numerical Solution. Suppose that the initial values satisfy $g(y_0)=0$ and $g_y(y_0)f(y_0,z_0)=\mathcal{O}(\delta)$ with some sufficiently small $\delta>0$ (we have to admit small perturbations in the hidden constraint, because in general the approximation z_1 of (6.5e) does not satisfy $g_y(y_1)f(y_1,z_1)=0$). By an induction argument we assume that the values (Y_j, Z_j) are already known for $j=1,\ldots,i-1$, and satisfy $Y_j = y_0+\mathcal{O}(h)$, $Z_j = z_0+\mathcal{O}(h+\delta)$. Then, Y_i is explicitly given by (6.5b), and we have $Y_i = y_0 + \mathcal{O}(h)$. As in (3.13) we now write the condition $0 = g(\widehat{Y}_i)$ as

$$0 = \int_0^1 g_y\big(y_0 + \tau(\widehat{Y}_i - y_0)\big)\, d\tau \cdot \sum_{j=1}^{i} \widehat{a}_{ij} f(Y_j, Z_j), \tag{6.6}$$

where $\widehat{Y}_i$ has to be replaced by (6.5c). This is a nonlinear equation of the form $F(Z_i, h)=0$. Since $F(z_0, 0) = \mathcal{O}(\delta)$ and

$$\frac{\partial F}{\partial z}(z_0, 0) = \widehat{a}_{ii} \cdot g_y(y_0) f_z(y_0, z_0),$$

it follows from the Implicit Function Theorem that (6.6) has a locally unique solution, if (6.3) and the condition

$$\widehat{a}_{ii} \neq 0 \qquad \text{for all } i \tag{6.7}$$

hold. Moreover we have $Z_i = z_0 + \mathcal{O}(h+\delta)$.

Error Propagation and Convergence. For inconsistent initial values we replace the nonlinear equation in (6.5c) by $g(\widehat{Y}_i) = g(y_0)$, so that the method is well-defined in a whole neighbourhood of the solution manifold (observe that the above existence result is still valid). Such an extension has the advantage that differentiation with respect to initial values is possible. The method (6.5) with z_1 from (6.5e), can thus be written as

$$\begin{aligned} y_{n+1} &= y_n + h\Phi(y_n, z_n, h) \\ z_{n+1} &= \Psi(y_n, z_n, h) \end{aligned} \tag{6.8}$$

with smooth functions Φ and Ψ. For the study of convergence and, in particular, of the order conditions the triangular matrix

$$W = \big(w_{ij}\big)_{i,j=1}^{s+1} = \begin{pmatrix} 1 & & & \\ \widehat{a}_{21} & \widehat{a}_{22} & & \\ \vdots & \vdots & \ddots & \\ \widehat{a}_{s+1,1} & \widehat{a}_{s+1,2} & \cdots & \widehat{a}_{s+1,s+1} \end{pmatrix}^{-1} \tag{6.9}$$

will play an important role.

Lemma 6.1. *Suppose that the method (6.5), satisfying (6.7), is written in the form (6.8). If* $g(y_0)=0$ *and* $g_y(y_0)f(y_0,z_0)=\mathcal{O}(h)$, *it holds*

$$\frac{\partial \Phi}{\partial z}(y_0,z_0,h)=\mathcal{O}(h), \qquad \frac{\partial \Psi}{\partial z}(y_0,z_0,h)=w_{s+1,1}\cdot I+\mathcal{O}(h),$$

where $w_{s+1,1}$ *is given by (6.9).*

Proof. From (6.5b) it follows that $\partial Y_i/\partial z_0=\mathcal{O}(h)$. Differentiation of (6.5c) with respect to z_0 thus yields

$$\frac{\partial \widehat{Y}_i}{\partial z_0}=h\sum_{j=1}^{i}\widehat{a}_{ij}f_z(y_0,z_0)\frac{\partial Z_j}{\partial z_0}+\mathcal{O}(h^2), \tag{6.10a}$$

$$0=g_y(y_0)\frac{\partial \widehat{Y}_i}{\partial z_0}+\mathcal{O}(h^2). \tag{6.10b}$$

Inserting (6.10a) into (6.10b) and multiplying with the inverse of the matrix $g_y(y_0)f_z(y_0,z_0)$, gives the relation

$$\sum_{j=1}^{i}\widehat{a}_{ij}\frac{\partial Z_j}{\partial z_0}=\mathcal{O}(h) \qquad \text{for } i=2,\ldots,s+1.$$

The statement now follows from $Z_1=z_0$, i.e., $\partial Z_1/\partial z_0=I$. □

Consider two pairs of initial values (y_0,z_0), $(\widetilde{y}_0,\widetilde{z}_0)$, satisfying $g(y_0)=0$, $g(\widetilde{y}_0)=0$, $g_y(y_0)f(y_0,z_0)=\mathcal{O}(h)$, $g_y(\widetilde{y}_0)f(\widetilde{y}_0,\widetilde{z}_0)=\mathcal{O}(h)$. It follows from Lemma 6.1 that the differences $\Delta y_0=y_0-\widetilde{y}_0$, ... satisfy the recursion

$$\begin{pmatrix}\|\Delta y_1\|\\ \|\Delta z_1\|\end{pmatrix}\le\begin{pmatrix}1+\mathcal{O}(h) & \mathcal{O}(h^2)\\ \mathcal{O}(1) & |w_{s+1,1}|+\mathcal{O}(h)\end{pmatrix}\begin{pmatrix}\|\Delta y_0\|\\ \|\Delta z_0\|\end{pmatrix}. \tag{6.11}$$

The local error of the method (6.5) is defined as usual. We let (y_1,z_1) be the numerical approximation for initial values $\big(y(x),z(x)\big)$ on the exact solution of (6.1), and denote it by $\delta y_h(x)=y_1-y(x+h)$, $\delta z_h(x)=z_1-z(x+h)$.

Theorem 6.2 (Murua 1995). *Consider the problem (6.1) with consistent initial values. Suppose that (6.7) holds and that*

$$|w_{s+1,1}|<1, \tag{6.12}$$

where $w_{s+1,1}$ *is given in (6.9). If the local error satisfies*

$$\delta y_h(x)=\mathcal{O}(h^{r+1}), \qquad \delta z_h(x)=\mathcal{O}(h^m), \tag{6.13}$$

then we have for $x_n-x_0\le Const$

$$y_n-y(x_n)=\mathcal{O}(h^{\min(r,m+1)}), \qquad z_n-z(x_n)=\mathcal{O}(h^{\min(r,m)}).$$

Proof. The recursion (6.11) allows us to apply Lemma VI.3.9 with $\varepsilon = h^2$ and $\alpha = |w_{s+1,1}| + \mathcal{O}(h)$. This shows that the contribution of the local error at x_i to the global error at x_n is bounded by

$$C\big(\|\delta y_h(x_i)\| + h^2\|\delta z_h(x_i)\|\big), \qquad C\big(\|\delta y_h(x_i)\| + (h^2 + \alpha^{n-1-i})\|\delta z_h(x_i)\|\big)$$

for the y- and z-component, respectively. Summing up these contributions proves the statement. □

Order Conditions. The order conditions for method (6.5) can be derived in the same way as for Runge-Kutta methods (previous section). The only difference is that at some places the coefficients a_{ij} have to be replaced by $\widehat{a}_{ij}$. Since $z_1 = Z_{s+1}$, the order conditions for the z-component can be directly obtained from Theorem 5.6. The result is the following:

Forming the Order Condition for a Given Tree. Attach to each vertex one summation index. Then the left-hand side of the order condition is a sum over all indices of a product with factors

$\widehat{a}_{si}$	if "i" is the index of a meagre root;
$w_{s+1,i}$	if "i" is the index of a fat root;
a_{ij}	if the meagre vertex "j" lies directly above the meagre vertex "i";
$\widehat{a}_{ij}$	if the meagre vertex "j" lies directly above the fat vertex "i";
w_{ij}	if the fat vertex "j" lies directly above the meagre vertex "i";

The right-hand side of the order condition is the inverse of the rational number γ, defined in Theorem 5.6.

In order to satisfy the assumption (6.13) of the convergence theorem, the order conditions have to be satisfied for trees $t \in DAT2_y$ with $\varrho(t) \le r$, and for trees $u \in DAT2_z$ with $\varrho(u) \le m-1$.

Construction of Methods. The trees of Sect. II.2 form a subset of the "index 2 trees" to be considered here. From the above construction principle it is clear that the coefficients $a_{ij}, b_i := \widehat{a}_{si}$ have to satisfy the classical order conditions of Sect. II.2. It is therefore natural to take a known, explicit Runge-Kutta method of a certain order and to determine $\widehat{a}_{ij}$ in such a way that the remaining order conditions are satisfied. Arnold (1995) and Murua (1995) have shown how half-explicit methods, based on the Dormand & Prince pair of Table II.5.2, can be constructed. Let us outline the main idea.

A significant simplification of the order conditions is obtained by requiring

$$\sum_{j=1}^{i} \widehat{a}_{ij} c_j^{q-1} = \frac{\widehat{c}_i^{\,q}}{q} \qquad \text{for} \quad i = 1, \ldots, s+1, \tag{6.14}$$

where $c_i = \sum_j a_{ij}$ and $\widehat{c}_i = \sum_j \widehat{a}_{ij}$. For $i = 1$, the relation (6.14) is automatically fulfilled because of $\widehat{a}_{1j} = 0$. For $i > 1$, it can be satisfied for $q = 1$ (definition of $\widehat{c}_i$), $q=2$, and $q=3$. The simplification in the order conditions is similar to that illustrated in Fig. II.5.2. By the definition of the matrix W, the relations of Eq. (6.14) are equivalent to

$$\sum_{j=1}^{i} w_{ij}\widehat{c}_j^{\,q} = q\, c_i^{q-1} \qquad \text{for } \; i=1,\ldots,s+1. \tag{6.15}$$

This implies further reductions in the set of order conditions. The few remaining ones can be treated in a straight-forward manner. For further details and for the coefficients of the resulting method we refer to the original article of Murua (1995). They have been incorporated in the code PHEM56 (see Sect. VII.7).

Application to Constrained Mechanical Systems. Consider the system

$$\begin{aligned} q' &= u && (6.16a)\\ M(q)u' &= f(q,u) - G^T(q)\lambda && (6.16b)\\ 0 &= g(q), && (6.16c) \end{aligned}$$

where $G(q)=g_q(q)$. Differentiating the constraint (6.16c) yields

$$0 = G(q)u. \tag{6.16d}$$

If $M(q)$ is invertible, the system (6.16a,b,d) is of the form (6.1) with $y=(q,u)$ and $z=\lambda$. The assumption (6.3) is equivalent to (1.47).

For this particular system the method (6.5) can be applied as follows: assume that Q_j, U_j, Λ_j, and $U_j' = M(Q_j)^{-1}\big(f(Q_j,U_j) - G^T(Q_j)\Lambda_j\big)$ are already given for $j=1,\ldots,i-1$. We then put

$$Q_i = q_0 + h\sum_{j=1}^{i-1} a_{ij}U_j, \qquad\qquad U_i = u_0 + h\sum_{j=1}^{i-1} a_{ij}U_j',$$

and compute Λ_i, U_i' from the system

$$\begin{pmatrix} M(Q_i) & G^T(Q_i) \\ G(\widehat{Q}_i) & 0 \end{pmatrix} \begin{pmatrix} U_i' \\ \Lambda_i \end{pmatrix} = \begin{pmatrix} f(Q_i,U_i) \\ R_i \end{pmatrix}, \tag{6.17}$$

where $\widehat{Q}_i = q_0 + h\sum_{j=1}^{i}\widehat{a}_{ij}U_j$ and $R_i = -G(\widehat{Q}_i)\big(u_0 + h\sum_{j=1}^{i-1}\widehat{a}_{ij}U_j'\big)/(h\widehat{a}_{ii})$ are known quantities. Hence, only linear systems of type (6.17) have to be solved. This makes half-explicit methods very attractive for the numerical solution of constrained mechanical systems. If necessary, this method can be combined with projections as explained in Sect. VII.2, so that also the position constraint is satisfied by the numerical approximation.

We remark that the methods proposed by Arnold (1995) satisfy $\widehat{Q}_i = Q_{i+1}$ for $i\geq 2$, so that some G evaluations can be saved in the computation of (6.17).

Extrapolation Methods

For nonstiff ordinary differential equations, the most efficient extrapolation algorithm is the GBS method (see Sect. II.9). Lubich (1989) extends this method to differential-algebraic equations of index 2.

Consider an initial value y_0 satisfying $g(y_0)=0$. Then, an approximation $S_h(x)$ to $y(x)$ (with $x=x_0+2mh$) is defined by

$$\begin{aligned} y_1 &= y_0 + hf(y_0,z_0), & g(y_1)&=0 & & \qquad (6.18a)\\ y_{i+1} &= y_{i-1} + 2hf(y_i,z_i), & g(y_{i+1})&=0, & i&=1,\ldots,2m \qquad (6.18b)\\ S_h(x) &= (y_{2m-1}+2y_{2m}+y_{2m+1})/4. & & & & \qquad (6.18c) \end{aligned}$$

The starting step is identical to the half-explicit Euler method, considered at the beginning of this section. It is implicit in z_0 and explicit in y_1. For the case that Eq. (6.1) in linear in z, i.e.,

$$f(y,z)=f_0(y)+f_z(y)z, \qquad (6.19)$$

we shall show below that the numerical approximations $S_h(x_0+2mh)$ and z_{2m} possess an h^2-expansion. Hence, these values can be used as the basis of an extrapolation method. The implementation is completely analogue to that for the GBS method (choice of the step number sequence, order and step size control, dense output, ...). Since the extrapolated values do not satisfy the constraint $g(y)=0$, it is recommended to project them onto this manifold (as explained in Sect. VII.2) after every accepted step.

The assumption (6.19) is satisfied for many interesting problems, e.g., for the constrained mechanical system (6.16a,b,d), where $z=\lambda$ plays the role of a Lagrange multiplier.

Theorem 6.3 (Lubich 1989). *Under the assumptions (6.3) and (6.19) the numerical solution of method (6.18) possesses an asymptotic h^2-expansion*

$$\begin{aligned} y_{2m}-y(x_{2m}) &= a_2(x_{2m})h^2+a_4(x_{2m})h^4+\ldots+a_{2N}(x_{2m})h^{2N}+\mathcal{O}(h^{2N+2})\\ z_{2m}-z(x_{2m}) &= b_2(x_{2m})h^2+b_4(x_{2m})h^4+\ldots+b_{2N}(x_{2m})h^{2N}+\mathcal{O}(h^{2N+2}) \end{aligned}$$

and another h^2-expansion for the error of $S_h(x_{2m})$.

The numerical solution $\{y_i\}$ of method (6.18) lies on the manifold defined by $g(y)=0$. In order to be able to apply the results and ideas of Sects. II.8 and II.9, we extend the method (6.18) to arbitrary initial values as follows:

$$\begin{aligned} y_1 &= y_0+hf(y_0,z_0), & g(y_1)&=g(y_0) & & \qquad (6.20a)\\ y_{i+1} &= y_{i-1}+2hf(y_i,z_i), & g(y_{i+1})&=g(y_{i-1}), & i&=1,\ldots,2m \qquad (6.20b) \end{aligned}$$

We further eliminiate the z-variables: using the identity

$$g(y_{i+1})-g(y_{i-1})=\int_{-1}^{1} g_y\Big(\frac{y_{i+1}+y_{i-1}}{2}+\sigma\,\frac{y_{i+1}-y_{i-1}}{2}\Big)\,d\sigma\cdot\Big(\frac{y_{i+1}-y_{i-1}}{2}\Big),$$

Eq. (6.20b) becomes

$$0=\int_{-1}^{1} g_y\Big(\frac{y_{i+1}+y_{i-1}}{2}+\sigma h f(y_i,z_i)\Big)\,d\sigma\cdot f(y_i,z_i). \tag{6.21}$$

By assumption (6.3) and the Implicit Function Theorem, Eq. (6.21) can be solved for z_i as a smooth function of $(y_{i+1}+y_{i-1})/2$, y_i, and h. Inserted into (6.20b) we obtain a recursion of the type

$$y_{i+1}=y_{i-1}+2h\Phi\big(y_i,(y_{i+1}+y_{i-1})/2,h\big). \tag{6.22}$$

The starting step (6.20a) can be rewritten in a similar way. We consider the more general system

$$w=v+hf(u,z), \qquad g(w)=g(v), \tag{6.23}$$

where u,v, and h are given. It can be written in the equivalent form

$$0=\int_0^1 g_y\big(v+\tau h f(u,z)\big)\,d\tau\cdot f(u,z),$$

which yields z as a smooth function of u,v, and h (again by the Implicit Function Theorem). Hence, the solution of (6.23) can be written as

$$w=v+h\Phi_0(u,v,h), \tag{6.24}$$

and the starting step (6.20a) becomes

$$y_1=y_0+h\Phi_0(y_0,y_0,h). \tag{6.25}$$

The crucial point of these reformulations is that the two-step method (6.22) and the starting step (6.25) are not only defined on the manifold $g(y)=0$, but on an open neighbourhood of it. Therefore, the standard ODE theory can be applied. Results for the method (6.22), (6.25) immediately carry over to the method (6.18), because both methods are identical for initial values satisfying $g(y_0)=0$.

Asymptotic Expansion for Symmetric Two-Step Methods. Motivated by the above reformulations we consider the method

$$y_1=y_0+h\Phi_0(y_0,y_0,h) \tag{6.26a}$$
$$y_{i+1}=y_{i-1}+2h\Phi\big(y_i,(y_{i+1}+y_{i-1})/2,h\big), \tag{6.26b}$$

where Φ_0 and Φ are arbitrary, smooth increment functions. We assume that $\Phi_0(y,y,0)=\Phi(y,y,0)=f(y)$, so that both methods are consistent with the ordinary differential equation $y'=f(y)$. In order to get an h^2-expansion of the error, the starting step (6.26a) has to be compatible with (6.26b) in the following sense: for arbitrary u_k,v_k, the three values

$$\begin{aligned} y_{2k-1}&:=v_k-h\Phi_0(u_k,v_k,-h), \qquad y_{2k}:=u_k,\\ y_{2k+1}&:=v_k+h\Phi_0(u_k,v_k,h)\end{aligned} \tag{6.27}$$

satisfy the recursion (6.26b).

Theorem 6.4. *If the method (6.26) satisfies the compatibility condition (6.27), the numerical approximations*

$$y_{2m}, \qquad (y_{2m+1}+y_{2m-1})/2$$

have an asymptotic expansion in even powers of h.

Proof. Inspired by Stetter's proof of Theorem II.9.2 we put $u_k := y_{2k}$, and let v_k be the solution of

$$y_{2k+1} := v_k + h\Phi_0(u_k, v_k, h). \tag{6.28}$$

We thus get the one-step method in doubled dimension

$$\begin{pmatrix} u_{k+1} \\ v_{k+1} \end{pmatrix} = \begin{pmatrix} u_k \\ v_k \end{pmatrix} + h^* \begin{pmatrix} \Phi(y_{2k+1}, (u_{k+1}+u_k)/2, h^*/2) \\ \frac{1}{2}(\Phi_0(u_k, v_k, h^*/2) + \Phi_0(u_{k+1}, v_{k+1}, -h^*/2)) \end{pmatrix},$$

where $h^* = 2h$, and y_{2k+1} is given by (6.28). The assumption (6.27) implies that this one-step method is symmetric. Therefore, $y_{2m} = u_m$ and v_m have an asymptotic h^2-expansion (see Theorem II.8.10). From

$$(y_{2m+1}+y_{2m-1})/2 = y_{2m} + h(\Phi_0(u_m, v_m, h) - \Phi_0(u_m, v_m, -h))$$

it follows that the same is true for $(y_{2m+1}+y_{2m-1})/2$. □

Proof of Theorem 6.3. We have already seen that the method (6.20) can be written in the form (6.26). All that remains to do is to check the compatibility condition (6.27). By definition of $\Phi_0(u, v, h)$ (see the equivalence of Eqs. (6.23) and (6.25)) we have

$$\begin{aligned} y_{2k-1} &= v_k - hf(u_k, z^-), \qquad & g(y_{2k-1}) &= g(v_k) \\ y_{2k+1} &= v_k + hf(u_k, z^+), \qquad & g(y_{2k+1}) &= g(v_k). \end{aligned}$$

Since f is linear in z, this implies (6.20b) with $z_{2k} = (z^- + z^+)/2$. The asymptotic h^2-expansion of y_{2m} and $S_h(x_{2m})$ thus follows from Theorem 6.4. From (6.21) we then see that also z_{2m} has an h^2-expansion. □

β-Blocked Multistep Methods

The convergence analysis of Sect. VII.3 shows that all roots of the σ-polynomial of a multistep method must lie inside the unit disc in order to get a convergent method of order p. This is a severe restriction and excludes, for example, all explicit and implicit Adams methods. Arévalo, Führer & Söderlind (1995) suggest a modification which allows the use of "nonstiff" multistep methods. The idea is to treat different parts of the problem by different discretizations.

For the index 2 problem

$$y' = f_0(y) + f_z(y)z, \qquad 0 = g(y), \tag{6.29}$$

where $f(y,z) = f_0(y) + f_z(y)z$ depends linearly on z, we consider the discretization

$$\sum_{i=0}^{k} \alpha_i y_{n+i} = h \sum_{i=0}^{k} \beta_i f(y_{n+i}, z_{n+i}) - h f_z(y_{n+k}) \sum_{i=0}^{k} \gamma_i z_{n+i}, \tag{6.30}$$

and denote the generating polynomials by

$$\varrho(\zeta) = \sum_{i=0}^{k} \alpha_i \zeta^i, \qquad \sigma(\zeta) = \sum_{i=0}^{k} \beta_i \zeta^i, \qquad \tau(\zeta) = \sum_{i=0}^{k} \gamma_i \zeta^i.$$

Theorem 6.5 (Arévalo, Führer & Söderlind 1995). *Let the index 2 problem (6.29) satisfy (6.3). Assume that the multistep method (ϱ, σ) is stable and of order p ($p = k$ or $p = k+1$), that $\tau(\zeta) = \gamma_k(\zeta - 1)^k$, and that all roots of $\sigma(\zeta) - \tau(\zeta)$ lie inside the unit disc $|\zeta| < 1$. Then the global error satisfies for $x_n - x_0 \le Const$*

$$y_n - y(x_n) = \mathcal{O}(h^p), \qquad z_n - z(x_n) = \mathcal{O}(h^k).$$

Proof. The special form of $\tau(\zeta)$ is equivalent to

$$\sum_{i=0}^{k} \gamma_i z(x_n + ih) = \mathcal{O}(h^k),$$

so that the newly added term in (6.30) is small. Moreover, this term is premultiplied by $f_z(y_{n+k})$, so that the local error satisfies

$$\delta y_h(x) = \mathcal{O}(h^{k+1}), \qquad P(x)\delta y_h(x) = \mathcal{O}(h^{p+1}),$$

where $P(x)$ is the projector of Definition 4.3.

With these observations in mind, the convergence result is obtained along the lines of the proof of Theorem 3.6. The only difference is that the coefficients β_i have to be replaced by $\beta_i - \gamma_i$ in Eqs. (3.43) and (3.44). □

In principle, one can take any convergent multistep method (ϱ, σ) of order $p = k$ or $p = k+1$, and try to optimize the parameter γ_k in $\tau(\zeta)$ in such a way that the roots of $\sigma(\zeta) - \tau(\zeta)$ become small. The result, for the implicit Adams methods, is rather disappointing. Only for $k \le 3$ it is possible to obtain convergent β-blocked Adams methods (Arévalo, Führer & Söderlind (1995), see also Exercise 3).

Difference Corrected BDF. Consider the $(k+1)$-step BDF method, defined in Eq. (III.1.22'), and replace $\nabla^{k+1} y_{n+1}$ by $\nabla^k f_{n+1}$. This leads to the so-called difference corrected BDF

$$\sum_{j=1}^{k} \frac{1}{j} \nabla^j y_{n+1} = h\Big(f_{n+1} - \frac{1}{k+1} \nabla^k f_{n+1}\Big), \tag{6.31}$$

introduced by Söderlind (1989). Method (6.31) is a k-step method of order $p = k+1$. Its ϱ-polynomial is identical to that of the BDF method and $\sigma(\zeta) = \zeta^k - (\zeta-1)^k/(k+1)$. With $\tau(\zeta) = -(\zeta-1)^k/(k+1)$ the difference $\sigma(\zeta) - \tau(\zeta)$ has all roots equal to zero. This is therefore an ideal candidate for a method of type (6.30).

Exercises

1. Construct all half-explicit methods (6.5) of order 3 ($r = m = 3$ in Eq. (6.13)) with $s = 3$ stages. You can take $c_2, c_3, \alpha, \widehat{c}_2, \widehat{c}_4$ as free parameters.

   Hint. Start with a classical Runge-Kutta method of order 3 (Exercise 4 of Sect. II.1), and show that the order conditions imply (6.14) for $q = 2$.

2. Show that the method (IV.9.15) of Bader & Deuflhard (1983) is of the form (6.26) with

$$\Phi(u, v, h) = f(u) - Ju + Jv$$
$$\Phi_0(u, v, h) = (I - hJ)^{-1}\big(f(u) - Ju + Jv\big).$$

   Check the assumption (6.27).

3. Let (ϱ_k, σ_k) be the generating polynomials of the k-step implicit Adams methods (Sect. III.1). For $k = 1, 2, \ldots, 10$ study numerically the function

$$R_k(\gamma) := \max\big\{|\zeta^*|\,;\ \ \zeta^* \text{ is root of } \sigma_k(\zeta) - \gamma(\zeta-1)^k = 0\big\}.$$

   For which values of k is it possible to find γ with $R_k(\gamma) < 1$?

VII.7 Computation of Multibody Mechanisms

> Dynamics of multibody systems is of great importance in the fields of robotics, biomechanics, spacecraft control, road and rail vehicle design, and dynamics of machinery.
>
> (W. Schiehlen 1990)

After having seen several different approaches for the numerical solution of constrained mechanical systems, we are interested in their efficiency when applied to a concrete situation. We consider two particular multibody mechanisms with constraints, one nonstiff and one stiff. General references for the computation of mechanical systems are Haug (1989) and Roberson & Schwertassek (1988).

Description of the Model

We first consider "Andrews' squeezer mechanism", which has become prominent through the work of Giles (1978) and Manning (1981), who promoted it as a test example for numerical codes; see also Ormrod & Andrews (1986). It consists of 7 rigid bodies connected by joints without friction in plane motion. It is represented in Fig. 7.1, which we have copied (with permission) from the book of Schiehlen (1990). The numerical constants, also taken from Schiehlen (1990), are displayed in Tables 7.1 and 7.2. The arrows in the right picture of Fig. 7.1 indicate the positions of the centres of gravity $C_1, \ldots, C_7$. In Table 7.1 the spring coefficient of the spring connecting the point D with C is denoted by c_0 and the unstretched length is ℓ_0. We suppose that the mechanism is driven by a motor, located at O, whose constant drive torque is given by $mom = 0.033$. The coordinate origin is the point O in Fig. 7.1 and the coordinates of the other fixed points A, B and C are given by

$$\begin{pmatrix} xa \\ ya \end{pmatrix} = \begin{pmatrix} -0.06934 \\ -0.00227 \end{pmatrix}, \quad \begin{pmatrix} xb \\ yb \end{pmatrix} = \begin{pmatrix} -0.03635 \\ 0.03273 \end{pmatrix}, \quad \begin{pmatrix} xc \\ yc \end{pmatrix} = \begin{pmatrix} 0.014 \\ 0.072 \end{pmatrix}. \tag{7.1}$$

Table 7.1. Geometrical parameters

$d = 0.028$	$da = 0.0115$	$e = 0.02$
$ea = 0.01421$	$zf = 0.02$	$fa = 0.01421$
$rr = 0.007$	$ra = 0.00092$	$ss = 0.035$
$sa = 0.01874$	$sb = 0.01043$	$sc = 0.018$
$sd = 0.02$	$zt = 0.04$	$ta = 0.02308$
$tb = 0.00916$	$u = 0.04$	$ua = 0.01228$
$ub = 0.00449$	$c_0 = 4530$	$\ell_0 = 0.07785$

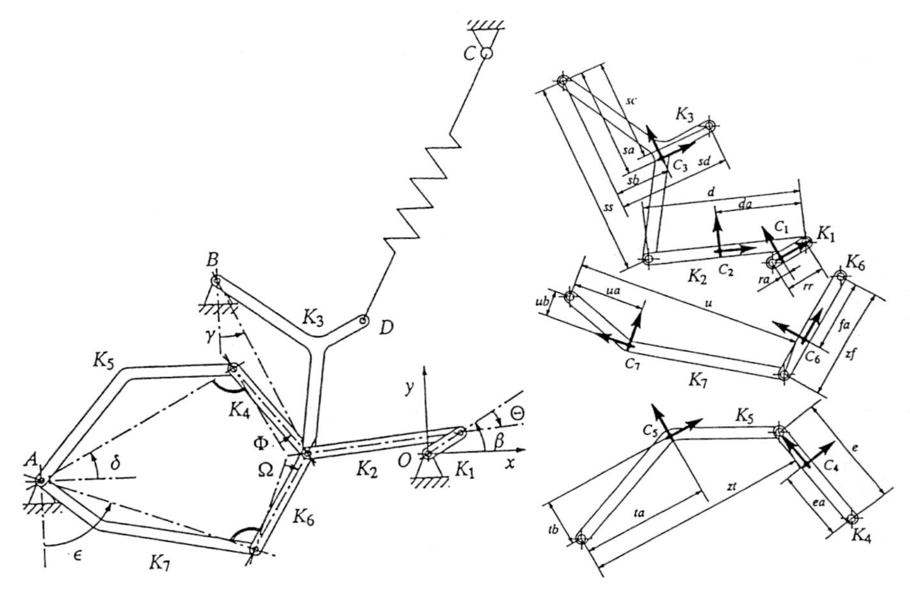

Fig. 7.1. Seven body mechanism (Schiehlen 1990, with permission)

Table 7.2. Parameters of the 7 bodies

No.	masses m_1 to m_7	inertias I_1 to I_7
1	0.04325	$2.194 \cdot 10^{-6}$
2	0.00365	$4.410 \cdot 10^{-7}$
3	0.02373	$5.255 \cdot 10^{-6}$
4	0.00706	$5.667 \cdot 10^{-7}$
5	0.07050	$1.169 \cdot 10^{-5}$
6	0.00706	$5.667 \cdot 10^{-7}$
7	0.05498	$1.912 \cdot 10^{-5}$

In order to derive the equations of motion we use the angles (see Fig. 7.1)

$$q_1 = \beta, \quad q_2 = \Theta, \quad q_3 = \gamma, \quad q_4 = \Phi, \quad q_5 = \delta, \quad q_6 = \Omega, \quad q_7 = \varepsilon, \tag{7.2}$$

as position coordinates for the mechanical system. If (x_j, y_j) are the cartesian coordinates of the centre of gravity C_j $(j = 1, \ldots, 7)$, the *kinetic energy* of the multibody system is

$$T = \sum_{j=1}^{7} m_j \frac{\dot{x}_j^2 + \dot{y}_j^2}{2} + \sum_{j=1}^{7} I_j \frac{\dot{\omega}_j^2}{2} \tag{7.3}$$

where ω_j is the total angle of rotation of the jth body and m_j, I_j are constants given in Table 7.2. The values of x_j, y_j, $\dot{x}_j^2 + \dot{y}_j^2$ and $\dot{\omega}_j$ can be obtained in terms of (7.2) by simple geometry (see Fig. 7.1):

$$
\begin{aligned}
C_1: \quad & x_1 = ra \cdot \cos\beta \\
& y_1 = ra \cdot \sin\beta \\
& \dot{x}_1^2 + \dot{y}_1^2 = ra^2 \cdot \dot{\beta}^2 \\
& \dot{\omega}_1 = \dot{\beta}
\end{aligned}
$$

$$
\begin{aligned}
C_2: \quad & x_2 = rr \cdot \cos\beta - da \cdot \cos(\beta + \Theta) \\
& y_2 = rr \cdot \sin\beta - da \cdot \sin(\beta + \Theta) \\
& \dot{x}_2^2 + \dot{y}_2^2 = \left(rr^2 - 2 \cdot da \cdot rr \cdot \cos\Theta + da^2\right) \cdot \dot{\beta}^2 \\
& \qquad + 2 \cdot \left(-rr \cdot da \cdot \cos\Theta + da^2\right) \cdot \dot{\beta} \cdot \dot{\Theta} + da^2 \cdot \dot{\Theta}^2 \\
& \dot{\omega}_2 = \dot{\beta} + \dot{\Theta}
\end{aligned}
$$

$$
\begin{aligned}
C_3: \quad & x_3 = xb + sa \cdot \sin\gamma + sb \cdot \cos\gamma \\
& y_3 = yb - sa \cdot \cos\gamma + sb \cdot \sin\gamma \\
& \dot{x}_3^2 + \dot{y}_3^2 = \left(sa^2 + sb^2\right) \cdot \dot{\gamma}^2 \\
& \dot{\omega}_3 = \dot{\gamma}
\end{aligned}
$$

$$
\begin{aligned}
C_4: \quad & x_4 = xa + zt \cdot \cos\delta + (e - ea) \cdot \sin(\Phi + \delta) \\
& y_4 = ya + zt \cdot \sin\delta - (e - ea) \cdot \cos(\Phi + \delta) \\
& \dot{x}_4^2 + \dot{y}_4^2 = (e - ea)^2 \cdot \dot{\Phi}^2 + 2 \cdot \left((e - ea)^2 + zt \cdot (e - ea) \cdot \sin\Phi\right) \cdot \dot{\Phi} \cdot \dot{\delta} \\
& \qquad + \left(zt^2 + 2 \cdot zt \cdot (e - ea) \cdot \sin\Phi + (e - ea)^2\right) \cdot \dot{\delta}^2 \\
& \dot{\omega}_4 = \dot{\Phi} + \dot{\delta}
\end{aligned}
$$

$$
\begin{aligned}
C_5: \quad & x_5 = xa + ta \cdot \cos\delta - tb \cdot \sin\delta \\
& y_5 = ya + ta \cdot \sin\delta + tb \cdot \cos\delta \\
& \dot{x}_5^2 + \dot{y}_5^2 = \left(ta^2 + tb^2\right) \cdot \dot{\delta}^2 \\
& \dot{\omega}_5 = \dot{\delta}
\end{aligned}
$$

$$
\begin{aligned}
C_6: \quad & x_6 = xa + u \cdot \sin\varepsilon + (zf - fa) \cdot \cos(\Omega + \varepsilon) \\
& y_6 = ya - u \cdot \cos\varepsilon + (zf - fa) \cdot \sin(\Omega + \varepsilon) \\
& \dot{x}_6^2 + \dot{y}_6^2 = (zf - fa)^2 \cdot \dot{\Omega}^2 + 2 \cdot \left((zf - fa)^2 - u \cdot (zf - fa) \cdot \sin\Omega\right) \cdot \dot{\Omega} \cdot \dot{\varepsilon} \\
& \qquad + \left((zf - fa)^2 - 2 \cdot u \cdot (zf - fa) \cdot \sin\Omega + u^2\right) \cdot \dot{\varepsilon}^2 \\
& \dot{\omega}_6 = \dot{\Omega} + \dot{\varepsilon}
\end{aligned}
$$

$$
\begin{aligned}
C_7: \quad & x_7 = xa + ua \cdot \sin\varepsilon - ub \cdot \cos\varepsilon \\
& y_7 = ya - ua \cdot \cos\varepsilon - ub \cdot \sin\varepsilon \\
& \dot{x}_7^2 + \dot{y}_7^2 = \left(ua^2 + ub^2\right) \cdot \dot{\varepsilon}^2 \\
& \dot{\omega}_7 = \dot{\varepsilon}
\end{aligned}
$$

The *potential energy* of the system is due to the motor at the origin and to the spring connecting the point D with C. By Hooke's law it is

$$U = -mom \cdot \beta + c_0 \frac{(\ell - \ell_0)^2}{2}, \tag{7.4}$$

where ℓ is the distance between D and C, namely

$$\begin{aligned} \ell &= \sqrt{(xd - xc)^2 + (yd - yc)^2} \\ xd &= xb + sc \cdot \sin\gamma + sd \cdot \cos\gamma \\ yd &= yb - sc \cdot \cos\gamma + sd \cdot \sin\gamma. \end{aligned}$$

Finally, we have to formulate the *algebraic constraints*. The mechanism contains three loops. The first loop connects O with B via K_1, K_2, K_3; the other two loops connect O with A, one via K_1, K_2, K_4, K_5, the other via K_1, K_2, K_6, K_7. For each loop we get two algebraic conditions:

$$\begin{aligned} rr \cdot \cos\beta - d \cdot \cos(\beta + \Theta) - ss \cdot \sin\gamma &= xb \\ rr \cdot \sin\beta - d \cdot \sin(\beta + \Theta) + ss \cdot \cos\gamma &= yb \\ rr \cdot \cos\beta - d \cdot \cos(\beta + \Theta) - e \cdot \sin(\Phi + \delta) - zt \cdot \cos\delta &= xa \\ rr \cdot \sin\beta - d \cdot \sin(\beta + \Theta) + e \cdot \cos(\Phi + \delta) - zt \cdot \sin\delta &= ya \\ rr \cdot \cos\beta - d \cdot \cos(\beta + \Theta) - zf \cdot \cos(\Omega + \varepsilon) - u \cdot \sin\varepsilon &= xa \\ rr \cdot \sin\beta - d \cdot \sin(\beta + \Theta) - zf \cdot \sin(\Omega + \varepsilon) + u \cdot \cos\varepsilon &= ya. \end{aligned} \tag{7.5}$$

With the position coordinates q from (7.2) the equations (7.5) represent the constraint $g(q) = 0$ where $g : \mathbb{R}^7 \to \mathbb{R}^6$. Together with the kinetic energy T of (7.3) the potential energy U of (7.4) and $L = T - U - \lambda_1 g_1 - \ldots - \lambda_6 g_6$ the equations of motion (1.46) are fully determined.

Fortran Subroutines

For the reader's convenience we include the essential parts of the FORTRAN subroutines describing the differential-algebraic problem. The equations of motion are of the form

$$M(q)\ddot{q} = f(q, \dot{q}) - G^T(q)\lambda \tag{7.6a}$$

$$0 = g(q) \tag{7.6b}$$

where $q \in \mathbb{R}^7$ is the vector defined in (7.2) and $\lambda \in \mathbb{R}^6$. In the following description the variables `Q(1),...,Q(7)` correspond to $\beta, \ldots, \varepsilon$ (exactly as in (7.2)) and `QP(1),...,QP(7)` to their derivatives $\dot{\beta}, \ldots, \dot{\varepsilon}$. In all subroutines we have used the abbreviations

```
SIBE = SIN (Q(1))            COBE = COS (Q(1))
SITH = SIN (Q(2))            COTH = COS (Q(2))
SIGA = SIN (Q(3))            COGA = COS (Q(3))
```

```
SIPH = SIN (Q(4))                COPH = COS (Q(4))
SIDE = SIN (Q(5))                CODE = COS (Q(5))
SIOM = SIN (Q(6))                COOM = COS (Q(6))
SIEP = SIN (Q(7))                COEP = COS (Q(7))

SIBETH = SIN (Q(1)+Q(2))         COBETH = COS (Q(1)+Q(2))
SIPHDE = SIN (Q(4)+Q(5))         COPHDE = COS (Q(4)+Q(5))
SIOMEP = SIN (Q(6)+Q(7))         COOMEP = COS (Q(6)+Q(7))

BEP = QP(1)                      THP = QP(2)
PHP = QP(4)                      DEP = QP(5)
OMP = QP(6)                      EPP = QP(7)
```

The remaining parameters XA,YA,...,D,DA,E,EA,...,M1,I1,M2,... are those of (7.1) and Tables 7.1 and 7.2. They usually reside in a COMMON block. The elements of $M(q)$ in (7.6) are given by

$$m_{ij} = \frac{\partial^2 L}{\partial \dot{q}_i \partial \dot{q}_j} = \frac{\partial^2 T}{\partial \dot{q}_i \partial \dot{q}_j}.$$

This matrix is symmetric and (due to the special arrangement of the coordinates) tridiagonal. The non-zero elements (on and below the diagonal) are

```
 M(1,1) = M1*RA**2 + M2*(RR**2-2*DA*RR*COTH+DA**2) + I1 + I2
 M(2,1) = M2*(DA**2-DA*RR*COTH) + I2
 M(2,2) = M2*DA**2 + I2
 M(3,3) = M3*(SA**2+SB**2) + I3
 M(4,4) = M4*(E-EA)**2 + I4
 M(5,4) = M4*((E-EA)**2+ZT*(E-EA)*SIPH) + I4
 M(5,5) = M4*(ZT**2+2*ZT*(E-EA)*SIPH+(E-EA)**2) + M5*(TA**2+TB**2)
+         + I4 + I5
 M(6,6) = M6*(ZF-FA)**2 + I6
 M(7,6) = M6*((ZF-FA)**2-U*(ZF-FA)*SIOM) + I6
 M(7,7) = M6*((ZF-FA)**2-2*U*(ZF-FA)*SIOM+U**2) + M7*(UA**2+UB**2)
+         + I6 + I7
```

The ith component of the function f in (7.6) is defined by

$$f_i(q,\dot{q}) = \frac{\partial(T-U)}{\partial q_i} - \sum_{j=1}^{7} \frac{\partial^2(T-U)}{\partial \dot{q}_i \partial q_j} \cdot \dot{q}_j.$$

Written as FORTRAN statements we have

```
 XD = SD*COGA + SC*SIGA + XB
 YD = SD*SIGA - SC*COGA + YB
 LANG = SQRT ((XD-XC)**2 + (YD-YC)**2)
 FORCE = - CO * (LANG - LO)/LANG
 FX = FORCE * (XD-XC)
 FY = FORCE * (YD-YC)
 F(1) = MOM - M2*DA*RR*THP*(THP+2*BEP)*SITH
 F(2) = M2*DA*RR*BEP**2*SITH
 F(3) = FX*(SC*COGA - SD*SIGA) + FY*(SD*COGA + SC*SIGA)
 F(4) = M4*ZT*(E-EA)*DEP**2*COPH
 F(5) = - M4*ZT*(E-EA)*PHP*(PHP+2*DEP)*COPH
 F(6) = - M6*U*(ZF-FA)*EPP**2*COOM
 F(7) = M6*U*(ZF-FA)*OMP*(OMP+2*EPP)*COOM
```

The algebraic constraints $g(q) = 0$ are given by the following six equations (see (7.5))

```
G(1) = RR*COBE - D*COBETH - SS*SIGA - XB
G(2) = RR*SIBE - D*SIBETH + SS*COGA - YB
G(3) = RR*COBE - D*COBETH - E*SIPHDE - ZT*CODE - XA
G(4) = RR*SIBE - D*SIBETH + E*COPHDE - ZT*SIDE - YA
G(5) = RR*COBE - D*COBETH - ZF*COOMEP - U*SIEP - XA
G(6) = RR*SIBE - D*SIBETH - ZF*SIOMEP + U*COEP - YA
```

And here is the Jacobian matrix $G(q) = g_q(q)$. The non-zero entries of this 6×7 array are

```
GQ(1,1) = - RR*SIBE + D*SIBETH
GQ(1,2) = D*SIBETH
GQ(1,3) = - SS*COGA
GQ(2,1) = RR*COBE - D*COBETH
GQ(2,2) = - D*COBETH
GQ(2,3) = - SS*SIGA
GQ(3,1) = - RR*SIBE + D*SIBETH
GQ(3,2) = D*SIBETH
GQ(3,4) = - E*COPHDE
GQ(3,5) = - E*COPHDE + ZT*SIDE
GQ(4,1) = RR*COBE - D*COBETH
GQ(4,2) = - D*COBETH
GQ(4,4) = - E*SIPHDE
GQ(4,5) = - E*SIPHDE - ZT*CODE
GQ(5,1) = - RR*SIBE + D*SIBETH
GQ(5,2) = D*SIBETH
GQ(5,6) = ZF*SIOMEP
GQ(5,7) = ZF*SIOMEP - U*COEP
GQ(6,1) = RR*COBE - D*COBETH
GQ(6,2) = - D*COBETH
GQ(6,6) = - ZF*COOMEP
GQ(6,7) = - ZF*COOMEP - U*SIEP
```

If we apply a numerical method to the index 1 formulation of the system, we also need the expression $g_{qq}(q)(\dot{q}, \dot{q})$. It is given by

```
 GQQ(1) = - RR*COBE*V(1)**2 + D*COBETH*(V(1)+V(2))**2 +
+           SS*SIGA*V(3)**2
 GQQ(2) = - RR*SIBE*V(1)**2 + D*SIBETH*(V(1)+V(2))**2 -
+           SS*COGA*V(3)**2
 GQQ(3) = - RR*COBE*V(1)**2 + D*COBETH*(V(1)+V(2))**2 +
+           E*SIPHDE*(V(4)+V(5))**2 + ZT*CODE*V(5)**2
 GQQ(4) = - RR*SIBE*V(1)**2 + D*SIBETH*(V(1)+V(2))**2 -
+           E*COPHDE*(V(4)+V(5))**2 + ZT*SIDE*V(5)**2
 GQQ(5) = - RR*COBE*V(1)**2 + D*COBETH*(V(1)+V(2))**2 +
+           ZF*COOMEP*(V(6)+V(7))**2 + U*SIEP*V(7)**2
 GQQ(6) = - RR*SIBE*V(1)**2 + D*SIBETH*(V(1)+V(2))**2 +
+           ZF*SIOMEP*(V(6)+V(7))**2 - U*COEP*V(7)**2
```

Computation of Consistent Initial Values

We first compute a solution of $g(q) = 0$. Since g consists of 6 equations in 7 unknowns we can fix one of them arbitrarily, say $\Theta(0) = 0$, and compute the remaining coordinates by Newton iterations. This gives

$$\begin{aligned}
\beta(0) &= -0.0617138900142764496358948458001 \\
\gamma(0) &= 0.455279819163070380255912382449 \\
\Phi(0) &= 0.222668390165885884674473185609 \\
\delta(0) &= 0.487364979543842550225598953530 \\
\Omega(0) &= -0.222668390165885884674473185609 \\
\varepsilon(0) &= 1.23054744454982119249735015568.
\end{aligned} \tag{7.7}$$

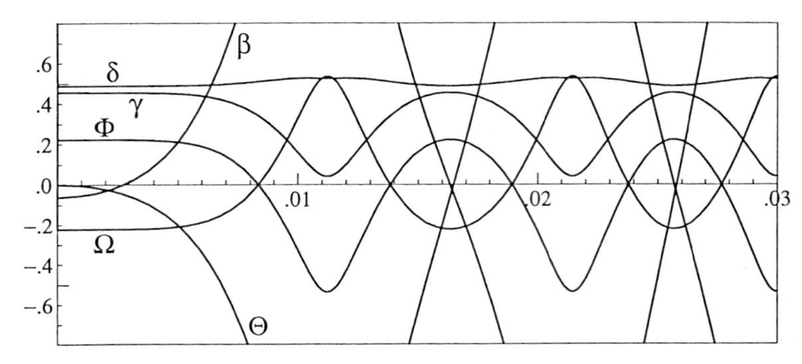

Fig. 7.2. Solution of 7 body mechanism

The condition $G(q)\dot{q} = 0$ is satisfied if we put

$$\dot{\beta}(0) = \dot{\Theta}(0) = \dot{\gamma}(0) = \dot{\Phi}(0) = \dot{\delta}(0) = \dot{\Omega}(0) = \dot{\varepsilon}(0) = 0. \tag{7.8}$$

The values of $\lambda(0)$ and $\ddot{q}(0)$ are then uniquely determined by (7.6a) and the twice differentiated constraint $0 = g_{qq}(q)(\dot{q},\dot{q}) + G(q)\ddot{q}$. We just have to solve a linear system with the matrix

$$\begin{pmatrix} M(q) & G^T(q) \\ G(q) & 0 \end{pmatrix}. \tag{7.9}$$

Observe that g_{qq} need not be evaluated, because $\dot{q}(0) = 0$. Due to the choice $\Theta(0) = 0$ most components of $\lambda(0)$ and $\ddot{q}(0)$ vanish. Only the first two of these are different from zero and given by

$$\begin{aligned} \ddot{\beta}(0) &= 14222.4439199541138705911625887 \\ \ddot{\Theta}(0) &= -10666.8329399655854029433719415 \\ \lambda_1(0) &= 98.5668703962410896057654982170 \\ \lambda_2(0) &= -6.12268834425566265503114393122. \end{aligned} \tag{7.10}$$

The solution of this seven body mechanism is plotted (mod 2π) in Fig. 7.2 for $0 \le t \le 0.03$.

Numerical Computations

We first transform (7.6) into a first order system by introducing the new variable $v = \dot{q}$. Our codes apply only to problems where the derivative is multiplied by a constant matrix. We therefore also consider $w = \ddot{q}$ as a variable so that (7.6a) becomes an algebraic relation. The various formulations of the problem, as discussed in Sect. VII.1, are now as follows:

Index 3 Formulation. With $v = \dot{q}$ and $w = \ddot{q}$ the system (7.6) can be written as

$$\dot{q} = v \tag{7.11a}$$
$$\dot{v} = w \tag{7.11b}$$
$$0 = M(q)w - f(q,v) + G^T(q)\lambda \tag{7.11c}$$
$$0 = g(q). \tag{7.11d}$$

Index 2 Formulation. If we differentiate $0 = g(q)$ once and replace (7.11d) by

$$0 = G(q)v, \tag{7.11e}$$

we get an index 2 problem which is mathematically equivalent to (7.6).

Index 1 Formulation. One more differentiation of (7.11e) yields

$$0 = g_{qq}(q)(v,v) + G(q)w, \tag{7.11f}$$

so that (7.11a,b,c,f) constitutes an index 1 problem.

We have applied several codes with many different tolerances between 10^{-2} and 10^{-10} to these formulations. The results are given in Fig. 7.3. We have plotted the computing time (on a SUN Spark 20 workstation) against the error of the (q,v)-components at $x_{\text{end}} = 0.03$ (in double logarithmic scale).

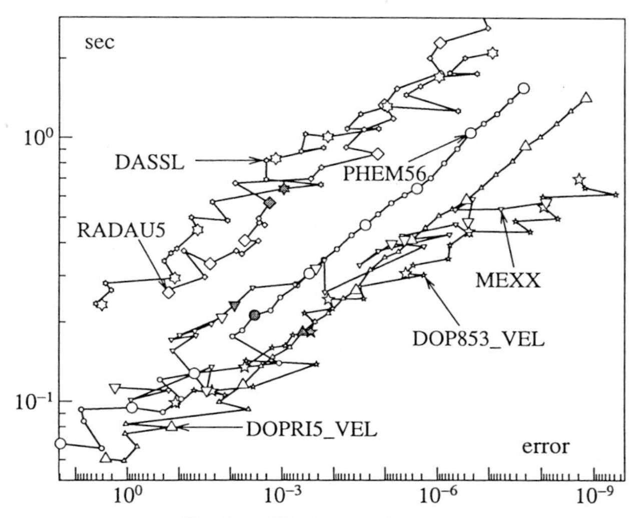

Fig. 7.3. Work-precision diagram

Explicit Runge-Kutta Methods. The index 1 formulation allows us to apply explicit methods such as DOPRI5 or DOP853 of Volume I. For this we have written a function subroutine which solves in each call the linear system (7.11c,f) for w and λ and inserts the result into (7.11a,b). Since there is no stiffness in the obtained differential equation for (q, v), it is not surprising that here the explicit codes work very efficiently (Fig. 7.3).

In order to avoid the drift-off phenomenon (see Sect. VII.2), we have also combined this method with projections onto the solution manifold. This can be implemented conveniently with help of the subroutine SOLOUT, which is called by DOPRI5 after every successful step (set IRTRN = 2 in order to indicate that the numerical approximation has been altered). The full projection (on position and velocity level, (7.11d) and (7.11e)) is slightly more expensive than velocity stabilization alone (denoted by DOPRI5_VEL in Fig. 7.3) and does not give improved results. The first picture of Fig. 7.4 shows the results of the three different implementations: the 'standard' approach is without any projection, 'velocity' means that we perform only velocity stabilization, and 'position' indicated that we do consecutive projections on the position and velocity level. We see that velocity stabilization gives the best results concerning achieved accuracy and computing time.

Half-Explicit Methods. These methods (discussed in Sect. VII.6) are especially adapted to the numerical solution of (nonstiff) constrained mechanical systems. Only linear systems with the matrix (7.9) have to be solved, otherwise the methods are explicit. Since they are applied directly to the index 2 formulation, the velocity constraint (7.11e) is automatically satisfied, and no subroutine for the computation of $g_{qq}(q)(v, v)$ is required.

The extrapolation code MEXX of Lubich (1989) (see also Lubich, Nowak, Poehle & Engstler 1992) implements the half-explicit mid-point rule (6.18). The existence of an h^2-expansion (Theorem 6.3) justifies extrapolation and thus yields methods of arbitrarily high order. It is not surprising that this code gives excellent results for high precision computations.

The first code implementing half-explicit Runge-Kutta methods is HEM5 of Brasey (1994). It has been modified and improved by Arnold (1995, code HEX5) and Murua (1995, code PHEM56). We have also included the results of the latter code (Fig. 7.3). It is slightly less efficient than DOPRI5_VEL in this particular example, because the evaluation of $g_{qq}(q)(v, v)$ is cheap. Arnold (1995) and Murua (1995) report about experiments (with expensive $g_{qq}(q)(v, v)$), where the half-explicit methods are superiour to explicit Runge-Kutta methods with velocity projection.

BDF. The famous code DASSL of Petzold (1982), see also Brenan, Campbell & Petzold (1989), is a realization of the BDF multistep formulas. It is written for problems of the general form $F(u, u', x) = 0$, so that it is not necessary to introduce $\ddot{q}$ of (7.6) as new variable. We applied it using default values for all parameters except for the scaling of the error estimation. We put INFO(2)=1 and

$$\text{ATOL(I)} = \text{RTOL(I)} = \begin{cases} Tol & \text{for } \text{I} = 1, \dots, 14, \\ 1.0\text{D0} & \text{for } \text{I} \geq 15, \end{cases}$$

which means that we control the accuracy for q and v, but not for the Lagrange

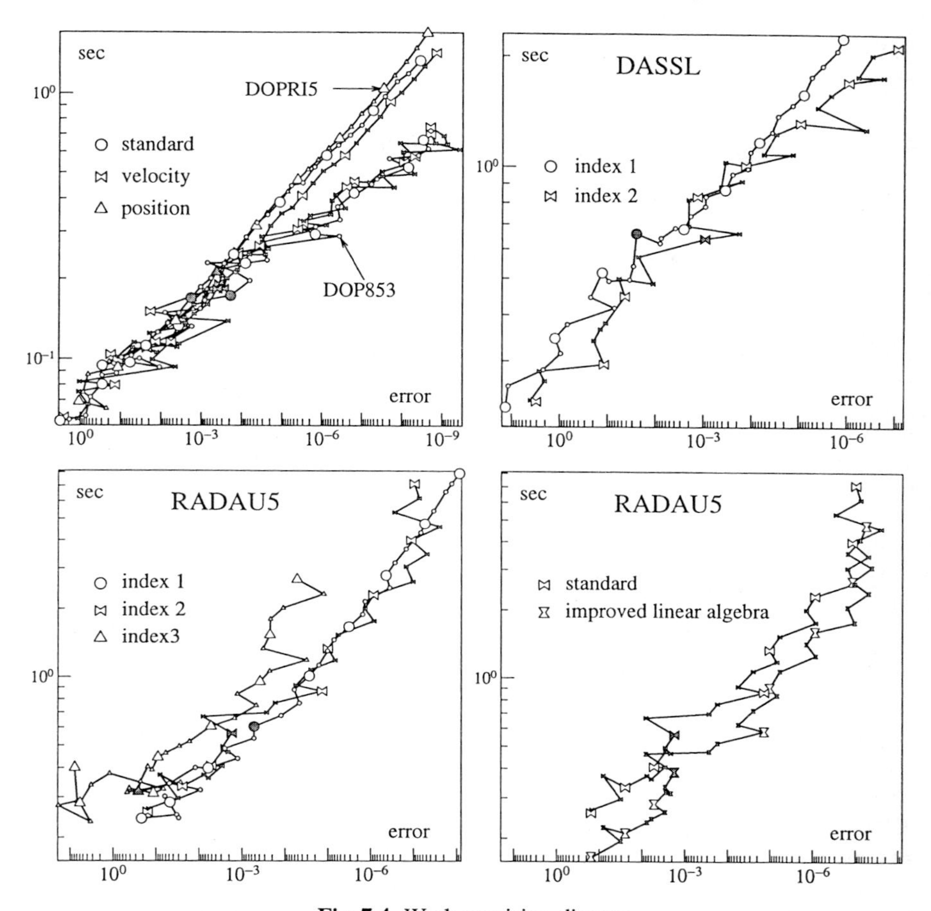

Fig. 7.4. Work-precision diagram

multipliers λ. In the comparisons of Fig. 7.3 (index 2 formulatin) and Fig. 7.4 we used the full Jacobian of the problem, obtained by numerical differentiation. This turned out to be more efficient than providing an analytic approximation, where the derivatives of f, M and G are neglected.

Implicit Runge-Kutta Methods. Our code RADAU5 is written for problems of the form $By' = f(x, y)$ with constant, possibly singular matrix B. It can therefore be applied to all three of the above formulations. Convergence is guaranteed by Theorem VI.1.1 for the index 1 formulation, by Theorems 4.5 and 4.6 for the index 2 formulation, and by the results of HLR89 for the index 3 case. However, the higher the index, the more difficult is it to solve the nonlinear Runge-Kutta equations. We have applied the code with the options IWORK(5) = 14, IWORK(6) = 0 and IWORK(7) = 13 (IWORK(5) = 7 and IWORK(6) = 7 for the index 3 formulation), so that the acceleration w and the Lagrange multiplier λ are scaled by h^2

in the error estimation. This guarantees the convergence of the simplified Newton iterations (see HLR89, Chapter 7 for a justification). Furthermore, we have exploited the special structure $\dot{q} = v$, $\dot{v} = w$ of our system by setting IWORK(9) = 14 and IWORK(10) = 7. This speeds up the computation of the arising linear systems. The results are given in Fig. 7.3 (index 2 formulation) and in the lower left picture of Fig. 7.3 for all three formulations of the problem. We have used an analytical approximation to the Jacobian (neglecting the derivatives of f, M and G) and did not apply any projection onto the solution manifold.

Savings in Linear Algebra. If the problem is nonstiff, one can use a reduced Jacobian for the solution of the nonlinear Runge-Kutta equations. Neglecting the derivatives of f, M and G (what we have done for the above calculations), we are led to linear systems of the form (in the index 2 case)

$$\begin{pmatrix} -\alpha I & I & 0 & 0 \\ 0 & -\alpha I & I & 0 \\ 0 & 0 & M & G^T \\ 0 & G & 0 & 0 \end{pmatrix} \begin{pmatrix} \Delta q \\ \Delta v \\ \Delta w \\ \Delta \lambda \end{pmatrix} = \begin{pmatrix} a \\ b \\ c \\ d \end{pmatrix} \tag{7.12}$$

where $\alpha = (h\gamma)^{-1}$, h the step size and γ an eigenvalue of the Runge-Kutta matrix. The evaluation of the matrix in (7.12) is free, because $M(q)$ and $G(q)$ have to be evaluated anyway for the right-hand side of the differential-algebraic system. Eliminating the variable Δv in the last row of (7.12) yields the smaller system

$$\begin{pmatrix} M & G^T \\ G & 0 \end{pmatrix} \begin{pmatrix} \Delta w \\ \Delta \lambda \end{pmatrix} = \begin{pmatrix} c \\ \alpha d + Gb \end{pmatrix} \tag{7.13}$$

which is of the same type as those for the explicit methods. Once a solution to (7.13) is known the values of Δv and Δq are easily obtained from the first two rows of (7.12). We observe that the matrix in (7.13) does not depend on $\alpha = (h\gamma)^{-1}$. Hence only *one* LU decomposition is necessary for a step, independently of the number of distinct eigenvalues of the Runge-Kutta matrix. An implementation of these ideas reduced considerably the work for solving the nonlinear systems (see last picture of Fig. 7.4).

A similar reduction of the linear algebra was first proposed by Gear, Gupta & Leimkuhler (1985) for the BDF schemes. The above idea is not restricted to the index 2 case, and extends straightforwardly to the index 1 and index 3 situations. We finally remark that one has the possibility of retaining the decomposed matrix of (7.13) over several steps even in the case when the step size is changed.

A Stiff Mechanical System

We now want to introduce some "stiffness" into the above mechanical system. To this end we take into account the elasticity of one of these bodies (K_6 appears to be the simplest one) and replace it by a spring with very large spring constant c_1. Thus the length of this spring will become an additional unknown variable q_8. We let the unstretched length be zf (of Table 7.1), and assume that the centre of gravity C_6 has constant distance fa from the upper joint (see Fig. 7.1). We further simplify the problem by assuming that the inertia of this body remains constant. Obviously the algebraic constraints (7.5) remain unchanged; we only have to replace the constant zf in (7.5) by the new variable q_8. The derivative matrix $G(q) = g'(q)$ has to be changed accordingly. It is now a 6×8 matrix.

The equations of motion for this modified problem are obtained as follows: in the *kinetic energy* (7.3) only the contribution of the 6th body (the new spring) changes, namely

$$
\begin{aligned}
C_6: \quad & x_6 = xa + u \cdot \sin\varepsilon + (q_8 - fa) \cdot \cos(\Omega + \varepsilon) \\
& y_6 = ya - u \cdot \cos\varepsilon + (q_8 - fa) \cdot \sin(\Omega + \varepsilon) \\
& \dot{x}_6^2 + \dot{y}_6^2 = (q_8 - fa)^2 \cdot \dot{\Omega}^2 + 2 \cdot \big((q_8 - fa)^2 - u \cdot (q_8 - fa) \cdot \sin\Omega\big) \cdot \dot{\Omega} \cdot \dot{\varepsilon} \\
& \qquad + \big((q_8 - fa)^2 - 2 \cdot u \cdot (q_8 - fa) \cdot \sin\Omega + u^2\big) \cdot \dot{\varepsilon}^2 \\
& \qquad + 2 \cdot u \cdot \cos\Omega \cdot \dot{\varepsilon} \cdot \dot{q}_8 + \dot{q}_8^2 \\
& \dot{\omega}_6 = \dot{\Omega} + \dot{\varepsilon}
\end{aligned}
$$

In the *potential energy* we have to add a term which is due to the new spring. We thus get (compare (7.4))

$$
U = -mom \cdot \beta + c_0 \cdot \frac{(\ell - \ell_0)^2}{2} + c_1 \cdot \frac{(q_8 - zf)^2}{2}, \tag{7.21}
$$

where the spring constant c_1 of the new spring is large. The resulting system is again of the form (7.6), but with $q \in \mathbb{R}^8$. The initial values (7.7), (7.9), (7.12) for the 7 angles (7.2) are consistent for the new problem, if we require in addition

$$
q_8(0) = zf, \qquad \dot{q}_8(0) = 0. \tag{7.22}
$$

This then implies $\ddot{q}_8(0) = 0$. For the choice $c_1 = 10^{10}$ we applied the implicit codes RADAU5 and DASSL to the above *stiff* mechanical system. The behaviour of these methods was nearly identical to that for the original problem (Fig. 7.4). So there was no need to draw another picture. Obviously, the explicit codes DOPRI5, PHEM56 and MEXX do not work any longer.

It should be remarked that for $Tol \le 1/c_1$ the efficiency of the implicit codes suddenly decreases. This is due to the fact that the exact solution of the problem (with the *initial* values described above) is highly oscillatory with frequency $\mathcal{O}(\sqrt{c_1})$ and amplitude $\mathcal{O}(1/c_1)$ about a smooth solution. A general theory for such situations has been elaborated by Ch. Lubich (1993). For very stringent tolerances any code is forced to follow the oscillations and the step sizes become small.

Exercises

1. Consider the differential equation (so-called "Kreiss problem")

$$y' = U^T(x)\begin{pmatrix} -1 & 0 \\ 0 & -1/\varepsilon \end{pmatrix} U(x)y, \qquad U(x) = \begin{pmatrix} \cos x & \sin x \\ -\sin x & \cos x \end{pmatrix} \tag{7.23}$$

and apply the Runge-Kutta code RADAU5 to this stiff problem. You will observe that, for a fixed tolerance, the number of function evaluations increases with decreasing $\varepsilon > 0$. Then apply the method to the equivalent system

$$y' = z, \qquad 0 = \begin{pmatrix} 1 & 0 \\ 0 & \varepsilon \end{pmatrix} U(x)z + U(x)y \tag{7.24}$$

and show that the number of function evaluations does not increase for $\varepsilon \to 0$.

a) Explain this phenomenon by studying the convergence of the simplified Newton iterations.

b) Prove that the index of the system (7.24) with $\varepsilon = 0$ is two.

VII.8 Symplectic Methods for Constrained Hamiltonian Systems

In principle, all approaches discussed in Sect. VII.2 can be employed for the numerical solution of constrained Hamiltonian systems. A disadvantage of these index reduction methods is, as we shall see below, that the symplectic structure of the flow is destroyed by the discretization.

In Sect. I.6 we have seen that the equations of motion for conservative mechanical systems can be written either in terms of position and velocity coordinates (Lagrangian formulation) or in terms of position and momentum coordinates (Hamiltonian formulation). For *constrained* mechanical systems the situation is exactly the same. In the present section we consider the Hamiltonian formulation

$$
\begin{aligned}
q' &= H_p(p,q) &\quad (8.1a)\\
p' &= -H_q(p,q) - G^T(q)\lambda &\quad (8.1b)\\
0 &= g(q). &\quad (8.1c)
\end{aligned}
$$

Here, $H : \mathbb{R}^n \times \mathbb{R}^n \to \mathbb{R}$ is the Hamiltonian function, H_p and H_q denote partial derivatives, $g : \mathbb{R}^n \to \mathbb{R}^m$ (with $m < n$) are the constraints, and $G(q) = g_q(q)$. If $T(q,\dot q) = \frac{1}{2}\dot q^T M(q)\dot q$ (with invertible $M(q)$) is the kinetic energy of a mechanical system and $U(q)$ its potential energy, we have $p = M(q)\dot q$ and

$$
H(p,q) = \frac{1}{2}p^T M(q)^{-1} p + U(q), \tag{8.2}
$$

(see Eq. (I.6.26)) in contrast to the Lagrange function, which is given by $\mathcal{L}(q,\dot q) = T(q,\dot q) - U(q)$. If $M(q) = I$ (the identity), we have $p = \dot q$ and both formulations, (1.46) and (8.1), are identical. If $M(q)$ depends on q, the formulation (8.1) may be numerically more advantageous than (1.46) (see Exercise 1).

Differentiating the constraint in (8.1) twice, we get

$$
\begin{aligned}
0 &= G(q)H_p(p,q), &\quad (8.3a)\\
0 &= \frac{d}{dq}\Big(G(q)H_p(p,q)\Big)H_p(p,q) - G(q)H_{pp}(p,q)\Big(H_q(p,q) + G^T(q)\lambda\Big), &\quad (8.3b)
\end{aligned}
$$

and we see that λ can be expressed in terms of p and q, if

$$
G(q)H_{pp}(p,q)G^T(q) \qquad \text{is invertible} \tag{8.4}
$$

in a neighbourhood of the considered solution. Therefore, (8.1) is a differential-algebraic system of index 3. If $H(p,q)$ is given by (8.2), condition (8.4) is the same as (1.47).

Properties of the Exact Flow

Every solution of the system (8.1) satisfies (8.1c) and (8.3a). It therefore lies on the manifold

$$\mathcal{M} = \{(p,q) \mid g(q)=0,\ G(q)H_p(p,q)=0\}. \tag{8.5}$$

Extracting λ from (8.3b) (this is possible, if (8.4) is satisfied), and inserting the resulting expression into (8.1b), yields a differential equation on the manifold $\mathcal{M}$. The situation here is completely analogous to that of (1.22) of Sect. VII.1.

Symplecticity. Our next aim is to extend the result of Theorem I.14.12 to constrained Hamiltonian systems. We consider the differential 2-form

$$\omega^2 = \sum_{I=1}^{n} dp^I \wedge dq^I \tag{8.6}$$

(p^I and q^I denote the components of the vectors p and q, respectively). The flow of the system (8.1), mapping an initial value $(p_0,q_0)\in\mathcal{M}$ onto $(p(t),q(t))\in\mathcal{M}$, is denoted by φ_t. For a differentiable function $g:\mathcal{M}\to\mathcal{M}$ we further denote by $g^*\omega^2$ the differential 2-form, defined by

$$(g^*\omega^2)(\xi_1,\xi_2) = \omega^2\big(g'(p,q)\xi_1, g'(p,q)\xi_2\big).$$

This is formally identical to Definition I.14.11, but here we are only interested in the case where ξ_1 and ξ_2 lie in the tangent space

$$T_{(p,q)}\mathcal{M} = \Big\{(u,v) \mid G(q)v=0,\ \frac{d}{dq}\big(G(q)H_p(p,q)\big)v + G(q)H_{pp}(p,q)u = 0\Big\}$$

of the manifold (8.5).

Theorem 8.1. *The flow $\varphi_t:\mathcal{M}\to\mathcal{M}$ of the system (8.1) is a symplectic transformation on $\mathcal{M}$, i.e.,*

$$(\varphi_t^*\omega^2)(\xi_1,\xi_2) = \omega^2(\xi_1,\xi_2)$$

for all t, for all (p,q), and for all ξ_1,ξ_2 lying in the tangent space $T_{(p,q)}\mathcal{M}$.

Proof. For $\xi\in T_{(p,q)}\mathcal{M}$ the tangent vector $\xi^t = \varphi_t'(p,q)\xi \in T_{(p(t),q(t))}\mathcal{M}$ is a solution of the variational equation

$$\begin{aligned}
\dot{\delta p^I} &= -\sum_{J=1}^{n}\frac{\partial^2 H}{\partial q^I\partial p^J}(p,q)\cdot\delta p^J - \sum_{J=1}^{n}\frac{\partial^2 H}{\partial q^I\partial q^J}(p,q)\cdot\delta q^J\\
&\quad -\sum_{K=1}^{m}\lambda^K\sum_{J=1}^{n}\frac{\partial^2 g^K}{\partial q^I\partial q^J}(p,q)\cdot\delta q^J - \sum_{K=1}^{m}\frac{\partial g^K}{\partial q^I}(q)\cdot\delta\lambda^K\\
\dot{\delta q^I} &= \sum_{J=1}^{n}\frac{\partial^2 H}{\partial p^I\partial p^J}(p,q)\cdot\delta p^J + \sum_{J=1}^{n}\frac{\partial^2 H}{\partial p^I\partial q^J}(p,q)\cdot\delta q^J,
\end{aligned}$$

where the $\delta\lambda^K$ (for $K=1,\ldots,m$) are obtained by differentiation of (8.3b). We now compute the time derivative of $\omega^2(\xi_1^t,\xi_2^t)$. The terms, not depending on λ or $\delta\lambda$, vanish by Theorem I.14.12. We therefore get

$$\begin{aligned}\frac{d}{dt}\omega^2(\xi_1^t,\xi_2^t) = -\Big(\sum_{K=1}^m \lambda^K \sum_{I,J=1}^n \frac{\partial^2 g^K(q)}{\partial q^I \partial q^J} dq^J \wedge dq^I \\ + \sum_{K=1}^m d\lambda^K \wedge \Big(\sum_{I=1}^n \frac{\partial g^K(q)}{\partial q^I} dq^I\Big)\Big)(\xi_1^t,\xi_2^t).\end{aligned} \tag{8.7}$$

Due to the symmetry of the second partial drivatives, the first expression of the right-hand side of Eq. (8.7) vanishes. The second expression also vanishes, because ξ_2^t lies in the tangent space $T_{(p(t),q(t))}\mathcal{M}$. Hence, $\omega^2(\xi_1^t,\xi_2^t)$ is constant, what proves the statement of the theorem. □

Preservation of the Hamiltonian. Differentiation of $H(p(t),q(t))$ with respect to time yields

$$-H_p^T H_q - H_p^T G^T \lambda + H_q^T H_p,$$

with all expressions evaluated at $(p(t),q(t))$. The first term cancels with the last one, and the remaining term vanishes, because $G(q)H_p(p,q)=0$ on the solution manifold. Consequently, the Hamiltonian function $H(p,q)$ is constant along solutions of (8.1).

First Order Symplectic Method

We shall now discuss in some detail the feasibility, the convergence, and the symplecticity of a simple first order method. The presented ideas will be useful for a better understanding of the later discussion of higher order methods.

Inspired by (II.16.54), we consider the following discretization of (8.1):

$$\widehat{p}_1 = p_0 - h\big(H_q(\widehat{p}_1,q_0) + G^T(q_0)\lambda_1\big) \tag{8.8a}$$
$$q_1 = q_0 + hH_p(\widehat{p}_1,q_0) \tag{8.8b}$$
$$0 = g(q_1). \tag{8.8c}$$

The numerical approximation $(\widehat{p}_1,q_1)$ satisfies the constraint (8.1c), but not (8.3a). Therefore, we append the projection

$$p_1 = \widehat{p}_1 - hG^T(q_1)\mu \tag{8.8d}$$
$$0 = G(q_1)H_p(p_1,q_1), \tag{8.8e}$$

so that method (8.8a-e) yields approximations that stay in the manifold $\mathcal{M}$ of Eq. (8.5).

Existence of the Numerical Solution. We consider a slightly more general system than (8.8). If the initial values are not consistent, we replace the relations (8.8c) and (8.8e) by

$$g(q_1) = g(q_0) + hG(q_0)H_p(p_0, q_0) \tag{8.9a}$$
$$G(q_1)H_p(p_1, q_1) = G(q_0)H_p(p_0, q_0). \tag{8.9b}$$

We shall show that the nonlinear system (8.8a,b), (8.9a) has a locally unique solution. Inspired by the proof of Theorem 3.1 we write

$$g(q_1) - g(q_0) = \int_0^1 g_q\big(q_0 + \tau(q_1 - q_0)\big)\, d\tau \cdot (q_1 - q_0).$$

Inserting $g(q_1)$ from (8.9a) and q_1 from (8.8b) and dividing by h yields

$$G(q_0)H_p(p_0, q_0) = \int_0^1 g_q\big(q_0 + \tau(q_1 - q_0)\big)\, d\tau \cdot H_p(\widehat{p}_1, q_0). \tag{8.10}$$

We next develop $H_p(\widehat{p}_1, q_0)$ as

$$H_p(\widehat{p}_1, q_0) = H_p(p_0, q_0) - h\int_0^1 H_{pp}\big(p_0 + \sigma(\widehat{p}_1 - p_0), q_0\big)\, d\sigma \big(H_q(\widehat{p}_1, q_0) + G^T(q_0)\lambda_1\big).$$

Inserting this formula into (8.10), an integration by parts shows that (8.9a) is equivalent to

$$0 = \int_0^1 (1-\tau) g_{qq}\big(q_0 + \tau(q_1 - q_0)\big)\, d\tau \cdot \big(H_p(p_0, q_0), H_p(\widehat{p}_1, q_0)\big) \tag{8.11}$$
$$- \int_0^1 g_q\big(q_0 + \tau(q_1 - q_0)\big)\, d\tau \int_0^1 H_{pp}\big(p_0 + \sigma(\widehat{p}_1 - p_0), q_0\big)\, d\sigma \big(H_q(\widehat{p}_1, q_0) + G^T(q_0)\lambda_1\big).$$

This is a linear system for λ_1 and allows us to express λ_1 smoothly in terms of $\widehat{p}_1, q_1$, and of the initial values p_0, q_0. We insert the resulting expression for λ_1 into (8.8a). Hence, (8.8a,b) becomes a nonlinear system for $\widehat{p}_1, q_1$, which, for sufficiently small h, has a unique solution close to p_0, q_0 (Implicit Function Theorem). It is interesting to note that, for $h \to 0$, the value λ_1 from (8.11) does not converge to $\lambda(0)$, given by (8.3b), but to the solution λ_0 of

$$0 = \frac{1}{2} g_{qq}(H_p, H_p) - GH_{pp}(H_q + G^T\lambda_0).$$

Here, all functions are evaluated at the initial value (p_0, q_0).

The existence of the solution (p_1, μ) to the system (8.8d), (8.9b) follows from the Newton-Kantorovich Theorem (Ortega & Rheinboldt 1970) with initial approximation $p_1 := \widehat{p}_1$, and $\mu = 0$, or also from the Implicit Function Theorem.

We have not only shown that the system (8.8) possesses a locally unique solution, but we have also seen that the replacement of (8.8c,e) by (8.9) extends the definition of the method to arbitrary initial values (close to $\mathcal{M}$). We thus have found a one-step method

$$\begin{pmatrix} p_1 \\ q_1 \end{pmatrix} = \begin{pmatrix} p_0 \\ q_0 \end{pmatrix} + h\Phi\left(\begin{pmatrix} p_0 \\ q_0 \end{pmatrix}, h\right) \tag{8.12}$$

in $\mathbb{R}^{2n}$, which reduces to (8.8) on the manifold $\mathcal{M}$. For smooth functions g and H also Φ is smooth, and the classical theory (convergence, asymptotic expansions, ...) can be applied to this method.

Convergence of Order 1. It is sufficient to show that the local error is of size $\mathcal{O}(h^2)$. The convergence then follows from Theorem II.3.6 applied to (8.12). From the above investigation on the existence of the numerical solution we know that $\widehat{p}_1 = p_0 + \mathcal{O}(h)$, $q_1 = q_0 + \mathcal{O}(h)$, and $\lambda_1 = \lambda_0 + \mathcal{O}(h)$. Consequently, we have from (8.8a,b) that

$$q_1 = q(t_0+h) + \mathcal{O}(h^2), \qquad \widehat{p}_1 = p(t_0+h) - hG^T(q_0)\,\delta\lambda + \mathcal{O}(h^2) \tag{8.13}$$

with $\delta\lambda = \lambda_0 - \lambda(t_0)$. The disturbing term $hG^T(q_0)\delta\lambda$ is eliminated by the projection (8.8d,e). This can be seen as follows: from (8.13) and (8.8d) we know that $p_1 = p(t_0+h) - G^T(q_0)\nu + \mathcal{O}(h^2)$, so that

$$G\big(q(t_0+h)\big)H_p\big(p(t_0+h) - G^T(q_0)\nu, q(t_0+h)\big) = \mathcal{O}(h^2).$$

By (8.4) and the Implicit Function Theorem this implies $\nu = \mathcal{O}(h^2)$, and the local error for both components (p and q) is of size $\mathcal{O}(h^2)$.

Symplecticity. Differentiation of the relations (8.8a,b) shows that (we use upper indices for the components)

$$\begin{aligned} d\widehat{p}_1^I &= dp_0^I - h\sum_{J=1}^{n}\frac{\partial^2 H}{\partial q^I \partial p^J}(\widehat{p}_1, q_0)d\widehat{p}_1^J - h\sum_{J=1}^{n}\frac{\partial^2 H}{\partial q^I \partial q^J}(\widehat{p}_1, q_0)dq_0^J \\ &\quad - h\sum_{K=1}^{m}\lambda_1^K\sum_{J=1}^{n}\frac{\partial^2 g^K}{\partial q^I \partial q^J}(q_0)dq_0^J - h\sum_{K=1}^{m}\frac{\partial g^K}{\partial q^I}(q_0)d\lambda_1^K \\ dq_1^I &= dq_0^I + h\sum_{J=1}^{n}\frac{\partial^2 H}{\partial p^I \partial p^J}(\widehat{p}_1, q_0)d\widehat{p}_1^J + h\sum_{J=1}^{n}\frac{\partial^2 H}{\partial p^I \partial q^J}(\widehat{p}_1, q_0)dq_0^J. \end{aligned}$$

Taking the exterior product of the first formula with dq_0^I, and of the second formula with $d\widehat{p}_1^I$, several terms cancel out (as in the proof of Theorem 8.1) and we obtain

$$\begin{aligned} \sum_{I=1}^{n} d\widehat{p}_1^I \wedge dq_0^I &= \sum_{I=1}^{n} dp_0^I \wedge dq_0^I - h\sum_{I,J=1}^{n}\frac{\partial^2 H}{\partial q^I \partial p^J}(\widehat{p}_1, q_0)d\widehat{p}_1^J \wedge dq_0^I \\ \sum_{I=1}^{n} d\widehat{p}_1^I \wedge dq_1^I &= \sum_{I=1}^{n} d\widehat{p}_1^I \wedge dq_0^I + h\sum_{I,J=1}^{n}\frac{\partial^2 H}{\partial p^I \partial q^J}(\widehat{p}_1, q_0)d\widehat{p}_1^I \wedge dq_0^J. \end{aligned}$$

Summing up both formulas yields

$$\sum_{I=1}^{n} d\widehat{p}_1^I \wedge dq_1^I = \sum_{I=1}^{n} dp_0^I \wedge dq_0^I, \tag{8.14}$$

what proves that the method (8.8a-c) is symplectic. In order to show that also the projection (8.8d,e) is symplectic, we compute

$$dp_1^I = d\widehat{p}_1^I - h\sum_{K=1}^{m} \mu^K \sum_{J=1}^{n} \frac{\partial^2 g^K}{\partial q^I \partial q^J}(q_1)dq_1^J - h\sum_{K=1}^{m} \frac{\partial g^K}{\partial q^I}(q_1)d\mu^K,$$

and we obtain as above (using $g(q_1)=0$) that

$$\sum_{I=1}^{n} dp_1^I \wedge dq_1^I = \sum_{I=1}^{n} d\widehat{p}_1^I \wedge dq_1^I. \tag{8.15}$$

Equations (8.14) and (8.15) together show that the complete procedure (8.8a-e) is symplectic.

SHAKE and RATTLE

These algorithms have been designed for problems with separable Hamiltonian

$$H(p,q) = \frac{1}{2}p^T M^{-1} p + U(q) \tag{8.16}$$

(constant matrix M), and are very popular in molecular dynamics simulation. Observe that for this Hamiltonian the problem (8.1) becomes the second order differential equation $Mq'' = -U_q(q) - G^T(q)\lambda$ with constraint (8.1c).

SHAKE. This method, due to Ryckaert, Ciccotti & Berendsen (1977), is given by

$$q_{n+1} - 2q_n + q_{n-1} = -h^2 M^{-1}\big(U_q(q_n) + G^T(q_n)\lambda_n\big) \tag{8.17a}$$

$$0 = g(q_{n+1}). \tag{8.17b}$$

In the absence of constraints it is identical to Störmer's method (Sect. III.10), which in molecular dynamics applications is often referred the Verlet method (Verlet 1967). The p-components are approximated by $p_n = M(q_{n+1} - q_{n-1})/2h$. For an implementation of this 2-step method a stabilized version is recommended (see the end of Sect. III.10).

RATTLE. Denoting $p_{n+1/2} := p_n - (h/2)(U_q(q_n) + G^T(q_n)\lambda_n)$, the SHAKE algorithm can be rewritten in the form

$$p_{n+1/2} = p_n - \frac{h}{2}\big(U_q(q_n) + G^T(q_n)\lambda_n\big) \tag{8.18a}$$

$$q_{n+1} = q_n + hM^{-1}p_{n+1/2} \tag{8.18b}$$

$$0 = g(q_{n+1}). \tag{8.18c}$$

The definition of p_{n+1} as in the SHAKE method requires the knowledge of q_{n+2}. In order to avoid this difficulty, Andersen (1983) suggests to define p_{n+1} by

$$\begin{aligned} p_{n+1} &= p_{n+1/2} - \frac{h}{2}\big(U_q(q_{n+1}) + G^T(q_{n+1})\mu_n\big) && (8.18d)\\ 0 &= G(q_{n+1})M^{-1}p_{n+1}, && (8.18e) \end{aligned}$$

so that also the hidden constraint (8.3a) is satisfied. These two equations constitute a linear system for (p_{n+1}, μ_n).

Extension to General Hamiltonian Functions. It was observed by Jay (1994) that the RATTLE algorithm can be extended to general Hamiltonian functions as follows: for consistent values $(p_n, q_n) \in \mathcal{M}$ define

$$\begin{aligned} p_{n+1/2} &= p_n - \frac{h}{2}\big(H_q(p_{n+1/2}, q_n) + G^T(q_n)\lambda_n\big) && (8.19a)\\ q_{n+1} &= q_n + \frac{h}{2}\big(H_p(p_{n+1/2}, q_n) + H_p(p_{n+1/2}, q_{n+1})\big) && (8.19b)\\ 0 &= g(q_{n+1}). && (8.19c)\\ p_{n+1} &= p_{n+1/2} - \frac{h}{2}\big(H_q(p_{n+1/2}, q_{n+1}) + G^T(q_{n+1})\mu_n\big) && (8.19d)\\ 0 &= G(q_{n+1})H_p(p_{n+1}, q_{n+1}). && (8.19e) \end{aligned}$$

This is the special case $s = 2$ of the Lobatto IIIA-IIIB pair to be discussed below.

The equations (8.19a-c) constitute a nonlinear system for the unknowns $p_{n+1/2}$, q_{n+1}, and λ_n. In the same way as for the method (8.8) we can reformulate Eq. (8.19c) in such a way that λ_n can be expressed smoothly in terms of p_n, q_n, $p_{n+1/2}$, q_{n+1}, and h. Hence, the numerical solution exists, is locally unique, and depends smoothly on h and on the initial values (p_n, q_n). The same is true for the system (8.19d,e). If the equations (8.19c,e) are replaced by (8.9), we get a smooth extension of the method (8.19), defined on a neighbourhood of $\mathcal{M}$ in $\mathbb{R}^{2n}$.

Theorem 8.2. *The numerical method (8.19) is symmetric, convergent of order 2, and symplectic.*

Proof. a) We consider the more general situation, where (8.19c,e) is replaced by (8.9). Replacing then h by $-h$, and exchanging (p_n, q_n) with (p_{n+1}, q_{n+1}) and

λ_n with μ_n, we obtain

$$\begin{aligned} p_{n+1/2} &= p_{n+1} + \frac{h}{2}\Big(H_q(p_{n+1/2},q_{n+1}) + G^T(q_{n+1})\mu_n\Big) \\ q_n &= q_{n+1} - \frac{h}{2}\Big(H_p(p_{n+1/2},q_{n+1}) + H_p(p_{n+1/2},q_n)\Big) \\ g(q_n) &= g(q_{n+1}) - hG(q_{n+1})H_p(p_{n+1},q_{n+1}) \\ p_n &= p_{n+1/2} + \frac{h}{2}\Big(H_q(p_{n+1/2},q_n) + G^T(q_n)\lambda_n\Big) \\ G(q_n)H_p(p_n,q_n) &= G(q_{n+1})H_p(p_{n+1},q_{n+1}). \end{aligned}$$

These are exactly the same equations as those of (8.19a,b,d) and (8,9), proving that even the extension of the method to a neighbourhood of $\mathcal{M}$ is symmetric.

b) We consider the method (8.19) as a mapping $(p_n,q_n)\mapsto(p_{n+1},q_{n+1})$ on the manifold $\mathcal{M}$ of Eq. (8.5). The same considerations as for (8.8) show that (8.19) is a method of order at least one. Since it is symmetric, its order has to be even (Sect. II.8). This proves that (8.19) is a convergent method of order 2.

c) The fact that the method (8.19) defines a symplectic transformation on $\mathcal{M}$ can be proved as for (8.8) (see Leimkuhler & Skeel (1994) for the case of a separable Hamiltonian (8.16)). We do not give details here, because the symplecticity of (8.19) also follows from Theorem 8.5 below. □

Remark 8.3. In a step by step application of method (8.19) the projection (8.19d,e) can be avoided at those points, where the value p_{n+1} is not needed for output. Indeed, from the second step on we can replace (8.19a) by

$$p_{n+1/2} = p_{n-1/2} - \frac{h}{2}\Big(H_q(p_{n+1/2},q_n) + H_q(p_{n-1/2},q_n) + G^T(q_n)(\lambda_n + \mu_{n-1})\Big)$$

without changing the numerical approximations q_n and $p_{n+1/2}$. The same trick is possible for method (8.8).

The Lobatto IIIA-IIIB Pair

Partitioned Runge-Kutta methods are well suited for unconstrained Hamiltonian systems (see Sect. II.16). We shall investigate here, how these methods can be extended to the constrained system (8.1). We consider

$$P_i = p_0 + h\sum_{j=1}^{s} a_{ij}k_j, \qquad Q_i = q_0 + h\sum_{j=1}^{s} \widehat{a}_{ij}\ell_j, \tag{8.20a}$$

$$p_1 = p_0 + h\sum_{i=1}^{s} b_i k_i, \qquad q_1 = q_0 + h\sum_{i=1}^{s} \widehat{b}_i\ell_i, \tag{8.20b}$$

$$k_i = -\frac{\partial H}{\partial q}(P_i,Q_i) - G^T(Q_i)\Lambda_i, \qquad \ell_i = \frac{\partial H}{\partial p}(P_i,Q_i), \tag{8.20c}$$

where b_i, a_{ij} and $\widehat{b}_i, \widehat{a}_{ij}$ are the coefficients of two Runge-Kutta schemes (c.f., Eq. (II.16.26)). For the moment, the values Λ_i $(i=1,\dots,s)$ are not yet specified. There are several possibilities to do this. One can either define them by $\Lambda_i = \lambda(P_i, Q_i)$, where $\lambda(p,q)$ is the function given by (8.3b), or one can define them implicitly by adding the conditions $G(Q_i)H_p(P_i,Q_i)=0$ or $g(Q_i)=0$.

We are interested in symplectic schemes. Therefore it is natural to consider methods satisfying the conditions of Theorem II.16.10.

Lemma 8.4. *If the coefficients of (8.20) satisfy*

$$b_i = \widehat{b}_i, \qquad i=1,\dots,s \tag{8.21}$$

$$b_i\widehat{a}_{ij} + \widehat{b}_j a_{ji} - b_i\widehat{b}_j = 0, \qquad i,j=1,\dots,s, \tag{8.22}$$

then we have the following relation for the expressions in (8.20):

$$\sum_{I=1}^{n} dp_1^I \wedge dq_1^I - \sum_{I=1}^{n} dp_0^I \wedge dq_0^I = h\sum_{i=1}^{s} b_i \sum_{K=1}^{m}\Big(\sum_{I=1}^{n}\frac{\partial g^K}{\partial q^I}(Q_i)dQ_i^I\Big)\wedge d\Lambda_i^K .$$

If the Hamiltonian is separable (i.e., $H(p,q)=T(p)+U(q)$), then the condition (8.22) alone implies the above relation.

Proof. We compute the expression $D=\sum_I dp_1^I\wedge dq_1^I - \sum_I dp_0^I\wedge dq_0^I$ following the lines of the proof of Theorem II.16.6 (see also the proof of Theorem II.16.10). All terms cancel with exception of those originating from the presence of $G^T(Q_i)\Lambda_i$ in (8.20c). We thus obtain

$$D=-h\sum_{i=1}^{s} b_i\sum_{K=1}^{m}\Big(\Lambda_i^K\sum_{I,J=1}^{n}\frac{\partial^2 g^K}{\partial q^J\partial q^I}(Q_i)dQ_i^J\wedge dQ_i^I + \sum_{I=1}^{n}\frac{\partial g^K}{\partial q^I}(Q_i)d\Lambda_i^K\wedge dQ_i^I\Big).$$

Due to the symmetry of the second derivative of g^K the term involving $dQ_i^J\wedge dQ_i^I$ vanishes identically. This proves the statement of the lemma. □

We are interested in partitioned Runge-Kutta methods that satisfy:

- the numerical solution stays on the manifold $\mathcal{M}$ of Eq. (8.5);
- the numerical flow $(p_0,q_0)\mapsto(p_1,q_1)$ is a symplectic transformation on $\mathcal{M}$;
- the order of convergence is higher than 2.

If the values Λ_i are determined by the condition

$$g(Q_i)=0 \qquad \text{for} \quad i=1,\dots,s, \tag{8.23}$$

then we have $\sum_I \partial g^K/\partial q^I(Q_i)dQ_i^I=0$, and it follows from Lemma 8.4 that the method (8.20) is symplectic, if (8.21) and (8.22) are satisfied. Hence, the second item holds. Here we see the importance of the conditions (8.23). Solving the index reduced system (8.1a,b), (8.3b) by a symplectic method would in general not result in a symplectic numerical flow on $\mathcal{M}$.

How can we achieve the first item, in particular the condition $g(q_1)=0$? The idea is to require the method $\widehat{b}_i, \widehat{a}_{ij}$ to be stiffly accurate, i.e.,

$$\widehat{a}_{sj}=\widehat{b}_j \qquad \text{for} \quad j=1,\ldots,s. \tag{8.24}$$

In this case we have $q_1 = Q_s$, and $g(q_1)=0$ is automatically satisfied by (8.23). The condition (8.24) together with (8.22) implies that (assuming nonzero $\widehat{b}_i$)

$$a_{is}=0 \qquad \text{for} \quad i=1,\ldots,s, \tag{8.25}$$

and the nonlinear system (8.20a,c), (8.23) no longer depends on Λ_s. This parameter, however, appears in the definition of p_1 in Eq. (8.20b) via k_s. There it can be used to impose the constraint $G(q_1)H_p(p_1,q_1)=0$.

Due to the condition (8.25) a new difficulty arises. If we consider (8.20b,c) as definition of the quantities p_1, q_1, k_i, ℓ_i, the remaining equations (8.20a) and (8.23) are a nonlinear system for $P_1,\ldots,P_s, Q_1,\ldots,Q_s, \Lambda_1,\ldots,\Lambda_{s-1}$. Counting the number of equations of this system $(2sn+sm)$ and the number of unknowns $(2sn+(s-1)m)$, one is readily convinced that this nonlinear system will usually not have a solution. The idea (Jay 1994, 1996) is to require

$$\widehat{a}_{1j}=0 \qquad \text{for} \quad j=1,\ldots,s, \tag{8.26}$$

so that $Q_1 = q_0$, and the condition (8.23) is automatically verified for $i=1$ (we always assume consistent initial values). By (8.22) this implies (for nonzero $\widehat{b}_i$)

$$a_{i1}=b_1 \qquad \text{for} \quad i=1,\ldots,s. \tag{8.27}$$

The Runge-Kutta matrices $\widehat{A}$ and A are both singular. Let $\widehat{A}_0$ be the $(s-1)\times s$ submatrix of $\widehat{A}$ obtained by deleting its first row, and let A_0 be the $s\times(s-1)$ submatrix of A formed by the first $s-1$ columns of A. In order to be able to prove the existence of a numerical solution of (8.20), (8.23), we require that the $(s-1)\times(s-1)$ matrix

$$\widehat{A}_0 A_0 \qquad \text{is invertible.} \tag{8.28}$$

We now extend the method to arbitrary initial values as follows: we replace condition (8.23) by

$$g(Q_i)=g(q_0)+\widehat{c}_i hG(q_0)H_p(p_0,q_0) \qquad \text{for} \quad i=1,\ldots,s,$$

($\widehat{c}_i=\sum_j \widehat{a}_{ij}$) and the condition $G(q_1)H_p(p_1,q_1)=0$ by (8.9b). Similar to Equation (8.10) we use

$$g(Q_i)-g(q_0)=h\int_0^1 g_q\big(q_0+\tau(Q_i-q_0)\big)\,d\tau\cdot\sum_{j=1}^{s}\widehat{a}_{ij}H_p(P_j,Q_j). \tag{8.29}$$

Then we develop

$$\begin{aligned} H_p(P_j,Q_j) &= H_p(p_0,Q_j) \\ &- h\int_0^1 H_{pp}(p_0+\sigma(P_j-p_0),Q_j)\,d\sigma\cdot\sum_{r=1}^{s-1}a_{jr}\big(H_q(P_r,Q_r)+G^T(Q_r)\Lambda_r\big), \end{aligned}$$

and insert this relation into (8.29). As in Eq. (8.11) we get a linear system for $\Lambda_1, \ldots, \Lambda_{s-1}$ which, for $h = 0$, has the solution Λ_r^0 given by

$$0 = \frac{\widehat{c}_i^2}{2} g_{qq}(H_p, H_p) + \Big(\sum_{j=1}^{s} \widehat{a}_{ij}\widehat{c}_j\Big) GH_{pq}H_p$$

$$- \sum_{r=1}^{s-1}\Big(\sum_{j=1}^{s} \widehat{a}_{ij}a_{jr}\Big) GH_{pp}(H_q + G^T\Lambda_r^0).$$

Here all functions are evaluated at (p_0, q_0). Due to (8.28) and (8.4) this system can be solved for Λ_r^0. The Implicit Function Theorem then guarantees the existence of a locally unique solution of the method (8.20), (8.23), and the existence of a smooth extension to a neighbourhood of $\mathcal{M}$.

The question is now: do there exist high order methods having all these properties?

Theorem 8.5. *The s-stage Lobatto IIIA-IIIB pair (Lobatto IIIA in the role of $\widehat{b}_i, \widehat{a}_{ij}$, and Lobatto IIIB in the role of b_i, a_{ij}; see Sect. IV.5 for their definition) satisfies (8.21), (8.22), (8.24), (8.25), (8.26), (8.27), and (8.28).*

Proof. Properties (8.21), (8.24), (8.25), (8.26), and (8.27) follow immediately from the definition of the methods. The symplecticity condition (8.22) has first been proved by Sun Geng (1993). We let $d_{ij} = b_i\widehat{a}_{ij} + \widehat{b}_j a_{ji} - b_i\widehat{b}_j$ and compute for $k = 1, \ldots, s$

$$\sum_{j=1}^{s} d_{ij}c_j^{k-1} = b_i\frac{c_i^k}{k} + \frac{b_i}{k}(1 - c_i^k) - b_i\frac{1}{k} = 0.$$

Here we have exploited the fact that the Lobatto IIIA method satisfies $C(s)$ and the Lobatto IIIB method satisfies $D(s)$ (see Table IV.5.13). Since the abscissae $c_1, \ldots, c_s$ of the Lobatto quadrature are distinct, the above Vandermonde type system has a unique solution $d_{ij} = 0$. This proves (8.22).

We next show that

$$\sum_{k=1}^{s-1}\Big(\sum_{j=1}^{s} \widehat{a}_{ij}a_{jk}\Big)c_k^{q-2} = \frac{c_i^q}{q(q-1)} \qquad \text{for } i, q = 2, \ldots, s. \tag{8.30}$$

This means that $\widehat{A}_0 A_0 V = W$, where V and W are nonsingular Vandermonde type matrices. This obviously implies (8.28). For $q = 2, \ldots, s-1$ Eq. (8.30) follows from the fact that the methods Lobatto IIIA and IIIB satisfy $C(s)$ and $C(s-2)$, respectively. It remains to show that the coefficients $\delta_i := \sum_k \sum_j \widehat{a}_{ij}a_{jk}c_k^{s-2} - c_i^s/s(s-1)$ vanish for all i. By (8.26) and $c_1 = 0$ we have $\delta_1 = 0$. Because of $\widehat{a}_{sj} = \widehat{b}_j = b_j$ and $c_s = 1$, the condition $\delta_s = 0$ is nothing else than an order condition (order s), which is satisfied (Sect. IV.5). Since the Lobatto IIIA and IIIB methods satisfy $D(s-2)$ and $D(s)$, respectively, it holds $\sum_i b_i c_i^{m-1}\delta_i = 0$ for

$m = 1, \ldots, s-2$. This proves that also $\delta_2, \ldots, \delta_{s-1}$ vanish, so that all relations of (8.30) are established. □

It still remains to discuss the order of convergence of the Lobatto IIIA-IIIB pair. Since we have succeeded in embedding the method into a one-step method that is defined in a whole neighbourhood of $\mathcal{M}$, the convergence theory of Sect. II.3 can be applied. We only have to investigate the local error of the method. Each of the methods has classical order $2s-2$ (Sect. IV.5), and it follows from Exercise 4 that, considered as partitioned Runge-Kutta method, the pair has also order $2s-2$. It has been shown in Jay (1994) that the presence of constraints (8.1c) does not reduce the order. The proof of this superconvergence result is very technical and long. Therefore we do not reproduce it here.

Composition Methods

Another possibility for obtaining high order symplectic methods for the system (8.1) is by composition of low order methods. The idea goes back to Yoshida (1990), and has been extended to constrained systems by Reich (1996).

Consider the second order symmetric method (8.19) and denote its extension to a neighbourhood of $\mathcal{M}$ by Φ_h. We shall study the following composition

$$\Phi_{c_1 h} \circ \Phi_{c_2 h} \circ \Phi_{c_1 h}. \tag{8.31}$$

The method (8.31) represents a one-step method, defined in a neighbourhood of $\mathcal{M}$. For initial values on $\mathcal{M}$, the numerical solution stays on $\mathcal{M}$. Moreover, the composition (8.31) is symplectic and symmetric. Observe that the projections (8.19d,e) can be avoided in an implementation of this method (see Remark 8.3). Concerning its order we have the following result.

Theorem 8.6. *Let Φ_h be the mapping $(p_0, q_0) \mapsto (p_1, q_1)$, defined by (8.19). If*

$$2c_1 + c_2 = 1, \qquad 2c_1^3 + c_2^3 = 0, \tag{8.32}$$

the composition method (8.31) is of order 4.

If Φ_h represents a one-step method that is symmetric, of order $p = 2k$, and defined in a neighbourhood of $\mathcal{M}$, then the relations

$$2c_1 + c_2 = 1, \qquad 2c_1^{p+1} + c_2^{p+1} = 0, \tag{8.33}$$

imply that the composition (8.31) is of order $p+2$.

Proof. We let $y_0 = (p_0, q_0)^T$ and $y(t) = \big(p(t), q(t)\big)^T$. The local error of the method (8.19) satisfies

$$y(t_0 + h) - \Phi_h(y_0) = d(y_0)h^3 + \mathcal{O}(h^4).$$

Since the basic method is of the form $\Phi_h(y_0) = y_0 + h\Psi(y_0, h)$, we have that

$$y\big(t_0 + (2c_1 + c_2)h\big) - \Phi_{c_1 h} \circ \Phi_{c_2 h} \circ \Phi_{c_1 h}(y_0) = (2c_1^3 + c_2^3)d(y_0)h^3 + \mathcal{O}(h^4).$$

The conditions (8.32) then imply that the method (8.31) is at least of order 3. Since it is symmetric, it has to be of order 4. The proof is easily adapted to the higher order situation. □

A solution of (8.32) is given by

$$c_1 = \frac{1}{2 - \sqrt[3]{2}}, \qquad c_2 = -\frac{\sqrt[3]{2}}{2 - \sqrt[3]{2}},$$

which shows that the intermediate step in the composition (8.31) is a 'back step' (negative step size $c_2 h$).

The result of Theorem 8.6 allows us to construct symplectic integrators for (8.1) of an arbitrary even order. However, the resulting method of order $p = 2k$ requires 3^{k-1} applications of the basic method (8.19).

In the case of unconstrained Hamiltonian systems it is known that better methods can be obtained by compositions of the form

$$\Phi_{c_1 h} \circ \Phi_{c_2 h} \circ \ldots \circ \Phi_{c_{s-1} h} \circ \Phi_{c_s h} \circ \Phi_{c_{s-1} h} \circ \ldots \circ \Phi_{c_2 h} \circ \Phi_{c_1 h} \tag{8.34}$$

(see Yoshida 1990, McLachlan 1995, Sanz-Serna & Calvo 1994). Reich (1996) studies the extension of these methods to constrained Hamiltonian systems and finds that additional order conditions are necessary. His investigation relies on a "backward error analysis" for integrators on manifolds.

Backward Error Analysis (for ODEs)

> Although backward analysis is a perfectly straightforward concept there is strong evidence that a training in classical mathematics leaves one unprepared to adopt it.
>
> (J.H. Wilkinson, NAG Newsletter 2/85)

In Sect. II.16 we have briefly explained the idea of backward error analysis for the symplectic Euler method. Here we present an extension to general one-step methods for ordinary differential equations. Consider

$$y' = f(y), \qquad y(0) = y_0, \tag{8.35}$$

and let $y_0 \mapsto y_1$ be an arbitrary one-step method for (8.35). We assume that $f(y)$ and the method are sufficiently often differentiable, so that the local error can be expanded into a Taylor series as

$$y_1 - y(h) = d_{p+1}(y_0)h^{p+1} + \ldots + d_N(y_0)h^N + \mathcal{O}(h^{N+1}). \tag{8.36}$$

Theorem 8.7. *Consider a one-step method of order* p, *and assume the local error to be given by (8.36). Then there exist functions* $f_j(y)$ *(for* $j=p,\dots,N$*), such that*

$$y_1 - \widetilde{y}(h) = \mathcal{O}(h^{N+1}), \tag{8.37}$$

where $\widetilde{y}(t)$ *is the solution of the perturbed differential equation*

$$\widetilde{y}' = f(\widetilde{y}) + h^p f_p(\widetilde{y}) + \ldots + h^{N-1} f_{N-1}(\widetilde{y}), \qquad \widetilde{y}(0) = y_0, \tag{8.38}$$

Remark. If the function $f(y)+h^p f_p(y)+\ldots+h^{N-1}f_{N-1}(y)$ satisfies a Lipschitz condition, the proof of Theorem II.3.4 shows that $y_n - \widetilde{y}(nh) = \mathcal{O}(h^N)$ on bounded intervals. This implies that the numerical approximation y_n is much closer to the solution of (8.38) than to that of (8.35). Hence, the study of the system (8.38) yields new insight into the behaviour of the numerical solution.

Proof. As a consequence of the nonlinear variation-of-constants formula (Theorem I.14.5) we have

$$\widetilde{y}(h) = y(h) + \int_0^h \frac{\partial y}{\partial y_0}(h, s, \widetilde{y}(s)) \cdot \Big(h^p f_p(\widetilde{y}(s)) + \ldots + h^N f_N(\widetilde{y}(s)) \Big) ds,$$

where $y(t, t_0, y_0)$ denotes the solution of (8.35) corresponding to initial values $y(t_0) = y_0$. Expanding the above integral into a Taylor series we obtain

$$\widetilde{y}(h) - y(h) = h^{p+1} f_p(y_0) + h^{p+2}\Big(f_{p+1} + \frac{1}{2} f_p' f + \frac{1}{2} f' f_p \Big)(y_0) + \ldots . \tag{8.39}$$

The condition (8.37) implies that the coefficients of (8.39) have to agree with those of (8.36) up to a certain order. We thus get $f_p(y) = d_{p+1}(y)$, $f_{p+1}(y) = d_{p+2}(y) - (f_p'(y)f(y) + f'(y)f_p(y))/2$, etc. The essential observation is that the coefficient of h^{j+1} in (8.39) contains $f_j(y)$ as linear term and further expressions that only depend on $f_i(y)$ with $i<j$. Hence, the functions $f_j(y)$ are recursively determined by the above comparison. □

Example 8.8. For an illustration of the above theorem we consider the Volterra-Lotka differential equation

$$u' = u(v-1), \qquad v' = v(2-u). \tag{8.40}$$

This system possesses the first integral

$$I(u,v) = 2\ln u - u + \ln v - v, \tag{8.41}$$

implying that the solutions are all periodic. Some of them are plotted in the left upper picture of Fig. 8.1.

We apply three different numerical methods to this differential equation. The first one is the well-known explicit Euler method $y_{n+1} = y_n + hf(y_n)$. The right upper picture of Fig. 8.1 shows the numerical solution and the exact solution (solid

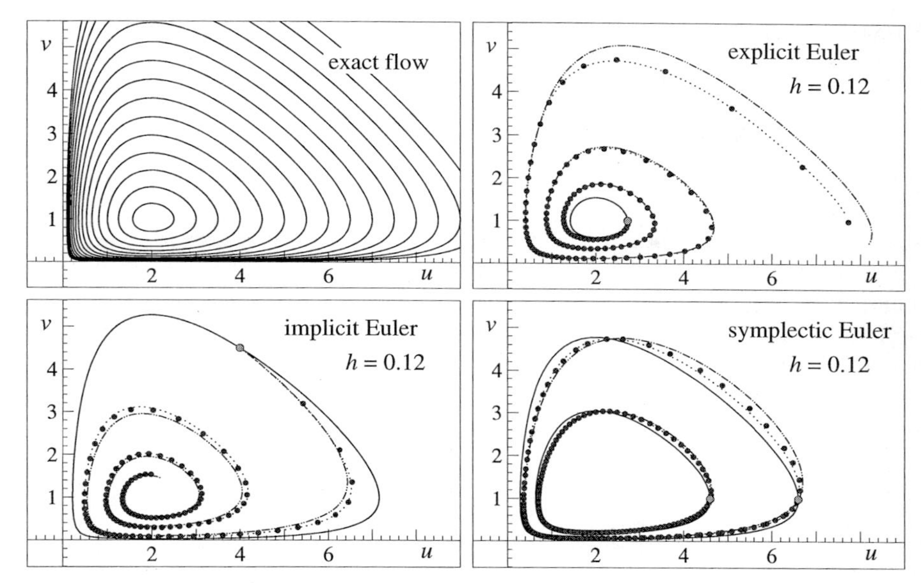

Fig. 8.1. Solutions of the perturbed differential equation for various methods

line) for the initial value $u_0 = 2.725$, $v_0 = 1$. Moreover, we have included the solutions of the perturbed differential equation (8.38) for $N = 1$ (dashed-dotted line) and for $N = 2$ (dotted line). For the explicit Euler method, Eq. (8.38) reads

$$\widetilde{y}\,' = f(\widetilde{y}) - \frac{h}{2}(f'f)(\widetilde{y}) + \frac{h^2}{12}\big(f''(f,f) + 4f'f'f\big)(\widetilde{y}). \tag{8.42}$$

We nicely observe the good agreement of the numerical solution with the exact solution of the perturbed system, even for the rather large step size $h = 0.12$.

The left lower picture shows the same experiment for the implicit Euler method $y_{n+1} = y_n + hf(y_{n+1})$. The perturbed differential equation is obtained from (8.42) by replacing h by $-h$ (this is, because the explicit Euler method is the adjoint method of the implicit Euler method).

The third method is the symplectic Euler method (see Eq. (8.45) below), which for the problem (8.40) is defined by

$$u_{n+1} = u_n + hu_n(v_{n+1} - 1), \qquad v_{n+1} = v_n + hv_{n+1}(2 - u_n).$$

The first term of the perturbed differential equation is

$$\begin{aligned} \widetilde{u}\,' &= \widetilde{u}(\widetilde{v} - 1) - h\widetilde{u}(\widetilde{u}\widetilde{v} - 4\widetilde{v} + \widetilde{v}^2 + 1)/2 \\ \widetilde{v}\,' &= \widetilde{v}(2 - \widetilde{u}) + h\widetilde{v}(\widetilde{u}\widetilde{v} - 5\widetilde{u} + \widetilde{u}^2 + 4)/2. \end{aligned} \tag{8.43}$$

The qualitative behaviour of this method is quite different from that of the previous methods. One can prove that the system (8.43) has a first integral close to $I(u,v)$ (Exercise 5). Hence the solutions are periodic, as it is the case for the original unperturbed system.

Example 8.9. For the Hamiltonian system (without constraints)

$$q' = H_p(p,q), \qquad p' = -H_q(p,q) \tag{8.44}$$

the method (8.8) becomes

$$q_1 = q_0 + hH_p(p_1,q_0), \qquad p_1 = p_0 - hH_q(p_1,q_0). \tag{8.45}$$

A similar method (implicit in q and explicit in p) has been considered in Sect. II.16, Formula (II.16.54). There we have computed the first terms of the perturbed differential equation (8.38), and we have noticed with surprise that it is also Hamiltonian. The same computation can be done here. We find that the perturbed differential equation for (8.45) is of the form

$$\widetilde{q}' = \widetilde{H}_p(\widetilde{p},\widetilde{q}), \qquad \widetilde{p}' = -\widetilde{H}_q(\widetilde{p},\widetilde{q}) \tag{8.46}$$

with (for $N=2$)

$$\widetilde{H} = H - \frac{h}{2}H_pH_q + \frac{h^2}{12}\Big(H_{pp}H_q^2 + H_{qq}H_p^2 + 4H_{pq}H_pH_q\Big).$$

For notational convenience we have assumed that p and q are scalars. However, with a suitable interpretation of the appearing expressions, the formula is also valid for problems with more than one degree of freedom.

Example 8.10. The second order method (8.19), when applied to the unconstrained system (8.44), becomes

$$\begin{aligned} q_1 &= q_0 + \frac{h}{2}\Big(H_p(p_{1/2},q_0) + H_p(p_{1/2},q_1)\Big) \\ p_1 &= p_0 - \frac{h}{2}\Big(H_q(p_{1/2},q_0) + H_q(p_{1/2},q_1)\Big), \end{aligned} \tag{8.47}$$

where $p_{1/2} = p_0 - (h/2)H_q(p_{1/2},q_0)$. Computing the dominant term of its local error, we see that the perturbed differential equation (8.38) is, for $N=2$, given by

$$\begin{aligned} \widetilde{q}' = H_p(\widetilde{p},\widetilde{q}) + \frac{h^2}{24}\Big(&-H_{ppp}H_q^2 + 2H_{ppq}H_pH_q + 2H_{pqq}H_p^2 \\ &+ 2H_{pq}H_{pq}H_p + 4H_{pp}H_{qq}H_p\Big)(\widetilde{p},\widetilde{q}) \\ \widetilde{p}' = -H_q(\widetilde{p},\widetilde{q}) + \frac{h^2}{24}\Big(&H_{ppq}H_q^2 - 2H_{pqq}H_pH_q - 2H_{qqq}H_p^2 \\ &- 2H_{pq}H_{pq}H_q + 2H_{pp}H_{qq}H_q - 6H_{pq}H_{qq}H_p\Big)(\widetilde{p},\widetilde{q}). \end{aligned}$$

One easily verifies that this is a Hamiltonian system (8.46) with

$$\widetilde{H} = H + \frac{h^2}{24}\Big(2H_{qq}H_p^2 - H_{pp}H_q^2 + 2H_{pq}H_pH_q\Big).$$

A Short Survey on Further Results. A further elaboration of backward error analysis for ordinary differential equations would take us beyond the scope of this chapter. We therefore collect some interesting results without going into details.

First of all, the mystery of the foregoing examples is well understood. In the situation, where the differential equation (8.35) is a Hamiltonian system, and where a symplectic integration method is applied, the perturbed system (8.38) is again Hamiltonian for all N. This result is proved by Hairer (1994), where explicit formulas for the functions $f_j(y)$ in terms of elementary differentials are provided, and where an explicit formula for the perturbed Hamiltonian is given. This explicit representation guarantees that $\widetilde{H}(p,q)$ is uniquely defined on regions where $H(p,q)$ is defined. Different proofs of this result can be found in Reich (1996) and Benettin & Giorgilli (1994).

If the function f in (8.35) is infinitely differentiable, then the truncation index N in Theorem 8.7 is arbitrary. In general, the series (8.38) diverges as $N \to \infty$ and the constants hidden in the $\mathcal{O}(h^{N+1})$ bounds of (8.37) tend to infinity with N, even if f is analytic. Therefore, it is interesting to find rigorous bounds on $y_1 - \widetilde{y}(h)$ for an optimally chosen N. Such results have been found independently by Benettin & Giorgilli (1994) and Hairer & Lubich (1996). As a consequence, one can show that for symplectic integrations the Hamiltonian remains bounded (with error of size $\mathcal{O}(h^p)$) over exponentially long times. Moreover, KAM theory can be applied to get more insight into the long-time behaviour of symplectic numerical schemes.

Backward Error Analysis on Manifolds

Consider the constrained Hamiltonian system (8.1), and a numerical one-step method which yields approximations (p_n, q_n) staying on the manifold $\mathcal{M}$ of Eq. (8.5). Can we extend the above backward error analysis for ODEs to this situation?

There are at least two ways to achieve this goal. The first one is to introduce local coordinates in order to obtain an unconstrained Hamiltonian system. The backward analysis for ODEs can then be applied to the one-step method written in local coordinates.

The second approach allows us to construct the perturbed Hamiltonian directly in the original coordinates. For the special case of separable Hamiltonians, this approach is due to Reich (1996). We shall explain it for the first and second order methods (8.8) and (8.19).

Backward Error Analysis for the Method (8.8). Consider first the subsystem (8.8a-c). The projection step (8.8d,e) will be treated later. In Eq. (8.11) the value λ_1 has been expressed in terms of $\widehat{p}_1, q_1, p_0, q_0$, even for inconsistent initial values. Inserting this function into (8.8a), the Eqs. (8.8a,b) represent two relations between the variable $\widehat{p}_1, q_1, p_0, q_0$, and h. By the Implicit Function Theorem these two relations allow us to express (p_0, q_1) in terms of $(\widehat{p}_1, q_0)$, and h. Consequently, the solution λ_1 of Eq. (8.11) can be written as a function of $(\widehat{p}_1, q_0, h)$. We denote it by

$$\lambda_1 = \lambda(\widehat{p}_1, q_0, h), \tag{8.48}$$

so that the system (8.8a,b) becomes

$$\begin{aligned} \widehat{p}_1 &= p_0 - h\big(H_q(\widehat{p}_1, q_0) + G^T(q_0)\lambda(\widehat{p}_1, q_0, h)\big) \\ q_1 &= q_0 + hH_p(\widehat{p}_1, q_0), \end{aligned} \tag{8.49}$$

and the constraint (8.9a) is automatically satisfied by the definition of $\lambda(\widehat{p}_1, q_0, h)$. We now consider the Hamiltonian function

$$\mathcal{H}(p,q) = H(p,q) + g(q)^T \lambda(p,q,h), \tag{8.50}$$

where $\lambda(p,q,h)$ is the function defined in (8.48). The corresponding Hamiltonian system is

$$\begin{aligned} q' &= H_p(p,q) + g(q)^T \lambda_p(p,q,h) \\ p' &= -H_q(p,q) - G^T(q)\lambda(p,q,h) - g(q)^T \lambda_q(p,q,h). \end{aligned} \tag{8.51}$$

The main observation is now that, for initial values satisfying $g(q_0) = 0$, the numerical solution $(\widehat{p}_1, q_1)$ of (8.49) is exactly the same as the numerical solution of the symplectic Euler method (8.45) applied to the (unconstrained) Hamiltonian system (8.51). Therefore, Example 8.9 shows that the numerical solution $(\widehat{p}_1, q_1)$ is $\mathcal{O}(h^4)$-close to the exact solution of (8.46), where in the definition of $\widetilde{H}$ the function H has to be replaced by $\mathcal{H}$ od Eq. (8.50).

The projection step (8.8d,e) can be treated similarly. The solution μ of (8.8d), (8.9b) depends on $\widehat{p}_1, q_1$, and h (the dependence on p_0, q_0 can be omitted, because the relations (8.8a,b) allow us to express them in terms of $\widehat{p}_1, q_1$, and h). Due to the relation (8.8d) we can also consider μ as a function of p_1, q_1, h, i.e., $\mu = \mu(p_1, q_1, h)$. We now consider the Hamiltonian

$$\mathcal{G}(p,q) = g(q)^T \mu(p,q,h), \tag{8.52}$$

and the corresponding Hamiltonian system

$$\begin{aligned} q' &= g(q)^T \mu_p(p,q,h) \\ p' &= -G^T(q)\mu(p,q,h) - g(q)^T \mu_q(p,q,h). \end{aligned} \tag{8.53}$$

If $g(q_1) = 0$, the numerical approximation p_1, computed from (8.8d), i.e., $p_1 = \widehat{p}_1 - hG^T(q_1)\mu(p_1, q_1, h)$, is identical to the numerical solution of (8.45), applied to the system (8.53) with initial values $(\widehat{p}_1, q_1)$. Again, we obtain from Example 8.9 that the numerical solution (p_1, q_1) is $\mathcal{O}(h^4)$-close to the exact solution of (8.46), where in the definition of $\widetilde{H}$ the function H has to be replaced by $\mathcal{G}$ of Eq. (8.52). We summarize our findings in the following theorem.

Theorem 8.11. *Consider the one-step method (8.8) and assume that the initial values are consistent, i.e.,* $(p_0, q_0) \in \mathcal{M}$. *Then it holds*

$$p_1 - \widetilde{p}(h) = \mathcal{O}(h^4), \qquad q_1 - \widetilde{q}(h) = \mathcal{O}(h^4),$$

where $\widetilde{p}(t), \widetilde{q}(t)$ *is the solution of the Hamiltonian system (8.46) with*

$$\widetilde{H} = \widehat{H} + \widehat{G} + \frac{h}{2}\{\widehat{H}, \widehat{G}\} + \frac{h^2}{12}\Big(\{\widehat{H}, \{\widehat{H}, \widehat{G}\}\} + \{\widehat{G}, \{\widehat{G}, \widehat{H}\}\}\Big)$$

where

$$\widehat{H} = \mathcal{H} - \frac{h}{2}\mathcal{H}_p\mathcal{H}_q + \frac{h^2}{12}\Big(\mathcal{H}_{pp}\mathcal{H}_q^2 + \mathcal{H}_{qq}\mathcal{H}_p^2 + 4\mathcal{H}_{pq}\mathcal{H}_p\mathcal{H}_q\Big)$$
$$\widehat{G} = \mathcal{G} - \frac{h}{2}\mathcal{G}_p\mathcal{G}_q + \frac{h^2}{12}\Big(\mathcal{G}_{pp}\mathcal{G}_q^2 + \mathcal{G}_{qq}\mathcal{G}_p^2 + 4\mathcal{G}_{pq}\mathcal{G}_p\mathcal{G}_q\Big),$$

and $\mathcal{H}$ *and* $\mathcal{G}$ *are given by (8.50) and (8.52), respectively. Here, the poisson bracket* $\{H, G\}$ *of two functions* $H, G : \mathbb{R}^n \times \mathbb{R}^n \to \mathbb{R}$ *is given by* $\{H, G\} := H_pG_q - H_qG_p$ *(see Eq. (II.16.65)).*

Proof. We consider the one-step method (8.8) as a composition of the mappings $(p_0, q_0) \mapsto (\widehat{p}_1, q_1)$ and $(\widehat{p}_1, q_1) \mapsto (p_1, q_1)$. Neglecting terms of size $\mathcal{O}(h^4)$, both mappings can be interpreted as the h-flow of Hamiltonian systems. The statement thus follows from the Campbell-Baker-Hausdorff Formula (II.16.83). □

Backward Error Analysis for the Method (8.19). We consider the solution λ_0 of (8.19a,b), (8.9a) as a function of $p_{1/2}, q_0$, and h, i.e., $\lambda_0 = \lambda(p_{1/2}, q_0, h)$, and the solution μ_0 of (8.19d), (8.9b) as a function of p_1, q_1 and h, i.e., $\mu_0 = \mu(p_1, q_1, h)$. The method (8.19) can therefore be written as the composition of

$$\begin{aligned} p_{1/2} &= p_0 - \frac{h}{2}\Big(H_q(p_{1/2}, q_0) + G^T(q_0)\lambda(p_{1/2}, q_0, h)\Big) \\ q_1 &= q_0 + \frac{h}{2}\Big(H_p(p_{1/2}, q_0) + H_p(p_{1/2}, q_1)\Big) \\ \widehat{p}_1 &= p_{1/2} - \frac{h}{2}\Big(H_q(p_{1/2}, q_1) + G^T(q_1)\lambda(p_{1/2}, q_1, h)\Big) \end{aligned} \tag{8.54}$$

with the projection step

$$p_1 = \widehat{p}_1 - \frac{h}{2}G^T(q_1)\nu(p_1, q_1, h), \tag{8.55}$$

where $\nu(p_1, q_1, h) = \mu(p_1, q_1, h) - \lambda(p_{1/2}, q_1, h)$. We see that, for consistent initial values $(p_0 . q_0) \in \mathcal{M}$, (8.54) is identical to (8.47) with $H(p, q)$ replaced by

$$\mathcal{H}(p, q) = H(p, q) + g(q)^T\lambda(p, q, h), \tag{8.56}$$

and the projection step (8.55) can be interpreted as method (8.45) with Hamiltonian function

$$\mathcal{G}(p, q) = \frac{1}{2}g(q)^T\nu(p, q, h). \tag{8.57}$$

In the same way as for the first order method we get:

Theorem 8.12. *Consider the method (8.19) and assume consistent initial values* $(p_0, q_0) \in \mathcal{M}$. *Then it holds*

$$p_1 - \widetilde{p}(h) = \mathcal{O}(h^4), \qquad q_1 - \widetilde{q}(h) = \mathcal{O}(h^4),$$

where $\widetilde{p}(t), \widetilde{q}(t)$ *is the solution of the Hamiltonian system (8.46) with*

$$\widetilde{H} = \widehat{H} + \widehat{G} + \frac{h}{2}\{\widehat{H}, \widehat{G}\} + \frac{h^2}{12}\Big(\{\widehat{H}, \{\widehat{H}, \widehat{G}\}\} + \{\widehat{G}, \{\widehat{G}, \widehat{H}\}\}\Big)$$

where

$$\widehat{H} = \mathcal{H} + \frac{h^2}{24}\Big(2\mathcal{H}_{qq}\mathcal{H}_p^2 - \mathcal{H}_{pp}\mathcal{H}_q^2 + 2\mathcal{H}_{pq}\mathcal{H}_p\mathcal{H}_q\Big)$$
$$\widehat{G} = \mathcal{G} - \frac{h}{2}\mathcal{G}_p\mathcal{G}_q + \frac{h^2}{12}\Big(\mathcal{G}_{pp}\mathcal{G}_q^2 + \mathcal{G}_{qq}\mathcal{G}_p^2 + 4\mathcal{G}_{pq}\mathcal{G}_p\mathcal{G}_q\Big),$$

and $\mathcal{H}$ *and* $\mathcal{G}$ *are given by (8.56) and (8.57), respectively.* □

The above two theorems show that, for consistent initial values, the numerical solution of the considered methods is (up to a certain order) the exact solution of an unconstrained perturbed Hamiltonian system. The perturbed Hamiltonian is defined in a neighbourhood of the manifold, so that all backward error analysis results for ODEs can be applied.

Exercises

1. (Jay 1995). The system (1.46) is equivalent to

$$\begin{aligned} q' &= u \\ (M(q)u)' &= M_q(q)(u,u) + f(q,u) - G^T(q)\lambda \\ 0 &= g(q). \end{aligned} \tag{8.58}$$

   In the case where (1.46) is obtained from the Lagrangian function $\mathcal{L}(q,\dot q) = \frac{1}{2}\dot q^T M(q)\dot q - U(q)$, show that $f(q,u)$ always contains the term $-M_q(q)(u,u)$ (Coriolis forces), which thus cancels out in the formulation (8.58).

2. Show that the example (2.1a-c) is of the form (8.1a-c) with Hamiltonian

$$H(p,q) = (p_1^2 + p_2^2)/2 + q_2.$$

   If we compute λ from (2.3), and insert it into (2.1a,b), the resulting differential equation is no longer Hamiltonian.

3. Give a second proof of Theorem 8.1 by applying Theorem I.14.12.

   Hint (Reich 1996). Let $\lambda = \lambda(p,q)$ be defined by (8.3b) and consider the unconstrained Hamiltonian system with Hamiltonian

$$H(p,q) + g(q)^T \lambda(p,q),$$

   whose flow reduces to that of (8.1) along the constraint manifold $\mathcal{M}$.

4. Consider a partitioned Runge-Kutta method applied to a partitioned ordinary differential equation (without constraints). Suppose that both methods are based on the same quadrature formula of order p, that the first method satisfies $C(\eta), D(\xi)$, and that the second method satisfies $C(\widehat{\eta}), D(\widehat{\xi})$. Prove that the pair has order

$$\min\Big(p, 2\min(\eta,\widehat{\eta})+2, \min(\eta,\widehat{\eta})+\min(\xi,\widehat{\xi})+2, \min(\eta+\xi,\widehat{\eta}+\widehat{\xi})+1\Big).$$

   Conclude that the Lobatto IIIA-IIIB pair has order $2s-2$.

   Hint. Apply the ideas of the proof of Theorem II.7.4 for the verification of the order conditions (Sect. II.15).

5. Compute a first integral of the differential equation (8.43). What is the reason for the existence of such an invariant?

   Hint. With the transformation $u = e^p$, $v = e^q$ you will get a Hamiltonian system.

   Result. $\widetilde{I}(u,v) = I(u,v) + h\big((u+v)^2 - 10u - 8v + 8\ln u + 2\ln v\big)/4.$

lso geht alles zu Ende allhier:
Feder, Tinte, Tobak und auch wir.
Zum letztenmal wird eingetunkt,
Dann kommt der große
schwarze

(W. Busch, Bilder zur Jobsiade 1872)

Appendix. Fortran Codes

During the preparation of this book several programs have been developed for solving stiff and differential-algebraic problems of the form

$$My' = f(x, y), \qquad y(x_0) = y_0, \tag{A.1}$$

where M is a constant square matrix. If M is singular, the problem is differential-algebraic. In this case the initial values have to be consistent.

The implicit Runge-Kutta code RADAU5 and its extension RADAUP can be applied to higher index (≥ 2) problems as well, whereas the Rosenbrock code RODAS and the exptrapolation code SEULEX are suited for explicit stiff differential equations and index 1 problems. The codes SDIRK4, ROS4, and SODEX are still available, but have not been updated.

In the case where M is not a constant matrix, suitable transformations and/or introduction of new variables allow us to bring every implicit differential equation to the form (A.1). If the problem is originally in one of the following forms

$$B(y)y' = f(x, y), \qquad y'' = f(x, y, y'), \qquad B(y)y'' = f(x, y, y'),$$

or the like, then the efficiency of the code can be increased by setting some parameters. This will be explained later in this appendix.

Communication with the code during integration can be done with help of the user-supplied subroutine SOLOUT. This is illustrated in the driver below. Further applications of this subroutine are discussed at the end of this appendix.

Experiences with all of our codes are welcome. The programs can be obtained by anonymous ftp (from "ftp.unige.ch" in the directory "pub/doc/math" or from "http://www.unige.ch/math/") or from the authors, if you send an IBM diskette.

Address: Section de Mathématiques, Case postale 240, CH-1211 Genève 24, Switzerland

E-mail: hairer@divsun.unige.ch wanner@divsun.unige.ch

Driver for the Code RADAU5

"The van der Pol equation problem is so much harder than the rest ..." (L.F. Shampine 1987)

We consider the van der Pol equation

$$\begin{aligned} y_1' &= y_2 & \qquad y_1(0) &= 2 \\ y_2' &= \big((1-y_1^2)y_2 - y_1\big)/\varepsilon & \qquad y_2(0) &= -0.66 \end{aligned}$$

with $\varepsilon = 10^{-6}$ on the interval [0,2]. The subroutines FVPOL, JVPOL compute the right-hand side of this differential equation and its Jacobian. The subroutine SOLOUT is used to print the solution at equidistant points.

```
C --------------------------------
C link driver radau5 decsol dc-decsol   or
C link driver radau5 lapack lapackc dc-lapack
C --------------------------------
        IMPLICIT REAL*8 (A-H,O-Z)
C --- PARAMETERS FOR RADAU5 (FULL JACOBIAN)
        PARAMETER (ND=2,LWORK=4*ND*ND+12*ND+20,LIWORK=3*ND+20)
        DIMENSION Y(ND),WORK(LWORK),IWORK(LIWORK)
        EXTERNAL FVPOL,JVPOL,SOLOUT
C --- PARAMETER IN THE DIFFERENTIAL EQUATION
        RPAR=1.0D-6
C --- DIMENSION OF THE SYSTEM
        N=2
C --- COMPUTE THE JACOBIAN ANALYTICALLY
        IJAC=1
C --- JACOBIAN IS A FULL MATRIX
        MLJAC=N
C --- DIFFERENTIAL EQUATION IS IN EXPLICIT FORM
        IMAS=0
C --- OUTPUT ROUTINE IS USED DURING INTEGRATION
        IOUT=1
C --- INITIAL VALUES
        X=0.0D0
        Y(1)=2.0D0
        Y(2)=-0.66D0
C --- ENDPOINT OF INTEGRATION
        XEND=2.0D0
C --- REQUIRED TOLERANCE
        RTOL=1.0D-4
        ATOL=1.0D0*RTOL
        ITOL=0
C --- INITIAL STEP SIZE
        H=1.0D-6
C --- SET DEFAULT VALUES
        DO I=1,20
           IWORK(I)=0
           WORK(I)=0.D0
        END DO
C --- CALL OF THE SUBROUTINE RADAU5
        CALL RADAU5(N,FVPOL,X,Y,XEND,H,
     +                  RTOL,ATOL,ITOL,
     +                  JVPOL,IJAC,MLJAC,MUJAC,
     +                  FVPOL,IMAS,MLMAS,MUMAS,
     +                  SOLOUT,IOUT,
     +                  WORK,LWORK,IWORK,LIWORK,RPAR,IPAR,IDID)
```

```
C --- PRINT FINAL SOLUTION
      WRITE (6,99) X,Y(1),Y(2)
 99   FORMAT(1X,'X =',F5.2,'    Y =',2E18.10)
C --- PRINT STATISTICS
      WRITE (6,90) RTOL
 90   FORMAT('         rtol=',D8.2)
      WRITE (6,91) (IWORK(J),J=14,20)
 91   FORMAT(' fcn=',I5,' jac=',I4,' step=',I4,' accpt=',I4,
     +      ' rejct=',I3,' dec=',I4,' sol=',I5)
      STOP
      END
C
      SUBROUTINE SOLOUT (NR,XOLD,X,Y,CONT,LRC,N,RPAR,IPAR,IRTRN)
C --- PRINTS SOLUTION AT EQUIDISTANT OUTPUT-POINTS BY USING "CONTR5"
      IMPLICIT REAL*8 (A-H,O-Z)
      DIMENSION Y(N),CONT(LRC)
      COMMON /INTERN/XOUT
      IF (NR.EQ.1) THEN
         WRITE (6,99) X,Y(1),Y(2),NR-1
         XOUT=0.2D0
      ELSE
 10      CONTINUE
         IF (X.GE.XOUT) THEN
C --- CONTINUOUS OUTPUT FOR RADAU5
            WRITE (6,99) XOUT,CONTR5(1,XOUT,CONT,LRC),
     +                   CONTR5(2,XOUT,CONT,LRC),NR-1
            XOUT=XOUT+0.2D0
            GOTO 10
         END IF
      END IF
 99   FORMAT(1X,'X =',F5.2,'    Y =',2E18.10,'    NSTEP =',I4)
      RETURN
      END
C
      SUBROUTINE FVPOL(N,X,Y,F,RPAR,IPAR)
C --- RIGHT-HAND SIDE OF VAN DER POL'S EQUATION
      IMPLICIT REAL*8 (A-H,O-Z)
      DIMENSION Y(N),F(N)
      F(1)=Y(2)
      F(2)=((1-Y(1)**2)*Y(2)-Y(1))/RPAR
      RETURN
      END
C
      SUBROUTINE JVPOL(N,X,Y,DFY,LDFY,RPAR,IPAR)
C --- JACOBIAN OF VAN DER POL'S EQUATION
      IMPLICIT REAL*8 (A-H,O-Z)
      DIMENSION Y(N),DFY(LDFY,N)
      DFY(1,1)=0.0D0
      DFY(1,2)=1.0D0
      DFY(2,1)=(-2.0D0*Y(1)*Y(2)-1.0D0)/RPAR
      DFY(2,2)=(1.0D0-Y(1)**2)/RPAR
      RETURN
      END
```

The result, obtained on a Sun SPARKstation 20, is the following:

```
X = 0.00    Y =  0.2000000000E+01 -0.6600000000E+00    NSTEP =   0
X = 0.20    Y =  0.1858210825E+01 -0.7575052373E+00    NSTEP =  10
X = 0.40    Y =  0.1693217727E+01 -0.9068995621E+00    NSTEP =  11
X = 0.60    Y =  0.1484573110E+01 -0.1233017457E+01    NSTEP =  13
X = 0.80    Y =  0.1083921362E+01 -0.6195010714E+01    NSTEP =  21
```

```
X = 1.00    Y = -0.1863641256E+01  0.7535196392E+00    NSTEP = 144
X = 1.20    Y = -0.1699715970E+01  0.8997232240E+00    NSTEP = 145
X = 1.40    Y = -0.1493380698E+01  0.1213958018E+01    NSTEP = 147
X = 1.60    Y = -0.1120822309E+01  0.4373266499E+01    NSTEP = 153
X = 1.80    Y =  0.1869064482E+01 -0.7496053261E+00    NSTEP = 275
X = 2.00    Y =  0.1706171005E+01 -0.8928020961E+00    NSTEP = 276
X = 2.00    Y =  0.1706171005E+01 -0.8928020961E+00
      rtol=0.10D-03
fcn= 2263 jac= 182 step= 293 accpt= 276 rejct=  9 dec= 251 sol=  662
```

Subroutine RADAU5

Implicit Runge-Kutta code based on the 3-stage Radau IIA method, given in Table IV.5.6. Details on the implementation are described in Section IV.8.

```
      SUBROUTINE RADAU5(N,FCN,X,Y,XEND,H,
     +                  RTOL,ATOL,ITOL,
     +                  JAC ,IJAC,MLJAC,MUJAC,
     +                  MAS ,IMAS,MLMAS,MUMAS,
     +                  SOLOUT,IOUT,
     +                  WORK,LWORK,IWORK,LIWORK,RPAR,IPAR,IDID)
C ----------------------------------------------------------
C     NUMERICAL SOLUTION OF A STIFF (OR DIFFERENTIAL ALGEBRAIC)
C     SYSTEM OF FIRST ORDER ORDINARY DIFFERENTIAL EQUATIONS
C                     M*Y'=F(X,Y).
C     THE SYSTEM CAN BE (LINEARLY) IMPLICIT (MASS-MATRIX M .NE. I)
C     OR EXPLICIT (M=I).
C     THE METHOD USED IS AN IMPLICIT RUNGE-KUTTA METHOD (RADAU IIA)
C     OF ORDER 5 WITH STEP SIZE CONTROL AND CONTINUOUS OUTPUT.
C     C.F. SECTION IV.8
C
C     AUTHORS: E. HAIRER AND G. WANNER
C              UNIVERSITE DE GENEVE, DEPT. DE MATHEMATIQUES
C              CH-1211 GENEVE 24, SWITZERLAND
C              E-MAIL:  HAIRER@DIVSUN.UNIGE.CH,  WANNER@DIVSUN.UNIGE.CH
C
C     THIS CODE IS PART OF THE BOOK:
C         E. HAIRER AND G. WANNER, SOLVING ORDINARY DIFFERENTIAL
C         EQUATIONS II. STIFF AND DIFFERENTIAL-ALGEBRAIC PROBLEMS.
C         SPRINGER SERIES IN COMPUTATIONAL MATHEMATICS 14,
C         SPRINGER-VERLAG 1991, SECOND EDITION 1996.
C
C     VERSION OF SEPTEMBER 30, 1995
C
C     INPUT PARAMETERS
C     ----------------
C     N           DIMENSION OF THE SYSTEM
C
C     FCN         NAME (EXTERNAL) OF SUBROUTINE COMPUTING THE
C                 VALUE OF F(X,Y):
C                    SUBROUTINE FCN(N,X,Y,F,RPAR,IPAR)
C                    REAL*8 X,Y(N),F(N)
C                    F(1)=...     ETC.
C                 RPAR, IPAR (SEE BELOW)
C
C     X           INITIAL X-VALUE
C
C     Y(N)        INITIAL VALUES FOR Y
```

```
C
C     XEND        FINAL X-VALUE (XEND-X MAY BE POSITIVE OR NEGATIVE)
C
C     H           INITIAL STEP SIZE GUESS;
C                 FOR STIFF EQUATIONS WITH INITIAL TRANSIENT,
C                 H=1.D0/(NORM OF F'), USUALLY 1.D-3 OR 1.D-5, IS GOOD.
C                 THIS CHOICE IS NOT VERY IMPORTANT, THE STEP SIZE IS
C                 QUICKLY ADAPTED. (IF H=0.D0, THE CODE PUTS H=1.D-6).
C
C     RTOL,ATOL   RELATIVE AND ABSOLUTE ERROR TOLERANCES. THEY
C                 CAN BE BOTH SCALARS OR ELSE BOTH VECTORS OF LENGTH N.
C
C     ITOL        SWITCH FOR RTOL AND ATOL:
C                   ITOL=0: BOTH RTOL AND ATOL ARE SCALARS.
C                     THE CODE KEEPS, ROUGHLY, THE LOCAL ERROR OF
C                     Y(I) BELOW RTOL*ABS(Y(I))+ATOL
C                   ITOL=1: BOTH RTOL AND ATOL ARE VECTORS.
C                     THE CODE KEEPS THE LOCAL ERROR OF Y(I) BELOW
C                     RTOL(I)*ABS(Y(I))+ATOL(I).
C
C     JAC         NAME (EXTERNAL) OF THE SUBROUTINE WHICH COMPUTES
C                 THE PARTIAL DERIVATIVES OF F(X,Y) WITH RESPECT TO Y
C                 (THIS ROUTINE IS ONLY CALLED IF IJAC=1; SUPPLY
C                 A DUMMY SUBROUTINE IN THE CASE IJAC=0).
C                 FOR IJAC=1, THIS SUBROUTINE MUST HAVE THE FORM
C                    SUBROUTINE JAC(N,X,Y,DFY,LDFY,RPAR,IPAR)
C                    REAL*8 X,Y(N),DFY(LDFY,N)
C                    DFY(1,1)= ...
C                 LDFY, THE COLUMN-LENGTH OF THE ARRAY, IS
C                 FURNISHED BY THE CALLING PROGRAM.
C                 IF (MLJAC.EQ.N) THE JACOBIAN IS SUPPOSED TO
C                    BE FULL AND THE PARTIAL DERIVATIVES ARE
C                    STORED IN DFY AS
C                       DFY(I,J) = PARTIAL F(I) / PARTIAL Y(J)
C                 ELSE, THE JACOBIAN IS TAKEN AS BANDED AND
C                    THE PARTIAL DERIVATIVES ARE STORED
C                    DIAGONAL-WISE AS
C                       DFY(I-J+MUJAC+1,J) = PARTIAL F(I) / PARTIAL Y(J).
C
C     IJAC        SWITCH FOR THE COMPUTATION OF THE JACOBIAN:
C                    IJAC=0: JACOBIAN IS COMPUTED INTERNALLY BY FINITE
C                       DIFFERENCES, SUBROUTINE "JAC" IS NEVER CALLED.
C                    IJAC=1: JACOBIAN IS SUPPLIED BY SUBROUTINE JAC.
C
C     MLJAC       SWITCH FOR THE BANDED STRUCTURE OF THE JACOBIAN:
C                    MLJAC=N: JACOBIAN IS A FULL MATRIX. THE LINEAR
C                       ALGEBRA IS DONE BY FULL-MATRIX GAUSS-ELIMINATION.
C                    0<=MLJAC<N: MLJAC IS THE LOWER BANDWITH OF JACOBIAN
C                       MATRIX (>= NUMBER OF NON-ZERO DIAGONALS BELOW
C                       THE MAIN DIAGONAL).
C
C     MUJAC       UPPER BANDWITH OF JACOBIAN  MATRIX (>= NUMBER OF NON-
C                 ZERO DIAGONALS ABOVE THE MAIN DIAGONAL).
C                 NEED NOT BE DEFINED IF MLJAC=N.
C
C     ----   MAS,IMAS,MLMAS, AND MUMAS HAVE ANALOG MEANINGS      -----
C     ----   FOR THE "MASS MATRIX" (THE MATRIX "M" OF SECTION IV.8):  -
C
C     MAS         NAME (EXTERNAL) OF SUBROUTINE COMPUTING THE MASS-
C                 MATRIX M.
C                 IF IMAS=0, THIS MATRIX IS ASSUMED TO BE THE IDENTITY
C                 MATRIX AND NEEDS NOT TO BE DEFINED;
```

```
C                    SUPPLY A DUMMY SUBROUTINE IN THIS CASE.
C                    IF IMAS=1, THE SUBROUTINE MAS IS OF THE FORM
C                       SUBROUTINE MAS(N,AM,LMAS,RPAR,IPAR)
C                       REAL*8 AM(LMAS,N)
C                       AM(1,1)= ....
C                       IF (MLMAS.EQ.N) THE MASS-MATRIX IS STORED
C                       AS FULL MATRIX LIKE
C                            AM(I,J) = M(I,J)
C                       ELSE, THE MATRIX IS TAKEN AS BANDED AND STORED
C                       DIAGONAL-WISE AS
C                            AM(I-J+MUMAS+1,J) = M(I,J).
C
C     IMAS          GIVES INFORMATION ON THE MASS-MATRIX:
C                       IMAS=0:  M IS SUPPOSED TO BE THE IDENTITY
C                          MATRIX, MAS IS NEVER CALLED.
C                       IMAS=1:  MASS-MATRIX  IS SUPPLIED.
C
C     MLMAS          SWITCH FOR THE BANDED STRUCTURE OF THE MASS-MATRIX:
C                       MLMAS=N: THE FULL MATRIX CASE. THE LINEAR
C                          ALGEBRA IS DONE BY FULL-MATRIX GAUSS-ELIMINATION.
C                       0<=MLMAS<N: MLMAS IS THE LOWER BANDWITH OF THE
C                          MATRIX (>= NUMBER OF NON-ZERO DIAGONALS BELOW
C                          THE MAIN DIAGONAL).
C                    MLMAS IS SUPPOSED TO BE .LE. MLJAC.
C
C     MUMAS          UPPER BANDWITH OF MASS-MATRIX (>= NUMBER OF NON-
C                    ZERO DIAGONALS ABOVE THE MAIN DIAGONAL).
C                    NEED NOT BE DEFINED IF MLMAS=N.
C                    MUMAS IS SUPPOSED TO BE .LE. MUJAC.
C
C     SOLOUT         NAME (EXTERNAL) OF SUBROUTINE PROVIDING THE
C                    NUMERICAL SOLUTION DURING INTEGRATION.
C                    IF IOUT=1, IT IS CALLED AFTER EVERY SUCCESSFUL STEP.
C                    SUPPLY A DUMMY SUBROUTINE IF IOUT=0.
C                    IT MUST HAVE THE FORM
C                       SUBROUTINE SOLOUT (NR,XOLD,X,Y,CONT,LRC,N,
C                                          RPAR,IPAR,IRTRN)
C                       REAL*8 X,Y(N),CONT(LRC)
C                       ....
C                    SOLOUT FURNISHES THE SOLUTION "Y" AT THE NR-TH
C                       GRID-POINT "X" (THEREBY THE INITIAL VALUE IS
C                       THE FIRST GRID-POINT).
C                    "XOLD" IS THE PRECEEDING GRID-POINT.
C                    "IRTRN" SERVES TO INTERRUPT THE INTEGRATION. IF IRTRN
C                       IS SET <0, RADAU5 RETURNS TO THE CALLING PROGRAM.
C
C          -----     CONTINUOUS OUTPUT: -----
C                    DURING CALLS TO "SOLOUT", A CONTINUOUS SOLUTION
C                    FOR THE INTERVAL [XOLD,X] IS AVAILABLE THROUGH
C                    THE FUNCTION
C                           >>>   CONTR5(I,S,CONT,LRC)   <<<
C                    WHICH PROVIDES AN APPROXIMATION TO THE I-TH
C                    COMPONENT OF THE SOLUTION AT THE POINT S. THE VALUE
C                    S SHOULD LIE IN THE INTERVAL [XOLD,X].
C                    DO NOT CHANGE THE ENTRIES OF CONT(LRC), IF THE
C                    DENSE OUTPUT FUNCTION IS USED.
C
C     IOUT           SWITCH FOR CALLING THE SUBROUTINE SOLOUT:
C                       IOUT=0:  SUBROUTINE IS NEVER CALLED
C                       IOUT=1:  SUBROUTINE IS AVAILABLE FOR OUTPUT.
C
C     WORK           ARRAY OF WORKING SPACE OF LENGTH "LWORK".
```

```
C                 WORK(1), WORK(2),.., WORK(20) SERVE AS PARAMETERS
C                 FOR THE CODE. FOR STANDARD USE OF THE CODE
C                 WORK(1),..,WORK(20) MUST BE SET TO ZERO BEFORE
C                 CALLING. SEE BELOW FOR A MORE SOPHISTICATED USE.
C                 WORK(21),..,WORK(LWORK) SERVE AS WORKING SPACE
C                 FOR ALL VECTORS AND MATRICES.
C                 "LWORK" MUST BE AT LEAST
C                             N*(LJAC+LMAS+3*LE+12)+20
C                 WHERE
C                    LJAC=N              IF MLJAC=N (FULL JACOBIAN)
C                    LJAC=MLJAC+MUJAC+1  IF MLJAC<N (BANDED JAC.)
C                 AND
C                    LMAS=0              IF IMAS=0
C                    LMAS=N              IF IMAS=1 AND MLMAS=N (FULL)
C                    LMAS=MLMAS+MUMAS+1  IF MLMAS<N (BANDED MASS-M.)
C                 AND
C                    LE=N               IF MLJAC=N (FULL JACOBIAN)
C                    LE=2*MLJAC+MUJAC+1 IF MLJAC<N (BANDED JAC.)
C
C                 IN THE USUAL CASE WHERE THE JACOBIAN IS FULL AND THE
C                 MASS-MATRIX IS THE INDENTITY (IMAS=0), THE MINIMUM
C                 STORAGE REQUIREMENT IS
C                             LWORK = 4*N*N+12*N+20.
C                 IF IWORK(9)=M1>0 THEN "LWORK" MUST BE AT LEAST
C                          N*(LJAC+12)+(N-M1)*(LMAS+3*LE)+20
C                 WHERE IN THE DEFINITIONS OF LJAC, LMAS AND LE THE
C                 NUMBER N CAN BE REPLACED BY N-M1.
C
C     LWORK       DECLARED LENGHT OF ARRAY "WORK".
C
C     IWORK       INTEGER WORKING SPACE OF LENGHT "LIWORK".
C                 IWORK(1),IWORK(2),...,IWORK(20) SERVE AS PARAMETERS
C                 FOR THE CODE. FOR STANDARD USE, SET IWORK(1),..,
C                 IWORK(20) TO ZERO BEFORE CALLING.
C                 IWORK(21),...,IWORK(LIWORK) SERVE AS WORKING AREA.
C                 "LIWORK" MUST BE AT LEAST 3*N+20.
C
C     LIWORK      DECLARED LENGHT OF ARRAY "IWORK".
C
C     RPAR, IPAR  REAL AND INTEGER PARAMETERS (OR PARAMETER ARRAYS) WHICH
C                 CAN BE USED FOR COMMUNICATION BETWEEN YOUR CALLING
C                 PROGRAM AND THE FCN, JAC, MAS, SOLOUT SUBROUTINES.
C
C ----------------------------------------------------------------------
C
C     SOPHISTICATED SETTING OF PARAMETERS
C     -----------------------------------
C              SEVERAL PARAMETERS OF THE CODE ARE TUNED TO MAKE IT WORK
C              WELL. THEY MAY BE DEFINED BY SETTING WORK(1),...
C              AS WELL AS IWORK(1),...  DIFFERENT FROM ZERO.
C              FOR ZERO INPUT, THE CODE CHOOSES DEFAULT VALUES:
C
C    IWORK(1)  IF IWORK(1).NE.0, THE CODE TRANSFORMS THE JACOBIAN
C              MATRIX TO HESSENBERG FORM. THIS IS PARTICULARLY
C              ADVANTAGEOUS FOR LARGE SYSTEMS WITH FULL JACOBIAN.
C              IT DOES NOT WORK FOR BANDED JACOBIAN (MLJAC<N)
C              AND NOT FOR IMPLICIT SYSTEMS (IMAS=1).
C
C    IWORK(2)  THIS IS THE MAXIMAL NUMBER OF ALLOWED STEPS.
C              THE DEFAULT VALUE (FOR IWORK(2)=0) IS 100000.
C
C    IWORK(3)  THE MAXIMUM NUMBER OF NEWTON ITERATIONS FOR THE
```

```
C              SOLUTION OF THE IMPLICIT SYSTEM IN EACH STEP.
C              THE DEFAULT VALUE (FOR IWORK(3)=0) IS 7.
C
C    IWORK(4)  IF IWORK(4).EQ.0 THE EXTRAPOLATED COLLOCATION SOLUTION
C              IS TAKEN AS STARTING VALUE FOR NEWTON'S METHOD.
C              IF IWORK(4).NE.0 ZERO STARTING VALUES ARE USED.
C              THE LATTER IS RECOMMENDED IF NEWTON'S METHOD HAS
C              DIFFICULTIES WITH CONVERGENCE (THIS IS THE CASE WHEN
C              NSTEP IS LARGER THAN NACCPT + NREJCT; SEE OUTPUT PARAM.).
C              DEFAULT IS IWORK(4)=0.
C
C       THE FOLLOWING 3 PARAMETERS ARE IMPORTANT FOR
C       DIFFERENTIAL-ALGEBRAIC SYSTEMS OF INDEX > 1.
C       THE FUNCTION-SUBROUTINE SHOULD BE WRITTEN SUCH THAT
C       THE INDEX 1,2,3 VARIABLES APPEAR IN THIS ORDER.
C       IN ESTIMATING THE ERROR THE INDEX 2 VARIABLES ARE
C       MULTIPLIED BY H, THE INDEX 3 VARIABLES BY H**2.
C
C    IWORK(5)  DIMENSION OF THE INDEX 1 VARIABLES (MUST BE > 0).  FOR
C              ODE'S THIS EQUALS THE DIMENSION OF THE SYSTEM.
C              DEFAULT IWORK(5)=N.
C
C    IWORK(6)  DIMENSION OF THE INDEX 2 VARIABLES. DEFAULT IWORK(6)=0.
C
C    IWORK(7)  DIMENSION OF THE INDEX 3 VARIABLES. DEFAULT IWORK(7)=0.
C
C    IWORK(8)  SWITCH FOR STEP SIZE STRATEGY
C              IF IWORK(8).EQ.1  MOD. PREDICTIVE CONTROLLER (GUSTAFSSON)
C              IF IWORK(8).EQ.2  CLASSICAL STEP SIZE CONTROL
C              THE DEFAULT VALUE (FOR IWORK(8)=0) IS IWORK(8)=1.
C              THE CHOICE IWORK(8).EQ.1 SEEMS TO PRODUCE SAFER RESULTS;
C              FOR SIMPLE PROBLEMS, THE CHOICE IWORK(8).EQ.2 PRODUCES
C              OFTEN SLIGHTLY FASTER RUNS
C
C       IF THE DIFFERENTIAL SYSTEM HAS THE SPECIAL STRUCTURE THAT
C            Y(I)' = Y(I+M2)   FOR  I=1,...,M1,
C       WITH M1 A MULTIPLE OF M2, A SUBSTANTIAL GAIN IN COMPUTERTIME
C       CAN BE ACHIEVED BY SETTING THE PARAMETERS IWORK(9) AND IWORK(10).
C       E.G., FOR SECOND ORDER SYSTEMS P'=V, V'=G(P,V), WHERE P AND V ARE
C       VECTORS OF DIMENSION N/2, ONE HAS TO PUT M1=M2=N/2.
C       FOR M1>0 SOME OF THE INPUT PARAMETERS HAVE DIFFERENT MEANINGS:
C       - JAC: ONLY THE ELEMENTS OF THE NON-TRIVIAL PART OF THE
C              JACOBIAN HAVE TO BE STORED
C              IF (MLJAC.EQ.N-M1) THE JACOBIAN IS SUPPOSED TO BE FULL
C                 DFY(I,J) = PARTIAL F(I+M1) / PARTIAL Y(J)
C                FOR I=1,N-M1 AND J=1,N.
C              ELSE, THE JACOBIAN IS BANDED ( M1 = M2 * MM )
C                 DFY(I-J+MUJAC+1,J+K*M2) = PARTIAL F(I+M1) / PARTIAL Y(J+K*M2)
C                FOR I=1,MLJAC+MUJAC+1 AND J=1,M2 AND K=0,MM.
C       - MLJAC: MLJAC=N-M1:  IF THE NON-TRIVIAL PART OF THE JACOBIAN IS FULL
C                0<=MLJAC<N-M1:  IF THE (MM+1) SUBMATRICES (FOR K=0,MM)
C                     PARTIAL F(I+M1) / PARTIAL Y(J+K*M2),  I,J=1,M2
C                    ARE BANDED, MLJAC IS THE MAXIMAL LOWER BANDWIDTH
C                    OF THESE MM+1 SUBMATRICES
C       - MUJAC: MAXIMAL UPPER BANDWIDTH OF THESE MM+1 SUBMATRICES
C                NEED NOT BE DEFINED IF MLJAC=N-M1
C       - MAS: IF IMAS=0 THIS MATRIX IS ASSUMED TO BE THE IDENTITY AND
C              NEED NOT BE DEFINED. SUPPLY A DUMMY SUBROUTINE IN THIS CASE.
C              IT IS ASSUMED THAT ONLY THE ELEMENTS OF RIGHT LOWER BLOCK OF
C              DIMENSION N-M1 DIFFER FROM THAT OF THE IDENTITY MATRIX.
C              IF (MLMAS.EQ.N-M1) THIS SUBMATRIX IS SUPPOSED TO BE FULL
C                 AM(I,J) = M(I+M1,J+M1)     FOR I=1,N-M1 AND J=1,N-M1.
```

```
C              ELSE, THE MASS MATRIX IS BANDED
C                 AM(I-J+MUMAS+1,J) = M(I+M1,J+M1)
C        - MLMAS: MLMAS=N-M1:  IF THE NON-TRIVIAL PART OF M IS FULL
C                 0<=MLMAS<N-M1:  LOWER BANDWIDTH OF THE MASS MATRIX
C        - MUMAS: UPPER BANDWIDTH OF THE MASS MATRIX
C                 NEED NOT BE DEFINED IF MLMAS=N-M1
C
C    IWORK(9)  THE VALUE OF M1.    DEFAULT M1=0.
C
C    IWORK(10) THE VALUE OF M2.    DEFAULT M2=M1.
C
C ----------
C
C    WORK(1)   UROUND, THE ROUNDING UNIT, DEFAULT 1.D-16.
C
C    WORK(2)   THE SAFETY FACTOR IN STEP SIZE PREDICTION,
C              DEFAULT 0.9D0.
C
C    WORK(3)   DECIDES WHETHER THE JACOBIAN SHOULD BE RECOMPUTED;
C              INCREASE WORK(3), TO 0.1 SAY, WHEN JACOBIAN EVALUATIONS
C              ARE COSTLY. FOR SMALL SYSTEMS WORK(3) SHOULD BE SMALLER
C              (0.001D0, SAY). NEGATIV WORK(3) FORCES THE CODE TO
C              COMPUTE THE JACOBIAN AFTER EVERY ACCEPTED STEP.
C              DEFAULT 0.001D0.
C
C    WORK(4)   STOPPING CRITERION FOR NEWTON'S METHOD, USUALLY CHOSEN <1.
C              SMALLER VALUES OF WORK(4) MAKE THE CODE SLOWER, BUT SAFER.
C              DEFAULT 0.03D0.
C
C    WORK(5) AND WORK(6) :  IF WORK(5) < HNEW/HOLD < WORK(6), THEN THE
C              STEP SIZE IS NOT CHANGED. THIS SAVES, TOGETHER WITH A
C              LARGE WORK(3), LU-DECOMPOSITIONS AND COMPUTING TIME FOR
C              LARGE SYSTEMS. FOR SMALL SYSTEMS ONE MAY HAVE
C              WORK(5)=1.D0, WORK(6)=1.2D0, FOR LARGE FULL SYSTEMS
C              WORK(5)=0.99D0, WORK(6)=2.D0 MIGHT BE GOOD.
C              DEFAULTS WORK(5)=1.D0, WORK(6)=1.2D0 .
C
C    WORK(7)   MAXIMAL STEP SIZE, DEFAULT XEND-X.
C
C    WORK(8), WORK(9)   PARAMETERS FOR STEP SIZE SELECTION
C              THE NEW STEP SIZE IS CHOSEN SUBJECT TO THE RESTRICTION
C                 WORK(8) <= HNEW/HOLD <= WORK(9)
C              DEFAULT VALUES: WORK(8)=0.2D0, WORK(9)=8.D0
C
C-----------------------------------------------------------------------
C
C     OUTPUT PARAMETERS
C     -----------------
C     X           X-VALUE FOR WHICH THE SOLUTION HAS BEEN COMPUTED
C                 (AFTER SUCCESSFUL RETURN X=XEND).
C
C     Y(N)        NUMERICAL SOLUTION AT X
C
C     H           PREDICTED STEP SIZE OF THE LAST ACCEPTED STEP
C
C     IDID        REPORTS ON SUCCESSFULNESS UPON RETURN:
C                   IDID= 1  COMPUTATION SUCCESSFUL,
C                   IDID= 2  COMPUT. SUCCESSFUL (INTERRUPTED BY SOLOUT)
C                   IDID=-1  INPUT IS NOT CONSISTENT,
C                   IDID=-2  LARGER NMAX IS NEEDED,
C                   IDID=-3  STEP SIZE BECOMES TOO SMALL,
C                   IDID=-4  MATRIX IS REPEATEDLY SINGULAR.
```

```
C
C   IWORK(14)  NFCN     NUMBER OF FUNCTION EVALUATIONS (THOSE FOR NUMERICAL
C                       EVALUATION OF THE JACOBIAN ARE NOT COUNTED)
C   IWORK(15)  NJAC     NUMBER OF JACOBIAN EVALUATIONS (EITHER ANALYTICALLY
C                       OR NUMERICALLY)
C   IWORK(16)  NSTEP    NUMBER OF COMPUTED STEPS
C   IWORK(17)  NACCPT   NUMBER OF ACCEPTED STEPS
C   IWORK(18)  NREJCT   NUMBER OF REJECTED STEPS (DUE TO ERROR TEST),
C                       (STEP REJECTIONS IN THE FIRST STEP ARE NOT COUNTED)
C   IWORK(19)  NDEC     NUMBER OF LU-DECOMPOSITIONS OF BOTH MATRICES
C   IWORK(20)  NSOL     NUMBER OF FORWARD-BACKWARD SUBSTITUTIONS, OF BOTH
C                       SYSTEMS; THE NSTEP FORWARD-BACKWARD SUBSTITUTIONS,
C                       NEEDED FOR STEP SIZE SELECTION, ARE NOT COUNTED
C-----------------------------------------------------------------------
```

Subroutine RADAUP

With the option `IWORK(11) = 3` this code is mathematically equivalent to RADAU5. The only difference is that explicit sums have been replaced by loops, and that the coefficients of the method have been put into arrays. This makes the code a little bit slower (in particular for small problems), but has the advantage that the coefficients of the method can be easily changed. At the moment, the coefficients of the Radau IIA methods of orders 5, 9, and 13 are available by setting `IWORK(11)` equal to 3, 5, and 7, respectively. The calling list is the same as for RADAU5.

```
      SUBROUTINE RADAUP(N,FCN,X,Y,XEND,H,
     +                  RTOL,ATOL,ITOL,
     +                  JAC ,IJAC,MLJAC,MUJAC,
     +                  MAS ,IMAS,MLMAS,MUMAS,
     +                  SOLOUT,IOUT,
     +                  WORK,LWORK,IWORK,LIWORK,RPAR,IPAR,IDID)
```

Subroutine RODAS

This is an implementation of the Rosenbrock method described in Section VI.3. It also satisfies the algebraic order conditions and can thus be applied to differential-algebraic problems of index 1. The calling list is:

```
      SUBROUTINE RODAS(N,FCN,IFCN,X,Y,XEND,H,
     +                  RTOL,ATOL,ITOL,
     +                  JAC ,IJAC,MLJAC,MUJAC,DFX,IDFX,
     +                  MAS ,IMAS,MLMAS,MUMAS,
     +                  SOLOUT,IOUT,
     +                  WORK,LWORK,IWORK,LIWORK,RPAR,IPAR,IDID)
```

Compared to RADAU5 we have three additional parameters. `IFCN` indicates whether the right-hand side $f(x,y)$ of the problem (A.1) is independent of x or not. In the case that f depends on x, the code needs the partial drivative $\partial f/\partial x$. This can be provided numerically (set `IDFX = 0` and supply a dummy subroutine for `DFX`) or analytically. In the latter case, one has to set `IDFX = 1` and one has to supply a subroutine computing $\partial f/\partial x$. Of course, the meaning of the WORK and IWORK parameters are not all the same as for RADAU5. They are decscibed in the comments of the code.

Subroutine SEULEX

This is an extrapolation code based on the linearly implicit Euler method (Sections IV.9 and VI.4). A dense output has been included in cooperation with A. Ostermann. The meaning of the input parameters is the same as for RODAS. The WORK and IWORK parameters are decscibed in the comments of the code.

```
      SUBROUTINE SEULEX(N,FCN,IFCN,X,Y,XEND,H,
     +                  RTOL,ATOL,ITOL,
     +                  JAC ,IJAC,MLJAC,MUJAC,
     +                  MAS ,IMAS,MLMAS,MUMAS,
     +                  SOLOUT,IOUT,
     +                  WORK,LWORK,IWORK,LIWORK,RPAR,IPAR,IDID)
```

Problems with Special Structure

If the first m_1 equations of (A.1) are of the form

$$y_i' = y_{i+m_2} \qquad \text{for} \quad i = 1, \ldots, m_1 \tag{A.2}$$

with m_1 being an integer multiple of m_2, and the remaining equations do not depend explicitly on $y_{m_1+1}', \ldots, y_n'$, it is recommended to set the parameters IWORK(9) and IWORK(10) equal to m_1 and m_2, respectively. This implies a more efficient treatment of the arising linear systems and is, in particular, advantageous for a large value of m_1.

If IWORK(9) is set to a nonzero value, care has to be taken with the definition of the subroutines JAC and MAS. Only the nontrivial part of the Jacobian (i.e., the rows with indices $m_1 + 1, \ldots, n$) have to be computed and stored in an array of dimension $(n - m_1) \times n$. Similarly, only the right lower block (of dimension $n - m_1$) of the matrix M has to be defined in the subroutine MAS. However, the subroutine FCN must contain the definition of all components of $f(x, y)$, in particular also the statement F(I) = Y(I+M2) for I=1,...,M1. Banded options are still possible. Typical situations, where (A.2) arises, are the following:

$\boldsymbol{y'' = f(x, y, y')}$. With the new variable $z = y'$ the system becomes

$$\begin{aligned} y' &= z \\ z' &= f(x, y, z), \end{aligned}$$

which is of the form (A.1). If $y \in \mathbb{R}^m$, both parameters IWORK(9) and IWORK(10) have to be set equal to m. Banded option can be used, if both $\partial f/\partial y$ and $\partial f/\partial y'$ are banded.

$C(x,y)y' = f(x,y)$. Again we introduce $z = y'$, so that this problem becomes equivalent to

$$\begin{aligned} y' &= z \\ 0 &= C(x,y)z - f(x,y). \end{aligned}$$

Both parameters `IWORK(9)` and `IWORK(10)` have to be set equal to the dimension of y. If only a few components of y' are multiplied by non-constant terms, then it may be more efficient to introduce new variables only for these components.

$C(x,y)y'' = f(x,y,y')$. With the new variables $z = y'$ and $u = z' = y''$, this problem can be written in the form (A.1) as follows

$$\begin{aligned} y' &= z \\ z' &= u \\ 0 &= C(x,y)u - f(x,y,z). \end{aligned}$$

Here m_2 is equal to the dimension of y, and $m_1 = 2m_2$.

Use of SOLOUT and of Dense Output

The subroutine SOLOUT, supplied by the user, is called after every accepted step and provides the solution over the whole step (dense output). This possibility can be used for tabulating the solution at prescribed output points (see the driver for RADAU5 above) or for graphical presentation of the solution. Further applications are the following:

Event location. Suppose we want to determine x such that $g\big(x, y(x)\big) = 0$, where $y(x)$ is the solution of (A.1). During integration one can check in the subroutine SOLOUT whether the values $g(x_{i-1}, y_{i-1})$ and $g(x_i, y_i)$ change sign. If this occurs, the dense output (which is available for all of our codes) can be used to localize the zero of $g\big(x, y(x)\big)$. This procedure is very useful for problems with discontinuous right-hand side (see Sect. II.6).

Projection. An efficient way for solving higher index differential-algebraic equations is index-reduction combinded with projection. If one applies a stiff (or non-stiff) code straightforwardly to an index-reduced problem, the obtained numerical solution will suffer from the so-called "drift-off" effect. In order to avoid this drift-off, it is recommended to project the numerical solution after every step onto the solution manifold of the problem. This can be conveniently done with help of the subroutine SOLOUT.

Bibliography

This bibliography includes the publications referred to in the text. Italic numbers in square brackets following a reference indicate the sections where the reference is cited.

R. Abraham, J.E. Marsden & T. Ratiu (1983): *Manifolds, Tensor Analysis, and Applications*. Applied Mathematical Series vol. 75, Springer-Verlag 1983; second edition 1988, 654 pp. *[VII.1]*

M. Abramowitz & I.A. Stegun (1964): *Handbook of mathematical functions*. Dover, 1000 pages. *[IV.2], [IV.4], [IV.12], [IV.13], [IV.14]*

C.A. Addison (1979): *Implementing a stiff method based upon the second derivative formulas*. Techn. Rep. 130/79, Dept. of Comput. Sc., Univ. of Toronto, Canada. *[V.3], [V.5]*

R.C. Aiken ed. (1985): *Stiff computation*. Oxford, Univ. Press, 462pp. *[IV.1], [IV.3], [IV.8], [V.5]*

G. Akilov, see L. Kantorovich & G. Akilov.

R. Alexander (1977): *Diagonally implicit Runge-Kutta methods for stiff O.D.E.'s*. SIAM J. Numer. Anal., vol. 14, pp. 1006-1021. *[IV.3], [IV.6]*

R. Alexander (1997): *Reliability of software for stiff initial value problems*. To appear in SIAM J. Sci. Comput. *[IV.10]*

T. Alishenas (1992): *Zur numerischen Behandlung, Stabilisierung durch Projektion und Modellierung mechanischer Systeme mit Nebenbedingungen und Invarianten*. Dissertation, Stockholm, TRITA-NA-9202. *[VII.2]*

T. Alishenas & Ö. Ólafsson (1994): *Modeling and velocity stabilization of constrained mechanical systems*. BIT, vol. 34, pp. 455-483. *[VII.2]*

R. Alt (1971): *Méthodes A-stables pour l'intégration de systèmes différentielles mal conditionnés*. Thèse, Univ. Paris VI. *[IV.6]*

H.C. Andersen (1983): *Rattle: a "velocity" version of the Shake algorithm for molecular dynamics calculations*. J. Comput. Phys., vol. 52, pp. 24-34. *[VII.8]*

G.C. Andrews, see also M.K. Ormrod & G.C. Andrews.

C. Arévalo, C. Führer & G. Söderlind (1996): *Stabilized multistep methods for index 2 Euler-Lagrange DAEs*. BIT, vol. 36, pp. 1-13. *[VII.6]*

S. Arimoto, see J. Nagumo, S. Arimoto & S. Yoshizawa.

M. Arnold (1993): *Stability of numerical methods for differential-algebraic equations of higher index*. Applied Numerical Mathematics, vol. 13, pp. 5-14. *[VII.1]*

M. Arnold (1995): *Half-explicit Runge-Kutta methods with explicit stages for differential-algebraic systems of index 2.* Submitted for publication. *[VII.6], [VII.7]*

V.I. Arnol'd (1979): *Matematičeskie metody klassičeskoi mechaniki.* Nauka, Moskva; English translation: Springer Verlag 1984, 1989. *[VII.1]*

W.E. Arnoldi (1951): *The principle of minimized iterations in the solution of the matrix eigenvalue problem.* Quart. Appl. Math., vol. 9, pp. 17-29. *[IV.10]*

U. Ascher (1989): *On numerical differential algebraic problems with application to semi-conductor device simulation.* SIAM J. Numer. Anal., vol. 26, pp. 517-538. *[VII.4]*

U. Ascher & G. Bader (1986): *Stability of collocation at Gaussian points.* SIAM J. Numer. Anal., vol. 23, pp. 412-422. *[IV.13]*

U.M. Ascher, H. Chin & S. Reich (1994): *Stabilization of DAEs and invariant manifolds.* Numer. Math., vol. 67, pp. 131-149. *[VII.2]*

U. Ascher & L.R. Petzold (1991): *Projected implicit Runge-Kutta methods for differential-algebraic equations.* SIAM J. Numer. Anal., vol. 28, pp. 1097-1120. *[VII.4]*

M. Athans & P.L. Falb (1966): *Optimal Control.* McGraw-Hill Book Company, New York, 879pp. *[VII.1]*

W. Auzinger, R. Frank, & F. Macsek (1990): *Asymptotic error expansions for stiff equations: the implicit Euler scheme.* SIAM J. Numer. Anal., vol. 27, pp. 67-104. *[VI.5]*

O. Axelsson (1969): *A class of A-stable methods.* BIT, vol. 9, pp. 185-199. *[IV.3], [IV.5]*

O. Axelsson (1972): *A note on a class of strongly A-stable methods.* BIT, vol. 12, pp. 1-4. *[IV.5]*

G. Bader & P. Deuflhard (1983): *A semi-implicit mid-point rule for stiff systems of ordinary differential equations.* Numer. Math., vol. 41, pp. 373-398. *[IV.9], [IV.10], [VII.6]*

G. Bader, see also U. Ascher & G. Bader; E. Hairer, G. Bader & Ch. Lubich.

C. Baiocchi & M. Crouzeix (1989): *On the equivalence of A-stability and G-stability.* Appl. Numer. Math., vol.5, pp. 19-22. *[V.6]*

M. Bakker (1971): *Analytical aspects of a minimax problem* (Dutch), Technical Note TN 62, Mathematical Centre, Amsterdam.

L.A. Bales, O.A. Karakashian & S.M. Serbin (1988): *On the A_0-acceptability of rational approximations to the exponential function with only real poles.* BIT, vol. 28, pp. 70-79. *[IV.4]*

G.P. Barker, A. Berman & R.J. Plemmons (1978): *Positive diagonal solutions to the Lyapunov equations.* Linear and Multilinear Algebra, vol. 5, pp. 249-256. *[IV.14]*

J. Baumgarte (1972): *Stabilization of constraints and integrals of motion in dynamical systems.* Comp. Meth. Appl. Mech. Eng., vol. 1, pp. 11-16. *[VII.2]*

G. Benettin & A. Giorgilli (1994): *On the Hamiltonian interpolation of near to the identity symplectic mappings with application to symplectic integration algorithms.* J. Statist. Phys., vol. 74, pp. 1117-1143. *[VII.8]*

H.J.C. Berendsen, see also J.-P. Ryckaert, G. Ciccotti & H.J.C. Berendsen.

A. Berman, see G.P. Barker, A. Berman & R.J. Plemmons.

S. Bernstein (1914): *Sur la définition et les propriétés des fonctions analytiques d'une variable réelle.* Math. Annalen, vol. 75, pp. 449-468. *[IV.11]*

S. Bernstein (1928): *Sur les fonctions absolument monotones.* Acta Mathematica, vol. 51, pp. 1-66. *[IV.11]*

M. Berzins & R.M. Furzeland (1985): *A user's manual for SPRINT – a versatile software package for solving systems of algebraic, ordinary and partial differential equations: part 1 – algebraic and ordinary differential equations.* Thornton Research Centre, Shell Research Ltd. TNER.85.058. *[V.5], [VII.3]*

T.A. Bickart (1977): *An efficient solution process for implicit Runge-Kutta methods.* SIAM J. Numer. Anal., vol. 14, 1022-1027. *[IV.8]*

T.A. Bickart & W.B. Rubin (1974): *Composite multistep methods and stiff stability.* In: Stiff Differential Systems, R.A. Willoughby (ed.), Plenum Press, New York. *[V.3]*

T.A. Bickart, see also H.M. Sloate & T.A. Bickart.

G. Birkhoff & R.S. Varga (1965): *Discretization errors for well-set Cauchy problems, I.* J. Math. Phys., vol. 44, pp. 1-23. *[IV.5]*

Å. Björck (1983): *A block QR algorithm for partitioning stiff differential systems.* BIT, vol. 23, pp. 329-345. *[IV.10]*

Å. Björck (1984): *Some methods for separating stiff components in initial value problems.* In: Numerical Analysis, Dundee 1983, D.F. Griffiths, ed., Lecture Notes in Math. 1066, Springer Verlag, pp. 30-43. *[IV.10]*

C. Bolley & M. Crouzeix (1978): *Conservation de la positivité lors de la discrétisation des problèmes d'évolution paraboliques.* R.A.I.R.O. Analyse numérique, vol. 12, pp. 237-245. *[IV.11]*

V.G. Boltyanskii, see L.S. Pontryagin, V.G. Boltyanskii, R.V. Gamkrelidze & E.F. Mishchenko.

V. Brasey (1992): *A half-explicit Runge-Kutta method of order 5 for solving constrained mechanical systems.* Computing, vol. 48, pp. 191-201. *[VII.6]*

V. Brasey (1994): *Half-explicit method for semi-explicit differential-algebraic equations of index 2.* Thèse N° 2664, Sect. Math., Univ. de Genève. *[VII.7]*

V. Brasey & E. Hairer (1993): *Half-explicit Runge-Kutta methods for differential-algebraic systems of index 2.* SIAM J. Numer. Anal., vol. 30, pp. 538-552. *[VII.6]*

K.E. Brenan (1983): *Stability and convergence of difference approximations for higher-index differential-algebraic systems with applications in trajectory control.* Doctoral thesis, Dep. Math., Univ. of California, Los Angeles. *[VII.1]*

K.E. Brenan, S.L. Campbell & L.R. Petzold (1989): *Numerical solution of initial-value problems in differential-algebraic equations.* North Holland, New York, 210pp. *[VII.1], [VII.3], [VII.7]*

K.E. Brenan & L.R. Engquist (1988): *Backward differentiation approximations of nonlinear differential/algebraic equations*, and Supplement. Math. Comp., vol. 51, pp. 659-676, pp. S7-S16. *[VII.3]*

K.E. Brenan & L.R. Petzold (1989): *The numerical solution of higher index differential-/algebraic equations by implicit Runge-Kutta methods.* SIAM J. Numer. Anal., vol. 26, pp. 976-996. *[VII.4]*

P.N. Brown, G.D. Byrne & A.C. Hindmarsh (1989): *VODE: a variable coefficient ODE solver.* SIAM J. Sci. Stat. Comput., vol. 10, pp. 1039-1051. *[V.5]*

T.D. Bui, see P. Kaps, S.W.H. Poon & T.D. Bui.

J.M. Burgers (1948): *A mathematical model illustrating the theory of turbulence.* Advances in appl. mech., vol. 1, pp. 171-199. *[V.8], [VI.6]*

K. Burrage (1978): *High order algebraically stable Runge-Kutta methods.* BIT, vol. 18, pp. 373-383. *[IV.5], [IV.13]*

K. Burrage (1978): *A special family of Runge-Kutta methods for solving stiff differential equations.* BIT, vol. 18, pp. 22-41. *[IV.5], [IV.6], [IV.8]*

K. Burrage (1982): *Efficiently implementable algebraically stable Runge-Kutta methods.* SIAM J. Numer. Anal., vol. 19, pp. 245-258. *[IV.13]*

K. Burrage (1987): *High order algebraically stable multistep Runge-Kutta methods.* SIAM J. Numer. Anal., vol. 24, pp. 106-115. *[V.9]*

K. Burrage (1988): *Order properties of implicit multivalue methods for ordinary differential equations.* IMA J. Numer. Anal., vol. 8, pp. 43-69. *[V.9]*

K. Burrage & J.C. Butcher (1979): *Stability criteria for implicit Runge-Kutta methods.* SIAM J. Numer. Anal., vol. 16, pp. 46-57. *[IV.12]*

K. Burrage & J.C. Butcher (1980): *Non-linear stability of a general class of differential equation methods.* BIT, vol. 20, pp. 185-203. *[IV.12], [V.9]*

K. Burrage, J.C. Butcher & F.H. Chipman (1980): *An implementation of singly-implicit Runge-Kutta methods.* BIT, vol. 20, pp. 326-340. *[IV.8]*

K. Burrage & W.H. Hundsdorfer (1987): *The order of B-convergence of algebraically stable Runge-Kutta methods.* BIT, vol. 27, pp. 62-71. *[IV.15]*

J.C. Butcher (1964): *Implicit Runge-Kutta processes.* Math. Comput., vol. 18, pp. 50-64. *[IV.5]*

J.C. Butcher (1964): *Integration processes based on Radau quadrature formulas.* Math. Comput., vol. 18, pp. 233-244. *[IV.5]*

J.C. Butcher (1975): *A stability property of implicit Runge-Kutta methods.* BIT, vol. 15, pp. 358-361. *[IV.12]*

J.C. Butcher (1976): *On the implementation of implicit Runge-Kutta methods.* BIT, vol. 6, pp. 237-240. *[IV.8]*

J.C. Butcher (1977): *On A-stable implicit Runge-Kutta methods.* BIT, vol. 17, pp. 375-378. *[IV.5]*

J.C. Butcher (1979): *A transformed implicit Runge-Kutta method.* J. Assoc. Comput. Mach., vol. 26, pp. 731-738. *[IV.8]*

J.C. Butcher (1981): *A generalization of singly-implicit methods.* BIT, vol. 21, pp. 175-189. *[V.3]*

J.C. Butcher (1982): *A short proof concerning B-stability.* BIT, vol. 22, pp. 528-529. *[IV.12]*

J.C. Butcher (1987): *Linear and non-linear stability for general linear methods.* BIT, vol. 27, pp. 182-189. *[V.9]*

J.C. Butcher (1987): *The equivalence of algebraic stability and AN-stability.* BIT, vol. 27, pp. 510-533. *[V.9]*

J.C. Butcher (1987): *The numerical analysis of ordinary differential equations. Runge-Kutta and general linear methods.* John Wiley & Sons, 512pp. *[IV.12]*

J.C. Butcher (1990): *Order, stepsize and stiffness switching.* Computing. vol 44, p. 209-220. *[IV.2]*

J.C. Butcher, see also K. Burrage & J.C. Butcher; K. Burrage, J.C. Butcher & F.H. Chipman.

G.D. Byrne & A.C. Hindmarsh (1975): *A polyalgorithm for the numerical solution of ordinary differential equations*. ACM Trans. Math. Software, vol. 1, pp. 71-96. *[V.5]*

G.D. Byrne & A.C. Hindmarsh (1987): *Stiff ODE solvers: a review of current and coming attractions*. J. of Comput. Physics, vol. 70, pp. 1-62. *[IV.10]*

G.D. Byrne, see also P.N. Brown, G.D. Byrne & A.C. Hindmarsh.

D.A. Calahan (1968): *A stable, accurate method of numerical integration for nonlinear systems*. Proc. IEEE, vol. 56, p. 744. *[IV.7]*

A. Callender, D.R. Hartree & A. Porter (1936): *Time-lag in a control system*. Phil. Trans. of the Royal Society (London), Series A, vol. 235, pp. 415-444. *[IV.2]*

M.P. Calvo, see also J.M. Sanz-Serna & M.P. Calvo.

S.L. Campbell (1982): *Singular Systems of Differential Equations II*. Pitman, London. *[VII.1]*

S.L. Campbell (1989): *A computational method for general higher index singular systems of differential equations*. IMACS Transactions Scientific Computing, vol. 1.2, pp. 555-560. *[VII.2]*

S.L. Campbell (1993): *Least squares completions for nonlinear differential algebraic equations*. Numer. Math., vol. 65, pp. 77-94. *[VII.2]*

S..L Campbell (1995): *High index differential algebraic equations*. J. Mech. Struct. & Machines, vol. 23, pp. 199-222. *[VII.1]*

S.L. Campbell & C.W. Gear (1995): *The index of general nonlinear DAEs*. Numer. Math., vol. 72, pp. 173-196. *[VII.1]*

S.L. Campbell & E. Moore (1995): *Constraint preserving integrators for general nonlinear higher index DAEs*. Numer. Math., vol. 69, pp. 383-399. *[VII.2]*

S.L. Campbell, see also K.E. Brenan, S.L. Campbell & L.R. Petzold.

J. Carr, D.B. Duncan & C.H. Walshaw (1995): *Numerical approximation of a metastable system*. IMA J. Numer. Anal., vol. 15, pp. 505-521. *[IV.10]*

J.R. Cash (1976): *Semi-implicit Runge-Kutta procedures with error estimates for the numerical integration of stiff systems of ordinary differential equations*. JACM, vol. 23, pp. 455-460. *[IV.7]*

J.R. Cash (1979): *Diagonally implicit Runge-Kutta formulae with error estimates*. J. Inst. Math. Applics, vol. 24, pp. 293-301. *[IV.6]*

J.R. Cash (1979): *Stable recursions, with applications to the numerical solution of stiff systems*. Academic Press, 223 pp. *[V.2]*

J.R. Cash (1980): *On the integration of stiff systems of O.D.E.s using extended backward differentiation formulae*. Numer. Math., vol. 34, pp. 235-246. *[V.3]*

J.R. Cash (1981): *Second derivative extended backward differentiation formulas for the numerical integration of of stiff systems*. SIAM J. Numer. Anal. vol. 18, pp. 21-36. *[V.3]*

J.R. Cash (1983): *The integration of stiff initial value problems in ODEs using modified extended backward differentiation formulas*. Comp. & Maths. with Appls., vol. 9, No. 5, pp. 645-657. *[V.3], [V.5]*

J.R. Cash & S. Considine (1992): *An* MEBDF *code for stiff initial value problems*. ACM Tans. Math. Software, vol. 18, No. 2, pp. 142-158. *[V.5]*

P.E. Chase (1962): *Stability properties of Predictor-Corrector methods for ordinary differential equations,* J. Assoc. Comput. Mach., vol. 9, pp.457-468. *[V.1]*

P.L. Chebyshev (Tchébychef) (1854): *Théorie des mécanismes connus sous le nom de parallélogrammes.* Mém. de l'Acad. Imp. St.-Pétersbourg, tome VII (1854), pp.539-568; Oeuvres Tome I, pp.111-143. *[IV.2]*

H. Chin, see also U.M. Ascher, H. Chin & S. Reich.

F.H. Chipman (1971): *A-stable Runge-Kutta processes.* BIT, vol. 11, pp. 384-388. *[IV.5]*

F.H. Chipman (1976): *A note on implicit A-stable RK methods with parameters.* BIT, vol. 16, pp. 223-227. *[IV.5]*

F.H. Chipman, see also K. Burrage, J.C. Butcher & F.H. Chipman.

G. Ciccotti, see also J.-P. Ryckaert, G. Ciccotti & H.J.C. Berendsen.

K. Clark (1988): *A structural form for higher index semistate equations I: Theory and applications to circuit and control theory.* Linear Alg. Appl., vol. 98, pp. 169-197. *[VII.1]*

L. Collatz (1950): *Numerische Behandlung von Differentialgleichungen.* Grundlehren, Springer Verlag, Band LX (later editions and translations). *[IV.10], [IV.15]*

P. Collet, J.-P. Eckmann, H. Epstein & J. Stubbe (1993): *Analyticity for the Kuramoto-Sivashinsky equation.* Physica D, vol. 67, pp. 321-326. *[IV.10]*

S. Considine, see also J.R. Cash & S. Considine.

G.J. Cooper (1985): *Reducible Runge-Kutta methods.* BIT, vol. 25, pp. 675-680. *[IV.12]*

G.J. Cooper (1986): *On the existence of solutions for algebraically stable Runge-Kutta methods.* IMA J. Numer. Anal., vol. 6, pp. 325-330. *[IV.14]*

G.J. Cooper & A. Sayfy (1979): *Semiexplicit A-stable Runge-Kutta methods.* Math. of Comp., vol. 33, pp. 541-556. *[IV.6]*

G.J. Cooper & A. Sayfy (1983): *Additive Runge-Kutta methods for stiff ordinary differential equations.* Math. of Comp., vol. 40, pp. 207-218. *[IV.7]*

R. Courant, K. Friedrichs & H. Lewy (1928): *Ueber die partiellen Differenzengleichungen der mathematischen Physik.* Math. Ann., vol. 100, pp. 32-74. *[IV.2]*

R. Courant, see A. Hurwitz & R. Courant.

G. Cramer (1750): *Introduction à l'analyse des lignes courbes algébriques.* Genève, 1750. *[IV.3]*

R.L. Crane & R.W. Klopfenstein (1965): *A predictor-corrector algorithm with an increased range of absolute stability.* J. ACM, vol. 12, pp.227-241. . *[V.1]*

M. Crouzeix (1975): *Sur l'approximation des équations différentielles opérationnelles linéaires par de méthodes de Runge-Kutta.* Thèse, Univ. Paris VI. *[IV.6]*

M. Crouzeix (1979): *Sur la B-stabilité des méthodes de Runge-Kutta.* Numer. Math., vol. 32, pp. 75-82. *[IV.12]*

M. Crouzeix, W.H. Hundsdorfer & M.N. Spijker (1983): *On the existence of solutions to the algebraic equations in implicit Runge-Kutta methods.* BIT, vol. 23, pp. 84-91. *[IV.14]*

M. Crouzeix & P.A. Raviart (1976): *Approximation des équations d'évolution linéaires par des méthodes à pas multiples.* C. R. Acad. Sc. Paris, Ser. A 283, pp. 367-370. *[V.7]*

M. Crouzeix & P.A. Raviart (1980): *Approximation des problèmes d'évolution.* Unpublished Lecture Notes, Université de Rennes. *[IV.6], [IV.14], [V.7]*

M. Crouzeix & F. Ruamps (1977): *On rational approximations to the exponential.* R.A.I.R.O. Analyse Numérique, vol. 11, pp. 241-243. *[IV.4]*

M. Crouzeix, see also C. Baiocchi & M. Crouzeix; C. Bolley & M. Crouzeix.

C.W. Cryer (1973): *A new class of highly stable methods. A_0-stable methods.* BIT, vol. 13, pp.153-159. *[V.2]*

A.R. Curtis (1983): *Jacobian matrix properties and their impact on choice of software for stiff ODE systems.* IMA J. Numer. Anal., vol. 3, pp. 397-415. *[IV.10]*

C.F. Curtiss & J.O. Hirschfelder (1952): *Integration of stiff equations.* Proc. Nat. Acad. Sci., vol. 38, pp.235-243. *[IV.1]*

G. Dahlquist (1951): *Fehlerabschätzungen bei Differenzenmethoden zur numerischen Integration gewöhnlicher Differentialgleichungen.* ZAMM, vol. 31, pp. 239-240. *[V.1]*

G. Dahlquist (1956): *Convergence and stability in the numerical integration of ordinary differential equations.* Math. Scand., vol. 4, pp. 33-53. *[V.7]*

G. Dahlquist (1963): *A special stability problem for linear multistep methods.* BIT, vol. 3, pp. 27-43. *[IV.3], [IV.9], [IV.12], [V.1], [V.6]*

G. Dahlquist (1975): *Error analysis for a class of methods for stiff nonlinear initial value problems.* Numerical Analysis, Dundee 1975, Lecture Notes in Math., No. 506, pp. 60-74. *[IV.12], [V.6]*

G. Dahlquist (1978): *G-stability is equivalent to A-stability.* BIT, vol. 18, pp. 384-401. *[IV.13], [V.6]*

G. Dahlquist (1978): *Positive functions and some applications to stability questions for numerical methods.* In: Recent Advances in Numerical Analysis, C. de Boor & G.H. Golub (eds.), Academic Press, New York, pp. 1-19. *[IV.5]*

G. Dahlquist (1983): *On one-leg multistep methods.* SIAM J. Numer. Anal., vol. 20, pp. 1130-1138. *[V.6], [V.7], [V.9]*

G. Dahlquist & R. Jeltsch (1979): *Generalized disks of contractivity for explicit and implicit Runge-Kutta methods.* TRITA-NA Report 7906. *[IV.12], [IV.13]*

G. Dahlquist & R. Jeltsch (1987): *Reducibility and contractivity of Runge-Kutta methods revisited.* Report Nr. 46, Inst. f. Geometrie u. Prakt. Math., RWTH Aachen. *[IV.12]*

G. Dahlquist, H. Mingyou & R. LeVeque (1983): *On the uniform power-boundedness of a family of matrices and the applications to one-leg and linear multistep methods.* Numer. Math., vol. 42, pp. 1-13. *[V.7]*

G. Dahlquist & G. Söderlind (1982): *Some problems related to stiff nonlinear differential systems.* In: Computing Methods in Applied Sciences and Engineering, V.R. Glowinski & J.L. Lions (eds.), North-Holland, INRIA *[V.7]*

G. Dahlquist, see also G. Söderlind & G. Dahlquist.

J.W. Daniel & R.E. Moore (1970); *Computation and theory in ordinary differential equations,* W.H. Freeman and Company, 172 pp. *[V.4]*

P.J. Davis (1963): *Interpolation and approximation.* Blaisdell 1963; Dover 1975. *[V.3]*

K. Dekker (1981): *Stability of linear multistep methods on the imaginary axis.* BIT, vol. 21, pp. 66-79. *[V.4]*

K. Dekker (1982): *On the iteration error in algebraically stable Runge-Kutta methods.* Report NW 138/82, Math. Centrum, Amsterdam. *[IV.14]*

K. Dekker (1984): *Error bounds for the solution to the algebraic equations in Runge-Kutta methods*. BIT, vol. 24, pp. 347-356. *[IV.14]*

K. Dekker & E. Hairer (1985): *A necessary condition for BSI-stability*. BIT, vol. 25, pp. 285-288. *[IV.14]*

K.Dekker, J.F.B.M. Kraaijevanger & J. Schneid (1990): *On the relation between algebraic stability and B-convergence for Runge-Kutta methods*. Numer. Math., vol. 57, pp.249-262. *[IV.15]*

K. Dekker & J.G. Verwer (1984): *Stability of Runge-Kutta methods for stiff nonlinear differential equations*. North-Holland, Amsterdam-New-York-Oxford. *[IV.12], [IV.14], [IV.15]*

K. Dekker, see also M.Z. Liu, K. Dekker & M.N. Spijker.

P. Deuflhard (1983): *Order and stepsize control in extrapolation methods*. Numer. Math., vol. 41, pp. 399-422. *[IV.9]*

P. Deuflhard (1985): *Recent progress in extrapolation methods for ordinary differential equations*. SIAM Review, vol. 27, pp. 505-535. *[IV.9]*

P. Deuflhard, E. Hairer & J. Zugck (1987): *One-step and extrapolation methods for differential-algebraic systems*. Numer. Math., vol. 51, pp. 501-516. *[VI.5]*

P. Deuflhard & U. Nowak (1987): *Extrapolation integrators for quasilinear implicit ODEs*. In P. Deuflhard & B. Engquist (eds.), Large-Scale Scientific Computing. Birkhäuser, Boston. *[VI.5], [VI.6]*

P. Deuflhard, see also G. Bader & P. Deuflhard.

G.A. Di Marzo (1992): *RODAS5(4), méthodes de Rosenbrock d'ordre 5(4) adaptées aux problèmes différentiels-algébriques*. Mémoire de diplôme en Mathématiques, Université de Genève 1992. *[IV.10], [VI.4]*

J.R. Dormand & P.J. Prince (1980): *A family of embedded Runge-Kutta formulae*. J. Comp. Appl. Math., vol. 6, pp. 19-26. *[IV.2]*

A.A. Dorodnicyn (1947): *Asymptotic solution of the van der Pol equation*. Prikl. Mat. i Meh., vol. 11, pp. 313-328; Translations AMS, Ser. 1, vol. 4, pp. 1-23. *[VI.1]*

B.L. Ehle (1968) :*High order A-stable methods for the numerical solution of systems of DEs*. BIT, vol. 8, pp. 276-278. *[IV.3], [IV.4], [IV.5]*

B.L. Ehle (1969) :*On Padé approximations to the exponential function and A-stable methods for the numerical solution of initial value problems*. Research Report CSRR 2010, Dept. AACS, Univ. of Waterloo, Ontario, Canada. *[IV.3], [IV.5]*

B.L. Ehle (1973): *A-stable methods and Padé approximations to the exponential*. SIAM J. Math. Anal., vol. 4, pp. 671-680. *[IV.4], [IV.5]*

B.L. Ehle & Z. Picel (1975): *Two-parameter, arbitrary order, exponential approximations for stiff equations*. Math. Comput., vol. 29, pp.501-511. *[IV.5]*

E. Eich (1993): *Convergence results for a coordinate projection method applied to mechanical systems with algebraic constraints*. SIAM J. Numer. Anal., vol. 30, pp. 1467-1482. *[VII.2]*

R. England (1982): *Some hybrid implicit stiffly stable methods for ordinary differential equations*. In: Numerical Analysis, Proc. Mexico, (ed. J.P. Hennart), Lecture Notes in Math., No. 909, Springer Verlag, pp. 147-158. *[V.3]*

L.R. Engquist, see K.E. Brenan & L.R. Engquist.

W.H. Enright (1974): *Optimal second derivative methods for stiff systems.* In: Stiff Differential Systems, ed. by R.A. Willoughby, Plenum Press, New York. *[V.3]*

W.H. Enright (1974): *Second derivative multistep methods for stiff ordinary differential equations,* SIAM J. Numer. Anal., vol. 11, pp. 321-331. *[V.3]*

W.H. Enright (1978): *Improving the efficiency of matrix operations in the numerical solution of stiff ordinary differential equations.* ACM Trans. on Math. Software, vol. 4, pp. 127-136. *[IV.8]*

W.H. Enright & T.E. Hull (1976): *Comparing numerical methods for the solution of stiff systems of ODEs arising in chemistry.* In: Numerical methods for differential systems, recent developments in algorithms, software and applications, L. Lapidus & W.E. Schiesser, Eds., Academic Press, New York, 1976, pp. 45-66. *[IV.10]*

W.H. Enright, T.E. Hull & B. Lindberg (1975): *Comparing numerical methods for stiff systems of ODEs.* BIT, vol. 15, pp. 10-48. *[IV.10]*

W.H. Enright & M.S. Kamel (1979): *Automatic partitioning of stiff systems and exploiting the resulting structure.* ACM TOMS, vol. 5, pp. 374-385. *[IV.10]*

M.A. Epton, see R.F. Sincovec, A.M. Erisman, E.L. Yip & M.A. Epton.

A.M. Erisman, see R.F. Sincovec, A.M. Erisman, E.L. Yip & M.A. Epton.

L. Euler (1737): *De fractionibus continuis dissertatio.* Comm. acad. sc. Petrop., vol. 9, pp. 98-137; Opera Omnia vol. XIV, pp. 187-215 (vide §7). *[IV.13]*

L. Euler (1752): *Elementa doctrinae solidorum.* Nov. comm. acad. sci. Petropolitanae vol. 4, p. 109-140; Opera Omnia vol. XXVI, pp. 71-93. *[IV.4]*

P.L. Falb, see M. Athans & P.L. Falb.

L. Fejér (1933): *Mechanische Quadraturen mit positiven Coteschen Zahlen.* Math. Zeitschrift, vol. 37, pp. 287-309. *[IV.13]*

A. Feng, C.D. Holland & S.E. Gallun (1984): *Development and comparison of a generalized semi-implicit Runge-Kutta method with Gear's method for systems of coupled differential and algebraic equations.* Comp. & Chem. Eng., vol. 8, pp. 51-59. *[VI.4]*

J. Field & R.M. Noyes (1974); *Oscillations in chemical systems. IV: Limit cycle behavior in a model of a real chemical reaction.* J. Chem. Phys., vol. 60, pp. 1877-1884. *[IV.10]*

R. Frank, J. Schneid & C.W. Ueberhuber (1981): *The concept of B-convergence.* SIAM J. Numer. Anal., vol. 18, pp. 753-780. *[IV.15]*

R. Frank, J. Schneid & C.W. Ueberhuber (1985): *Stability properties of implicit Runge-Kutta methods.* SIAM J. Numer. Anal., vol. 22, pp. 497-514. *[IV.14], [IV.15]*

R. Frank, J. Schneid & C.W. Ueberhuber (1985): *Order results for implicit Runge-Kutta methods applied to stiff systems.* SIAM J. Numer. Anal., vol. 22, pp. 515-534. *[IV.14], [IV.15]*

R. Frank, see also W. Auzinger, R. Frank, & F. Macsek.

J.N. Franklin (1959): *Numerical stability in digital and analogue computation for diffusion problems.* J. Math. Phys., vol 37, pp. 305-315. *[IV.2]*

A. Friedli (1978): *Verallgemeinerte Runge-Kutta Verfahren zur Lösung steifer Differentialgleichungssysteme.* Oberwolfach Conference 1976, Lecture Notes in Math. 631, pp. 35-50. *[IV.11]*

K. Friedrichs, see R. Courant, K. Friedrichs & H. Lewy.

C. Führer (1988): *Differential-algebraische Gleichungssysteme in mechanischen Mehrkörpersystemen: Theorie, numerische Ansätze und Anwendungen*. Doctoral thesis, Technische Universität München *[VII.2]*.

C. Führer & B.J. Leimkuhler (1991): *Numerical solution of differential-algebraic equations for constrained mechanical motion*. Numer. Math., vol. 59, pp. 55-69. *[VII.2]*

C. Führer, see also C. Arévalo, C. Führer & G. Söderlind.

H. Fujita & T. Kato (1964): *On the Navier-Stokes initial value problem. I.* Arch. Rat. Mech. Anal., vol. 16, pp. 269-315. *[V.8]*

R.M. Furzeland, see M. Berzins & R.M. Furzeland.

B.G. Galerkin (1915): *Series expansions for some cases of equilibria of plates and beams* (Russian). Vestnik Ingenerov Petrograd, H.10. *[IV.10]*

S.E. Gallun, see A. Feng, C.D. Holland & S.E. Gallun.

R.V. Gamkrelidze, see L.S. Pontryagin, V.G. Boltyanskii, R.V. Gamkrelidze & E.F. Mishchenko.

F.R. Gantmacher (1954): *Teorya Matrits*. Two volumes, Gosudarstv. Izdat. Techn.-Teor. Lit., Moscva 1953; translations: Chelsea NY 1959, Interscience NY and London 1959, D. Verl. d. Wiss. Berlin 1958/59, Dunod Paris 1966. *[VII.1]*

C.W. Gear (1971): *Numerical initial value problems in ordinary differential equations*, Prentice Hall, 253 pp. *[V.2], [V.5]*

C.W. Gear (1971): *Simultaneous numerical solution of differential-algebraic equations*. IEEE Trans. Circuit Theory, vol. CT-18, pp. 89-95. *[VI.2]*

C.W. Gear (1982): *Automatic detection and treatment of oscillatory and/or stiff ordinary differential equations*. In: Numerical integration of differential equations, Lecture Notes in Math., vol. 968, pp. 190-206. *[IV.1]*

C.W. Gear (1988): *Differential-algebraic equation index transformations*. SIAM J. Sci. Stat. Comput., vol. 9, pp. 39-47. *[VII.4]*

C.W. Gear (1990): *Differential-algebraic equations, indices, and integral algebraic equations*. SIAM J. Numer. Anal., vol. 27. *[VII.1]*

C.W. Gear, G.K. Gupta & B. Leimkuhler (1985): *Automatic integration of Euler-Lagrange equations with constraints*. J. Comp. Appl. Math., vol. 12 & 13, pp. 77-90. *[VII.1], [VII.3], [VII.7]*

C.W. Gear, H.H. Hsu & L. Petzold (1981): *Differential-algebraic equations revisited*. Proc. Numerical Methods for Solving Stiff Initial Value Problems, Oberwolfach, BRD. *[VII.3]*

C.W. Gear & L.R. Petzold (1983): *Differential/algebraic systems and matrix pencils*. In: Matrix Pencils, B. Kagstrom & A. Ruhe (eds.), Lecture Notes in Math. 973, Springer Verlag, pp. 75-89. *[VII.1]*

C.W. Gear & L.R. Petzold (1984): *ODE methods for the solution of differential/algebraic systems*. SIAM J. Numer. Anal., vol. 21, pp. 716-728. *[VII.1], [VII.3]*

C.W. Gear & Y. Saad (1983): *Iterative solution of linear equations in ODE codes*. SIAM J. Sci. Stat. Comput., vol. 4, pp. 583-601. *[IV.10]*

C.W. Gear, see also S.L. Campbell & C.W. Gear.

E. Gekeler (1979): *Uniform stability of linear multistep methods in Galerkin prodecures for parabolic problems*. J. Math. Sciences, vol. 2, pp. 651-667. *[V.7]*

E. Gekeler (1984): *Discretization Methods for Stable Initial Value Problems.* Lecture Notes in Math., No. 1044, Springer Verlag. *[V.7]*

Y. Genin (1974): *An algebraic approach to A-stable linear multistep-multiderivative integration formulas.* BIT, vol. 14, pp. 382-406. *[V.4]*

D.R.A. Giles (1978): *A comparison of three problem-oriented simulation programs for dynamic mechanical systems.* Thesis, Univ. Waterloo, Ontario. *[VII.7]*

A. Giorgilli, see also G. Benettin & A. Giorgilli.

G.H. Golub & C.F. Van Loan (1989): *Matrix Computations.* Second edition, John Hopkins Univ. Press, Baltimore and London. *[VII.1]*

B.A. Gottwald (1977): *MISS — Ein einfaches Simulations-System für biologische und chemische Prozesse*, EDV in Medizin und Biologie, vol. 3, pp. 85-90. *[IV.10]*

A.R. Gourlay (1970): *A note on trapezoidal methods for the solution of initial value problems.* Math. of Comp., vol. 24, pp. 629-633. *[IV.3]*

J.A. van de Griend & J.F.B.M. Kraaijevanger (1986): *Absolute monotonicity of rational fonctions occuring in the numerical study of initial value problems.* Numer. Math., vol. 49, pp. 413-424. *[IV.11]*

E. Griepentrog & R. März (1986): *Differential-algebraic equations and their numerical treatment.* Teubner Texte zur Math., Band 88. *[VI.1], [VII.1], [VII.3]*

R.D. Grigorieff (1977): *Numerik gewöhnlicher Differentialgleichungen, Bd. 2, Mehrschrittverfahren.* Teubner Studienbücher, 411 Seiten "mit 49 Figuren, 32 Tabellen und zahlreichen Beispielen". *[V.1]*

R.D. Grigorieff & J. Schroll (1978): *Über $A(\alpha)$-stabile Verfahren hoher Konsistenzordnung.* Computing, vol. 20, pp. 343-350. *[V.2]*

A. Guillou & B. Lago (1961): *Domaine de stabilité associé aux formules d'intégration numérique d'équations différentielles, à pas séparés et à pas liés. Recherche de formules à grand rayon de stabilité.* ler Congr. Assoc. Fran. Calcul, AFCAL, Grenoble, Sept. 1960, pp. 43-56. *[IV.2]*

A. Guillou & J.L. Soulé (1969): *La résolution numérique des problèmes différentiels aux conditions initiales par des méthodes de collocation.* R.I.R.O., vol. R-3, pp. 17-44. *[V.3]*

G.K. Gupta, see C.W. Gear, G.K. Gupta & B. Leimkuhler.

K. Gustafsson (1991): *Control theoretic techniques for stepsize selection in explicit Runge-Kutta methods.* ACM Trans. Math. Soft., vol. 17, pp. 533-554. *[IV.2]*

K. Gustafsson (1994): *Control-theoretic techniques for stepsize selection in implicit Runge-Kutta methods.* ACM Trans. Math. Soft., vol. 20, pp. 496-517. *[IV.8]*

K. Gustafsson, M. Lundh & G. Söderlind (1988): *A PI stepsize control for the numerical solution of ordinary differential equations.* BIT, vol. 28, pp. 270-287. *[IV.2]*

E. Hairer (1980: *Highest possible order of algebraically stable diagonally implicit Runge-Kutta methods.* BIT, vol. 20, pp. 254-256. *[IV.13]*

E. Hairer (1982): *Constructive characterization of A-stable approximations to* $\exp z$ *and its connection with algebraically stable Runge-Kutta methods.* Numer. Math., vol. 39, pp. 247-258. *[IV.5]*

E. Hairer (1986): *A- and B-stability for Runge-Kutta methods - characterizations and equivalence.* Numer. Math., vol. 48, pp. 383-389. *[IV.13]*

E. Hairer (1994): *Backward analysis of numerical integrators and symplectic methods.* Annals of Numer. Math., vol. 1, pp. 107-132. *[VII.8]*

E. Hairer, G. Bader & Ch. Lubich (1982): *On the stability of semi-implicit methods for ordinary differential equations.* BIT, vol. 22, pp. 211-232. *[IV.9], [IV.11]*

E. Hairer & Ch. Lubich (1988): *Extrapolation at stiff differential equations.* Numer. Math., vol. 52, pp. 377-400. *[VI.5]*

E. Hairer & Ch. Lubich (1988b): *On extrapolation methods for stiff and differential-algebraic equations.* Teubner Texte zur Mathematik, Band 104, Teubner, Leipzig, pp. 64-73. *[VI.5]*

E. Hairer & Ch. Lubich (1996): *The life-span of backward error analysis for numerical integrators.* Numer. Math. *[VII.8]*

E. Hairer, Ch. Lubich & M. Roche (1988): *Error of Runge-Kutta methods for stiff problems studied via differential algebraic equations.* BIT, vol. 28, pp. 678-700. *[VI.3]*

E. Hairer, Ch. Lubich & M. Roche (1989): *Error of Rosenbrock methods for stiff problems studied via differential algebraic equations.* BIT, vol. 29, pp. 77-90. *[VI.3]*

E. Hairer, Ch. Lubich & M. Roche (1989): *The numerical solution of differential-algebraic systems by Runge-Kutta methods* (abbreviated as HLR89). Lecture Notes in Math. 1409, Springer Verlag. *[VI.1], [VII.1], [VII.3], [VII.4], [VII.5], [VII.7]*

E. Hairer & A. Ostermann (1990): *Dense output for extrapolation methods.* Numer. Math., vol. 58, pp. 419-439. *[VI.5]*

E. Hairer & H. Türke (1984): *The equivalence of B-stability and A-stability.* BIT, vol. 24, pp. 520-528. *[IV.5], [IV.13]*

E. Hairer & G. Wanner (1981): *Algebraically stable and implementable Runge-Kutta methods of high order.* SIAM J. Numer. Anal., vol. 18, pp. 1098-1108. *[IV.5], [IV.13]*

E. Hairer & G. Wanner (1982): *Characterization of non-linearly stable implicit Runge-Kutta methods.* In: Numerical integration of differential equations, Lecture Notes in Math., vol. 968, pp. 207-219. *[IV.5], [IV.13]*

E. Hairer & G. Wanner (1995): *Analysis by its history.* Undergraduate Texts in Mathematics, Springe-Verlag New York. *[IV.4],*

E. Hairer & G. Wanner (1996): *On a generalization of a theorem of von Neumann.* To appear in ZAMM. *[IV.12]*

E. Hairer & M. Zennaro (1996): *On error growth functions of Runge-Kutta methods.* To appear in Appl. Numer. Math. *[IV.11], [IV.12]*

E. Hairer, see also V. Brasey & E. Hairer; K. Dekker & E. Hairer; P. Deuflhard, E. Hairer & J. Zugck; G. Wanner, E. Hairer & S.P. Nørsett.

G. Hall (1985): *Equilibrium states of Runge-Kutta schemes.* ACM Trans. Math. Software, vol. 11, pp. 289-301. *[IV.1], [IV.2]*

G. Hall (1986): *Equilibrium states of Runge-Kutta schemes, part II.* ACM Trans. Math. Software, vol. 12, pp. 183-192. *[IV.2]*

G. Hall & D.J. Higham (1988): *Analysis of stepsize selection schemes for Runge-Kutta codes.* IMA J. Numer. Anal., vol. 8, pp. 305-310. *[IV.2]*

G. Hall, see also D.J. Higham & G. Hall.

R.W.Hamming (1959): *Stable predictor-corrector methods for ordinary differential equations.* J. ACM, vol. 6, pp. 37-47. *[V.1]*

R.W. HansonSmith, see D.S. Watkins & R.W. HansonSmith.

D.R. Hartree, see A. Callender, D.R. Hartree & A. Porter.

E.J. Haug (1989): *Computer-aided Kinematics and Dynamics of Mechanical Systems.* Allyn & Bacon, Boston. *[VII.7]*

E.J. Haug, see also R.A. Wehage & E.J. Haug.

F. Hausdorff (1921): *Summationsmethoden und Momentfolgen.* Math. Zeitschrift, vol. 9, pp. 74-109 and pp. 280-299. *[IV.11]*

P. Henrici (1962): *Discrete Variable Methods in Ordinary Differential Equations.* Wiley, New York. *[V.7]*

D. Henry (1981): *Geometric Theory of Semilinear Parabolic Equations.* Springer Lecture Notes in Mathematics 840. *[V.8]*

Ch. Hermite (1873): *Sur la fonction exponentielle.* Comptes rendus de l'Acad. Sciences, vol. 77, pp. 18-24, 74-79, 226-233, 285-293. Œuvres, tome III, pp. 150-181. *[IV.3]*

K.L. Hiebert, see L.F. Shampine & K.L. Hiebert.

D.J. Higham & G. Hall (1990): *Embedded Runge-Kutta formulae with stable equilibrium states.* J. of Comp. and Appl. Math., vol. 29, pp. 25-33. *[IV.2]*

D.J. Higham (1989): *Analysis of the Enright-Kamel partitioning method for stiff ordinary differential equations.* IMA J. Numer. Anal., vol. 9, pp. 1-14. *[IV.10]*

D.J. Higham, see also G. Hall & D.J. Higham.

A.C. Hindmarsh (1980): LSODE *and* LSODI, *two new initial value ordinary differential equation solvers.* ACM-SIGNUM Newsletter 15, pp. 10-11. *[IV.10], [V.5], [VII.3]*

A.C. Hindmarsh (1983): ODEPACK, *a systematized collection of ode solvers.* In Scientific Computing, R.S. Stepleman et al. (eds.), North-Holland, Amsterdam, pp. 55-64. *[V.5]*

A.C. Hindmarsh, see also P.N. Brown, G.D. Byrne & A.C. Hindmarsh; G.D. Byrne & A.C. Hindmarsh.

J.O. Hirschfelder, see C.F. Curtiss & J.O. Hirschfelder.

E. Hofer (1976): *A partially implicit method for large stiff systems of ODE's with only few equations introducing small time-constants.* SIAM J. Numer. Anal., vol. 13, pp. 645-663. *[IV.10]*

C.D. Holland, see A. Feng, C.D. Holland & S.E. Gallun.

E. Hopf (1950): *The partial differential equation* $u_t + uu_x = \mu u_{xx}$. Comm. on Pure and Appl. Math., vol. 3, pp. 201-230. *[VI.5], [VI.6]*

P.J. van der Houwen (1968): *Finite difference methods for solving partial differential equations.* MC Tract 20, Math. Centrum, Amsterdam. *[IV.2]*

P.J. van der Houwen (1973): *One-step methods with adaptive stability functions for the integration of differential equations.* Lecture Notes in Mathematics No. 333, Springer-Verlag, Berlin, pp. 164-174. *[IV.7]*

P.J. van der Houwen (1977): *Construction of integration formulas for initial value problems.* North Holland series in Applied Math. and Mech., 269 pp. *[IV.2], [IV.11]*

P.J. van der Houwen & B.P. Sommeijer (1980): *On the internal stability of explicit,* m*-stage Runge-Kutta methods for large* m*-values.* Z. Angew. Math. Mech., vol. 60, pp. 479-485. *[IV.2]*

H.H. Hsu, see C.W. Gear, H.H. Hsu & L. Petzold.

T.E. Hull, see W.H. Enright & T.E. Hull; W.H. Enright, T.E. Hull & B. Lindberg.

W.H. Hundsdorfer (1985): *The numerical solution of nonlinear stiff initial value problems: an analysis of one step methods.* CWI Tract, Nr. 12, Mathematisch Centrum, Amsterdam. *[IV.11], [IV.12], [IV.14]*

W.H. Hundsdorfer (1986): *Stability and B-convergence of linearly implicit Runge-Kutta methods.* Numer. Math., vol. 50, pp. 83-95. *[IV.15]*

W.H. Hundsdorfer & M.N. Spijker (1981): *A note on B-stability of Runge-Kutta methods.* Numer. Math., vol. 36, pp. 319-331. *[IV.12]*

W.H. Hundsdorfer & M.N. Spijker (1987): *On the algebraic equations in implicit Runge-Kutta methods.* SIAM J. Numer. Anal., vol. 24, pp. 583-594. *[IV.14]*

W.H. Hundsdorfer & B.I. Steininger (1991): *Convergence of linear multistep and one-leg methods for stiff nonlinear initial value problems.* BIT vol. 31, p.124-143. *[V.6], [V.7]*

W.H. Hundsdorfer, see also K. Burrage & W.H. Hundsdorfer; M. Crouzeix, W.H. Hundsdorfer & M.N. Spijker; J.G. Verwer, W.H. Hundsdorfer & B.P. Sommeijer.

A. Hurwitz & R. Courant (1925): *Funktionentheorie.* 2. Aufl., Verlag von Julius Springer, Berlin. *[V.4]*

A.F. Huxley, see A.L. Hodgkin & A.F. Huxley.

A. Iserles (1981): *Generalized order star theory, in : Padé approximations and its applications.* Amsterdam 1980, ed. M.G. de Bruin & H. van Rossum, Lecture Notes in Math. #888. *[IV.4]*

A. Iserles & S,P. Nørsett (1984): *A proof of the first Dahlquist barrier by order stars.* BIT, vol. 24, pp. 529-537. *[V.4]*

A. Iserles & G. Strang (1983): *The optimal accuracy of difference schemes.* Trans. Am. Math. Soc., vol. 277, pp. 779-803. *[IV.4]*

A. Iserles & R.A Williamson (1983): *Stability and accuracy of semi-discretized finite difference methods.* IMA J. Numer. Anal., vol. 4, pp. 289-307. *[IV.4]*

C.G.J. Jacobi (1826): *Ueber Gauss' neue Methode die Werthe der Integrale näherungsweise zu finden.* Journ. f. reine u. angew. Math., vol. I, pp. 301-308; Werke Vol. VI (1981), pp. 1-11. *[IV.5]*

L. Jay (1993): *Convergence of a class of Runge-Kutta methods for differential-algebraic systems of index 2*, BIT, vol. 33, pp. 137-150. *[VII.4]*

L. Jay (1994): *Runge-Kutta type methods for index three differential-algebraic equations with applications to Hamiltonian systems.* Thesis No. 2658, Univ. Genève. *[VII.8]*

L. Jay (1995): *Structure-preserving integrators.* Submitted for publication. *[VII.8]*

L. Jay (1996): *Symplectic partitioned Runge-Kutta methods for constrained Hamiltonian systems.* SIAM J. Numer. Anal., vol. 33, pp. 368-387. *[VII.8]*

R. Jeltsch (1976): *Stiff stability and its relation to A_0- and $A(0)$-stability*, SIAM J. Numer. Anal., vol. 13, pp. 8-17. *[V.2]*

R. Jeltsch (1976): *Note on A-stability of multistep multiderivative methods.* BIT, vol. 16, pp. 74-78. *[V.4]*

R. Jeltsch (1978): *Stability on the imaginary axis and A-stability of linear multistep methods.* BIT, vol. 18, pp. 170-174. *[V.4]*

R. Jeltsch (1988): *Order barriers for difference schemes for linear and nonlinear hyperbolic problems.* In: Numerical Analysis 1987, D.F. Griffiths & G.A. Watson (eds.), Pitman Research Notes in Math., No. 170, pp. 157-175. *[IV.4]*

R. Jeltsch & O. Nevanlinna (1978): *Largest disk of stability of explicit Runge-Kutta methods.* BIT, vol. 18, pp. 500-502. *[IV.4]*

R. Jeltsch & O. Nevanlinna (1981): *Stability of explicit time discretizations for solving initial value problems.* Numer. Math., vol. 37, pp. 61-91; Corrigendum: Numer. Math., vol. 39, p.155. *[IV.4]*

R. Jeltsch & O. Nevanlinna (1982): *Stability and accuracy of time discretizations for initial value problems.* Numer. Math., vol. 40, pp. 245-296. *[IV.4], [V.2], [V.4]*

R. Jeltsch, see also G. Dahlquist & R. Jeltsch.

M.S. Kamel, see W.H. Enright & M.S. Kamel.

L. Kantorovich & G. Akilov (1959): *Functional Analysis in Normed Spaces.* Fizmatgiz, Moscow (German translation: Akademic-Verlag, Berlin, 1964). *[VI.3]*

P. Kaps (1977): *Modifizierte Rosenbrockmethoden der Ordnungen 4,5 und 6 zur numerischen Integration steifer Differentialgleichungen.* Dissertation, Univ. Innsbruck. *[IV.7]*

P. Kaps & A. Ostermann (1989): *Rosenbrock methods using few LU-decompositions.* IMA J. Numer. Anal., vol. 9, pp. 15-27. *[IV.7]*

P. Kaps & A. Ostermann (1990): *$L(\alpha)$-stable variable order Rosenbrock-methods.* in: K. Strehmel, ed., *Numerical treatment of differential equations*, Teubner Texte zur Mathematik, Band 121, p. 80-91. *[IV.7]*

P. Kaps, S.W.H. Poon & T.D. Bui (1985): *Rosenbrock methods for stiff ODEs: a comparison of Richardson extrapolation and embedding technique.* Computing, vol. 34, pp. 17-40. *[IV.7]*

P. Kaps & P. Rentrop (1979): *Generalized Runge-Kutta methods of order four with stepsize control for stiff ordinary differential equations.* Numer. Math., vol. 33, pp. 55-68. *[IV.7]*

P. Kaps & G. Wanner (1981): *A study of Rosenbrock-type methods of high order.* Numer. Math., vol. 38, pp. 279-298. *[IV.7]*

O.A. Karakashian, see L.A. Bales, O.A. Karakashian & S.M. Serbin.

T. Kato (1960): *Estimation of iterated matrices, with application to the von Neumann condition.* Numer. Math., vol. 2, pp. 22-29. *[V.7]*

T. Kato (1966): *Perturbation Theory for Linear Operators.* Grudlehren der math. Wissenschaften, Bd. 132, Springer Verlag, Berlin. *[V.7]*

T. Kato, see H. Fujita & T. Kato.

S.L. Keeling (1989): *On implicit Runge-Kutta methods with a stability function having distinct real poles.* BIT, vol. 29, pp. 91-109. *[IV.4]*

M.D. Kirszbraun (1934): *Ueber die zusammenziehenden und Lipschitzschen Transformationen.* Fund. Math., vol. 23, pp. 77-108. *[IV.12]*

R.W. Klopfenstein, see R.L. Crane & R.W. Klopfenstein.

A.K. Kong, see R.D. Skeel & A.K. Kong.

J.F.B.M. Kraaijevanger (1985): *B-convergence of the implicit midpoint rule and the trapezoidal rule.* BIT, vol. 25, pp. 652-666. *[IV.15]*

J.F.B.M. Kraaijevanger (1986): *Absolute monotonicity of polynomials occuring in the numerical solution of initial value problems.* Numer. Math., vol. 48, pp. 303-322. *[IV.11]*

J.F.B.M. Kraaijevanger (1991): *A characterization of Lyapunov diagonal stability using Hadamard products.* Linear Alg. Appl., vol. 151, pp. 245-254. *[IV.14]*

J.F.B.M. Kraaijevanger & J. Schneid (1991): *On the unique solvability of the Runge-Kutta equations.* Numer. Math., vol. 59, pp. 129-157. *[IV.14], [IV.15]*

J.F.B.M. Kraaijevanger, see also K.Dekker, J.F.B.M. Kraaijevanger & J. Schneid; J.A. van de Griend & J.F.B.M. Kraaijevanger; M.Z. Liu & J.F.B.M. Kraaijevanger.

H.O. Kreiss (1962): *Über die Stabilitätsdefinition für Differenzengleichungen die partielle Differentialgleichungen approximieren.* BIT, vol. 2, pp. 153-181. *[V.7]*

F.T. Krogh (1966): *Predictor-Corrector methods of high order with improved stability characteristics.* J. Assoc. Comput. Mach., vol. 13, pp. 374-385. *[V.1]*

L. Kronecker (1874): *Über Schaaren von quadratischen und bilinearen Formen.* Akad. der Wiss. Berlin 19. Jan. 1874, Werke vol. I, pp. 351-413. *[VII.1]*

L. Kronecker (1890): *Algebraische Reduction der Schaaren bilinearer Formen.* Akad. der Wiss. Berlin 27. Nov. 1890, Werke vol. III^2, pp. 141-155. *[VII.1]*

V.I. Krylov (1959): *Priblizhennoe Vychislenie Integralov.* Goz. Izd. Fiz.-Mat. Lit., Moscow. English translation: Approximate calculation of integrals. Macmillan, New York, 1962. *[V.3]*

P. Kunkel & V. Mehrmann (1995): *Canonical forms for linear differential-algebraic equations with variable coefficients.* J. Comp. Appl. Math., vol. 56, pp. 225-251. *[VII.1]*

P. Kunkel & V. Mehrmann (1996): *Regular solutions of nonlinear differential-algebraic equations and their numerical determination.* Preprint, TU Chemnitz-Zwickau. *[VII.2]*

M.A. Kurdi (1974): *Stable high order methods for time discretization of stiff differential equations.* Thesis, Univ. of California. *[IV.6]*

A. Kværnø (1990): *Runge-Kutta methods applied to fully implicit differential-algebraic equations of index* 1. Math. Comp., vol. 54, pp. 583-625. *[VII.5]*

B. Lago, see A. Guillou & B. Lago.

J.L. Lagrange (1776): *Sur l'usage des fractions continues dans le calcul intégral.* Nouv. Mém. de l'Acad. royale du Sc. et Belles-Lettres de Berlin, Oeuvres Tome quatrième, pp. 301-332. *[IV.3]*

J.L. Lagrange (1788): *Méchanique analitique.* Paris, chez la Veuve Desaint, Libraire, MDCCLXXXVIII, avec approbation et privilège du Roi. Oeuvres vol. 11 et 12. *[IV.1]*

S. Lang (1962): *Introduction to differentiable manifolds.* John Wiley 1962; third and enlarged edition: *Differential and Riemannian manifolds.* Graduate Texts in Mathematics, Springer 1995. *[VII.1]*

J.D. Lawson (1967): *Generalized Runge-Kutta processes for stable systems with large Lipschitz constants.* SIAM J. Numer. Anal., vol. 4, pp. 372-380. *[IV.9]*

V.I. Lebedev (1989): *Explicit difference schemes with time-variable steps for solving stiff systems of Equations.* Sov. J. Numer. Anal. Math. Modelling 1989, vol. .4, N2, pp. 111-135. *[IV.2]*

V.I. Lebedev (1994): *How to solve stiff systems of differential equations by explicit methods.* In: *Numerical methods and applications*, ed. by G.I. Marchuk, pp. 45-80, CRC Press 1994. *[IV.2]*

V.I. Lebedev (1995): *Extremal polynomials with restrictions and optimal algorithms.* Manuscript, Russian Academy of Science, Moscow. *[IV.2]*

V.I. Lebedev & S.I. Finogenov (1976): *On the utilization of ordered Tchebychef parameters in iterative methods.* Zh. Vychisl. Mat. Mat Fiziki vol. 16, Nr. 4 pp. 895-910, (in Russian). *[IV.2]*

V.I. Lebedev & A.A. Medovikov (1994): *Explicit methods of second order for the solution of stiff systems of ordinary differential equations* (russian). Manuscript, Russian Academy of Science, Moscow. *[IV.2]*

B. van Leer, see P. Sonneveld & B. van Leer.

B. Leimkuhler, see C. Führer & B. Leimkuhler; C.W. Gear, G.K. Gupta & B. Leimkuhler.

B.J. Leimkuhler & R.D. Skeel (1994): *Symplectic numerical integrators in constrained Hamiltonian systems.* J. Comput. Phys., vol. 112, pp. 117-125. *[VII.8]*

M.-N. Le Roux (1980): *Méthodes multipas pour des équations paraboliques non linéaires.* Numer. Math., vol. 35, pp. 143-162. *[V.8]*

R.J. LeVeque & L.N. Trefethen (1984): *On the resolvent condition in the Kreiss matrix theorem.* BIT, vol. 24, pp. 584-591. *[V.7]*

R. LeVeque, see also G. Dahlquist, H. Mingyou & R. LeVeque.

H. Lewy, see R. Courant, K. Friedrichs & H. Lewy.

I. Lie (1990): *The stability function for multistep collocation methods.* Numer. Math., vol. 57, pp. 779-787. *[V.3]*

I. Lie & S.P. Nørsett (1989): *Superconvergence for multistep collocation.* Math. of Comput., vol. 52, pp. 65-79. *[V.3]*

B. Lindberg (1971): *On smoothing and extrapolation for the trapezoidal rule.* BIT, vol. 11, pp. 29-52. *[IV.9]*

B. Lindberg (1972): *A simple interpolation algorithm for improvement of the numerical solution of a differential equation.* SIAM J. Numer. Anal., vol. 9, pp. 662-668. *[VI.5]*

B. Lindberg (1974): *On a dangerous property of methods for stiff differential equations.* BIT, vol. 14, pp. 430-436. *[IV.3]*

B. Lindberg, see also W.H. Enright, T.E. Hull & B. Lindberg.

W. Liniger (1956): *Zur Stabilität der numerischen Integrationsmethoden für Differentialgleichungen.* Thèse, Université de Lausanne, 95 p. *[V.6]*

W. Liniger & R.A. Willoughby (1970): *Efficient integration methods for stiff systems of ordinary differential equations.* SIAM J. Numer. Anal., vol. 7, pp. 47-66. *[IV.8]*

W. Liniger, see also O. Nevanlinna & W. Liniger; F. Odeh & W. Liniger.

M.Z. Liu, K. Dekker & M.N. Spijker (1987): *Suitability of Runge-Kutta methods.* J. Comp. Appl. Math., vol. 91, pp. 53-63. *[IV.14]*

M.Z. Liu & J.F.B.M. Kraaijevanger (1988): *On the solvability of the systems of equations arising in implicit Runge-Kutta methods.* BIT, vol. 28, pp. 825-838. *[IV.14]*

C.F. Van Loan, see G.H. Golub & C.F. Van Loan.

L. Lopez & D. Trigiante (1989): *A projection method for the numerical solution of linear systems in separable stiff differential equations.* Intern. J. Computer Math., vol. 30, pp. 191-206. *[IV.10]*

P. Lötstedt (1985): *Discretization of singular perturbation problems by BDF methods.* Report No.99, Uppsala Univ., Dept. of Comp. Sci. *[VI.2]*

P. Lötstedt (1985): *On the relation between singular perturbation problems and differential-algebraic equations.* Report No.100, Uppsala Univ., Dept. of Comp. Sci. *[VI.2]*

P. Lötstedt & L. Petzold (1986): *Numerical solution of nonlinear differential equations with algebraic constraints I: Convergence results for backward differentiation formulas.* Math. Comput., vol 46, pp. 491-516. *[VII.3]*

Ch. Lubich (1988): *Convolution quadrature and discretized operational calculus I.* Numer. Math., vol. 52, pp. 129-145. *[V.7]*

Ch. Lubich (1989): *Linearly implicit extrapolation methods for differential-algebraic systems.* Numer. Math., vol. 55, pp. 197-211. *[VI.6] [VII.1]*

Ch. Lubich (1989): h^2 *-extrapolation methods for differential-algebraic systems of index 2.* Impact Comput. Sc. Eng., vol. 1, pp. 260-268. *[VII.7], [VII.6]*

Ch. Lubich (1991): *On the convergence of multistep methods for nonlinear stiff differential equations.* Numer. Math., vol. 58, pp. 839-853, and Erratum (Numer. Math., vol. 61, pp. 277-279) *[V.7], [V.8], [VI.2]*

Ch. Lubich (1991): *Extrapolation integrators for constrained multibody systems.* Impact Comp. Sci. Eng., vol. 3, pp. 213-234. *[VII.2]*

Ch. Lubich (1991): *On projected Runge-Kutta methods for differential-algebraic equations.* BIT, vol. 31, pp. 545-550. [VII.5]

Ch. Lubich (1993): *Integration of stiff mechanical systems by Runge-Kutta methods.* ZAMP, vol. 44, pp. 1022-1053. [VII.7]

Ch. Lubich, see also E. Hairer, G. Bader & Ch. Lubich; E. Hairer & Ch. Lubich; E. Hairer, Ch. Lubich & M. Roche.

Ch. Lubich, U. Nowak, U. Pöhle & Ch. Engstler (1992): *MEXX – numerical software for the integration of constrained mechanical multibody systems.* Preprint SC 92-12, Konrad-Zuse-Zentrum, Berlin. *[VI.7]*

Ch. Lubich & A. Ostermann (1993): *Runge-Kutta methods for parabolic equations and convolution quadrature.* Math. Comp., vol. 60, pp. 105-131. *[V.8]*

Ch. Lubich & M. Roche (1990): *Rosenbrock methods for differential-algebraic systems with solution-dependent singular matrix multiplying the derivative.* Computing, vol. 43, pp. 325-342. *[VI.6]*

M. Lundh, see K. Gustafsson, M. Lundh & G. Söderlind.

F. Macsek, see W. Auzinger, R. Frank, & F. Macsek.

D.W. Manning (1981): *A computer technique for simulating dynamic multibody systems based on dynamic formalism.* Thesis, Univ. Waterloo, Ontario. *[VII.7]*

M. Marden (1966): *Geometry of polynomials.* Mathematical Surveys, American Mathematical Society, Providence, Rhode Island, 2nd edition, 243 p. *[V.7]*

A.A. Markov (1890): *On a question of Mendeleiev.* Petersb. Proceedings LXII, 1-24 (Russian). *[IV.2]*

J.E. Marsden, see R. Abraham, J.E. Marsden & T. Ratiu.

R. März (1989): *Index-2 differential-algebraic equations.* Results in Mathematics, vol. 15, pp. 149-171. *[VII.1]*

R. März (1990): *Higher index differential-algebraic equations: Analysis and numerical treatment.* Banach Center Publ., 24, Numer. Anal. and Math. Modelling, pp. 199-222. *[VII.1], [VII.3]*

R. März, see also E. Griepentrog & R. März.

W.S. Massey (1980): *Singular homology theory.* Graduate Texts in Mathematics 70, Springer Verlag, 265 pp. *[IV.4]*

R.I. McLachlan (1995): *On the numerical integration of ordinary differential equations by symmetric composition methods.* SIAM J. Sci. Comput., vol. 16, pp. 151-168. *[VII.8]*

V. Mehrmann, see also P. Kunkel & V. Mehrmann.

M.L. Michelsen (1976): *Semi-implicit Runge-Kutta methods for stiff systems, program description and application examples.* Inst. f. Kemiteknik, Danmarks tekniske Højskole, Lyngby. *[VI.4]*

K. Miller & R.N. Miller (1981): *Moving finite elements. I.* SIAM J. Numer. Anal., vol. 18, pp. 1019-1032. *[VI.6]*

H. Mingyou, see G. Dahlquist, H. Mingyou & R. LeVeque.

G.J. Minty (1962): *On a simultaneous solution of a certain system of linear inequalities.* Proc. Amer. Math. Soc., vol. 13, pp. 11-12. *[IV.12]*

E.F. Mishchenko, see L.S. Pontryagin, V.G. Boltyanskii, R.V. Gamkrelidze & E.F. Mishchenko.

J.I. Montijano (1983): *Estudio de los metodos SIRK para la resolucion numérica de ecuaciones diferenciales de tipo stiff.* Thesis, Univ. Zaragoza. *[IV.14]*

E. Moore, see also S.L. Campbell & E. Moore.

R.E. Moore, see J.W. Daniel & R.E. Moore.

K.W. Morton, see R.D. Richtmyer & K.W. Morton.

H.N. Mülthei (1982): *Maximale Konvergenzordnung bei der numerischen Lösung von Anfangswertproblemen mit Splines.* Numer. Math., vol. 39, pp. 449-463. *[V.3]*

H.N. Mülthei (1982): *A-stabile Kollokationsverfahren mit mehrfachen Knoten.* Computing, vol. 29, pp. 51-61. *[V.3]*

S. Müller, A. Prohl, R. Rannacher & S. Turek (1994): *Implicit time-discretization of the nonstationary incompressible Navier-Stokes equations.* Proc. 10th GAMM-Workshop, Kiel, W. Hackbusch & G. Wittum eds., Vieweg. *[IV.3]*

A. Murua (1995): *Partitioned half-explicit Runge-Kutta methods for differential-algebraic systems of index 2.* Submitted for publication. *[VII.6], [VII.7]*

C.L. Navier (1823): *Mémoire sur les lois du mouvement des fluides* (lu à l'Acad. le 18 mars 1822). Paris, Mém. de l'Acad. Royale des Sciences, Tome VI, pp. 389-440. *[V.8]*

J. von Neumann (1951): *Eine Spektraltheorie für allgemeine Operatoren eines unitären Raumes.* Math. Nachrichten, vol. 4, pp. 258-281. *[IV.11]*

O. Nevanlinna (1976): *On the logarithmic norms of a matrix.* Report HTKK–MAT–A94, Helsinki Univ. of Tech. *[VI.3]*

O. Nevanlinna (1976): *On error bounds for G-stable methods.* BIT, vol. 16, pp. 79-84. *[V.6]*

O. Nevanlinna (1977): *On the numerical integration of nonlinear initial value problems by linear multistep methods.* BIT, vol. 17, pp. 58-71. *[V.8]*

O. Nevanlinna (1985): *Matrix valued versions of a result of von Neumann with an application to time discretization.* J. Comput. Appl. Math., vol. 12& 13, pp. 475-489. *[V.7]*

O. Nevanlinna & W. Liniger (1978): *Contractive methods for stiff differential equations, I.* BIT, vol. 18, pp. 457-474. *[V.7]*

O. Nevanlinna & W. Liniger (1979): *Contractive methods for stiff differential equations, II.* BIT, vol. 19, pp. 53-72. *[V.7]*

O. Nevanlinna & F. Odeh (1981): *Multiplier techniques for linear multistep methods.* Numer. Funct. Anal. Optim., vol. 3, pp. 377-423. *[V.8]*

O. Nevanlinna, see also R. Jeltsch & O. Nevanlinna.

K. Nipp & D. Stoffer (1995): *Invariant manifolds and global error estimates of numerical integration schemes applied to stiff systems of singular perturbation type – Part I: RK-methods.* Numer. Math., vol. 70, pp. 245-257. *[VI.3]*

S.P. Nørsett (1974): *Multiple Padé approximations to the exponential function.* Report No. 4/74, Dept. of Math., Univ. of Trondheim, Norway. *[IV.4]*

S.P. Nørsett (1974): *Semi-explicit Runge-Kutta methods.* Report No. 6/74, Dept. of Math., Univ. of Trondheim, Norway. *[IV.6]*

S.P. Nørsett (1975): *Runge-Kutta methods with coefficients depending on the Jacobian.* Report No. 1/75, Dept. of Math., Univ. of Trondheim, Norway. *[IV.7]*

S.P. Nørsett (1975): *C-polynomials for rational approximations to the exponential function.* Numer. Math., vol. 25, pp.39-56. *[IV.3]*

S.P. Nørsett (1976): *Runge-Kutta methods with a multiple real eigenvalue only.* BIT, vol. 16, pp. 388-393. *[IV.8]*

S.P. Nørsett & G. Wanner (1979): *The real-pole sandwich for rational approximations and oscillation equations.* BIT, vol. 19, pp. 79-94. *[IV.3], [IV.4]*

S.P. Nørsett & G. Wanner (1981): *Perturbed collocation and Runge-Kutta methods.* Numer. Math., vol. 38, pp. 193-208. *[IV.5], [IV.13]*

S.P. Nørsett & A. Wolfbrandt (1977): *Attainable order of rational approximations to the exponential function with only real poles.* BIT, vol. 17, pp. 200-208. *[IV.4]*

S.P. Nørsett & A. Wolfbrandt (1979): *Order conditions for Rosenbrock types methods.* Numer. Math., vol. 32, pp. 1-15. *[IV.7]*

S,P. Nørsett, see also A. Iserles & S,P. Nørsett; I. Lie & S.P. Nørsett; G. Wanner, E. Hairer & S.P. Nørsett.

U. Nowak, see P. Deuflhard & U. Nowak.

R.M. Noyes, see J. Field & R.M. Noyes.

F. Odeh & W. Liniger (1977): *Non-linear fixed-h stability of linear multistep formulae.* J. Math. Anal. Appl., vol. 61, pp. 691-712. *[V.8]*

F. Odeh, see also O. Nevanlinna & F. Odeh.

Ö. Ólafsson, see also T. Alishenas & Ö. Ólafsson.

R.E. O'Malley (1974): *Introduction to Singular Perturbations.* Academic Press, New York. *[VI.3]*

M.K. Ormrod & G.C. Andrews (1986): *Advent: a simulation program for constrained planar kinematic and dynamic systems.* Publications of the Amer. Soc. of Mech. Eng., 86-DET-97. *[VII.7]*

J.M. Ortega & W.C. Rheinboldt (1970): *Iterative Solution of Nonlinear Equations in Several Variables.* Academic Press, NewYork. *[VI.3], [VII.3], [VII.4], [VII.8]*

A. Ostermann (1988): *Ueber die Wahl geeigneter Approximationen an die Jacobimatrix bei linear-impliziten Runge-Kutta-Verfahren.* Dissertation, Univ. Innsbruck, pp. 66. *[IV.11]*

A. Ostermann (1990): *Continuous extensions of Rosenbrock-type methods.* Computing, vol. 44, pp. 59-68. *[VI.4]*

A. Ostermann, see also E. Hairer & A. Ostermann; P. Kaps & A. Ostermann, Ch. Lubich & A. Ostermann.

H. Padé (1892): *Sur la représentation approchée d'une fonction par des fractions rationnelles.* Première Thèse ("A Monsieur Hermite"), Ann. Ec. Norm. Sup. (3), vol. 9, Supp. 3-93, Oeuvres pp. 72-165. *[IV.3]*

H. Padé (1899): *Mémoire sur les développements en fractions continues de la fonction exponentielle pouvant servir d'introduction à la théorie des fractions continues algébriques.* Ann. Ec. Norm. Sup. (3), vol. 16, pp. 395-426; Oeuvres pp. 231-262. *[IV.3]*

M.A. Parseval (1799): Private communication to S.F. Lacroix. See: Lacroix, *Traité des différences et des séries*, Paris 1800, p. 377, or *Traité du calcul diff. et du calcul int.*, 2e éd, vol. 3, p. 394, Paris 1819. Also published in Paris Mémoires présentés par divers savants à l'acad. d. sc., vol 1, (1806), p. 639.

A. Pazy (1983): *Semigroups of Linear Operators and Applications to Partial Differential Equations.* Appl. Math. Sciences 44, Springer Verlag. *[V.7]*

F. Peherstorfer (1981): *Characterization of positive quadrature formulas.* SIAM J. Math. Anal., vol. 12, pp. 935-942. *[IV.13]*

O. Perron (1913): *Die Lehre von den Kettenbrüchen.* Teubner, 520 pp., 3rd ed., repr. 1977. *[IV.13]*

L.R. Petzold (1982): *A description of* DASSL: *A Differential/Algebraic System Solver.* Proceedings of IMACS World Congress, Montreal, Canada. *[VII.3], [VII.7]*

L.R. Petzold (1983): *Automatic selection of methods for solving stiff and nonstiff systems of ordinary differential equations.* SIAM J. Sci. Stat. Comp., vol. 4, pp. 136-148. *[IV.2]*

L.R. Petzold (1986): *Order results for implicit Runge-Kutta methods applied to differential/algebraic systems.* SIAM J. Numer. Anal., vol. 23, pp. 837-852. *[VI.1], [VII.4]*

L.R. Petzold, see also U. Ascher & L.R. Petzold; K.E. Brenan, S.L. Campbell & L.R. Petzold; K.E. Brenan & L.R. Petzold; C.W. Gear, H.H. Hsu & L. Petzold; C.W. Gear & L.R. Petzold; P. Lötstedt & L. Petzold.

Z. Picel, see B.L. Ehle & Z. Picel.

R.J. Plemmons, see G.P. Barker, A. Berman & R.J. Plemmons.

B. van der Pol (1926): *On "Relaxation Oscillations".* Phil. Mag., vol. 2, pp. 978-992; reproduced in: B. van der Pol, Selected Scientific Papers, vol. I, North-Holland Publ. Comp. Amsterdam (1960). *[VI.1]*

G. Pólya & G. Szegö (1925): *Aufgaben und Lehrsätze aus der Analysis.* Two volumes, Grundlehren Band XX, Springer Verlag, many later editions and translations. *[IV.4]*

L.S. Pontryagin, V.G. Boltyanskii, R.V. Gamkrelidze & E.F. Mishchenko (1961): *The mathematical theory of optimal processes.* Fizmatgiz Moscow, english translations: Wiley 1962, Pergamon Press 1964; german translation: Oldenbourg 1964. *[VII.1]*

S.W.H. Poon, see P. Kaps, S.W.H. Poon & T.D. Bui.

A. Porter, see A. Callender, D.R. Hartree & A. Porter.

F.A. Potra (1995): *Runge-Kutta integrators for multibody dynamics.* Mechanics of Structures and Machines, vol. 23, pp. 181-197. *[VII.2]*

F.A. Potra & W.C. Rheinboldt (1990): *Differential-geometric techniques for solving differential algebraic equations.* In E.J. Haug & R.C. Deyo, eds, Real-Time Integration of Mechanical System Simulation, Springer-Verlag, Berlin, pp. 155-191. *[VII.2]*

F.A. Potra & W.C. Rheinboldt (1991): *On the numerical solution of Euler-Lagrange equations.* Mech. Struct. & Mech., vol. 19(1), pp. 1-18. *[VII.2]*

W.H. Press, B.P. Flannery, S.A. Teukolsky & W.T. Vetterling (1986,1989): *Numerical Recipes, the art od scientific computing (FORTRAN version).* Cambridgre University Press, 702 pp. *[IV.10]*

P.J. Prince, see J.R. Dormand & P.J. Prince.

A. Prothero & A. Robinson (1974): *On the stability and accuracy of one-step methods for solving stiff systems of ordinary differential equations.* Math. of Comput., vol. 28, pp. 145-162. *[IV.3], [IV.15]*

V. Puiseux (1850): *Recherches sur les fonctions algébriques.* Journal de Math. vol 15, pp. 365-480. *[V.4]*

T. Ratiu, see R. Abraham, J.E. Marsden & T. Ratiu.

P.A. Raviart, see M. Crouzeix & P.A. Raviart.

S. Reich (1996): *Symplectic integration of constrained Hamiltonian systems by composition methods.* SIAM J. Numer. Anal., vol. 33, pp. 475-491. *[VII.8]*

S. Reich (1996): *On higher-order semi-explicit symplectic partitioned Runge-Kutta methods for constrained Hamiltonian systems.* Numer. Math. *[VII.8]*

S. Reich, see also U.M. Ascher, H. Chin & S. Reich.

M. Reimer (1967): *Zur Theorie der linearen Differenzenformeln.* Math. Zeitschr., vol. 95, pp. 373-402. *[V.4]*

E.Ya. Remez (1957): *General computation methods of Chebyshev approximation.* UkSSR Acad. Sci. Publ., Kiev 1957 (in Russian).

P. Rentrop, M. Roche & G. Steinebach (1989): *The application of Rosenbrock-Wanner type methods with stepsize control in differential-algebraic equations.* Numer. Math., vol. 55, pp. 545-563. *[VI.1], [VI.4]*

P. Rentrop, see also P. Kaps & P. Rentrop.

J.D. Reymond (1989): *Implementation des méthodes Radau IIA d'ordre 7 et 9.* Diploma thesis, Univ. Geneva. *[IV.10]*

W.C. Rheinboldt (1984): *Differential-algebraic systems as differential equations on manifolds.* Math. Comp., vol. 43, pp. 473-482. *[VII.1]*

W.C. Rheinboldt, see J.M. Ortega & W.C. Rheinboldt; F.A. Potra & W.C. Rheinboldt.

R.D. Richtmyer & K.W. Morton (1967): *Difference Methods for Initial-Value Problems.* Wiley-Interscience. *[V.7]*

B. Riemann (1857): *Allgemeine Voraussetzungen und Hülfsmittel für die Untersuchung von Functionen unbeschränkt veränderlicher Größen.* J. f. d. r. u. angew. Math., vol. 54, pp. 101-104; Werke pp. 81-84. *[V.4]*

R.E. Roberson & R. Schwertassek (1988): *Dynamics of Multibody Systems*. Springer Verlag. *[VII.7]*

B.C. Robertson (1987): *Detecting stiffness with explicit Runge-Kutta formulas*. Rep. 193/87, Dept. Comp. Sci., University of Toronto. *[IV.2]*

H.H. Robertson (1966): *The solution of a set of reaction rate equations*. In: J. Walsh ed.: Numer. Anal., an Introduction, Academ. Press, pp. 178-182. *[IV.1], [IV.10]*

A. Robinson, see A. Prothero & A. Robinson.

M. Roche (1988): *Rosenbrock methods for differential algebraic equations*. Numer. Math., vol. 52, pp. 45-63. *[VI.4]*

M. Roche (1988): *Runge-Kutta and Rosenbrock methods for differential-algebraic equations and stiff ODEs*. Doctoral thesis, Université de Genève. *[VII.5]*

M. Roche (1989): *Runge-Kutta methods for differential algebraic equations*. SIAM J. Numer. Anal., vol. 26, pp. 963-975. *[VI.4]*

M. Roche, see also E. Hairer, Ch. Lubich & M. Roche; Ch. Lubich & M. Roche; P. Rentrop, M. Roche & G. Steinebach.

H.H. Rosenbrock (1962/63): *Some general implicit processes for the numerical solution of differential equations*. Computer J., vol. 5, pp. 329-330. *[IV.7]*

F. Ruamps, see M. Crouzeix & F. Ruamps.

W.B. Rubin, see T.A. Bickart & W.B. Rubin.

J.-P. Ryckaert, G. Ciccotti & H.J.C. Berendsen (1977): *Numerical integration of the cartesian equations of motion of a system with constraints: molecular dynamics of n-alkanes*. J. Comput. Phys., vol. 23, pp. 327-341. *[VII.8]*

Y. Saad (1981): *Krylov subspace methods for solving large unsymmetric linear systems*. Math. Comp., vol. 37, pp. 105-126. *[IV.10]*

Y. Saad (1982): *The Lanczos biorthogonalization algorithm and other oblique projection methods for solving large unsymmetric systems*. SIAM J. Numer. Anal., vol. 19, pp. 485-506. *[IV.10]*

Y. Saad, see also C.W. Gear & Y. Saad.

I.W. Sandberg & H. Sichman (1968): *Numerical integration of systems of stiff nonlinear differential equations*. The Bell System Technical Journal, vol. 47, pp. 511-527. *[IV.12]*

J.M. Sanz-Serna & M.P. Calvo (1994): *Numerical Hamiltonian Problems*. Appl. Math. and Math. Comput. 7, Chapman & Hall, 207pp. *[VII.8]*

V.K. Saul'ev (1960): *Integration of parabolic type equations with the method of nets* (in Russian). Moscow, Fizmatgiz 1960. *[IV.2]*

A. Sayfy, see G.J. Cooper & A. Sayfy.

E. Schäfer (1975): *A new approach to explain the "High Irradiance Responses" of photomorphogenesis on the basis of phytochrome*. J. of Math. Biology, vol. 2, pp. 41-56. *[IV.10]*

W. Schiehlen, ed. (1990): *Multibody systems handbook*. Springer Verlag, Berlin. *[VII.7]*

R. Scherer (1979): *A necessary condition for B-stability*. BIT, vol. 19, pp. 111-115. *[IV.3], [IV.12]*

J. Schneid (1987): *B-convergence of Lobatto IIIC formulas*. Numer. Math., vol. 51, pp. 229-235. *[IV.15]*

J. Schneid, see also K.Dekker, J.F.B.M. Kraaijevanger & J. Schneid; R. Frank, J. Schneid & C.W. Ueberhuber; J.F.B.M.Kraaijevanger & J. Schneid.

C. Schneider (1991): *ROW-methods adapted to differential-algebraic systems.* Math. Comp., vol. 56, pp. 201-213. *[VI.4]*

C. Schneider (1991b): Private communication. *[VI.4]*

C. Schneider (1993): *Analysis of the linearly implicit mid-point rule for differential-algebraic equations.* Electronic Transactions on Numerical Analysis, vol. 1, pp. 1-10. *[VI.5]*

I.J. Schoenberg (1953): *On a Theorem of Kirszbraun and Valentine.* Amer. Math. Monthly, vol. 60, pp. 620-622. *[IV.12]*

S. Scholz (1989): *Order barriers for the B-convergence of ROW methods.* Computing, vol. 41, pp. 219-235. *[IV.15]*

S. Scholz, see also J.G. Verwer & S. Scholz.

J. Schroll, see R.D. Grigorieff & J. Schroll.

J. Schur (1918): *Über Potenzreihen, die im Innern des Einheitskreises beschränkt sind.* J. Reine u. angew. Math., vol. 147, pp. 205-232. *[V.3]*

R. Schwertassek, see R.E. Roberson & R. Schwertassek.

S.M. Serbin, see L.A. Bales, O.A. Karakashian & S.M. Sebin.

L.F. Shampine (1977): *Stiffness and nonstiff differential equation solvers, II: detecting stiffness with Runge-Kutta methods.* ACM TOMS, vol. 3, pp. 44-53. *[IV.2]*

L.F. Shampine (1980): *Implementation of implicit formulas for the solution of ODEs.* SIAM J. Sci. Stat. Comput., vol. 1, pp. 103-118. *[IV.8]*

L.F. Shampine (1981): *Evaluation of a test set for stiff ODE solvers.* ACM Trans. Math. Soft., vol. 7, pp. 409-420. *[IV.10]*

L.F. Shampine (1982): *Implementation of Rosenbrock methods.* ACM Trans. Math. Soft., vol. 8, pp. 93-113. *[IV.7]*

L.F. Shampine (1986): *Conservation laws and the numerical solution of ODEs.* Comp. Maths. Appls., vol. 12B., pp. 1287-1296. *[VII.2]*

L.F. Shampine (1987): *Control of step size and order in extrapolation codes.* J. Comp. Appl. Math., vol. 18, pp. 3-16. *[IV.9]*

L.F. Shampine & K.L. Hiebert (1977): *Detecting stiffness with the Fehlberg (4,5) formulas.* Comp. & Maths. with Appls., vol. 3, pp. 41-46. *[IV.2]*

L.F. Shampine & H.A. Watts (1979): *DEPAC — design of a user oriented package of ODE solvers.* Report SAND-79-2374, Sandia Nat. Lab., Albuquerque, New Mexico. *[V.5]*

H. Sichman, see I.W. Sandberg & H. Sichman.

R.F. Sincovec, A.M. Erisman, E.L. Yip & M.A. Epton (1981): *Analysis of descriptor systems using numerical algorithms.* IEEE Trans. Aut. Control, AC-26, pp. 139-147. *[VII.3]*

R.D. Skeel, see also B.J. Leimkuhler & R.D. Skeel.

R.D. Skeel & A.K. Kong (1977): *Blended linear multistep methods.* ACM TOMS, vol. 3, pp. 326-343. *[V.2], [V.3], [V.5]*

H.M. Sloate & T.A. Bickart (1973): *A-stable composite multistep methods.* J. ACM, vol. 20, pp. 7-26. *[V.3]*

P.E. Sobolevskiĭ (1959): *On non-stationary equations of hydrodynamics for viscous fluid.* Doklady Akad. Nauk USSR, vol. 128, pp. 45-48. *[V.8]*

G. Söderlind (1981): *On the efficient solution of nonlinear equations in numerical methods for stiff differential systems.* Report TRITA-NA-8114, Royal Inst. of Tech., Stockholm. *[IV.10]*

G. Söderlind (1989): *A multi-purpose system for the numerical integration of ODEs.* Appl. Math. Comp., vol. 31, pp. 346-360. *[VII.6]*

G. Söderlind & G. Dahlquist (1981): *Error propagation and stiff differential systems of singular perturbation type.* Rep. TRITA-NA-8108, Royal Inst. of Tech., Stockholm. *[VI.2]*

G. Söderlind, see also C. Arévalo, C. Führer & G. Söderlind; G. Dahlquist & G. Söderlind; K. Gustafsson, M. Lundh & G. Söderlind.

B.P. Sommeijer & J.G. Verwer (1980): *A performance evaluation of a class of Runge-Kutta-Chebyshev methods for solving semi-discrete parabolic differential equations.* Report NW91/80, Mathematisch Centrum, Amsterdam. *[IV.2]*

B.P. Sommeijer (1991): *RKC, a nearly-stiff ODE solver.* Available from netlib@ornl.gov, send rkc.f from ode. *[IV.2], [IV.10]*

B.P. Sommeijer, see P.J. van der Houwen & B.P. Sommeijer; J.G. Verwer, W.H. Hundsdorfer & B.P. Sommeijer.

A. Sommerfeld (1942): *Vorlesungen über theoretische Physik.* Bd.1., Mechanik; translated from the 4th german ed.: Acad. Press. *[IV.1], [VII.1]*

P. Sonneveld & B. van Leer (1985): *A minimax problem along the imaginary axis.* Nieuw Archief V. Wiskunde (4), vol. 3, pp. 19-22. *[IV.2]*

G. Sottas (1984): *Dynamic adaptive selection between explicit and implicit methods when solving ODE's.* Report, Sect. de math., Univ. Genève. *[IV.2]*

G. Sottas & G. Wanner (1982): *The number of positive weights of a quadrature formula.* BIT, vol. 22, pp. 339-352. *[IV.13]*

J.L. Soulé, see A. Guillou & J.L. Soulé.

M.N. Spijker (1983): *Contractivity in the numerical solution of initial value problems.* Numer. Math., vol. 42, pp. 271-290. *[IV.11]*

M.N. Spijker (1985): *Feasibility and contractivity in implicit Runge-Kutta methods.* J. Comp. Appl. Math., vol. 12 et 13, pp. 563-578. *[IV.14]*

M.N. Spijker (1985): *Stepsize restrictions for stability of one-step methods in the numerical solution of initial value problems.* Math. Comp., vol. 45, pp. 377-392. *[IV.11]*

M.N. Spijker (1986): *The relevance of algebraic stability in implicit Runge-Kutta methods.* Teubner Texte zur Mathematik 82 (K. Strehmel, ed.), pp. 158-164. *[IV.15]*

M.N. Spijker (1991): *On a conjecture by LeVeque and Trefethen related to the Kreiss matrix theorem.* BIT, vol. 31, pp. 551-555. *[V.7]*

M.N. Spijker, see also M. Crouzeix, W.H. Hundsdorfer & M.N. Spijker; W.H.Hundsdorfer & M.N. Spijker; M.Z. Liu, K. Dekker & M.N. Spijker.

I.A. Stegun, see M. Abramowitz & I.A. Stegun.

T. Steihaug & A. Wolfbrandt (1979): *An attempt to avoid exact Jacobian and nonlinear equations in the numerical solution of stiff differential equations.* Math. Comp., vol. 33, pp. 521-534. *[IV.7]*

G. Steinebach (1995): *Order-reduction of ROW-methods for DAEs and method of lines applications*. Preprint, TH Darmstadt. *[VI.4]*

G. Steinebach, see P. Rentrop, M. Roche & G. Steinebach.

B.I. Steininger, see W.H. Hundsdorfer & B.I. Steininger.

V. Steklov (1916): *On the approximate computation of definite integrals with the help of so-called mechanical quadrature I. Convergence of mechanical quadrature formulas*. Petrograd, Bull. Acad. Sciences, ser. VI, vol. 10, pp. 169-186 (russian). See also same Journal vol. 11 (1917), pp. 557-558 for a french explanation. *[IV.13]*

H.J. Stetter (1968): *Improved absolute stability of predictor-corrector schemes*. Computing, vol. 3, pp. 286-296. *[V.1]*

H.J. Stetter (1973): *Analysis of discretization methods for ordinary differential equations*. Springer, Berlin. *[IV.3], [IV.9], [IV.12]*

G.W. Stewart (1972): *On the sensitivity of the eigenvalue problem* $Ax=\lambda Bx$. SIAM J. Numer. Anal., vol. 9, pp. 669-686. *[VII.1]*

T.J. Stieltjes (1884): *Quelques recherches sur la Théorie des quadrature dites mécaniques*. Annales Scientif. de l'Ecole Norm. Sup., troisième série, tome I, pp. 409-426. *[IV.12]*

G.G. Stokes (1845): *On the theories of the internal friction of fluids in motion, and the equilibrium and motion of elastic solids*. Cambr. Phil. Soc. Trans., vol. 8. Republished in: G.G. Stokes, Mathematical and Physical Papers, vol. 1, Cambridge 1880. *[V.8]*

D. Stoffer, see also K. Nipp & D. Stoffer.

G. Strang, see A. Iserles & G. Strang.

K. Strehmel & R. Weiner (1982): *Behandlung steifer Anfangswertprobleme gewöhnlicher Differentialgleichungen mit adaptiven Runge-Kutta Methoden*. Computing, vol. 29, pp. 153-165. *[IV.11]*

K. Strehmel & R. Weiner (1987): *B-convergence results for linearly implicit one step methods*. BIT, vol. 27, pp. 264-281. *[IV.11], [IV.15]*

Sun Geng (1993): *Symplectic partitioned Runge-Kutta methods*. J. Comput. Math., vol. 11, pp. 365-372. *[VII.8]*

A.G. Sveshnikov, see A.N. Tikhonov, A.B. Vasil'eva & A.G. Sveshnikov.

G. Szegö (1939): *Orthogonal Polynomials*. AMS Coll. Publ., vol. XXIII, 403pp. *[IV.13]*

G. Szegö, see also G. Pólya & G. Szegö.

E. Tadmor (1981): *The equivalence of L_2-stability, the resolvent condition, and strict H-stability*. Lin. Alg. and its Applics., vol. 41, pp. 151-159. *[V.7]*

P.G. Thomsen, see S.P. Nørsett & P.G. Thomsen.

A.N. Tikhonov (1952): *Systems of differential equations containing small parameters in the derivatives*. Mat. Sb. (Russian), vol. 31 (73), pp. 575-586. *[VI.3]*

A.N. Tikhonov, A.B. Vasil'eva & A.G. Sveshnikov (1985): *Differential Equations*. Trans. from the Russian by A.B. Sossinskij. Springer Verlag, 238pp. *[VI.3]*

L.N. Trefethen, see R.J. LeVeque & L.N. Trefethen.

D. Trigiante, see L. Lopez & D. Trigiante.

H. Türke, see E. Hairer & H. Türke.

C.W. Ueberhuber, see R. Frank, J. Schneid & C.W. Ueberhuber.

R. Vanselow (1979): *Stabilitäts-und Fehleruntersuchungen bei numerischen Verfahren zur Lösung steifer nichtlinearer Anfangswertprobleme.* Diplomarbeit, Sektion Mathematik, TU-Dresden. *[IV.12]*

J.M. Varah (1979): *On the efficient implementation of implicit Runge-Kutta methods.* Math. Comp., vol. 33, pp. 557-561. *[IV.8]*

R.S. Varga, see G. Birkhoff & R.S. Varga.

A.B. Vasil'eva (1963): *Asymptotic behaviour of solutions to certain problems involving nonlinear differential equations containing a small parameter multiplying the highest derivatives.* Usp. Mat. Nauk (Russian), vol. 18, pp.15-86. English translation: Russian Math. Surveys, vol.18, Nr. 3, pp. 13-84. *[VI.3]*

A.B. Vasil'eva, see also A.N. Tikhonov, A.B. Vasil'eva & A.G. Sveshnikov.

M.V. van Veldhuizen (1984): *D-stability and Kaps-Rentrop methods.* Computing vol. 32, pp. 229-237. *[IV.7], [VI.4]*

L. Verlet (1967): *Computer "experiments" on classical fluids. I. Thermodynamical properties of Lennard-Jones molecules.* Physical Review, vol. 159, pp. 98-103. *[VII.8]*

J.G. Verwer (1980): *On generalized Runge-Kutta methods using an exact Jacobian at a non-step point.* ZAMM, vol. 60, pp. 263-265. *[IV.7]*

J.G. Verwer (1996): *Explicit Runge-Kutta methods for parabolic partial differential equations.* To appear in Applied Numerical Mathematics. *[IV.2]*

J.G. Verwer, W.H. Hundsdorfer & B.P. Sommeijer (1990): *Convergence properties of the Runge-Kutta-Chebyshev method.* Numer. Math., vol. 57, pp. 157-178. *[IV.2]*

J.G. Verwer & S. Scholz (1983): *Rosenbrock methods and time-lagged Jacobian matrices.* Beiträge zur Numer. Math., vol. 11, pp. 173-183. *[IV.7]*

J.G. Verwer, see also K. Dekker & J.G. Verwer.

P.P. Wakker (1985): *Extending monotone and non-expansive mappings by optimization.* Cahiers du C.E.R.O., vol. 27, pp. 141-149. *[IV.12]*

G. Wanner (1976): *A short proof on nonlinear A-stability.* BIT, vol. 16, pp. 226-227. *[IV.12]*

G. Wanner (1980): *Characterization of all A-stable methods of order* $2m-4$. BIT, vol. 20, pp. 367-374. *[IV.5]*

G. Wanner, E. Hairer & S.P. Nørsett (1978): *Order stars and stability theorems.* BIT, vol. 18, pp. 475-489. *[IV.4], [IV.6], [V.4]*

G. Wanner, E. Hairer & S.P. Nørsett (1978): *When I-stability implies A-stability.* BIT, vol. 18, p. 503. *[IV.4]*

G. Wanner, see also E. Hairer & G. Wanner; P. Kaps & G. Wanner; S.P. Nørsett & G. Wanner; G. Sottas & G. Wanner.

W. Wasow (1965): *Asymptotic expansions for ordinary differential equations.* Interscience, John Wiley & Sons, New York, 263pp. *[VI.3]*

D.S. Watkins & R.W. HansonSmith (1983): *The numerical solution of sparably stiff systems by precise partitioning.* ACM trans. Math. Soft., vol. 9, pp. 293-301. *[IV.10]*

H.A. Watts, see L.F. Shampine & H.A. Watts.

R.A. Wehage & E.J. Haug (1982): *Generalized coordinate partitioning for dimension reduction in analysis of constrained dynamic systems.* J. Mechanical Design, vol. 104, pp. 247-255. *[VII.2]*

K. Weierstrass (1868): *Zur Theorie der bilinearen und quadratischen Formen.* Akad. der Wiss. Berlin 18. Mai. 1868, Werke vol. II, pp. 19-44. *[VII.1]*

R. Weiner, see K. Strehmel & R. Weiner.

D.V. Widder (1946): *The Laplace Transform.* Princeton University Press, London. *[IV.11]*

O.B. Widlund (1967): *A note on unconditionally stable linear multistep methods.* BIT, vol. 7, pp. 65-70. *[IV.3], [V.1], [V.2]*

J.H. Wilkinson (1965): *The Algebraic Eigenvalue Problem.* Clarendon Press, Oxford, 662 p. *[IV.2]*

R.A Williamson, see A. Iserles & R.A Williamson.

R.A. Willoughby (ed.) (1974): *Stiff Differential Systems.* Plenum Press, New York. *[IV.1]*

R.A. Willoughby, see also W. Liniger & R.A. Willoughby.

A. Wolfbrandt (1977): *A study of Rosenbrock processes with respect to order conditions and stiff stability.* Thesis, Chalmers Univ. of Techn., Göteborg, Sweden. *[IV.4], [IV.7]*

A. Wolfbrandt, see also S.P. Nørsett & A. Wolfbrandt; T. Steihaug & A. Wolfbrandt.

K. Wright (1970): *Some relationships between implicit Runge-Kutta, collocation and Lanczos τ methods, and their stability properties.* BIT, vol. 10, pp.217-227. *[IV.3]*

J. Yen (1993): *Constrained equations of motion in multibody dynamics as ODEs on manifolds.* SIAM J. Numer. Anal., vol. 30, pp. 553-568. *[VII.2]*

E.L. Yip, see R.F. Sincovec, A.M. Erisman, E.L. Yip & M.A. Epton.

H. Yoshida (1990): *Construction of higher order symplectic integrators.* Phys. Lett. A, Vol.150, p.262-268. *[VII.8]*

S. Yoshizawa, see J. Nagumo, S. Arimoto & S. Yoshizawa.

Yuan Chzao Din (1958): *Some difference schemes of solution of first boundary problem for linear differential equations with partial derivatives,* (in Russian) Thesis cand.phys. math. Sc., Moscov MGU 1958.

E.C. Zeeman (1972): *Differential equations for the heartbeat and nerve impulse.* Published in *Towards a theoretical biology* (Edited C.H. Waddington) Edinburgh University Press, Volume 4, pp. 8-67. Reprinted in *Catastrophe theory, Selected papers 1972-1977*, Addison-Wesley 1977, pp. 81-140. *[IV.10]*

J. Zugck, see P. Deuflhard, E. Hairer & J. Zugck.

Symbol Index

Subject Index

Druck: STRAUSS OFFSETDRUCK, MÖRLENBACH
Verarbeitung: SCHÄFFER, GRÜNSTADT